Lecture Notes in Artificial Intell

T0238662

Subseries of Lecture Notes in Comput.

LNAI Series Editors

Randy Goebel
University of Alberta, Edmonton, Canada
Yuzuru Tanaka
Hokkaido University, Sapporo, Japan
Wolfgang Wahlster
DFKI and Saarland University, Saarbrücken, Germany

LNAI Founding Series Editor

Joerg Siekmann
DFKI and Saarland University, Saarbrücken, Germany

Henrik Legind Larsen
Maria J. Martin-Bautista
María Amparo Vila
Troels Andreasen
Henning Christiansen (Eds.)

Flexible Query Answering Systems

10th International Conference, FQAS 2013
Granada, Spain, September 18-20, 2013
Proceedings

 Springer

Volume Editors

Henrik Legind Larsen
Aalborg University, Department of Electronic Systems
6700 Esbjerg, Denmark
E-mail: hll@es.aau.dk

Maria J. Martin-Bautista
María Amparo Vila
University of Granada
Department of Computer Science and Aritificial Intelligence
18071 Granada, Spain
E-mail: {mbautis; vila}@decsai.ugr.es

Troels Andreasen
Henning Christiansen
Roskilde University, CBIT
4000 Roskilde, Denmark
E-mail: {troels; henning}@ruc.dk

ISSN 0302-9743   e-ISSN 1611-3349
ISBN 978-3-642-40768-0   e-ISBN 978-3-642-40769-7
DOI 10.1007/978-3-642-40769-7
Springer Heidelberg New York Dordrecht London

Library of Congress Control Number: 2013946851

CR Subject Classification (1998): I.2, H.3, H.2, H.4, H.5, C.2, F.1

LNCS Sublibrary: SL 7 – Artificial Intelligence

Typesetting: Camera-ready by author, data conversion by Scientific Publishing Services, Chennai, India

Printed on acid-free paper

Springer is part of Springer Science+Business Media (www.springer.com)

Preface

This volume constitutes the Proceedings of the 10th International Conference on Flexible Query-Answering Systems, FQAS 2013, held in Granada, Spain, during September 18–20, 2013. This biennial conference series has been running since 1994, starting in Roskilde, Denmark, where it was also held in 1996, 1998, and 2009; in 2000 it was held in Warsaw, Poland, in 2002 in Copenhagen, Denmark, in 2004 in Lyon, France, in 2006 in Milan, Italy, and in 2011 in Ghent, Belgium.

FQAS is the premier conference concerned with the very important issue of providing users of information systems with flexible querying capabilities, and with an easy and intuitive access to information. More specifically, the overall theme of the FQAS conferences is the modelling and design of innovative and flexible modalities for accessing information systems. The main objective is to achieve more expressive, informative, cooperative, and productive systems that facilitate retrieval from information repositories such as databases, libraries, heterogeneous archives, and the Web.

With these aims, FQAS is a multidisciplinary conference drawing on several research areas, including information retrieval, database management, information filtering, knowledge representation, computational linguistics and natural language processing, artificial intelligence, soft computing, classical and non-classical logics, and human-computer interaction.

The sessions were organized in a general session track and a parallel special session track with a total of 59 original papers contained in these proceedings. The general track covers the current main stream fields: querying-answering systems, semantic technology, patterns and classification, personalization and recommender systems, searching and ranking, and Web and human-computer interaction. The special track covers some specific and, typically, newer fields, namely: environmental scanning for strategic early warning, generating linguistic descriptions of data, advances in fuzzy querying and fuzzy databases: theory and applications, fusion and ensemble techniques for online learning on data streams, and intelligent information extraction from texts.

We wish to thank all authors for their excellent papers and the referees, publisher, sponsors, and local organizers for their efforts. Special thanks to the organizers of the special sessions, the invited speakers, members of the Advisory Board, and members of the Program Committee. All of them made the success of FQAS 2013 possible.

September 2013

Henrik Legind Larsen
Maria J. Martin-Bautista
María Amparo Vila
Troels Andreasen
Henning Christiansen

Organization

FQAS 2013 was organized by the Department of Computer Science and Artificial Intelligence of the University of Granada, and by the Department of Electronic Systems, Aalborg University. The local organization was done in cooperation with the High Technical School of Computer Science and Telecommunications of the University of Granada.

General and Program Co-chairs

Henrik Legind Larsen	Aalborg University, Denmark
Maria J. Martin-Bautista	University of Granada, Spain
María Amparo Vila	University of Granada, Spain

Local Organizing Committee

Maria J. Martin-Bautista	University of Granada, Spain (Chair)
Juan Miguel Medina	University of Granada, Spain
Daniel Sánchez	University of Granada, Spain
Nicolás Marín	University of Granada, Spain
Ignacio Blanco	University of Granada, Spain
José M. Serrano	University of Jaén, Spain
Carlos Molina	University of Jaén, Spain
Belén Prados	University of Granada, Spain
Carmen Martínez-Cruz	University of Jaén, Spain
Carlos D. Barranco	Pablo de Olavide University, Spain
M. Dolores Ruiz	University of Granada, Spain
Jesús R. Campaña	University of Granada, Spain
Miguel Molina	University of Granada, Spain
María Ros	University of Granada, Spain
Rita Castillo	University of Granada, Spain
Jose Enrique Pons	University of Granada, Spain
Sergio Jaime	University of Granada, Spain
Irene Díaz-Valenzuela	University of Granada, Spain
María Martínez-Rojas	University of Granada, Spain

Steering Committee

Henrik Legind Larsen	Aalborg University, Denmark
Troels Andreasen	Roskilde University, Denmark
Henning Christiansen	Roskilde University, Denmark

International Advisory Board

Guy De Tré, Belgium
Miguel Delgado, Spain
Hassane Essafi, France
Janusz Kacprzyk, Poland
Donald Kraft, USA
Amihai Motro, USA

Gabriella Pasi, Italy
Olivier Pivert, France
Zbigniew W. Ras, USA
Nicolas Spyratos, France
Sławomir Zadrożny, Poland

Program Committee

Troels Andreasen, Denmark
Carlos Barranco, Spain
Ignacio Blanco, Spain
Igor Boguslavsky, Spain
Gloria Bordogna, Italy
Mohand Boughanem, France
Piotr Brodka, Poland
Robert Burduk, Poland
Jesús R. Campaña, Spain
Jesús Cardeñosa, Spain
Paola Carrara, Italy
Gladys Castillo Jordán, Portugal
Panagiotis Chountas, UK
Henning Christiansen, Denmark
Fabio Crestani, UK
Bruce Croft, USA
Juan Carlos Cubero, Spain
Alfredo Cuzzocrea, Italy
Boguslaw Cyganek, Poland
Ireneusz Czarnowski, Poland
Ernesto Damiani, Italy
Agnieszka Dardzinska, Poland
Bernard De Baets, Belgium
Estelle De Marco, France
Guy De Tré, Belgium
Miguel Delgado, Spain
Marcin Detyniecki, France
Mike Dillinger, USA
Jørgen Fischer Nilsson, Denmark
William Fleischman, USA
José Galindo, Spain
Carolina Gallardo, Spain
Ana Garcia-Serrano, Spain
Anne Gerdes, Denmark

Manuel Graña, Spain
Maria Grineva, USA
Allel Hadjali, France
Sven Helmer, UK
Enrique Herrera-Viedma, Spain
David Hicks, Denmark
Tzung-Pei Hong, Taiwan
Luis Iraola, Spain
Janusz Kacprzyk, Poland
Tomasz Kajdanowicz, Poland
Ioannis Katakis, Greece
Etienne Kerre, Belgium
Murat Koyuncu, Turkey
Donald Kraft, USA
Marzena Kryszkiewicz, Poland
Mounia Lalmas, UK
Henrik Legind Larsen, Denmark
Anne Laurent, France
David Losada, Spain
Edwin Lughofer, Austria
Zongmin Ma, China
Nicolas Marín, Spain
Palmira Marrafa, Portugal
Christophe Marsala, France
Maria J. Martin-Bautista, Spain
Carmen Martínez-Cruz, Spain
Juan M. Medina, Spain
Ernestina Menasalvas, Spain
Sadaaki Miyamoto, Japan
Carlos Molina, Spain
Miguel Molina, Spain
Amihai Motro, USA
Joachim Nielandt, Belgium
Andreas Nuernberger, Germany

Peter Oehrstroem, Denmark
Jose A. Olivas, Spain
Daniel Ortiz-Arroyo, Denmark
Gabriella Pasi, Italy
Fred Petry, USA
Olivier Pivert, France
Olga Pons, Spain
Henri Prade, France
Belén Prados, Spain
Zbigniew W. Ras, USA
Guillaume Raschia, France
Francisco P. Romero, Spain
Maria Ros, Spain
Jacobo Rouces, Denmark
M. Dolores Ruiz, Spain
Daniel Sánchez, Spain
Florence Sedes, France
Jose M. Serrano, Spain
Miguel A. Sicilia, Spain
Dragan Simić, Serbia
Andrzej Skowron, Poland

Nicolas Spyratos, France
Jerzy Stefanowski, Poland
Zbigniew Telec, Poland
Vicenc Torra, Spain
Farouk Toumani, France
Bogdan Trawinski, Poland
Gracian Triviño, Spain
Maurice Van Keulen, The Netherlands
Maarten Van Someren,
   The Netherlands
Suzan Verberne, The Netherlands
María Amparo Vila, Spain
Yana Volkovich, Spain
Peter Vojtas, Slovakia
Jef Wijsen, Belgium
Michal Wozniak, Poland
Adnan Yazici, Turkey
Sławomir Zadrożny, Poland
Shuigeng Zhou, China

Additional Reviewers

Alberto Alvarez, Spain
Alberto Bugarin, Spain
Manuel P. Cuéllar, Spain
Irene Diaz-Valenzuela, Spain
Yun Guo, China

Marie-Jeanne Lesot, France
Andrea G.B. Tettamanzi, Italy
Daan Van Britsom, Belgium
Ronald R. Yager, USA
Man Zhu, China

Special Session Organizers

Babak Akhgar, UK
Carlos Barranco, Spain
Jesús Cardeñosa, Spain
Henning Christiansen, Denmark
Guy De Tré, Belgium
Hassane Essafi, France
Carolina Gallardo, Spain
Anne Gerdes, Denmark
Manuel Graña, Spain
Maria Grineva, USA
Luis Iraola, Spain
Maria J. Martin-Bautista, Spain
Henrik Legind Larsen, Denmark

Edwin Lughofer, Austria
Nicolas Marín, Spain
Juan M. Medina, Spain
Daniel Ortiz-Arroyo, Denmark
Daniel Sánchez, Spain
Bogdan Trawinski, Poland
Gracian Triviño, Spain
Yana Volkovich, Spain
Michal Wozniak, Poland
Ronald R. Yager, USA
Sławomir Zadrożny, Poland
Shuigeng Zhou, China

Sponsoring Institutions

- University of Granada
- Aalborg University
- High Technical School of Computer Science and Telecommunications of the University of Granada
- European Union's Seventh Framework Programme (FP7/2007–2013) under grant agreement n. 312651 (ePOOLICE project)

Table of Contents

Query-Answering Systems

Conceptual Pathway Querying of Natural Logic Knowledge Bases
from Text Bases 1
  *Troels Andreasen, Henrik Bulskov, Jørgen Fischer Nilsson,
  Per Anker Jensen, and Tine Lassen*

Query Rewriting for an Incremental Search in Heterogeneous Linked
Data Sources 13
  *Ana I. Torre-Bastida, Jesús Bermúdez, Arantza Illarramendi,
  Eduardo Mena, and Marta González*

FILT – Filtering Indexed Lucene Triples – A SPARQL Filter Query
Processing Engine– 25
  Magnus Stuhr and Csaba Veres

Improving Range Query Result Size Estimation Based on a New
Optimal Histogram 40
  Wissem Labbadi and Jalel Akaichi

Has FQAS Something to Say on Taking Good Care of Our Elders? 57
  María Ros, Miguel Molina-Solana, and Miguel Delgado

Question Answering System for Dialogues: A New Taxonomy
of Opinion Questions 67
  Amine Bayoudhi, Hatem Ghorbel, and Lamia Hadrich Belguith

R/quest: A Question Answering System 79
  Joan Morrissey and Ruoxuan Zhao

Answering Questions by Means of Causal Sentences 91
  C. Puente, E. Garrido, and J.A. Olivas

Ontology-Based Question Analysis Method 100
  Ghada Besbes, Hajer Baazaoui-Zghal, and Antonio Moreno

Fuzzy Multidimensional Modelling for Flexible Querying of Learning
Object Repositories 112
  Gloria Appelgren Lara, Miguel Delgado, and Nicolás Marín

Environmental Scanning for Strategic Early Warning

Special Session Organized by: Henrik Legind Larsen, Babak Akhgar,
Hassane Essafi, Daniel Ortiz-Arroyo, Maria J. Martin-Bautista,
Ronald R. Yager, Anne Gerdes

Using Formal Concept Analysis to Detect and Monitor Organised
Crime   124
   Simon Andrews, Babak Akhgar, Simeon Yates, Alex Stedmon, and
   Laurence Hirsch

Analysis of Semantic Networks Using Complex Networks Concepts   134
   Daniel Ortiz-Arroyo

Detecting Anomalous and Exceptional Behaviour on Credit Data
by Means of Association Rules   143
   Miguel Delgado, Maria J. Martin-Bautista, M. Dolores Ruiz, and
   Daniel Sánchez

Issues of Security and Informational Privacy in Relation
to an Environmental Scanning System for Fighting Organized Crime . . .   155
   Anne Gerdes, Henrik Legind Larsen, and Jacobo Rouces

Semantic Technology

Algorithmic Semantics for Processing Pronominal Verbal Phrases   164
   Roussanka Loukanova

Improving the Understandability of OLAP Queries by Semantic
Interpretations   176
   Carlos Molina, Belen Prados-Suárez, Miguel Prados de Reyes, and
   Carmen Peña Yañez

Semantic Interpretation of Intermediate Quantifiers and Their
Syllogisms   186
   Petra Murinová and Vilém Novák

Ranking Images Using Customized Fuzzy Dominant Color
Descriptors   198
   J.M. Soto-Hidalgo, J. Chamorro-Martínez,
   P. Martínez-Jiménez, and Daniel Sánchez

Generating Linguistic Descriptions of Data

Special Session Organized by: Daniel Sánchez, Nicolás Marín, and
Gracian Triviño

Linguistic Descriptions: Their Structure and Applications   209
   Vilém Novák, Martin Štěpnička, and Jiří Kupka

Landscapes Description Using Linguistic Summaries and
a Two-Dimensional Cellular Automaton   221
   Francisco P. Romero and Juan Moreno-García

Comparing f_β-Optimal with Distance Based Merge Functions   233
   Daan Van Britsom, Antoon Bronselaer, and Guy De Tré

Flexible Querying with Linguistic F-Cube Factory   245
   *R. Castillo-Ortega, Nicolás Marín, Daniel Sánchez, and
   Carlos Molina*

Mathematical Morphology Tools to Evaluate Periodic Linguistic
Summaries   257
   *Gilles Moyse, Marie-Jeanne Lesot, and
   Bernadette Bouchon-Meunier*

Automatic Generation of Textual Short-Term Weather Forecasts
on Real Prediction Data   269
   A. Ramos-Soto, A. Bugarin, S. Barro, and J. Taboada

Increasing the Granularity Degree in Linguistic Descriptions
of Quasi-periodic Phenomena   281
   Daniel Sanchez-Valdes and Gracian Trivino

A Model-Based Multilingual Natural Language Parser — Implementing
Chomsky's X-bar Theory in ModelCC   293
   Luis Quesada, Fernando Berzal, and Juan-Carlos Cubero

Patterns and Classification

Correlated Trends: A New Representation for Imperfect and Large
Dataseries   305
   Miguel Delgado, Waldo Fajardo, and Miguel Molina-Solana

Arc-Based Soft XML Pattern Matching   317
   *Mohammedsharaf Alzebdi, Panagiotis Chountas, and
   Krassimir Atanassov*

Discrimination of the Micro Electrode Recordings for STN Localization
during DBS Surgery in Parkinson's Patients   328
   Konrad Ciecierski, Zbigniew W. Raś, and Andrzej W. Przybyszewski

Image Classification Based on 2D Feature Motifs   340
   Angelo Furfaro, Maria Carmela Groccia, and Simona E. Rombo

Advances in Fuzzy Querying and Fuzzy Databases: Theory and Applications

Special Session Organized by: Guy De Tré, Slawomir Zadrozny, Juan Miguel Medina, and Carlos D. Barranco

Wildfire Susceptibility Maps Flexible Querying and Answering......... 352
    Paolo Arcaini, Gloria Bordogna, and Simone Sterlacchini

Enhancing Flexible Querying Using Criterion Trees 364
    Guy De Tré, Jozo Dujmović, Joachim Nielandt, and
    Antoon Bronselaer

A Possibilistic Logic Approach to Conditional Preference Queries 376
    Didier Dubois, Henri Prade, and Fayçal Touazi

Bipolar Conjunctive Query Evaluation for Ontology Based Database
Querying 389
    Nouredine Tamani, Ludovic Liétard, and Daniel Rocacher

Bipolar Querying of Valid-Time Intervals Subject to Uncertainty 401
    Christophe Billiet, José Enrique Pons, Olga Pons, and Guy De Tré

Declarative Fuzzy Linguistic Queries on Relational Databases 413
    Clemente Rubio-Manzano, Pascual Julián-Iranzo,
    Esteban Salazar-Santis, and Eduardo San Martín-Villarroel

Finding Similar Objects in Relational Databases — An
Association-Based Fuzzy Approach.................................. 425
    Olivier Pivert, Grégory Smits, and Hélène Jaudoin

M2LFGP: Mining Gradual Patterns over Fuzzy Multiple Levels 437
    Yogi S. Aryadinata, Arnaud Castelltort, Anne Laurent, and
    Michel Sala

Building a Fuzzy Valid Time Support Module on a Fuzzy
Object-Relational Database 447
    Carlos D. Barranco, Juan Miguel Medina, José Enrique Pons, and
    Olga Pons

Personalization and Recommender Systems

Predictors of Users' Willingness to Personalize Web Search 459
    Arjumand Younus, Colm O'Riordan, and Gabriella Pasi

Semantic-Based Recommendation of Nutrition Diets for the Elderly
from Agroalimentary Thesauri..................................... 471
    Vanesa Espín, María V. Hurtado, Manuel Noguera, and
    Kawtar Benghazi

Enhancing Recommender System with Linked Open Data   483
   Ladislav Peska and Peter Vojtas

Making Structured Data Searchable via Natural Language
Generation   495
   Jochen L. Leidner and Darya Kamkova

Searching and Ranking

On Top-k Retrieval for a Family of Non-monotonic Ranking
Functions   507
   Nicolás Madrid and Umberto Straccia

Using a Stack Decoder for Structured Search   519
   Kien Tjin-Kam-Jet, Dolf Trieschnigg, and Djoerd Hiemstra

On Cosine and Tanimoto Near Duplicates Search among Vectors
with Domains Consisting of Zero, a Positive Number and a Negative
Number   531
   Marzena Kryszkiewicz

L2RLab: Integrated Experimenter Environment for Learning to Rank ...   543
   *Óscar José Alejo, Juan M. Fernández-Luna, Juan F. Huete, and
   Eleazar Moreno-Cerrud*

Fusion and Ensemble Techniques for Online Learning on Data Streams

Special Session Organized by: Manuel Graña, Edwin Lughofer,
Bogdan Trawinski, and Michal Wozniak

Heuristic Classifier Chains for Multi-label Classification   555
   Tomasz Kajdanowicz and Przemyslaw Kazienko

Weighting Component Models by Predicting from Data Streams Using
Ensembles of Genetic Fuzzy Systems   567
   *Bogdan Trawiński, Tadeusz Lasota, Magdalena Smętek, and
   Grzegorz Trawiński*

Weighted Aging Classifier Ensemble for the Incremental Drifted Data
Streams   579
   Michał Woźniak, Andrzej Kasprzak, and Piotr Cal

An Analysis of Change Trends by Predicting from a Data Stream Using
Neural Networks   589
   *Zbigniew Telec, Tadeusz Lasota, Bogdan Trawiński, and
   Grzegorz Trawiński*

Web and Human-Computer Interaction

An Autocompletion Mechanism for Enriched Keyword Queries to RDF
Data Sources   601
  Grégory Smits, Olivier Pivert, Hélène Jaudoin, and François Paulus

Querying Sentiment Development over Time   613
  Troels Andreasen, Henning Christiansen, and Christian Theil Have

SuDoC: Semi-unsupervised Classification of Text Document Opinions
Using a Few Labeled Examples and Clustering   625
  František Dařena and Jan Žižka

Efficient Visualization of Folksonomies Based on «Intersectors»   637
  A. Mouakher, S. Heymann, S. Ben Yahia, and B. Le Grand

Intelligent Information Extraction from Texts
Special Session Organized by: Jesús Cardeñosa, Carolina Gallardo, and
Luis Iraola

Ukrainian WordNet: Creation and Filling   649
  Anatoly Anisimov, Oleksandr Marchenko, Andrey Nikonenko,
  Elena Porkhun, and Volodymyr Taranukha

Linguistic Patterns for Encyclopaedic Information Extraction   661
  Jesús Cardeñosa, Miguel Ángel de la Villa, and Carolina Gallardo

Contextualization and Personalization of Queries to Knowledge Bases
Using Spreading Activation   671
  Ana B. Pelegrina, Maria J. Martin-Bautista, and Pamela Faber

Utilizing Annotated Wikipedia Article Titles to Improve a Rule-Based
Named Entity Recognizer for Turkish   683
  Dilek Küçük

Author Index   693

Conceptual Pathway Querying of Natural Logic Knowledge Bases from Text Bases

Troels Andreasen[1], Henrik Bulskov[1], Jørgen Fischer Nilsson[2],
Per Anker Jensen[3], and Tine Lassen[3]

[1] CBIT, Roskilde University
{troels,bulskov}@ruc.dk
[2] IMM, Technical University of Denmark
jfn@imm.dtu.dk
[3] IBC, Copenhagen Business School
{paj.ibc,tla.ibc}@cbs.dk

Abstract. We describe a framework affording computation of conceptual pathways between a pair of terms presented as a query to a text database. In this framework, information is extracted from text sentences and becomes represented in natural logic, which is a form of logic coming much closer to natural language than predicate logic. Natural logic accommodates a variety of scientific parlance, ontologies and domain models. It also supports a semantic net or graph view of the knowledge base. This admits computation of relationships between concepts simultaneously through pathfinding in the knowledge base graph and deductive inference with the stored assertions. We envisage use of the developed pathway functionality, e.g., within bio-, pharma-, and medical sciences for calculating bio-pathways and causal chains.

1 Introduction

This paper addresses the problem of retrieving conceptual pathways in large text databases. A conceptual pathway is a sequence of propositions or terms semantically linking two given terms according to the principles described below. For instance, one may query a bio-science knowledge base about any connections between diabetes and infectious diseases in order to get sequences of propositions from the knowledge base connecting these two terms. Such a functionality would draw on a formal ontology and presumably additional formalized commonsense and domain-specific background knowledge.

In our approach, this pathfinding functionality is sought achieved by computationally extracting as much as possible of the textual meaning into a logical language. The applied target logic is a version of natural logic. The pathfinding computation is then pursued as a computational inference process. Accordingly, the system supporting the intended functionality comes in two main parts: A component for extracting information from text sentences using available lexical, linguistic and domain-specific resources and a component for computationally resolving pathway queries to the natural logic knowledge base resulting from information extraction from a text database.

H.L. Larsen et al. (Eds.): FQAS 2013, LNAI 8132, pp. 1–12, 2013.

The knowledge base is augmented by formal ontologies and other auxiliary resources also formulated in natural logic. This systems architecture is to favor robustness rather than completeness in the text analysis while still relying on a rigorous use of logic-based representations. We focus on text sources within scientific domains, in particular (molecular) bio-science and bio-technology, and pharma-medical areas. These areas as well as many other scientific and technological areas tend to conform to stereotypical forms of natural language in order to achieve precision and avoid ambiguities. Moreover, these areas rely on a rather fixed body of common background knowledge for the specialist reader.

This paper focusses on the applied natural logic and the conceptual pathfinding in a natural-logic knowledge base. An approach to extraction from the text database is addressed in [2].

The present paper is structured as follows: In section 2, we discuss general principles for deducing pathways in logical knowledge bases. In section 3, we describe the complementary graph view of logical knowledge bases. In section 4 and 5, we explain our use of natural logic in knowledge bases. In section 6, we explain how we re-shape natural logic in Datalog clauses, with a supporting inference engine as described in section 7. Finally, section 8 concludes.

2   Deductive Querying of Knowledge Bases

Querying of knowledge bases and databases is commonly handled as a deductive reasoning process. Given a knowledge base, KB, conceived of as a collection of logical propositions, query answering is conducted as computing of constructive solutions to an existential assertion

$$KB \vdash \exists x_1, x_2, ..., x_n \ p(x_1, x_2, ..., x_n)?$$

where the computable instantiations of the existential variables form the answer extension relation. It is well-known that computing of relational database queries with common declarative query languages is subsumed, at least to a large extent, by this deductive querying principle.

2.1   Deducing Relationships

Below, we present and develop an alternative, innovative query functionality for logical knowledge bases stated in natural logic. We consider knowledge bases KB comprising concept- or class terms entering into (binary) relationships forming propositions in natural logic. The prominent relationship is the concept inclusion (subsumption) relationship conventionally denoted by isa. As such, the KB supports formal ontologies. In addition, there are $ad\ hoc$ causal, partonomic, temporal, locative and other relationships according to needs.

The key functionality logically conforms with deduction of relevant relationships r between two or more given concept terms, say a and b:

$$KB \vdash \exists r \ r(a, b)?$$

Formally, with a variable ranging over relations, the query assertion is thus higher order.

Relationships are to be understood in a general logical sense transcending the ones which are explicitly present in the knowledge base propositions: As a simple example, consider a KB simply consisting of propositions $p(a, c)$ and $q(c, b)$. The relation r can be instantiated to $\lambda x, y(p(x, c) \wedge q(c, y))$ yielding a pathway from a to b via c. The property of transitivity (possessed, *e.g.*, by the inclusion relation) would obviously play a crucial rôle for such inferences.

From a logical point of view, recognition of the existence of relationships may be due to appeal to appropriate, restricted comprehension principles. Such principles would appear as special cases of the general logical comprehension principle for binary relations:

$$\exists r \forall x, y \ (r(x, y) \leftrightarrow \Phi[x, y])$$

where $\Phi[x, y]$ is, at the outset, any formula within the applied logic having the sole free variables x and y. Thus, the general principle says that any binary relation expressible within the logic is recognized as existing logically. We wish to abstain from applying higher order logic (type theory) with quantification over relations. This is accomplished by adopting a metalogic framework as to be described below in section 6. The above general comprehension principle clearly provides a vast supply of relations being irrelevant from a query answer point of view. Therefore, the applied comprehension should be constrained so as to ensure, ideally, only relevant query answers.

Heuristically, one may favor transitivity and reversal of relationships, supported by cost priority policies. This leads next to the alternative "semantic net" conception of the logical knowledge base as a labeled graph.

3   The Graph Knowledge Base View

We assume so far that the KB consists of simple propositions comprising a relator connecting two relata in the form of concepts or classes as for example: *betacell produce insulin*. Below we augment with more complex propositions and describe these from the point of view of natural logic.

As it appears, the KB may be conceived of as a finite, directed labeled graph with concept terms as nodes and with directed arcs labeled with relationships. Linguistically, the simple propositions take the form *CN Vt CN*, where *CN* denotes common nouns and *Vt* a transitive verb.

3.1   Deducing Pathways

In the graph ("semantic network") view of KB, the deduction of relationships between nodes conforms with calculating paths between the nodes. As in route finding in maps, priority might be given to shortest paths.

In calculating a heuristic path length measure, one may rely on certain transitivity properties, notably the transitivity of *isa* and certain other relations such as causal and partonomic ones. Inverse relationships may also enter into paths. Moreover, for natural logic the so-called monotonicity rules apply. According to these rules, the subject term in a proposition may be specialized whereas the object term may be generalized.

Given in the *KB*, for instance,

> *betacell produce insulin* and
> *betacell isa cell* and
> *insulin isa hormone* and
> *hormone isa protein*

as depicted in figure 1, for queries concerning relationships, say, between *betacell* and *protein* one may deduce the path via *produce insulin, which is a hormone, which is a protein*. By monotonicity it follows from this example that betacells produce protein. Moreover, given the terms *cell* and *insulin*, one may return the path via *betacell*. Notice that in the latter example the *isa* relationship is used inversely.

4   Natural Logic

Logical knowledge bases usually apply description logic or logical rule clauses (e.g. DATALOG). Description logic forces all sentences effectively into copular forms as discussed in [7]. Description logic specifications therefore tend to be far from natural language formulations containing e.g. transitive verbs. Similarly, logical rule clauses are far removed from natural language formulations due to their use of quantified variables.

Natural logic is a generic term used for forms of symbolic logic coming closer to natural language forms than modern predicate logic. The natural logic tradition has roots back to Aristotelian logic, and is currently pursued by research groups e.g. at the University of Amsterdam [3,6,4] and at Stanford University [11] spurred by developments in computational logic and envisioned applications in IT.

Natural logic stresses the use of reasoning rules which are more natural and intuitive than the ones applied in mathematical logic at the expense of achieving complete coverage as sought in mathematical logic. The present project adopts a form of natural logic comprising propositions such as the ones in the left-hand column of the table below, where the right-hand column shows corresponding user-readable forms:

Fig. 1. A *KB* including a pathway connecting *betacell* and *protein*

Internal natural logic representation	User-readable form
cell (produce a) stimulate b	*cell that produce a stimulate b*
cell (partof y) stimulate (u partof v)	*cell in y stimulate u in v*
cell (produce x (contain z)) stimulate y	*cell that produce x*
	that contain z stimulate y

The relative pronoun *"that"* is added for readability; it has no bearings on the semantics. Parentheses are often added for disambiguation, e.g. *cell (that produce x (that contain z)) stimulate y.*

From the point of view of traditional logic, natural logic is a subset of predicate logic dealing with monadic and binary predicates. As an example the proposition *cell (that produce a) stimulate b* corresponds to the following expression in predicate logic:

$$\forall x (cell(x) \rightarrow \exists y(produce(x,y) \wedge a(y)) \rightarrow \exists z(b(z) \wedge stimulate(x,z)))$$

However, the natural logic propositions are not translated into predicate logic in the system, since the computational reasoning takes place at the natural logic level as to be explained below.

4.1 Non-monotonic Logic in Natural Logic

Unlike traditional natural logic and description logic, we are going to appeal to the closed world assumption (CWA) for handling negation in certain cases. Notably, by default, all classes (coming from nouns and more generally noun phrases) in the ontology are conceived of as disjoint unless otherwise stated, in accordance with common conventions for ontologies and in contrast to description logic.

Consider the following example: Suppose that the knowledge base has *betacell produce insulin*. Suppose further that alpha cells and beta cells are assumed to be disjoint by the default convention in absence of any overlapping class of cells in the knowledge base. Then the truth value of the proposition *alphacell produce insulin* cannot be determined so far in our suggested approach to negative information with the constrained appeal to negation as nonprovability. Thus, the answer is "don't know" or "open", rather than a plain "no" from the CWA as customary in database querying. This limitation is motivated by the expected partiality of the logical knowledge bases. However, if the knowledge base also contains *cell (that produce insulin) isa betacell* (extracted, say, as contribution from a natural language sentence *only betacells produce insulin*), then the above question is to be answered in the negative.

Thus, there is no notion of empty or null class in the applied logic. As stated, it is assumed that classes like alphacells and betacells are disjoint as long as there is no common subclass given explicitly in the ontology (and no inclusion between the said classes). More generally, the dismissal of classical sentential negation may be replaced by limited forms of negation as in *betacell (unless abnorm-betacell) produce insulin* – stating a non-monotonic exception from the normative case without logical inconsistencies. This approach to denials is known from deductive databases. However, it seems to be innovative in the combination with natural logic proposed here. This departure from traditional natural logic (as well as description logic) has positive bearings on the computational complexity of the deduction process: The computational complexity in our framework does not exceed that of a deductive query process in DATALOG, which is bound to be of low polynomial complexity, cf. [5].

5   From Natural Language to Natural Logic

Complete computerized translation of text into a natural logic would be an over-ambitious goal far beyond current theories. Therefore, as part of our framework we develop an extractor module which builds on more modest but principled approaches for computational conversion of the main textual content into the applied natural logic. The extractor establishes a correlation between the text and admissible natural logic forms, natural logic propositions, in order to capture as much propositional information as possible from the text. The conversion into natural logic propositions progresses heuristically by applying a variety of linguistic transformations and reductions of the text.

By way of example, coordinated constructions are decomposed into un-coordinated propositions. The translation of restrictive relative clauses is tentatively dealt with as a subtype of anaphor resolution in that the relative pronoun is substituted by its antecedent, which would also typically be the strategy for resolving anaphoric reference. Basically, the extractor is to process one sentence at a time and convert the meaning content into a number of propositions in natural logic. This implies that inter-sentential anaphora might be left unresolved. Even intra-sentential anaphora might also be given up as unresolvable, particularly if they violate the logical limitations of the natural logic, cf. the infamous "donkey sentences" (for instance with a back referring "it"). On the other hand, common anaphor patterns (with pronouns) may well be accommodated by dedicated natural logic schemes tailored to anaphor schemes.

Passive constructions are transformed into their active counterparts, from which a predicate-argument structure is regularly deducible. Predicative adjective constructions, which are full propositions, may be transformed into concepts by placing the adjective as an attributive modifier of the subject. A number of inflectional, lexical, and phrasal elements are deleted from the natural language texts. These elements include inflectional affixes, determiners, adverbs, non-restrictive relative clauses, modals and auxiliary verbs, and non-clausal adverbials.

The extractor further has to handle complex sentential and phrasal forms such as topicalization, apposition, and nominalization, to name a few.

5.1   Example

Consider the following miniature corpus on the role of the enzyme amylase and the influence of pancreas disease.[1]

> Amylase is an enzyme that helps digest carbohydrates. It is produced in the pancreas and the glands that make saliva. When the pancreas is diseased or inflamed, amylase releases into the blood.

> Pancreatitis usually develops as a result of gallstones or moderate to heavy alcohol consumption over a period of years.

From the 4 sentences above the following natural logic propositions can be extracted:

- amylase isa enzyme
- amylase cause (digestion of carbohydrate)
- pancreas produce amylase
- (gland that produce saliva) produce amylase
- (diseased pancreas) cause (amylase release into blood)
- (inflamed pancreas) cause (amylase release into blood)
- gallstone cause pancreatitis
- (alcohol consumption) cause pancreatitis

The target for querying is the content of the corpus or more specifically the knowledge that can be drawn from this – that is, propositions like the ones listed above. However, rather than restricting the system to what can be extracted directly from the corpus, our approach is to embed extracted propositions into a background domain ontology and to target the enriched ontology for query evaluation. Figure 2 shows an example: A subontology for an enriched background ontology corresponding to ontological context for the extracted propositions given above. The background ontology used is UMLS [12].

Various types of queries can be developed for targeting an enriched ontology – one obviously involves starting with a free form textual query and to analyze this in the same manner as the corpus. However, to go directly to the core of conceptual pathway queries, we will here only consider queries specified by two concepts and query answers given as possible ways to relate these concepts in the enriched ontology. A sample two-concept query that can be evaluated within the subontology shown in figure 2 is the following.

Query: gallstone, blood
Answer:
- gallstone cause (pancreas with pancreatitis) isa (diseased pancreas) cause (amylase release into blood) release-into blood

As is evident in figure 2, a single pathway connecting the two query concepts can be identified, and thus an answer to the query can be this pathway. The subgraph corresponding to the answer is shown in figure 3.

---

[1] The two paragraphs are taken from National Library of Medicine's *MedlinePlus* [13] and Wolters Kluwer Health's physician-authored clinical decision support resource *UpToDate* [14] respectively.

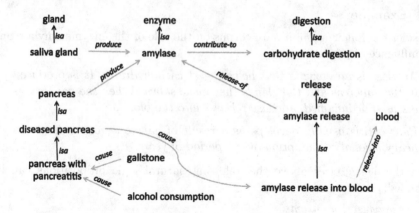

Fig. 2. Ontological context for given miniature example corpus

Obviously, for many queries, multiple connecting pathways may contribute to the answer, as in the following example.

Query: *pancreas, digestion*
Answer:
- *pancreas produce amylase contribute-to (digestion of carbohydrate) isa digestion*
- *pancreas has-specialization (diseased pancreas) cause (amylase release into blood) isa (amylase release) release-of amylase contribute-to (digestion of carbohydrate) isa digestion*

The answer is also shown as a subgraph in figure 4. Multiple connecting pathways can be dealt with in different ways. A simple approach is to eliminate all but the shortest path, which in this case would provide only the first pathway including the edge from pancreas to amylase. Pathways can also be ranked and listed in the according order in the result. Different principles for measuring rank weight can be applied. Obviously a simple path length principle can be applied, but additional properties such as relation importance (where *isa* properly would be among the most important) and node weight (discriminating concepts that are central to an explanation of a pathway connection from concepts that are not) can be taken into account. In addition, in case of multiple connections, the choice

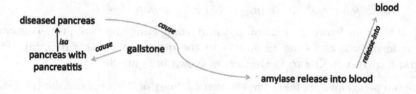

Fig. 3. A pathway answer to the query *gallstone, blood*

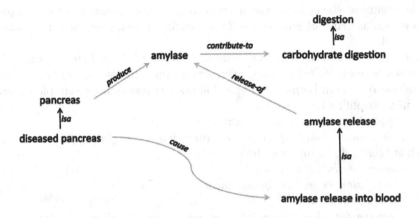

Fig. 4. A multiple pathway answer to the query *pancreas, digestion*

of pathways to present can also be subject to query framework parameterization or to query specification through special constructs (such as SHORTEST, ALL etc.).

6 Flattening Natural Logic into Clauses

The natural logic forms applied in the knowledge base are embedded in a deductive database language (DATALOG with extensions). This means that we adopt a two-level or meta-logical form of knowledge base logic in which assertions like the above are embedded in an outer logical layer taking care of the computational deduction (computation of conceptual pathways). For example, the sentence *"betacell produce insulin"* is logically conceived of as the natural logic proposition *every betacell produce some insulin* as an instance of the full natural logic form $Q_1 \ A \ R \ Q_2 \ B$, where Q_1 and Q_2 are quantifiers (e.g. *every, some*), cf. [10]. With the main case Q_1 being *every* and Q_2 being *some*, we get $\forall x(A(x) \rightarrow \exists y(R(x,y) \wedge B(y)))$ yielding at the meta level *everysome(A, R, B)* with the distinguished quantifier predicate *everysome*[2]. Accordingly, the natural logic assertion *betacells produce insulin* at our meta level re-appears as the atomic formula *everysome(betacell, produce, insulin)* representing a quantifier-annotated triple, as appearing in the graphic visualizations.

The latter can be reasoned within a deductive database language with variables ranging over classes such as betacell and binary relations such as produce, cf [8]. Accordingly, the inference rules are to be shaped as DATALOG clauses. The clauses may in turn be re-expressed in a common database query language with the assertions stored in database relations. Superficially, the applied triple form resembles RDF representations, but is here logically supported by the deductive inference engine.

---

[2] Correspondingly, we have *everyevery, someevery,* and *somesome*.

The recursive structure of the natural logic calls for generation of complex classes such as *cell-that-produce-insulin* appearing as constants at the database tuple level.

Consider the following sentence in natural language[3], "*The hypothalamus also secretes a hormone called somatostatin, which causes the pituitary gland to stop the release of growth hormone*". Below follows a translation into simple natural logic in a "triplification" process:

> *hypothalamus secrete somatostatin (a hormone)*
> *(that cause stopping (of release (of growth hormone) (from pituitary gland))*

which is broken down into the three assertions

> *somatostatin isa hormone*
> *hypothalamus secrete somatostatin*
> *somatostatin cause*
> *stopping (of release (of growth hormone) (from pituitary gland)) ...*

A further decomposition is carried out in order to arrive at the simple natural logic propositions for the KB.

7   Inference Engine for Conceptual Pathfinding

> "*You see, my dear Watson*"– *[...]* –"*it is not really difficult to construct a series of inferences, each dependent upon its predecessor and each simple in itself. If, after doing so, one simply knocks out all the central inferences and presents one's audience with the starting-point and the conclusion, one may produce a startling, though possibly a meretricious, effect.*"
>
> – Sherlock Holmes *(The Adventure of the Dancing Men)*

The meta-level logic comprises clauses for performing logical deduction on the natural logic KB contents by means of reflexivity and transitivity of *isa* and the monotonicity properties of natural logic, cf. [8,9].

The inference engine clauses in a nutshell:

$$isa^+(C, D) \leftarrow isa(C, X)$$
$$isa^+(C, D) \leftarrow isa^+(C, X), isa(X, D)$$
$$isa^*(C, C)$$
$$isa^*(C, D) \leftarrow isa^+(C, D)$$
$$path(C, [isa], D) \leftarrow isa^+(C, D)$$
$$path(C', [R], D) \leftarrow isa^*(C', C) \wedge everysome(C, R, D)$$
$$path(C, [R], D') \leftarrow everysome(C, R, D) \wedge isa^*(D, D')$$
$$path(C, [R1, R2], D) \leftarrow path(C, R1, X) \wedge path(X, R2, D)$$

The mid trace argument transcends DATALOG; the argument here serves merely to explain the principle for accumulating path information. Ascription of path costs is not shown. In the actual implementation, path computation is to be conducted by an appropriate graph search algorithm guided by heuristic costs for ranking of paths.

---

[3] From: *emedicinehealth* [15].

In addition, there are clauses expressing properties of applied relations such as causal and partonomic transitivity facilitating short cuts with respect to cost. There may also be appropriate clauses for reverse paths (with accompanying costs) for the various relations including *isa*. Thus, the explicit knowledge from the text base is represented at the natural logic level, whereas common sense knowledge as well as pure logical reasoning capabilities are formalized as clauses at the metalogic level.

As a distinguished feature of our approach, the logical reasoning is supported by the meta-level logic and is to be implemented as an "inference engine" taking advantage of the efficient storage and access provided by current database technology. This is in order to obtain the proof-of-concept of a technology which is capable of computationally processing large amounts of information. In other words, the inference machine eventually is to be realized on top of a database platform. This motivates decomposing natural logic assertions into atomic propositions admitting storage in database relations. The pathway-finding inference process may then be controlled by efficient search algorithms.

Unlike traditional natural logic, the presently applied version of natural logic omits denials with classical logic in favor of limited use of CWA and non-monotonic logic as common in databases and logic programming and formal ontologies. This departure from traditional natural logic (as well as description logic) has positive bearings on the computational complexity of the deduction process: The computational complexity in our framework does not exceed that of a deductive query process in DATALOG, which is bound to be of low polynomial complexity.

8 Summary and Conclusion

We have presented a framework for deductive conceptual pathfinding in knowledge bases derived from text databases. Our approach entertains a dual view on the knowledge bases: Firstly, as a collection of assertions in natural logic supported by appropriate inference rules stated at the metalogic level as DATALOG clauses. Secondly, as a labeled directed graph where the nodes are concepts. The graph representation facilitates the pathfinding view exploited in our framework.

References

1. Andreasen, T., Bulskov, H., Zambach, S., Lassen, T., Madsen, B.N., Jensen, P.A., Thomsen, H.E., Fischer Nilsson, J.: A semantics-based approach to retrieving biomedical information. In: Christiansen, H., De Tré, G., Yazici, A., Zadrozny, S., Andreasen, T., Larsen, H.L. (eds.) FQAS 2011. LNCS, vol. 7022, pp. 108–118. Springer, Heidelberg (2011)
2. Andreasen, T., Bulskov, H., Jensen, P.A., Lassen, T.: Extracting Conceptual Feature Structures from Text. In: Kryszkiewicz, M., Rybinski, H., Skowron, A., Raś, Z.W. (eds.) ISMIS 2011. LNCS, vol. 6804, pp. 396–406. Springer, Heidelberg (2011)
3. van Benthem, J.: Essays in Logical Semantics, Studies in Linguistics and Philosophy, vol. 29. D. Reidel Publishing Company (1986)

4. van Benthem, J.: Natural Logic, Past And Future, Workshop on Natural Logic, Proof Theory, and Computational Semantics 2011, CSLI Stanford (2011), http://www.stanford.edu/~icard/logic&language/index.html

5. Benjamin, N.G., Horrocks, I., Volz, R., Decker, S.: Description logic programs: combining logic programs with description logic. In: Proceedings of the 12th International Conference on World Wide Web (WWW 2003), pp. 48–57. ACM, New York (2003)

6. Muskens, R.: Towards Logics that Model Natural Reasoning, Program Descrip-tion Research program in Natural Logic (2011), http://lyrawww.uvt.nl/~rmuskens/natural/

7. Fischer Nilsson, J.: On Reducing Relationships to Property Ascriptions, Information Modelling and Knowledge Bases XX. In: Kiyoki, Y., Tokuda, T., Jaakkola, H., Chen, X., Yoshida, N. (eds.). Frontiers in Artificial Intelligence and Applications, vol. 190 (2009) ISBN: 978-1-58603-957-8

8. Fischer Nilsson, J.: Querying class-relationship logic in a metalogic framework. In: Christiansen, H., De Tré, G., Yazici, A., Zadrozny, S., Andreasen, T., Larsen, H.L. (eds.) FQAS 2011. LNCS, vol. 7022, pp. 96–107. Springer, Heidelberg (2011)

9. Fischer Nilsson, J.: Diagrammatic Reasoning with Classes and Relationships. In: Moktefi, A., Shin, S.-J. (eds.) Visual Reasoning with Diagrams. Studies in Universal Logic. Birkhäuser, Springer (March 2013)

10. Sanchez Valencia, V.: The Algebra of Logic. In: Gabbay, D.M., Woods, J. (eds.) Handbook of the History of Logic. The Rise of Modern Logic: From Leibniz to Frege, vol. 3. Elsevier (2004)

11. MacCartney, B., Manning, C.: An Extended Model of Natural Logic. In: Bunt, H., Petukhova, V., Wubben, S. (eds.) Proceedings of the 8th IWCS, Tilburg, pp. 140–156 (2009)

12. Bodenreider, O.: The Unified Medical Language System (UMLS): integrating biomedical terminology. Nucleic Acids Research 32, D267–D270 (2004)

13. Amylase, National Library of Medicine's MedlinePlus (2013), http://www.nlm.nih.gov/medlineplus/ency/article/003464.htm (retrieved March 1, 2013)

14. Pancreatitis, Wolters Kluwer Health's physician-authored clinical decision support resource UpToDate (2013), http://www.uptodate.com/contents/acute-pancreatitis-beyond-the-basics (retrieved March 1, 2013)

15. Hypothalamus, emedicinehealth (2013), http://www.emedicinehealth.com/anatomy_of_the_endocrine_system/page2_em.htm#hypothalamus (retrieved March 1, 2013)

Query Rewriting for an Incremental Search in Heterogeneous Linked Data Sources

Ana I. Torre-Bastida[1], Jesús Bermúdez[2], Arantza Illarramendi[2], Eduardo Mena[3], and Marta González[1]

[1] Tecnalia Research & Innovation
{isabel.torre,marta.gonzalez}@tecnalia.com
[2] Departamento de Lenguajes y Sistemas Informáticos, UPV-EHU
{a.illarramendi,jesus.bermudez}@ehu.es
[3] Departamento de Informática e Ingeniería de Sistemas, Univ. Zaragoza
emena@unizar.es

Abstract. Nowadays, the number of linked data sources available on the Web is considerable. In this scenario, users are interested in frameworks that help them to query those heterogeneous data sources in a friendly way, so avoiding awareness of the technical details related to the heterogeneity and variety of data sources. With this aim, we present a system that implements an innovative query approach that obtains results to user queries in an incremental way. It sequentially accesses different datasets, expressed with possibly different vocabularies. Our approach enriches previous answers each time a different dataset is accessed. Mapping axioms between datasets are used for rewriting the original query and so obtaining new queries expressed with terms in the vocabularies of the target dataset. These rewritten queries may be semantically equivalent or they could result in a certain semantic loss; in this case, an estimation of the loss of information incurred is presented.

Keywords: Semantic Web, Linked Open Data Sources, SPARQL query, vocabulary mapping, query rewriting.

1 Introduction

In recent years an increasing number of RDF open data sources are emerging, partly due to the existence of techniques to convert non RDF datasources into RDF ones, supported by initiatives like the Linking Open Data (LOD)[1] with the aim of creating a "Web of Data". The Linked Open Data cloud diagram[2] shows datasets that have been published in Linked Data Format (around 338 datasets by 2013[3]), and this diagram is continuosly growing. Moreover, although those datasets follow the same representation format, they can deal with heterogeneous

---

[1] http://www.w3.org/wiki/SweoIG/TaskForces/
CommunityProjects/LinkingOpenData

[2] http://lod-cloud.net/state/

[3] http://datahub.io/lv/group/lodcloud?tags% 3Dno-vocab-mappings

H.L. Larsen et al. (Eds.): FQAS 2013, LNAI 8132, pp. 13–24, 2013.

vocabularies to name the resources. In that scenario, users find difficulties in taking advantage of the contents of many of those datasets because they get lost with the quantity and the variety of them. For example, a user that is only familiar with BNE (Biblioteca Nacional de España)[4] dataset vocabularies could be interested in accessing BNB (Bristh National Bibliography)[5], DBpedia[6], or VEROIA (Public Library of Veroia)[7] datasets in order to find more information. However, not being familiar with the vocabularies of those datasets may dissuade him from trying to query them.

So, taking into account the significant volume of Linked Data being published on the Web, numerous research efforts have been oriented to find new ways to exploit this Web of Data. Those efforts can be broadly classified into three main categories: Linked Data browsers, Linked Data search engines, and domain-specific Linked Data appplications [1]. The proposal presented in this paper can be considered under the category of Linked Data search engines and more particularly, under human-oriented Linked Data search engines, where we can find other approaches such as Falcons[8] and SWSE[9], amongst others. Nevertheless, the main difference of our proposal with respect to existing engines lies in the fact that it provides the possibility of obtaining a broader response to a query formulated by a user by combining the following two aspects: (1) an automatic navigation through different datasets, one by one, using mappings defined among datasets; and (2) a controlled rewriting (generalization/specialization) of the query formulated by the user according to the vocabularies managed by the target dataset.

In summary, the novel contribution of this paper is the development of a system that provides the following main advantages:

- *A greater number of datasets at the users disposal.* Using our system the user can gain access to more datasets without bothering to know the existence of those datasets or the heterogeneous vocabularies that they manage. The system manages the navigation into different datasets.
- *Incremental answer enrichment.* By accessing different datasets the user can obtain more information of interest. For that, the system manages existing mapping axioms between datasets.
- *Exact or approximate answers.* If the system is not capable of obtaining a semantically equivalent rewriting for the query formulated by the user it will try to obtain a related query by generalizing/specializing that query and it will provide information about the loss in precision and/or recall with respect to the original query.

In the rest of the paper we present first some related works in section 2. Then, we introduce an overview of the query processing approach in section 3.

---

[4] BNE - (http://datos.bne.es/sparql)

[5] BNB - (http://bnb.data.bl.uk/sparql)

[6] DBpedia - (http://wiki.dbpedia.org/Datasets)

[7] VEROIA - (http://libver.math.auth.gr/sparql)

[8] http://ws.nju.edu.cn/falcons/objectsearch/index.jsp

[9] http://swse.deri.org/

We follow with a detailed explanation of the query rewriting algorithm and with a brief presentation of how the information loss is measured in sections 4 and 5. Finally we end with some conclusions in section 6.

2   Related Works

According to the growth of the Semantic Web, SPARQL query processing over heterogeneous data sources is an active research field. Some systems (such as DARQ [10], FedX [12]) consider federated approaches over distributed data sources with the ultimate goal of virtual integration. One main difference with our proposal is that they focus on top-down strategies where the relevant sources are known while in our proposal the sources are discovered during the query processing. Nevertheless one main drawback for query processing over heterogeneous data is that existing mapping axioms are scarce and very simples (most of them are of the owl:sameAs type)

Even closer to our approach are the works related to SPARQL query rewriting. Some of them, such as [7] and [2], support the query rewriting with mapping axioms described by logic rules that are applied to the triple patterns that compose the query; [7] uses a quite expressive specific mapping language based on Description Logics and [2] uses less expressive Horn clause-like rules. In both cases, the mapping language is much more expressive than what is usually found in datasets metadata (for instance, VoID[10] linksets) and the approach does not seem to scale up well due to the hard work needed to define that kind of mapping.

Another approach to query rewriting is query relaxation, which consists of reformulating the triple patterns of a query to retrieve more results without excessive loss in precision. Examples of that approach are [6] and [3]. Each work presents a different methodology for defining some types of relaxation: [6] uses vocabulary inference on triple patterns and [3] uses a statistical language modeling technique that allows them to compute the similarity between two entities. Both of them define a ranking model for the presentation of the query results. Although our proposal shares with them the goal of providing more results to the user, they are focused more on generalizing the query while we are focused on rewriting the query trying to preserve the meaning of the original query as much as possible, and so generalizing or specializing parts of the query when necessary in the new context.

The query rewriting problem is also considered [5], but we differ in the way to face it. In our proposal we cope with existing mapping axioms, that relate different vocabularies, and we make the most of them in the query rewriting process. In contrast, [5] disregards such mapping axioms and looks for results in the target dataset by evaluating similarity with an Entity Relevance Model (ERM) calculated with the results of the original query. The calculation of the ERM is based on the number of word occurrences in the results obtained, which are later used as keywords for evaluating similarity. The strength of this method turns into its weakness in some scenarios because there are datasets that make

---

[10] http://vocab.deri.ie/void

abundant use of codes to identify their entities and those strings do not help as keywords.

Finally, query rewriting has also been extensively considered in the area of ontology matching [4]. A distinguising aspect of our system is the measurement of information loss. In order to compute it we adapt the approach presented in [11] and further elaborated in [9] to estimate the information loss when a term is substituted by an expression.

3   An Overview of the Query Processing Approach

In this section we present first some terminology that will be used throughout the rest of the paper. Then we show the main steps followed by the system to provide an answer to one query formulated by the user. Finally, we present a brief specification of the main components of the system that are involved in the process of providing the answer.

With respect to the terminology used, we consider datasets that are modeled as RDF graphs. An RDF *graph* is a set of RDF triples. An RDF *triple* is a statement formed by a *subject*, a *predicate* and an *object* [8]. Elements in a triple are represented by IRIs and objects may also be represented by literals. We use *term* for any element in a triple. Each dataset is described with terms from a declared vocabulary set. Let us use *target* dataset for the dataset over which the query is going to be evaluated, and we use *target* vocabulary set for its declared vocabulary set.

SPARQL queries[11] are made of graph patterns. A *graph pattern* is a query expression made of a set of triple patterns. A *triple pattern* is a triple where any of its elements may be a variable. When a triple pattern of a query is expressed with terms of the target vocabulary set we say that the triple pattern is *adequate* for the target dataset. When every triple pattern of a query is *adequate* for a target dataset, we say that the query is *adequate* for that target dataset.

The *original* query is expressed with terms from a *source* vocabulary set. Let us call $\mathcal{T}$ the set of terms used by the original query. As long as any term in $\mathcal{T}$ belongs to a vocabulary in the target vocabulary set, the original query is adequate for the target dataset and the query can be properly processed over that dataset. However, if there were terms in $\mathcal{T}$ not appearing in the target vocabulary set, triple patterns of the original query including any such term should be rewritten into appropriate graph patterns, with terms taken from the target vocabulary set, in order to become an adequate query for the target dataset.

Terms in $\mathcal{T}$ appearing in synonymy mapping axioms (i.e. expressed with any of the properties owl:sameAs, owl:equivalentClass, owl:equivalentProperty) with a term in the target vocabulary set can be directly replaced by the synonym term. Those terms in $\mathcal{T}$ not appearing in the target vocabulary set and not appearing in synonymy mapping axioms with terms in the target vocabulary set are called *conflicting* terms. Since there is no guarantee for enough

---

[11] http://www.w3.org/TR/rdf-sparql-query/

synonym mapping axioms between source and target vocabulary sets that allow a semantic preserving rewriting of the original query into an adequate query for the target vocabulary, we must cope with query rewritings with some loss of information. The goal of the query rewriting algorithm is to replace every triple pattern including conflicting terms with a graph pattern adequate for the target dataset.

3.1 Main Query Processing Steps

The query which we will use as a running example is "Give me resources whose author is Tim Berners-Lee". The steps followed to answer that query are presented next:

1. The user formulates the query dealing with a provided GUI. For that, he uses terms that belong to a vocabulary that he is familiar with (for example, DBLP and FOAF vocabularies in this case). Notice that it is not required that the user knows the SPARQL language for RDF, he should only know the terms dblp:Tim_Berners-Lee, and foaf:maker from the DBLP and FOAF vocabularies. The system produces the following query:

```
PREFIX foaf: <http://xmlns.com/foaf/0.1/>
PREFIX dblp: <http://dblp.13s.de/d2r/resource/authors/>
{?resource foaf:maker dblp:Tim_Berners-Lee>}
```

2. The system asks the user for a name of a dataset in which he is interested in finding the answer. If the user does not provide any specific name, then the system shows the user different possible datasets that belong to the same domain (e.g., bibliographic domain). If the user does not select any of them then the system selects one. Following the previous example, we assume that the user selects DBpedia dataset among those presented by the system.

3. The system first tries to find the query terms in the selected dataset. If it finds them, it runs the query processing. Otherwise the system tries to rewrite the query formulated by the user into another equivalent query using mapping axioms. At this point two different situations may happen:

   (a) The system finds synonymy mapping axioms, defined between the source and target vocabularies, that allows it to rewrite each term of the query into an equivalent term in the target vocabulary (for instance, mapping axioms of the type dblp:Tim_Berners-Lee owl:sameAs dbpedia: Tim_Berners-Lee). Following the previous example, the property foaf:maker is replaced with dbpedia-owl:author. The rewritten query is the following:

   ```
   PREFIX dbpedia: <http://dbpedia.org/resource/>
   PREFIX dbpedia-owl: <http://dbpedia.org/ontology/>
   {?resource dbpedia-owl:author dbpedia:Tim_Berners-Lee>}
   ```

   Then the system obtains the answer querying the DBpedia dataset and shows the answer to the user through the GUI. The results obtained by the considered query are:

```
http://dbpedia.org/resource/Tabulator
http://dbpedia.org/resource/Weaving_the_Web:_The_Original_Design_
and_Ultimate_Destiny_of_the_World_Wide_Web_by_its_inventor
```

(b) The system does not find synonymy mapping axioms for every term in the original query. In this case, the triple including the conflicting term is replaced with a graph pattern until an adequate query is obtained. In sections 4 and 5 we present the algorithm used for the rewriting and an example that illustrates the behaviour, respectively.

4. The system asks the user if he is interested in querying another dataset. If the answer is *No* the process ends. If the answer is *Yes* the process returns to step 2.

3.2 System Modules

In order to accomplish the steps presented in the previous subsection the system handles the following modules:

- *Input/Output Module.* This module manages a GUI that facilitates, on the one hand, the task of querying the datasets using some predefined forms; and, on the other hand, presents the obtained answer with a friendly appearance.
- *Rewriting Module.* This module is in charge of two main tasks: *Query analysis* and *Query rewriting*. The *Query analysis* consists of parsing the query formulated by the user and obtaining a tree model. For this task, the *Query Analyzer* module implemented with ARQ[12] is used. In this task the datasets that belong to the domain considered in the query are also selected. Concerning *Query rewriting*, we have developed an algorithm (explained in section 4) that rewrites the query expressed using a source vocabulary into an adequate query. The algorithm makes use of mapping axioms expressed as RDF triples and which can be obtained through SPARQL endpoints or RDF dumps.

  The mapping axioms we are considering in this paper are those triples whose subject and object are from different vocabularies and the predicate is one of the following terms: `owl:sameAs`, `rdfs:subClassOf`, `rdfs:subPropertyOf`, `owl:equivalentClass`, and `owl:equivalentProperty`. Future work will consider a broader set of properties for the mapping axioms.
- *Evaluation Module.* Taking into account that different rewritings could be possible for a query, the goal of this module is to evaluate those different rewritings and to select the one that incurs the least information loss. For that it handles some defined metrics (see section 5.1) and the information stored in the VoID statistics of the considered datasets.
- *Processing Module.* Once the best query rewriting is selected, this module is in charge of obtaining the answer for the query by accessing the corresponding dataset.

---

[12] Apache Jena/ARQ (`http://jena.apache.org/documentation/query/index.html`)

4 Query Rewriting Algorithm

In this section we present the query rewriting algorithm. Its foundation is a graph traversing algorithm looking for the nearest terms (that belong to the target dataset) of a conflicting term.

We follow two guiding principles for the replacement of conflicting terms: (1) a term can be replaced with the conjunction of its directly subsuming terms. (2) a term can be replaced with the disjunction of its directly subsumee terms. These guiding principles are recursively followed until adequate expressions are accomplished.

A distinguishing feature of our working scenario is that source and target vocabulary sets are not necessarily fully integrated. Notice that datasets are totally independent from one another and our system is only allowed to access them by their particular web services (SPARQL endpoint or programmatic interface). Therefore, our system depends only on the declared vocabulary sets and the published mapping axioms. Infered relationships between terms are not taken into account unless the target system provides them. We are aware of the limitations of that consideration, but we think that it is quite a realistic scenario nowadays.

In the following, we present the algorithm that obtains an adequate query expression for the target dataset with the minimum loss of information with respect to the original query Q measured by our proposed metrics.

First of all, the original query Q is decomposed into triple patterns which in turn are decomposed into the collection of terms $\mathcal{T}$. This step is represented in line 4 in the displayed listing of the algorithm. Notice that variables are not included in $\mathcal{T}$. Variables are maintained unchanged in the rewritten query. Neither literal values are included in $\mathcal{T}$. Literal values are processed by domain specific transformer functions that take into account structure, units and measurement systems.

Then, for each term in $\mathcal{T}$, a collection of expressions is constructed and gathered with the term. Each expression represents a possible substitution of the triple pattern including the conflicting term for a graph pattern adequate for the target dataset. See lines 5 to 10 in the algorithm. Considering these expressions associated with each term, the set of all possible adequate queries is constructed (line 12) and the information loss of each query is measured and the query with the least loss is selected (line 14).

The core of the algorithm is the REWRITE routine (line 7) which examines source and target vocabularies, with their respective mapping axioms, in order to discover possible substitutions for a given term in a source vocabulary. Let us consider terms in a vocabulary as nodes in a graph and relationships between terms (specifically rdfs:subClassOf, rdf:subPropertyOf, owl:equivalentClass, owl:equivalentProperty, and owl:sameAs) as directed labeled edges between nodes. Notice that, due to mapping axioms between two vocabularies, we can consider those vocabularies as parts of the same graph. REWRITE routine performs a variation of a Breadth First Search traverse from a conflicting term, looking for its *frontier* of terms that belong to a target vocabulary. A term f belongs to the *frontier* of a term t if it satisfies the following three conditions: (a) f belongs to a target vocabulary, (b) there is a *trail* from t to f, and (c) there is not another term g (different from f) belonging to a target vocabulary in that trail. A *trail* from a node t to a node f is a sequence of edges that connects t and f independent of the direction of the edges. For instance, t `rdfs:subClassOf` r, f `rdfs:subClassOf` r is a trail from t to f.

Although a trail admits the traversing of edges in whatever direction, our algorithm keeps track of the pair formed by each node in the trail and the direction of the edge followed during the traverse since that is crucial information for producing the adequate expressions for substitution. Notice that we are interested in obtaining a conjunction expression with the directly subsuming terms, and a disjunction expression with the directly subsumee terms. For that reason, different routines are used to traverse the graph. In line 28 of the algorithm, directSuper(t) is the routine in charge of traversing the edges leaving t. In line 30 of the algorithm, directSub(t) is the routine in charge of traversing the edges entering t. Whenever a synonym to a term in a target vocabulary is found (line 25), such information is added to a queue (line 26) that stores the result of the REWRITE routine.

Termination of our algorithm is guaranteed because the graph traverse prevents the processing of a previously visited node (avoiding cycles) and furthermore a natural threshold parameter is established in order to limit the maximum distance from the conflicting term of a visited node in the graph.

```
1   //Returns an adequate query for onto_target,
2   //produced by a rewriting of Q with the least loss of information
3   QUERY_SELECTION(Q, ontoSource, ontoTarget) return Query
4     terms = DecomposeQuery(Q); // terms is the set of terms in Q
5     for each term in terms do
6     {
7       rewritingExpressions = REWRITE(term, ontoSource, ontoTarget);
8       //stores the term together with its adequate rewriting expressions
9       termsRewritings.add(term, rewritingExpressions);
10    }
11    //Constructs queries from the expressions obtained for each term
12    possibleQueries = ConstructQuery(Q, termsRewritings);
13    //Selects and returns the query that provides less loss of information
14    return LeastLoss(Q, possibleQueries);
15
16
17  //Constructs a queue of adequate expressions for term in onto_target
18  REWRITE(term, ontoSource, ontoTarget) return Queue<Expression>
19    resultQueue = new Queue();
20    traverseQueue = new Queue();
21    traverseQueue.add(term);
22    while not traverseQueue.isEmpty() do
23    {
24      t = traverseQueue.remove();
25      if has_synonym(t, ontoTarget) then
26         resultQueue.add(map(t, ontoTarget));
27      else //t is a conflicting term
28         {          ceiling = directSuper(t);
29                    traverseQueue.enqueueAll(ceiling);
30                    floor = directSub(t);
31                    traverseQueue.enqueueAll(floor);        }
32    }
33    return resultQueue;
```

5   Estimation of Information Loss

In this section we describe how we measure the loss of information caused by the rewriting of the original query. Also we explain in detail a use case that needs these rewritings to achieve an adequate query.

5.1   Measuring the Loss of Information

The system measures the loss of information using a composite measure adapted from [11]. This measure is based on the combination of the metrics *precision* and *recall* from Information Retrieval literature. We measure the proportion of retrieved data that is relevant (*precision*) and the proportion of relevant data that is retrieved (*recall*).

To calculate these metrics, we use datasets metadata published as VoID statistics. There are VoID statements that inform us of the number or entities of a class or the number of pairs of resources related by a property in a certain dataset. For instance, in :DBpedia dataset, the class dbpedia:Book has 26198 entities and there are 4102 triples with the property dbpedia:notableWorks.

```
:DBpedia a void:Dataset;
    void:classPartition [              void:propertyPartition [
        void:class dbpedia:Book;           void:property dbpedia:notableWorks;
        void:entities 26198;    ];        void:triples 4102;    ];
```

Given a conflicting term *ct*, we define *Ext(ct)* as the extension of *ct*; that is the collection of relevant instances for that term. Let us call *Rewr(ct)* to an expression obtained by the rewriting of a conflicting term *ct*, and *Ext(Rewr(ct))* to the extension of the rewritten expression, that is the retrieved instances for that expression.

We define ♯*Ext(ct)* as the number of entities (resp. triples) registered for *ct* in the dataset (this value should be obtained from the metadata statistics). In the case of *Ext(Rewr(ct))*, we cannot expect a registered value in the metadata. Instead we calculate an estimation for an interval of values [♯*Ext(Rewr(ct).low)*, ♯*Ext(Rewr(ct).high)*] which bound the minimum and the maximum cardinality of the expression extension. Those values are used for the calculation of our measures of precision and recall. However, due to the lack of space and the intricacy of the different cases that must be taken into account, we will not to present a detailed explanation for the calculation here.

Allow us to say that precision and recall of a rewriting of a conflicting term *ct* will be measured with an interval [*Precision(ct).low*, *Precision(ct).high*] where *Precision(ct).low* = $\mathcal{L}$(♯*Ext(ct)*, ♯*Ext(Rewr(ct).low)*, ♯*Ext(Rewr(ct).high)*) and *Precision(ct).high* = $\mathcal{H}$(♯*Ext(ct)*, ♯*Ext(Rewr(ct).low)*, ♯*Ext(Rewr(ct).high)*) are functional values calculated after a careful analysis of the diverse semantic relationships between *ct* and *Rewr(ct)*. Offered only as a hint, consider that the functions are variations on the following formulae, presented in [9]:

$$Precision(ct) = \frac{\sharp(Ext(ct) \cap Ext(Rewr(ct)))}{\sharp Ext(Rewr(ct))}; Recall(ct) = \frac{\sharp(Ext(ct) \cap Ext(Rewr(ct)))}{\sharp Ext(ct)}$$

In order to provide the user with a certain capacity for expressing preferences on precision or recall, we introduce a real value parameter α $(0 \leq \alpha \leq 1)$ for tuning the function to calculate the loss of information due to the rewriting of a conflicting term. Again, this measure is expressed as an interval of values:

$$Loss(ct).low = 1 - \frac{1}{\alpha(\frac{1}{Precision(ct).high}) + (1 - \alpha)(\frac{1}{Recall(ct).high})} \tag{1}$$

$$Loss(ct).high = 1 - \frac{1}{\alpha(\frac{1}{Precision(ct).low}) + (1 - \alpha)(\frac{1}{Recall(ct).low})} \tag{2}$$

Finally, many functions can be considered for the calculation of the loss of information incurred for the rewriting of the entire original query Q. We are aware that more research and experimentation is needed to select the most appropriate ones for our task. Nevertheless, for the sake of this paper, let us use a very simple and effective one such as the maximum among the set of values that represent the losses.

$$Loss(Q).low = max\{Loss(ct).low \mid ct \text{ conflicting term in } Q\} \tag{3}$$

$$Loss(Q).high = max\{Loss(ct).high \mid ct \text{ conflicting term in } Q\} \tag{4}$$

5.2   Rewriting Example

This section describes in detail an example of the process followed by our system in the case that loss of information is produced during the rewriting process. Consider that the system is trying to answer the original query shown in figure 1, which is expressed with terms in the proprietary *bdi* vocabulary, and that the user decides to commit the query to the DBpedia dataset. Some of the mapping axioms at the disposal of the system are as follows:

```
bdi:Document rdfs:subClassOf dbpedia:Work .
bdi:Publication rdfs:subClassOf bdi:Document .
dbpedia:WrittenWork rdfs:subClassOf bdi:Publication .
dbpedia:Website rdfs:subClassOf bdi:Publication .
dbpedia:Miguel_de_Cervantes owl:sameAs bdi:Miguel_de_Cervantes.
dbpedia:notableWork owl:sameAs bdi:isAuthor .
```

During the process, two possible rewritings are generated, as shown in figure 1. The one on the left is due to the pair of mapping axioms that specify that `dbpedia:Work` is a superclass of the conflicting term `bdi:Publication`; and, the one on the right is due to a pair of mapping axioms that specify that `dbpedia:WritenWork` and `dbpedia:Website` are subclasses of `bdi:Publication` (see those terms in the shaded boxes of figure 1).

The calculation of the loss information for each rewriting is as follows. Notice that the only conflicting term, in this case, is $(bdi : Publication)$. Firstly, the extension of the conflicting term and the rewriting expresssions are calculated.

$Ext(bdi:Publication) = 503;$
$Ext(dbpedia:Work) = 387599;$

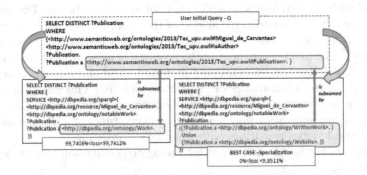

Fig. 1. Rewriting expressions generated

$Ext(dbpedia:WrittenWork \cup dbpedia:Website).low = min[40016, 2438] = 2438;$

$Ext(dbpedia:WrittenWork \cup dbpedia:Website).high = 40016+2438 = 42454.$

Secondly, precision and recall taking into account the relationships between the conflicting term and its rewriting expressions are calculated.

$With\ respect\ to\ Rewr(bdi:Publication) = dbpedia:Work$

$[Precision.low = 0,0012960;\ Precision.high = 0,0012977;\ Recall = 1]$

$With\ respect\ to\ Rewr(bdi:Publication) = db:WrittenWork \cup db:Website$

$[Precision = 1;\ Recall.low = 0,828969;\ Recall.high=1]$

Then, the loss of information interval for $bdi : Publication$ with a parameter $\alpha = 0.5$ (meaning equal preference on precision and recall) is calculated.

$With\ respect\ to\ Rewr(bdi:Publication) = dbpedia:Work$

$[Loss(bdi:Publication).low = 0,997408;\ Loss(bdi:Publication).high= 0,997412\ ]$

$With\ respect\ to\ Rewr(bdi:Publication) = db:WrittenWork \cup db:Website$

$[Loss(bdi:Publication).low = 0;\ Loss(bdi:Publication).high= 0.093511]$

Considering the above information loss intervals, the system will choose the second option (replacing `bdi:Publication` with `db:WrittenWork` ∪ `db:Website`) as the loss of information is estimated to be between 0% and 9% (i.e., very low even with the posibility of being 0%, that is no loss of information). However, the first option (replacing `bdi:Publication` with `dbpedia:Work`) is estimated to incur in a big loss of information (about 99.7%), which is something that could be expected: `dbpedia:Work` references many works that are not publications. Anyway, in absence of the second option, the first one (despite returning many references to works that are not publications) also returns the publications included in `dbpedia:Work` which could satisfy the user. The alternative, not dealing with imprecise answers, would return nothing when a semantic preserving query into a new dataset cannot be achieved.

6   Conclusions

For this new era of Web of Data we present in this paper a proposal that offers the users the possibility of querying heterogeneous Linked Data sources in a friendly way. That means that users do not need to take notice of technical details associated with the heterogeneity and variety of existing datasets. The proposal gives the opportunity to enrich the answer of the query incrementally, by visiting different datasets one by one, without needing to know the particular features of each dataset. The main component of the proposal is an algorithm that rewrites queries formulated by the users, using preferred vocabularies, into other ones expressed using the vocabularies of the datasets visited. This algorithm makes an extensive use of mapping axioms already defined in the datasets. To rewrite preserving query semantics may be difficult many times, for that reason the algorithm also handles rewritings with some loss of information.

Experiments are being carried out for tuning the estimation loss formulae.

Acknowledgements. This work is together supported by the TIN2010-21387-CO2-01 project and the Iñaki Goenaga (FCT-IG) Technology Centres Foundation.

References

1. Bizer, C., Heath, T., Berners-Lee, T.: Linked data-the story so far. International Journal on Semantic Web and Information Systems (IJSWIS) 5(3), 1–22 (2009)
2. Correndo, G., Salvadores, M., Millard, I., Glaser, H., Shadbolt, N.: Sparql query rewriting for implementing data integration over linked data. In: Proceedings of the 2010 EDBT/ICDT Workshops, p. 4. ACM (2010)
3. Elbassuoni, S., Ramanath, M., Weikum, G.: Query relaxation for entity-relationship search. In: The Semantic Web: Research and Applications, pp. 62–76 (2011)
4. Euzenat, J., Shvaiko, P.: Ontology matching, vol. 18. Springer, Heidelberg (2007)
5. Herzig, D., Tran, T.: One query to bind them all. In: COLD 2011, CEUR Workshop Proceedings, vol. 782 (2011)
6. Hurtado, C., Poulovassilis, A., Wood, P.: Query relaxation in rdf. Journal on Data Semantics X, 31–61 (2008)
7. Makris, K., Gioldasis, N., Bikakis, N., Christodoulakis, S.: Ontology mapping and sparql rewriting for querying federated rdf data sources. In: Meersman, R., Dillon, T., Herrero, P. (eds.) OTM 2010. LNCS, vol. 6427, pp. 1108–1117. Springer, Heidelberg (2010)
8. Manola, F., Miller, E., McBride, B.: Rdf primer w3c recommendation (February 10, 2004)
9. Mena, E., Kashyap, V., Illarramendi, A., Sheth, A.: Imprecise answers on highly open and distributed environments: An approach based on information loss for multi-ontology based query processing. International Journal of Cooperative Information Systems (IJCIS) 9(4), 403–425 (2000)
10. Quilitz, B., Leser, U.: Querying distributed rdf data sources with sparql. In: Bechhofer, S., Hauswirth, M., Hoffmann, J., Koubarakis, M. (eds.) ESWC 2008. LNCS, vol. 5021, pp. 524–538. Springer, Heidelberg (2008)
11. Salton, G.: Automatic Text Processing: The Transformation, Analysis, and Retrieval of. Addison-Wesley (1989)
12. Schwarte, A., Haase, P., Hose, K., Schenkel, R., Schmidt, M.: FedX: Optimization techniques for federated query processing on linked data. In: Aroyo, L., Welty, C., Alani, H., Taylor, J., Bernstein, A., Kagal, L., Noy, N., Blomqvist, E. (eds.) ISWC 2011, Part I. LNCS, vol. 7031, pp. 601–616. Springer, Heidelberg (2011)

FILT – Filtering Indexed Lucene Triples
– A SPARQL Filter Query Processing Engine–

Magnus Stuhr[1] and Csaba Veres[2]

[1] Computas AS, Norway
Magnus.Stuhr@computas.com
[2] University of Bergen, Bergen, Norway
Csaba.Veres@infomedia.uib.no

Abstract. The Resource Description Framework (RDF) is the W3C recommended standard for data on the semantic web, while the SPARQL Protocol and RDF Query Language (SPARQL) is the query language that retrieves RDF triples. RDF data often contain valuable information that can only be queried through filter functions. The SPARQL query language for RDF can include filter clauses in order to define specific data criteria, such as full-text searches, numerical filtering, and constraints and relationships between data resources. However, the downside of executing SPARQL filter queries is the frequently slow query execution times. This paper presents a SPARQL filter query-processing engine for conventional triplestores called FILT (Filtering Indexed Lucene Triples), built on top of the Apache Lucene framework for storing and retrieving indexed documents, compatible with unmodified SPARQL queries. The objective of FILT was to decrease the query execution time of SPARQL filter queries. This aspect was evaluated by performing a benchmark test of FILT compared to the Joseki triplestore, focusing on two different use-cases; SPARQL regular expression filtering in medical data, and SPARQL numerical/logical filtering of geo-coordinates in geographical locations.

Keywords: RDF full-text search, SPARQL filter queries, SPARQL regex filtering, SPARQL numerical filtering, RDF data indexing, Lucene.

1   Introduction

RDF (Resource Description Framework) is a language for describing things or entities on the World Wide Web [8]. RDF data is structured as connected graphs, and is composed of triples. A triple is a statement consisting of three components: a subject, a predicate and an object. The World Wide Web Consortium (W3C) standard query language for looking up RDF data is the SPARQL Protocol and RDF Query Language, referred to as SPARQL [13]. SPARQL makes it possible to retrieve and manipulate RDF data, whether the data is stored in a native RDF store, or expressed as RDF through middleware conversion mechanisms. SPARQL queries are expressed in the same syntax as RDF, namely as triples.

H.L. Larsen et al. (Eds.): FQAS 2013, LNAI 8132, pp. 25–39, 2013.

As the Web evolves into one enormous database, locating and searching for specific information poses a challenge. RDF data consists of graphs defined by triples, meaning that there are many more relationships and connections between data resources, compared to the traditional Web structure consisting of clear text documents. The RDF data structure offers a more flexible and accurate way of retrieving information, as specific relationships between data resources can be looked up. Moreover, the architecture of the Semantic Web poses a need for another search design opposed to the traditional Web. However, full-text searches will also be important when searching the Semantic Web, as there usually exist a great deal of textual descriptions and numerical values stored as literals in most RDF data sets. Moreover, full-text searches in RDF data are important, because users often do not know to a full extent what information exists. SPARQL is a good way of searching for explicit data relationships and occurrences in RDF data sets, also offering the possibility of performing full-text searches and filtering terms and phrases through SPARQL filter clauses. These filter clauses enables the filtering of logical expressions and variables expressed in the general SPARQL query. Examples of SPARQL clauses are filtering string values, regular expressions, logical expressions and language metadata. Unfortunately, SPARQL filter clauses pose a major challenge when it comes to query-execution time. When applying filter clauses in SPARQL queries, the queries have to perform matching of logical expressions or terms and phrases, meaning that the SPARQL queries will execute slower than general SPARQL queries. As SPARQL filter queries can discover data relationships that general SPARQL queries cannot, they play an important role in retrieving RDF data. However, because SPARQL filter queries in most cases have a much slower query-execution time than general SPARQL queries; it is easy to shy away from applying filter clauses to the queries. Minack et al. [9] argue that literals are what connect humans to the Semantic Web, giving meaning and an understanding to all the data that exist on the Web. If literals are taken away from RDF data, the directed graphs that amount to the Web of Data will merely be a set of interconnected nodes that are to a certain extent name- and meaningless. This argument suggests that discovering efficient ways of filtering literals in RDF data will be of great value to the information retrieval aspect of the Semantic Web.

This paper presents a technique for optimizing the query-execution times of SPARQL filter queries. A prototype solution called FILT (Filtering Indexed Lucene Triples) has been built in order to show that a general SPARQL filter query processor can decrease the query-execution time of SPARQL filter queries, thus enhancing the value of integrating full-text searches with the SPARQL query language. The paper is divided into six sections apart from the introduction: section 2 presents the implementation and features of FILT, section 3 presents previous related work, section 4 presents the framework for evaluating FILT through a benchmark test, section 5 presents the results of the benchmark test, section 6 discuss the results, and finally section 7 presents conclusions and further work.

2   Implementation of FILT with Apache Lucene

FILT is a SPARQL filter-processing engine and enables storing and querying of RDF data through the Apache Lucene framework [3]. It is supports unmodified SPARQL queries, meaning that users do not have to re-write their SPARQL queries in order to execute them. The main purpose of FILT is to decrease the query-execution time of SPARQL queries containing filter clauses, thus optimizing the efficiency of semantic information retrieval. FILT currently provides storing of triples, a SPARQL endpoint, and a SPARQL querying user-interface. FILT can store any data set stated as triples. The data set must be expressed in one of the three most common syntaxes for RDF triples: N-Triples, Turtle or RDF/XML. Moreover, FILT will supplement a traditional triplestore by stripping filter queries away from the SPARQL query during a pre-processing phase. It then passes the set of triples that match the filter conditions back to the Jena SPARQL query engine. General SPARQL queries without filter clauses are sent directly to an external triplestore SPARQL endpoint, or to a local RDF model of the entire data set. This means that a SPARQL endpoint URL of a triplestore, or the raw RDF data set file, has to be specified in FILT in order for any type of SPARQL query to execute properly. The architecture of FILT is shown in Figure 1. This figure illustrates how SPARQL queries are executed through FILT. There are several steps in this process: first, the user issues a SPARQL query. If the query does not contain filter clauses, the query is immediately executed through an external RDF store, either a triplestore or a local RDF model loaded into the Jena framework. If the SPARQL query contains filter clauses, it is sent to the query-rewriting module that performs two processes: extracting the filter clauses from the query and transforming them into Lucene queries, and stripping the filter clauses from the SPARQL query, leaving only the general SPARQL query. The Lucene queries, constructed based on the filter clauses in the query, are executed through the Lucene index consisting of the indexed data of the entire RDF data set. The output of the Lucene queries executed through the index consists of triples that will be the foundation of building an internal RDF model. This RDF model contains the triples corresponding to the filter clauses of the SPARQL query, and the general SPARQL query stripped of filter clauses will be executed over this local model. Finally, the output returned from the general SPARQL query is the final query output that is returned to the user that issued the SPARQL query.

As mentioned, FILT is built on top of the Apache Lucene framework. Apache Lucene is a free open-source high-performance information retrieval engine written in the Java Programming language. It offers full-featured text search, based on indexing mechanisms. Lucene is a vital part of storing and querying data in FILT. A Lucene index contains a set of documents that contain one or more fields. These fields can be stored as text or numerical values, and can either be analyzed or not analyzed by the Lucene library, which will later affect how the given information can be retrieved. Moreover, a Lucene Document Field is a separated part of a document that can be indexed so that terms in the field can be used to retrieve the document through Lucene queries. The index structure in FILT is based on a dynamic index structure that

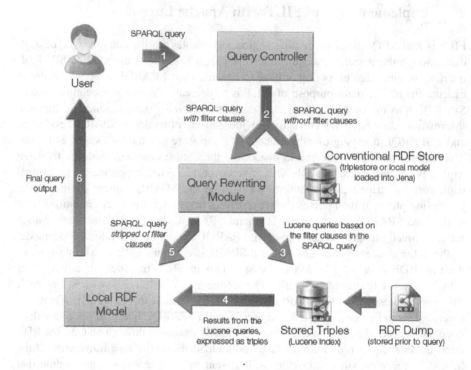

SPARQL query

User

Query Controller

SPARQL query
with filter clauses

SPARQL query
without filter clauses

Conventional RDF Store
(triplestore or local model
loaded into Jena)

Final query
output

Query Rewriting
Module

SPARQL query
stripped of filter
clauses

Lucene queries based on
the filter clauses in the
SPARQL query

Local RDF
Model

Results from the
Lucene queries,
expressed as triples

Stored Triples
(Lucene Index)

RDF Dump
(stored prior to query)

Fig. 1. The architecture of FILT

contains a default "graph" and "subject" fields, which contain the RDF graph
locations (paths to data set files) and the subject URI of the data entity being indexed.
Apart from this, the index structure is a dynamic structure that names each field in a
document by its predicate URI and giving it the object-value of the given triple as its
input. Moreover, this means that apart from the static field named "subject", the other
document-field names will vary depending on what the predicate URI is. This makes
it easy to query the index by specifying predicate names for the field names in the
Lucene queries. The overall index structure can be described in a more formal way
like this:

for each sub-graph in the superior graph {
  new Document
  *add field to document(FieldName: graph, FieldValue: <The filename of the data
set file>)*
  add field to document(FieldName: subject, FieldValue: <subject-URI>)
  for each predicate and object in graph {
    add field to document(FieldName: predicate, FieldValue: <object-value>)
  }
}

FILT translates SPARQL queries into Lucene queries in order to retrieve information from the pre-stored index. Only SPARQL queries with filter clauses run through the index. All other queries run through the local model or the SPARQL endpoint specified by the data set owner. FILT has mainly focused on implementing compatibility with SPARQL regex filter clauses and SPARQL logical/numerical expression filter clauses. In FILT, the regex filter clause is executed through the RegexQuery class in Lucene. This query class allows regular expression to be matched against text stored in the index documents. To illustrate how FILT deals with the aspect of number filtering, look at this SPARQL query containing a "logical expression" filter clause for filter numbers: *SELECT * WHERE {?s geo:lat ?lat; geo:long ?long. Filter(xsd:double(?lat) > 50 && ?long = 60)}*. The objective of this filter clause is to find all data entities where the latitude is above 50 and the longitude equals 60. These expressions can easily be translated into existing Lucene queries, namely the NumericRangeQuery and the RegexQuery classes. The first expression "xsd:double(?lat) > 50" is translated into the NumericRangeQuery "geo:lat:[50 TO *]" and the second expression "?long = 60" is transformed into the RegexQuery "geo:long:60". In this case, the NumericRangeQuery "geo:lat:[50 TO *]" has defined the lower term in the query to be exclusive, meaning that only data entities with a latitude over 50 returns true. If the lower term was set to be inclusive, data entities with a latitude equaling 50 would also return true. This would be correct to apply if the filter expression rather stated "xsd:double(?lat) >= 50". The same principles apply to any NumericRangeQuery, whether the query contain only a lower term or an upper term, or both. Any expression containing the EQUAL expression operator ("=") or the NOT EQUAL expression operator ("!="), regardless of filter value, is translated into the RegexQuery. If the query is based on the equal operator, it will only include the filter value itself as the query input, such as the query just mentioned: "geo:long:60". However, if the filter expression stated "?long != 60" instead of "?long = 60", the RegexQuery would have to generate a regular expression with a "negative look-ahead" condition, in order to find data entities with a latitude not matching the value "60". This RegexQuery would look like this: *geo:long^(?!.*60).*$)*. The built-in Lucene query library offers the possibility of easily translating simple number filtering into different queries. However, more complex number filtering cannot be directly translated into Lucene queries. This can be demonstrated through this query: *SELECT ?subject WHERE {?subject geo:lat ?lat; geo:long ?long . FILTER ((xsd:double(?lat) - 37.785834 <= 0.040000) && (37.785834 - xsd:double(?lat) <= 0.040000) &&(xsd:double(?long) - -122.406417 <= 0.040000) && (-122.406417 - xsd:double(?long) <= 0.040000))}*.

The filter clause expressions in this query is tricky to filter by using Lucene queries, as none of the built-in Lucene query classes can execute mathematical expressions containing numeric operators. This means that in order to execute the number filtering expressions in the filter clause, the mathematical expressions have to be simplified in order to meet the requirements of the Lucene query libraries. FILT translates complex numeric expressions into more simple expressions in order to meet the requirements of the built-in Lucene query library. The rules for simplifying the numerical expressions are based on the standard mathematical rules for equations and inequalities. The Lucene queries are built based on the filter clauses in the given SPARQL query that is being executed, and each specific filter clause is converted to one or more separate Lucene queries. When every filter clause have been divided into distinct Lucene queries, these different Lucene queries will be joined as one large query and finally executed over the index.

3   Previous Work

Interesting research has been conducted within the area of semantic searching and indexing of RDF data. Sindice is a lookup-index over data entities crawled on the Semantic Web [12]. SIREn is a semantic information retrieval engine plugin to Lucene [7], and is the search engine that Sindice is based on. SIREn includes a node-based indexing scheme for semi-structured data, based on the Entity-Attribute value model [6]. As the Sindice project focuses mostly on storing and querying decentralized, heterogeneous data sources as a semantic search-engine on the Web of Data, FILT heads in the direction of storing and querying pre-defined data sets where the data schema is fully known. FILT does not analyze or tokenize the data being indexed so that all data values are stored as their full value, meaning that they also have to be queried by denoting their entire data values. As FILT is mainly a SPARQL filter query processing engine, this indexing approach supports the idea behind SPARQL queries, where the data schema is fully known to the user executing the query.

SEMPLORE [14] also offers full-text searches through indexed RDF data. SEMPLORE treats any data value that has a data type property as a virtual keyword of concepts, meaning it will be available for full-text searches. These virtual keywords of concepts can be combined with concepts in an ontology using Boolean operators. Opposed to SPARQL queries, where a query can have multiple query targets, the querying capabilities of SEMPLORE restrict the queries to have a single query target. This supports conventional ways of retrieving information on the Web, but FILT differs from this solution in terms of letting the users query multiple targets through SPARQL queries. In addition, FILT is a database solution opposed to SEMPLORE, which is mainly a web solution.

Castillo et al. [5] present a solution called RDFMatView for decreasing the query processing time of SPARQL queries containing multiple graph patterns. As several implemented SPARQL processors are built on top of relational databases, SPARQL queries are translated into one or more SQL queries. If queries have more than one graph pattern, the query processing requires roughly as many joins as the query has graph patterns. Castillo et al. [5] argue that optimizing these joins is vital in order to achieve scalable SPARQL systems. In order to avoid the computation of several join queries RDFMatView indexes fractions of queries that occur frequently in executed queries. Only graph patterns that are used together regularly in queries are indexed. RDFMatView matches FILT in terms of indexing data in order to decrease the query-execution time of SPARQL queries, but it only focuses on decreasing the query execution time of SPARQL queries with multiple graph patterns, disregarding the complications of SPARQL filter queries regarding query-execution time.

There exist several solutions trying to implement efficient full-text searches through the SPARQL query language. Apache Jena LARQ [2] is a querying solution based on Lucene and the Jena SPARQL query engine Apache Jena ARQ [1]. NEPOMUK [10] also offers the translation of full-text searches from the regex filter clause in SPARQL queries into Lucene queries. FILT differs from LARQ and NEPOMUK in terms of not just implementing full-text searches, but also implementing the filtering of logical expressions and several other SPARQL filter clauses. In addition, LARQ and NEPOMUK do not translate SPARQL queries into customized query solutions for the users, but rather offer the possibility for the users to rewrite the queries themselves. Moreover, LARQ and NEPOMUK offer extensions for performing full-text searches on literals, whereas FILT propose a solution for executing full-text searches and logical expression filtering on any triple-component through an index, directly translated from user-generated SPARQL queries. Minack et al. [9] present the Sesame LuceneSail solution, a part of the NEPOMUK project. Sesame LuceneSail is a solution for performing full-text search on RDF data by storing the data in a Lucene index and executing keyword queries through the index. FILT differs from this in terms of not being dependent on an external triplestore when executing SPARQL filter queries, as the general graph pattern SPARQL query stripped from filter clauses is executed over the relevant triples extracted from the Lucene query. In addition, Sesame LuceneSail has certain restrictions on its query expressiveness in terms of not offering the possibility of querying more than one keyword query on each subject of a triple. FILT offers the same flexibilities and expressiveness as defined in the SPARQL query language, as FILT directly translates SPARQL filter queries into Lucene queries, obtaining the exact same results as executing the SPARQL queries through a conventional triplestore.

Many triplestores contain built-in mechanisms for coping with queries containing filtering functions. For instance, the Jena and Joseki (http://www.joseki.org/) SPARQL engines provide a possibility of executing full-text queries through LARQ. The difference between the full-text search-engine in LARQ compared to FILT is that LARQ requires the SPARQL queries to include different syntaxes that do not correspond with the general SPARQL syntax. FILT does not require any additional statements or functions in the SPARQL queries and executes regular SPARQL queries with filter clauses. Full-text searches through FILT are simply run by adding a regex filter clause in the SPARQL query based on the standard SPARQL syntax. Another example of a built-in mechanism for executing specific filtering functions is the SQL MM function for executing geospatial queries in the Virtuoso triplestore (http://virtuoso.openlinksw.com/). The SQL MM function in Virtuoso makes it more efficient to execute geospatial queries [11]. However, just as Joseki and Jena combined with LARQ, the built-in SQL MM filtering function in Virtuoso is dependent on another query-syntax than SPARQL filter queries, meaning that the SPARQL queries have to be modified from their original syntax in order to benefit from the built-in filtering mechanisms. FILT is not dependent on additional filter statements or different query syntaxes in order to execute filter queries, as FILT simply execute queries of the standard SPARQL syntax.

4   Benchmark Evaluation

In this project, an extensive benchmark evaluation of FILT has been performed. The objective of the benchmark test was to compare the features of FILT to the Joseki triplestore by evaluating several metrics regarding the speed of query execution. The benchmark evaluation included executing two pre-defined sets of SPARQL filter queries over two separate data sets. The two different data sets that the queries were executed over were the DrugBank data set and the Geographic Coordinates RDF graph of the DBpedia data set. The DrugBank data set contains 766,920 triples, whereas the Geographic Coordinates data set contains 1,771,100 triples. For this benchmark evaluation, both the DrugBank data set and the Geographical Coordinates (DBpedia) data sets were divided into three data sets; each with a distinct amount of triples. The data sets were split into one sub-set containing 1/7 of the total amount of triples and one sub-set containing 1/2 of the total amount of triples. Finally, the entire data set was tested. These data sets were loaded into two different data stores: FILT and Joseki. Joseki is a triplestore for Jena, developed by W3C RDF Data Access Working Group. It supports the SPARQL protocol and the SPARQL RDF Query Language. The version of FILT that will be applied in the benchmark evaluation is v1.0, and the Joseki version used is v3.4.4. The query mixes were executed over each of the divided data sets, both through the Joseki triplestore and FILT, in order to illustrate the scalability performance of a conventional triplestore opposed to FILT. The DrugBank data set can be downloaded from: http://dl.dropbox.com/u/21236338/drugbank.zip. The Geographical Coordinates of DBpedia data set can be downloaded from: http://downloads.dbpedia.org/ 3.7/en/geo_coordinates_en.nt.bz2.

The metrics of this benchmark evaluation are based on the performance metrics specified by Bizer & Shultz [4]. The metrics are "Milliseconds per Query (MSpQ)", "Average Query Execution Time (aQET)", "Overall Runtime (oaRT)" and "Average Query Execution Time over all Queries (aQEToA)". However, the benchmark evaluation in this paper will only evaluate and present the aQET. The aQET will be calculated by the average time it takes to execute a single query multiple times. The aQET of each query will then be combined with the aQET of the queries of the same query form. Moreover, this means that the aQET of all SELECT queries will be calculated into a combined aQET for SELECT queries. The same procedure will be repeated with all query forms. This way it is possible to analyze the performance of the two data stores based on different query forms. The query mixes of both the regex use-case and the numerical filtering use-case contained 24 queries; six queries of each SPARQL query form (SELECT, DESCRIBE, CONSTRUCT and ASK). This way, the performance of all the query forms isolated could be analyzed. The query mixes were executed three times for each data set sizes. Prior to each execution of the query mixes, the data sets were re-loaded along with executing a warm-up query-mix.

5   Results

This section will refer to each of the data set sizes of the DrugBank and Geographical Coordinates data sets as "S" for the smallest data set version, "M" for the medium data set version, and "L" for the large data set, consisting of the entire data set. The results from the DrugBank data set and the Geographical Coordinates data set were each analyzed in a separate, two way analysis of variance (ANOVA) with the factors Size (S, M, L) and Store (FILT, Joseki). The critical values for F will be reported in the results with the signifiers "*" where the probability number is less than 0.05, "**" where the probability number is less than 0.01, and "***" where the probability is less than 0.001.

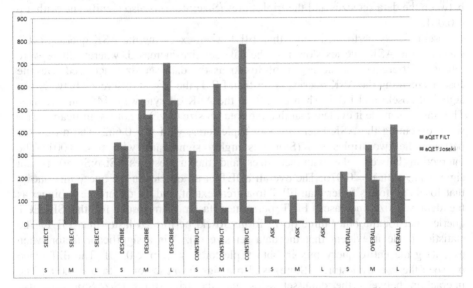

Fig. 2. The overall benchmark results of the DrugBank regular expression filtering use-case

The overall results of the DrugBank regular expression filtering use-case are shown in Figure 2. The results of the DrugBank use-case indicate that the SELECT queries of the query mix had a significant difference in the results of FILT and Joseki. FILT performs faster than Joseki with SELECT regex queries for all data set sizes. The results indicate that the larger the data set is, Joseki performs significantly worse, as opposed to FILT that more or less performs in the same way regardless of data set size, with small differences in the aQET. The probability numbers showed that the data set size (Size) is a significant factor when executing the SELECT queries in both triplestores, with $p < 0.01$. The difference between the two triplestores (Store) is also a significant factor, with $p < 0.001$. The interaction between the data set sizes and the triplestores (Size:Store) is not significant, with $p < 0.10$. Further, the chart shows that, as opposed to the results of the SELECT queries, FILT and Joseki performed almost similar on the small data set size (S) when executing the DESCRIBE queries, with

Joseki having a slight advantage. However, as the data set size increased Joseki performed faster than FILT. The statistics made it evident that the data set size (Size) is a significant factor when executing the DESCRIBE queries in both triplestores, with $p < 0.001$. The difference between the two triplestores (Store) is also a significant factor, with $p < 0.001$. The interaction between the data set sizes and the triplestores (Size:Store) is also significant, with $p < 0.01$. The results also show that Joseki performed better than FILT when executing the CONSTRUCT queries, regardless of the data set size. As the data set size increased FILT performed worse, whereas Joseki performed more or less the same for all data set sizes. The statistics made it evident that the data set size (Size) is a significant factor when executing the CONSTRUCT queries in both triplestores, with $p < 0.001$. The difference between the two triplestores (Store) is also a significant factor, with $p < 0.001$. The interaction between the data set sizes and the triplestores (Size:Store) is also significant, with a p < 0.001.

Joseki clearly performed better than FILT when executing the ASK queries. FILT executed the ASK queries slower as the data set size increased, whereas there were minimal differences in the aQET of Joseki as the data set size increased. Despite Joseki executing the ASK queries faster than FILT, the largest difference between the aQET of Joseki and FILT when executing the ASK queryieswere 145 milliseconds. The statistics made it evident that that the data set size (Size) is not a significant factor when executing the ASK query in both triplestores, with $p = 0.662$. The difference between the two triplestores (Store) is highly significant, with $p < 0.001$. The interaction between the data set sizes and the triplestores (Size:Store) is not significant, with $p = 0.076$. The overall aQET of all queries in the query mix shows that Joseki performs faster than FILT to a great extent, and the difference is bigger as the data set size increases. FILT performed faster than Joseki for the SELECT queries, but for the other three query forms Joseki performed faster than FILT. The statistics made it evident that the data set size (Size) is a significant factor when executing the entire query mix in both triplestores, with $p < 0.001$. The difference between the two triplestores (Store) is also a significant factor, with $p < 0.001$. The interaction between the data set sizes and the triplestores (Size:Store) is also significant, with $p < 0.001$.

To summarize the SPARQL regex use-case, FILT outperforms Joseki when it comes to SELECT queries. The results also show that Joseki performs faster than FILT with the other query forms: DESCRIBE, CONSTRUCT and ASK.

The results of the Geographical Coordinates use-case clearly show that the SELECT queries of the query mix had a significant difference in the results of FILT and Joseki. Figure 3 shows that FILT performed remarkably faster than Joseki for the six SELECT queries in the query mix. The difference between FILT and Joseki for the small data set (S), consisting of 250,000 triples, were noteworthy, and as the data set size increased FILT performs significantly faster than Joseki. The biggest difference in the aQET of the SELECT queries occurred when executing the queries over the large data set (L), consisting of 1,700,000 triples, where FILT executed the SELECT queries more than 35,000 milliseconds (35 seconds) faster than Joseki. The statistics made it evident that the data set size (Size) is a significant factor when

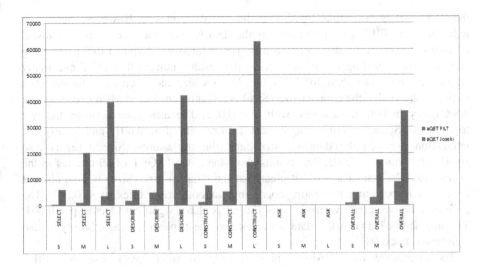

Fig. 3. The overall benchmark results of the Geographical Coordinates numerical/logical filtering use-case

executing the SELECT queries in both triplestores, $p < 0.001$. The difference between the two triplestores (Store) is also a significant factor, $p < 0.001$. The interaction between the data set sizes and the triplestores (Size:Store) is also significant, $p < 0.001$.

Further, the chart shows that there is a similarity between the aQET of SELECT queries and DESCRIBE queries in both FILT and Joseki. However, both FILT and Joseki performed faster when executing the SELECT queries compared to DESCRIBE queries. The difference of the aQET between FILT and Joseki were significant when executing the DESCRIBE queries. The biggest difference in the aQET of the DESCRIBE queries occurred when executing the DESCRIBE queries over the large data set (L), consisting of 1,700,000 triples, with a time difference of 27,000 milliseconds (27 seconds). The statistics made it evident that the data set size (Size) is a significant factor when executing the DESCRIBE queries in both triplestores, $p < 0.001$. The difference between the two triplestores (Store) is also a significant factor, $p < 0.001$. The interaction between the data set sizes and the triplestores (Size:Store) is also significant, $p < 0.001$.

The results clearly indicate that FILT performed better than Joseki when executing the CONSTRUCT queries, regardless of the data set size. The biggest difference in the aQET of the two CONSTRUCT queries occurred when executing the CONSTRUCT queries over the large data set (L), consisting of 1,700,000 triples, with a time difference of 46,000 milliseconds (46 seconds). The statistics made it evident that the data set size (Size) is a significant factor when executing the CONSTRUCT queries in both triplestores, $p < 0.001$. The difference between the two triplestores (Store) is also a significant factor, $p < 0.001$. The interaction between the data sizes and the triplestores (Size:Store) is also significant, $p < 0.001$. Joseki executed the ASK queries faster than FILT, regardless of data set size. However, there is an indication that FILT performs faster as the data set size increases, whereas Joseki performs slower as the data set size increases. Moreover, despite FILT performing slower when executing the ASK queries, the results indicate that FILT eventually would perform faster than Joseki as the data set size increased even further. The statistics shows that the data set size (Size) is a significant factor when executing the ASK query in both triplestores, with $p < 0.001$. The difference between the two triplestores (Store) is also a significant factor, with $p < 0.001$, and finally the interaction between the data set sizes and the triplestores (Size:Store) is also significant, with $p < 0.001$. The results show the overall aQET of all queries in the query mix. The statistics made it clear that the data set size (Size) is a significant factor when executing the entire query mix in both triplestores, $p < 0.001$. The difference between the two triplestores (Store) is also a significant factor, $p < 0.001$. The interaction between the data set sizes and the triplestores (Size:Store) is also significant, $p < 0.001$.

To summarize the SPARQL numerical/logical filter query use-case, FILT outperforms Joseki to a great extent for all query forms, except ASK queries. The biggest difference in the aQET between FILT and Joseki occurred when executing the query mix over the large data set (L), where FILT performed 28 milliseconds (28 seconds) faster than Joseki. The biggest difference for any of the query forms occurred when executing the CONSTRUCT queries, where FILT executed the queries 46 seconds faster than Joseki for the large data set.

6   Discussion

The results of the benchmark evaluation show that FILT outperforms Joseki on SELECT queries in both use cases. In addition, every query form apart from the ASK queries was performed significantly faster with FILT than by Joseki in the SPARQL numerical/logical filter query use-case. However, this was not the case with the with the SPARQL regular expression filter query use-case, as Joseki performed faster than FILT with the DESCRIBE, CONSTRUCT and ASK query forms. The results of the ASK, CONSTRUCT and DESCRIBE queries in the query mix of the SPARQL regular expression filter use-case affected the overall results of the use-case to a great extent, despite the aQET of the SELECT queries being faster in FILT than Joseki. It is worth mentioning that even though Joseki performs better than FILT for the CONSTRUCT, DESCRIBE and ASK query forms in the SPARQL regex filter query use-case; the differences in the aQET between Joseki and FILT are so small that they are hardly noticeable in a real-world querying scenario unless the times are actually recorded. This means that it is hard to locate any noticeable factors in the architecture of FILT that can lead to the aQET of the three query forms being slower than Joseki. However, certain aspects of how FILT returns query results are worth discussing in light of the different outcomes of the four SPARQL query forms.

FILT executes all query forms in the exact same manner; the SPARQL filter clauses are being executed through Lucene, and the general SPARQL query is being executed through the Jena SPARQL processing engine. However, the difference in the way FILT returns query results from SELECT queries on one hand, and DESCRIBE and CONSTRUCT queries on the other hand, is that the results of the DESCRIBE and CONSTRUCT queries are converted from a Jena RDF model to a text string containing the raw RDF data, whereas SELECT queries are merely returned a SPARQL XML result set. Converting the Jena RDF model to a text string containing the raw RDF data is necessary in order to send the result object across the HTTP protocol, as a raw Jena RDF model cannot be sent through the HTTP protocol. This process is not time-consuming, but in many cases the time being spent by this conversion procedure is enough for FILT to return the results of the DESCRIBE and CONSTRUCT queries slower than Joseki, meaning that the aQET will be slower. It is likely that this conversion process is a major cause to the disadvantage FILT has compared to Joseki when executing DESCRIBE and CONSTRUCT regex queries. For the SPARQL numerical/logical filter query case, the conversion process would not have a significant outcome on the results, because Joseki was already executing the queries several seconds slower than FILT.

Moreover, a couple of hundred milliseconds spent on converting the results are not noticeable in the SPARQL numerical/logical filter query use-case. Optimizing the process of returning results from DESCRIBE and CONSTRUCT queries in FILT are worth having a closer look at if FILT should be developed further. ASK queries are constructed to check if the graph patterns and functions in the queries exists or do not exists in the data set. FILT copes with ASK queries the same way it copes with all the other query forms; the filter clauses are executed through Lucene and the general SPARQL query is executed through the Jena SPARQL processing engine. FILT does not retrieve all the entities that match the filter clauses executed through Lucene, but merely one of the entities. This is because as long as one entity corresponds to the filter clauses in the ASK query, this is enough for the filter clauses to be true. The entity is then being loaded into a local RDF model where the general SPARQL query is being executed. The results are finally returned as a SPARQL XML result set with a true or false binding. In FILT this is the most obvious and efficient way to deal with ASK queries discovered in this project, and it is difficult to say why Joseki outperforms FILT when it comes to all ASK queries, regardless of the two different use-cases. Finally, it is still worth mentioning that the highest time difference between FILT and Joseki with all ASK queries is only 145 milliseconds, which is hardly noticeable in a real-world querying scenario. Also, the results of the ASK queries executed in the SPARQL numerical/logical filter use-case indicate that FILT will eventually execute the ASK queries faster if the data set size increases further. A final aspect worth discussing is the index structure of FILT and the variety of Lucene queries that are executed depending on what the SPARQL filter clauses of a query represent. The index structure in terms of document field analyzers and the entire indexer itself (Lucene provides several different indexing classes) may be factors that to some extent can provide answers as to why there are significant differences between the two use-cases. Also, the SPARQL regular expression filter clauses are executed through the Lucene RegexQuery class, whereas SPARQL numerical/logical filter clauses are mainly executed through the NumericRangeQuery, meaning that it is possible that the two Lucene query types have entirely different ways of filtering through data, and that one of them may be considerably faster than the other.

The fact that Joseki struggles largely with SPARQL numerical/logical filter queries compared to SPARQL regex filter queries suggests that the major strength of Joseki lies in coping with SPARQL regex filter queries. FILT however, copes much better with SPARQL regular expression queries than Joseki does with SPARQL numerical/logical filter queries. This means that the weakness of FILT is much less significant and noticeable than the weakness of Joseki. Additionally, if the results of both use-cases were combined into one huge result set, FILT would outperform Joseki to a great extent. This is because even though FILT performs slightly slower than Joseki in the SPARQL regex use-case the query execution times are still very low (in most cases the aQET does not even reach a whole second). Finally, a conclusion can be drawn stating that FILT is a solution that should be used for executing SPARQL SELECT regex filter queries and SPARQL numerical/logical filter queries of all query forms.

7   Conclusions and Future Work

This paper has demonstrated the practical advantages of using a text-indexing platform in conjunction with a regular triplestore, for executing certain kinds of SPARQL queries. Our implementation of FILT, based on Lucene, demonstrated that in the most successful cases, FILT returned results 46 seconds faster than Joseki. In usability terms, a18 second response from FILT is far more acceptable than a 64-second response from Joseki. The aim now is to implement FILT as a general architecture that can be deployed by any triplestore maintainer. The advantage of our approach is that it is agnostic about the companion triplestore, and does not require any special syntax. In other words, it can be transparently deployed alongside any triplestore.

A number of outstanding issues need to be resolved. First, we need to solve the puzzling limitations in CONSTRUCT and DESCRIBE regex filter queries, as well as ASK queries of both regex and numerical SPARQL filter queries. Second, we need to include more rewrite rules to cope with the full range of FILTER queries. Finally, we need to ensure that the solution is scalable to any required implementation. Once these issues are resolved, FILT will be distributed as a simple package that will handle the indexing of RDF data in the triplestore, and be deployed as a seamless layer that passes non-FILTER queries onto the regular triplestore, but executes FILTER queries through its own speedy execution engine.

References

[1] Apache Jena ARQ, ARQ - A SPARQL Processor for Jena (2012),
    http://incubator.apache.org/jena/documentation/larq/
    index.html

[2] Apache Jena LARQ (2012), LARQ - adding free text searches to SPARQL,
    `http://incubator.apache.org/jena/documentation/query/index.html`

[3] Apache Lucene Core, Apache Lucene Core (2011),
    `http://lucene.apache.org/core/`

[4] Bizer, C., Schultz, A.: The Berlin SPARQL Benchmark. The Proceedings of the International Journal on Semantic Web and Information Systems (IJSWIS) 5(2), 24 (2009),
    `http://www.igi-global.com/article/berlin-sparql-benchmark/4112`, doi:10.4018/jswis.2009040101

[5] Castillo, R., Rothe, C., Leser, U.: RDFMatView: Indexing RDF Data Using Materialized SPARQL Queries. In: Proceedings of the 6th International Workshop on Scalable Semantic Web Knowledge Base Systems (SSWS 2010), vol. 669, pp. 80–95 (2010),
    `http://ceur-ws.org/Vol-669`

[6] Delbru, R., Campinas, S., Tummarello, G.: Searching Web Data: an Entity Retrieval and High-Performance Indexing Model. Web Semantics: Science, Services and Agents on the World Wide Web, Web-Scale Semantic Information Processing 10, 33–58 (2012),
    `http://www.sciencedirect.com/science/article/pii/S1570826811000230`, doi:10.1016/j.websem.2011.04.004

[7] Delbru, R., Toupikov, N., Catasta, M., Tummarello, G.: A Node Indexing Scheme for Web Entity Retrieval. In: Aroyo, L., Antoniou, G., Hyvönen, E., ten Teije, A., Stuckenschmidt, H., Cabral, L., Tudorache, T. (eds.) ESWC 2010, Part II. LNCS, vol. 6089, pp. 240–256. Springer, Heidelberg (2010),
    `http://dx.doi.org/10.1007/978-3-642-13489-0_17`, doi:10.1007/978-3-642-13489-0_17.

[8] Manola, F., Miller, E.: RDF Primer, W3C Recommendation (2004),
    `http://www.w3.org/TR/rdf-primer/`

[9] Minack, E., Sauermann, L., Grimnes, G., Fluit, C., Broekstra, J.: The Sesame LuceneSail: RDF Queries with Full-text Search. NEPOMUK Technical Report 2008-1 (2008),
    `http://www.dfki.uni-kl.de/~sauermann/papers/Minack%2B2008.pdf`

[10] NEPOMUK, NEPOMUK - The Social Semantic Desktop - FP6-027705 (2008),
    `http://nepomuk.semanticdesktop.org/nepomuk/`

[11] OpenLink Software, OpenLink Virtuoso Universal Server: Documentation. RDF and Geometry (2009),
    `http://docs.openlinksw.com/virtuoso/rdfsparqlgeospat.html`
    (retrieved May 13, 2012)

[12] Oren, E., Delbru, R., Catasta, M., Cyganiak, R., Stenzhorn, H., Tummarello, G.: Sindice.com: A Document-oriented Lookup Index for Open Linked Data. Proceedings of the International Journal of Metadata, Semantics and Ontologies 3(1/2008), 37–52 (2008),
    `http://inderscience.metapress.com/content/3518208222365647`, doi:10.1504/IJMSO.2008.021204

[13] Prud'hommeaux, E., Seaborne, A.: SPARQL Query Language for RDF. W3C working draft, 4 (January 2008), `http://www.w3.org/TR/rdf-sparql-query`

[14] Wang, H., Liu, Q., Penin, T., Fu, L., Zhang, L., Tran, T., Yu, Y., Pan, Y.: Semplore: A scalable IR approach to search the Web of Data. Web Semantics: Science, Services and Agents on the World Wide Web 7(3), 177–188 (2009),
    `http://www.sciencedirect.com/science/article/pii/S1570826809000262`, doi:10.1016/j.websem.2009.08.001

Improving Range Query Result Size Estimation
Based on a New Optimal Histogram

Wissem Labbadi and Jalel Akaichi

Computer Science Department, ISG-University of Tunis, Le Bardo, Tunisia
wissem.labbadi@yahoo.fr, jalel.akaichi@isg.rnu.tn

Abstract. Many commercial relational Data Base Management Systems (DBMSs) maintain histograms to approximate the distribution of values in the relation attributes and based on them estimate query result sizes. A histogram approximates the distribution by grouping data into buckets. The estimation-errors resulting from the loss of information during the grouping process affect the accuracy of the decision, made by query optimizers, about choosing the most economical evaluation plan for a query. In front of this challenging problem, many histogram-based estimation techniques including the equi-depth, the v-optimal, the max-diff and the compressed histograms have well contributed to approximate the cost of a query evaluation plan. But, most of the times the obtained estimates have much error. Motivated by the fact that inaccurate estimations can lead to wrong decisions, we propose in this paper an efficient algorithm, called Compressed-V2, for accurate histogram constructions. Both theoretical and effective experiments are done using benchmark data set showing the promising results obtained using the proposed algorithm. We think that this algorithm will significantly contribute for helping to solve the problem of Multi-Query Optimization (MQO) resulting from queries interactions especially in Relational Data Warehouses (RDW) which represent the ideal environment in which complex OLAP queries interact with each other.

Keywords: Optimal histograms, Query result size estimation, Intermediate query result distribution, DBMS, Estimation error, Multi-query optimization, Query interaction.

1 Introduction

Many commercial DBMSs maintain a variety of types of histograms to summarize the contents of the database relation by approximating the distribution of values in the relation attributes and based on them estimate sizes and value distributions in query results [1, 2, 3, 4]. Different techniques for constructing histograms are described in [5]. The simplest approach for constructing a histogram on attribute X is by partitioning the domain D of X into β ($\beta > 1$) mutually disjoint subsets called buckets. A histogram approximates the distributions by grouping the data values into buckets. This grouping into buckets loses information. This loss of information engenders errors in estimates based on these histograms. The resulting estimation-sizes errors

H.L. Larsen et al. (Eds.): FQAS 2013, LNAI 8132, pp. 40–56, 2013.
© Springer-Verlag Berlin Heidelberg 2013

directly or transitively affect the accuracy of the decision, made by query optimizers, about choosing the most efficient access plan for a query [6] and undermine the validity of the optimizers' decisions.

The problem of query optimization consists in choosing, among many different query evaluation plans, the most economical one for a given query. Since the number of query evaluation plans increases exponentially with the number of relations involving the query [7], query optimization was becoming a worthwhile problem. A query can be performed by means of different intermediate operations such as join. A simple sequence of join operations that leads to the same final result is called a query evaluation plan [7]. Each query evaluation plan has an associated cost which depends on the number of operations performed in the intermediate joins. In [8], it has shown that errors in query result size estimates may increase exponentially with the number of operations performed in the intermediate joins. In worse of the cases, the chance of choosing the optimal query evaluation plan decreases since the query optimizer uses erroneous data to accomplish its task [9]. In that case, the query optimizer must estimate various parameters for the intermediate results of the operations and then use the obtained values to estimate the corresponding parameters of the results of subsequent operations [9]. Even if the original errors are small, their transitive effect on estimates derived for the final result may be devastating and so leading query optimizers to wrong decisions. For multi-join queries that are processed as a sequence of many join operations, the transitive effect of error propagation among the intermediate results on the estimates derived for the complete query may be destructive. This problem has been solved by approximating the cost of a query evaluation plan using histogram-based estimation techniques including the equi-width [10], the equi-depth [11], the v-optimal [1, 9, 12, 13], the max-diff [3] and the compressed histograms [3]. The idea was to estimate the query result sizes of the intermediate results and based on them selecting the most efficient and economical query evaluation plan.

Another important problem in which query result estimation techniques may be very useful is the phenomenon of query interaction which raises the problem of multiple queries optimization (MQO) especially in the relational data warehouse context (RDW). Relational data warehouses represent the ideal environment in which complex OLAP queries interact with each other. The problem of MQO combines the problem of efficient buffer management and the problem of query scheduling [14, 15]. It consists in finding an optimal scenario of queries processing that permits a total benefits from the buffered intermediate results which represents a major cause of performance problems in database systems. In fact, before executing a given query, it may get benefit from the actual content of the buffer if it has some intermediate results with previous queries [16]. Based on this scenario, if the query scheduler has a snapshot of the buffer content (intermediate results), it may reorder the queries to allow them getting benefit from the buffer [16].

Motivated by the fact that inaccurate estimations can lead to wrong decisions, our contribution can be summarized on preparing an experimental comparative study of the effectiveness of the different optimal histograms reported in the literature in order to identify the best one for reducing error in the estimations of sizes and value distributions especially in the results of queries with high complexity, e.g., multi-join queries. We envisage by this study to determine the main features of a good histogram in order to take them into account when developing our algorithm called Compressed-v2 algorithm for accurate histogram construction. Both theoretical and effective experiments are done using real data sets.

This paper is organized as follows. Section 2 provides an overview of several earlier and some more recent classes of histograms that are close to optimal and effective in many estimation problems. In section 3, we propose a new technique based on an effective algorithm called HistConst to construct a very promising histogram called compressed-v2 in terms of query result size estimation accuracy. In section 4, we propose a running example to show the efficiency of our algorithm. Section 5 presents a set of experiments to compare the effectiveness of the different histogram. Finally, Section 6 concludes and outlines some of the open problems in this area.

2   State of the Art

The buckets in a histogram are determined according to a partitioning rule and are limited by the disk space. We classify, in this section, the histograms listed in the literature into two classes based on two partitioning constraints. The first constraint consists in partitioning the attribute domain based on trivial rules and it concerns earlier histograms like the equi-width [10] and the equi-depth [11] histograms. The second constraint aims to avoid grouping vastly different values into the same bucket and it covers relative recent histograms like v-optimal [1, 9, 12, 13], max-diff [3] and compressed histograms [3].

2.1   Earlier Histograms

Trivial Histograms. This kind of histograms has a single bucket where all the attribute values fall into the same bucket. Frequencies approximated based on this histogram are identical for all attribute values [17]. This histogram assumes the uniform distribution over the entire attribute domain [6] and this assumption, however, didn't hold in real data. That's why trivial histograms usually have large error rate in query result estimation [8, 18].

Example 1. Let us consider the histogram maintaining, over one single bucket, the approximated frequencies of the attribute SALARY in a relation R with information on 100 employees (see Fig. 1). The domain of SALARY is the interval from 1000 \$ to 5000 \$.

According to the histogram in Fig. 1, the number of employees having for example a salary equal to 1500 \$, denoted SEL (SALARY=1500 \$) [11], is approximated by the average frequency of all salaries which is $S(R)/v(R, SALARY)$ where $S(R)$ is the size of the relation R and $V(R, SALARY)$ is the number of distinct values present in the attribute SALARY. Three different approaches were proposed in the literature to approximate the number of distinct values within a bucket [1, 6, 11].

The accurate number of tuples satisfying the above query is anywhere from 0 to 100. So, this approximation can be wrong at least by 50% (for $V(R, SALARY) = 2$) and in general usually by more than 50% (for $V(R, SALARY) > 2$) which represents a very large error rate.

Equi-Width Histograms. This kind of histograms consists in dividing the domain of the attribute values into K equal-width buckets and counting the number of tuples falling into each bucket [11]. Typically, equi-width histograms have 10 to 20 buckets [10].

Example 2. Let us consider the histogram maintaining the distribution of the attribute SALARY from example 1 (see Fig. 2). For reasons of simplicity, let this histogram divide the domain of the attribute into 3 equal-width buckets.

Continuing with the same query from example1, the accurate number of the employees having **SALARY = 1500\$** is anywhere from 0 to 48. So the true percentage of the employees with **SALARY = 1500 \$ is** anywhere from 0 to 0.48 ($0 \le$ SEL ($=1500$ \$) ≤ 0.48). An estimation, on average, of SEL ($=1500$ \$) from the histogram in Fig. 2 corresponds to the mid-point in this range which is 0.24. So, this estimate can be wrong by 0.24. In general, the maximum error in estimating SEL ($= Const$) on average, denoted SEL~ ($=Const$) [11], is half the height of the bucket in which $Const$ falls.

In [11], it has been shown that estimations of histograms belonging to the class of *equi-width* histograms are often better than *trivial* ones. They have frequently large errors since they force buckets to have equal width without controlling the height of each bucket. In such a histogram, we may find too high buckets and too low other ones. This huge disparity is due to the unexpected distribution of values over the entire attribute. In general, the distributions of values in the attributes of relations rarely follow any functional description, such as Zipf distribution [19] which leads to an inequitable distribution of values over the different buckets. In that case, if the bucket in which $Const$ falls is too high, the range in which SEL ($=Const$) belongs will be very large (the superior limit of the range is close to 100%) and a selectivity estimate will be wrong by 50% (mid-point in this range).

We can conclude that in order to control the maximum estimation error, the height of each bucket in the histogram should be controlled. Hence, the idea of creating histograms having buckets with equal height instead of equal width.

Equi-Depth Histograms. The maximum error in estimating from a histogram the selectivity of comparison, based on relational operators, is half the height of the bucket in which the comparison constant falls into. This error can be very close to 0.5, with an unlucky distribution of attributes values, where the tallest bucket contains almost 100% of the tuples in the relation. Creating a histogram where the attribute values are equally distributed over the different buckets will avoid having, in all cases, large errors in selectivity estimates. Such a histogram is called equi-depth [11]. In an equi-depth histogram, called also equi-height, the sum of the frequencies in each bucket is the same. This kind of histograms guarantees estimation with small error (usually < 0.5) and the maximum error can be reduced to an arbitrarily small value by increasing sufficiently the number of buckets in a way that half the height of a bucket will be negligible. For the construction of this histogram, we must first sort the attribute values in an ascending order to obtain a height balanced histogram.

Example 3. The distribution of the salaries of 100 employees in an equi-sum fashion over 3 equal height buckets is represented in Fig. 3.

Again choosing the maximum error in selectivity estimates as the half of the bucket, the estimation of SEL (=1500$) can be wrong at most by 0.16 (half the height of the first bucket in which 1500 falls). This error is 1 time and a half less than the error that can be present in the selectivity estimate obtained using an equi-width histogram. But the difficulty in this type of histograms consists in how to determine the required boundaries of the buckets in order to guarantee the equality of height between the different buckets.

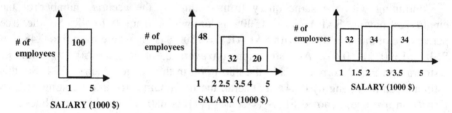

Fig. 1. Distribution of salaries over one singleton bucket   **Fig. 2.** Distribution of salaries over equal-width buckets   **Fig. 3.** Distribution of salaries over equal-height buckets

2.2   Relative Recent Histograms

V-Optimal Histograms. The v-optimal histograms [9, 12, 13], called also variance-optimal try to avoid grouping vastly different values into a bucket by reducing the weighted variance between the actual and the approximate distribution over all the approximated values within each bucket [13]. This variance is defined as $\sum_{j=1}^{\beta} p_j V_j$, where p is the number of frequencies, V is the variance of frequencies in the jth bucket and β is the maximum number of buckets.

The v-optimal histogram is optimal for estimating on average the result sizes of equality join and selection queries [1]. In order to approximating the number of distinct values with in a bucket, contrary to the previous histograms which instead of storing the actual number of distinct values in each bucket, they make assumptions about it such as continuous values assumption and point value assumption and both can lead to significant estimation errors, V-optimal histograms record every distinct attribute value that appeared in each bucket. Since bucket groups close frequencies and under the above assumption, all frequencies will be close to the average of frequencies so that estimations will be close to the actual results.

Definition 1. *Let H1 and H2 be two different histograms partitioning the values of an attribute X into the same number β (≥1) of buckets. The v-optimal histogram on X, among H1 and H2, is the histogram with the least variance.*

Example 4. Fig. 4 illustrates the above definition by the meaning of two different histograms H1 and H2 on the attribute SALARY with 3 buckets in each one. The frequencies of the different attribute values are listed in Fig. 4. a Fig. 4. b and Fig. 4. c show the partitioning technique employed for grouping the frequencies into buckets respectively for H1 and H2.

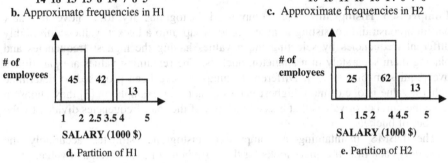

(1, 12) (1.5, 20) (2, 16) (2.5, 10) (3, 8) (3.5, 14) (4, 10) (4.5, 6) (5, 4)
a.Pairs of actual values and corresponding frequencies

14 18 13 13 8 14 7 8 5
b. Approximate frequencies in H1

10 15 11 15 16 14 6 7 6
c. Approximate frequencies in H2

d. Partition of H1

e. Partition of H2

Fig. 4. Example of optimal histograms

The cumulated variances V_{H1} and V_{H2} respectively of H1 and H2 are calculated as follows. $V_{H1} = 14+74+16.6667 = \mathbf{104.6667}$ and $V_{H2} = 24.5+52.8+0.5 = \mathbf{77.8}$. Based on Definition 1, the v-optimal histogram on the attribute SALARY, defined previously, between H1 and H2 is H2 since it has the least cumulated variance.

Maxdiff Histograms. The maxdiff histograms try to avoid grouping vastly different values within a bucket by inserting a boundary between two adjacent values v_i and v_{i+1} if the difference between the area of v_{i+1} and v_i is one of the β-1 largest such differences [3]. The area a_i of v_i is defined as $a_i = f(v_i).s_i$ where s_i is the spread of v_i and is defined as $s_i = v_{i+1} - v_i$ [3, 20]. Continuing with the set of values shown in Fig. 4a, the differences between the areas of the different successive values, noted Δ area, are calculated in Table 1. So, according to this table the bucket boundaries of a maxdiff histogram, approximating the distribution of values in Table 1 over 3 buckets, are inserted respectively between the two pairs of adjacent values (2, 2.5) and (3, 3.5) since they differ the most than the other pairs of adjacent values. The corresponding Maxdiff histogram is illustrated in Fig. 5.

This histogram estimates the number of tuples having the value 1500 in the attribute SALARY to be 22 engendering then an error on average that can reach 22%.

The comparison between the different histograms based on the error obtained in the estimates provided by each one for the same query (SEL (SALARY = 1500)) shows that v-optimal and max-diff are significantly more accurate and practical than earlier histograms.

Table 1. Computing the spread, area and Δ area

Value	1	1.5	2	2.5	3	3.5	4	4.5	5
Frequency	12	20	16	10	8	14	10	6	4
Spread	0.5	0.5	0.5	0.5	0.5	0.5	0.5	0.5	-
Area	6	10	8	5	4	7	5	3	-
Δ Area	4	2	3	1	3	2	2	-	-

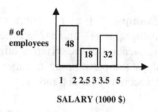

Fig. 5. Distribution of salaries in a Maxdiff histogram

Compressed Histograms. The compressed histograms try to achieve the new partition constraint consisting at avoiding to group, into a bucket, values with highly different frequencies by selecting the n values having the highest frequencies and placing them separately in n singleton buckets. The remaining values are partitioned over equi sum buckets [3]. Different techniques have been proposed to determine either a value is one of the n highest values or not. For example, in [3] they choose n to be the number of values that exceed the sum of the total frequencies divided by the number of buckets.

The DBMS maintaining a compressed histogram estimates accurately the selectivity each time the query looks for the periodicity of a high frequent value.

Example 5. Let's consider a compressed histogram approximating the distribution of the salaries over 5 buckets (see Fig. 6). According to [3] to choose the highest values, 1500$ is considered a high frequent value and is stored separately in a singleton bucket.

The compressed histograms by keeping values with high frequencies in singleton buckets and grouping contiguous values into buckets, they achieve great accuracy in estimating selectivity in databases [3]. That's, this histogram provides an accurate estimation on average (with a null error) of the same previous query SEL (SALARY = 1500$).

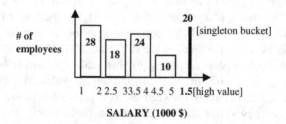

Fig. 6. Distribution of salaries in a compressed histogram

3   Compressed-V2 Histogram

The problem of constructing a good histogram and maintaining it well is primordial for the validity of the query optimizers' decisions [8, 18]. Due to their typically low-error estimates and simplicity in representing data distributions in low costs, there has been considerable work on identifying good histograms for estimating the result sizes of various query operators with reasonable accuracy [1, 3, 11, 21, 22]. The proposed histograms differ in how the attribute values are assigned to buckets to achieve good estimates and especially by the error rate in their estimates. In this work, we propose an efficient algorithm for constructing an improved version, called *compressed-v2*, of existing *compressed* histogram [3].We developed both theoretical and effective experiments to underline the effectiveness and the accuracy of our algorithm and to prove that the new version of *compressed* histogram generates the lowest estimation error among the existing techniques.

In a *compressed-v2* histogram, the n highest attribute values are stored separately in n singleton buckets. In our algorithm, we choose n to be the number of values that exceed the sum of all values divided by the number of buckets. The rest of values are partitioned over *maxdiff* buckets [11] instead of being partitioned over *equi-depth* ones [3]. An optimization phase is applied to the exceptional buckets in order to guarantee they generate good estimations. An exceptional bucket is a *maxdiff* bucket taller than the *equi-depth* bucket(s) approximating both the same distribution of values.

The problem of multi-query processing consists in finding an optimal scenario of query processing that permits a total benefits from the buffered intermediate results which represents a major cause of performance problems in database systems. The effectiveness of our histogram in estimating the size and the distribution of the intermediate results helps to well ordering the queries in order to allow them to get benefit form the buffer.

3.1   Definitions and Problem Formulation

In this section we define the accuracy of a histogram and formulate the problem studied in this paper.

Definition 2. *Let H1, respectively H2 be a compressed, respectively compressed-v2 histograms approximating the frequency distribution of an attribute X. We say that H2 is more optimal than H1 if and only if the error of H2 in approximating the frequency of each infrequent value of X is strictly less than the error of H1 in approximating the frequency of the same value.*

Theorem 1. Given a frequency distribution of a data set, a max-difference bucket with a height h1 provides estimation on average more accurate than an equi-depth bucket with a height h2 for all $h1 \le h2$.

Proof. Consider a relation R containing an attribute X. The value set V of X is the set of values of X that are present in R and F the set of their corresponding frequencies. Let M and E be respectively a maxdiff and an equi-depth histograms constructed by partitioning the values of V into β (≥ 1) buckets.

Let $(h_i^M)_{i=1..N}$ and $(h_i^E)_{i=1..k}$ be the respective heights of the buckets $(B_i^M)_{i=1..N}$ and $(B_i^E)_{i=1..k}$ that compose respectively the histograms M and E.

Let's take a maxdiff bucket B_i^M and an equi-depth bucket B_j^E, having common values that lie in their ranges, such that $h_i^M \leq h_j^E$ for a given $1 \leq i \leq N$ and $1 \leq j \leq K$. To prove that the frequency approximations on average of the common values based on the bucket B_i^M are more accurate than those based on the bucket B_j^E, it suffices to prove that: Error $(B_i^M) \leq$ Error (B_j^E), where Error(B_i^M), respectively Error (B_j^E) represents the total error of the approximation of B_i^M, respectively of B_j^E.

This inequality is verified since M, the max-diff histogram, is already constructed by minimizing the difference between the grouped values, whereas equi-depth permits vastly different values to be stored in the same bucket. Thus, the values grouped in B_i^M are close to the average of frequencies in B_i^M while those in B_j^E are dispersed from the average of frequencies in B_j^E. Hence, Error $(B_i^M) \leq$ Error (B_j^E).

The case where $h_i^M > h_j^E$ and there are common approximated values between B_i^M and B_j^E for $1 \leq i \leq N$ and $1 \leq j \leq K$ represents the main problem we focus in this paper. We try to improve the accuracy of these$(B_i^M)_{i=1..N}$, called exceptional buckets, using the proposed *HistConst algorithm*.

3.2   HistConst Algorithm

In general, the construction of a histogram on an attribute is performed on two steps. The first consists on partitioning the frequencies of the attribute into buckets, and the second step is to approximate the frequencies and values in each bucket in some technique [2]. We suggest in this section a naïve algorithm called **HistConst** which gives an accurate histogram with respect to the estimation error specified for the given sequence of values and number of buckets in O (n) time. The **HistConst** algorithm is illustrated in Fig. 7.

HistConst Algorithm. This algorithm takes in input the approximate frequencies of the attribute values and the number of permitted buckets. The **HistConst** algorithm proceeds as follow. First, there will be a call to the procedure **Find( )** to determine the highest values to store them separately in singleton buckets. Then, the procedure **maxdiff( )** takes care to partition the rest of values, over the remaining buckets in a maxdiff fashion, by inserting a bucket boundary between two adjacent values that differ the max. In the optimization phase, we try to reduce the height of the exceptional buckets to guarantee accurate estimations. This phase proceeds as follows:

We consider the height of an equi-depth bucket as a threshold.

Migrate, from each *exceptional bucket*, the *minimum values* in their order in the bucket range to the *previous bucket* while the height of this latter is lower than the threshold and the height of the *exceptional bucket* remains greater than the threshold. Once the previous bucket reach the threshold and the exceptional bucket is still higher than the threshold, then migrate all possible *maximum values* in the bucket range to the *next bucket* without that this latter exceeds the threshold.

Values from the previous bucket (respectively next bucket) can be migrated, if necessary, in their turn to its previous (respectively its next) bucket in order to respect the maximum tolerated height for a bucket.

Algorithm HistConst

Objective: Construct an optimal histogram with respect to the estimation error specified for the given sequence of values and number of buckets.

Input:B:Number of permitted buckets $(b_1, b_2, ...,b_B)$

threshold: maximum tolerated height for a bucket

check: booleanvariable that receive True if the actual bucket is an exceptional one.

Output: compressed-v2 histogram

begin

1. Find (F, V,B, V')
2. Maxdiff(L, F, B', maxdiff)

Optimization phase

3. **Repeat**
     check :=false
4. **Fori** := 1 toB'do {
5. If (exceptional_bucket(b_i)) then {
6. check:=true
7. **While** (h(prev_bucket(b_i)) < threshold) and (h(b_i) > threshold) **do** { // $h(b_i)$:determines the height of b_i
8. migrate (min_val(b_i), b_i, prev_bucket(b_i)) // $min\_val(b_i)$:determinesminimum value in the range of b_i
9. }
10. If h(prev_bucket(b_i) ≥ threshold) then // $prev\_bucket(b_i)$: determines the previous bucket of b_i
11. **While** (h(next_bucket(b_i)) < threshold) and (h(b_i) > threshold) **do** { // $next\_bucket(b_i)$: determines the successive bucket of b_i
12. migrate (max_val($_{bi}$), $_{bi}$, next_bucket($_{bi}$)) //$max\_val(b_i)$:determines maximum value in the range of b_i
13. }
14. }
15. }
16. **Until** (check = false)
17. Result:return compressed-v2

end

Fig. 7. The HistConst algorithm

Finding Highest Values. We present in the Fig. 8 a pseudo code to find the high frequent values among those actually present in the relation.

Procedure Find (F, V, B, **V'**)

Inputs:V: set of values of the attribute that are present in the relation, V= {v_i | $1 \le i \le N$}
     F: frequency vector of the attribute, F= {f(v_i) | $1 \le i \le N$}

Output: V': set of the high frequent values, V' = {v_i | f(v_i) > $\frac{\Sigma f(vi)}{B}$, $1 \le i \le N$}

begin

1. **fori**: = 1to N**do**{
2.   if (f(v_i) >$\frac{\Sigma f(vi)}{B}$) then
3.   Add(v_i, V')
4. }
5. Result: return V'

end

Fig. 8. Code of the procedure Find

Having the approximated values and the corresponding approximated frequencies, the procedure **Find( )** takes care to determine the highest values to store them separately in singleton buckets. Each value is compared to the sum of all source values divided by the number of total buckets. If the value exceeds this quotient, then is considered a high frequent value.

Constructing Maxdiff Histogram. After finding the highest values and storing each one separately in a singleton bucket, we propose an algorithm of the procedure maxdiff presented in Fig. 9 to partition the remaining values following the technique that consists in separating vastly different values into different buckets.

Procedure Maxdiff(L,F, B',**maxdiff**)

Inputs:L:set of the remaining values (low frequent values) that are present in the relation, $L = \{v_j \mid f(v_j) \leq \frac{\sum f(v_j)}{B}, 1 \leq j \leq M < N\}$

B':number of the remaining buckets (non singleton buckets) for grouping the low frequent values

Output:maxdiff:max-diff histogram partitioning the remaining values

begin

1. **for** i:= 1 **to** (B'-1) **do** {
2.   [max_area := 0]**for** j :=1 **to** (M-1) **do** {
3. Δ Area := [f(v_{j+1})*S_{j+1}] – [f(v_j)*S_j]
4.   If (Δ Area >max_area) then {
5.   max_area := Δarea
6.   bound :=j
7.   }
8.   }
9.   *Insert bucket_boundary* (v_{bound}, $v_{bound+1}$)
10. }
11. Result: return maxdiff

end

Fig. 9. Code of the procedure maxdiff

The procedure **Maxdiff** begins first by calculating the differences between all the adjacent values. Then, it inserts bucket boundaries between the pairs of adjacent values that differ the most in their frequencies with respect to the number of buckets permitted to partition the remaining values.

Procedure Exceptional_bucket(b^M)

Input:b^M:maxdiff bucket

begin

1. If (h (b^M) > threshold) then
2. Exceptional_bucket:= True
3. ElseExceptional_bucket := False
4. Result: return Exceptional_bucket

end

Fig. 10. Code of the function Exceptional_bucket

Finding Exceptional Buckets. We propose in this section a boolean function to determine the exceptional buckets (Fig. 10). We remind that the threshold is the height of an equi-depth bucket partitioning the same values which is approximately equal to $\frac{\sum f(vi)}{B'}$ with $1 \leq i \leq M$. We remind also that B' is the number of buckets used to partition the remaining values.

After approximating the low frequent values and their frequencies in each bucket according to the maxdiff partitioning technique, this function returns true for each exceptional bucket, false else. In the affirmative, the corresponding bucket undergoes the change of the optimization phase described above in order to adjust its height with respect to the height of an equi-depth bucket.

In Table 2., we describe the complexity of the HistConst main components.

Table 2. HistConst time complexity

Algorithm	Time Complexity
Procedure Find	O(N)
Procedure Maxdiff	O(M)
Algorithm HistConst	O(N)
Procedure Exceptional_bucket	O(B')

Where N is the number of attribute values, M is the number of low frequent values and B' is the number of non-singleton buckets.

4   Running Example

Suppose we have the following values from an integer-valued attribute with their corresponding low frequencies (see Table 3).

Table 3. Set of integer values with their frequencies

Value	1	2	3	4	5	6	7	8	9
Frequency	2	1	2	3	4	1	2	3	2
ΔFrequency	1	1	1	1	3	1	1	1	

Following the *HistConst* algorithm steps to partition these values over four buckets as in a maxdiff histogram, the three pairs of adjacent values that differ the most in their frequencies are (5, 6), (2, 3) and (7, 8). Thus, the bucket boundaries are placed between these adjacent values (see Fig. 11).

According to *HistConst* algorithm, the bucket with a range [3, 5] and a height equal to 9 is considered an exceptional bucket. In the context of approving exceptional buckets in a compressed-v2 histogram, the procedure Exeptional_bucket( )migrates the value 3 from the second bucket to the first bucket, as shown in Fig. 12, since the two adjacent values 2 and 3 are contiguous and grouping them into the same bucket doesn't affect the accuracy of frequency approximation inside a bucket.

The result of the optimization phase is a histogram that discards vastly different values and then partitions them like in an equi-depth histogram such that the sum of frequencies in each bucket is approximately the same. This is in contrast to the equi-depth histograms that permit vastly different values to be stored in the same bucket. The resulting histogram is illustrated in Fig. 13.

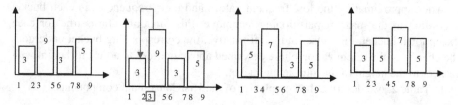

Fig. 11. Partitioning of low frequent values in a maxdiff fashion

Fig. 12. Improvement in phase

Fig. 13. Partitioning the rest of values after optimization

Fig. 14. Partitioning infrequent values in equi-depth fashion

4.1   Selectivity Estimation of Low Frequent Values with Compressed and Compressed-v2 Histograms

The accuracy of estimates of range query result sizes obtained through maintained histograms depends heavily on the partitioning rules used to group attribute values into buckets [9, 11]. Here, we compare the average errors incurred when estimating the selectivity only of low frequent values based on compressed and compressed-v2 histograms. The same highest values are chosen for the two types of histograms and hence their frequencies are similarly approximated in both histograms.

The compressed histogram approximating, over four buckets, the frequencies of these values is illustrated in Fig. 14. To investigate the accuracy of query result size estimates obtained from compressed and compressed-v2 histograms, we choose to compare the accuracy of the selectivity estimation of the values 5 and 6 obtained from the two types of histograms described as follows:

- *Compressed*: The value 5 falls in the third bucket (Fig. 14). Then, SEL (SALARY = 5) is estimated by the average of frequencies in this bucket which is 2. The true fraction of tuples with salary equal to 5 is 4, then this estimate is wrong by 0.5. Similarly, the value 6 falls into the same bucket and its selectivity is estimated on average by 2. The true fraction of tuples with salary equal to 6 is 1 and hence this estimate is wrong by 0.5.
- **Compressed-v2:** The value 5 falls in the second bucket (Fig. 13) and SEL (SALARY = 5) is estimated by 3. The true fraction of tuples with salary equal to 5 is 4 which mean that the estimate is wrong by 0.25. Contrary to the compressed histogram, the value 6 is separated from the value 5 and is stored in the third bucket (Fig. 13) since they are judged as two large different values. SEL (SALARY = 6) is estimated on average by 1 where the real frequency is 1. The error in the estimate in this case is equal to 0.

Comparing the errors in the estimates based on the two types of histograms, we see clearly that the compressed-v2 approximates much better the frequency of the values 5 and 6 than the compressed histogram. The frequencies of the other values are almost equally approximated in the two histograms since each group contains contiguous values which are stored in approximately equal height buckets in both histograms.

5 Experimentation Results

We investigated the effectiveness of the different histogram types cited above for estimating range query result sizes. The average errors due to the different histograms, as a function of the number of buckets, are computed each time when estimating based on histograms the result size of a selection query where the selectivity conditions are related each time to values with different frequencies like infrequent values, balanced values and very frequent values.

The experiments were conducted on six different histogram algorithms including equi-width, equi-depth, v-optimal, maxdiff, compressed and compressed-v2 in three specified histogram data-frequency category including low, balanced and very high, while using two different distributions of the attribute Salary from the *American League Baseball Salaries (Albb)* and *National League Baseball Salaries (Nlbb)* databases respectively for the years 2003 and 2005. The values in the attribute Salary vary from 300 000$ to 22 000 000$ in the two databases.

The frequencies of the values in the attribute Salary in the database of *Albb* (respectively Nlbb) vary from 1 to 44 (respectively from 1 to 25). We consider [1..9] (respectively [1..8]) to be the range of low frequencies, [10..30] (respectively [9..15]) the range of balanced frequencies and [31..44] (respectively [16..25]) to be the range of very high frequencies.

In the following experiment, we studied the performance of the different histograms by comparing through several graphs the typical behavior of the histograms errors in approximating the frequencies of different values with varying the number of buckets. For an efficient study of the effectiveness of these histograms, we select randomly three values from each database: an infrequent value, a balanced value and a very frequent value. The errors in approximating the frequency of a given value are represented in a graph separately. The x-axis of each graph shows the number of buckets and the y-axis shows the average error of each histogram for different number of buckets.

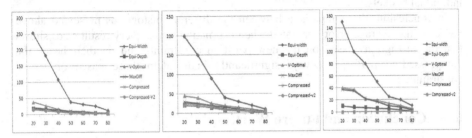

Fig. 15. Average error as a function of the number of buckets of approximating, in Nlbb, the frequency of a) an infrequent value, b) a balanced value, and c) a very frequent value

We select from the database of Nlbb the following values with their corresponding real frequencies (1000000, 1), (500000, 10) and (316000, 25). The errors of the six histograms in approximating the frequencies of these values are illustrated respectively in Fig. 15.a, Fig.15.b and Fig. 15.c.

We select from the database of Albb the following values with their corresponding real frequencies (7500000, 1), (600000, 10) and (300000, 25). The errors of the six histograms in approximating the frequencies of these values are illustrated respectively in Fig. 16.a, Fig.16.b and Fig. 16.c.

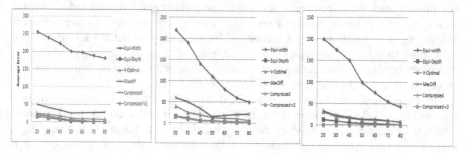

Fig. 16. Average error as a function of the number of buckets of approximating, in Albb, the frequency of a) an infrequent value, b) a balanced value, and c) a very frequent value

Looking at the different figures, we observe that the error generated is monotony proportional to the number of buckets. As shown in the two figures, the accuracy can be reached when increasing the number of buckets for all histogram types and the compressed-v2, compressed, max-diff and v-optimal histograms are significantly better than the others that they show the least error for different number of buckets. Moreover, the equi-width histogram exhibits the worst accuracy.

Based on the different figures, we distinguish clearly, by comparing the average errors generated by the entire set of histograms when estimating the selectivity of infrequent values, balanced or very frequent values, a set of effective histograms, i.e. compressed-v2, compressed, V-optimal and Max-diff, where the compressed-v2 presents each time the least approximation error for different number of buckets. The same behavior of compressed-v2 errors in all the figures improves the victory of this histogram over the other ones that it gives 100% accurate approximation of the frequencies the highest values since their actual frequencies are stored separately in individual buckets.

In conclusion, the comparison between the different histograms presented above based on the average error generated when estimating range query result sizes shows that the histograms based on the new partition constraints and on their heads the compressed-v2 performs always significantly better than those based on trivial constraints.

6   Conclusion and Future Work

The problem of minimizing the error in estimating range query result sizes remains a real challenge despite the serious research done on identifying classes of optimal histograms that generate least errors in the estimations of sizes and value distributions especially in the results of queries with high complexity, e.g., multi-join queries.

In this paper, we provided an overview of several earlier and some more recent classes of histograms that are close to optimal and effective in many estimation problems. In addition to that, we have introduced a new algorithmic technique, HistConst, for constructing a more accurate histogram called compressed-v2. An experimental comparative study was proposed to study the effectiveness of the different classes of optimal histograms reported in the literature and our proposed histogram in estimating sizes and value distributions especially in the results of complex queries, e.g., multi-join queries. The experiments show that estimations based on our histogram are always better than those based on the other remaining types of histograms.

The identification of the optimal histogram remains an open field. As several new research opportunities appear, we will try to identify optimal histograms for different types of queries to limit not only the average estimation error but also other metrics of error, to determine the appropriate number of buckets to build the optimal histogram and to find the histogram that can handle uncertain data.

An important direction for research is to focus on the problem of data stream which is the transmission of the flow of data that changes over time. Existing database systems do not process data streams efficiently and this makes this area a popular search field [23, 24].

References

1. Ioannidis, Y., Poosala, V.: Balancing histogram optimality and practicality for query result size estimation. In: Proceedings of the 1995 ACM SIGMOD International Conference on Management of Data, pp. 233–244 (1995)
2. Jagadish, H.V., Koudas, N., Muthukrishnan, S., Poosala, V., Sevcik, K., Suel, T.: Optimal histograms with quality guarantees. In: Proceedings of the 24th International Conference on Very Large Data Bases (VLDB), New York, USA, pp. 275–286 (1998)
3. Poosala, V., Ioannidis, Y.E., Haas, P.J., Shekita, E.J.: Improved histograms for selectivity estimation of range predicates. In: Proceedings of the 1996 ACM SIGMOD International Conference on Management of Data, pp. 294–305 (1996)
4. Jagadish, H.V., Jin, H., Ooi, B.C., Tan, K.-L.: Global optimization of histograms. In: Proceedings of the 1998 ACM SIGMOD International Conference on Management of Data, pp. 223–234 (2001)
5. Yu, C., Philip, G., Meng, W.: Distributed top-N query processing with possibly uncooperative local systems. In: Proc. 29th VLDB Conf., Berlin, Germany, pp. 117–128 (2003)
6. Selinger, P.G., Astrahan, M.M., Chamberlin, D.D., Lorie, R.A., Price, T.G.: Access path selection in a relational database management system. In: Proceedings of the ACM SIGMOD International Symposium on Management of Data, Boston, Mass., pp. 23–34 (June 1979)
7. John Oommen, B., Rueda, L.G.: An empirical comparison of histogram-like techniques for query optimization. In: Proceedings of the 2nd International Conference on Entreprise Information Systems, Stafford, UK, July 4-7, pp. 71–78 (2000)

8. Ioannidis, Y., Christodoulakis, S.: On the propagation of errors in the size of join results. In: Proceedings of the 1991 ACM SIGMOD Conference, Denver, CO, pp. 268–277 (May 1991)

9. Ioannidis, Y., Christodoulakis, S.: Optimal histograms for limiting worst-case error propagation in the estimates of query optimizers. To appear in ACM-TODS (1992)

10. Kooi, R.P.: The optimization of queries in relational databases. PhD thesis, Case Western Reserver University (September 1980)

11. Shapiro, G.P., Connell, C.: Accurate Estimation of the Number of Tuples Satisfying a Condition. In: Proceedings of ACM-SIGMOD Conference, pp. 256–276 (1984)

12. Ioannidis, Y.: Universality of serial histograms. In: Proceedings of the 19th Int. Conf. on Very Large Databases, pp. 256–267 (December 1993)

13. Poosala, V., Ioannidis, Y.: Estimation of query-result distribution and its application in parallel-join load balancing. In: Proceedings of the 22nd Int. Conf. on Very Large Databases, pp. 448–459 (1996)

14. Gupta, A., Sudarshan, S., Viswanathan, S.: Query scheduling in multi query optimization. In: IDEAS, pp. 11–19 (2001)

15. Thomas, D., Diwan, A.A., Sudarshan, S.: Scheduling and caching in multi query optimization. In: COMAD, pp. 150–153 (2006)

16. Kerkad, A., Bellatreche, L., Geniet, D.: Queen-Bee: Query interaction- aware for buffer allocation and scheduling problem. In: Cuzzocrea, A., Dayal, U. (eds.) DaWaK 2012. LNCS, vol. 7448, pp. 156–167. Springer, Heidelberg (2012)

17. Ioannidis, Y.: Query optimization. In: ACM Computing Surveys, Symposium Issue on the 50th Anniversary of ACM, vol. 28, pp. 121–123 (1996)

18. Christodoulakis, S.: Implications of certain assumptions in database performance evaluation. ACM TODS 9(2), 163–186 (1984)

19. Zipf, G.K.: Human Behavior and the Principle of Least Effort: an Introduction to Human Ecology. Addison-Wesley, Cambridge (1949)

20. Liu, Y.: Data preprocessing. Department of Biomedical, Industrial and Human Factors Engineering Wright State University (2010)

21. Ioannidis, Y., Poosala, V.: Histogram-based solutions to diverse database estimation problems. IEEE Data Engineering Bulletin 18(3), 10–18 (1995)

22. Muralikrishna, M., Dewitt, D.J.: Equi-depth histograms for estimating selectivity factors for multi-dimensional queries. In: Proceedings of ACM SIGMOD Conference, pp. 28–36 (1988)

23. Mousavi, H., Zaniolo, C.: Fast and Accurate Computation of Equi-Depth Histograms over Data Streams. In: Proceedings of EDBT, Uppsala, Sweden, March 22-24 (2011)

24. Gomes, J.S.: Adaptive Histogram Algorithms for Approximating Frequency Queries in Dynamic Data Streams. In: 12th International Conference on Internet Computing, ICOMP 2011, Las Vegas, NV, July 18-21 (2011)

Has FQAS Something to Say on Taking Good Care of Our Elders?

María Ros*, Miguel Molina-Solana, and Miguel Delgado

Department Computer Science and Artificial Intelligence,
Universidad de Granada,
18071 Granada, Spain
marosiz@decsai.ugr.es, {miguelmolina,mdelgado}@ugr.es

Abstract. The increasing population of elders in the near future, and their expectations for a independent, safe and in-place living require new practical systems and technologies to fulfil their demands in sustainable ways. This paper presents our own reflection on the great relevance of FQAS' main topics for recent developments on the context of Home Assistance. We show how those developments employ several techniques from the FQAS conference scope with the aim of encouraging researchers to test their systems in this field.

1   Introduction

Flexible Query Answering Systems (FQAS) is a conference focusing on the key issue in the information society of providing easy, flexible, and intuitive access to information to everybody. In targeting this issue, the conference draws on several research areas, such as information retrieval, database management, information filtering, knowledge representation, soft computing, management of multimedia information, and human-computer interaction.

The theoretical developments on those areas have traditionally be applied to a wide range of fields with the aim of improving performance, data representation, and interaction with users, among other aspects.

With the occasion of the tenth edition of the FQAS conference, we feel the necessity of reflecting on how the conference has contributed to several areas of interest for both the research community and the general population, with the aim of putting in value some of the practical applications that the topics of the conference might have. In particular, we focus in this paper on assisting technologies for elderly people at their own home.

Successful ageing has indeed become one of the most important problems in our society in recent years. According to a recent study from Eurostat [1], by 2060, most European countries are likely to have a proportion of oldest-old of more than 10%, against the 1-2% from 1960 (a hundred years before).

Past, and still current, approaches to provide these people with an appropriate quality of life are mainly based on increasing resources such as nursing homes

---

* Corresponding author.

H.L. Larsen et al. (Eds.): FQAS 2013, LNAI 8132, pp. 57–66, 2013.
© Springer-Verlag Berlin Heidelberg 2013

and caregivers. To the date, those efforts have been proved insufficient, and the approach is hardly scalable and manageable, as more and more people demand assistance. What is more important, elders are not very receptive to the fact of moving out to a new home, leaving their previous life behind.

In light of this situation, new solutions have been proposed to handle it by means of developing suitable tools for home assistance, that is, by assisting elders at their own homes with the help of technologies. In devising those tools, we believe that the aforementioned FQAS' research areas have a lot to say.

In fact, these developments are not only aimed at elders enjoying a good health and conditions, but also to a wide range of people who suffer from cognitive impairments, memory deficiencies, or other diseases such as Alzheimer's. Without the aid of those systems, those people could hardly live independently at their own homes.

Therefore, the present paper highlights some of the most relevant proposals found in the recent literature in the field of home assistance, specially focusing on elders and ageing in place. As said, we believe this is a topic of growing interest that should be carefully addressed by several agents (Governments, families, caregivers, patients) involving people from several disciplines (Medicine, Robotics, Artificial Intelligence, Sociology). The European Commission thinks also in this way and considers this as one of its research priorities [2].

The paper describes how several of those systems are related with the topics addressed at the FQAS conference; in particular, we will focus on Information Retrieval, Knowledge Representation, Domain and User Context Modelling, Approximate Reasoning, and Human Computer Interaction.

By means of the present review, we also aim to encourage researchers to apply their developments into the field of Home Assistance. Surprisingly for us, contributions to the conference focusing on Home Assistants are very scarce. In fact, we were only able to find the work by Tablado *et al.* on Tele-assistance[3].

2   Knowledge Representation and Information Retrieval

Within the broad area of Assisted Living, systems aimed at monitoring patients are the most numerous. Their general goal is to maintain registers of the vital constants of the patients, their movements and other variables, taking samples at small time intervals. This way, users are continuously monitored and if something goes wrong, and alarm could be fired and emergency bodies, or family, be notified.

To do so, the systems need an appropriate process of data gathering, information retrieval and knowledge representation. All of those, are topics of interest for the FQAS community. In Assisted Living applications, those processes are general applied in controlled and known environments called *Smart Homes* [4,5], houses equipped with special structured wiring to provide occupants with different applications to control their own environment. For instance, remotely controlling or interacting with different devices, monitoring their activities, or offering guidelines to improve them, are a few examples of the kind of applications and services that can be easily found in a Smart Home.

Miskelly [6] presented an interesting overview on different devices that can be embedded in a home environment, such as electronic sensors, fall detectors, door monitors or pressure mats in order to gather the data.

Several projects have been proposed to build Smart Homes for people with some kind of impairment. One of those is the Center for Eldercare and Rehabilitation Technology (CERT) [7] focused on the TigerPlace facility [8,9]. TigerPlace is an elderly residence where multiple sensing technologies has been installed in order to monitor the residents. This project implied an interdisciplinary collaboration between electrical and computer engineers, gerontological nurses, social workers and physical therapists at the University of Missouri.

With the availability of intelligent mobile devices, more and more projects are incorporating them as the way of gathering sensing data [10]. Winkler and colleagues [11] have studied the feasibility of a remote monitoring system based on telemonitoring via a mobile phone network. Several portable home devices for ECG, body weight, blood pressure and self-assessment measurements are all connected to a PDA by means of a local Bluetooth network. Users are then continuously monitored, and their vital constants forwarded to central servers where electronic records are available to telemedical centres. Additionally, the system is able to provide efficient detection of home emergencies, contacting with the corresponding service (ambulance, general practitioner, specialist or local hospital depending on the case).

Also, there are projects that take advantage of mobile devices as the tool to interface with users. For instance, *MASPortal* is a grid portal application for the assistance of people for medical advice at their homes. It provides remote access to medical diagnostic and treatment advice system via Personal Digital Assistants (PDA).

Other studies have employed the monitored data to extract behavioural patterns that could be extrapolated to other people with the aim of detecting potential changes in health status. In one of such studies [12], the authors reported the extraction of a dozen of behavioural patterns after monitoring 22 residents in an assisted living facility. They were able to model circadian activity rhythms (CARs) and their deviations, and to use these data, together with caregiver information about monitored resident, to detect deviations in activity patterns. The system would warn the caregivers, if necessary, allowing them to intervene in a timely manner.

Other efforts of building Smart Homes include the *iDorm* [13,14], CASAS [15,16] and the Georgia Tech Aware Home [17], as well as another eight projects reviewed by Droes *et al.* [18].

These projects increase their utility and relevance when we refer to houses in which elders live alone. In those situations, the virtual assistant acquires the role of caregiver and should be able to respond to special situations and emergencies, such as falls or dismays, as soon as they are identified from the available data.

3   Domain and User Context Modelling

Most systems we have already cited make use of proper knowledge representation. However, there are some which specifically focus on representing the user and his context in an accurate way. Not in vain, Aging in Place applications make extensive use of Context-aware services, in order to better represent the surrounding conditions of users and to personalise their functioning by means of adjusting their behaviour to changes.

Sanchez-Pi et al. [19] proposed a set of categories that might be adapted to the user's preferences and to the environmental conditions. Elders are then expected to specify their relevant activities according to three categories: comfort (temperature control, light control or music control), autonomy enhancement (medication, shopping or cooking) and emergency assistant (assistance, prediction, and prevention of emergencies). The authors also proposed a prototype of system, taking advantage of the context.

A different system, a context-aware pervasive healtcare for chronic conditions system, H-SAUDE, was proposed in [20]. It was based on a decision-level data fusion technique and aimed for monitoring and reporting critical health conditions of hypertensive patients while at home.

Helal et al. [21] described an indoor precision tracking system taking advantage of available context-aware services. It was developed as an OSGi-based framework, integrating the different tracking services into an standard platform. They also employed ultrasonic sensor technology to monitor the environment. The system has been tested at House of Matilda (an in-laboratory mock up house) in the Pervasive Computing Laboratory at the University of Florida.

The MavHome (Managing An Intelligent Versatile Home) project [4] at the University of Texas at Arlington aimed to develop a home capable of offering personalized services to their inhabitants. Authors also control the ambiance of the environment, including temperature, ventilation and lighting and the efficiency of the cost of utilities, such as gas and electricity. To achieve such a goal, an appropriate representation of those concepts and the knowledge involved were needed.

The CARA (Context Aware Real-time Assistant) project [22] is a pervasive healthcare system specifically oriented to elders with chronic diseases. It aims at remotely monitoring patients and notifying caregivers if necessary. To do that, it employs context-aware data fusion and inference mechanisms using both fuzzy representation and fuzzy reasoning. The context is described and modelled as a set of fuzzy rules, and processed by a related context-aware reasoning middleware.

Graf et al. [23] described a robotic assistant able to lead users to specific rooms in the house and to manipulate objects. In particular, it had the capability to navigate autonomously in indoor environments, be used as an intelligent walking support, and execute manipulation tasks. The system was tested with elderly users from an assisted living facility and a nursery home, and in a sample home environment. The authors reported good results achieving those tasks.

Another interesting system that takes full advantage of user context modelling is the one presented in [24]. *Autominder* assists the elderly in creating daily plans, decision making and execution of plans. It is able to cover user's preferences and adjust its knowledge accordingly to user's history. Besides, temporal information is also included as constraints that the system should follow to provide appropriate reminders.

In [25], authors propose a health care web information system for assisting living of elderly people at Nursing Homes. They developed a tool with the aim of helping doctors and nurses to have a more realistic health status of patients and their contexts. With such information, doctors could also be provided with reports about management outputs such as medical supplies.

In this kind of projects, besides the employment of a domain and user context model, an appropriate Human-Computer Interaction is an issue of particular importance. As seen, both are within the scope of the FQAS conference.

4 Approximate Reasoning Models

When talking about systems that monitor users' lives, one of the areas involved is that of Approximate Reasoning. Probabilistic systems offer great flexibility to control alternatives of behaviour realisation in different environments. A study about the advantages and drawbacks of employing stochastic techniques for recognizing human behaviours was done by Naeem and Bigham [26]. On the other hand, Singla *et al.* [27] tested various probabilistic modelling methods that evaluate situations in which user has not performed the sequence correctly, and the activities are interrupted and interwoven. The conclusion was that models should be flexible but, at the same time fixed enough, to be able to recognize activities in which users make some irregular actions.

Even though, other techniques such as Data mining and Fuzzy systems have been used in several projects obtaining good results in the detection of abnormal ADL [28].

Our previous experience in the issue involve a method for detecting and recognizing behaviours by means of inductive learning and based on the detection of temporal relations between actions using Data Mining techniques[29,30]. Additionally, in [31] we presented some learning and matching methods for models of behaviours based on Learning Automata. They were tested at the *iSpace*, a real test-bed developed by the University of Essex.

Project *CARA* [22], already mentioned, uses the continuous contextualization of the user activities as input to a reasoning system based on Fuzzy logic to predict possibly risky situations.

4.1 Activity Recognition

One of the most challenging issues within Ambient Assisted Living (AAL) is, without doubt, the activity recognition process[18]. The goal here consists of designing models and systems to label the different activities that inhabitants perform in a controlled environment, and providing some personalized support services to those activities.

In general, there are two main trends in the field of Activity Recognition. Firstly, a *low level* activity recognition process, in which researchers focus their attention on basic activities, such as, walking, running, or crouching [32]. Secondly, an *abstract level* activity recognition process. In this level, authors manage high level activities known as Activities of Daily Living (ADL) [33,34], a sequence of atomic actions within user's daily routine, such as toileting, making a meal, or leaving home.

In the literature, a variety of studies can be found proposing solutions to the different aspects within this area. For example, the underlying sensor network for user data acquisition is the main focus of works such as [33,35]; whereas the representation and modelling is studied in [36,13,14]. Precisely, Hagras *et al.* propose a fuzzy control system to detect user activities in a ambient intelligent environment known as iDorm[14]. Concretely, they propose a type-2 rule based system to control lighting and temperature accordingly to multiple users profiles.

5   Human Computer Interaction

Within the area of Home Assistance, an appropriate Human-Computer Interaction is an issue of particular importance, due to the particular characteristics of their target users: mainly elderly or disabled people. Natural interfaces and spoken language should not be nice additions, but required components, in these systems.

Dingli and Lia [37] proposed a home butler system to help with common activities at home. What makes this project special is its ability of generating dialogues. The system was able to create them by using television series scripts, looking for syntactical repetitive structure. The dialogues were structured as networks, with nodes representing possible situations/questions, and arcs representing potential answers.

However, what we can extensively find in literature are works focusing on robotics, manly embedding computational capabilities to common appliances. Acting this way, the capabilities of objects already known by users are extended by means of reasoning and communication. In any case, users do not need to learn how to use new objects, because the old ones keep their traditional behaviour.

The overview by Pineau and colleagues [38] is still a relevant document despite the time passed since its publication. In it, the authors introduced the challenges that robotic assistants pose in nursing homes, and described several attempts on developing such systems. In general, on top of those elements, a central and virtual assistant is often build as interface and manager.

Service robotics cover a wide range of applications and tasks such as: vacuuming and cleaning, gardening and lawnmowers, personal robotic assistants, telepresence, teleassistance and health, entertainment, and home security and privacy [39].

Among those, there are some specially aimed at people with some type of disabilities such as the portable five-degree-of-freedom manipulator arm (ASI-BOT) [40]. It is useful for helping in successfully deal with Activities of Daily Living (ADL) (eating, shaving, drinking) performance. The arm is able to climb to different surfaces, providing a portable and friendly interface.

5.1 Social Networks

Apart from connecting users with the computer, one of the transversal goals of the systems we are describing is the ability to connect users with other users, family and caregivers. To achieve such aims, the concept of Social Network is of relevance; and in particular, there is one application in which Social Networks are heavily employed: conferencing.

The ACTION project [41] was a pioneer on using information and communication technology to support frail elders and their family carers, by connecting homes and call centres through a videoconferencing system. The aims were to enhance elder's quality of life, independence and preparedness, and to break social isolation. Reduction on both the sense of loneliness and isolation were reported by users.

6 Conclusions

The increasing population of elders in the near future, and their expectations for a independent, safe and in-place living, require new practical systems and technologies to fulfil their demands in sustainable ways. Research and projects in the context of Home Assistance are experimenting a incredible growth, and many funding bodies have them among their priorities.

On the other hand, the Flexible Query Answering Systems (FQAS) conference focuses on providing easy, flexible, and intuitive access to information to everybody. To do so, the conference draws on several research areas, such as information retrieval, database management, information filtering, knowledge representation, soft computing, management of multimedia information, and human-computer interaction.

With the occasion of its Tenth edition and the aim of highlighting the valuable contributions that FQAS could be offering to the Home Assistance field, our paper presented an overview on how different projects in this field are related with the main topics of the FQAS conference.

Paradoxically, few contributions to the FQAS conference have directly addressed the area of Home Assistance so far. Therefore, we encourage researchers to test their theoretical developments on this interesting field for future editions of the conference.

From our perspective, we can only anticipate a even deeper integration in the following years between the research contributions within the scope of FQAS and the developments in home assistance, as the later gain in maturity and intelligent behaviours.

Acknowledgements. The authors would like to thank the Spanish Ministry of Education for their funding under the project TIN2009-14538-C02-01. Miguel Molina-Solana and María Ros are also funded by the Research Office at the University of Granada.

References

1. Eurostat: Eurostat: Population projections,
   http://epp.eurostat.ec.europa.eu/statistics_explained/index.php/
   Population_projections#Grandparent_boom_approaching (accessed April 1, 2013)
2. Commission, E.: Ambient assisted living: Joint programme (2013),
   http://www.aal-europe.eu/wp-content/uploads/2012/04/
   AALCatalogue_onlineV4.pdf (accessed April 1, 2013)
3. Tablado, A., Illarramendi, A., Bagüés, M.I., Bermúdez, J., Goñi, A.: A flexible data processing technique for a tele-assistance system of elderly people. In: Christiansen, H., Hacid, M.-S., Andreasen, T., Larsen, H.L. (eds.) FQAS 2004. LNCS (LNAI), vol. 3055, pp. 270–281. Springer, Heidelberg (2004)
4. Cook, D.J., Youngblood, M., Heierman III, E.O., Gopalratnam, K., Rao, S., Litvin, A., Khawaja, F.: Mavhome: an agent-based smart home. In: Procs. of the First IEEE Int. Conf. onPervasive Computing and Communications (PerCom 2003), pp. 521–524 (2003)
5. Sadri, F.: Ambient intelligence: A survey. ACM Comput. Surv. 43(4), 1–66 (2011)
6. Miskelly, F.: Assistive technology in elderly care. Age Ageing 30(6), 455–458 (2001)
7. Rantz, M.J., Dorman-Marek, K., Aud, M., Tyrer, H.W., Skubic, M., Demiris, G., Hussam, A.: A technology and nursing collaboration to help older adults age in place. Nursing Outlook 53(1), 40–45 (2005)
8. Rantz, M., Aud, M., Alexander, G., Oliver, D., Minner, D., Skubic, M., Keller, J., He, Z., Popescu, M., Demiris, G., Miller, S.: Tiger place: An innovative educational and research environment. In: AAAI in Eldercare: New Solutions to Old Problems (2008)
9. Skubic, M., Rantz, M.: Active elders: Center for eldercare and rehabilitation technology (accessed April 1, 2013)
10. Abbate, S., Avvenuti, M., Bonatesta, F., Cola, G., Corsini, P., Vecchio, A.: A smartphone-based fall detection system. Pervasive and Mobile Computing 8(6), 883–899 (2012)
11. Winkler, S., Schieber, M., Luecke, S., Heinze, P., Schweizer, T., Wegertseder, D., Scherf, M., Nettlau, H., Henke, S., Braecklein, M., Anker, S.D., Koehler, F.: A new telemonitoring system intended for chronic heart failure patients using mobile telephone technology - feasibility study. International Journal of Cardiology 153(1), 55–58 (2011)
12. Virone, G., Alwan, M., Dalal, S., Kell, S.W., Turner, B., Stankovic, J.A., Felder, R.: Behavioral patterns of older adults in assisted living. Trans. Info. Tech. Biomed. 12(3), 387–398 (2008)
13. Doctor, F., Hagras, H., Callaghan, V.: A fuzzy embedded agent-based approach for realizing ambient intelligence in intelligent inhabited environments. IEEE Transactions on Systems, Man, and Cybernetics, Part A 35(1), 55–65 (2005)
14. Hagras, H., Callaghan, V., Colley, M., Clarke, G., Pounds-Cornish, A., Duman, H.: Creating an ambient-intelligence environment using embedded agents. IEEE Intelligent Systems 19(6), 12–20 (2004)

15. Rashidi, P., Cook, D.J.: Keeping the resident in the loop: adapting the smart home to the user. Trans. Sys. Man Cyber. Part A 39(5), 949–959 (2009)
16. Rashidi, P., Cook, D.J., Holder, L.B., Schmitter-Edgecombe, M.: Discovering activities to recognize and track in a smart environment. IEEE Trans. on Knowledge and Data Engineering 23(4), 527–539 (2011)
17. Kientz, J.A., Patel, S.N., Jones, B., Price, E., Mynatt, E.D., Abowd, G.D.: The georgia tech aware home. In: CHI 2008 Extended Abstracts on Human Factors in Computing Systems, CHI EA 2008, pp. 3675–3680. ACM (2008)
18. Droes, R.M., Mulvenna, M., Nugent, C., Finlay, D., Donnelly, M., Mikalsen, M., Walderhaug, S., van Kasteren, T., Krose, B., Puglia, S., Scanu, F., Migliori, M.O., Ucar, E., Atlig, C., Kilicaslan, Y., Ucar, O., Hou, J.: Healthcare systems and other applications. IEEE Pervasive Computing 6(1), 59–63 (2007)
19. Sánchez-Pi, N., Molina, J.M.: A smart solution for elders in ambient assisted living. In: Mira, J., Ferrández, J.M., Álvarez, J.R., de la Paz, F., Toledo, F.J. (eds.) IWINAC 2009, Part II. LNCS, vol. 5602, pp. 95–103. Springer, Heidelberg (2009)
20. Copetti, A., Loques, O., Leite, J.C.B., Barbosa, T.P.C., da Nobrega, A.C.L.: Intelligent context-aware monitoring of hypertensive patients. In: 1st Int. ICST Workshop on Situation Recognition and Medical Data Analysis in Pervasive Health Environments (2009)
21. Helal, S., Winkler, B., Lee, C., Kaddoura, Y., Ran, L., Giraldo, C., Kuchibhotla, S., Mann, W.: Enabling location-aware pervasive computing applications for the edlerly. In: Procs. 1st IEEE Int. Conf. on Pervasive Computing and Communications, pp. 531–536 (2003)
22. Yuan, B., Herbert, J.: Fuzzy CARA - a fuzzy-based context reasoning system for pervasive healthcare. Procedia Computer Science 10, 357–365 (2012)
23. Graf, B., Hans, M., Schraft, R.D.: Care-O-bot II - Development of a next generation robotic home assistant. Autonomous Robots 16(2), 193–205 (2004) Times Cited: 60
24. Pollack, M., Brown, L., Colbry, D., McCarthy, C., Orosz, C., Peintner, B., Ramakrishnan, S., Tsamardinos, I.: Autominder: an intelligent cognitive orthotic system for people with memory impairment. Robotics and Autonomous Systems 44(3-4), 273–282 (2003)
25. Stefanos, N., Vergados, D.D., Anagnostopoulos, I.: Health care information systems and personalized services for assisting living of elderly people at nursing home. In: Procs. of the 2008 3rd Int. Workshop on Semantic Media Adaptation and Personalization, pp. 122–127. IEEE Computer Society (2008)
26. Naeem, U., Bigham, J.: A comparison of two hidden markov approaches to task identification in the home environment. In: Procs. 2nd Int. Conf. on Pervasive Computing and Applications, pp. 624–634 (2007)
27. Singla, G., Cook, D.J., Schmitter-Edgecombe, M.: Tracking activities in complex settings using smart environment technologies. International Journal of Bio-Sciences, Psychiatry and Technology 1(1) (2009)
28. Acampora, G., Loia, V.: A proposal of an open ubiquitous fuzzy computing system for ambient intelligence. In: Lee, R., Loia, V. (eds.) Computational Intelligence for Agent-based Systems. SCI, vol. 72, pp. 1–27. Springer, Heidelberg (2007)
29. Ros, M., Cuéllar, M., Delgado, M., Vila, A.: Online recognition of human activities and adaptation to habit changes by means of learning automata and fuzzy temporal windows. Information Sciences 220, 86–101 (2013)
30. Ros, M., Delgado, M., Vila, A.: Fuzzy method to disclose behaviour patterns in a tagged world. Expert Systems with Applications 38(4), 3600–3612 (2011)

31. Ros, M., Delgado, M., Vila, A., Hagras, H., Bilgin, A.: A fuzzy logic approach for learning daily human activities in an ambient intelligent environment. In: IEEE Int. Conf. on Fuzzy Systems, pp. 1–8 (2012)

32. Ermes, M., Parkka, J., Mantyjarvi, J., Korhonen, I.: Detection of daily activities and sports with wearable sensors in controlled and uncontrolled conditions. IEEE Transactions on Information Technology in Biomedicine 12(1), 20–26 (2008)

33. Philipose, M., Fishkin, K.P., Perkowitz, M., Patterson, D.J., Fox, D., Kautz, H., Hahnel, D.: Inferring activities from interactions with objects. IEEE Pervasive Computing 3(4), 50–57 (2004)

34. Liao, L., Patterson, D.J., Fox, D., Kautz, H.: Learning and inferring transportation routines. Artif. Intell. 171(5-6), 311–331 (2007)

35. Park, S., Kautz, H.: Hierarchical recognition of activities of daily living using multi-scale, multi-perspective vision and rfid. In: 2008 IET 4th International Conference on Intelligent Environments, pp. 1–4 (2008)

36. Boger, J., Hoey, J., Poupart, P., Boutilier, C., Fernie, G., Mihailidis, A.: A planning system based on markov decision processes to guide people with dementia through activities of daily living. Trans. Info. Tech. Biomed. 10(2), 323–333 (2006)

37. Dingli, A., Lia, S.: Home butler creating a virtual home assistant. In: Czachórski, T., Kozielski, S., Stańczyk, U. (eds.) Man-Machine Interactions 2. AISC, vol. 103, pp. 251–258. Springer, Heidelberg (2011)

38. Pineau, J., Montemerlo, M., Pollack, M., Roy, N., Thrun, S.: Towards robotic assistants in nursing homes: Challenges and results. Robotics and Autonomous Systems 42(3-4), 271–281 (2003) Times Cited: 101

39. Gonzalez Alonso, I.: Service Robotics. In: Service Robotics within the Digital Home: Applications and Future Prospects, vol. 53, pp. 89–114. Springer (2011)

40. Jardon Huete, A., Victores, J.G., Martinez, S., Gimenez, A., Balaguer, C.: Personal autonomy rehabilitation in home environments by a portable assistive robot. IEEE Transactions on Systems Man and Cybernetics Part C - Applications and Reviews 42(4), 561–570 (2012)

41. Savolainen, L., Hanson, E., Magnusson, L., Gustavsson, T.: An internet-based videoconferencing system for supporting frail elderly people and their carers. Journal of Telemedicine and Telecare 14(2), 79–82 (2008)

Question Answering System for Dialogues: A New Taxonomy of Opinion Questions

Amine Bayoudhi[1], Hatem Ghorbel[2], and Lamia Hadrich Belguith[1]

[1] ANLP Group, MIRACL Laboratory, University of Sfax, B.P. 1088, 3018, Sfax Tunisia
bayoudhi.amine@gmail.com, l.belguith@fsegs.rnu.tn
[2] ISIC Lab, HE-Arc Ingénierie, University of Applied Sciences, CH-2610 St-Imier Switzerland
hatem.ghorbel@he-arc.ch

Abstract. Question analysis is an important task in Question Answering Systems (QAS). To perform this task, the system must procure fine-grained information about the question types. This information is defined by the question taxonomy. In the literature, factual question taxonomies were the object of many research works. However, opinion question taxonomies did not get the same attention because they are more complicated. Besides, most QAS were focusing on monologal texts, while dialogues have rarely been explored by information retrieval tools. In this paper, we investigate the use of dialogue data as an information source for opinion QAS. Hence, we propose a new opinion question taxonomy in the context of an Arabic QAS for political debates and we propose then an approach to classify these questions. Obtained results were relevant with a precision of around 91.13% for the opinion classes' classification.

Keywords: question taxonomy, opinion question classification, sentiment analysis, Question Answering Systems.

1 Introduction

Nowadays, information sources are becoming much larger. As a result, finding the appropriate piece of information using the least effort is becoming more difficult. Question Answering Systems (QAS) are information retrieval tools designed to make this task easier; they offer the user the possibility to formulate his queries in natural language and to get concise and precise answers.

Question analysis is considered as an important task in QAS. To perform this task, the system must procure fine-grained information about the question types. This information is defined by the question taxonomy. In the literature, factual question taxonomies were the object of many research works. However, opinion question taxonomies did not get the same attention.

Dialogues, as the main modality of communication in human interaction, make an essential part of information sources. They occur either directly (i.e. professional meetings, TV programmes and political debates) or virtually (i.e. social networks or blogs). During dialogues, interlocutors perform different interactive actions: they

H.L. Larsen et al. (Eds.): FQAS 2013, LNAI 8132, pp. 67–78, 2013.
© Springer-Verlag Berlin Heidelberg 2013

exchange information, express opinions, make decisions, etc. Nevertheless, dialogues have rarely been explored by information retrieval tools such as QAS. This is due to the lack of linguistic resources, in particular annotated oral corpora, as well as the complexity of processing related to the specific aspects of oral conversations.

The current research is part of a framework aiming to implement an Arabic QAS for political debates. In this paper, we investigate the use of dialogue data as an information source for a QAS and we propose a new opinion question taxonomy in this context. We propose also an approach to classify these questions, based on opinion extraction and machine learning techniques. The rest of this paper is organized as follows. In section 2, we review a selection of previous works related to the opinion question classification. In section 3, we propose our opinion question taxonomy in QAS for dialogues. In section 4, we illustrate our classification approach, report and discuss the obtained results. Finally, we conclude and provide some perspectives in section 5.

2   Related Works

In this section, we present a brief overview of the question type taxonomies, opinion extraction techniques and automatic question classification in the QAS.

2.1   Question Type Taxonomies

Question type taxonomy refers to the set of categories into which questions have to be classified [1]. In the literature, most of the proposed taxonomies concern factual questions. Their architecture can be flat [2] or hierarchical [3]. The taxonomy of Hovy et al. [4] and that of Li and Roth [5] are the most used ones for factual questions. On the other hand, we find few works proposing taxonomies for opinion questions. We cite in this context the works of Ku et al. [6] which deal with the analysis of questions and the retrieval of answer passages for opinion QAS. The training corpus is gathered from conferences question data and Internet Polls, and includes the authors' own corpus called OPQ corpus (created using the NTCIR-2 and NTCIR-3[1] topic data collected from news article).The proposed taxonomy classifies the questions into factual and opinion questions, and then subdivides the opinion questions into six fine-grained types: Holder, Target, Attitude, Reason, Majority and Yes/No. Besides, we cite the works of Moghaddam and Easter [7] addressing the problem of answering opinion questions about products by using reviewers' opinions. The adopted taxonomy, inspired from the works of Ku et al. [6], has dropped out the type Holder since it is irrelevant in mining product reviews domain. Moreover, the type Majority has been replaced by the question form attribute.

---

[1] http://research.nii.ac.jp/ntcir/permission/
perm-en.html#ntcir-3-qa

2.2   Opinion Extraction Techniques

Opinion extraction is an emerging research area in the opinion mining domain. It aims at extracting the main components of a subjective expression such as the opinion holder, the target towards whom or which the opinion is expressed, and the opinion polarity. The used techniques are based on supervised learning [8] and unsupervised learning [9]. In this context, we cite the model of Paroubek et al. [10] proposed for the evaluation of the opinion mining annotations performed in the industrial context of the DOXA project[2]. This model represents opinion expression within eight attributes:

— *Opinion marker*: the linguistic elements which express an opinion.
— *Opinion polarity*: the more or less positive impression felt while reading an opinion expression.
— *Source*: the opinion holder.
— *Target*: the object, issue or person towards which the opinion is expressed.
— *Intensity*: the strength of the expression.
— *Theme/Topic*: reference of the addressed topic in the document containing the opinion expression.
— *Information*: the more or less factual aspect of the opinion expression.
— *Engagement*: the relative implication that the opinion holder is supposed to have to support his opinion expression.

2.3   Question Classification Approaches

In the literature, we distinguish three different approaches for the question classification: rule based approach, machine learning approach and hybrid approach.

— *Rule based approach*: it consists in associating to the question a number of manually defined rules, called hand-crafted rules [11]. This approach is generally based on interrogative words used in questions. The disadvantages of this approach are linked to the overabundance of the rules to define.
— *Machine learning approach*: it consists of extracting a set of features from the questions themselves and using them to build a classifier that allows predicting the adequate type of the question. In effect, works adopting this approach differ according to: *i)* the type of the classifier in use such as Naive Bayes [12], SVM classifiers [13], and decision trees [14], *ii)* the selected classification features that can be symbolic [15], morpho-syntactic using Part-of-speech tags [16], semantic using hypernyms relations of WordNet [17] or statistical [6].
— *Hybrid approach*: it consists in combining the two previous approaches using as learning features manually defined rules [1], [18].

---

[2] DOXA is a project supported by the numeric competitiveness center CAP DIGITAL of Ile-de-France region which aims at defining and implementing an OSA semantic model for opinion mining in an industrial context.

3   Proposed Taxonomy for Opinion Question Classification

In order to identify the different question types in a QAS for dialogues, we have built a study corpus of questions. In this section, we start by explaining the construction steps of the study corpus and presenting its specification details. Then, we argue some question specificities in QAS designed for dialogue data. Finally, we describe our proposed taxonomy for the question classification and we provide some discussion notes.

3.1   Building the Study Corpus

We have built the study corpus COPARQ (Corpus of OPinion ARabic Questions) in order to determine the question types that can be asked in the QAS for political debates. The corpus (Table 1) was built after collecting 14 episodes of political debates broadcast on Aljazeera satellite channel. Starting from these manually transcribed episodes, we have prepared 14 questionnaires containing for every episode: the title, the subtitles, the date, the interlocutors and their affiliations. The questionnaires are distributed to 14 volunteers of different profiles (students, teachers, workers, etc.) so that they contribute with questions relative to the discussed topics. The total number of gathered questions is 620.

Table 1. Specifications of the study corpus COPARQ

Corpus specification	Value
Number of episodes	14
Total number of words in episodes	80,151
Number of participants	14
Total number of questions	620
Average number of questions per episode	44.28
Total number of words in gathered questions	7,549
Average number of words per question	12.176

3.2   Question Specificities of QAS for Dialogues

In a QAS for dialogues, question types differ from those in a QAS for texts in many issues.

First, users in QAS for dialogues tend to ask questions especially about the subjective aspect of the utterances. These questions, generally identified through opinion markers (i.e. opinion verbs, adjectives and adverbs), are hard to classify if the subjective aspect of the question is implicit. That's why, using the existing taxonomies (originally designed for QAS for texts) may not provide efficient results in the classification of questions in QAS for dialogues. For example, the question " من المسؤول عن جرائم القناصة إبان الثورة" ("Who is responsible for the snipers' crimes during the revolution?") will be most likely classified as a factual question since it does not contain any explicit subjective information. However, the user asked this question to know the feedbacks of all the dialogue participants in the question issue. He wanted to say "حسب الضيوف من المسؤول عن جرائم القناصة إبان الثورة ؟" (According to the guests, who is responsible for the snipers' crimes during the revolution?), and this is an opinion question. Therefore, a question taxonomy in QAS for dialogues must have flexible type definitions and must be provided with adaptation techniques to support implicit subjective questions.

Second, Yes/No questions are classified according to the existing taxonomies as objective or subjective questions. But in our case, Yes/No questions, which have the form of "Q X according to P?" where Q is a Yes/No question form (i.e. is, do, have), X is a statement and P is a person, can be written into the form of "Does P believes that X?". For example, the question "هل نجحت الحكومة الحالية في قيادة البلاد في المرحلة الانتقالية؟" (Did the current government succeed in leading the country during the transitional period?) can be written into the form of "هل يعتبر الضيوف أن الحكومة الحالية نجحت في قيادة البلاد في المرحلة الانتقالية؟" (Do the guests consider that the current government succeeded in leading the country during the transitional period?). Therefore, we consider that all Yes/No questions, in our QAS for dialogues, are opinion questions.

Third, asked questions in a QAS for dialogues concern the interlocutors' feelings and attitudes as well as their beliefs and arguments (i.e. How does a specific action happen according to a given interlocutor). They may directly query information about what an interlocutor feels or thinks such as the question "ما هو رأي قيس سعيد في تسليم البغدادي المحمودي إلى ليبيا؟" (What does Qais Saiyed think about the extradition of Baghdadi Mahmoudi to Libya?). Also, they may query information about the discussed topic according to the opinion of a given interlocutor such as the question "كيف تم تسليم البغدادي المحمودي إلى ليبيا حسب رأي قيس سعيد؟" (How was Baghdadi Mahmoudi extradited to Libya according to Qais Saiyed?). However, the existing taxonomies do not make distinction between these two types of opinion questions despite the fact that they do not share the same answering strategies.

3.3   Proposed Taxonomy for Opinion Questions

Since the existing taxonomies are not completely convenient for questions in QAS for dialogues, we propose, after a deep study of the corpus, a new question taxonomy within the framework of an Arabic QAS for political debates. This taxonomy, unlike the Ku et al. taxonomy for instance (addressed to news articles which are more structured than dialogues), allows us to solve the issues raised from the specificities of dialogue data by making some reformulations to the questions and setting more precise question type definitions. To define this taxonomy, we are essentially inspired from the model proposed by Paroubek et al. [10], since this model gives a synthetic view of the main opinion mining models (20 models) listed in the literature. It was also successfully used to automatically detect and annotate topics, feelings and opinions in English and French texts.

In order to differentiate between objectivity/subjectivity levels and fine-grained opinion information, we propose a two level hierarchical taxonomy. The first level namely question categories describes high level classes depending on the degree of the subjective aspect; the second level namely opinion question classes describes opinion question types according to the requested information and the expected answer.

First Level: Question Categories. With reference to the Paroubek et al. model for annotating opinion expressions [10], we propose, at a first stage, three main categories to classify questions within the framework of an Arabic QAS for political debates. The categories are: Thematic, Informational and Opinionated (Table 2).

— *Thematic*: it is the category of questions asking for the discussed topics and the involved interlocutors. These questions can be answered using classic techniques of information extraction (e.g. bag of words, TF-IDF) and they do not require deep semantic analysis. This category contains questions asking whether a given interlocutor has participated in the discussion of a given topic, or asking to report the communication of a given interlocutor in a given topic.
— *Informational*: it is the category of questions in which the factual aspect dominates the subjective aspect. It contains questions asking information about an event, a person, an object or an issue according to a given interlocutor. This information can be named entities or any other type of non factual questions such as reason, manner or definition.
— *Opinionated*: it is the category of questions asking for an opinion expression attribute such as attitude, opinion holder or target. The extraction of these attributes is one of the issues dealt with in the opinion extraction domain [8].

Table 2. Examples for the question categories

Question category	Example
Thematic	هل قال نور الدين البحيري شيئا عن قانون الأحزاب الجديد؟
	Did Noureddine Beheiri say anything about the new political parties Act?
Informational	كيف استطاع الرئيس المخلوع مغادرة البلاد حسب رأي محمد الغنوشي؟
	According to Mohamed Ghannouchi, How did the ousted president manage to leave the country?
Opinionated	ما هو موقف راشد الغنوشي من السعودية بعد استقبالها الرئيس المخلوع؟
	What does Rached Ghannouchi think of Saudi Arabia after it received the ousted president?

Second Level: Opinion Question Classes. Since our QAS is designed for opinion questions, we are interested in the current work in the Opinionated category. In fact, giving the opinion QAS more fine-grained information about the question types will improve its performance more than simply distinguishing between subjective and factual information [19]. That's why, we have proceeded, at a second stage, with a supplementary level of classification that concerns the category Opinionated. Inspired by the model of Paroubek et al. [10] and the classification of Ku et al. [6], we define seven opinion classes for this category (Table 3).

- Attitude: asks about the attitude of the given holder towards the given target.
- Yes/No: asks whether the given holder has adopted the specified attitude towards the given target.
- Holder: asks about who expressed the specified attitude towards the given target.
- Target: asks about toward whom or what the given holder has the given attitude.
- Reason: asks about the reasons for which the given holder has expressed the specified attitude towards the given target.
- Majority: asks about which of the opinions (listed or not) is the one of the given holder toward the given target.
- Intensity: Asks about how far the given holder has the specified attitude toward the given target.

Table 3. Examples of the opinion classes

Opinion class	Example
Attitude	ما هو موقف سهام بن سدرين من وضعية حقوق الإنسان في تونس قبل الثورة ؟ What's the attitude of Sihem Bensedrine about human rights situation in Tunisia before the revolution?
Yes/No	هل يظن عصام الشابي أنه بالفعل هناك أياد خفية تعبث بالثورة ؟ Does Issam Chebbi think that there are actually unknown forces trying to sabotage the revolution?
Holder	من من الحاضرين يعتقد أن الثورة التونسية هي بداية ثورة عربية عامة ستمتد لبقية الدول العربية؟ Among those present people, who thinks that the Tunisian revolution is the beginning of a general Arab revolution that will be widespread in the remaining Arab countries?
Target	فيمن تشك سهام بن سدرين أن يكون المسؤول عن جرائم القناصة إبان الثورة ؟ Who does Sihem Ben Sedrine suspect for being responsible of the snipers' crimes during the revolution?
Reason	لماذا يرى منصف المرزوقي أن التمويل الخارجي يهدد مستقبل الديمقراطية في تونس ؟ Why does Moncef Marzougui think that foreign funding can threaten the future of democracy in Tunisia?
Majority	هل يعتبر محمد الأحمري أن التجربة الديمقراطية الغربية تجربة مثالية نموذجية أم أنها تشكو العديد من النقائص رغم ما حققته من إنجازات ؟ Does Muhammad Alahmari consider that the occidental democratic experience is a perfect and typical one, or that it is suffering from a number of flaws in spite of its accomplishments?
Intensity	إلى أي حد يعتبر أمان الله المنصوري أن شباب الثورة في تونس قادر على المشاركة في إدارة البلاد في المرحلة القادمة ؟ How far does Amen-Allah Almansouri believe that the revolution youth in Tunisia are able to take part in running the country during the coming period?

3.4 Discussions about the Proposed Taxonomy

In the context of QAS for text data, most researches on opinion question classification, similarly to sentence classification, addressed the problem of subjectivity classification.

Nevertheless, seeing that the subjective aspect is quite dominant over the factual aspect in the questions of QAS for dialogues, we consider that this problem should be differently addressed in our context. Indeed, factual aspect exists only as minor information parts of the question. Therefore, we have proposed, instead of the factual class, the Informational category to include opinionated question which have a more or less factual aspect. Besides, the category Informational allows discriminating opinion questions asking about beliefs or arguments among those asking about attitudes or feelings. Previous researches [6] [7] omitted this distinction despite the fact that answering strategies to these questions are completely divergent. This was stated especially by Somasundaran et al. [19] who developed an automatic classifier for recognizing sentiment and arguing attitudes.

In the matter of the opinion classes, we note that the class Intensity was not taken into account in the classification of Ku et al. [6]. But, we have noticed after observing our study corpus that users' questions in political debates focus sometimes on the opinion intensity of an interlocutor (5% of opinionated questions in the study corpus). Thus, we have added the class Intensity to our taxonomy.

In addition, Moghaddam et al. [7] considered that questions belonging to the class Majority defined in [6] can be expressed as Target, Reason, Attitude and Yes/No, and therefore it is not an independent class of question. In this way, they did not consider the class Majority and add instead an attribute called question form. This attribute is an additional description defined for every class and it allows distinguishing between the simple form and the comparative form of questions. In accordance with this hypothesis, we enrich our taxonomy with the attribute question form (Table 4). Despite the fact that values of this attribute do not still cover all question forms, we believe that they are sufficient to resolve most of cases we are dealing with.

Table 4. Examples of the two question forms for the class Attitude

Question form	Example
Simple	ما هو رأي الحزب الديمقراطي التقدمي في قانون الأحزاب الجديد ؟
	What does the Democratic Progressive Party think of the new Political Parties Act?
Comparative	ما هو رأي سهام بن سدرين في أن انتهاكات حقوق الإنسان في حكم بن علي أكثر بكثير منها في حكم بورقيبة ؟
	What does Sihem Bensedrine think of the assertion that the violation of human rights was far greater during the rule of Ben-Ali than during the rule of Bourguiba?

Nevertheless, we maintain the class Majority because we believe that this class has an independent answer type, conversely to Moghaddam et al. [7]. In fact, Moghaddam et al. consider that, for example, in the question "Why is Canon X better than Samsung Y?", there is confusion between the class Majority and the class Reason. However, we consider that the question belongs only to the class Reason, since it asks about reason and does not list options as recommended by the class majority such as the question "What do you prefer better, Canon X or Samsung Y?". In addition, we confirm that the illustrated question is of a comparative form, in accordance with the proposition of Moghaddam et al. [7] of using the question form attribute. This attribute would be very useful in the information extraction task.

4 Proposed Approach for the Opinion Question Classification

Our approach of question classification is inspired from the techniques of opinion extraction and it is based on supervised machine learning methods. It consists of two main phases: the extraction of learning features and the automatic question classification. This approach requires different resources and linguistic tools such as a morphosyntactic tagger, lexical resources and an annotated training corpus.

4.1 Extraction of Classification Features

With reference to the works presented in section 2.2, we have chosen to adopt lexical, morpho-syntactic and statistic features. Extraction of these features is performed in four steps:

1. Extraction of POS tags and verbs tense: extracts POS tags by using an Arabic POS tagger. This step enables also to detect the tense of the verb if the question contains a verbal phrase.
2. Extraction of interrogative words: extracts interrogative words (lexical features) by using exhaustive lists of interrogative words such as "من" (who), interrogative words attached to prepositions such as "لأي" (for what), and imperative verbs used in an interrogation context of as "اذكر" (list).
3. Extraction of opinion markers: extracts question opinion markers (lexical features) by using lists of opinion verbs, nouns, adjectives or adverbs such as "اعتقد" (think), "رأي" (opinion), "إيجابي" (positive) and "أفضل" (better).
4. Extraction of statistic features: extracts statistic features by calculating the number of words in the question. This extraction is performed after removing punctuation and stop words such as "و" (and), "في" (in) and "من" (from). In addition, this step allows calculating the probabilities of unigrams and bigrams such as "أكد" (confirm) "علق" (comment), "حول موضوع" (about the subject of). Unigrams and bigrams are used mainly to identify the Thematic category.

4.2 Training Corpus

Our training corpus (Table 5) is collected from three sources: i) the COPARQ corpus (see section 4.1); ii) extracts from Polls created by some TV channels (Aljazeera, Al-Alam, Russia Today); iii) Selected questions from international conferences corpus (TREC, TAC and CLEF) after their translation to Arabic. The training corpus was annotated by two linguistic experts according to our proposed taxonomy.

To evaluate disagreement degree between the two annotators, we have calculated the kappa coefficient which allows measuring agreement between the annotators. The

Average kappa value obtained is around **0.97** (**0.96** for the question categories annotations and **0.99** for the opinion classes annotations), which allows to judge that our training corpus is quite homogenous.

Table 5. Specifications of the training corpus

Source	Total number of questions	Total size (number of words)	Average question length (number of words)
COPARQ	620	7,531	12.146
Conferences	723	6,000	8.298
Polls	596	5,915	9.942
Total	1,939	19,446	10.028

4.3   Results and Discussions

We have evaluated our classification approach in terms of precision (2) which measures the ability to classify the question into the appropriate category or class. The precision is calculated after applying the 10-fold cross validation evaluation method.

$$\text{Precision} = \frac{\text{Number of well classified questions}}{\text{Total number of questions}} \tag{1}$$

Table 6 illustrates the results of the classification into question categories and into opinion classes according to four algorithms: the three most common learning algorithms Naïve Bayes, decision trees and SVM, and the Zero-R as a baseline algorithm. In particular, SVM provided the best performance with a rate of **87.9%** for the question categories' classification and **91.13%** for the opinion classes' classification.

Table 6. Results of the question classification

Algorithms	Precision of the question categories' classification (%)	Precision of the opinion classes' classification (%)
Rule based	75.65	63.34
Naïve Bayes	81.05	90.67
Decision tree	86.58	90.03
SVM	**87.9**	**91.13**

Concerning the opinion classes' classification, the results are good and show that the selected classification features are relevant. Hence, we consider that shallow features that we have used are sufficient to get a good opinion question classification for Arabic. We note that Ku et al. [6] have also used, to classify Chinese opinion questions, shallow features compound of heuristic rules and scores calculated based on unigrams and bigrams. They obtained a nearly similar average performance around of 92.5%. The little difference might be due to the nature of the selected topics. While they used news articles data, we have used political debates data which have much more fuzzy and irregular structure.

Besides, precision obtained for the classification into question categories reached **87.9%** (**87.8%** by Ku et al. using a sentiment lexicon of over 10,000 words). The main difficulty encountered in our classifier is due to the ambiguity in recognizing factual information in the question to discriminate between Informational and Opinionated categories. Indeed, this task, already considered difficult for texts, is more for a question whose content is shorter and therefore contains less lexical information. In addition, the limits of used Arabic linguistic tools reduced the performance of the classifier. For example, the ambiguity due to non-vowel words causes confusion between the preposition "مِن" (from) and the interrogative word "مَن" (who).

5   Conclusion and Perspectives

In this paper, we have proposed a new taxonomy for the question classification in an opinion QAS for political debates, inspired by opinion mining and sentiment analysis models. This taxonomy, composed of two classification levels, provides a wider and more comprehensive description of opinion questions. In addition, we have proposed an approach for the automatic classification of opinion questions based on different shallow features. To evaluate the proposed approach, we have developed a classification tool using four different learning algorithms. The results were encouraging and reached an average accuracy of 91.31% for the opinion classes' classification. These results show that the shallow features are sufficient enough to build a satisfactorily accurate classifier for opinion question.

As perspectives, we intend to evaluate our question classification tool within each training corpus source separately. The aim is to compare the obtained results per corpus source dataset in order to evaluate the affect of the question topic domain on the classification performance. Moreover, we intend to build a sentiment lexicon to collect opinion markers and to assign degrees of subjectivity to them. The lexicon will allow us to solve the problem of detecting the subjective nature of the questions and subsequently to improve the results obtained in the question category classification. In addition, it can be used to define polarity and calculate its intensity in the information extraction of our opinion QAS.

References

1. Silva, J., Coheur, L., Mendes, A.C., Wichert, A.: From symbolic to sub-symbolic information in question classification. Artificial Intelligence Review 35, 137–154 (2011)
2. Lehnert, W.G.: A conceptual theory of question answering. In: Grosz, B.J., Sparck Jones, K., Webber, B.L. (eds.) Natural Language Processing, Kaufmann, Los Altos, CA, pp. 651–657 (1986)
3. Singhal, A., Abney, S., Bacchiani, M., Collins, M., Hindle, D., Pereira, F.: AT&T at TREC-8. In: Voorhees, E. (ed.) Proceedings of TREC8. NIST (2000)
4. Hovy, E.H., Hermjakob, U., Ravichandran, D.: A Question/Answer Typology with Surface Text Patterns. In: Poster in Proceedings of the Human Language Technology Conference, San Diego, CA (2002)

5. Li, X., Roth, D.: Learning question classifiers. In: Proceedings of the 19th International Conference on Computational Linguistics, Morristown, NJ, USA, pp. 1–7 (2002)
6. Ku, L.W., Liang, Y.T., Chen, H.H.: Question Analysis and Answer Passage Retrieval for Opinion Question Answering Systems. International Journal of Computational Linguistics and Chinese Language Processing 13(3), 307–326 (2008)
7. Moghaddam, S., Ester, M.: AQA : Aspect-based Opinion Question Answering. In: International Conference on Data Mining, ICDM Workshops 2011, Vancouver, Canada, December 11-14, pp. 89–96 (2011)
8. Elarnaoty, M., AbdelRahman, S., Fahmy, A.: A machine learning approach for opinion holder extraction in Arabic. International Journal of Artificial Intelligence & Applications (IJAIA) 3(2) (2012)
9. Kim, S.M., Hovy, E.: Determining the Sentiment of Opinions. In: COLING (2004)
10. Paroubek, P., Pak, A., Mostefa, D.: Annotations for Opinion Mining Evaluation in the Industrial Context of the DOXA project. In: LREC 2010, Malta, May 17-23 (2010)
11. Hull, D.A.: Xerox TREC-8 question answering track report. In: Voorhees and Harman (1999)
12. Yadav, R., Mishra, M.: Question Classification Using Naïve Bayes Machine Learning Approach. International Journal of Engineering and Innovative Technology (IJEIT) 2(8), 291–294 (2013)
13. Wang, Y., Yu, Z., Lin, H., Guo, J., Mao, C.: Chinese Question Classification Transfer Learning Method based on Feature Mapping. Journal of Computational Information Systems 9(6), 2261–2267 (2013)
14. Tomuro, N.: Question terminology and representation of question type classification. In: Second International Workshop on Computational Terminology, vol. 14 (2002)
15. Razmara, M., Fee, A., Kosseim, L.: Concordia University at the TREC 2007 QA track. In: Proceedings of the TREC 2007, Gaithersburg, USA (2007)
16. Khoury, R.: Question Type Classification Using a Part-of-Speech Hierarchy. In: Kamel, M., Karray, F., Gueaieb, W., Khamis, A. (eds.) AIS 2011. LNCS, vol. 6752, pp. 212–221. Springer, Heidelberg (2011)
17. Cai, D., Sun, J., Zhang, G., Lv, D., Dong, Y., Song, Y., Yu, C.: HowNet Based Chinese Question Classification. In: Proceedings of the 20th Pacific Asia Conference on Language, Information and Computation (2006)
18. Kuchmann-Beauger, N., Aufaure, M.-A.: Structured Data-Based Q&A System Using Surface Patterns. In: Christiansen, H., De Tré, G., Yazici, A., Zadrozny, S., Andreasen, T., Larsen, H.L. (eds.) FQAS 2011. LNCS, vol. 7022, pp. 37–48. Springer, Heidelberg (2011)
19. Somasundaran, S., Wilson, T., Wiebe, J., Stoyanov, V.: QA with Attitude: Exploiting Opinion Type Analysis for Improving Question Answering in On-line Discussions and the News. In: International Conference on Weblogs and Social Media, ICWSM (2007)

R/quest: A Question Answering System

Joan Morrissey and Ruoxuan Zhao

University of Windsor, School of Computer Science, 401 Sunset Ave, Windsor
N9B 3P4, Canada
{joan,zhao11t}@uwindsor.ca

Abstract. In this paper, we discuss our novel, open-domain question answering (Q/A) system, R/quest. We use web page snippets from Google™ to extract short paragraphs that become candidate answers. We performed an evaluation that showed, on average, 1.4 times higher recall and a slightly higher precision by using a question expansion method. We have modified the Cosine Coefficient Similarity Measure to take into account the rank position of a candidate answer and its length. This produces an effective ranking scheme. We have a new question refinement method that improves recall. We further enhanced performance by adding a Boolean NOT operator. R/quest on average provides an answer within the top 2 to 3 paragraphs shown to the user. We consider this to be a considerable advance over search engines that provide millions of ranked web pages which must be searched manually to find the information needed.

Keywords: Question answering systems, candidate answers, information retrieval, web crawling, question expansion, question refinement, modified cosine similarity measure, system evaluation.

1   Introduction and Related Work

Suppose that you want to know the ingredients contained in Coca Cola. If you use Google™ to find the answer you will get over 5 million ranked web pages in less than a second. However, you will have to search through the pages to find the actual answer. In contrast, with our question answering system, R/quest, the first answer returned to the user is "The primary ingredients of Coca-Cola syrup include either high fructose corn syrup or sucrose derived from cane sugar, caramel color, caffeine, phosphoric acid, coca extract, lime extract, vanilla and glycerin". R/quest saves users' time and effort by giving them a correct, short answer rather than a ranked list of web pages. In this paper, we describe how R/quest works and comment on its novel features.

A great deal of research is being done in the area of Question Answering. Most systems are based on either Nature Language Processing (NLP) or Information Retrieval (IR) techniques. Using these technologies, various non open-domain Q/A systems have been developed for different purposes. In [1-5], researchers present some Q/A systems which process medical information and knowledge. They assert that their systems produce more accurate answers in comparison to Google™. Depending on the system, knowledge was grouped by such categories as symptoms, causes,

H.L. Larsen et al. (Eds.): FQAS 2013, LNAI 8132, pp. 79–90, 2013.
© Springer-Verlag Berlin Heidelberg 2013

treatments, and other relevant areas. The queries were subjected to a semantic analysis in order to determine the answer. There are also interactive Q/A systems. Here, dialogue is used to help users build more specific queries and, thus, get better answers. In [6-7], some dialogue is introduced. Users are limited in that they can only ask certain types of questions. The systems will recommend some existing question patterns to them. Next, users can fill in the blanks with their own words to create system-readable questions. In [8-13], researchers focus their work on monolingual and multilingual Q/A systems. They came up with different strategies to process users' questions written in languages such as Chinese, German, French, Arabic and others, based on the specific language structure. Monolingual Q/A systems usually work with some syntactic and semantic analysis functions, thesauri and other knowledge. The multilingual Q/A systems discussed in [14-16] profess to work with more than one individual language. They make use of machine translation systems to convert the users' questions into the same language used in their databases. The answer is also translated back into the original language, if necessary. Thus, users can obtain information from a wider collection of data and are more likely to get a correct answer.

Essentially, most IR systems are based on matching index terms in the question with those in the stored documents. This works very well, for the most part. However, such systems lack any type of semantic information. Thus, the question "what is AI" will not match a document containing the phrase "Artificial intelligence is the intelligence of machines and robots and the branch of computer science that aims to create it". This happens because the document shares no terms with the question. The semantic connection between "AI" and "Artificial intelligence" is missed. This is known as the vocabulary mismatch problem.

Working on this issue, Sahami and Heilman [17] came up with an improved retrieval strategy. They utilized results from Google™ to expand their questions and documents to increase the probability that the question and document would share some common terms. The documents and the question were sent as independent queries to Google™. Then they made use of webpage snippets [17] to expand the original documents and question. Snippets are text segments selected from web pages returned by Google™. The authors analyzed a number of snippets and saved the 50 terms with the highest TF-IDF weights [18]. These new terms were added to each question and every document. Hence, the probability that the document and question would share terms was greatly increased. In addition, the problem of missing semantic information was decreased to some extent.

Sahami and Heilman [17] also claim that their retrieval strategy is suitable for a query suggestion system. Here, a number of short existing queries are saved for future use. Those stored queries are small enough to be sent to Google™ for expansion. Once the system finishes expanding the entire query collection, it only needs to expand each new query by sending it to Google™ once.

However, their strategy is extremely expensive for IR systems which have dynamic databases. For instance, if an IR system imports 200 documents to answer a particular question, then it has to perform 201 expansions and Google™ searches. The time needed is prohibitive. Moreover, their retrieval strategy [17] relies on systems only having short queries. Long queries cannot be used to perform their required expansions. In our work, one focus is to deal with the deficiencies in their methodology.

However, our main focus is to improve information retrieval from the Internet. Our contribution is to develop an open-domain Q/A system based on traditional IR techniques with some improvements.

The rest of paper is organized as follows: In section 2 we describe our work, with emphasis on the methodology used in our Q/A system; evaluation and testing of our system is discussed in section 3; we discuss our conclusions in section 4; finally, in section 5 we propose some future work.

2   R/quest Methodology

2.1   Answer Retrieval Using Question Expansion

Below is the system diagram for R/quest.

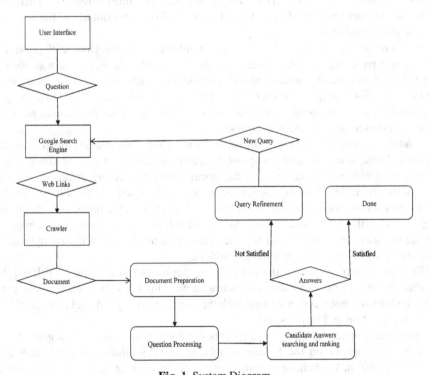

Fig. 1. System Diagram

Once R/quest receives a question from a user through its interface, it will send the question to Google™ to obtain related web pages. Then, R/quest's web crawler goes to each web page, follows the links and identifies their textual content. Paragraphs are selected and each is saved as an independent document. Note that we often refer to the paragraphs as *documents* in this paper, because they are saved as *documents*. In truth, they are short paragraphs. The two terms are used interchangeably in this paper.

All documents are then indexed using traditional Information Retrieval (IR) techniques [18]. This includes the elimination of very high frequency words, stemming and conflation. We also used TF-IDF [18] term weighting.

After getting a question, R/quest retrieves a small number of candidate answers for the user to view. If the user is not satisfied with the answers then they can refine their question. This new question will be sent to Google™ with the goal of obtaining a better answer. This process may be repeated until the user is satisfied.

Traditional term-based IR systems, in simple words, detect the terms shared by the question and documents and generally use weights to rank documents to present to the user. However, one important disadvantage is that the semantic information is missing. Hence, important answers may be missed because the question does not have any terms in common with a document. Using the previous example, the question "What is AI" will not retrieve the answer "Artificial intelligence is the intelligence of machines and robots and the branch of computer science that aims to create it". There are no common terms so the perfect answer is missed. This is an example of the vocabulary mismatch problem.

To overcome this issue, we make use of information that Google™ provides, namely web page snippets. A snippet gives the reason why the web page was chosen as a potential answer. It contains the web page text segments which share terms with the question. For example, if a user enters the question "what is AI" the first snippet returned is "what is artificial intelligence (AI)? ... This article for the layman answers basic questions about artificial intelligence".

In addition, snippets also contain other terms which are strongly related to the question. These terms are used to expand the question in order to retrieve similar documents from our collection, namely, the documents retrieved while processing the original question. Based on the snippet shown above, the question "what is AI" can be expanded with the phrase "artificial intelligence". Therefore, the document "Artificial intelligence is the intelligence of machines and robots and the branch of computer science that aims to create it" will now be retrieved as an answer. With this approach, we overcome the vocabulary mismatch problem.

We expand each question with the top 50 snippet terms based on their TF-IDF [18] weights. We match the expanded question with our initial results. Hence, documents that originally did not share any terms with the question are a good deal more likely to match the question and be retrieved.

The use of snippets in query expansion is not uncommon. However, a goal of our work was to improve on the Sahami and Heilman [17] solution to the vocabulary mismatch problem. We believe that, in their snippet based approach, too much time is required to send more than 200 documents to Google™ in order to answer one question. However, the documents stored by R/quest are collected from the internet by Google™ based on each question received. Thus, the documents and the question are absolutely connected from the beginning. As a consequence, if R/quest were to expand its documents using terms from Google™ then more non-relevant documents would be returned. For example: Consider document d_5 that is a paragraph from the web page w_1 although it is not relevant to the question q. However, w_1 was retrieved by Google™ based on q. Therefore, q will get expanded with snippets from w_1. We assume that w_1 has several snippets. However, snippet s_1 is relevant to q and snippet s_2, which shares some terms with s_1, is not. If we expand d_5 by sending it back to Google™ then d_5 will almost certainly be expanded by s_2. Consequently, it is very likely that the non-relevant document d_5 will be returned as an answer although it should not be. As a result of our methodology, our snippet based expansion technique retrieves less non-relevant documents than the Sahami and Heilman method [17].

2.2   Candidate Answer Ranking

R/quest ranks retrieved documents based on the similarity between each document and the question. To perform ranking, a document is assigned a numeric value indicating its similarity to the question. Documents with the highest rank are expected to be the most relevant and most likely to contain the correct answer. Documents are presented to the user in decreasing rank order.

We use the Cosine Coefficient Similarity Measure (CCSM) [18] to calculate the similarity between the question and the retrieved text documents. CCSM is defined as:

$$similarity~(j, q) = \frac{\sum_{i=1}^{n} w_{ij} * w_{iq}}{\sqrt{\sum_{i=1}^{n} w_{ij}^2} * \sqrt{\sum_{i=1}^{n} w_{iq}^2}} \tag{1}$$

In the formula, q represents the question and j is document. Furthermore, i indicates a term shared by the document j and the question q. There are n common terms. Therefore, w_{ij} is the weight of the term i in the document j and w_{iq} is the weight of the term i in the question q. In R/quest, before refinement, each term in the question has a weight of 1 during the first ranking of candidate answers.

However, R/quest is not dealing with traditional IR text documents such as journal papers. It ranks paragraphs returned by Google™. Therefore, we have modified the original CCSM to work with these paragraphs.

The quality of the web page is treated as an important factor in our Modified Cosine Coefficient Similarity Measure (MCCSM). Since R/quest receives a list of ranked links from Google™, R/quest treats that rank information as an indication of the web page's quality. A highly ranked web page is more likely to contain good quality text to construct a high-quality answer. If R/quest retrieves the top 10 web pages from Google™, then there will be 10 sets of paragraphs saved in its document collection. The rank information for each document is also stored. The rank information is incorporated into the modified similarity measure as follows:

$$similarity~(j, q) = \frac{\sum_{i=1}^{n} w_{ij} * w_{iq}}{\sqrt{\sum_{i=1}^{n} w_{ij}^2} * \sqrt{\sum_{i=1}^{n} w_{iq}^2}} * \frac{1}{\sqrt{r_j}} \tag{2}$$

In this formula, r_j is the rank of the web page from which the document j was extracted. Because the range of the original CCSM is from 0 to 1, the rank information is first normalized as

$$\frac{1}{\sqrt{r_j}} \tag{3}$$

Thus, if the web page rank is 1, then the original similarity will be multiplied by 1. However, if the document is from the second ranked page then the similarity will be multiplied by 0.87. Thus, the higher the rank, the higher the similarity.

The rank position is only one factor we consider. We also take into account the length of the retrieved paragraph from a web page. Based on our experimental observations, a longer paragraph from a web page typically contains more valuable information. In the same way as the rank information was incorporated, we normalize the length factor to effectively integrate it into our similarity measure.

$$\text{similarity } (j, q) = \frac{\sum_{i=1}^{n} w_{ij}*w_{iq}}{\sqrt{\sum_{i=1}^{n} w_{ij}{}^2}*\sqrt{\sum_{i=1}^{n} w_{iq}{}^2}}*\frac{1}{\sqrt{r_j}}*(1-\frac{1}{l_j}) \tag{4}$$

Here l_j is the length of the document j. However, duplicate terms are counted as one term. Hence, taking into account both the rank position and the length of a document, R/quest calculates similarity as:

$$\text{similarity } (j, q) = \frac{(l_j-1)*(\sum_{i=1}^{n} w_{ij}*w_{iq})}{\sqrt{\sum_{i=1}^{n} w_{ij}{}^2}*\sqrt{\sum_{i=1}^{n} w_{iq}{}^2}*\sqrt{r_j}*l_j} \tag{5}$$

2.3   Question Refinement

In R/quest, after a user obtains a ranked list of candidate answers, they will be asked if they are satisfied with the result. If the answer is yes, that means they have found a high quality answer in an acceptable time. If the answer is no, then the user may refine their question. This is resubmitted to Google™ and new candidate answers are displayed to the user. These are likely to answer the user's question.

There are three steps in refinement. No step is mandatory. All three steps are independent and affect the result differently. First we ask the user to edit their question. They can add and eliminate terms. R/quest is different to other search engines in that users of our Q/A system have the previous search results visible on the page where they refine their questions. We believe this helps them understand the effect of each term in their last question and helps them to improve it. If the question has not been revised, R/quest will not send it to Google™ again. If the user did opt to make some changes then the new question will be sent to Google™ and a new set of candidate answers will be presented to the user.

The second step is to allow the user to add weights to the terms in their question. Weights vary from 1 to 10. The weight is the user's estimation of how important a term is in the question. R/quest will use these new weights to re-rank the documents. Term weighting does not change the list of candidate answers; it re-ranks them.

The third and last step in refinement is the addition of the Boolean operator "NOT". Traditional IR systems just match terms using a similarity measure to rank documents. However, users can not say that they do not want documents that contain a particular term. IR systems do not filter out documents that users do not want to see. Hence, we incorporated the Boolean operator "NOT" to allow users to indicate the terms that they do not want to see in the answer. Question refinement may be executed repeatedly, until a user is satisfied that they have a correct answer to their query.

3   Evaluation

To formally evaluate R/quest's overall performance, we submitted 50 English questions from the Text REtrieval Conference (TREC) [19]. Based on the standard answers TREC provided, we recorded the rank position of the correct answer produced by R/quest for each test question. The x-axis shows the number of the TREC question; the y-axis shows the shows the rank of the paragraph where the answer was found.

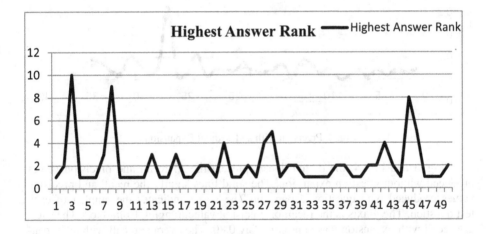

Fig. 2. Answer Ranking

The average rank was 2.18. This implies R/quest finds the correct answer in the top 2 to 3 ranked documents. This means that R/quest users only need to look at 2 to 3 text paragraphs on average to find a correct answer. We consider this to be a considerable improvement over search engines, such as Google™, where users often have to visit a large number of web pages to find an answer.

We also evaluated the effectiveness of our question expansion methodology using the ADI SMART test collection [20]. It consists of 82 documents, 35 questions and the correct answers. During evaluation, we submitted the 35 questions to R/quest, once with expansion and once without. In each case, we compared our answers with the answers provided with the test collection. We then produced precision and recall graphs [18] for the 35 questions. The results are shown below.

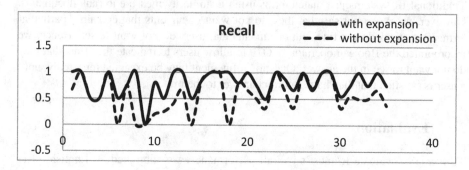

Fig. 3. Recall with and without Expansion

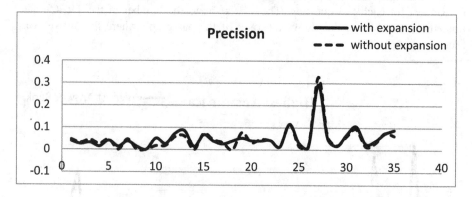

Fig. 4. Precision with and without Expansion

In Fig. 3 and Fig. 4, the dotted lines represent the 35 recall and precision values produced without question expansion. The solid lines signify the recall and precision values produced with question expansion. The x-axis indicates the number of each test question. The y-axis is the recall or precision value, ranging from 0 to 1.The average recall with expansion was approximately 0.80. The average recall without expansion was around 0.56. Hence, expansion means that the system retrieves about 1.4 times more relevant documents. Furthermore, the average precision with expansion was roughly 0.05 and with none was around 0.048. This shows that our expansion retrieves less non-relevant documents. Therefore we get both higher recall and higher precision. This result is counter-intuitive. Classically, higher recall correlates with lower precision. However, in our expansion method we almost always retrieve relevant documents. However, without expansion sometimes no relevant documents are retrieved. This explains why we get both higher recall and precision. By and large, we can say that expansion retrieves more relevant documents and less non-relevant documents than with no expansion. We conclude that query expansion is important and that our method works well.

With R/quest, users use less effort to find the answer to their question. They have less candidate answers to read. Hence, we redefine precision as follows:

IF all the relevant documents in the collection have been retrieved by the system then

    o Record the rank number of the lowest ranked retrieved document

    o Calculate precision as:

$$Precision = \frac{number\ of\ relevant\ documents\ retrieved}{the\ lowest\ rank\ of\ the\ relevant\ documents}$$

ELSE

    o Calculate precision as:

$$Precision = \frac{number\ of\ relevant\ documents\ retrieved}{number\ of\ documents\ retrieved}$$

Fig. 5. Calculating Precision

With this new method, we obtain the precision values shown in Fig. 6. The average new precision was about 0.081. The average precision produced with no refinement was around 0.07. The difference is not very significant but shows promise for our query refinement method.

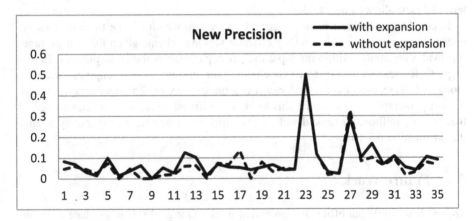

Fig. 6. New Precision Values

4   Conclusion

In this paper, we presented our open-domain Q/A system, R/quest. We used web page snippets from Google™ to expand users' questions. We addressed the vocabulary mismatch problem in section 2.1. The use of snippets to expand queries is not uncommon, however we chose to focus on improving the method used by Sahami and

Heilman [17]. Our expansion technique retrieves less non-relevant documents than theirs and is very efficient in comparison. Thus, our work is an improvement on that of Sahami and Heilman [17]. We performed an evaluation, using the ADI SMART test collection [20], that showed, on average, our query expansion technique resulted in a 1.4 times higher recall and a slightly higher precision. Before performing an objective overall evaluation of R/quest, we submitted approximately 200 very different queries. In the vast majority of cases we retrieved a short, correct answer within the top three ranked answers returned. This was very encouraging and led us to compare R/quest, as a Q/A system, with 50 English questions from a Text REtrieval Conference (TREC) test collection [19]. This work shows that R/quest finds the correct answer in the top 2 to 3 ranked documents. This means that R/quest users only need to look at 2 to 3 text paragraphs on average to find a correct answer. We modified the Cosine Coefficient Similarity Measure [18] to take into account the rank position and length of a document. This produces an improved ranking of our candidate answers. Because our query expansion method retrieves more relevant documents and less non-relevant documents than when no expansion is used, we redefined the definition of precision when query expansion is used. Precision values are improved, reflecting the fact that users' have less paragraphs to review with query expansion. Thus, literally, the retrieval of an answer is a more precise process.

We further enhanced performance by adding a question refinement process that includes the addition of a Boolean NOT operator. This allows users to indicate the terms that they do not want to see in the answer. Traditional, non-Boolean IR systems do not allow users to say that they do not want documents that contain a particular term. R/quest allows a user to do so.

In conclusion, R/quest is an effective Q/A system which is able to answer users' questions on any topic. It has a very efficient ranking scheme given that, on average, it provides an answer within the top 2 to 3 short paragraphs shown to the user. In this respect, R/quest is a great improvement over current search engines, such as Google™. Users can now find an answer to a question by reading, on average, at most 3 short paragraphs. We consider this to be a significant advance over search engines that provide millions of ranked web pages which must be searched manually to find the information needed.

5   Future Work

R/quest is a simple but efficient system that works. Our goal was to produce a system that would effectively and quickly provide a user with a short and correct answer to their query. We also wanted this answer to be one of the first retrieved by our ranking method. We believe that we have achieved this goal.

However, work remains to be done. Our next step is to evaluate R/quest against an existing similar, open-domain Q/A system. The TREC [19] and ADI [20] collections are small. Nevertheless, evaluating R/quest against them suggests we are progressing in the right direction. However, proper, large scale benchmarking is a very important part of this research area. This is our next important goal.

The Cosine Coefficient Similarity Measure [18] is a measure used to evaluate the similarity between two text documents. Although many other similarity measures can be used [18], we found that in our initial testing of R/quest, with Google™, it produced good results and chose not to investigate other similarity measures. After proper benchmarking of R/quest, we may find we have to investigate other similarity measures such as Dice and Jaccard [18] to explore if improvements can be found.

The use of snippets is not uncommon. We chose to concentrate on their use by Sahami and Heilman [17]. A future goal is to study other uses of snippets, specifically in the area of passage retrieval in IR, again with the objective of obtaining improvements in R/quest.

We have already introduced the Boolean NOT operator into R/quest. We are currently implementing a full Boolean interface that would offer a choice of interfaces to users. Our motivation is our understanding that Boolean interfaces can be extremely efficient in practice. An example is Westlaw [21], one of the biggest legal databases available that supports the use of a Boolean interface.

Acknowledgement. This research was fully funded by a grant from the University of Windsor to Joan Morrissey. The authors thank the reviewers for their very helpful comments.

References

1. Athentikos, S.J., Han, H., Brooks, A.D.: A Framework of a Logic-based Question-Answering System for the Medical Domain (LOQAS-Med). In: ACM Symposium on Applied Computing, pp. 847–851 (2009)
2. Cao, Y.G., Liu, F., Simpson, P., Antieau, L., Bennett, A., Cimino, J., Ely, J., Yu, H.: Ask-HERMES An Online Question Answering System for Complex Clinical Questions. Journal of Biomedical Informatics 44, 277–288 (2011)
3. Liang, C.: Improved Intelligent Answering System Research and Design. In: 3rd International Conference on Teaching and Computational Science, pp. 583–589 (2009)
4. Cairns, B.L., Nielsen, R.D., Masanz, J.J., Martin, J.H., Palmer, M.S., Ward, W.H., Savova, G.K.: The MiPACQ Clinical Question Answering System. In: AMIA Annual Symposium, pp. 171–180 (2011)
5. Tao, C., Solbrig, H.R., Sharma, D.K., Wei, W.-Q., Savova, G.K., Chute, C.G.: Time-oriented question answering from clinical narratives using semantic-web techniques. In: Patel-Schneider, P.F., Pan, Y., Hitzler, P., Mika, P., Zhang, L., Pan, J.Z., Horrocks, I., Glimm, B. (eds.) ISWC 2010, Part II. LNCS, vol. 6497, pp. 241–256. Springer, Heidelberg (2010)
6. Leuski, A., Traum, D.: NPCEditor: A Tool for Building Question-Answering Characters. In: Seventh International Conference on Language Resources and Evaluation (LREC 2010), pp. 2463–2470 (2010)
7. Hao, T., Liu, W., Zhu, C.: Semantic pattern-based user interactive question answering: User interface design and evaluation. In: Huang, D.-S., Gan, Y., Gupta, P., Gromiha, M.M. (eds.) ICIC 2011. LNCS, vol. 6839, pp. 363–370. Springer, Heidelberg (2012)

8. Xia, L., Teng, Z., Ren, F.: Question Classification for Chinese Cuisine Question Answering System. IEEE Transactions on Electrical and Electronic Engineering 4, 689–695 (2009)
9. Dong, T., Furbach, U., Glöckner, I., Pelzer, B.: A Natural Language Question Answering System as a Participant in Human Q&A Portals. In: 22nd International Joint Conference on Artificial Intelligence, pp. 2430–2435 (2011)
10. Trigui, O., Belguith, L.H., Rosso, P.: DefArabic QA: Arabic Definition Question Answering System. In: Workshop on LR & HLT for Semitic Languages, pp. 40–44 (2011)
11. Čeh, I., Ojsteršek, M.: Developing a Question Answering System for the Slovene Language. Transactions on Information Science and Applications 6, 1533–1543 (2009)
12. Tannier, X., Moriceau, V.: FIDJI: Web Question-Answering. In: Seventh International Language Resources and Evaluation, pp. 2375–2379 (2010)
13. Bouma, G., Kloosterman, G., Mur, J., van Noord, G., van der Plas, L., Tiedemann, J.: Question Answering with Joost. In: Peters, C., Jijkoun, V., Mandl, T., Müller, H., Oard, D.W., Peñas, A., Petras, V., Santos, D. (eds.) CLEF 2007. LNCS, vol. 5152, pp. 257–260. Springer, Heidelberg (2008)
14. Ansa, O., Arregi, X., Otegi, A., Soraluze, A.: Ihardetsi: a Basque Question Answering System. In: Peters, C., Deselaers, T., Ferro, N., Gonzalo, J., Jones, G.J.F., Kurimo, M., Mandl, T., Peñas, A., Petras, V. (eds.) CLEF 2008. LNCS, vol. 5706, pp. 369–376. Springer, Heidelberg (2009)
15. Bowden, M., Olteanu, M., Suriyentrakorn, P., D'Silva, T., Moldovan, D.: Multilingual Question Answering through Intermediate Translation: Lcc's PowerAnswer. In: Peters, C., Jijkoun, V., Mandl, T., Müller, H., Oard, D.W., Peñas, A., Petras, V., Santos, D. (eds.) CLEF 2007. LNCS, vol. 5152, pp. 273–283. Springer, Heidelberg (2008)
16. Olteanu, M., Davis, C., Volosen, I., Moldovan, D.: Phramer: An Open Source Statistical Phrase-based Translator. In: Workshop on Statistical Machine Translation, pp. 146–149. Association for Computational Linguistics, USA (2006)
17. Sahami, M., Heilman, T.D.: A Web-based Kernel Function for Measuring the Similarity of Short Text Snippets. In: 15th International Conference on World Wide Web, pp. 377–386 (2006)
18. Manning, C.D., Raghavan, P., Schütze, H.: An Introduction to Information Retrieval. Cambridge University Press, New York (2008)
19. National Institute of Standards and Technology, http://trec.nist.gov/overview.html
20. Spark-Jones, K., van Rijsbergen, C.J.: Information Retrieval Test Collections. Journal of Documentation 32, 59–75 (1976)
21. Westlaw, http://www.westlaw.com/

Answering Questions by Means of Causal Sentences

C. Puente[1], E. Garrido[1], and J.A. Olivas[2]

[1] Advanced Technical Faculty of Engineering ICAI
Pontifical Comillas University, Madrid, Spain
{Cristina.Puente,Eduardo.Garrido}@upcomillas.es
[2] Information Technologies and Systems Dept
University of Castilla-La Mancha, Ciudad Real, Spain
Joseangel.olivas@uclm.es

Abstract. The aim of this paper is to introduce a set of algorithms able to configure an automatic answer from a proposed question. This procedure has two main steps. The first one is focused in the extraction, filtering and selection of those causal sentences that could have relevant information for the answer. The second one is focused in the composition of a suitable answer with the obtained information in the previous step.

Keywords: Causal questions, Causality, Causal Sentences, Causal Representation.

1 Introduction

Causality is a fundamental notion in every field of science, in fact there is a great presence of causal sentences in scientific language [1]. Causal links are the basis for the scientific theories as they permit the formulation of laws. In experimental sciences, such as physics or biology, causal relationships only seem to be precise in nature. But in these sciences, causality also includes imperfect or imprecise causal chains.

There have been many works related to causal extraction in text documents, like Khoo and Kornfilt [2], who developed an automatic method for extracting causal information from Wall Street Journal texts, or Khoo and Chan [3] who developed a method to identify and extract cause-effect information explicitly expressed without knowledge-based inference in medical abstracts using the Medline database.

On the other hand, the ability to synthesize information losing as little information as possible from a text has been studied in various disciplines throughout history. In this area the work of linguistics, logic and statistics [4] are the most relevant issues to take into consideration to propose a suitable answer to a given question.

Taking these works as inspiration, this paper deals with two algorithms. The first one is focused on the extraction of causal sentences from texts belonging to different genres or disciplines, using them as a database of knowledge about a given topic. Once the information has been selected, a question is proposed to choose those sentences where this concept is included. These statements are treated automatically in order to achieve a graphical representation of the causal relationships with nodes labeled with linguistic hedges that denote the intensity with which the causes or effects happen, and the arcs marked with fuzzy quantifiers that show the strength of

H.L. Larsen et al. (Eds.): FQAS 2013, LNAI 8132, pp. 91–99, 2013.

the causal link. In turn, nodes are divided into cells denoting location, sub location or other contextual issues that must permit a better understanding of the meaning of the cause. The node 'cause' includes also a special box indicating the intensity with which the cause occurs.

The second algorithm is in charge of the generation of an answer by reading the information represented by the causal graph obtained in the previous step. Redundant information is removed, and the most relevant information is classified using several algorithms such as collocation algorithms like SALSA or classical approaches like keywords depending on the context, TF-IDF algorithm.

The answer is generated in natural language thanks to another algorithm which is able to build phrases using a generative grammar.

2 Selection of the Input Information

In [5], Puente, Sobrino, Olivas & Merlo described a procedure to automatically display a causal graph from medical knowledge included in several medical texts.

A Flex and C program was designed to analyze causal phrases denoted by words like 'cause', 'effect' or their synonyms, highlighting vague words that qualify the causal nodes or the links between them. Another C program receives as input a set of tags from the previous parser and generates a template with a starting node (cause), a causal relation (denoted by lexical words), possibly qualified by fuzzy quantification, and a final node (effect), possibly modified by a linguistic hedge showing its intensity. Finally, a Java program automates this task. A general overview of the extraction of causal sentences procedure is the following:

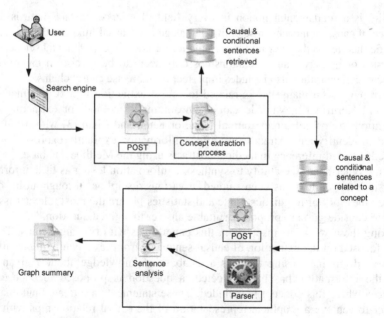

Fig. 1. Extraction and representation of causal sentences

Once the system was developed, an experiment was performed to answer the question *What provokes lung cancer?*, obtaining a set of 15 causal sentences related to this topic which served as input for a causal graph representation. The whole process was unable to answer the question directly, but was capable of generating a causal graph with the topics involved in the proposed question as shown in figure 2.

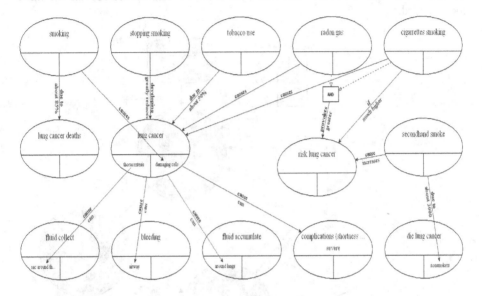

Fig. 2. Causal representation related to the question *What provokes lung cancer?*

With this causal graph, we want to go a step further in this paper to generate the answer to the proposed question by means of a summary, processing the information contained in the causal nodes and the relationships among them.

3   Summarizing the Content of the Causal Graph

The ideal representation of the concepts presented in Fig. 2 would be a natural language text. This part of the article presents the design of a possible approach to do so. Main problems are discussed and the general ways to solve them are introduced.

The size of the graph could be bigger than the presented one as not all the causal sentences are critical to appear in the final summary. It is necessary to create a summary of the information of the graph in order to be readable by a human as if it was a text created by other human.

The causal graph presented in Fig. 2 has the problem that the concepts represented could have a similar meaning in comparison to other concepts. For example, *"smoking"* and *"tobacco use"* have the similar meaning in the graph so one of these concepts could be redundant.

Not only is this relationship of synonymy but other semantic relations such as hyperonymy or meronymy are important as well.

It is possible to create a process to read the concepts of the graph sending them to an ontology like Wordnet or UMLS and retrieving different similarity degrees according to each relation [6].

Several analysis and processes need to be executed to obtain a summary by means of an automatic algorithm. The following diagram shows the design of the summary system that is created to solve this issue, including the main processes and the main tools needed.

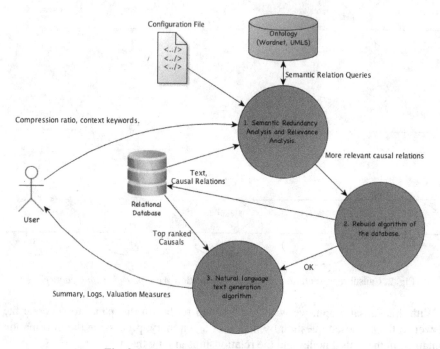

Fig. 3. Basic design of the summarization process

This procedure is not simple, due to the fact that the meaning of the words has to be discovered with help of the context. A polysemic word is one that has several meanings referring to its written representation.

A redundancy analysis process is created to solve this problem, taking into account the multiple synsets of every word of the concepts that is been analyzed. It is also taken into account the context of the text having keywords of every context and other measures.

To do so, Wordnet synsets are queried from Java thanks to *Jwnl* and *RiWordnet* tools to find out the meaning of these terms. The output of the process will consist of the possible relations between all pairs of entities compared, declaring the type and intensity of this relationship.

The degree of their similarity with other concepts is computed as well, being a measure to take into account in the relevance analysis. Different algorithms of similarity between concepts such as Path Length, Leacock & Chodorow [7] or Wu Palmer [8], are executed through platforms like Wordnet::Similarities.

A comparison matrix is then created with all this information, showing the similarity between terms according to different semantic relations. Those concepts with higher similarity degrees with others are the redundant ones.

$$M = \begin{pmatrix} 0 & m_{12} & m_{13} & . & . & . & m_{1n} \\ 0 & 0 & m_{23} & . & . & . & m_{2n} \\ . & . & . & . & . & . & m_{3n} \\ . & . & . & . & m_{ij} & . & m_{in} \\ . & . & . & . & . & . & . \\ . & . & . & . & . & 0 & m_{n-1n} \\ 0 & 0 & 0 & . & . & 0 & 0 \end{pmatrix}$$

Fig. 4. Comparison matrix built by the semantic redundancy algorithm

After running the whole process, a list of semantic relations between entities is obtained. This list contains all the information of the relations and a list of semantic entities containing the entities which are going to be deleted. This is the entry for the graph reconstruction algorithm. Additionally, a report is obtained on this first version with the final results:

```
Final results
====================
Synonyms: 6
Hypernymy/Hyponymy: 13
Meronymy/Holonymy: 0
Entailment: 0
Verb groups: 0
Non related: 72
Total compared concepts: 91
Percentage of reduction of the graph: 79.12088 %
=======================================
Concepts to review:
-> lung cancer deaths
-> risk lung cancer
-> die lung cancer
-> stopping smoking
-> tobacco use
-> cigarettes smoking
-> secondhand smoke
-> fluid collect
-> fluid accumulate
=======================================
```

Fig. 5. Final results

Once a relation has been found, the problem is choosing which term is the most relevant. In the example mentioned above, the question would be, what is the most important concept, "smoking" or "tobacco use". In [9] we proposed a mechanism that gives the answer to that question performing an analysis of the relevance of each concept. To do so, classical measures that analyses the appearance of the concepts in the text like TF-Algorithm are used. Connective algorithms that analyses the graph as SALSA or HITS [10] are also used.

When a causal relation is going to be moved to other concept due to the fact that this concept is going to be erased according to the semantic redundancy or relevance ranking algorithm, if the causal relation also exists in the concept which is not going to be erased then two different grades exists. In order to see which implication degree is the resultant one an expression is proposed:

$$NGa = (1-s)*Ga + s*(relA/(relA+relB)*Ga + relB/(relA+relB)*Gb);$$
$$/ NGa [0,1] \forall \{s,relA,relB,Ga,Gb\} [0,1]$$

Being NGa the new degree of the concept A and being the concept B the one which is going to be erased, s [0,1] the semantic similarity between the two concepts, Ga and Gb [0,1] the implication degree of the concepts, relA and relB [0,1] the relevance of both terms according to the relevance ranking algorithm. Using this expression the new implication degree is calculated in function of all of the parameters of both nodes.

If the implication does not exist in the node which is not going to be erased then the expression of the new degree is the following one:

$$NGa = s*(relB/(relA+relB)*Gb);$$
$$/ NGa [0,1] \forall \{s,relA,relB,Gb\} [0,1]$$

After these analysis, the information of the graph has been summarized obtaining the following graph:

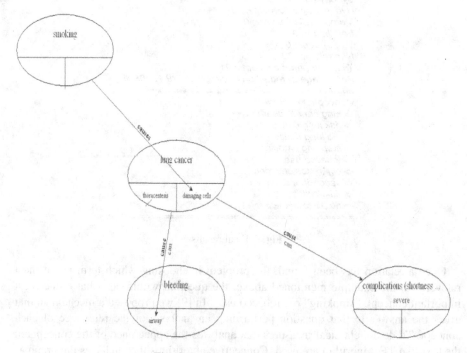

Fig. 6. Causal graph summarized

The summary process has a configuration module depending on the user's preferences and the nature and context of the text to be analyzed. All the modules and measures can be parameterized by means of a weight-value algorithm. In order to have a better reading of the graph, the information needs to be expressed in natural language. The last process consists of an algorithm that generates natural language given the top ranked causals by the semantic redundancy analysis and relevance analysis. We have performed two experiments varying the compression rate to evaluate the obtained results and check the configuration of the algorithm. In the first experiment, we used a compression rate of 0.3, obtaining as a result the following summary:

"Cigarettes smoking causes die lung cancer occasionally and lung cancer normally.Tobacco use causes lung cancer constantly and die lung cancer infrequently.Lung cancer causes die lung cancer seldom and fluid collect sometimes. It is important to end knowing that lung cancer sometimes causes severe complication."

The original text length is 1497 characters and the summary length is 311 so the system has been able to achieve the compression rate, being the summary less than the 30% of the original text. In this case, the main information has been included, removing redundant information. The system has chained sentences with the same causes to compose coordinate sentences and reduce the length of the final summary. As seen, the grammatical and semantic meaning is quite precise and accurate, without losing relevant information. In the second experiment, the compression rate was the lowest, to remove all the redundant and irrelevant information, it was set so the summary represents a 10% of the original text, so the result was the following:

"Lung cancer is frequently caused by tobacco use. In conclusion severe complication is sometimes caused by lung cancer."

In this case, the system just takes the information of the three most relevant nodes, one cause, one intermediate node, and an effect node, creating a summary with the most relevant nodes included in the graph. As it can be seen, the length of the summary is of 118 characters, what represents less than a 10% of the length of the original text, the system is able to modify its behaviour according to different configurations of the weights of the redundance and relevance algorithms and the compression rate. We made experiments with other texts which passes essential quality tests such as measuring the syntax of the texts or assuring the compression ratio, precision and recall are generally good in these experiments. Having this original text we can see the logic of the summary:

"This is a text inspired by the famous case discussed in Ethic class Ford Pinto. When a CEO introduces a new product in the industry it has several options, new product options rarely are doing no quality test and unfrequently a fast manufacturing test is done. New product options normally imply doing a normal manufacturing process. If the organization is not meeting standards then incidents are occasionally caused by this behavior. No quality tests may produce incidents but no quality tests often imply

being the first in the market. A fast manufacturing process rarely produces incidents but there are cases. Fast manufacturing processes often implies being first in the market. A fast manufacturing process is hardly ever the cause of losses because of things that are not done. Incidents are constantly the cause of jail or prison by the CEO"s that do not follow the security standards, but the temptation is the following condition: If company is being the first in the market then it will always earn profits for a short time. But jail or prison is always the cause of losses and human lifes, CEO"s have to be aware of the ethics of not following the security standards."

Having a compression rate of 0.2 and using the configuration by default the obtained summary is the following one:

"What is discussed is that being first in the market is implied hardly ever by new product options.Loss is never caused by being first in the market. Prison is constantly implied by incident.Eventually, prison always produces loss."

The original length of this text was of 1180 characters, and using a compression rate of 0.2, the length of the obtained summary has been 231 characters, which is actually a 19,58% of the original text. The degrees are logical according to the original text and the most important semantic content of the text is contained in the summary.

4   Conclusions and Future Works

The massive amount of information, growing constantly, is a problem that should be treated using systems like the one proposed in this paper. This system is able to extract the most relevant knowledge contained in texts to create a causal database related to a given topic.

Using this database, the representation algorithm is able to create a causal graph containing the main concepts of a proposed question. With this graph, we have developed a procedure to remove the irrelevant information for an automatic answer. By removing the redundancy, we are able to compose an answer suitable to the proposed question, with different levels of compression.

As future works, we would like answer more complex questions, like how questions, which require for a more complex mechanism to be answered.

Acknowledgments. Partially supported by TIN2010-20395 FIDELIO project, MEC-FEDER, Spain.

References

1. Bunge, M.: Causality and modern science. Dover (1979)
2. Khoo, C., Kornfilt, J., Oddy, R.N., Myaeng, S.H.: Automatic extraction of cause-effect information from newspaper text without knowledge-based inferencing. Literary and Linguistic Computing 13(4), 177–186 (1998)

3. Khoo, C.S.G., Chan, S., Niu, Y.: Extracting causal knowledge from a medical database using graphical patterns. In: ACL Proceedings of the 38th Annual Meeting on Association for Computational Linguistics, Morristown, USA, pp. 336–343 (2000)
4. Basagic, R., Krupic, D., Suzic, B.: Automatic Text Summarization, Information Search and Retrieval. In: WS 2009, Institute for Information Systems and Computer. Graz University of Technology, Graz (2009)
5. Puente, C., Sobrino, A., Olivas, J.A., Merlo, R.: Extraction, Analysis and Representation of Imperfect Conditional and Causal sentences by means of a Semi-Automatic Process. In: Proceedings IEEE International Conference on Fuzzy Systems (FUZZ-IEEE 2010), Barcelona, Spain, pp. 1423–1430 (2010)
6. Varelas, G., Voutsakis, E., Raftopoulou, P., Petrakis, G.M.E., Milios, E.: Semantic Similarity Methods in WordNet and their Application to Information Retrieval on the Web. In: WIDM 2005, Bremen, Germany (November 5, 2005)
7. Leacock, C., Chodorow, M.: Combininglocal context and WordNet similarity for word sense identification. In: Fellbaum 1998, pp. 265–283 (1998)
8. Wu, Z., Palmer, M.: Verb semantics and lexical selection. In: 32nd Annual Meeting of the Association for Computational Linguistics, pp. 133–138, Resnik (1994)
9. Puente, C., Garrido, E., Olivas, J.A., Seisdedos, R.: Creating a natural language summary from a compressed causal graph. In: Proc. of the Ifsa-Nafips 2013, Edmonton, Canada (2013)
10. Najork, M.: Comparing The Effectiveness of HITS and SALSA. In: Proc. of the Sixteenth ACM Conference on Information and Knowledge Management, CIKM 2007, Lisboa, Portugal, November 6-8 (2007)

Ontology-Based Question Analysis Method

Ghada Besbes[1], Hajer Baazaoui-Zghal[1], and Antonio Moreno[2]

[1] Riadi-GDL, ENSI Campus Universitaire de la Manouba, Tunis, Tunisie
ghada.besbes@gmail.com, hajer.baazaouizghal@riadi.rnu.tn
[2] ITAKA Research Group, Departament d'Enginyeria Informatica i Matematiques,
Universitat Rovira i Virgili, Av. Paisos Catalans, 26. 43007, Tarragona, Spain
antonio.moreno@urv.cat

Abstract. Question analysis is a central component of Question Answering systems. In this paper we propose a new method for question analysis based on ontologies (*QAnalOnto*). *QAnalOnto* relies on four main components: (1) Lexical and syntactic analysis, (2) Question graph construction, (3) Query reformulation and (4) Search for similar questions. Our contribution consists on the representation of generic structures of questions and results by using typed attributed graphs and on the integration of domain ontologies and lexico-syntactic patterns for query reformulation. Some preliminary tests have shown that the proposed method improves the quality of the retrieved documents and the search of previous similar questions.

Keywords: Question-Answering systems, ontology, lexico-syntactic patterns, typed attributed graphs.

1 Introduction

With the rapid growth of the amount of online electronic documents, the classic search techniques based on keywords have become inadequate. Question Answering systems are considered as advanced information retrieval systems, allowing the user to ask a question in natural language (NL) and returning the precise answer instead of a set of documents. The search process in a Question Answering system is composed of three main steps: question analysis, document search and answer extraction from relevant documents. Generally, Question Answering (QA) systems aim at providing answers to NL questions in an open domain context and can provide a solution to the problem of response accuracy. This requirement has motivated researchers in the QA field to incorporate knowledge-processing components such as semantic representation, ontologies, reasoning and inference engines. Our work hypothesis is that, if the user starts with a well-formulated question, answers will be more relevant; this is why, in this work, we focus on question analysis. So, the aim of this paper is to design and implement a new method dedicated to question analysis in a QA system. Indeed, our goals consist on improving the representation of the question's structure by using typed attributed graphs and improving the results of query reformulation by using domain ontologies and lexico-syntactic patterns.

H.L. Larsen et al. (Eds.): FQAS 2013, LNAI 8132, pp. 100–111, 2013.

In this method, first of all, lexical and syntactic analyses are applied to the user's question. Second, a question graph, containing all the information about the question, is constructed based on a generic question graph using knowledge from WordNet and from a question ontology. Then the question is reformulated based on lexico-syntactic patterns and the domain knowledge represented in an ontology. Finally, the method stores the question graph and the reformulated question in a question base in order to extract analysis results for similar questions later. Our method is dedicated to QA systems as it deals with NL queries asked in a question form, considered as a particular case of information retrieval systems. The evaluation is conducted using information retrieval metrics such as precision and MAP.

The remaining of this paper is organized as follows. Section 2 presents an overview of works related to question analysis techniques. Section 3 describes our method of question analysis based on ontologies. Section 4 presents and discusses some experimental results of our proposal. Finally, section 5 concludes and proposes directions for future research.

2   Related Works

The question analysis component is the first step of the search process in a question-answering system. This analysis aims to determine the question's structure as well as the significant information (expected answer type, terms' grammatical functions, etc.) that are considered as clues for identifying the precise answer. Question analysis methods can be classified depending on their level of linguistic analysis: (i) **Lexical analysis:** The lexical level of NL processing is centered on the concept of a word, and the techniques used for lexical analysis are generally a pre-treatment for the following analysis. The most used techniques are the following: tokenization (division of the question into words) and keyword extraction [1], lemmatization [2](considering the root to group words of the same family) and removing stop words [1] (the elimination of common words that do not affect the meaning of the question to reduce the number of words to be analyzed). (ii) **Syntactic analysis:** Information extracted from the question analysis component is the basis for answer extraction. This component constructs a representation of the question, which differs from one system to the other and contains various types of information and knowledge. The purpose of this analysis is to preserve the syntactic structure of the question by exploiting the syntactic functions of words in the questions [3]. Question-answering systems use different techniques of NL processing, including the following: Part-of-speech tagging or POS tagging [4] (giving each word a tag that represents information about its class and morphological features), named entity recognition (identifying objects as classes that can be categorized as locations, quantity, names, etc.) and the use of a syntactic parser. (iii) **Semantic analysis:** In some question-answering systems, analyzing the question goes beyond vocabulary and syntax up to semantics and query reformulation. This phase includes the extraction of semantic relations between the question words [5] to make a semantic representation as in the Javelin system [6]. The purpose of semantic analysis is to detect and represent semantic knowledge in order to use it for inference or matching when extracting the answer. To do this, several systems rely on semantic techniques in order to have a better analysis of the question. In the case of query reformulation and enrichment, most systems use tools and semantic knowledge such as WordNet [7] or ontologies to extract other semantic forms for the question keywords. In fact, ontology-based question-answering systems such as QuestIO [8], AquaLog [9] and QASYO [12] use an internal representation of knowledge in the form of an ontology. The purpose of using an ontology as a knowledge representation is either to extract the answer directly as in Querix [10] or to reformulate the query by rewriting the user's question using the ontology concepts.

In general, the purpose of question analysis is to collect information on the subject of the question, to represent it and to formally submit a request to the search engine. The previous study allows identifying the following limits on the different levels of linguistic analysis: (i) **Lexical analysis:** Question analysis in many question-answering systems is reduced to the lexical analysis, and extracted keywords are used as search queries for the information retrieval system without any reformulation. This method does not represent the question and does not extract the terms' grammatical functions. (ii) **Syntactic analysis:** with only a syntactic analysis, the query reformulation problem is still not resolved. In addition, the question's representation has only the terms used in it and their morpho-syntactic classes; therefore, it does not represent the question's semantic knowledge.(iii) **Semantic analysis:** Query reformulation at this level focuses only on retrieving potentially relevant documents, not answer-bearing ones.

Our main objective is to improve the question analysis component in QA systems in order to improve their performance. During the study of the state of the art we identified the following items to address: finding similar questions from a question base, representing analyzed questions and reformulating the queries.

1. The process of finding similar questions in a question base is a computationally expensive, and most similarity measures are designed to deal with concepts not with questions. We therefore applied a filtering on the question base in order to lighten the process and we combined statistic and semantic similarity measures suitable for questions.

2. During the question analysis process, we are confronted with the problems of determining its structure and the lack of expressiveness of representation formalisms that do not respect the granularity of the concepts used in the question. Therefore, we used a generic graph (in fact, a typed attributed graph) to represent the structure of the question

3. Query reformulation is not rich enough. It lacks external knowledge such as ontologies to bring new concepts and terms. It is also oriented towards relevant documents not answer-bearing ones. Through our method, we tried to solve the problems of query reformulation by using a domain ontology combined with lexico-syntactic patterns.

3   Question Analysis Method

The proposed method relies on four main components: (1) Lexical and syntactic analysis, (2) Question graph construction, (3) Query reformulation and (4) Search for similar questions. These components will be detailed in the following sections.

3.1   Proposed Method's Description

Figure 1 provides a general view of the proposed method for analyzing NL questions. The goal is to identify all the terms of a question and their grammatical functions in the question and to obtain useful information for the answer's extraction. First, the user submits a question in NL. The method performs a lexical and syntactic analysis (1). The syntactic analysis is based on POS tagging in order to identify the grammatical morpho-syntactic class for each term used in the question. These results are interpreted by the question ontology, synonyms for each term are extracted from WordNet and the structure of the question is defined using the generic graph that contains all the general structures of questions. The method builds a typed attributed graph (2) that contains all the information available in the question. Then, the question is reformulated (3) using lexico-syntactic patterns and the concepts of a domain ontology. The patterns required in this reformulation process are retrieved from the question ontology. Concepts that are semantically related to the terms of the question are extracted from a domain ontology to enrich the question. Using a question base, all questions are recorded along with their analysis results, that is to say, the typed attributed graph of the question and the result of query reformulation. Thus the method can search for similar questions (4) and the user has the option to extract directly the results of analysis of a stored similar question. The output of the method is a reformulated query ready to be submitted to a search engine and a set of useful, well-structured questions that will be used by the answer extraction component.

3.2   Lexical and Syntactic Analysis Component

The first step is lexical analysis which includes the following two processes:

- Tokenization: It's the division of the text into words that can be managed in the next steps of the analysis.
- Lemmatization: This process considers the root of a word. For example, all verbs are reduced to the infinitive (eaten, ate -> eat), plural nouns are reduced to singular, etc. In this way, a search using any of the word's variants will lead to the same result.

The second step is syntactic analysis. At this level of analysis, POS tagging is applied to the question. This is the process of associating a tag to each word in the question that represents information about its class and morphological features.

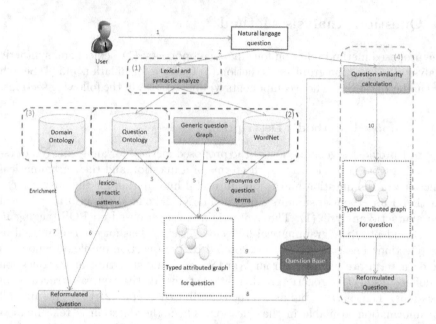

Fig. 1. General architecture of QAnalOnto

3.3 Question Graph Construction Component

The question graph is a representation of the user's question in an intuitive and understandable form that contains all the information included in the question necessary to search for its answer.

Generic Question Graph. The main advantage of using graphs resides in its capability to represent relations, even multiple ones, between objects. The generic question graph contains all forms of predefined questions. It is used to identify the question's structure. It is a typed attributed graph. This type of graph is a pair $(NG; EG)$ where NG is a set of attributed nodes and EG is a set of attributed edges. An attributed node $n \in NG = (Tn, AVn)$ has a type Tn and a set of attribute values AVn. An attributed edge $e \in EG = (Te; AVe; Oe, De)$ has a type Te, a set of attribute values AVe, an attributed node that represents the origin of the edge Oe and an attributed node that represents the destination of the edge De.

Figure 2 shows a generic graph of a simple question: WH + Verb + Subject.

The nodes "WH","Verb" and "Subject" are subgraphs composed of "Term" nodes that represent, respectively, the kind of a WH-question, the main verb in the question and its subject.

A node "Term" (Ti) is the smallest conceptual unit representing a term in a question. The node "Term" consists of the following attributes: type ("Term"), value (question term), POS tag, lemma, category (WH, verb, subject) and synonyms (extracted from WordNet).

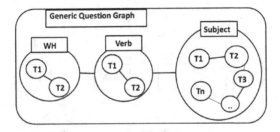

Fig. 2. Example of generic question graph

Each node "WH", "Verb" and "Subject" is itself a typed attributed graph $C = (TC; RTC)$ where TC is the list of attributed "Term" nodes and RTC is a set of typed edges between terms which represent the relation "followed_by" which specifies the order of the different terms of the question.

Construction Steps. In the first component, we performed a lexical and syntactic analysis on the question in order to extract the terms used in the question and their tags. Using these results, the system constructs the question graph. The construction process is divided on three steps:

1. Detection of the question's structure: Using the parsed question and the question ontology we can extract the question's structure from the generic question graph (that contains the structures of all types of questions allowed in the system). The system passes the parsed question by the question ontology in order to interpret the tags and determine the answer type. The question ontology is a manually constructed ontology that contains all the tags classified by category, so, tags are recognized and returned to the generic graph to identify and extract the question's structure. In fact, in the question ontology each kind of question has different answer types. From the results of the tagging, the ontology defines the expected answer type for the question. The ontology also contains lexico syntactic patterns for each type of question, that can be used to reformulate it.

   Part of the question ontology focused on the question "where" is represented in Figure 3.

   The ellipsis boxes represent classes, the rectangular ones represent the tags returned by POS tagger. Their super classes represent their grammatical functions (WH, verb, subject, etc.) and their subclasses represent the NL terms used in the question (When, Where, etc.). The solid edges represent the relation "subClassOf" and the dotted lines represent object properties. NL concepts are linked to their types through the "has_type" property and to their patterns through the "has_pattern" property. These elements are themselves subclasses of the concepts Types and Patterns respectively.

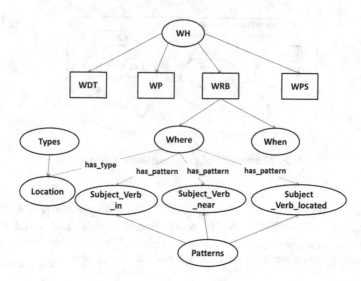

Fig. 3. Part of the question ontology

2. Instantiation of the generic graph: Using the parsed question, we instantiate the part of the generic question graph that contains the structure determined in the previous step. The result is a question graph that has the determined structure and contains the question's information. In fact, this graph is an instantiation of the generic graph that contains filled nodes of type "Term" containing the question's words. The terms of the same category form a graph and belong to the same type node: "WH", "Verb" or "Subject" (according to the example shown in Figure 2).Edges between these nodes are of type "followed_by" which specify the order of words in the user's question.

Example: Figure 4 is a question graph applied to the question "where is the tallest monument in the world?". This graph is an instantiation of the generic question graph shown in figure 2.

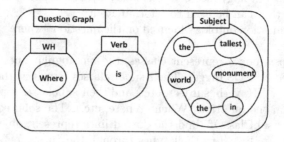

Fig. 4. Example of question graph

3. Synonym detection: WordNet is used in this step to extract the terms' semantics. We complete the question graph with the terms' synonyms in order to create a complete graph which contains the words, their grammatical functions, the structure of the question and synonyms. Adding synonyms to the graph is crucial for future search in the question base. In fact, the same question can be asked in several ways or expressed with different words and have the same meaning, in which case the system must be able to identify the different forms using the various synonyms stored in the question graph.

3.4 Query Reformulation Component

The analysis process requires query reformulation which consists on adding terms related to the question's keywords and expanding it. The resulting reformulated query will be submitted to the search engine that will return a set of documents from which the answer is extracted. The query reformulation is based, in our method, on two techniques which are the use of lexico-syntactic patterns and of a domain ontology. The aim is to guide the search engine to relevant documents for the search topic (using a domain ontology) and to answer-bearing documents (using patterns that define the answer's structure).

Query Reformulation Based on Patterns. The patterns used in this method are intended to reconstruct the user's question in order to guide it to the answer. Therefore, these answer patterns are applied to extract the candidate passage and locate the correct answer.

For each question type (what, where, who) there is an associated set of answer patterns. According to the question type of the submitted query, answer patterns are retrieved from the question ontology and instantiated with question terms. For instance, for the question: "where is the tallest monument in the world?", the method identifies from the question ontology the following patterns: Subject_Verb_in, Subject_Verb_near et Subject_Verb_located. The method reformulates the query using these patterns and obtains the following questions: "the tallest monument in the world is in", "the tallest monument in the world is near", "the tallest monument in the world is located".

Query Reformulation Based on a Domain Ontology. In order to add more semantic information to guide the search towards relevant documents, we use a domain ontology from which the method extracts, for the terms of the query that correspond to an ontology concept, its sub-classes and its related concepts. The method specializes the query by adding more specific concepts extracted from the ontology. This refinement increases the number of specific concepts and subsequently, increases the precision.

Let's take for example the question "where is the tallest monument in the world?" After the reformulation based on patterns in the previous section, we enrich the reformulated query with concepts related to the concept "monument" extracted from a domain ontology. We use the subclasses of this concept ("statue", "arch ", "memorial") to enrich the reformulated query. We obtain three final re-formulated queries: "the tallest monument/statue/arch/memorial in the world is in", "the tallest monument/statue/arch/memorial in the world is near" and "the tallest monument/statue/arch/memorial in the world is located".

3.5   Search for Similar Questions

This module of QAnalOnto retrieves similar questions stored in the question base. The method lists the similar questions ordered by similarity to the one asked by the user and, if the user chooses one, the corresponding analysis re-sult and the reformulated query will be returned. However, the question base can be large, and the direct application of similarity measures can slow down the search process. To overcome this problem, we apply a filtering process that selects candidate questions from the base and removes questions that have to-tal dissimilarity with the one asked by the user. In fact, the chosen questions from the base have the same expected answer type and at least one common keyword. On these questions, we will apply the similarity measures in order to classify them by their relevance to the user's question.

Several measures of semantic similarity, with different properties and results exist in the literature. The similarity measure we propose is based on the work of [11]. It combines the statistic similarity and the semantic similarity between the user's question and the questions stored in the question base. The statistic similarity is based on dynamically formed vectors: the two compared questions are represented with two vectors formed by their words instead of considering all the words in the question base and then their cosine product is computed to obtain the statistic similarity. The semantic similarity is calculated using the distance between two words w1 and w2 in WordNet as follows:

$$\frac{minDistToCommonParent}{DistFromCommonParentToRoot + minDistToCommonParent}$$

In this formula $minDistToCommonParent$ indicates the shortest path be-tween two words to the common parent and $DistFromCommonParentToRoot$ indicates the path length from the common parent to the root.

The overall similarity is an average of statistical and semantic similarities.

4   Experimental Evaluation

A prototype has been developed to show that the proposed method can improve the performance of the retrieval task. It provides a user interface that allows these main functionalities: search for similar questions from the question base, construction of the question graph and reformulation of the user's query. Since the proposed method provides an analysis of the question and reformulates the query to be submitted to the search engine, we experimentally evaluate its per-formance by testing its capacity for (1) retrieving relevant documents after query reformulation and (2) retrieving similar questions from the question base.

4.1 Search Results Evaluation

To evaluate the query reformulation component, we computed: (1) Exact precision measures P@10, P@30, P@50 and P@100 representing respectively, the mean precision values at the top 10, 30, 50 and 100 returned documents; (2) MAP representing the Mean Average Precision computed over all topics.

Two main scenarios have been tested:

- The first scenario represents the baseline which is a classic search using keywords without performing any query reformulation.
- The second scenario represents results obtained after reformulating using both lexico-syntactic patterns and ontologies.

The improvement value is computed as follows:

$$Improvement = \frac{\text{Reformulation-result} - \text{Baseline-result}}{\text{Baseline-result}}$$

Table 1. Improvement in average precision at top n documents and MAP

	P@10	P@20	P@30	P@50	P@100	MAP
Baseline	0.60	0.32	0.212	0,171	0.065	0,273
QAnalOnto	0.783	0.39	0.256	0,206	0.078	0,341
Improvement	30,5%	21,87%	20.75%	20,46%	20%	24,90%

The evaluation results are calculated using the LEMUR [1] tool for Information Retrieval evaluation. Besides, we rely on the INEX 2010 [2] collection of documents. We measured the precision for several queries using the INEX topics and then we averaged these results. The evaluation results shown in table 1 represent the precision obtained according to the number of retrieved documents (10, 20, 30, 50 and 100), and we observe a significant improvement of the relevance of the retrieved information. In Table 1, we outline the computed MAP and the average precision at the top n documents and their percentages of improvement. We observe that reformulating queries using both lexico-syntactic patterns and a domain ontology improves the retrieval precision by 24,9%. In fact, using lexico-syntactic patterns guides the search towards answer-bearing documents and specifying the question's keywords and enriching it using the domain ontology improves the precision.

4.2 Similar Question Search Evaluation

To evaluate the search for similar questions, we used a set of queries (20 WH-questions from different domains). For each of them we created manually:(1) a set of questions containing the same words with different meanings, (2) a set

---

[1] http://www.lemurproject.org/
[2] https://inex.mmci.uni-saarland.de/about.html

of questions with different words but with the same structure and answer type and (3) one question with different words and the same meaning. In fact, this question is the only one considered similar to the tested question.

This set of questions is inserted into the question base. During the experimentations, we calculate the similarities between the user's question and each question extracted from the question base after filtering. We extract the most similar questions to the user's question and we return an ordered set of questions. To evaluate our method, the statistic, semantic and overall similarities have been calculated. For performance evaluation, we use the measures:

- Success at n (S@n), which means the percentage of queries for which we return the correct similar question in the top n (1, 2, 5, and 10) returned results. For example, s@1=50% means that the correct answer is at rank 1 for 50% of the queries.
- Mean Reciprocal Rank (MRR) calculated over all tested questions. The reciprocal rank is 1 divided by the rank of the similar question. The MRR is the average of the reciprocal ranks of results for the tested questions.

Table 2. s@n and MRR

	s@1	s@2	s@5	s@10	MRR
Semantic Similarity	15%	30%	60%	70%	0,338
Statistic Similarity	40%	70%	85%	95%	0,604
Overall Similarity	55%	95%	100%	100%	0,76

Table 2 represents s@n and the MRR measures that consider the rank of the correct similar question. The experimental results show that the overall similarity gives the best results and achieves a good performance. In fact, s@2=95%, that is to say for 95% of the questions, the similar question is extracted in 55% of the cases in the first position and 40% of the cases in the second.

5   Conclusion

This paper presents a new question analysis method based on ontologies. Our contribution can be summarized in: (1) representing the questions' structures by a generic graph; (2) representing the question by a typed attributed graph to ensure the representation of knowledge based on different levels of granularity; and (3) using lexico-syntactic patterns and domain ontologies to improve the query reformulation process and guide the search towards relevant (using domain ontologies) and answer-bearing documents (using patterns that define the structure of the answer).

Experiments were conducted and showed an improvement of the precision of information returned after the query reformulation and good similar questions extraction results.

As perspectives, we plan to develop automatic learning techniques to update the generic question graph and complete this work by adding an answer extraction method to search for answers in documents automatically.

Acknowledgments. This work has been supported by the Spanish-Tunisian AECID project A/030058/10, A Frameworkfor the Integration of Ontology Learning and Semantic Search.

References

1. Liu, H., Lin, X., Liu, C.: Research and Implementation of Ontological QA System based on FAQ. Journal of Convergence Information Technology Vol. 5, N. 3 (2010)
2. Hammo, B., Abu-salem, H., Lytinen, S., Evens, M.: QARAB: A Question Answering System to Support the Arabic Language. In: Workshop on Computational Approaches to Semitic Languages, ACL (2002)
3. Monceaux, L., Robba, I.: Les analyseurs syntaxiques: atouts pour une analyse des questions dans un systme de question-rponse. Actes de Traitement Automatique des Langues Naturelles, Nancy (2002)
4. Gaizauskas, R., Greenwood, M.A., Hepple, M., Roberts, I., Saggion, H.: The University of Sheffields TREC 2004 Q&A Experiments. In: Proceedings of the 13th Text REtrieval Conference (2004)
5. Wang, Y., Wang, W., Huang, C., Chang, T., Yen, Y.: Semantic Representation and Ontology Construction in the Question Answering System. In: CIT 2007, pp. 241–246 (2007)
6. Nyberg, E., Mitamura, T., Carbonell, J., Callan, J., Collins-Thompson, K., Czuba, K., Duggan, M., Hiyakumoto, L., Hu, N., Huang, Y., Ko, J., Lita, L., Murtagh, S., Pedro, V., Svoboda, D.: The Javelin question answering system at TREC 2002. In: Proceedings of the 11th Text Retrieval Conference (2002)
7. Miller, G.A.: WordNet: A lexical database for English. Communications of the ACM 38(11), 39–41 (1995)
8. Tablan, V., Damljanovic, D., Bontcheva, K.: A Natural Language Query Interface to Structured Information. In: Bechhofer, S., Hauswirth, M., Hoffmann, J., Koubarakis, M. (eds.) ESWC 2008. LNCS, vol. 5021, pp. 361–375. Springer, Heidelberg (2008)
9. Lopez, V., Uren, V., Motta, E., Pasin, M.: AquaLog: An ontology-driven question answering system for organizational semantic intranets (2007)
10. Kaufmann, E., Bernstein, A., Zumstein, R.: Querix: A natural language interface to query ontologies based on clarification dialogs. In: Cruz, I., Decker, S., Allemang, D., Preist, C., Schwabe, D., Mika, P., Uschold, M., Aroyo, L.M. (eds.) ISWC 2006. LNCS, vol. 4273, pp. 980–981. Springer, Heidelberg (2006)
11. Song, W., Feng, M., Gu, N., Wenyin, L.: Question Similarity Calculation for FAQ Answering. In: Proceedings of the Third International Conference on Semantics, Knowledge and Grid, pp. 298–301 (2007)
12. Moussa, A.M., Abdel-Kader, R.F.: QASYO: A Question Answering System for YAGO Ontology. International Journal of Database Theory and Application 4(2), 99–112 (2011)

Fuzzy Multidimensional Modelling for Flexible Querying of Learning Object Repositories

Gloria Appelgren Lara[1], Miguel Delgado[2], and Nicolás Marín[2,*]

[1] LATAM Airlines Group, Santiago, Chile
gloria.appelgren@lan.com
[2] University of Granada, Department of Computer Science and AI,
18071 - Granada, Spain
mdelgado@ugr.es, nicm@decsai.ugr.es

Abstract. The goal of this research is to design a fuzzy multidimensional model to manage learning object repositories. This model will provide the required elements to develop an intelligent system for information retrieval on learning object repositories based on OLAP multidimensional modeling and soft computing tools. It will handle the uncertainty of this data through a flexible approach.

Keywords: Learning Object Metadata, Data Warehousing, Multidimensional Model, Fuzzy Set, Fuzzy Techniques.

1 Introduction

In recent years, one of the major challenges in e-learning is the standardization of content. The Learning Objects technology allows contents that comply with certain standards, such as those indicated in the rules of the Sharable Content Object Reference Model - SCORM[3,16], to be reused in different distance learning platforms, making interoperability possible. Unfortunately, there are still shortcomings in the management and evaluation of content.

In the literature, there are many definitions for the term Learning Object (LO). One of the most accepted is established by the IEEE Standards Committee: a LO is *any digital or other entity that can be used, reused or referenced during a learning process supported by technology*[9]. The focus of LO technology is the encapsulation of content, so that it becomes an autonomous unit, i.e. a LO is self-contained and devoted to present a concept or idea. This has been established to be structured as a combination of educational content: lecture notes, presentations, tutorials, etc., and their respective metadata: title, author, rank, age, etc. These LOs and their metadata are stored in Learning Object

---

* Work partially supported by project Representación y Manipulación de Objetos Imperfectos en Problemas de Integración de Datos, Junta de Andalucía, Consejería de Economía, Innovación y Ciencia(P07-TIC-03175).

H.L. Larsen et al. (Eds.): FQAS 2013, LNAI 8132, pp. 112–123, 2013.

Repositories (LORs), which correspond to *stores* that provide the mechanisms for searching, exchanging and reusing LOs.

The Learning Object Metadata (LOM) is a model formally approved by IEEE and widely accepted in the e-learning field[13]. LOM is based on previous efforts made to describe educational resources on projects ARIADNE, IMS and Dublin Core[6]. Its aim is to create structured descriptions of educational resources. Its data model specifies which aspects of learning object should be described and what vocabularies may be used in that description. The model consists of a hierarchical description of nine major categories that group the other fields: General, Lifecycle, Meta-metadata, Technical, Educational Use, Rights, Relation, Annotation and Classification. This standard aims to ensure interoperability between repositories from various sources.

In order to manage knowledge coming from LOs, we can take advantage of a Data Warehouse and techniques for online analytical processing (Online Analytical Processing - OLAP)[2] appropriately adapted for the management of LORs. Though OLAP systems and related intelligent management tools (Business Intelligence - BI) are really targeted to the business, in this paper, we propose their use in education, specifically to handle LO repositories.

Data Warehousing technology, due to their analytic orientation, proposes a different way of thinking within the information system area, which is supported by a specific data model, known as multi-dimensional data model, which seeks to provide the user with an interactive high-level vision of the business operation. In this context, the dimensional modeling is a technique for modeling understandable views to friendly support *end user* operations. The basic idea is that users easily visualize the relationships between the various components of the model.

In general, multidimensional models are oriented to the generation of ad-hoc reports that enable business decisions based on more accurate data[5]. However, data usually are incomplete and, in many repositories, they are expressed in natural language and are often affected by the inherent imprecision of this language. As Molina et al. [15] indicated, it is possible to model multidimensional data cubes based on fuzzy set theory and thus allow the management of uncertainty and imprecision in the data. In learning object repositories we can find the presence of uncertainty in the expressions of a LO metadata such as "Neuroscience for Kids"; in this expression we have the presence of the word *kid*, which is a categorization regarding the age of the target users of the LO, understood by any person but not easily managed by machines, that is what we do by applying fuzzy logic.

In this work, the idea is to provide an intelligent system that allows users to analyze the learning object type that best fits their way of learning or the learning of their students in any field of knowledge, developing and implementing a DW OLAP technique to integrate the fuzzy sets theory to ease the extraction of knowledge [12]. The main contribution is the flexible design of datacubes for the management of a Learning Object Repository (G-LOR).

The paper is organized as follows: Section 2 briefly describes the previous concepts related to learning objects, repositories and data warehouses, and states the problem we face in this paper. Section 3 is devoted to explain the LOM IEEE standard. Section 4 develops the proposed solution to the problem using a fuzzy multidimensional model of learning object repositories. We include a use case that exemplifies this proposal. Finally, we end the paper with some conclusions and guidelines for future work.

2   Background and Motivation

2.1   LOs, LORs and LO Metadata

There are many definitions of learning objects (Learning Object, LO). As we have mentioned in the introduction, we use as reference the IEEE standard, which defines a learning object as *any digital or other entity that can be used, reused or referenced during learning process supported by technology*[9]. LOs are usually organized in Learning Object Repositories (LORs), that are repositories that allow us to store, search, retrieve, view and download LOs from all areas of knowledge. Hence the object and the repository are complementary.

In particular, the Learning Object Repositories can be classified into the object repositories containing learning to download and incorporate into a learning platform, and metadata repositories that contain the object information and a link to its location on the Internet.

The search and retrieval of LOs is guided through the use of metadata that describe these learning objects. In this sense, LOM (Learning Object Metadata) is the IEEE standard e-learning, formally approved and widely accepted [9]. LOM is based on previous efforts made to describe educational resources on projects ARIADNE, IMS and Dublin Core[6]. The aim of LOM is the creation of structured descriptions of educational resources. Its data model specifies which aspects of a learning object should be described and what vocabularies may be used in that description. The model consists of a hierarchical description of nine major categories that group the other fields: General, Lifecycle, Metametadata, Technical, Educational Use, Rights, Relation, Annotation and Classification.

2.2   Datawarehousing, OLAP and OLTP

Data Warehousing is the design and implementation of processes, tools, and facilities to manage and deliver complete, timely, accurate, and understandable information for decision making[1].

The OLAP techniques (Online Analytical Processing) develop a multidimensional analysis, also called *analysis of data hypercubes*. The data handled by this technique are imported both from external sources and from production databases. These databases are feeding production systems based on OLTP (Online Transactional Processing).

For the sake of decision-making, it is necessary to have a large amount of information organized and with specific characteristics. Thus there is a consensus that data warehouses are the ideal structure for this. We recall the definition made by WH Inmon [10], a pioneer in the field, *A data warehouse is a set oriented data by topic, integrated, time-varying and non-volatile which is used to support decision making*. There are four main categories of OLAP tools. These are classified according to the architecture used to store and process multidimensional data, these are: Multidimensional OLAP (MOLAP), Relational OLAP (ROLAP), Hybrid OLAP (HOLAP) and Desktop OLAP (Dolap, Desktop OLAP).

2.3 Fuzzy Sets Theory and OLAP

In classical data warehouses (DWH), classification of values takes place in a sharp manner; because of this real world values are difficult to be measured and smooth transition between classes does not occur. According to [14] a Fuzzy Data Warehouse (FDWH) is a data repository which allows integration of fuzzy concepts on dimensions and facts. Then, it contains fuzzy data and allows the processing of these data. Data entry with *lack of clarity* in data storage systems, offers the possibility to process data at higher level of abstraction and improving the analysis of imprecise data. It also provides the possibility to express business indicators in natural language using terms such as high, low, about 10, almost all, etc., represented by appropriate membership functions.

Different approaches have been proposed for integrating fuzzy concepts in Data Warehouses. For example, Delgado et al. [8] present dimensions where some members can be modeled as fuzzy concepts. Additionally, different fuzzy aggregation functions are used in this approach, which have been developed for different OLAP (Online Analytical Processing) operations, such as roll-up, drill, dice, among others, applying fuzzy operations [7].

Castillo et al. [4] also use fuzzy multidimensional modelling, proposing two methods for linguistically describing time series data in a more natural way, based on the use of hierarchical fuzzy partition of time dimension. This approach introduces two alternative strategies to find the phrases that make up an aggregate. In summary, it is possible to develop a DW and OLAP techniques applied to integrate fuzzy set theory to ease the extraction of knowledge[12].

2.4 Need for DW Techniques in LORs

Learning Objects are developed and stored in various repositories on the web. They involve an enormous potential for the benefit of e-learning. However, there are many technical issues to be considered so that they can be reusable, interchangeable or manageable.

The most widely used standard for learning object repositories is called LOM (Learning Object Metadata) defined by the IEEE (Institute of Electrical and Electronics Engineers, Inc.). This standard, has different sections and fields to describe in detail a learning object and its characteristics through a series of metadata grouped into categories. The proposed metadata are of generic type:

title, description of the subject, format, language. Other LOM model elements, education related, are the least densely populated: the duration of the learning activity, difficulty, structure, granularity, etc. [11]. Therefore we are in a scenario with incomplete data, usually affected by imperfections of different types such as imprecision, inconsistencies, etc.

The information associated with these fields is usually stored in natural language. This feature provides *good expression* but does not facilitate or enable a setting for the automatic inference and reasoning on the metadata records[18]. In conclusion, current models suggest very rigid structures for the representation of the domains[15].

The storage and retrieval of learning objects can benefit from the use of Data Warehouse System and OLAP techniques [2] that allow flexible intelligent management on learning object repositories. The multidimensional model is highly appropriate to represent complex metadata models and OLAP operations could serve as basis for a friendly querying of the repositories. Additionally, the use of fuzzy subsets theory together with these tools, permits to handle data imperfections produced by the use of natural language when inserting LOs metadata.

This paper seeks to add flexibility and uncertainty management by designing a multidimensional model where we will apply fuzzy sets for that goal in fact and dimensions tables. The objective is to design a flexible model for the intelligent management of learning object repositories (G-LOR).

3   Learning Object Metadata Standards

As we have commented before, LOM (IEEE Learning Object Metadata) is the standard for e-learning formally approved and widely accepted [5]. LOM is based on previous efforts made to describe educational resources on projects ARIADNE, IMS and Dublin Core [2]. The aim of LOM is the creation of structured descriptions of educational resources. Its data model specifies which aspects of a learning object should be described and what vocabulary may be used in that description.

This model proposes a hierarchical description in nine major categories that group the other fields: General, Life Cycle, Meta-Metadata, Technical, Educational, Rights, Relation, Annotation and Classification. Table 1 describes each category of the LOM model.

We have used the LOM model to design a database to store the learning objects. The database is described in figure 1 presented below, where we can see that the general entity which controls access to the data for each learning object based in LOM.

This database was generated in the Database Management System (DBMS) Oracle express 11g. This is a repository of LO as proof of concepts. And that is the base of our proposal for fuzzy multidimensional modelling of learning object repositories.

Table 1. Categories of the LOM model

Category	Description
General	Information that describes the learning object as a whole. It describes the purpose of education. Identifier includes fields such as IT, title, description, etc.
Life Cycle	Characteristics related to the history and present state of the learning object and those who have affected this object during its evolution.
Meta-Metadata	About the metadata themselves, not regarding the learning object being described. It contains information such as who has contributed to the creation of metadata and the kind of contribution he has made.
Technical	Technical requirements and technical characteristics of the learning object, such as size, location or format in which it is located. Additionally, this element stored potential technical requirements to use the object referred to metadata.
Educational Uses	Policies educational use of the resource. This category includes different pedagogical features of the object. Typically, includes areas such as resource type - exercise, diagram, figure - and level of interactivity between the user and the object-high, medium, low-, or the context of resource use - college, primary education, doctorate- among others.
Rights	Terms rights that concern the resource exploitation. Details on the intellectual property of the resource. It also describes the conditions of use and price when applicable.
Value	Value of the resource described with other learning objects. Explains the type of relationship of the learning resource to other LOs. It has a name-value detailing the name of the LO-related and type of relationship, is part of, is based on, etc.
Annotation	Includes comments on the educational use of the learning object, as well as its author and date of creation.
Classification	Description subject of the appeal in a classification system. It reports if the LO belongs to some particular subject. For example, physics or history. It allows as much detail as you want by nesting of topics.

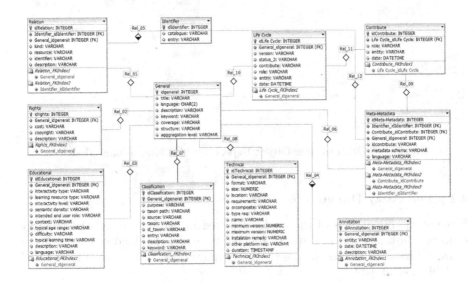

Fig. 1. The LOM model

4   A Fuzzy Multidimensional Model of Learning Object Repositories

The database described in the previous section is the basis for the design of a multidimensional model whose objective is to allow the flexible querying of learning objects from this repository.

A dimensional model can be expressed as a table with a composite primary key, called the fact table, and a set of additional tables called dimension tables.

In our application domain, the fact table is called Learning Objects (LO) and we propose a set of 12 dimension tables, namely: typical range of age, degree of difficulty, length, creation date, type of interactivity, level of aggregation, contribution and life cycle, localization, path taxonomic classification, context or scope, level of interactivity and finally, semantic density. Figure 2 describes this star model.

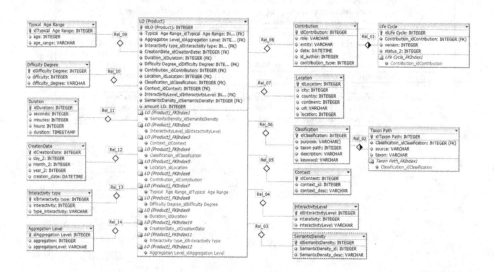

Fig. 2. M-LOR. Learning Object Repository Multidimensional Model.

4.1   The Fuzzy Dimensions

Some of the dimensions of the model depicted in figure 2 are related to fuzzy concepts. In this work, in order to be able to model this kind of fuzzy dimensions, we have considered to use the fuzzy multidimensional model introduced by Molina et al. [15].

The model proposed by Molina et al. is founded on the use of dimensions where the hierarchies of members are defined through the use of a fuzzy kinship relation. The use of this type of hierarchies make the modelling of dimensions related to fuzzy concepts possible, because membership functions of labels can be used to set out the mentioned kinship relation.

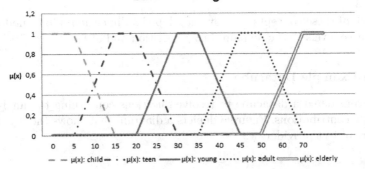

Fig. 3. Age partition

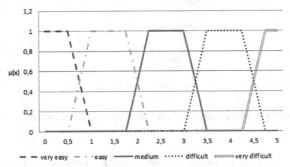

Fig. 4. Difficulty partition

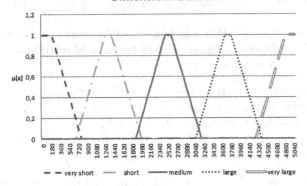

Fig. 5. Duration partition

In our fuzzy model for learning object repositories, we have used fuzzy concepts like Age, Difficulty, and Duration to define fuzzy hierarchies. See figures 3, 4, and 5.

For each of these concepts, we have developed a dimension in our model with at least three levels (basic domain, fuzzy partition and all).

4.2   An Example Datacube

We have considered a datacube to resolve questions concerning the analysis of community contributions. Figure 6 depicts a diagram that shows the star model of learning object provider.

Fig. 6. Star Model about Analysis Community Contribution

The datacube is build in order to solve queries aimed at the measurement of the productivity of a community that provides learning objects and study the relationships between these four variables of the LOM model. An example query could be *to find the amount of free radicals prepared in 2010 for large communities of adults, which difficulty degree is very difficult and duration is medium.*

The fuzzy scheme of the data warehouse may be more complicated than the crisp one due to additional dimensions, fact tables, and relationships[17].

The dimensions by extension, are:

- Contribution role type: short, medium, and large community
- Creation Date: day, month, and year
- Age: child, adolescent, young, adult, and elderly
- Difficulty: very easy, easy, medium, difficult, and very difficult
- Duration: very short, short, medium, long, and very long

See Figure 7 in order to see the levels of the hierarchy developed for each dimension.

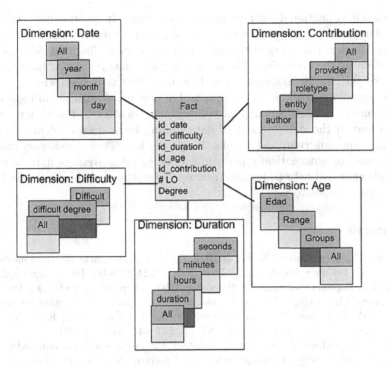

Fig. 7. Datacube Scheme for the Analysis of LO contributions

Table 2. Facts

#LO	Degree	Difficulty Degree	Contribution	Duration	Age Range
2	0,6	5 (very difficult)	Provider 1 (large community)	75 min (verylarge)	23 (Young)
3	0,7	4 (difficult)	Provider 2 (medium community)	20 min (short)	18 (Teen)
1	0,8	4 (difficult)	Provider 2 (medium community)	36 min (medium)	15 (Teen)
3	0,3	1 (very easy)	Provider 4 (short community)	10 min (very short)	7 (Child)
3	0,5	5 (very difficult)	Provider 5 (large community)	30 min (medium)	48 (Adult) t

Some example facts are shown in the table 2.

The operations to be performed to respond to the query are:

- Dice on Age dimension with the condition "adult" at the group level.
- Dice on Contribution role type on the condition "large community"
- Dice on the condition Difficulty Degree dimension "very difficult"
- Dice on Duration on the condition "medium".
- Roll-up on the scale Creation Date to define level Year 2010.

5 Conclusions and Future Work

In this paper we describe and apply concepts of Learning Objects, Repositories, Precise and Fuzzy Data Warehouses. We show how to use concepts and tools used in business intelligence to the area of management of learning object repositories widely used in educational communities and e-learning. It raised issues related to the standardization and management of learning object repositories and proposes a solution through the use of fuzzy multidimensional modelling that can ease the management at the e-learning. The main scientific contribution is the design of buckets for the flexible management of Learning Objects Repositories.

Currently we are working in the proposal of new multidimensional models to analyze content demand, quality of content, user profiles in the various communities and in the development of data mining techniques to define models containing grouping rules and regulations prediction. These prediction models could be used to automatically perform sophisticated analysis of data to identify trends that will help to identify new opportunities and choose the best for learning object repositories.

References

1. Ballard, C., Herreman, D., Schau, D., Bell, R., Kim, E., Valencic, A.: Data modeling techniques for data warehousing. Tech. Rep. SG24-2238-00, IBM - International Technical Support Organization (February 1998), http://www.redbooks.ibm.com
2. Baruque, C.B., Melo, R.N.: Using data warehousing and data mining techniques for the development of digital libraries for lms. In: Isaías, P.T., Karmakar, N., Rodrigues, L., Barbosa, P. (eds.) ICWI, pp. 227–234. IADIS (2004)
3. Bohl, O., Schellhase, J., Sengler, R., Winand, U.: The sharable content object reference model (scorm). a critical review. In: Proceedings of the International Conference on Computers in Education, ICCE 2002, p. 950. IEEE Computer Society Press, Los Alamitos (2002)
4. Castillo-Ortega, R., Marín, N., Molina, C., Sánchez, D.: Building linguistic summaries with f-cube factory. In: Actas del XVI Congreso Español sobre Tecnologás y Lógica Fuzzy, pp. 620–625 (2012)
5. Connolly, T.M., Begg, C.E.: Database Systems: A Practical Approach to Design, Implementation and Management, 5/E. Addison-Wesley (2010)
6. Currier, S.: Metadata for learning resources: An update on standards activity for 2008. Tech. Rep. 55, ARIADNE (April 2008),
   http://www.ariadne.ac.uk/issue55/currier/
7. Delgado, M., Molina, C., Rodríguez Ariza, L., Sánchez, D., Vila Miranda, M.Λ.: F-cube factory: a fuzzy olap system for supporting imprecision. International Journal of Uncertainty. Fuzziness and Knowledge-Based Systems 15(suppl.-1), 59–81 (2007)
8. Delgado, M., Molina, C., Sánchez, D., Vila, M.A., Rodriguez-Ariza, L.: A fuzzy multidimensional model for supporting imprecision in olap. In: Proceedings of the 2004 IEEE International Conference on Fuzzy Systems, vol. 3, pp. 1331–1336 (2004)
9. Hodgin, W.: Draft standard for learning object metadata. Tech. Rep. IEEE 1484.12.1-2002, Learning Technology Standards Committee of the IEEE (July 2002)
10. Inmon, W.: Building the Data Warehouse. Wiley (2005)
11. Kraan, W.: Learning object metadata use survey: sticking the short and wide in the long and thin. Tech. rep., CETIS - Centre for Educational Technology and Interoperability Standards (2004), http://zope.cetis.ac.uk/content2/20041015015543
12. Laurent, A.: Querying fuzzy multidimensional databases: unary operators and their properties. Int. J. Uncertain. Fuzziness Knowl. -Based Syst. 11(suppl.), 31–45 (2003), http://dx.doi.org/10.1142/S0218488503002259

13. López Guzmán, C.: Los Repositorios de Objetos de Aprendizaje como soporte a un entorno e-learning. Master's thesis, University of Salamanca (2005)
14. Małysiak-Mrozek, B.z., Mrozek, D., Kozielski, S.: Processing of crisp and fuzzy measures in the fuzzy data warehouse for global natural resources. In: García-Pedrajas, N., Herrera, F., Fyfe, C., Benítez, J.M., Ali, M. (eds.) IEA/AIE 2010, Part III. LNCS, vol. 6098, pp. 616–625. Springer, Heidelberg (2010)
15. Molina, C., Rodríguez Ariza, L., Sánchez, D., Vila, M.A.: A new fuzzy multidimensional model. IEEE T. Fuzzy Systems 14(6), 897–912 (2006)
16. Poltrack, J., Hruska, N., Johnson, A., Haag, J.: The next generation of scorm: Innovation for the global force. In: Proceedings of Interservice-Industry Training, Simulation, and Education Conference (I-ITSEC), pp. 1–9 (2012)
17. Sapir, L., Shmilovici, A., Rokach, L.: A methodology for the design of a fuzzy data warehouse. In: Proceedings of the 4th International IEEE Conference on Intelligent Systems (IS 2008), vol. 1, pp. 2.14–2.21. IEEE (2008)
18. Soto-Carrión, J., Garcia, E., Sánchez-Alonso, S.: Problemas de almacenamiento e inferencia sobre grandes conjuntos de metadatos en un repositorio semántico de objetos de aprendizaje. In: Actas del III Simposio Plurisciplinar sobre objetos y diseños de aprendizaje, Oviedo, Spain (2006)

Using Formal Concept Analysis to Detect and Monitor Organised Crime

Simon Andrews, Babak Akhgar, Simeon Yates,
Alex Stedmon, and Laurence Hirsch

CENTRIC*
Sheffield Hallam University, Sheffield, UK
{s.andrews,b.akhgar,s.yates,a.stedmon,l.hirsch}@shu.ac.uk

Abstract. This paper describes some possible uses of Formal Concept Analysis in the detection and monitoring of Organised Crime. After describing FCA and its mathematical basis, the paper suggests, with some simple examples, ways in which FCA and some of its related disciplines can be applied to this problem domain. In particular, the paper proposes FCA-based approaches for finding multiple instances of an activity associated with Organised Crime, finding dependencies between Organised Crime attributes, and finding new indicators of Organised Crime from the analysis of existing data. The paper concludes by suggesting that these approaches will culminate in the creation and implementation of an Organised Crime 'threat score card', as part of an overall environmental scanning system that is being developed by the new European ePOOLICE project.

1 Introduction

Efficient and effective scanning of the environment for strategic early warning of Organised Crime (OC) is a significant challenge due to the large and increasing amount of potentially relevant information that is accessible [5, 18]. The types of question and analysis required are not always clear-cut or of a straightforward numerical/statistical nature, but rather necessitate a more conceptual or semantic approach. New developments in computational intelligence and analytics, have opened up new solutions for meeting this challenge. A theoretical development of particular interest for this purpose is Formal Concept Analysis (FCA), with its faculty for knowledge discovery and ability to intuitively visualise hidden meaning in data [3, 19]. This is particular important in environmental scanning, where many of the signals are weak, with information that may be incomplete, imprecise or unclear.

The potential for FCA to reveal semantic information in large amounts of data is beginning to be realised by developments in efficient algorithms and their implementations [1, 13] and by the better appropriation of diverse data for FCA [2, 4].

---

* Centre of Excellence in Terrorism, Resilience, Intelligence and Organised Crime Research.

H.L. Larsen et al. (Eds.): FQAS 2013, LNAI 8132, pp. 124–133, 2013.
© Springer-Verlag Berlin Heidelberg 2013

This paper describes some possible approaches for detecting and monitoring OC using these advances in FCA.

2 Formal Concept Analysis

Formal Concept Analysis (FCA) was introduced in the 1990s by Rudolf Wille and Bernhard Ganter [8], building on applied lattice and order theory developed by Birkhoff and others in the 1930s. It was initially developed as a subsection of Applied Mathematics based on the mathematisation of concepts and concepts hierarchy, where a concept is constituted by its *extension*, comprising of all objects which belong to the concept, and its *intension*, comprising of all attributes (properties, meanings) which apply to all objects of the extension. The set of objects and attributes, together with their relation to each other, form a *formal context*, which can be represented by a cross table.

Airlines	Latin America	Europe	Canada	Asia Pacific	Middle east	Africa	Mexico	Caribbean	USA
Air Canada	×	×	×	×	×		×	×	×
Air New Zealand		×		×					×
Nippon Airways		×		×					×
Ansett Australia				×					
Austrian Airlines		×	×	×	×	×			×

2.1 Formal Contexts

The cross-table above shows a formal context representing destinations for five airlines. The elements on the left side are formal objects; the elements at the top are formal attributes. If an object has a specific property (formal attribute), it is indicated by placing a cross in the corresponding cell of the table. An empty cell indicates that the corresponding object does not have the corresponding attribute. In the Airlines context above, Air Canada flies to Latin America (since the corresponding cell contains a cross) but does not fly to Africa (since the corresponding cell is empty).

In mathematical terms, a formal context is defined as a triple $\mathbb{K} := (G, M, I)$, with G being a set of objects, M a set of attributes and I a relation defined between G and M. The relation I is understood to be a subset of the cross product between the sets it relates, so $I \subseteq G \times M$. If an object g has an attribute m, then $g \in G$ relates to m by I, so we write $(g, m) \in I$, or gIm. For a subset of objects $A \subseteq G$, a derivation operator $'$ is defined to obtain the set of attributes, common to the objects in A, as follows:

$$A' = \{m \in M \mid \forall g \in A : gIm\}$$

Similarly, for a subset of attributes $B \subseteq M$, the derivation operator $'$ is defined to obtain the set of objects, common to the attributes in B, as follows:

$$B' = \{g \in G \mid \forall m \in B : gIm\}$$

2.2   Formal Concepts

Now, a pair *(A, B)* is a *Formal Concept* in a given formal context (G, M, I) only if $A \subseteq G$, $B \subseteq M$, $A' = B$ and $B' = A$. The set A is the extent of the concept and the set B is the intent of the concept. A formal concept is, therefore, a closed set of object/attribute relations, in that its extension contains all objects that have the attributes in its intension, and the intension contains all attributes shared by the objects in its extension. In the Airlines example, it can be seen from the cross-table that Air Canada and Austrian Airlines fly to both USA and Europe. However, this does not constitute a formal concept because both airlines also fly to Asia Pacific, Canada and the Middle East. Adding these destinations completes (closes) the formal concept:

({Air Canada, Austrian Airlines}, {Europe, USA, Asia Pacific, Canada, Middle East}).

2.3   Galois Connections

Another central notion of FCA is a duality called a 'Galois connection', which is often observed between items that relate to each other in a given domain, such as objects and attributes. A Galois connection implies that "if one makes the sets of one type larger, they correspond to smaller sets of the other type, and vice versa" [15]. Using the formal concept above as an example, if Africa is added to the list of destinations, the set of airlines reduces to {*Austrian Airlines*}.

2.4   Concept Lattices

The Galois connections between the formal concepts of a formal context can be visualized in a *Concept Lattice* (Figure 1), which is an intuitive way of discovering hitherto undiscovered information in data and portraying the natural hierarchy of concepts that exist in a formal context.

A concept lattice consists of the set of concepts of a formal context and the subconcept-superconcept relation between the concepts. The nodes in Figure 1 represent formal concepts. It is conventional that formal objects are noted slightly below and formal attributes slightly above the nodes, which they label.

A concept lattice can provide valuable information when one knows how to read it. As an example, the node which is labeled with the formal attribute 'Asia Pacific' shall be referred to as *Concept A*. To retrieve the extension of Concept A (the objects which feature the attribute 'Asia Pacific'), one begins at the node where the attribute is labeled and traces all paths which lead down from the node. Any objects one meets along the way are the objects which have that

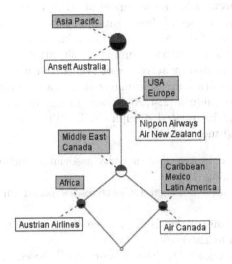

Fig. 1. A lattice corresponding to the Airlines context

particular attribute. Looking at the lattice in Figure 1, if one takes the attribute 'Asia Pacific' and traces all paths which lead down from the node, one will collect all the objects. Thus Concept A can be interpreted as 'All airlines fly to Asia Pacific'. Similarly, the node which is labeled with the formal object 'Air New Zealand' shall be referred to as *Concept B*. To retrieve the intension of Concept B (the attributes of 'Air New Zealand'), one begins at the node where the object is labeled and traces all paths which lead up from the node. Any attributes one meets along the way, are the attributes of that particular object. Looking at the lattice once again, if one takes the object 'Air New Zealand' and traces all paths which lead up from the node, one will collect the attributes 'USA', 'Europe', and 'Asia Pacific'. This can be interpreted as 'The Air New Zealand airline flies to USA, Europe and Asia Pacific'. The concept that we formed previously by inspecting the cross-table, is the node in the center of the lattice; the one labeled with 'Middle East' and 'Canada'. It becomes quite clear, for example, that although Air New Zealand and Nippon Airways also fly to Europe, USA and Asia Pacific, only Air Canada and Austrian Airlines fly to Canada and the Middle East as well.

Although the Airline context is a small example of FCA, visualising the formal context clearly shows that concept lattices provide richer information than from looking at the cross-table alone. This type of hierarchical intelligence that is gleaned from FCA is not so readily available from other forms of data analysis.

2.5   Representing Organised Crime with FCA

To represent Organised Crime (OC) with FCA it is necessary to consider what are suitable as the objects of study and what are attributes of those objects.

For example, the objects could be instances of crime or types of crime and the attributes could be properties of these crimes. A formal context can be created from recorded instances of crime or from domain knowledge regarding the types of OC. Alternatively, for horizon scanning or situation assessment purposes, objects could be represented by activities or events that may be associated with OC. From appropriate data sources, formal contexts can be created using existing software tools and techniques [2, 4, 20]. Then, using the formalisms and tools available in FCA and its related disciplines, it will be possible to carry out analyses to detect and monitor OC:

- Finding multiple instances of an activity associated with OC based on Frequent Itemset Mining [10].
- Finding dependencies between OC attributes based on association rules [11, 12].
- Finding new OC indicators from existing data based on Machine Learning/Classification methods [6].
- Developing and using an OC 'threat score card', based on association rules (strength of association between an indicator and an OC).

The following sections illustrate these possibilities using simple examples.

3   Detecting OC Activities

Let us say that the purchasing of a certain type of fluorescent light tube is common in the cultivation of cannabis plants. An OC gang does not want to make its presence known by making large numbers of purchases from the same location/web site, so they make an effort to spread their purchases over several locations/sites. However, it may still be possible to detect this activity using Frequent Itemset Mining (FIMI) [9, 10]. This uses the notion of frequency of occurrence of a group of items (the so-called item-set). It is akin to FCA with objects being represented by instances. If we monitor the purchasing of tubes, FIMI can be used to automatically highlight possible clusters, thus alerting the possibility of OC. The itemset (attributes) need to be carefully considered and may be a combination of quantity of tubes purchased, location of purchase (stores or towns/areas for delivery from web sites) and time frames. Thus we may be automatically alerted of a number of purchases occurring in a particular time frame and in a particular geographical area.

Although the computation required to carry out the analysis is intensive, recent developments in high-performance concept mining tools [1, 13] mean that this type of monitoring could be carried out in real-time situation assessment. The outputs of the analysis would be suitable for visualising on a map (see figure 2) and end-users can be provided with the ability to alter parameters such as geographical area size and time frame, as well as being able to select different OC activities to analyse.

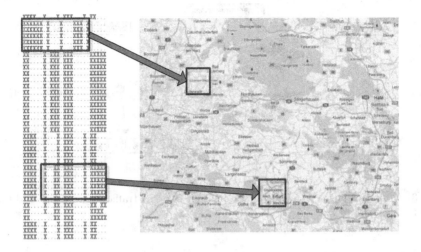

Fig. 2. Visualisaing OC Activities from Frequent Itemsets

4   Finding OC Dependencies

By creating a formal context of Organised Crime using information such as that from the EU survey [16] it may be possible to reveal hidden dependencies between types of OC or between certain OC activities. Such dependencies are often called association rules and are inherent in FCA, being the ratio of the number of objects in concepts that have Galois connections. Using the simple airlines example above, one can say that if an airline flies to the USA (Air New Zealand, Nippon Airways, Austrian Airlines and Air Canada) then there is a 50% chance that it will also fly to Canada (Austrian Airlines and Air Canada). Similarly, if an airline flies to Africa then there is 100% chance it will also fly to the Middle East.

Now, if we take OC, we could investigate the association between drugs trafficking and the use of violence by OC gangs. Taking information from the same survey, FCA produces the lattice in figure 3. The numbers represent the number of OC gangs. Thus, perhaps surprisingly, in this sample of gangs at least, there is little difference in the use of violence by gangs who traffic drugs and those who do not. The profile of violence use is similar in both cases.

Using FCA tools such as ConExp [20] (which was used to produce the lattices in this paper), it is possible to investigate all associations between attributes in a formal context by calculating and listing the association rules. Typically, if we are carrying out an exploratory analysis of attributes, we will be interested in rules that show a strong association and that involve a (statistically) significant number of objects. The list below is a number of association rules generated by the information from the same OC gang (OCG) survey. The numbers in between angle brackets are the number of gangs involved and the percentages show the

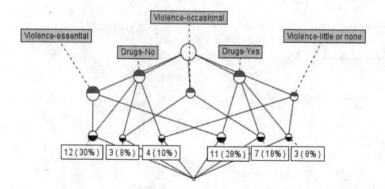

Fig. 3. A concept lattice showing the association between trafficking in drugs and the use of violence

strength of the association. Thus, for example from rule 1, the use of violence is usually essential for gangs with multiple criminal activities; from rules 2 and 3, there appears to be a strong link between a low level of trans-border activity and lack of cooperation between OC gangs; from rules 6 and 10, local/regional political influence and extensive penetration into the legitimate economy usually imply that OCGs find the use of both violence and corruption essential.

1. <13>Activity-multiple =[85%]=><11>Violence-essential;
2. <12>OCG Cooperation-none =[83%]=><10>Trans-border Activity-1-2 counties;
3. <12>Trans-border Activity-1-2 counties =[83%]=><10>OCG Cooperation-none;
4. <10>Economy Penetration-none/limited =[90%]=><9>Political Influence-none;
5. <10>Political Influence-local/regional Economy Penetration-extensive =[90%]=>
   <9>Violence-essential;
6. <11>Violence-essential Political Influence-local/regional =[82%]=>
   <9>Economy Penetration-extensive;
7. <9>Structure-Rigid hierarchy Economy Penetration-extensive =[89%]=>
   <8>Corruption-essential;
8. <9>Corruption-essential Political Influence-local/regional =[89%]=>
   <8>Economy Penetration-extensive;
9. <9>Structure-Rigid hierarchy Corruption-essential =[89%]=>
   <8>Economy Penetration-extensive;
10. <10>Political Influence-local/regional Economy Penetration-extensive =[80%]=>
    <8>Corruption-essential;
11. <10>Violence-occasional =[80%]=><8>Activity-1 primary plus others;
12. <10>Activity-2-3 activities =[80%]=><8>Economy Penetration-extensive;
13. <10>Structure-Devolved hierarchy =[80%]=><8>Violence-essential;

5   Finding New Indicators for OC

The notions of dependency and association can be taken a step further by analysing the links between situations, events and activities and the occurrence or emergence of OC. There are many known indicators of OC [5, 17] but FCA may provide a means of discovering new, less obvious ones. The problem may be considered akin to a classification problem, either classification instances as OC or not OC, or by classifying instances as particular types of OC. Whilst there exist several well-known techniques of classification (such as those automated in the field of Machine Learning [6]), FCA has shown potential in this area [7] and, with the evolving of high-performance algorithms and software [1,13] FCA may provide an approach that can be applied to large volumes of data in real time situation assessment. To illustrate the possibility an example is taken here using the well-known (in Machine Learning) data set of agaricus-lepiota mushroom, some of which are edible and some poisonous, with no obvious indicator for each. The data set contains a number a physical attributes of the mushrooms, such as stalk shape and cap colour and the issue is to find a reliable method of classifying the mushrooms as poisonous or edible. The concept lattices in figure 4 were produced from the data set and show some strong associations between various combinations of attributes and the classes edible and poisonous. The numbers are the number of mushrooms. The poisonous class also shows an interesting feature with zero mushrooms in the bottom concept - indicating that there appears to be two distinct and disjoint groups of poisonous mushroom, classified by two different sets of attributes. It is important to note that no single attribute (such as a foul odor) is a reliable indicator of a class. It is only in combination with other attributes (such as bulbous root and chocolate spore colour) that reliable sets of indicators are found.

With appropriate existing data (to be used as training data), a similar analysis should be possible for OC, to reveal possible sets of indicators for OC that can be used as part of a 'horizon scanning' system to detect or predict the emergence of OC.

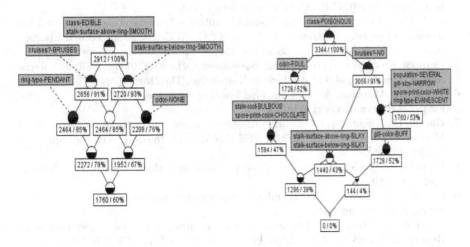

Fig. 4. Concept lattices showing indicators for edible and poisonous mushrooms

6 Conclusion: Developing an OC 'Threat Score Card'

Although the work presented here is mainly of a propositional nature, it shows potential for FCA to be applied in the domain of detecting and monitoring OC. The culmination of the FCA for OC may be in the creation and implementation of an OC 'threat score card'. Using known and newly discovered indicators, association rules can be used to provide a weighting of the indicators. The resulting 'score card' can be implemented as part of an horizon scanning system for the detection of OC and OC types and situation assessment of the possible emergence of OC (OC types) if certain environmental conditions (indicators) pertain. Indeed, this is the proposed role of FCA in the new European ePOO-LICE project [14] (grant agreement number: FP7-SEC-2012-312651), where it will play a part as one of several data analysis tools in a prototype pan-European OC monitoring and detection system.

Acknowledgment. The research leading to these results has received funding from the European Union's Seventh Framework Programme (FP7/2007-2013) under grant agreement n 312651.

References

1. Andrews, S.: In-close2, a high performance formal concept miner. In: Andrews, S., Polovina, S., Hill, R., Akhgar, B. (eds.) ICCS 2011. LNCS, vol. 6828, pp. 50–62. Springer, Heidelberg (2011)
2. Andrews, S., Orphanides, C.: FcaBedrock, a formal context creator. In: Croitoru, M., Ferré, S., Lukose, D. (eds.) ICCS 2010. LNCS, vol. 6208, pp. 181–184. Springer, Heidelberg (2010)
3. Andrews, S., Orphanides, C.: Knowledge discovery through creating formal contexts, pp. 455–460. IEEE Computer Society (2010)
4. Becker, P., Correia, J.H.: The ToscanaJ Suite for Implementing Conceptual Information Systems. In: Ganter, B., Stumme, G., Wille, R. (eds.) Formal Concept Analysis. LNCS (LNAI), vol. 3626, pp. 324–348. Springer, Heidelberg (2005)
5. Europol. Eu organised crime threat assessment: Octa 2011. file no. 2530-274. Technical report, Europol, O2 Analysis & Knowledge, The Hague (2011)
6. Frank, A., Asuncion, A.: UCI machine learning repository (2010), http://archive.ics.uci.edu/ml
7. Ganter, B., Kuzntesov, S.O.: Formalizing hypotheses with concepts. In: Ganter, B., Mineau, G.W. (eds.) ICCS 2000. LNCS (LNAI), vol. 1867, pp. 342–356. Springer, Heidelberg (2000)
8. Ganter, B., Wille, R.: Formal Concept Analysis: Mathematical Foundations. Springer (1998)
9. Goethals, B.: Frequent itemset mining implementations repository, http://fimi.ua.ac.be/
10. Goethals, B., Zaki, M.: Advances in frequent itemset mining implementations: Report on fimi'03. SIGKDD Explorations Newsletter 6(1), 109–117 (2004)
11. Imberman, S., Domanski, D.: Finding association rules from quantitative data using data booleanization (1999)

12. Kuznetsov, S.O.: Mathematical aspects of concept analysis. Journal of Mathematical Science 18, 1654–1698 (1996)
13. Outrata, J., Vychodil, V.: Fast algorithm for computing fixpoints of galois connections induced by object-attribute relational data. Inf. Sci. 185(1), 114–127 (2012), doi:10.1016/j.ins.2011.09.023
14. Pastor, R.: epoolice: Early pusuit against organised crime using environmental scanning, the law and intelligence systems (2013), https://www.epoolice.eu/
15. Priss, U.: Formal concept analysis in information science. Annual Review of Information Science and Technology (ASIST) 40 (2008)
16. United Nations: Global programme against transnational organized crime. Results of a pilot survey of forty selected organized criminal groups in sixteen countries. Technical report, United Nations: Offcie on Drugs and Crime (2002)
17. General Secretariat. Serious and organised crime threat assessment (socta) - methodology. Technical report, Council of the European Union (2012)
18. CISC Strategic Criminal Analytical Services. Strategic early warning for criminal intelligence. Technical report, Criminal Intelligence Service Canada (CISC), Central Bureau, Ottawa (2007)
19. Valtchev, P., Missaoui, R., Godin, R.: Formal concept analysis for knowledge discovery and data mining: The new challenges. In: Eklund, P. (ed.) ICFCA 2004. LNCS (LNAI), vol. 2961, pp. 352–371. Springer, Heidelberg (2004)
20. Yevtushenko, S.A.: System of data analysis "concept explorer". In: Proceedings of the 7th National Conference on Artificial Intelligence KII 2000, pp. 127–134 (2000) (in Russian)

Analysis of Semantic Networks Using Complex Networks Concepts

Daniel Ortiz-Arroyo

Computational Intelligence and Security Laboratory
Department of Electronic Systems, Aalborg University
Niels Bohrs Vej 8, Esbjerg, Denmark
do@es.aau.dk

Abstract. In this paper we perform a preliminary analysis of semantic networks to determine the most important terms that could be used to optimize a summarization task. In our experiments, we measure how the properties of a semantic network change, when the terms in the network are removed. Our preliminary results indicate that this approach provides good results on the semantic network analyzed in this paper.

Keywords: Complex Networks, Semantic Networks, Information Theory.

1 Introduction

Automatic text summarization is a computer processing task that consists in selecting those sentences within a text that best represent its contents. One way to perform summarization is by assigning a score to each of document's sentences, according to its importance.

Many approaches have been explored in the past to perform automatic text summarization. Among these are the application of TF-IDF[1] to assign importance scores or the use of more elaborated algorithms based on fuzzy logic, genetic algorithms, neural networks, semantic role labeling, and latent semantic analysis.

Automatic text summarization can be applied not only to full documents but also to a group of phrases or sentences contained in a document. The goal is to extract those keywords or terms that best summarize sentences' contents. After these sentences have been extracted from a document, they can be represented as a semantic network.

In general a semantic network is one form of knowledge representation that depicts how terms or concepts are inter-related. Different types of semantic networks are used for different purposes. For instance, semantic networks can be used in defining concepts, representing beliefs or causality, or in performing inferences.

---

[1] Term frequency-inverse document frequency.

H.L. Larsen et al. (Eds.): FQAS 2013, LNAI 8132, pp. 134–142, 2013.

We use a broad definition of what semantic networks are to represents not only relationships between concepts but how words or terms used in phrases or sentences are inter-related. In particular we use certain word properties, such as their position within a sentence or their frequency of co-occurrence with other words. Other properties that can be used to create a semantic network from sentences are its syntactical structure, or the grammatical category to which the words in them belong.

In these semantic networks the words within sentences are the nodes in the graph and the syntactical or grammatical relationship existing between words represent the edges. This type of semantic network is described in [5] and will be used in this paper.

Previous studies [6] have shown that semantic networks have some of the properties that complex networks possess.

Complex networks are networks that are neither random[2] nor regular. Complex networks have some non-trivial topological properties that differentiate them from random and regular networks.[3] The discovery of these properties has produced an exponential growth of interest in these networks during the last years.

Some of the well known properties of complex networks are *scale-free* degree distribution and *small-world effect*. In a scale-free network, the degree distribution of the nodes follows a power-law. This basically means that a few nodes in the network have connections to many other nodes, but most nodes in the network have just a few connections with the rest of the nodes. An example of a network with scale-free degree distribution is the Internet. Its scale-free property explains why the Internet network is resilient to the random failures that may occur in some of the nodes. The probability that a random failure occurs in one of the few of the nodes that have a large number of connections is smaller compared to the probability that a node with few connections fails.

The power-law describes probability distributions that also commonly occur in other phenomena in nature and society. An example is the Pareto distribution. This distribution describes how wealth is distributed within society i.e. that a few percentage of a population owns most of wealth of a country and that most population owns little of that wealth.

Another property that characterizes complex networks is the *small-world* effect. This effect characterizes complex networks that have a high global clustering coefficient. This means that nodes in a complex network tend to lay at relatively short geodesic distances[4] between each other, compared to how nodes are clustered in a random network. In social networks this property commonly occurs in the form of closing triads that describe fact that "the friends of my friends are also commonly my friends".

---

[2] Random Networks are also called Erdos-Renyi networks.

[3] In this paper we will use as synonyms the terms graph and network, node and vertex, and edge and link.

[4] Geodesic distances are also called shortest paths.

The small-world effect is also known as the "six-degrees" of separation, a metric that describes the average number of links that separates two persons in a social network. A similar effect has been observed in networks extracted from bibliographic cites in mathematical papers (called the Erdos number) or from movie actors (called the Bacon number). In these networks the average degree of separation between authors or actors is even smaller than six.

The scale-free power-law distribution can be used to build synthetic models of complex networks, using a preferential attachment process. In the preferential attachment process, a network is built iteratively by connecting new nodes with higher probability to nodes in the network that are already highly connected.

Complex networks have multiple applications in a wide variety of fields such as the Internet, energy, traffic, sociology, neural networks, natural language etc.

Interestingly, the distribution of words in natural languages show some of the known properties of complex networks. For instance, the well known Zipf's law, states that the frequency of words follows a power-law distribution. This fact has been used to compress text documents efficiently by assigning shortest codes to most frequently used words.

In this paper we use semantic networks extracted from sentences and methods from complex networks to find the terms within these sentences that best summarize its contents. We compare the experimental results obtained by applying two different methods from complex networks. Our preliminary results indicate that this approach shows good results in the experiment we have performed.

This paper is organized as follows. Section 2 presents a brief summary of related work. Section 3 describes the methods we used and the intuitions behind them. Section 4 presents the preliminary results of our experiments and section 5 concludes the paper and describes future work.

2   Related Work

There is a plethora of research work in automatic summarization systems and complex networks. In this section we will provide a brief summary of the research work that is directly related to the approach presented in this paper.

Many approaches have been proposed in the literature to perform automatic summarization. Among these are supervised and unsupervised machine learning-based methods.

In [7] both methods were applied to the summarization task. Classifiers were constructed using supervised methods such as J48, Naive Bayes, and SVM[5]. In the same work, classifiers were also induced using the HITS algorithm in an unsupervised way. Results of the experiments reported in [7] show that supervised methods work better when large labeled training sets are available, otherwise unsupervised methods should be used.

In [5] an approach to extract keyphrases from books is presented. Phrases are represented as semantic networks and centrality measures are applied to extract those phrases that are the most relevant. The method employs an unsupervised

---

[5] Support Vector Machine.

machine learning method and the concepts of *betweeness centrality* and *relation centrality* as feature weights to extract keyphrases. Relation centrality measures dynamically, the contribution of a node to the connectedness of the network. Relation centrality counts statically, how many routes betweeness centrality is actually shortening.

On the side of complex networks, the communication efficiency of a network is defined in [3] as a function that is inversely proportional to the length of the shortest path between any two nodes. The effect that one node has on the overall efficiency of a network is found by calculating how the network's efficiency changes when that node is removed. Those nodes that have a larger, detrimental effect on network's communication efficiency, are considered the most important since their removal will force network's communication to happen through larger paths. This approach was employed to find the importance of the members of a terrorist organization in [3].

In [1] an approach to find sets of key players within a social network was presented. The method consists in selecting simultaneously k players via combinatorial optimization.

In [6] it was shown that several types of semantic networks have a small-world structure with sparse connectivity. Authors found that these semantic networks have short average path lengths between words, and a strong local clustering that is typical in structures that have the small-world property. The distribution of the number of connections observed, indicates that these networks follow a scale-free pattern of connectivity.

In a related work described in [4], we found that the concept of entropy can be applied to find sets of key-players within a social network. This approach works well in networks that have a sparse number of edges. The reason is that the removal of a node in dense networks will still keep the network very dense, making the changes in entropy very small.

Shannon's definition of entropy used as a metric to identify important nodes in a network has been previously reported in a diversity of research work. For instance, in analyzing social networks extracted from a corpus of emails [2], in finding key players in social networks [4], and in other very different application domains such as city planning [8]. However, the definition of the probability distributions used in these works to calculate entropy changes slightly. For instance in [8] the probability distribution employs all shortest paths that pass through certain node and [4] includes all shortest paths that originate from a node.

3   Finding the Most Important Terms in a Semantic Network

The main objective of this paper is to determine if some of the concepts applied in complex networks and social network analysis are useful to find the most important terms within the phrases or sentences of a document, that best summarize its content.

In this approach, the terms used in phrases are represented as a semantic network. The semantic network may be obtained in different ways. One way is by using the relative position of words within a phrase or group of phrases. Other methods analyze the syntactic relation of the terms among each other and/or using the grammatical category to which they belong.

In our experiments we have used the semantic network that represents the phrases extracted from a book that best represent its content as is described in [5]. The method used in that work to generate the semantic network, employs neighboring relations and the co-occurrence of terms within phrases.

In our analysis we have used the concept of centrality entropy. Centrality entropy represents the uncertainty that nodes could be able to reach other nodes in the network through shortest paths when a node is removed from the network. Centrality entropy can be calculated using Shannon's definition of entropy:

$$C_e(G) = -\sum_{i=1}^{n} p_g(i) log(p_g(i)) \tag{1}$$

where $C_e(G)$ is the centrality entropy of graph G and $p_g(i)$ represents the probability distribution of the shortest paths from node i to all other nodes in the network. This probability distribution is defined as:

$$p_g(i) = \frac{g_p(i)}{\sum_{j=1}^{n} g_p(j)} \tag{2}$$

where the numerator $g_p(i)$ is the number of shortest paths that communicate node i with all other nodes in the network and the denominator is the total number of shortest paths that exist in the network. Note that the actual length of the shortest paths[6] is not used to calculate centrality entropy. Entropy, defined in this way, changes as nodes are deleted from the graph, disconnecting some nodes and reducing as a consequence, the number of shortest paths available in the network to communicate the rest of the nodes in the graph.

A similar method has been proposed in [3] to detect important nodes in a network. The method determines how the communication efficiency of a network changes when nodes are removed. In this case communication efficiency may be interpreted as how important a node is to establish a semantic link between the terms in the network. Communication efficiency is measured using the equation described in [3]:

$$E(G) = \frac{1}{n(n-1)} \sum_{i \neq j} \frac{1}{ls_{ij}} \tag{3}$$

where $E(G)$ is the efficiency of graph G, n the total number of nodes in the graph, and ls_{ij} is the length of the shortest path between nodes i and j. The equation shows that communication efficiency is inversely proportional to the length of the shortest path.

---

[6] Shortest paths are also called *geodesic paths*.

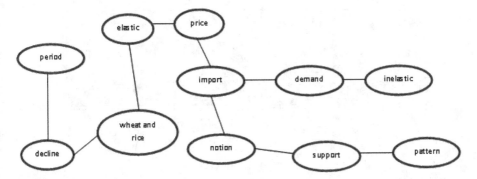

Fig. 1. An example of a syntactic semantic network extracted from 2 sentences

To procedure used in both cases, to measure the efficiency of a network and to find centrality entropy, consists in disconnecting nodes one by one and measuring the efficiency or entropy of the resulting network.

4   Experimental Results

In our preliminary experimental results we used two sentences that were extracted from a book and analyzed syntactically as is described in [5]. The sentences are:

"The import price elasticities remain less than one for both wheat and rice and decline over the entire period. This pattern again tends to support the notion that import demand is inelastic"

Arguably the main subject of these two sentences is *"the notion of how import of wheat and rice behaves"*. Therefore, we could conclude that the terms in the semantic network that may be used to summarize the main topic of these two sentences are {*notion, import, wheat and rice*}

The semantic network generated from these two sentences was taken from [5] and is shown in Fig. 1. As is described in [5], the sentences were pre-processed using stop word removal and stemming. Then, sentences were selected and the network was created using an unsupervised machine learning method that employs as feature weights, two different centrality measures.

We apply our method to analyze how the entropy of the semantic network changes when nodes in the semantic network are removed. First, we calculate the total entropy of the network using Eq. 1. Afterwards, nodes are removed one by one, recalculating in each iteration, the probability distributions and the total entropy of the graph.

Using this method we obtained a plot that shows how entropy changes in Fig. 2.

The entropy defined in Eq. 1 provides a measure of the probability that a node could be reached from any other node in the graph through shortest paths. In a

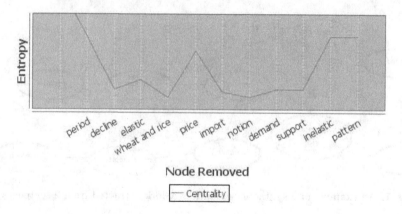

Fig. 2. Drop in total entropy when nodes in the semantic network are removed one by one

fully connected graph, the probability is 1 since a node can reach any other node in the graph through a single edge. Hence, no matter which node is removed the entropy will be the same since the remaining nodes will still keep the graph fully connected.

As graphs become more sparse, some nodes could be reached through shortest or non-shortest paths. However, in centrality entropy only shortest paths are used since we are interested in finding the nearest related terms. In the semantic network that we will analyze, the shortest paths represent how semantically close are the terms in the network.

When nodes are removed from the graph, these nodes that produce the largest drop in entropy are considered the most important since their removal will reduce the number of shortest paths that the remaining nodes in a graph could use to reach the rest of nodes in the graph.

A threshold value can be used to determine how many of these important terms will be included in the summarization task.

The centrality entropy drop graph obtained in Fig. 2, indicates that the nodes that have most effect, when removed from the network are {*notion, import, wheat and rice,* } and to a lesser degree {*decline, period, elastic, price, demand, support, inelastic, pattern*}. By changing the threshold value more or less terms could be included as the most important ones.

Interestingly, the {*price*} term was not detected as an important term by the centrality entropy calculation. This is firstly due to the fact that, as can be seen in Fig. 2, the {*import*} term works as a hub for terms {*notion*} and {*demand*}, making it important since its removal will reduce the number of shortest paths that will be available in the graph compared to the effect that the term {*price*} may produce on entropy when removed.

We could ask why our method finds that terms such as {*wheat and rice*} are more important than other terms such as {*elastic*} or {*decline*}. These terms seem to have similar importance judged by their position in the network.

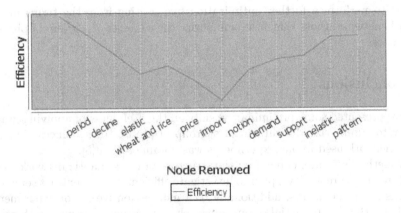

Node Removed

— Efficiency

Fig. 3. Drop in network's communication efficiency when nodes in the semantic network are removed one by one

The reason is that when the term {*decline*} is removed, the node {*period*} becomes isolated from the graph and the number of shortest paths available in the graph decreases proportionally for the rest of the nodes. However, that single isolated node does not contribute to the total shortest paths available.

When node {*elastic*} is removed, the original graph is split into two graphs. The one containing nodes { *period, decline, wheat and rice*} and the one containing {*price, import, demand, inelastic, notion, support, pattern*}. In this case when node {*elastic*} is removed, the number of shortest paths will be reduced since the larger graph will not be able to reach the smaller graph. However, the smaller graph with 3 nodes still provides some local shortest paths to reach these local nodes i.e. there will be 6 shortest paths within the smaller network.

Finally, when node {*wheat and rice*} is removed, the graph is again split into two graphs, but in this case the smaller graph consisting of only two nodes {*period, decline*}, provides only 2 shortest paths in the smaller network, decreasing the total amount of shortest paths available and with this the probability that some node in the network could reach any other node.

Fig. 3 shows how network's communication efficiency, defined in Eq. 3, changes when nodes are removed one by one from the network. The plot shows that the terms that produce the maximum drop in efficiency are firstly {*import, price*} and then {*elastic, wheat and rice, notion*} with the rest of terms having a lesser degree on the drop in efficiency. Arguably, these first two terms {*import, price*}, do not fully capture the *"the notion of how import of wheat and rice behaves"*. However, if we increase the threshold value, other terms such as {*elastic, wheat and rice, notion*} will be included in the set of most important terms.

Our preliminary results indicate that centrality entropy is a metric that produces good results when applied to select the most important terms in a semantic network. These terms can be used to summarize the content of the two sentences used in the example.

Given that the terms in the semantic network were selected as the most important ones in the phrase extraction phase described in [5], our method can be used to perform a further optimization by selecting from the terms in the semantic network, those that best summarize the contents of a group of phrases or sentences.

5   Conclusions

We have presented some preliminary results on the usefulness of applying graph entropy to summarize the subject of a group of phrases or sentences. The semantic network used in our experiments was obtained from [5].

Our method's results depend on the structure of the semantic network used. Therefore, in future work we plan to investigate efficient ways to extract semantic networks from documents, additionally to a more extensive set of experiments to evaluate the real potential of our approach, comparing its results with other summarization systems.

Finally, a more extensive analysis of semantic networks using other methods from complex networks is also planned in future work.

Acknowledgements. The research leading to the results has received funding from the European Union's Seventh Framework Programme (FP7/2007-2013) under grant agreement no. 312651.

References

1. Borgatti, S.P.: Identifying sets of key players in a network. Computational, Mathematical and Organizational Theory 12(1), 21–34 (2006)
2. Shetty, J., Adibi, J.: Discovering important nodes through graph entropy the case of enron email database. In: LinkKDD 2005: Proceedings of the 3rd International Workshop on Link Discovery, pp. 74–81. ACM, New York (2005)
3. Latora, V., Marchiori, M.: How the science of complex networks can help developing strategies against terrorism. Chaos, Solitons and Fractals 20(1), 69–75 (2003)
4. Ortiz-Arroyo, D.: Discovering Sets of Key Players in Social Networks. In: Abraham, A., Aboul-Ella, H., Vaclav, S. (eds.) Computational Social Network Analysis: Computer Communications and Networks, pp. 27–47. Springer, London (2010)
5. Huang, C., Tian, Y., Zhou, Z., Ling, C.X., Huang, T.: Keyphrase Extraction using Semantic Networks Structure Analysis. In: ICDM 2006, Sixth International Conference on Data Mining, pp. 275–284. IEEE (2006)
6. Steyvers, M., Tenenbaum, J.B.: The Large-Scale Structure of Semantic Networks: Statistical Analyses and a Model of Semantic Growth. In: Cognitive Science, vol. 29, pp. 41–78 (2005)
7. Litvak, M., Last, M.: Graph-Based Keyword Extraction for Single-Document Summarization. In: Coling 2008: Proceedings of the Workshop on Multi-Source Multilingual Information Extraction and Summarization, pp. 17–24 (2008)
8. Volchenkov, D., Blanchard, P.: Discovering Important Nodes Through Graph Entropy Encoded in Urban Space Syntax. In: eprint arXiv:0709.4415, ARXIV (2007)

Detecting Anomalous and Exceptional Behaviour on Credit Data by Means of Association Rules

Miguel Delgado[1], Maria J. Martin-Bautista[1], M. Dolores Ruiz[1], and Daniel Sánchez[1,2]

[1] Dpt. Computer Science and A.I., CITIC-UGR, University of Granada
C/Periodista Daniel Saucedo Aranda s/n , Granada 18071, Spain
mdelgado@ugr.es, {mbautis,mdruiz,daniel}@decsai.ugr.es
[2] European Centre for Soft Computing, Mieres, Spain
daniel.sanchezf@softcomputing.es

Abstract. Association rules is a data mining technique for extracting useful knowledge from databases. Recently some approaches has been developed for mining novel kinds of useful information, such us peculiarities, infrequent rules, exception or anomalous rules. The common feature of these proposals is the low support of such type of rules. Therefore, finding efficient algorithms for extracting them are needed.

The aim of this paper is three fold. First, it reviews a previous formulation for exception and anomalous rules, focusing on its semantics and definition. Second, we propose efficient algorithms for mining such type of rules. Third, we apply them to the case of detecting anomalous and exceptional behaviours on credit data.

Keywords: Data mining, association rules, exception rules, anomalous rules, fraud, credit.

1 Introduction

Association rules are one of the frequent used tools in data mining. They allow to identify novel, useful and comprehensive knowledge. The kind of knowledge they try to extract is the appearance of a set of items together in most of the transactions in a database. An example of association rule is "most of transactions that contain hamburger also contain beer", and it is usually noted $hamburger \rightarrow beer$. The intensity of the above association rule is frequently measured by the *support* and the *confidence* measures [1]. The support is the percentage of transactions satisfying both parts of the rule and the confidence measures the proportion of transactions that satisfying the antecedent, also satisfies the consequent. That is, the confidence gives an estimation of the conditional probability of the consequent given the antecedent [1]. There also exist many proposals imposing new quality measures for extracting semantically or even statistically different association rules [11]. In this line, the certainty factor [4] has some advantages over

H.L. Larsen et al. (Eds.): FQAS 2013, LNAI 8132, pp. 143–154, 2013.
© Springer-Verlag Berlin Heidelberg 2013

the confidence as it extracts more accurate rules and therefore, the number of mined rules is substantially reduced.

There are few approaches dealing with the extraction of unusual or exceptional knowledge that might be useful in some contexts. We focus in those proposals that allow to obtain some uncommon information, specially on exception and anomalous rules [20,3]. In general, these approaches are able to manage rules that, being infrequent, provide a specific domain information usually delimited by an association rule.

Previous approaches using data mining techniques for fraud detection try to discover the usual profiles of legitimate customer behaviour and then search the anomalies using different methodologies such us clustering [10]. The main scope of this paper is to apply such kind of "infrequent" rules to the case of detecting exceptional or anomalous behaviour automatically that could help for fraud detection, obtaining the common customer behaviour as well as some indicators (exceptions) that happen when the behaviour deviates from an usual one and the anomalous deviations (anomalies). For this purpose, we will perform several experiments in financial data concerning credits.

The structure of the paper is the following: next section offers a brief description of background concepts and related works on this topic. In section 3, we review previous proposals for mining exception and anomalous rules. Section 4 describes our proposal for mining exception and anomalous rules using the certainty factor. Section 5 presents the algorithm for extracting these kinds of rules and its application to the real dataset German-statlog about credits in a certain bank in section 6. Finally, section 7 contains the conclusions and some lines for future research.

2   Background Concepts and Related Work

2.1   Association Rules

Given a set I ("set of items") and a database D constituted by a set of transactions, each one being a subset of I, association rules [1] are "implications" of the form $A \rightarrow B$ that relate the presence of itemsets A and B in transactions of D, assuming $A, B \subseteq I$, $A \cap B = \emptyset$ and $A, B \neq \emptyset$.

The support of an itemset is defined as the probability that a transaction contains the itemset, i.e. $\text{supp}(A) = |\{t \in D \mid A \subseteq t\}| / |D|$.

The ordinary measures to assess association rules are the *support* (the joint probability $P(A \cup B)$)

$$\text{Supp}(A \rightarrow B) = \text{supp}(A \cup B) \tag{1}$$

and the *confidence* (the conditional probability $P(B|A)$)

$$\text{Conf}(A \rightarrow B) = \frac{\text{supp}(A \cup B)}{\text{supp}(A)}. \tag{2}$$

Given the minimum thresholds *minsupp* and *minconf*, that should be imposed by the user, we will say that $A \rightarrow B$ is *frequent* if $\text{Supp}(A \rightarrow B) \geq minsupp$, and *confident* if $\text{Conf}(A \rightarrow B) \geq minconf$.

Definition 1. *[4] An association rule $A \to B$ is strong if it exceeds the minimum thresholds minsupp and minconf imposed by the user, i.e. if $A \to B$ is frequent and confident.*

An alternative framework was proposed in [4] where the accuracy is measured by means of Shortliffe and Buchanan's certainty factors [17], as follows:

Definition 2. *[5] Let $\mathrm{supp}(B)$ be the support of the itemset B, and let $\mathrm{Conf}(A \to B)$ be the confidence of the rule. The certainty factor of the rule, denoted as $CF(A \to B)$, is defined as*

$$
\begin{cases}
\dfrac{\mathrm{Conf}(A \to B) - \mathrm{supp}(B)}{1 - \mathrm{supp}(B)} & \text{if } \mathrm{Conf}(A \to B) > \mathrm{supp}(B) \\
\dfrac{\mathrm{Conf}(A \to B) - \mathrm{supp}(B)}{\mathrm{supp}(B)} & \text{if } \mathrm{Conf}(A \to B) < \mathrm{supp}(B) \\
0 & \text{otherwise.}
\end{cases}
\tag{3}
$$

The certainty factor yields a value in the interval [-1, 1] and measures how our belief that B is in a transaction changes when we are told that A is in that transaction. Positive values indicate that our belief increases, negative values mean that our belief decreases, and 0 means no change. Certainty factor has better properties than confidence and other quality measures (see [6] for more details), and helps to solve some of the confidence drawbacks [4,5]. In particular, it helps to reduce the number of rules obtained by filtering those rules corresponding to statistical independence or negative dependence.

Analogously, we will say that $A \to B$ is *certain* if $\mathrm{Supp}(A \to B) \geq minCF$, where $minCF$ is the minimum threshold for the certainty factor given by the user. The definition for strong rules can be reformulated when using CF as a rule which must be frequent and certain.

Definition 3. *[4] An association rule $A \to B$ is very strong if both rules $A \to B$ and $\neg B \to \neg A$ are strong.*

In addition, the certainty factor has the following property $CF(A \to B) = CF(\neg B \to \neg A)$, which tell us that when using the certainty factor, a strong rule is also very strong [4].

2.2   Related Works

The common denominator when mining association rules is their high support. Usually the mining process, as for instance Apriori [1], uses a candidate generation function which exploits the downward closure property of support (also called anti-monotonicity) which guarantees that for a frequent itemset all its subsets are also frequent. The problem here is that exception and anomalous rules are infrequent rules, and therefore such property cannot be used. In the literature we can find different approaches utilizing infrequent rules for capturing a novel type of knowledge hidden in data.

Peculiarity rules are discovered from the data by searching the relevant data among the peculiar data [26]. Roughly speaking, peculiar data is given by the attributes which contain any peculiar value. A peculiar value will be recognized when it is very different from the rest of values of the attribute in the data set. Peculiarity rules are defined as a new type of association rule representing a kind of regularity hidden in a relatively small number of peculiar data.

Infrequent rules are rules that do not exceed the minimum support threshold. They have been studied mainly for intrusion detection joint with exceptions [25,27]. There exists some approaches for mining them: in [19] the authors modify the known *Lambda* measure for obtaining more interesting rules using some pruning techniques. In [27] infrequent items are obtained first and then some measures are used for mining the infrequent rules. In particular, they used correlation and interest measures together with an incremental ratio of conditional probabilities associated to pairs of items. In [8] the infrequent rules are extracted using a new structure called co-occurrence transactional matrix instead of new interest measures.

Exception rules were first defined as rules that contradict the user's common belief [20]. In other words, for searching an exception rule we have to find an attribute that changes the consequent of a strong rule [23,12,22].

We can find two different ways of mining exception rules: direct or indirect techniques. The formers are in most of the cases highly subjective as the set of user's beliefs is compared to the set of mined rules [18,15,13]. The indirect techniques use the knowledge provided by a set of rules (usually strong rules) and then the exception rules are those that contradict or deviate this knowledge [22,25]. Good surveys on this topic can be found in [9,24,7].

Anomalous rules are in appearance similar to exception rules, but semantically different. An anomalous association rule is an association rule that appears when the strong rule "fails". In other words, it is an association rule that complement the usual behaviour represented by the strong rule [3]. Therefore, the anomalous rules will represent the unusual behaviour, having in general low support.

3   Previous Approaches for Discovering Exception and Anomalous Rules

Exception rules were first defined as rules that contradict the user's common belief [20]. For mining this type of rules we will follow the notation by means of a set of rules which has been considered in [2]. An exception rule is defined joint with the strong rule that represents the common belief. Formally we have two rules noted by (*csr,exc*) where *csr* stands for *common sense rule* which is equivalent to the definition of strong rule; and *exc* represents the exception rule:

$$X \text{ strongly implies the fulfilment of } Y, \text{ (and not } E) \quad (csr)$$
$$\text{but, } X \text{ in conjunction of } E \text{ implies } \neg Y. \quad (exc)$$

For instance, if X represents antibiotics, Y recovery and E staphylococcus, it could be found the following exception rule [3]:

"with the help of *antibiotics*, the patient tends to *recover*,
unless *staphylococcus* appears",

in this case the combination of staphylococcus with antibiotics leads to death. This example shows how the presence of E changes the usual behaviour of rule $X \rightarrow Y$, where the value of Y is the patient recovery meanwhile $\neg Y$ is the patient death.

The problem description for exception rules extraction were first presented as obtaining a set of pairs of rules (common sense rule + exception rule) by Suzuki et al. in [21] composed by $(X \rightarrow y, X \wedge E \rightarrow y')$ where y and y' are two different values of the same item, and X, E are two itemsets. But for mining them they define a third rule for achieving more reliable results. This rule is called *reference rule*, *ref* for short, and setted as $E \rightarrow y'$ that must have low confidence.

Hussain et al. present a different approach also based on a triple (csr, ref, exc) as we show in Table 1 but instead of using the confidence for the exception rule $X \wedge E \rightarrow \neg Y$ they define a measure based on the difference of relative information of *exc* respect to *csr* and *ref*. Although the reference rule is defined in [12] as $E \rightarrow \neg Y$ with low support and/or low confidence, they check whether $E \rightarrow Y$ is a strong rule [12], which is an equivalent condition.

Table 1. Schema for mining exception rules given by Hussain et al.

$X \rightarrow Y$	Common Sense rule	(high *supp* and high *conf*)
$X \wedge E \rightarrow \neg Y$	Exception rule	(low *supp* and high *conf*)
$E \rightarrow \neg Y$	Reference rule	(low *supp* and/or low *conf*)

There are other proposals [24,13] that differ from those presented by Suzuki and Hussain et al. but we focus on these because their formulation are nearer to our proposal. In addition these two approaches not only find the unusual or contradictory behaviour of a strong rule, but also the 'agent' that causes it, represented by E.

Following the schema in Table 1, several types of knowledge can be discovered by adjusting the three involved rules in the triple (csr, exc, ref). This is the case of Berzal et al. approach in [3] and [2], where they capture anomalous knowledge.

An *anomalous* rule is an association rule that is verified when the common rule fails. In other words, it comes to the surface when the dominant effect produced by the strong rule is removed [3]. Table 2 shows its formal definition, where the more confident the rules $X \wedge \neg Y \rightarrow A$ and $X \wedge Y \rightarrow \neg A$ are, the stronger the anomaly is. In this approach, there is no imposition over the support of the anomalous and the reference rules.

An example of anomalous rule will be: *"if a patient have symptoms X then he usually has the disease Y; if not, he has the disease A"*. Anomalous rules have different semantics than exception rules, trying to capture the deviation from the common sense rule (i.e. from the usual behaviour). In other words: when X,

Table 2. Schema for mining anomalous rules given by Berzal et al.

$X \to Y$	Common Sense rule (high *supp* and high *conf*)
$X \land \neg Y \to A$ Anomalous rule	(high *conf*)
$X \land Y \to \neg A$ Reference rule	(high *conf*)

then we have either Y (usually) or A (unusually). In this case A is not an agent like E, but it is the alternative behaviour when the usual fails.

In both cases, exception and anomalous rules, the reference rule acts as a pruning criterion to reduce the high number of obtained exceptions or anomalies. On the contrary, our approach will reduce the number of exceptions and anomalies by means of a stronger measure than the confidence.

4 Our Proposal for Mining Exception and Anomalous Rules

This section presents alternative approaches for mining exception and anomalous rules.

4.1 Our Approach for Exception Rules

For the case of exceptions, we offer an alternative approach that does not need the imposition of the reference rule, and we use the certainty factor instead of the confidence for validating the pair of rules (*csr*,*exc*).

The first reason which motivates to reject the use of the reference rule is that it does not offer a semantic enrichment when defining exception rules. Second reason is that the reference rule should be defined in the *csr* antecedent's domain, because the definition of the exception rule does not make sense out of the dominance of X (the *csr* antecedent). Then, we reformulate the triple as follows.

Definition 4. *[7] Let X, Y and E be three non-empty itemsets in a database D. Let $D_X = \{t \in D : X \subset t\}$, that is, D_X is the set of transactions in D satisfying X. We define an* exception rule *as the pair of rules (csr, exc) satisfying the following two conditions:*

- $X \to Y$ *is frequent and certain in D*   (*csr*)
- $E \to \neg Y$ *is certain in D_X*   (*exc*)

where $\varphi \to \psi$ is a certain rule if it exceeds imposed threshold for the certainty factor.

With Definition 4 we achieve two important issues when mining exception rules: (1) to reduce the quantity of extracted pairs (*csr*, *exc*); (2) to obtain reliable exception rules.

We want to remark that we restrict to D_X when defining *exc* because we want that the exception rule is true in the dominance of the common sense rule antecedent. If we look again to the previous example, we can see that searching for exception rules is focused on finding the 'agent' E which, interacting with X, changes the usual behaviour of the common sense rule, that is, it changes the *csr* consequent. In addition, our definition can be formulated as the pair $(X \rightarrow Y, X \wedge E \rightarrow X \wedge \neg Y)$, but this choice for the *exc* is not allowed in usual definitions of association rules because antecedent and consequent are not disjoint. Nevertheless, by restricting to D_X our proposal coincides with the previous approach (without restricting to D_X) when using the confidence measure, i.e., $\text{Conf}(X \wedge E \rightarrow \neg Y) = \text{Conf}_X(E \rightarrow \neg Y)$.

4.2 Our Approach for Anomalous Rules

Our approach for extracting anomalous rules is based on the same two ideas we used for exception rules:

1. To define anomalous rules using the domain D_X.
2. To use the certainty factor instead of the confidence. The certainty factor reduces the number of common sense rules since it discards non-reliable rules and, as a consequence, the number of anomalous rules is also reduced.

In [7] there is an analysis of the reference rule taken in the approach of Berzal et al. This analysis concludes affirming that the increasing of $\text{Conf}(X \wedge Y \rightarrow \neg A)$ is higher as $\text{Supp}(X \rightarrow Y)$ increases. This leads to affirm that the reference rule condition depends on the following supports $\text{Supp}(X \rightarrow Y) = \text{supp}(X \cup Y)$ and $\text{supp}(X \cup Y \cup A)$. This gives reason to propose an alternative formulation for anomalous rules changing the reference rule for a stronger condition (as we prove in Theorem 1) than the one given in [3,2].

Definition 5. *Let X, Y be two non-empty itemsets and A an item. We define an anomalous rule by the triple (csr, anom, ref) satisfying the following conditions:*

- $X \rightarrow Y$ *is frequent and certain (csr).*
- $\neg Y \rightarrow A$ *is certain in D_X (anom).*
- $A \rightarrow \neg Y$ *is certain in D_X (ref).*

Comparing our formulation with the one of Berzal et al., our approach is equivalent to that from a formal point of view if *anom* and *ref* are defined in D_X, because $A \rightarrow \neg Y$ is equivalent to $\neg\neg Y \rightarrow \neg A \equiv Y \rightarrow \neg A$.

The following theorem shows a relation between our definition for anomalous rules and the definition given by Berzal et al. [3], in other words, it shows that our approach is more restrictive than the one proposed in [3].

Theorem 1. *[7] Let X, Y and A be arbitrary itemsets. The following inequality holds*

$$\text{Conf}(X \wedge A \rightarrow \neg Y) \leq \text{Conf}(X \wedge Y \rightarrow \neg A) \tag{4}$$

if and only if

$$\text{supp}(X \cup A) \leq \text{supp}(X \cup Y).$$

Our proposal is similar and logically equivalent to that of Berzal et al. but it does not have the disadvantage that the confidence of the rule $X \wedge Y \rightarrow \neg A$ is affected by an increment when the support of $X \cup Y$ is high (see [7] for more details).

It can be proven that $\mathrm{Conf}_X(A \rightarrow B) = \mathrm{Conf}(X \wedge A \rightarrow B)$, but this is not true when using the certainty factor. This is due to the appearance of the consequent's support in D or D_X in the computation of certainty factor:

$$
\begin{aligned}
CF(X \wedge \neg Y \rightarrow X \wedge A) \neq CF_X(\neg Y \rightarrow A) \\
CF(X \wedge A \rightarrow X \wedge \neg Y) \neq CF_X(A \rightarrow \neg Y)
\end{aligned}
\tag{5}
$$

because

$$
\mathrm{supp}(X \wedge A) = \frac{|X \cap A|}{|D|} \neq \frac{|X \cap A|}{|X|} = \mathrm{supp}_X(A).
\tag{6}
$$

5   Algorithm

We have proposed new approaches using the certainty factor for mining exception rules as well as anomalous rules. Mining exceptions and anomalies associated to a strong rule offers a clarification about the agents that perturbs the strong rule's usual behaviour, in the case of exceptions, or the resulting perturbation, if we find anomalies.

The algorithm 1, called **ERSA** (Exception Rule Search Algorithm), is able to mine together the set of common sense rules in a database with their associated exceptions. For anomalous rules, **ERSA** can be modified into **ARSA** (Anomalous Rule Search Algorithm) only by changing step 2.2.1. The process is very similar, in this case we take $A \in I$ (we do not impose not to have attribute in common with the items in the *csr*), and then we compute the CFs for the anomalous and the reference rule.

In our implementation we only consider exceptions and anomalies given by a single item, for a simpler comprehension of the obtained rules. To mine the association rules we have used an itemset representation by means of BitSets. Previous works [14,16] have implemented the Apriori algorithm using a bit-string representation of items. Both obtained quite good results with respect to time. One advantage of using a bit-string representation of items is that it speeds up logical operations such as conjunction or cardinality.

The algorithm complexity depends on the total number of transactions n and the number of obtained items i having in the first part a theoretical complexity of $O(n2^i)$, but in the second part it also depends on the number of *csr* obtained (r). So, theoretically both **ARSA** and **ERSA** have $O(nri2^i)$. Although this is a high complexity, in the performed experiments with several real databases, the algorithm takes reasonable times. In fact, the two influential factors in the execution time are the number of *csr* extracted.

The memory consumption in both algorithms, **ARSA** and **ERSA**, is high because the vector of BitSets associated to the database is stored in memory, but for standard databases this fact does not represent any problem. For instance,

---

Algorithm 1. ERSA (Exception Rule Search Algorithm)

Input: Transactional database, $minsupp$, $minconf$ or $minCF$
Output: Set of association rules with their associated exception rules.

1. **Database Preprocessing**
    1.1 Transformation of the transactional database into a boolean database.
    1.2 Database storage into a vector of BitSets.
2. **Mining Process**
    2.1 **Mining Common Sense Rules**
        Searching the set of candidates (frequent itemsets) for extracting the csr.
        Storing the indexes of BitSet vectors associated to candidates and their supports.
        csr extraction exceeding $minsupp$ and $minconf/minCF$ thresholds
        2.2.1 **Mining Exception Rules**
            For every common sense rule $X \to Y$ we compute the possible exceptions:
            For each item $E \subset I$ (except those in the common sense rule)
            Compute $X \wedge E \wedge \neg Y$ and its support
            Compute $X \wedge E$ and its support
            Using confidence:
                If $Conf(X \wedge E \to \neg Y) \geq minconf$ **then** we have an exception
            Using certainty factor:
                Compute $supp_X(\neg Y)$
                If $CF_X(E \to \neg Y) \geq minCF$ **then** we have an exception rule

---

database Barbora[1] used in the PKDD99 conference held in Prague [16] consists in 6181 transactions and 12 attributes (33 items). The required memory in this case for the vector of BitSets is 107 kb, and for 61810 transactions is 1.04 MB. More details about the algorithm can be found in [7].

6   Experimental Evaluation

The benchmark data set German-statlog, about credits and the clients having a credit in a German bank, from the UCI Machine Learning repository has been used to empirically evaluate the performance of **ERSA** and **ARSA** algorithms. It is composed of 1000 transactions and 21 attributes, from which 18 are categorical or numerical, and 3 of them are continuous. The numerical continuous attributes have been categorized into meaningful intervals.

For the experiments, we used a 1.73GHz Intel Core 2Duo notebook with 1024MB of main memory, running Windows XP using Java. Tables 3 and 4 show respectively the number of rules and the employed time when mining exception and anomalous rules using our algorithm. In this collection of experiments we impose as 3 the limit of the maximum number of items in the antecedent or the consequent of the csr in order to obtain more manageable rules.

Once the rules are obtained, an expert should clarify if some of them are really interesting. We highlight here some of them, that we think they are in some sense remarkable.

---

[1] http://lispminer.vse.cz/download

Table 3. Number of *csr*, *exc* and *anom* rules found for different thresholds in German-statlog database

minsupp	*minCF* = 0.8			*minCF* = 0.9			*minCF* = 0.95		
	csr	*exc*	*anom*	*csr*	*exc*	*anom*	*csr*	*exc*	*anom*
0.08	674	66	326	309	11	39	270	6	10
0.1	384	27	208	137	4	12	123	3	5
0.12	226	11	142	62	1	3	57	0	2

Table 4. Time in seconds for mining exception and anomalous rules for different thresholds in German-statlog database

minsupp	*minCF* = 0.8		*minCF* = 0.9		*minCF* = 0.95	
	ERSA	**ARSA**	**ERSA**	**ARSA**	**ERSA**	**ARSA**
0.08	137	139	116	116	115	116
0.1	73	71	64	63	63	64
0.12	43	43	38	38	38	38

"*IF* present employment since 7 years *AND* status & sex = single male *THEN* people being liable to provide maintenance for = 1($Supp = 0.105$ & $CF = 0.879$) *EXCEPT* when Purpose = business ($CF = 1$)".

Previous exception rule tell us that when the Purpose = business the previous *csr* changes its behaviour. We have also found anomalous rules as for instance

"*IF* property = real estate *AND* number of existing credits on this bank = 1 *THEN* age is in between 18 and 25 ($Supp = 0.082$ & $CF = 0.972$) *OR* property = car (unusually with $CF_1 = 1$, $CF_2 = 1$)".

This common sense rule has an anomalous rule introduced by the clause *OR* indicating that this is the unusual behaviour of the *csr*. Like in this example, we have observed that many anomalous rules contain items that are complementary to the common sense rule consequent, that is, A and Y has the attribute in common, but they differ in the value. This is very useful in order to see what is the usual behaviour (strong association) and their anomalous or unusual behaviours.

7   Conclusions and Future Research

Mining exception or anomalous rules can be useful in several domains. We have analysed their semantics and formulation, giving a new proposal that removes the imposition of the reference rule for the case of exceptions. Relative to anomalous rules our approach uses a more restrictive reference rule. Our approaches are also sustained in using the certainty factor as an alternative to confidence, achieving a smaller and a more accurate set of exceptions or anomalies. We also provide efficient algorithms for mining these kinds of rules. These algorithms have been run in a database about credits, obtaining a manageable set of interesting rules that should be analysed by an expert.

For future works we are interested in the development of a new approach for searching exceptional and anomalous knowledge with uncertain data. The first idea is to smooth the definitions presented here by means of fuzzy association rules. Other interesting task concerns the search of exception or anomalous rules in certain levels of action.

Acknowledgements. The research reported in this paper was partially supported by the Andalusian Government (Junta de Andalucía) under project P11-TIC-7460; from the Spanish Ministry for Science and Innovation by the project grants TIN2009-14538-C02-01, TIN2009-08296 and TIN2012-30939; and the project FP7-SEC-2012-312651, funded from the European Union in the Seventh Framework Programme [FP7/2007-2013] under grant agreement No 312651.

References

1. Agrawal, R., Manilla, H., Sukent, R., Toivonen, A., Verkamo, A.: Fast discovery of Association rule. In: Advances in Knowledge Discovery and Data Mining, pp. 307–328. AAA Press (1996)
2. Balderas, M.A., Berzal, F., Cubero, J.C., Eisman, E., Marín, N.: Discovering hidden association rules. In: KDD Workshop on Data Mining Methods for Anomaly Detection, Chicago, pp. 13–20 (2005)
3. Berzal, F., Cubero, J.C., Marín, N., Gámez, M.: Anomalous association rules. In: IEEE ICDM Workshop Alternative Techniques for Data Mining and Knowledge Discovery (2004)
4. Berzal, F., Delgado, M., Sánchez, D., Vila, M.A.: Measuring accuracy and interest of association rules: A new framework. Intelligent Data Analysis 66(3), 221–235 (2002)
5. Delgado, M., Marín, N., Sánchez, D., Vila, M.A.: Fuzzy association rules: General model and applications. IEEE Trans. on Fuzzy Systems 11(2), 214–225 (2003)
6. Delgado, M., Ruiz, M.D., Sánchez, D.: Studying interest measures for association rules through a logical model. International Journal of Uncertainty, Fuzziness and Knowledge-Based Systems 18(1), 87–106 (2010)
7. Delgado, M., Ruiz, M.D., Sánchez, D.: New approaches for discovering exception and anomalous rules. International Journal of Uncertainty, Fuzziness and Knowledge-Based Systems 19(2), 361–399 (2011)
8. Ding, J., Yau, S.S.T.: TCOM, an innovative data structure for mining association rules among infrequent items. Computers & Mathematics with Applications 57(2), 290–301 (2009)
9. Duval, B., Salleb, A., Vrain, C.: On the discovery of exception rules: A survey. Studies in Computational Intelligence 43, 77–98 (2007)
10. Fawcet, T., Provost, F.: Adaptative fraud detection. In: Data Mining and Knowledge Discovery, pp. 291–316 (1997)
11. Geng, L., Hamilton, H.J.: Interestingness measures for data mining: A survey. ACM Comput. Surv. 38(3), 9 (2006)
12. Hussain, F., Liu, H., Suzuki, E., Lu, H.: Exception rule mining with a relative interestingness measure. In: Terano, T., Liu, H., Chen, A.L.P. (eds.) PAKDD 2000. LNCS, vol. 1805, pp. 86–97. Springer, Heidelberg (2000)

13. Liu, H., Lu, H., Feng, L., Hussain, F.: Efficient search of reliable exceptions. In: Zhong, N., Zhou, L. (eds.) PAKDD 1999. LNCS (LNAI), vol. 1574, pp. 194–204. Springer, Heidelberg (1999)

14. Louie, E., Lin, T.Y.: Finding association rules using fast bit computation: Machine-oriented modeling. In: Ohsuga, S., Raś, Z.W. (eds.) ISMIS 2000. LNCS (LNAI), vol. 1932, pp. 486–494. Springer, Heidelberg (2000)

15. Padmanabhan, B., Tuzhilin, A.: A belief driven method for discovering unexpected patterns. In: Proceedings of the 4th International Conference on Knowledge Discovery and Data Mining, pp. 94–100 (1998)

16. Rauch, J., Šimunek, M.: An alternative approach to mining association rules. Studies in Computational Intelligence (SCI) 6, 211–231 (2005)

17. Shortliffe, E., Buchanan, B.: A model of inexact reasoning in medicine. Mathematical Biosciences 23, 351–379 (1975)

18. Silberschatz, A., Tuzhilin, A.: User-assisted knowledge discovery: how much should the user be involved. In: ACM-SIGMOD Workshop on Research Issues on Data Mining and Knowledge Discovery (1996)

19. Sim, A.T.H., Indrawan, M., Srinivasan, B.: Mining infrequent and interesting rules from transaction records. In: 7th WSEAS Int. Conf. on AI, Knowledge Engineering and Databases (AIKED 2008), pp. 515–520 (2008)

20. Suzuki, E.: Discovering unexpected exceptions: A stochastic approach. In: Proceedings of the Fourth International Workshop on RSFD, pp. 225–232 (1996)

21. Suzuki, E.: Undirected discovery of interesting exception rules. International Journal of Pattern Recognition and Artificial Intelligence 16(8), 1065–1086 (2002)

22. Suzuki, E.: Discovering interesting exception rules with rule pair. In: Proc. Workshop on Advances in Inductive Rule Learning at PKDD 2004, pp. 163–178 (2004)

23. Suzuki, E., Shimura, M.: Exceptional knowledge discovery in databases based on information theory. In: Proceedings of the Second International Conference on Knowledge Discovery and Data Mining, pp. 275–278. AAAI Press (1996)

24. Taniar, D., Rahayu, W., Lee, V., Daly, O.: Exception rules in association rule mining. Applied Mathematics and Computation 205, 735–750 (2008)

25. Yao, Y., Wang, F.Y., Zeng, D., Wang, J.: Rule + exception strategies for security information analysis. IEEE Intelligent Systems, 52–57 (2005)

26. Zhong, N., Ohshima, M., Ohsuga, S.: Peculiarity oriented mining and its application for knowledge discovery in amino-acid data. In: Cheung, D., Williams, G.J., Li, Q. (eds.) PAKDD 2001. LNCS (LNAI), vol. 2035, pp. 260–269. Springer, Heidelberg (2001)

27. Zhou, L., Yau, S.: Efficient association rule mining among both frequent and infrequent items. Computers & Mathematics with Applications 54(6), 737–749 (2007)

Issues of Security and Informational Privacy in Relation to an Environmental Scanning System for Fighting Organized Crime

Anne Gerdes, Henrik Legind Larsen, and Jacobo Rouces

Computational Intelligence and Security Lab.,
Department of Electronic Systems,
Aalborg University, Esbjerg, Denmark
{gerdes,hll,jrg}@es.aau.dk

Abstract. This paper clarifies privacy challenges related to the EU project, ePOOLICE, which aims at developing a particular kind of open source information filtering system, namely a so-called environmental scanning system, for fighting organized crime by improving law enforcement agencies opportunities for strategic proactive planning in response to emerging organized crime threats. The environmental scanning is carried out on public online data streams, focusing on modus operandi and crime trends, not on individuals. Hence, ethical and technical issues – related to societal security and potential privacy infringements in public online contexts – are being discussed in order to safeguard privacy all through the system design process.

Keywords: informational privacy, environmental scanning, security, ethics.

1 Introduction

In this paper we set out to analyze preliminary issues of informational privacy in relation to the development of an efficient and effective environmental scanning system [1], ePOOLICE, for early warning and detection of emerging organized crime threats. Our ambition here is to develop privacy enhancing security technology by incorporating privacy considerations from the outset of the system development process in order to safeguard both the technological implementation as well as the use procedures surrounding the system. In Sect. 2 we present the overall aims of the ePOOLICE project followed by a discussion of national security versus citizens' right to privacy (Sect. 3). Here, it is argued that core issues should not be addressed as a strict dichotomy of realms, formulated in a clash between citizens right to privacy as opposed to national security; rather we have to strike a balance between two dimensions of security at a national and individual level [2]. Likewise, ethical issues of privacy have traditionally been conceptualized in a dichotomy between public versus private or intimate spheres, and approached by implementing solutions, which protect personal sensitive information from public disclosure. However, the increasing use of open source

H.L. Larsen et al. (Eds.): FQAS 2013, LNAI 8132, pp. 155–163, 2013.

public data streams in flexible query-answering systems, calls for a reframing of privacy in order to account for privacy issues in public spheres (Sect. 4). Hence, this paper emphasizes privacy challenges in relation to environmental scanning of public accessible on line sources, and roughly outlines preliminary technical solutions as well as illustrates how a justificatory framework based on conceptual integrity [3] offers an adequate account for issues of informational privacy in public online accessible sources.

2   Aims of the ePOOLICE Project

In dealing with the challenges posed by organized crime, law enforcement agencies (LEAs) are faced with a field continuously progressing with widespread activities and means for easily adapting to new crime markets. Hence, the project aims at developing an efficient and effective environmental scanning system as part of an early warning system for the detection of emerging organized crime threats and changes in modus operandi, particularly focusing on illegal immigration, trafficking and cybercrime.

The environmental scanning takes departure in a structured framework including a number of societal domains, which divide the environment into political, economic, social, technical, environmental and legislative domains, coined with an acronym as PESTEL domains [4, 489 ff.]. Changes in PESTEL domains might lead to changes in organized crime modus operandi. Hence, ePOOLICE sets out to refine a methodology to monitor heterogeneous information sources in PESTEL domains, identifying and prioritizing indicators to outline a strategic early warning process. Central to the solution is the development of an environmental knowledge repository of all relevant information and knowledge, including scanned information and derived, learned or hypothesized knowledge, as well as the metadata needed for credibility and confidence assessment, traceability, and privacy protection management. For effective and efficient utilization, as well as for interoperability, the repository will apply a standard representation form for all information and knowledge. For effective and efficient scanning of the raw information sources, the project will develop an intelligent environmental radar that will utilize the knowledge repository for focusing the scanning. A key part of this process is semantic filtering for identification of data items that constitute weak signals of emerging organized crime threats, exploiting fully the concept of crime hubs, crime indicators, and facilitating factors, as understood by our user partners.

This way, the monitoring system has knowledge about a number of organized crime types, their facilitators, identifying signatures and indicators. For each crime type, it "knows" what to look for and the relevant information sources to scan. More important sources are scanned more frequently. For instance, from electricity consumption and medical treatment statistics combined with some information from police narratives and other sources, the system may recognize a pattern that is likely to be caused by emerging home-grown cannabis activity in some area. The analyst is alerted about the finding. If confirmed by the analyst, the system will start a more detailed scanning for known indicators and signals of this organized crime type. An indicator, like medical treatment statistics, may be a "necessary", but not "sufficient", indicator of several organized crime types; only certain patterns of indicators can provide a sufficiently strong recognition of an organized crime type. The system can also be set to alarm in cases where an abnormal and unexplainable behavior in some indicator, e.g., immigration or financial transactions, is observed.

Consequently, the overall aims of the system are to alert LEAs to potentially significant changes in organized crime modi operandi before they mature. This will improve the situation in fighting organized crime by ensuring that law enforcement agencies are well armed, use proactive planning for countering threats and are able to detect and deal with discontinuities or strategic new situations, i.e. discover "weak signals" [5] which can be sorted out as important discontinuities in the environment and interpreted as early signs of an emerging organized crime menace.

3   Balancing Security and Privacy

After the end of the Cold War, the classical state-oriented security concept has undergone a change towards a more individual-centered approach, emphasizing the integrity and security of the individual and protection from threats. Similarly, in the context of security technology, security can be defined as nonattendance of danger at a state level, as well as at a societal level with reference to the citizens forming the society. Likewise, the EU Security Strategy (2003) emphasizes the need to act proactively in dealing with key security threats, among which terrorism and organized crime are to be found [2, 16 ff.]. As such, it is generally acknowledged that trust is essential for a flourishing society and that relations of trust are easily maintained and better preserved in moral communities [6] [7]; or in Smith's sarcastic formulation thereof: "*if there is any society among robbers and murderers, they must at least... abstain from robbing and murdering one another.*"[8]. At the same time, societal trust basically rest on the ability of citizens to rely on that in interacting with others, including government authorities, their integrity and autonomy will be respected [9]; and to provide for this, privacy is a highly held value, which has to be properly protected. Hence, as citizens, we are reluctant towards any kinds of surveillance technologies, which may potentially restrict our privacy and thereby constrain our freedom and possibilities for acting as autonomous individuals in the formation of our identities. But privacy is not only important for individuals; privacy also has to be recognized as a societal good or collective value of crucial importance to economic and societal development in democratic liberal societies. If citizens fear intrusive agents of government in ordinary life contexts, they may start to adjust their actions in order not to contrast with mainstream behavior [10] [11, 221 ff.]. Within this kind of panoptic setting, creativity and drive in society may be hindered, if individuals feel an urge to carry out performance-oriented "as-if" behavior.

Accordingly, in order to balance societies' overall security needs without compromising citizens' right to privacy and democracy, different legal sources underscore the importance of protecting privacy against government intrusion. Hence, ePOOLICE has to be developed in legal compliance with EU member states privacy legislation. At the international level, the European Convention for the Protection of Human Rights and Fundamental Freedom (ECHR), which the European charter of Human Rights is based on, stresses the importance of the citizens' right to privacy and protection of personal data and feeds into the local laws of EU member states. Likewise, the Convention for the Protection of Individuals with right to Automatic processing of Personal Data (Council of Europe, 1981) positions data protection as a fundamental right, subsequently backed up by the Data Protection Directive (Directive 95/46/EC), to which member states national legislations are aligned. This directive is currently undergoing transformation and the status of the new directive is not yet settled. Also, the non-binding OECD Guidelines of Protection of Privacy and Transborder Flows of Personal Data (1980, revised in 1999) codifies eight internationally agreed upon principles related to fair information practices (regarding collection, use, purpose and disclosure of personal information). Consequently, from a technological perspective, privacy issues in ePOOLICE may to a certain extent be handled by employing techniques of anonymization of person names and identifiers, access control via logging of all access, as well as techniques for statistical privacy security in order to avoid identification of data-subjects through small and special statistical populations.

Nevertheless, there still seems to be a dichotomous clash between citizens' right to privacy as opposed to national security, implying that more security is necessarily followed by more surveillance, which may give raise to civil society concerns regarding privacy rights in ePOOLICE. But given that security is also an intrinsic value for human well-being at a fundamental level, we might move beyond the dichotomy between citizens' right to privacy and national security and instead conceptualize security in terms of interacting and mutual dependent dimensions of security; i.e., as individual security and national security. This is also reflected in the EU Security Strategy; and from this viewpoint, we are faced with the challenge of striking a balance between two sides of security; formulated as absence of organized crime threats and preservation of individual autonomy as a presumption for democracy.

From a public point of view, an example of European citizens' opinion on privacy and security issues can be found in a participatory technology assessment, which concludes that citizens are open to legitimate security measures for crime prevention, whereas reference to terror treats does not justify privacy limitations for most citizens [2, 26 ff.]. Consequently, it seems to be the case that people are prepared to value security over legitimate restrictions of informational privacy in specific contexts reflecting individual dimensions of security. To elaborate on this from a legal point of view, any limitation to fundamental rights of privacy and personal data protection has to respect some basic principles in order to be legitimate and ensure that privacy is not violated. Hence, limitations have to rest on a legal basis and must be formulated with such a degree of precision that it enables citizens to understand how their navigation and conduct in society are affected by the given limitation. Moreover, a restriction must pursue a legitimate aim, i.e., be in accordance with listed legitimate aims, formulated within each article of rights in the ECHR, as aims that justify interference. Furthermore, any limitation must correspond to a real need of society and must be seen as an efficient instrument (for instance in relation to crime reduction and security). Finally, the principle of proportionality seeks to guarantee that the limitation is balanced to the aim pursued. In order to minimize the infringement of privacy rights and to assess the proportionality of a restriction, the main issues to settle are whether the overall effect of the constraint is reasonable and whether it is the least intrusive mean available. Here, to ensure that privacy is not violated, the ePOOLICE project must see to that the requirement of proportionality of the privacy restriction is satisfied. Given these circumstances, the ePOOLICE project strives to enhance both privacy and security by introducing pro-active privacy enhancing design principles throughout all stages of the development process – for instance in relying on the well-established Privacy by Design principles by the Canadian information and privacy commissioner Cavoukian [12]. In this way, the project seeks to develop technological solutions that support privacy compliant use.

Yet, even in the presence of both legal and general public back-up to privacy restricting technologies such as ePOOLICE, a problem still resides in the fact that an assessment of proportionality is not easy to deal with in a precise manner. Judging whether the privacy interference caused by ePOOLICE is a suitable, necessary and adequate means for fighting organized crime on a strategic level, implies, among other things, a measurement of security gains. However, security advantages are not easy to calculate – neither ahead nor ex-post. Hence, from the fact that security technologies have proved to be effective, we cannot presuppose this outcome for ePOOLICE in advance. Also, if it turns out to be the case, ex-post, that we observe a decline in organized crime after the implementation of ePOOLICE, we still need to carry out a thorough evaluation to justify if and how ePOOLICE contributed to this outcome.

4 Privacy Issues in ePOOLICE

Within the last decade, advancements in data mining and environmental scanning techniques have exacerbated privacy concerns. As a consequence thereof, individual citizens have become more and more transparent to a variety of actors; including, amongst others, government authorities as well as fellow citizens, corporations and online data vendors. Furthermore, at the same time they have experienced a reduction in transparency with right to knowledge of what is being known about them, where and by whom. On top of this, as Web users, we contribute to our own potential de-privatization by spreading information about ourselves on the Web, i.e., by being present at social networking platforms, or by enjoying the convenience of seamless internet transactions based on personalized services in exchange for personal data. Needless to say, that this might raise privacy concerns associated with lack of autonomy in controlling the flow of information about oneself across different contexts, as well as lack of confidentiality and trust in relying on that intended or unintended information-based harm will not occur.

In ePOOLICE, environmental scanning is carried out as an ongoing process of monitoring various open source public data streams. Within PESTEL domains, key open sources are scanned for information – i.e. research reports, the Web, social media, news media, and national statistics, public online databases, and digital libraries. In principle, personal data are not relevant in the information collection context of ePOOLICE. As such, the system will not make use of or aggregate personal data or maintain a database for storing or managing personal data or other kinds of sensitive information, and the environmental scanning techniques developed cannot trace back to individuals. Moreover, by means of a broad-spectrum scan of open sources, the system functions as a tool for tactical planning focusing on modus operandi, hotspot locations, crime patterns and trends. Hence, the use context of the envisioned system is situated at the strategic level, implying that the system does not support the operational level at all, but serves a pure preventive purpose in scaffolding sense-making activities carried out by law enforcement agents and analysts engaged in countering threats and acting proactively in dealing with upcoming trends in organized crime. Within the overall framework of the ePOOLICE project, one might assume that privacy is well protected; both from a legal as well as from an ethical perspective, since – and in accordance with the acknowledged general view that privacy protection has to be applied to personal information – no data subject is identified or under surveillance, and no personal and intimate information per se is involved in the identification of relevant data and interpretation of relevant patterns of information and knowledge. However, ePOOLICE gives raise to privacy concerns precisely due to the scan of online open sources, which may introduce new ethical and legal issues.

Privacy is typically characterized as an instrumental value of great importance for the promotion of a variety of intrinsic values; particularly autonomy [13], [14], integrity and development of personality [13], and freedom from intrusion into intimate spheres [15], as well as friendship [16], and more broadly, intimate relationships [17]. In the context of ePOOLICE, we refer to informational privacy as individuals' ability to control the flow of personal information, including how information is exchanged and transferred [18] [19]. From this perspective, we are faced with two scenarios in which environmental scanning may raise informational privacy concerns:

1. Environmental scanning of open source documents.
2. Environmental scanning of social networking platforms and media, including reports relying on social network analyses in order to disclose trends in communication, which combined with other indicators, support prediction of developments in organized crime.

Since personal (and often also sensitive) information is highly accessible online, the inherent risk of unintentionally identifying data-subjects during the raw data scanning process of open source documents is fairly high. Here, we have to bear in mind that personal data include information, which may identify an individual indirectly by means of different fragments of sources. This challenge to privacy can be met by syntactic data protection techniques, such as de-identification of micro data from sources containing identifiers [20]. Hence, metadata needed for privacy protection should be included in the knowledge repository in order to ensure that personal data, as well as data streams, which enable indirect identification, are excluded from data streams used in the subsequent process of environmental monitoring, which, by means of relevant fusion approaches, allows for automated detection of relevant organized crime types and anomalies. Consequently there seem to be solutions at hand to ensure non-disclosure of data subjects. On the other hand, the environmental monitoring of the environment may come up with patterns of information and point to indicators that hold the potential to sort groups by race, belief, gender or sexual orientation, etc. Still, when based on objective statistical analysis, the use of criminal profiling, by LEAs, is legal. Nevertheless, following the precautionary principle, we need to stress the importance of avoiding potential discrimination, which affords categorization of people into damaging stereotypes.

From a legal point of view, personal information in social networking platforms is protected by the Data Protection Directive (Directive 95/46/EC). As such, the environmental scanning of social networking sites provides a systematic approach for exploring and mapping patterns of communication and relationships among networks at a general level without singling out actors, i.e. unique data subjects. Nevertheless, public environmental scanning may slip under the radar of privacy restrictions, but still imply privacy discomfort among people, due to privacy concerns regarding information traffic across contexts representing distinctive spheres in life. A justificatory conceptual framework, for the systematic exploration of people's reactions to technology can be found in Nissenbaum [3], who has coined the term "contextual integrity" in order to explain for and tie adequate protection against informational moral wrongdoing. Information flows always have to be seen according to context-sensitive norms, representing a function of: the types of information in case, the respectively roles of communicators, and principles for information distribution between the parties. Consequently, contextual integrity is defined, not as a right to control over information, but as a right to appropriate flows of personal information in contexts with right to two norms [3, 127 ff.]: Norms of "appropriateness" and norms of "distribution", i.e., the moment of transfer of information from part X to $Y_{1...n}$. Violations of one of these norms represent a privacy infringement [3].

In the case of ePOOLICE, new flows of information are established and may cause a potential violation of contextual integrity, since information gathering via environmental scanning of communication streams on social networking sites may possibly be judged inappropriate to that context and violate the ordinary governing norms of distribution within it. In this case, ePOOLICE would be framed as a pure panoptic technology, giving raise to surveillance concerns and self-censorship among citizens. On the other side, organized crime is a growing threat to society, which has to be proactively dealt with, and it is in fact possible that these new flows of information will not violate contextual integrity, since, as discussed above (Sect. 3), people might be willing to accept new flows of information caused by environmental scanning if these are judged to be valuable in the context of achieving safety and security against organized crime. Furthermore, whether civic society will embrace or reject new strategic intelligence practices depends on peoples' ability to gain insight into the working of environmental scanning technologies. Hence, it takes an effort to ensure dissemination of research results to the public in order to allow for a public dialogue on an informed background.

5   Concluding Remarks

Privacy issues in ePOOLICE may be adequately dealt with from a legal perspective and still yield privacy concerns due to the fact that alterations in flows of information may lead to violation of contextual integrity. Hence, the overall judgment of ethical implications related to ePOOLICE goes beyond the scope of a standalone privacy evaluation of the system, implying that the context-sensitive tradeoff between privacy and security has to be taken into consideration as well. Consequently, we have to ensure that the new flows of information effected by ePOOLICE respect the integrity of social life by representing adequate means to achieve values of security and safety in a balanced way that does not compromise citizens' right to privacy and democracy.

Acknowledgment. The research leading to these results has received funding from the European Union's Seventh Framework Programme (FP7/2007-2013) under grant agreement n° 312651.

References

[1] Choo, C.W.: The art of scanning the environment. Bulletin of the American Society for information Science and Technology 25(3), 21–24 (1999)
[2] Raguse, M., Meints, M., Langfeldt, O., Peissl, W.: Preparatory action on the enhancement of the european industrial potential in the field of security research. Technical report, PRISE (2008)
[3] Nissenbaum, H.F.: Privacy in context: Technology, policy, and the integrity of social life. Stanford Law & Politics (2010)
[4] Beken, T.V.: Risky business: A risk-based methodology to measure organized crime. Crime, Law and Social Change 41(5), 471–516 (2004)
[5] Ansoff, H.I.: Managing strategic surprise by response to weak signals. California Management Review 18(2), 21–33 (1975)
[6] Delhey, J., Newton, K.: Who trusts?: The origins of social trust in seven societies. European Societies 5(2), 93–137 (2003)
[7] Fukuyama, F.: Social capital and the global economy. Foreign Affairs, 89–103 (1995)

[8] Smith, A.: The theory of moral sentiments. In: Raphael, D.D., Macfie, A.L. (eds.). Clarendon Press, Oxford (1759)

[9] Løgstrup, K.: The ethical demand. University Notre Dame, USA (1997)

[10] Peissl, W.: Surveillance and security: A dodgy relationship. Journal of Contingencies and Crisis Management 11(1), 19–24 (2003)

[11] Regan, P.M.: Legislating privacy: Technology, social values, and public policy. Univ. of North Carolina Press (1995)

[12] Cavoukian, A., et al.: Privacy by design: The 7 foundational principles. Information and Privacy Commissioner of Ontario, Canada (2009)

[13] Benn, S.I.: Privacy, freedom, and respect for persons. In: Pennock, J.R., Chapman, J.W. (eds.) Nomos XIII: Privacy. Atherton Press, New York (1971)

[14] Johnson, D.G.: Computer ethics. Prentice Hall (1994)

[15] Warren, S.D., Brandeis, L.D.: The right to privacy [the implicit made explicit] (1984)

[16] Fried, C.: Privacy. The Yale Law Journal 77(3), 475–493 (1968)

[17] Rachels, J.: Why privacy is important. Philosophy & Public Affairs 4(4), 323–333 (1975)

[18] Tavani, H.T.: Informational privacy, data mining, and the internet. Ethics and Information Technology 1(2), 137–145 (1999)

[19] Van den Hoven, M.: Privacy and the varieties of moral wrong-doing in the information age. Computers and Society 27, 33–37 (1997)

[20] De Capitani di Vimercati, S., Foresti, S., Livraga, G., Samarati, P.: Data privacy: definitions and techniques. International Journal of Uncertainty, Fuzziness and Knowledge-Based Systems 20, 793–817 (2012)

Algorithmic Semantics for Processing Pronominal Verbal Phrases

Roussanka Loukanova

Independent Research, Uppsala, Sweden
rloukanova@gmail.com

Abstract. The formal language of acyclic recursion L_{ar}^λ (FLAR) has a distinctive algorithmic expressiveness, which, in addition to computational fundamentals, provides representation of underspecified semantic information. Semantic ambiguities and underspecification of information expressed by human language are problematic for computational semantics, and for natural language processing in general. Pronominal and elliptical expressions in human languages are ubiquitous and major contributors to underspecification in language and other information processing. We demonstrate the capacity of the type theory of L_{ar}^λ for computational semantic underspecification by representing interactions between reflexives, non-reflexive pronominals, and VP ellipses with type theoretic, recursion therms. We present a class of semantic underspecification that propagates and presents in question-answering interactions. The paper introduces a technique for incremental presentation of question-answer interaction.

1 Background and Recent Developments

1.1 Algorithmic Intensionality in the Type Theory L_{ar}^λ

Moschovakis developed a class of formal languages of recursion, as a new approach to the mathematical notion of algorithm, for computational semantics of artificial and natural languages (NLs), e.g., see Moschovakis [10]-[11]. In particular, the theory of acyclic recursion L_{ar}^λ in Moschovaki [11] models the concepts of meaning and synonymy. For initial applications of L_{ar}^λ to computational syntax-semantics interface in Constraint-Based Lexicalized Grammar (CBLG) of natural language, see Loukanova [8]. The formal system L_{ar}^λ is a higher-order type theory, which is a proper extension of Gallin's TY$_2$, (Gallin [3]), and thus, of Montague's Intensional Logic (IL). L_{ar}^λ extends Gallin's TY$_2$, by adding a second kind of variables, recursion variables, to its pure variables, and by formation of recursive terms with a recursion operator, which is denoted by the constant where, and used in infix notation. I.e, for any L_{ar}^λ-terms $A_0 : \sigma_0, \ldots, A_n : \sigma_n$ $(n \geq 0)$, and any pairwise different recursion variables (locations) of the corresponding types, $p_1 : \sigma_1, \ldots, p_n : \sigma_n$, such that the set of assignments $\{p_1 := A_1, \ldots, p_n := A_n\}$ is acyclic, the expression $(A_0 \text{ where } \{p_1 := A_1, \ldots, p_n := A_n\})$ is an L_{ar}^λ-term. The where-terms represent recursive computations by designating functional recursors: intuitively, the denotation of the term A_0 depends on the denotations

H.L. Larsen et al. (Eds.): FQAS 2013, LNAI 8132, pp. 164–175, 2013.

of $p_1, \ldots, p_n$, which are computed recursively by the system of assignments $\{p_1 := A_1, \ldots, p_n := A_n\}$. In an acyclic system of assignments, these computations close-off. The formal syntax of L_{ar}^{λ} allows only recursive terms with acyclic systems of assignments. The languages of recursion (e.g., FLR, L_r^{λ} and L_{ar}^{λ}) have two semantic layers: denotational semantics and referential intensions. The recursive terms of L_{ar}^{λ} are essential for encoding two-fold semantic information. **Denotational Semantics:** For any given semantic structure $\mathfrak{A}$, a denotation function, den, is defined compositionally on the structure of the L_{ar}^{λ}-terms. In any standard structure $\mathfrak{A}$, there is exactly one, well-defined denotation function, den, from terms and variable assignments to objects in the domain of $\mathfrak{A}$. Thus, for any variable assignment g, a L_{ar}^{λ}-term A of type σ *denotes* a uniquely defined object $den(A)(g)$ of the sub-domain $\mathfrak{A}_\sigma$ of $\mathfrak{A}$. L_{ar}^{λ} has a reduction calculus that reduces each term A to its canonical form $cf(A) \equiv A_0$ where $\{p_1 := A_1, \ldots, p_n := A_n\}$, which is unique modulo congruence, i.e., with respect to renaming bound variables and reordering of assignments. **Intensional Semantics:** The *referential intension*, $\mathsf{Int}(A)$, of a meaningful term A is the tuple of functions (a recursor) that is defined by the denotations $den(A_i)$ $(i \in \{0, \ldots n\})$ of the parts of its canonical form $cf(A) \equiv A_0$ where $\{p_1 := A_1, \ldots, p_n := A_n\}$. Intuitively, for each meaningful term A, the intension of A, $\mathsf{Int}(A)$, is the *algorithm* for computing its denotation $den(A)$. Two meaningful expressions are synonymous iff their referential intensions are naturally isomorphic, i.e., they are the same algorithms. Thus, the algorithmic meaning of a meaningful term (i.e., its sense) is the information about how to "compute" its denotation step-by-step: a meaningful term has sense by carrying instructions within its structure, which are revealed by its canonical form, for acquiring what they denote in a model. The canonical form $cf(A)$ of a meaningful term A encodes its intension, i.e., the algorithm for computing its denotation, via: (1) the basic instructions (facts), which consist of $\{p_1 := A_1, \ldots, p_n := A_n\}$ and the head term A_0, that are needed for computing the denotation $den(A)$, and (2) a terminating rank order of the recursive steps that compute each $den(A_i)$, for $i \in \{0, \ldots, n\}$, for incremental computation of the denotation $den(A) = den(A_0)$.

1.2 Restricted β-reduction in L_{ar}^{λ} and Pronominal Underspecification

In addition to Moschovakis [11], Loukanova [7] provides more evidence for restricted β-reduction, by using NL expressions with pronominal noun phrases (NPs), some of which can be in anaphora-antecedent relations with other NPs. In linguistic syntactic theories, the syntactic co-occurrence distribution of reflexive pronouns is handled by various versions of the so called, *Binding Theory*. An elegant version of a *Binding Theory* for handling anaphoric relations of reflexives, in contrast to non-reflexives, is given in Sag et al. [13], which assumes a grammar system with potentials for syntax-semantics interface, and is used in Loukanova [7]. Reflexive pronouns have genuine, strict co-reference semantic relations with their antecedent NPs, and are subject to special syntax-semantics restrictions: e.g., a reflexive pronoun has to be co-referential with a preceding syntactic argument of the same verb to which it is an argument. On the other hand, non-reflexive pronouns are typically open for ambiguity, depending on the context of using the expressions in which they occur. There are two distinctive cases for non-reflexives: (i) While a non-reflexive pronoun can have an *antecedent* expression that occurs in the larger expression, the non-reflexive and its antecedent, are not strictly co-referential: typically, the non-reflexive pronoun refers "secondarily" to the object that is already obtained (i.e., calculated) as the referent of its antecedent. Thus, in contrast to reflexives, the semantic relation between a non-reflexive pronoun and its antecedent is not reflexive. (ii) In addition, in a given context, a non-reflexive can be used to refer directly to some object, without the use of any antecedent, by an agent[1], technically called a speaker, that uses the entire, encompassing expression in the context.

Outside of any specific context of use, NL expressions that include non-reflexive pronouns, e.g., "John visits his GP and he honors her.", are ambiguous with respect to (i) and (ii), and do not have any specific interpretation. Nevertheless, such NL expressions are meaningful, by having *underspecified, abstract linguistic meaning*. Loukanova [7] demonstrates that the type theory L_{ar}^{λ}, by its two levels of denotational and intensional semantics, with the restricted β-reduction, is highly expressive and can represent simultaneously linguistic distinctions of pronominal NPs and semantic underspecification of NL expressions that have occurrences of pronominal NPs. The semantic phenomena of pronominal expressions exhibit: (1) the semantic distinctions between reflexives and non-reflexives; (2) the semantic underspecification of non-reflexives between usages with or without NP antecedents; (3) propagation of the interactions between pronominal NPs and VP ellipsis, in many modes of language usage, and in particular, in questions-answers.

Reflexives vs. Repeated Names: A sentence like (1a) is naturally rendered to the L_{ar}^{λ} term (1a), as in any typical λ-calculus, and distinctively for L_{ar}^{λ}, to the L_{ar}^{λ} term (1c). The terms (1a), (1b), and (1c) are referentially, i.e., algorithmically equivalent by the the Referential Synonymy Theorem of L_{ar}^{λ} (i.e., by the algorithmic synonymy criteria):

$$\text{Mary likes herself.} \xrightarrow{\text{render}} \lambda x \, like(x,x)(mary) \tag{1a}$$

$$\Rightarrow_{cf} \lambda x \, like(x,x)(m) \text{ where } \{m := mary\} \tag{1b}$$

$$\approx like(m,m) \text{ where } \{m := mary\} \tag{1c}$$

On the other hand, (2a), a simple example for a sentence with repeated occurrences of the same naming expression, is more naturally rendered to the L_{ar}^{λ} term (2a), or directly to its canonical form (2b):

$$\text{Mary likes Mary.} \xrightarrow{\text{render}} like(mary)(mary) \tag{2a}$$

$$\Rightarrow_{cf} like(m_1)(m_2) \text{ where } \{m_1 := mary, \, m_2 := mary\} \tag{2b}$$

---

[1] The agent can be an automatic system involving language processing.

Corollary 1. *The L_{ar}^λ-terms (1a), (1b), and (1c) are not referentially, i.e., algorithmically, equivalent to the terms (2b), i.e.:*

$$\lambda x\, like(x,x)(mary) \not\approx like(mary, mary) \tag{3}$$

Proof. The canonical terms (1b) and (2b) do not have corresponding parts that are denotationally equivalent. Therefore, (3) follows by the Referential Synonymy Theorem of L_{ar}^λ.

Thus the sentences (1a) and (2a) properly do not have the same meaning in L_{ar}^λ, as they do not convey the same information in NL.

2   VP Ellipsis and Underspecification

The systematic ambiguities of VP ellipses and their alternative interpretations were originally classified by Geach [4], who coined the terms *strict* and *sloppy* *readings*. For example,

John visits his GP, and Peter does too.	(4a)
John likes his wife, and Peter does too.	(4b)
John likes himself, and Peter does too.	(4c)

By limiting the sentence (4a) to readings where John visits his own GP, (4a) has a strict reading, where Peter visits the same person (i.e., John's GP), and a sloppy reading, where Peter visits his own GP. Similarly, the sentence (4c) has a strict reading, where Peter likes that same John, and a sloppy reading, where Peter likes himself.

Various type-logic grammars have been used to analyse VP ellipses in the above paradigm. For example, Carpenter (see [1]) exploits Lambek calculus for derivations of λ-terms, as representations of VP meanings that can be used as antecedent of the elided VP. The intensional (algorithmic) semantics of the type theory L_{ar}^λ provides semantic representation of VP ellipses and underspecification related to them, at its object level, which is further evidence for the restricted β-reduction rule. Recursive L_{ar}^λ-terms can represent the alternative semantic readings of NL sentences that have occurrences of VP ellipses. Even more importantly, L_{ar}^λ can be used to represent, at its object level language, the abstract linguistic meanings of such NL sentences, which, in absence of sufficient context information, are underspecified. Furthermore, the syntax-semantics characteristics of pronominal NPs interact with VP ellipses. We use the L_{ar}^λ facilities for semantic representation of VP ellipses, which incorporate the semantic distinctions between reflexives and non-reflexives, and semantic underspecification of non-reflexives.

By the linguistic Binding Theory, the non-reflexive pronoun "his" in (4a) is not constrained to strictly co-refer with the preceding, subject NP "John", as is the reflexive "himself" in (4c). Nevertheless, depending on specific contexts of usage of (4a), "his" can refer secondarily to the individual that is provided as the referent of the preceding NP "John". Alternatively, "his" can directly denote some object, via the speaker's references. Thus, in fact the sentence (4a) has two strict readings, in both John and Peter visit the same person: in one, John visits his own GP, and Peter visits the same person; and in another, John visits the GP of the person that is the speaker's denotation of "his" (and she is not necessarily his own GP), and Peter visits the same person.

In this work, we extend the technique for semantic analysis of pronominal NPs, introduced in Loukanova [7], to the semantic analysis of VP ellipses, by using L_{ar}^{λ}-terms that represent underspecified, abstract linguistic meanings of NL expressions having occurrences of VP ellipses. The alternative strict and sloppy renderings of VP ellipses, can be obtained from the underspecified renderings.

For example, the sentence (4a) provides a pattern for semantic underspecification, when it is out of any context of use. It can be rendered into the recursive L_{ar}^{λ}-term in (5b)-(5f), where h_1 and h_2 are free recursion variables:

$$\text{John visits his GP, and Peter does too.} \xrightarrow{\text{render}} \tag{5a}$$

$$\big[p_1 \ \& \ p_2\big] \text{ where } \{p_1 := L(h_1)(j), \tag{5b}$$
$$p_2 := L(h_2)(p), \tag{5c}$$
$$L := \lambda(x)visit(H(x)), \tag{5d}$$
$$H := his(D), D := GP, \tag{5e}$$
$$j := john, p := peter\} \tag{5f}$$

In (5b)-(5f), by the "currying" order, $\text{den}(his(D)(a)) = \text{den}([\lambda(x)his(D)(x)])(a)$ and expresses 'den(a) has the particular relationship to the value den(D), where the relationship is specified by the specification of the recursion variable D'. In this example, $D := GP$, with GP being an abbreviation for a more complex L_{ar}^{λ} term that renders the common noun "general practitioner". Thus,

$$\text{den}\Big(\big(his(D)\big)(h)\Big) = \text{den}\Big(the\big(provide(medical(care))(h)\big)\Big) \tag{6a}$$

$$\text{den}(D) = general(practitioner) \tag{6b}$$

In this paper, we ignore the additional semantic information that is carried by the pronouns for personification and gender, which will add more parts and components to the rendering terms. In (6a), the recursion variable *his* is assigned a specific denotation, which in $his(D)(h)$ expresses the relationship association of the value den$(his(D)(h)) = d$ to den(h) as the medical doctor assigned to provide medical care to den(h). I.e., the denotational value of the recursion variable *his* is parametric relationship between two objects. The pronominal "his" and its rendering can denote belonging, ownership, or other association relationships, which depend on the noun that is its complement and on the context. A term $his(X)(h)$ is parametric with respect to the association relationship of h with the unique object that would be the denotational value of the $his(X)(h)$ after specifying the property den(X). In addition, den(X), can be context dependent, even after specification of X. E.g., the NP "his book" may be rendered to $his(book)$

and depending on context, $\mathsf{den}(his(book))(a)$ may denote a book that is either owned by a or written by a.

From a computational perspective, for semantic rendering of the sentence (5a), it is better to take the sub-term $his(D)$ in (5b)-(5f) to be an abbreviation for the term (7a)-(7c), which is in canonical form.

$$H := his(D) \equiv \lambda(y)the(B(y)) \text{ where} \tag{7a}$$
$$\{\, B := \lambda(y)\lambda(x)(P(x)(y) \;\&\; D(x)), \tag{7b}$$
$$P := provide_{to}(A), \; A := D(C), \; C := care \,\} \tag{7c}$$

The canonical term (7a)-(7c) provides the same denotation $\mathsf{den}[his(D)(h)]$ as the terms in (6a)-(6b) and (8a)-(8b).

$$\mathsf{den}[his(D)(h)] = \mathsf{den}[the(\lambda(x)(provide_{to}(medical(care))(x)(h) \tag{8a}$$
$$\&\; general(practitioner)(d)))] \tag{8b}$$

It is an advantage to use the canonical term (7a)-(7c) because, unlike the terms in (6a)-(6b) and (8a)-(8b), it provides a computational pattern, i.e., an algorithms for computation of the denotations of a class of expressions of the form "his X", where X is a nominal expression of the noun (N) category, for various professions, e.g., medical specialists and other civil care. The recursion variable D in the canonical term (7a)-(7c) can be instantiated with specific term depending on X in "his X", with some appropriate adjustments of the terms in the assignments. E.g., by varying the assignments for the recursion variables P, A, C, $D := administration(assistant)$, $D := director$, etc., and as in (5b)-(5f), $D := GP$, $D := cardiologist$, etc. If, in an oversimplified manner, we would have taken his for pronominal constant $his(D) \approx \lambda(z)his(D)(z)$. On the contrary the explanation above provides justification that the possessive pronoun[2] his is not a constant even by ignoring personification from semantic information. In virtue of h_1 and h_2 being free recursion variables, (5b)-(5f) represents underspecified, abstract linguistic meaning of the sentence (5a). All the alternative readings of (5a) can be obtained by extending the system of assignments in (5b)-(5f), respectively by:

1. $h_1 := j$, $h_2 := h_1$
   The result is a term representing a strict reading, where John visits his own GP, and Peter visits the same person.
2. $h_2 := h_1$, and h_1 is a free recursion variable.
   The result is a term for another strict reading, where John visits the individual that would be denoted by the free recursion variable h_1, via the speaker's references, and Peter visits the same individual, denoted by h_2 not by direct denotation, but by picking it from the value of h_1.
3. $h_1 := j$, $h_2 := p$
   This is for a sloppy reading, where each of the men, John and Peter, visits his own GP.

---

[2] There are other possibilities for rending possessive pronouns and other possessives, which can not be covered by this paper for sake of space.

In the above cases, the assignment $h_1 := j$ is a computational step having the effect of non-reflexive co-denotation, which is not strictly algorithmic co-reference. We consider the notion of reference in L_{ar}^λ computationally, as the algorithmic steps prescribed by the assignments in order to compute, i.e., to "obtain" the denotation. This respects the non-reflexive lexical form of the pronoun "his".

Corollary 2. *The term (5b)-(5f) (and each term obtained from it by adding the additional assignments (1), (2), and (3)) is not intensionally (i.e., algorithmically) equivalent to any explicit[3] λ-term; and thus, it is not intensionally (i.e., algorithmically) equivalent to any term of Gallin's TY_2.*

Proof. It is by one of the fundamental theorems of L_{ar}^λ, Theorem §3.24 given in Moschovakis [11], because: (i) the term (5b)-(5f) (and each of the extended terms) is in a canonical form, and (ii) the recursion variable L occurs in more than one part of (5b)-(5f).

3   Question-Answer Interaction with Underspecified Answers

In this section we extend the application of the higher-order theory of algorithms L_{ar}^λ to interactive question-answer semantic representation. We demonstrate the interactive technique by using a representative example. This example has been selected because it combines several semantic phenomena (1) the nature of information accumulation; (2) the interactive contribution of question-answering to information update; (3) specification of underspecified semantic parameters; (4) answers that provide partial specification by leaving semantic parameters underspecified. In particular, the example shows that a prominent class of semantic underspecification, the elliptic verbal phrases, propagates and presents in question-answering interactions.

The example (5a) provides a pattern for a broad pronominal VP underspecification in question-answering interactions:

$$\text{John visits his GP.} \tag{9a}$$

$$\text{What about Peter?} \tag{9b}$$

$$\text{He does so too.} \tag{9c}$$

Now:

$$\text{John visits his GP.} \xrightarrow{\text{render}} \tag{10a}$$

$$p_1 \text{ where } \{p_1 := L(h_1)(j), \tag{10b}$$

$$L := \lambda(x)visit(H(x)), \tag{10c}$$

$$H := his(D), \ D := GP, \tag{10d}$$

$$j := john\} \tag{10e}$$

---

[3] A term A is explicit if the constant where does not occur in it.

The added question extends the rendering with prompting:

$$\text{John visits his GP. What about Peter? } \xrightarrow{\text{render}} \tag{11a}$$

$$[p_1 \ \& \ p_0?] \text{ where } \{p_1 := L(h_1)(j), \tag{11b}$$

$$p_0 := [what(V)](p) \tag{11c}$$

$$L := \lambda(x)visit(H(x)), \tag{11d}$$

$$H := his(D), \ D := GP, \tag{11e}$$

$$j := john, \ p := peter\} \tag{11f}$$

In terms like (11a)-eqrefGP-st-t5-qa, the sub-term p_0 in the head expression $[p_1 \ \& \ p_0?]$ is marked by "?" and is not per se L_{ar}^λ term interpreted as a statement. We call a marked sub-expression p_0? of an expression E a *prompt for instantiation of p_0*. Since $p_0 := [what(V)](p)$ is in (11f), an instantiation of the free recursion variable V can provide the prompt. We call p_0 and V *prompted recursion variables*. In this case, p_0 is called a *primary prompt*, and the assignment $V := L(h_2)(h)$ in (12e) an *instantiation of the prompted recursion variable V*.

$$\text{John visits his GP. What about Peter? He does so too.} \tag{12a}$$

$$\text{(What he does is the same.) } \xrightarrow{\text{render}} \tag{12b}$$

$$[p_1 \ \& \ p_0] \text{ where } \{p_1 := L(h_1)(j), \tag{12c}$$

$$p_0 := [what(V)](p) \tag{12d}$$

$$V := L(h_2)(h), \tag{12e}$$

$$L := \lambda(x)visit(H(x)), \tag{12f}$$

$$H := his(D), \ D := GP, \tag{12g}$$

$$j := john, \ p := peter\} \tag{12h}$$

In (12a)-(12h), the expression "does so too" is rendered as pronominal VP by "sloppy" co-reference with the term rendering the VP in the sentence "Peter [visits his GP]$_{VP}$". This is achieved via instantiation of the *prompted recursion variable* $V := L(h_2)(h)$ by the term $L(h_2)(h)$ that shares the same form pattern as that of the canonical form that renders the antecedent VP, $L(h_1)(h)$. This instantiation leaves the arguments of L fully underspecified because of the pronoun "He" adds ambiguity, with respect to its denotation, either as free for setting by the speaker's perceptual references, or by co-reference to the denotation of the antecedent NP *Peter*, by adding the assignment $h := p$ to the term (12a)-(12h).

4   Ongoing and Future Work

The above analysis demonstrates the elegance of L_{ar}^λ. By its new theory of intensionality, as algorithmic sense, the L_{ar}^λ offers, at its object level, L_{ar}^λ-terms that grasp interactions between semantic properties of different kinds of structures in artificial and natural languages, and combinations of such languages.

In human languages various lexical and syntactic components express and contribute to propagation of ambiguities and underspecification, e.g., pronominal NPs, VP ellipses, various syntactic categories expressing quantifiers and attitudes. The theory L_{ar}^{λ} offers representation of ambiguity and underspecification at the object level of its formal language, without the use of external recourses. Aside pronominal underspecification, possessive constructs in NL also involve various kinds of ambiguities on their own. We plan semantic analysis of the classes of possessives and their interaction with linguistic and extralinguistic factors, such as contexts and users. E.g., pronominal underspecification interacts with underspecification and ambiguities of possessives. Human language processing, in general, and especially in the environment of interactions should be based on more essential, intrinsically semantic, information, not only on pure lexical and phrasal syntax, for reliable language and information processing. In addition, while expressing core informational components, different languages, and different language communities withing the same languages, can express or suppress different additional components of semantic information. What kinds of intrinsically semantic information is at disposal to language users depends on the domain areas of information. When translating between languages, the canonical terms that present semantic information may need to be modified. For example, to express information that presents semantically in the lexical units of one language may need to be expressed by extra phrases in another language.

Generating human language expressions based on semantic information, uses techniques for presenting semantic components by expressions of some formal logic language, typically called *logic forms* (LFs).

Machine translation (MT) systems target presenting and preserving all semantic information expressed by the languages they handle. Presenting semantic information is most reliable when using logic for its model-theoretic and inference techniques. Semantic transfer approaches to MT, like generation systems, have been using varieties of logic forms. Classical semantic transfer approaches to MT, typically constitute stages that involve: (1) *parsing* a source NL expression, which produces (2) *semantic representation* of the input language expressions, e.g., in LFs, (3) a *transfer component*, which converts the source LF into a target semantic representation, also a LF. (4) a *generator*, which converts the target LFs into expression(s) of a target language. It has been understood that semantic approaches to translation and generation is better carried out algorithmically when logic forms consists of basic, atomic (or literal) formulae. The so called "flat semantic representation" has been proposed in MT systems that use lists of atomic formulas for representations of the parts of the human language. Such methods have been encountering problems of information distortion with respect to ambiguities, in particular for quantifier scopes. Various formal language have been used, from first order to varieties of higher-order logic languages. Spurious and real ambiguities have been exhibiting serious problems.

A technique, Minimal Recursion Semantics (MRS), has been implemented for semantic representation in HPSG grammares, see Copestake et al. [2]. MRS employs unification methods in feature-value structure descriptions, presented in Sag et al. [13]. While MRS has been extensively used in well-developed and rich large-scale systems, it has been lacking strict mathematical formalization. The theory of L_{ar}^λ provides MRS with the necessary formalization, see Loukanova [6,5]. Independently, L_{ar}^λ has been directly used in generalized CBLG approach, for syntax-semantics interfaces, see Loukanova [8].

The development of multilingual grammatical framework GF, see Ranta [12], has introduced a new approach for simultaneous translation between multiple languages, by using syntax based on type theory of dependent type. GF is development of reliable automatic system for language processing, which is based on solid formal mathematics for expressing syntax of human language. It captures in a uniquely expressive way the distinctions between *abstract*, corresponding to representation of core semantic information, and *concrete* syntax, which presents the details of "external" syntax, specific to different human language. These distinctions grasp in a novel, computational way the classic ideas of *deep* and *surface* structure levels. GF's abstract syntax grasps more adequately concepts of universal syntax of human languages. We consider that GF can be developed further by specialized components for semantic information, in particular by syntax-semantics interfaces.

Semantic information can be presented reliably in L_{ar}^λ terms. In addition, the generation process can be better carried on from logical forms presenting semantic information into human language expression on the basis of terms in canonical forms. The recursion terms in canonical forms present all the basic semantic components expressed lexically and by phrases, and in addition they provide the computational steps and information that compose together the basic facts. There is ongoing work to define render translations from NL to L_{ar}^λ, by computational syntax-semantics interface in Constraint-Based Lexicalized Grammar (CBLG), e.g., by the technique in Loukanova [8]. Furthermore, underspecification and ambiguity are signature features of natural languages and, in general, of information. The theory L_{ar}^λ offers representation of such phenomena in reliable, mathematically founded, way. The work in this paper is a part of extended work on computational syntax-semantics interface in information transfers between languages, which can be artificial, natural, or combined.

More specifically, VP ellipses, like that in (4c), are subject to restrictions by the Binding Theory, e.g., as expressed in Sag et al. [13], and, while having less alternative readings, are very interesting, especially for the combination of reflexives with VP-ellipsis underspecification. For example, "John likes himself", the reflexive pronoun "himself", which fills the complement argument of the verb "likes" is constrained to co-refer with the subject NP "John". Such constraint can be formalized by a version of the Binding Theory similar to that in Sag et al. [13]. An extensive classification of VP-ellipsis is outside of the space and scope of this paper and will be presented in an extended work. Syntax-semantics interface that covers respective formalization of Binding Theory is also a topic of upcoming work.

Further work is necessary to put the theory L_{ar}^λ in applications for representing and processing semantic information. E.g., various documentation forms are structured according to common patterns, e.g., in administrative records in education and health systems. Such forms may use predefined structures to be filled with specific information in human language, which is often domain restricted. Syntax-semantics interface technics can be developed for representing the information in such documents in reliable, mathematical form, by using L_{ar}^λ, without loss of information. Different health systems use document forms that vary in structure, but compile the same, or at least similar information. The formal language L_{ar}^λ and its theory can be used for reliable transfer and exchange of information between documents in the same or different systems. That can be done by a technique for syntax-semantics interfaces, as in Loukanova [8], specially formulated depending on the domains of applications.

The theory of L_{ar}^λ is strictly functional by using currying types and terms, as an extension of TY_2 Gallin [3]. Such a theory of recursion is beneficial to functional systems for language processing including for proving improvements of type-theoretic systems are based on Montagovian computational semantics. A prospective work is on relational variant of L_{ar}^λ closer to semantic models that use situation theory, see Loukanova [9]. Relational models are more native to many contemporary applications, e.g., visualization with incorporated language interactions, information systems with ontological basis, etc.

5   Conclusions

One of the fundamental features of human language, even when used in specific context, is that the information it expresses is parametric. Often human language expressions are semantically meaningful and carry partly known information with parametric components, i.e., the language expressions are well-formed and meaningful, but with underspecified meanings. Specific interpretations can be obtained by assigning values of certain types to the parameters. A distinctive cognitive feature of humans is that they tend to interpret such semantically parametric expressions depending on their state of mind and by being in specific contexts. Systems that use partly or fully human language can include components that are underspecified or partial due to ambiguous or underspecified human language, e.g., by being brief, missing informational pieces of context, or in other ways. Where information is underspecified, it should be kept as such, in clear form, without forcibly presented by specific interpretations, without sufficient facts and justification. The formal language of type theory of recursion provides typed terms, with free and bound recursion variables as components, to represent partly known information in an algorithmic way.

Human language is fundamentally prone to ambiguity and underspecification. Rendering such documents by a formal language such as that of the typed recursion L_{ar}^λ provides algorithmic way of representing underspecification and ambiguities. Recursion L_{ar}^λ terms come with recursion variables, which are restricted to be of certain types and to satisfy 'known' restrictions expressed by the recursion parts, in an algorithmic and faithful way. The acyclicity over recursive terms provide reliable computations that end and provide results. It is important to refrain from cyclic computations when they are unnecessary, and to add full recursion, for modeling genuinely partial information that requires it. Pronominal and elliptical expressions are a class of language constructions across interactions between humans. Their presentation by type-theoretic terms of L_{ar}^{λ}, given in this paper, introduces algorithmic handling of widely spread, and often inevitable, underspecification. The contribution of this paper belongs to new methodological and theoretical development with potentials for multiple applications, including question-answer systems.

References

1. Carpenter, B.: Type-Logical Semantics. The MIT Press, Cambridge (1998)
2. Copestake, A., Flickinger, D., Pollard, C., Sag, I.: Minimal recursion semantics: an introduction. Research on Language and Computation 3, 281–332 (2005)
3. Gallin, D.: Intensional and Higher-Order Modal Logic. North-Holland (1975)
4. Geach, P.T.: Reference and Generality. Cornell University Press, Ithaca (1962)
5. Loukanova, R.: From Montague's rules of quantification to minimal recursion semantics and the language of acyclic recursion. In: Bel-Enguix, G., Dahl, V., Jiménez-López, M.D. (eds.) Biology, Computation and Linguistics — New Interdisciplinary Paradigms, Frontiers in Artificial Intelligence and Applications, vol. 228, pp. 200–214. IOS Press, Amsterdam (2011)
6. Loukanova, R.: Minimal recursion semantics and the language of acyclic recursion. In: Bel-Enguix, G., Dahl, V., Puente, A.O.D.L. (eds.) AI Methods for Interdisciplinary Research in Language and Biology, pp. 88–97. SciTePress – Science and Technology Publications, Rome (2011)
7. Loukanova, R.: Reference, co-reference and antecedent-anaphora in the type theory of acyclic recursion. In: Bel-Enguix, G., Jiménez-López, M.D. (eds.) Bio-Inspired Models for Natural and Formal Languages, pp. 81–102. Cambridge Scholars Publishing (2011)
8. Loukanova, R.: Semantics with the language of acyclic recursion in constraint-based grammar. In: Bel-Enguix, G., Jiménez-López, M.D. (eds.) Bio-Inspired Models for Natural and Formal Languages, pp. 103–134. Cambridge Scholars Publishing (2011)
9. Loukanova, R.: Situated agents in linguistic contexts. In: Filipe, J., Fred, A. (eds.) Proceedings of the 5th International Conference on Agents and Artificial Intelligence, February 15-18, vol. 1, pp. 494–503. SciTePress – Science and Technology Publications, Barcelona (2013)
10. Moschovakis, Y.N.: Sense and denotation as algorithm and value. In: Oikkonen, J., Vaananen, J. (eds.) Lecture Notes in Logic, vol. (2), pp. 210–249. Springer (1994)
11. Moschovakis, Y.N.: A logical calculus of meaning and synonymy. Linguistics and Philosophy 29, 27–89 (2006)
12. Ranta, A.: Grammatical Framework: Programming with Multilingual Grammars. CSLI Publications, Stanford (2011)
13. Sag, I.A., Wasow, T., Bender, E.M.: Syntactic Theory: A Formal Introduction. CSLI Publications, Stanford (2003)

Improving the Understandability of OLAP Queries by Semantic Interpretations

Carlos Molina[1,*], Belen Prados-Suárez[2], Miguel Prados de Reyes[3], and Carmen Peña Yañez[3]

[1] Department of Computer Sciences, University of Jaen, Jaen, Spain
carlosmo@ujaen.es
[2] Department of Software Engineering, University of Granada, Granada, Spain
belenps@ugr.es
[3] Computer Science Department, San Cecilio Hospital, Granada, Spain
{prados,camenpy}@decsai.ugr.es

Abstract. Everyday methods providing managers with elaborated information making more comprehensible the results obtained of queries over OLAP systems are required. This problem is relatively recent due to the huge amount of information they store, but so far there are few proposals facing this issue, and they are mainly focused on presenting the information to the user in a comprehensible language (natural language). Here we go further and introduce a new mathematical formalism, the *Semantic Interpretations*, to supply the user not only understandable responses, but also semantically meaningful results.

Keywords: Queries interpretation, OLAP, Fuzzy Logic, Semantic Interpretation.

1 Introduction

Nowadays more enterprises and big organizations require advanced methods providing managers with elaborated and comprehensible information. It is especially relevant in the cases of organizations working over OLAP systems due to the immense amount of information that is stored using datacubes.

This problem is relatively recent so there are not a lot of proposals that face this problem. Most of the existing techniques are focused on presenting the information to the user in a comprehensible language for him/her; i.e. in natural language. This is the case of the linguistic summary methods, that analyze great amount of data to provide the user with results of the *"Q of the X verify the*

* The research reported in this paper was partially supported by the Andalusian Government (Junta de Andalucía) under project P07-TIC03175 "Representación y Manipulación de Objetos Imperfectos en Problemas de Integración de Datos: Una Aplicación a los Almacenes de Objetos de Aprendizaje", the Spanish Government (Science and Innovation Department) under project TIN2009-08296 and also by project UJA11/12/56 from the University of Jaen.

H.L. Larsen et al. (Eds.): FQAS 2013, LNAI 8132, pp. 176–185, 2013.
© Springer-Verlag Berlin Heidelberg 2013

property Y" structure, where Q is a quantifier. However this is not enough when the user needs the result of the query to be semantically meaningful.

An example of this situation takes place when a manager of a group of health centers has to evaluate the performance of the medical doctors. This manager may query about the number of patients that are attended by a given doctor, obtaining, as a result the number of 15 patients per day. This value does not show whether this doctor works a lot or, otherwise, attends to very few patients. Therefore, it would be necessary to perform the query comparing the results with the attendance values of the other medical doctors working at the same center. With it the manager may get the conclusion that all the staff at the same center have a similar productivity; however, he/she still does not know if it is a good productivity or not: the value obtained does not have the same meaning if the health center attends to a small population than if it is at a big crowded city. Hence, to know if this number is appropriated, it would be necessary to perform a query comparing this value with the ones of the medical doctors working at other health centers with similar characteristics. In other words: 15 patients/day may be a good rate in a small center (where the productivity uses to be medium) but a bad rate in a big hospital (where the average productivity uses to be high or very high).

In this example the same user has performed three different queries over the same data but with distinct purposes, each requiring a different interpretation according to the granularity of the information with which this data is compared.

The research field closer to the problem of the meaning of the queries is, as mentioned above, the linguistic summary field. According to Bouchon-Meunier, B. & Moyse [2], proposals in this scope can be categorized in two groups. On the one hand can be found the proposals using fuzzy logic quantifiers [14,17,15,7,9,1,12,11]. On the other hand, proposals base on nature languages generation (NLG) [16,13,6,8,4,5].

Nevertheless, all of these techniques doesn't take into account the granularity of the information and just tell the user "how many of the X verify Y", when what the user really wants is to analyze the same data item from different points of view (alone, compared with a small set or with a bigger set) each with a different meaning.

This is why in this paper we introduce the concept of Semantic Interpretation of the results of queries on a datacube. To this purpose in section 2 we present the multidimensional model used as reference, whereas in section 3 we describe then notion of semantic interpretations. Next section presetns the adapted multidimensional model including the Semantic Interpretations. In Section 5 an ilustrative example is shown. The last section presents the main conclusions.

2   Multidimensional Model

The base for the semantic interpretation is a multidimensional model to store the data and query it. In this section we briefly present the model. A detail definition can be found in [10].

2.1   Multidimensional Structure

In this section we present the structure of the fuzzy multidimensional model.

Definition 1. *A dimension is a tuple* $d = (l, \leq_d, l_\perp, l_\top)$ *where* $l = l_i, i = 1, ..., n$ *so that each* l_i *is a set of values* $l_i = \{c_{i1}, ..., c_{in}\}$ *and* $l_i \cap l_j = \emptyset$ *if* $i \neq j$, *and* $\leq_d$ *is a partial order relation between the elements of* l *so that* $l_i \leq_d l_k$ *if* $\forall c_{ij} \in l_i \Rightarrow \exists c_{kp} \in l_k / c_{ij} \subseteq c_{kp}$. $l_\perp$ *and* $l_\top$ *are two elements of* l *so that* $\forall l_i \in l \; l_\perp \leq_d l_i \leq_d l_\top$.

We denote level to each element l_i. To identify the level l of the dimension d we will use $d.l$. The two special levels $l_\perp$ and $l_\top$ will be called *base level* and *top level* respectively. The partial order relation in a dimension is what gives the hierarchical relation between levels.

Definition 2. *For each pair of levels* l_i *and* l_j *such that* $l_j \in H_i$, *we have the relation* $\mu_{ij} : l_i \times l_j \to [0, 1]$ *and we call this the* **kinship relation***.*

If we use only the values 0 and 1 and we only allow an element to be included with degree 1 by an unique element of its parent levels, this relation represents a crisp hierarchy. If we relax these conditions and we allow to use values in the interval [0,1] without any other limitation, we have a fuzzy hierarchical relation.

Definition 3. *We say that any pair* (h, α) *is a* **fact** *when* h *is an m-tuple on the attributes domain we want to analyze, and* $\alpha \in [0, 1]$.

The value α controls the influence of the fact in the analysis. The imprecision of the data is managed by assigning an α value representing this imprecision. Now we can define the structure of a fuzzy DataCube.

Definition 4. *A DataCube is a tuple* $C = (D, l_b, F, A, H)$ *such that* $D = (d_1, ..., d_n)$ *is a set of dimensions,* $l_b = (l_{1b}, ..., l_{nb})$ *is a set of levels such that* l_{ib} *belongs to* d_i, $F = R \cup \emptyset$ *where* R *is the set of facts and* $\emptyset$ *is a special symbol,* H *is an object of type history,* A *is an application defined as* $A : l_{1b} \times ... \times l_{nb} \to F$, *giving the relation between the dimensions and the facts defined.*

For a more detailed explication of the structure and the operations over then, see [10].

2.2   Operations

Once we have the structure of the multidimensional model, we need the operations to analyse the data in the datacube. In this section we present the elements needed to apply the normal operations (roll-up, drill-down, pivto and slice).

Definition 5. *An aggregation operator* G *is a function* $G(B)$ *where* $B = (h, \alpha)/(h, \alpha) \in F$ *and the result is a tuple* (h', α').

The parameter that operator needs can be seen as a fuzzy bag ([3]). In this structure there is a group of elements that can be duplicated, and each one has a degree of membership.

Definition 6. *For each value a belonging to d_i we have the set*

$$F_a = \begin{cases} \bigcup_{l_i \in H_{l_i}} F_b/b \in l_j \wedge \mu_{ij}(a,b) > 0 & \text{if } l_i \neq l_b \\ \{h/h \in H \wedge \exists a_1, ..., a_n A(a_1, ..., a_n) = h\} & \text{if } l_i = l_b \end{cases} \qquad (1)$$

The set F_a represents all the facts that are related to the value a.

With this structure, the basic operations over datacubes are defined: roll-up, drill-down, dice, slice and pivot (see [10] for definition and properties).

3 Semantic Interpretation

In this section we present the inclusion of semantic interpretations in the fuzzy multidimensional model and the query process using those. Next section presents the structure of the semantic interpretations. Section 3.2 studies the aggregations functions related to the semantic of the results. The last section presents the process of the query.

3.1 Structure

A Semantic Interpretation (SI) is a structure associated to each fact. Elements:

- $L = L_1, ..., L_m$: a set of linguistic labels over the basic domain. The set has not to be a partition but this characteristic is desirable.
- $f_a(L,c)$: a function to adapt the labels in L to a cardinality c. As c can be a fuzzy set the function has to be able to work with this kind of data. The function f_a has to be continue and monotone.
- $G = G_1, ..., G_n$: a set indicating the aggregation functions that keep unaltered the meaning.

Multiple SI can be associated to each measure. On each fact we have to store as a metadata the cardinality associated to the value. This value means the number of values that were aggregated to obtain this value but depends on the aggregation function used. Next section present the study about this value according the the type of aggregations applied.

When a value is going to be shown, the system applies the semantic interpretation to translate the value into a label. In this process we can differentiate two different approaches:

- *Independence interpretation.* In this situation each value is studied without considering the context (the rest of the values) so we obtained an independence interpretation of the value. In this case, the cardinality to adapt the labels is the one store in the value.
- *Relative interpretation.* In this case, the values are compared with the other facts in the query so the interpretation is relative to the complete query so the cardinality to adapt the labels depends on the complete set of values. In this case, the system calculates the average cardinality of all the values and uses this value to adapt the labels.

3.2   Aggregation Functions

Aggregation functions have an important role in the query process and in the semantic of the results. In this section we will study the different aggregation function type we can find according to the cardinality of the results and if a change of semantic occurs.

Let be a set of value $V = v_1, ..., v_m$, which set of cardinality is $C = n_1, ..., n_m$, considering these two factors, we can classify the functions in three categories:

- *Aggregators.* These functions aggregate the values and the cardinality is the sum of the cardinalities of each value

$$c = \sum_i^m n_i \qquad (2)$$

  The only aggregation function that satisfies this behaviour is the *sum*.
- *Summaries.* In that case, the functions take a set of values and obtain a value that summaries the complete set. Then the cardinality has to represent the average cardinality of the values.

$$c = \frac{\sum_i^m n_i}{m} \qquad (3)$$

  Most of the statistic indicators are in this category (*maximum, minimum, average, median, percentiles,* etc.).
- *Others.* These functions represent a complete change of the semantic of the values so the result has to be considered in a new domain. In that case, the cardinality should be established to 1.

$$c = 1 \qquad (4)$$

  In this category we found functions like the *variance* or the *count*.

Once we have studied the aggregation functions we have all the elements to show the query process with the SIs in datacubes

4   Semantic Interpretations for DataCubes

One we have define the formalism for SI, we introduce this concept in the multidimensional model previously define. To be able to use the SI we need to add information to each fact that represent the cardinality for the concrete value. So, we have to redefine the fact (definition 3) including the metadata.

Definition 7. *We say that any tuple (h, α, m) is a **fact** when h is an m-tuple on the attributes domain we want to analyze, $\alpha \in [0, 1]$ and m the metatada for these values.*

In the m element of the structure we introduce the cardinality c needed for the SI. This value is updated each time we query the datacube so, the aggregators have to work with this metadata.

Definition 8. *An aggregation operator G is a function $G(B)$ where $B = (h, \alpha, m)/(h, \alpha, m) \in F$ and the result is a tuple (h', α', m').*

4.1   Query Process

In this section we present the query process considering the use of the *SI*. We can differentiate two phases on the query process: the OLAP query over the DataCube and the report with the results. In both phases the *SI* are involved in a different way. Let show the process on each one:

- *Query over the DataCube.* In this phase is where the values are calculated. Inside this process we have to calculate the metatada of each one so, in the next phase, the values can be shown using the *SI*. The cardinality is calculated over each value considering the aggregation function used as shown in section 3.2. In this process the system has to control if the semantic has changed. On each value the system check if the aggregation function used is in the set G of each *SI*. If the function is not included, then this *SI* is deleted. In next phase (the report) the user can used only the *SI*s that satisfies this restriction.
- *Report.* Once the query has finished the result is shown to the user in a report. In this process the user has to choose the way to represent the values (the *SI* to use) and the interpretation (independence or relative as shown in section 3.1). After these steps, the system adapts the labels of each value using the f_a functions and the right cardinality (the absolute or the average).

4.2   Learning the f_a Functions

In previous section we have presented the query process using *SI*. One of the phase adapts the labels in L so the labels are fitted to the new cardinality. This process is carried out using the f_a function. The quality of the result will depend on this function, so an important point is the process to define it. Asking the user for that function is not always possible because most of the times the user is not able to use a mathematic expression to define his/her interpretations. So, we propose to learn the functions. The learn process will have the followings steps:

1. First we ask the user for an interpretation over the basic domain so we can define the set L of labels over it.
2. To learn the function now the system runs some queries over the DataCube showing the results.
3. For each query the system ask the user to associate a label of L to the value. The system stores the associations and the cardinalities of each value.
4. With these associations the system tries to fit a function that satisfies the interpretation with the corresponding cardinality. In this process, the system will try continue and monotone functions to adapt the labels. If the fitted function has good quality (the adapted labels correspond to the labels associated by the user) the process end. In other case, the process go but to point 2 to show more queries so the system have more data to fit the function.

5   Example

In this section we will present a small example to show in details the propose method. Let suppose we have a simple datacube only with two dimensions (time and centre) and only one measure (number of patients). The hierarchies for both dimensions are shown in Figure 1.

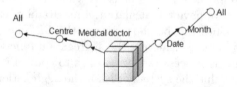

Fig. 1. Example datacube

For the measure we define a *SI* indicating if the number of patient attended by a doctor is Low, Normal or High. At base level (doctor and day) the fuzzy partition is shown in Figure 2.

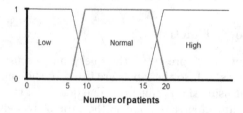

Fig. 2. Fuzzy partition over the measure Number of Patients

The *SI* is valid for aggregations like sum and average. The last aspect to define is the f_a function. In this example we suppose is lineal and just we multiply the points of the fuzzy label by the new cardinality (e.g. if Low is define as $(0, 0, 5, 10)$ for one doctor in a day, for two doctors the label will be $(2 \cdot 0, 2 \cdot 0, 2 \cdot 5, 2 \cdot 10) = (0, 0, 10, 20)$). Let suppose that we have two centres with different size. One (C1) is placed in a city and there are 500 medical doctors in the staff. The second one (C2) is placed in a small village and only 10 doctors are working in that centre. If a manager asks the system to calculate the number of patients attended by both centres each month we can get the Table 1.

The result shows very different results for each centre and it is not easy interpretable due to differences in the size of each one. Let apply our proposal and obtain the label that best represent each value. For centre C1 we have to calculate the cardinality of the result so we can adjust the labels. We have the datacube defined in the granularity doctor by day, so, for each month we have aggregate the values for 500 medical doctors a 20 working days for month, so

Table 1. Query results

Centre	Month	Patients
C1	January	125,000
C1	February	130,000
...	...	...
C2	January	4,200
C2	February	5,000
...	...	...

the cardinality is $500 \cdot 20 = 10.000$. We adjust the fuzzy partition for this new cardinality as shown in Figure 3. In the case of centre C2 then the cardinality is $10 \cdot 20 = 200$ (Figure 4).

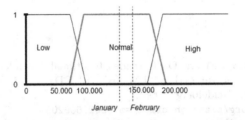

Fig. 3. Labels adaptation for query for centre C1

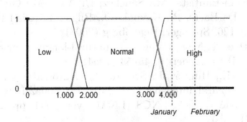

Fig. 4. Labels adaptation for query for centre C2

Table 2. Query results using *Semantic Interpretation*

Centre	Month	Patients	Label
C1	January	125,000	Normal
C1	February	130,000	Normal
...	...	...	...
C2	January	4,200	High
C2	February	5,000	High
...	...	...	...

In the figures we have indicated the values for the Table 1, so we have the labels associated to each result. In Table 2 we have added the label associated to each value.

If we use the *SI* in the example we see how the values are adapted so the user has the interpretation of the values directly.

6   Conclusions

In this paper we have introduced the new concept of Semantic Interpretation, that provides the OLAP systems with the new capability of querying about the same given item with different purposes obtaining in each case a result with a different meaning. With our proposal, the semantic of the results of the query can be distinct and adapted to the needs of the user, by taking into account the granularity of the information considered.

References

1. Bosc, P., Lietard, L., Pivert, O.: Extended functional dependencies as a basis for linguistic summaries. In: Żytkow, J.M. (ed.) PKDD 1998. LNCS, vol. 1510, pp. 255–263. Springer, Heidelberg (1998),
   http://dl.acm.org/citation.cfm?id=645802.669203
2. Bouchon-Meunier, B., Moyse, G.: Fuzzy linguistic summaries: Where are we, where can we go? In: 2012 IEEE Conference on Computational Intelligence for Financial Engineering Economics (CIFEr), pp. 1–8 (2012)
3. Delgado, M., Martín-Bautista, M., Sánchez, D., Vila, M.: On a characterization of fuzzy bags. In: De Baets, B., Kaynak, O., Bilgiç, T. (eds.) IFSA 2003. LNCS, vol. 2715, pp. 119–126. Springer, Heidelberg (2003)
4. Goldberg, E., Kittredge, N.D., Using, R.I.: natural-language processing to produce weather forecasts. IEEE Expert 9, 45–53 (1994)
5. Portet, F., Reiter, E., Hunter, J., Sripada, S.: Automatic generation of textual summaries from neonatal intensive care data. In: Bellazzi, R., Abu-Hanna, A., Hunter, J. (eds.) AIME 2007. LNCS (LNAI), vol. 4594, pp. 227–236. Springer, Heidelberg (2007)
6. Yu, J., Reiter, E., Sripada, J.H., Sumtime-turbine, S.: A knowledge-based system to communicate gas turbine time-series data. In: The 16th International Conference on Industrial & Engineering Applications of Artificial Intelligence and Expert Systems (2003)
7. Kacprzyk, J., Wilbik, A.: Linguistic summaries of time series using a degree of appropriateness as a measure of interestingness. In: Ninth International Conference on Intelligent Systems Design and Applications, ISDA 2009, pp. 385–390 (2009)
8. Danlos, L., Combet, F.M., Easytext, V.: an operational nlg system. In: ENLG 2011, 13th European Workshop on Natural Language Generation (2011)
9. Lietard, L.: A new definition for linguistic summaries of data. In: IEEE International Conference on Fuzzy Systems, FUZZ-IEEE 2008 (IEEE World Congress on Computational Intelligence), pp. 506–511 (2008)
10. Molina, C., Rodriguez-Ariza, L., Sanchez, D., Vila, A.: A new fuzzy multidimensional model. IEEE Transactions on Fuzzy Systems 14(6), 897–912 (2006)

11. Castillo-Ortega, R., Marín, N., Sánchez, D.: A fuzzy approach to the linguistic summarization of time series. Journal of Multiple-Valued Logic and Soft Computing 17(2,3), 157–182 (2011)
12. Rasmussen, D., Yager, R.R.: Finding fuzzy and gradual functional dependencies with summarysql. Fuzzy Sets Syst. 106(2), 131–142 (1999), http://dx.doi.org/10.1016/S0165-0114(97)00268-6
13. Sripada, S., Reiter, E., Davy, I.: Sumtime-mousam: Configurable marine weather forecast generator. Tech. rep. (2003)
14. Yager, R.R.: A new approach to the summarization of data. Information Sciences 28, 69–82 (1982)
15. Yager, R.R.: Fuzzy summaries in database mining. In: Proceedings the 11th Conference on Artificial Intelligence for Applications (1995)
16. Yseop: Faire parler les chiffres automatiquement, http://www.yseop.com/demo/diagFinance/FR/
17. Zadeh, L.: A computational approach to fuzzy quantifiers in natural languages. Computers & Mathematics with Applications 9, 149–184 (1983)

Semantic Interpretation of Intermediate Quantifiers and Their Syllogisms*

Petra Murinová and Vilém Novák

Centre of Excellence IT4Innovations, Division of the University of Ostrava
Institute for Research and Applications of Fuzzy Modeling
30. dubna 22, 701 03 Ostrava 1, Czech Republic
{petra.murinova,vilem.novak}@osu.cz
http://irafm.osu.cz/

Abstract. This paper is a contribution to the formal theory of *intermediate quantifiers* (linguistic expressions such as *most, few, almost all, a lot of, many, a great deal of, a large part of, a small part of*). The latter concept was informally introduced by P. L. Peterson in his book and formalized in the frame of higher-order fuzzy logic by V. Novák. The main goal of this paper is to demonstrate how our theory works in an intended model. We will also show, how validity of generalized intermediate syllogisms can be semantically verified.

Keywords: Generalized quantifiers, intermediate quantifiers, fuzzy type theory, evaluative linguistic expressions, generalized Aristotle's syllogisms.

1 Introduction

The linguist and philosopher P. L. Peterson in his book [13] introduced informally the concept of intermediate quantifier. This is a linguistic expression such as *most, few, almost all, a lot of, many, a great deal of, a large part of, a small part of*, etc. Moreover, using informal tools he also suggested several concrete quantifiers and demonstrated that 105 generalized Aristotelian syllogisms should be valid for them. This original and well written book inspired V. Novák in [10] to suggest a formalization of Peterson's concept in the frame of higher-order fuzzy logic. Namely, the intermediate quantifiers are modeled by means of a special formal theory T^{IQ} of the Łukasiewicz fuzzy type theory (L-FTT) introduced by in [8]. The formal theory of intermediate quantifiers was further developed in [7] where we formally proved that all the syllogisms analysed in [13] are valid in T^{IQ}. moreover, since all the proofs proceed syntactically, the syllogisms are valid *in all models*.

In view of the classical theory of generalized quantifiers, the intermediate quantifiers are special generalized quantifiers of type $\langle 1, 1 \rangle$ (cf. [6,14,15]) that

---

* The paper has been supported by the European Regional Development Fund in the IT4Innovations Centre of Excellence project (CZ.1.05/1.1.00/02.0070).

H.L. Larsen et al. (Eds.): FQAS 2013, LNAI 8132, pp. 186–197, 2013.

are *isomorphism-invariant* (cf. [4,5]). This property immediately follows from model theory of higher order fuzzy logic introduced in [12].

The semantics of intermediate quantifiers is modeled using the concept of *evaluative linguistic expressions* that are special expressions of natural language, for example, *small, medium, rather big, very short quite roughly strong*, etc. Each nontrivial intermediate quantifier is taken as a classical quantifier "all" or "exists" whose universe of quantification is modified using a proper evaluative expression. Since the semantics of the latter is modeled by means of a special theory T^{Ev} of Ł-FTT, the theory T^{IQ} is obtained as an extension of T^{Ev}.

Let us emphasize that the meaning of each evaluative expression is construed as a specific formula representing *intension* whose interpretation in a model is a function from the set of possible worlds[1] into a set of fuzzy sets. Intension thus determines in each context a corresponding extension that is a fuzzy set in a universe determined in the given context. The fuzzy set is constructed as a *horizon* that can be shifted along the universe. For the details of the theory T^{Ev} including its special axioms and the motivation — see [9]. From it follows that we can easily construct also intensions of the intermediate quantifiers.

The main goal of this paper is to give examples of syllogism in a specific intended model. After a brief introduction of the basic concepts of Ł-FTT we present the basic concepts of the theory of intermediate quantifiers and their syllogisms. In Section 4, we construct a model of the theory T^{IQ} and give examples of syllogisms for each of four possible figures.

2 Basic Concepts of the Fuzzy Type Theory

The basic syntactical objects of Ł-FTT are classical — see [1], namely the concepts of *type* and *formula*. The atomic types are ϵ (elements) and o (truth values). General types are denoted by Greek letters $\alpha, \beta, \ldots$. The set of all types is denoted by *Types*. The *language* of Ł-FTT denoted by J, consists of variables $x_\alpha, \ldots$, special constants $c_\alpha, \ldots$ ($\alpha \in \textit{Types}$), the symbol λ, and brackets.

The truth values form an MV-algebra (see [2]) extended by the delta operation. It can be seen as the residuated lattice

$$\mathcal{L} = \langle L, \vee, \wedge, \otimes, \rightarrow, \mathbf{0}, \mathbf{1}, \Delta \rangle, \tag{1}$$

where $\mathbf{0}$ is the least and $\mathbf{1}$ is the greatest element, $L = \langle L, \otimes, \mathbf{1} \rangle$ is a commutative monoid, the $\rightarrow$ is residuation fulfilling the *adjunction property*

$$a \otimes b \leq c \quad \text{iff} \quad a \leq b \rightarrow c,$$

and $a \vee b = (a \rightarrow b) \rightarrow b$ holds for all $a, b, c \in L$. The Δ is a unary operation fulfilling 6 special axioms (cf. [3,8]). A special case is the standard Łukasiewicz MV_Δ-algebra

$$\mathcal{L} = \langle [0,1], \vee, \wedge, \otimes, \rightarrow, 0, 1, \Delta \rangle \tag{2}$$

---

[1] In our theory, we prefer to speak about *contexts*.

where

$$\wedge = \text{minimum}, \qquad\qquad \vee = \text{maximum},$$
$$a \otimes b = \max(0, a + b - 1), \qquad a \rightarrow b = \min(1, 1 - a + b),$$
$$\neg a = a \rightarrow 0 = 1 - a, \qquad \Delta(a) = \begin{cases} 1 & \text{if } a = 1, \\ 0 & \text{otherwise.} \end{cases}$$

We will also consider the operation of Łukasiewicz disjunction $a \oplus b = \min(1, a + b)$.

For the precise definitions of formulas, axioms, inference rules and properties of the fuzzy type theory — see [8,11].

3   Theory of Intermediate Quantifiers

As mentioned, the formal theory of intermediate quantifiers T^{IQ} is a special theory of L-FTT extending the theory T^{Ev} of evaluative linguistic expressions.

Definition 1. *Let* $R \in \text{Form}_{o(o\alpha)(o\alpha)}$ *be a formula. Put*

$$\mu := \lambda z_{o\alpha} \, \lambda x_{o\alpha} \, (R z_{o\alpha}) x_{o\alpha}. \tag{3}$$

We say that the formula $\mu \in \text{Form}_{o(o\alpha)(o\alpha)}$ *represents a* measure on fuzzy sets *in the universe of type* $\alpha \in \text{Types}$ *if it has the following properties:*

(i) $\Delta(x_{o\alpha} \equiv z_{o\alpha}) \equiv ((\mu z_{o\alpha}) x_{o\alpha} \equiv \top)$ *(M1)*,

(ii) $\Delta(x_{o\alpha} \subseteq z_{o\alpha}) \& \Delta(y_{o\alpha} \subseteq z_{o\alpha}) \& \Delta(x_{o\alpha} \subseteq y_{o\alpha}) \Rightarrow$ [2]
   $((\mu z_{o\alpha}) x_{o\alpha} \Rightarrow (\mu z_{o\alpha}) y_{o\alpha})$ *(M2)*,

(iii) $\Delta(z_{o\alpha} \neq \emptyset_{o\alpha}) \& \Delta(x_{o\alpha} \subseteq z_{o\alpha}) \Rightarrow$
   $((\mu z_{o\alpha})(z_{o\alpha} - x_{o\alpha}) \equiv \neg (\mu z_{o\alpha}) x_{o\alpha})$ *(M3)*,

(iv) $\Delta(x_{o\alpha} \subseteq y_{o\alpha}) \& \Delta(x_{o\alpha} \subseteq z_{o\alpha}) \& \Delta(y_{o\alpha} \subseteq z_{o\alpha}) \Rightarrow$
   $((\mu z_{o\alpha}) x_{o\alpha} \Rightarrow (\mu y_{o\alpha}) x_{o\alpha})$ *(M4)*.

Note that the measure is normed with respect to a distinguished fuzzy set $z_{o\alpha}$. Using μ we will characterize size of the considered fuzzy sets.

For the following definition, we have to consider a set of selected types $\mathcal{S}$ to which our theory will be confined. The reason is to avoid possible difficulties with interpretation of the formula μ for complex types which may correspond to sets of very large, possibly non-measurable cardinalities. This means that our theory is not fully general. We do not see it as a limitation, though, because one can hardly imagine the meaning of "most X's" over a set of inaccessible cardinality. On the other hand, our theory works whenever there is a model in which we can define a measure in the sense of Definition 1. The theory T^{IQ} defined below is thus parametrized by the set $\mathcal{S}$. Below, we will work with formulas of types α

---

[2] This implication is interpreted using Łukasiewicz implication which was defined above.

(arbitrary objects), $o\alpha$ (fuzzy sets of objects of type α) and possibly formulas of more complicated types that will be explained in the text.

Let us introduce the following special formula representing a fuzzy set of all measurable fuzzy sets in the given type α:

$$\mathbf{M}_{o(o\alpha)} := \lambda z_{o\alpha}\,(\mathbf{\Delta}(\mu z_{o\alpha})z_{o\alpha}\,\&(\forall x_{o\alpha})(\forall y_{o\alpha})$$

$$((\text{M2})\,\&\,(\text{M4}))\,\&\,(\forall x_{o\alpha}))(\text{M3}) \quad (4)$$

where (M2)–(M4) are the respective axioms from Definition 1.

Definition 2. *Let $\mathcal{S} \subseteq$ Types be a distinguished set of types and $\{R \in Form_{o(o\alpha)(o\alpha)} \mid \alpha \in \mathcal{S}\}$ be a set of new constants. The theory of intermediate quantifiers $T^{IQ}[\mathcal{S}]$ w.r.t. $\mathcal{S}$ is a formal theory of L-FTT defined as follows:*

(i) The language of $T^{IQ}[\mathcal{S}]$ is $J^{Ev} \cup \{R_{o(o\alpha)(o\alpha)} \in Form_{o(o\alpha)(o\alpha)} \mid \alpha \in \mathcal{S}\}$.
(ii) Special axioms of $T^{IQ}[\mathcal{S}]$ are those of T^{Ev} and $(\exists z_{o\alpha})\mathbf{M}_{o(o\alpha)}z_{o\alpha}$, $\alpha \in \mathcal{S}$.

In the following definition, we introduce two kinds of intermediate quantifiers.

Definition 3. *Let $T^{IQ}[\mathcal{S}]$ be a theory of intermediate quantifiers in the sense of Definition 2 and $Ev \in Form_{oo}$ be intension of some evaluative expression. Furthermore, let $z \in Form_{o\alpha}$, $x \in Form_{\alpha}$ be variables and $A, B \in Form_{o\alpha}$ be formulas, $\alpha \in \mathcal{S}$, such that*

$$T^{IQ} \vdash \mathbf{M}_{o(o\alpha)}B_{o\alpha}$$

holds true. Then a type $\langle 1, 1 \rangle$ intermediate generalized quantifier interpreting the sentence "$\langle$Quantifier$\rangle$ B's are A" is one of the following formulas:

$$(Q^{\forall}_{Ev}\,x)(B, A) := (\exists z)((\mathbf{\Delta}(z \subseteq B)\,\&\,(\forall x)(z\,x \Rightarrow Ax)) \wedge Ev((\mu B)z)), \quad (5)$$

$$(Q^{\exists}_{Ev}\,x)(B, A) := (\exists z)((\mathbf{\Delta}(z \subseteq B)\,\&\,(\exists x)(zx \wedge Ax)) \wedge Ev((\mu B)z)). \quad (6)$$

In some cases, only non-empty subsets of B must be considered. This is a special case of intermediate quantifiers with *presupposition*.

Definition 4. *Let $T^{IQ}[\mathcal{S}]$ be a theory of intermediate quantifiers in the sense of Definition 2 and Ev, z, x, A, B be the same as in Definition 3. Then an intermediate generalized quantifier with presupposition is the formula*

$$(^{*}Q^{\forall}_{Ev}\,x)(B, A) \equiv (\exists z)((\mathbf{\Delta}(z \subseteq B)\,\&\,(\exists x)zx\,\&\,(\forall x)(z\,x \Rightarrow Ax)) \wedge Ev((\mu B)z)).$$
$$(7)$$

We will now introduce definitions of several specific intermediate quantifiers based on the analysis provided by Peterson in his book [13]. We can see that all the intermediate quantifiers are defined using specific evaluative expressions. Shapes of their extensions, i.e., of the corresponding fuzzy sets in the context $[0, 1]$ are depicted in Fig. 1 below.

A: All B are $A := Q^{\vee}_{Bi\Delta}(B, A) \equiv (\forall x)(Bx \Rightarrow Ax)$,

E: No B are $A := Q^{\vee}_{Bi\Delta}(B, \neg A) \equiv (\forall x)(Bx \Rightarrow \neg Ax)$,

P: Almost all B are $A := Q^{\vee}_{Bi\,Ex}(B, A) \equiv$
$$(\exists z)((\Delta(z \subseteq B) \,\&\, (\forall x)(zx \Rightarrow Ax)) \wedge (Bi\,Ex)((\mu B)z)),$$

B: Few B are A ($:=$ Almost all B are not A) $:= Q^{\vee}_{Bi\,Ex}(B, \neg A) \equiv$
$$(\exists z)((\Delta(z \subseteq B) \,\&\, (\forall x)(zx \Rightarrow \neg Ax)) \wedge (Bi\,Ex)((\mu B)z)),$$

T: Most B are $A := Q^{\vee}_{Bi\,Ve}(B, A) \equiv$
$$(\exists z)((\Delta(z \subseteq B) \,\&\, (\forall x)(zx \Rightarrow Ax)) \wedge (Bi\,Ve)((\mu B)z)),$$

D: Most B are not $A := Q^{\vee}_{Bi\,Ve}(B, \neg A) \equiv$
$$(\exists z)((\Delta(z \subseteq B) \,\&\, (\forall x)(zx \Rightarrow \neg Ax)) \wedge (Bi\,Ve)((\mu B)z)),$$

K: Many B are $A := Q^{\vee}_{\neg(Sm\,\bar{\nu})}(B, A) \equiv$
$$(\exists z)((\Delta(z \subseteq B) \,\&\, (\forall x)(zx \Rightarrow Ax)) \wedge \neg(Sm\,\bar{\nu})((\mu B)z)),$$

G: Many B are not $A := Q^{\vee}_{\neg(Sm\,\bar{\nu})}(B, \neg A) \equiv$
$$(\exists z)((\Delta(z \subseteq B) \,\&\, (\forall x)(zx \Rightarrow \neg Ax)) \wedge \neg(Sm\,\bar{\nu})((\mu B)z)),$$

I: Some B are $A := Q^{\exists}_{Bi\Delta}(B, A) \equiv (\exists x)(Bx \wedge Ax)$,

O: Some B are not $A := Q^{\exists}_{Bi\Delta}(B, \neg A) \equiv (\exists x)(Bx \wedge \neg Ax)$.

4   Generalized Intermediate Syllogisms

A *syllogism* (or *logical appeal*) is a special kind of logical argument in which the *conclusion* is inferred from two *premises*: the *major premise* (first) and *minor premise* (second). The syllogisms will be written as triples of formulas $\langle P_1, P_2, C \rangle$. The *intermediate syllogism* is obtained from any traditional syllogism (valid or not) when replacing one or more of its formulas by formulas containing intermediate quantifiers.

We say that a syllogism $\langle P_1, P_2, C \rangle$ is *valid* in a theory T if $T \vdash P_1 \,\&\, P_2 \Rightarrow C$, or equivalently, if $T \vdash P_1 \Rightarrow (P_2 \Rightarrow C)$. Note that this means that in every model $\mathcal{M} \models T$ the inequality

$$\mathcal{M}(P_1) \otimes \mathcal{M}(P_2) \leq \mathcal{M}(C) \tag{8}$$

holds true. Suppose that Q_1, Q_2, Q_3 are intermediate quantifiers and $X, Y, M \in Form_{o\alpha}$ are formulas representing properties. Then the following figures can be considered:

Figure I	Figure II	Figure III	Figure IV
Q_1 M is Y	Q_1 Y is M	Q_1 M is Y	Q_1 Y is M
Q_2 X is M	Q_2 X is M	Q_2 M is X	Q_2 M is X
Q_3 X is Y	Q_3 X is Y	Q_3 X is Y	Q_3 X is Y

4.1 Interpretation

In this section we will present examples of syllogisms of every figure and show their validity in a simple model with a finite set M_ϵ of elements. The frame of the constructed model is the following:

$$\mathcal{M} = \langle (M_\alpha, =_\alpha)_{\alpha \in Types}, \mathcal{L}_\Delta \rangle \tag{9}$$

where $M_o = [0,1]$ is the support of the standard Łukasiewicz MV_Δ-algebra. The fuzzy equality $=_o$ is the Łukasiewicz biresiduation $\leftrightarrow$. Furthermore, $M_\epsilon = \{u_1, \ldots, u_r\}$ is a finite set with a fixed numbering of its elements and $=_\epsilon$ is defined by

$$[u_i =_\epsilon u_j] = \left(1 - \min\left(1, \frac{|i-j|}{s}\right)\right)$$

for some fixed natural number $s \leq r$. This a separated fuzzy equality w.r.t. the Łukasiewicz conjunction $\otimes$. It can be verified that all the logical axioms of L-FTT are true in the degree 1 in $\mathcal{M}$ (all the considered functions are weakly extensional w.r.t. $\mathcal{M}(\equiv)$). Moreover, $\mathcal{M}$ is nontrivial because $1 - \frac{|i-j|}{s} \in (0,1)$ implies $\frac{|i-j|}{s} \in (0,1)$ and thus, taking the assignment p such that $p(x_\epsilon) = u_i$, $p(y_\epsilon) = u_j$ and considering $A_o := x_\epsilon \equiv y_\epsilon$, we obtain $\mathcal{M}_p(A_o \vee \neg A_o) \in (0,1)$.

To make $\mathcal{M}$ a model of T^{Ev} and T^{IQ}, we must define interpretation of $\sim$; namely, we put $\mathcal{M}(\sim) = \leftrightarrow^2$, $\mathcal{M}(\dagger) = 0.5$ and put $\mathcal{M}(\boldsymbol{\nu})$ equal to a function $\nu_{a,b,c}$ which is a simple partially quadratic function given in [9]. In Fig. 1, extensions of several evaluative expressions used below are depicted. It can be verified that $\mathcal{M} \models T^{\mathrm{Ev}}$.

The distinguished set $\mathcal{S} \subset Types$ is defined as follows: $\alpha \in \mathcal{S}$ iff α does not contain a subtype of the form βo. As follows from the analysis in [12], all sets M_α for $\alpha \in \mathcal{S}$ are finite.

Let $A \subseteq_\sim M_\alpha$, $\alpha \in \mathcal{S}$ be a fuzzy set. We put

$$|A| = \sum_{u \in \mathrm{Supp}(A)} A(u), \qquad u \in M_\alpha. \tag{10}$$

Furthermore, for fuzzy sets $A, B \subseteq_\sim M_\alpha$, $\alpha \in \mathcal{S}$ we define

$$F_R(B)(A) = \begin{cases} 1 & \text{if } B = \emptyset \text{ or } A = B, \\ \frac{|A|}{|B|} & \text{if } B \neq \emptyset \text{ and } A \subseteq B, \\ 0 & \text{otherwise.} \end{cases} \tag{11}$$

Interpretation of the constants $R \in Form_{o(o\alpha)(o\alpha)}$, $\alpha \in \mathcal{S}$ is defined by $\mathcal{M}(R) = F_R$ where $F_R : \mathcal{F}(M_\alpha) \times \mathcal{F}(M_\alpha) \to L$ is the function (11). It can be verified that axioms (M1)–(M4) are true in the degree 1 in $\mathcal{M}$. Thus, $\mathcal{M} \models T^{\mathrm{IQ}}$.

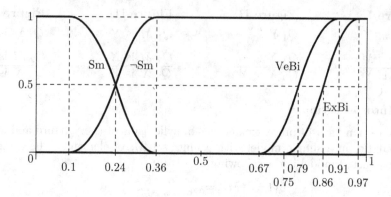

Fig. 1. Shapes of the extensions of evaluative expressions in the context $[0, 1]$ used in the examples below

4.2 Examples of Strongly Valid Syllogisms of Figure I

Below, we will demonstrate on concrete examples, how some of the syllogisms, whose validity was syntactically proved in [7], behave in the model constructed above. In the examples below we will suppose that the truth values of the major and minor premises in the above considered model are equal to 1. By (8), the truth value of the conclusion must also be equal to 1.

We will now analyze what kinds of sets fulfil these requirements. First, we will consider the following syllogism.

$$P_1\text{: All men are bigheaded.}$$
ATT-I: $\dfrac{P_2\text{: Most politicians are men.}}{C\text{: Most politicians are bigheaded.}}$

Let $\mathcal{M}$ be a model in (9). Let M_ϵ be a set of people. Let $\mathrm{Men}_{o\epsilon}$ be a formula representing "men". We will suppose that its interpretation is a classical set $\mathcal{M}(\mathrm{Men}_{o\epsilon}) = M \subseteq M_\epsilon$. Furthermore, let $\mathrm{Pol}_{o\epsilon}$ be a formula representing "politicians" interpreted by $\mathcal{M}(\mathrm{Pol}_{o\epsilon}) = P \subseteq M_\epsilon$ where P is a classical set. Finally, let $\mathrm{BH}_{o\epsilon}$ be a formula interpreted by a classical set $\mathcal{M}(\mathrm{BH}_{o\epsilon}) = H \subseteq M_\epsilon$.

Major premise: "All men are bigheaded." Note that this is the classical general quantifier and so, by the assumption

$$\mathcal{M}((\forall x_\epsilon)(\mathrm{Men}_{o\epsilon}(x_\epsilon) \Rightarrow \mathrm{Hum}_{o\epsilon}(x_\epsilon))) =$$
$$\bigwedge_{m \in M_\epsilon} (\mathcal{M}(\mathrm{Men}_{o\epsilon})(m) \to \mathcal{M}(\mathrm{BH}_{o\epsilon})(m)) = 1$$

we conclude that $M \subseteq H$.

Minor premise: *"Most politicians are men."* This is a non-classical intermediate quantifier. By the assumption, its interpretation in the above model

$$\mathcal{M}((\exists z_{o\epsilon})((\Delta(z_{o\epsilon} \subseteq \mathrm{Pol}_{o\epsilon}) \,\&\, (\forall x_\epsilon)(z_{o\epsilon}x_\epsilon \Rightarrow \mathrm{Men}_{o\epsilon}))$$
$$\wedge (Bi\ Ve)((\mu\mathrm{Pol}_{o\epsilon})z_{o\epsilon}))) = 1 \quad (12)$$

leads to the requirement to find the greatest subset $\mathcal{M}(z_{o\epsilon}) = M' \subseteq P$ such that:

$$\mathcal{M}(\Delta(z_{o\epsilon} \subseteq \mathrm{Pol}_{o\epsilon})) = 1, \quad (13)$$
$$\mathcal{M}((\forall x_\epsilon)(z_{o\epsilon}x_\epsilon \Rightarrow \mathrm{Men}_{o\epsilon}) = 1, \quad (14)$$
$$\mathcal{M}((Bi\ Ve)((\mu\mathrm{Pol}_{o\epsilon})z_{o\epsilon})) = 1. \quad (15)$$

One can verify that this holds if $M' = M$.

From (15) and the interpretation of evaluative expressions (see Figure 1) it follows that $\mathcal{M}((\mu\mathrm{Pol}_{o\epsilon})z_{o\epsilon}) = F_R(P, M) \geq 0.91$. Thus, for example, if $|P| = 100$ then $|M| \geq 91$.

Conclusion: *"Most politicians are bigheaded."* This is the non-classical intermediate quantifier. Its interpretation in the above model is

$$Q^{\forall}_{Bi\ Ve}(\mathrm{Men}_{o\epsilon}, \mathrm{BH}_{o\epsilon}) :=$$
$$(\exists z_{o\epsilon})((\Delta(z_{o\epsilon} \subseteq \mathrm{Men}_{o\epsilon}) \,\&\, (\forall x_\epsilon)(z_{o\epsilon}x_\epsilon \Rightarrow \mathrm{BH}_{o\epsilon})) \wedge (Bi\ Ve)((\mu\mathrm{Men}_{o\epsilon})z_{o\epsilon}). \quad (16)$$

Because we are dealing with classical sets, we conclude that to find a truth value of (16) requires to find a set $\mathcal{M}(z_{o\epsilon}) = H'$, where $H' \subseteq M$ and $H' \subseteq H$, which maximizes the truth value

$$\mathcal{M}((Bi\ Ve)((\mu\mathrm{Men}_{o\epsilon})z_{o\epsilon})). \quad (17)$$

But from the first premise we know that $M \subseteq H$. From the fact that $F_R(P, M)$ provides the truth value 1 in (15) and from $H' \subseteq H$ we conclude that $M \subseteq H'$. Hence, $\mathcal{M}((\mu\mathrm{Men}_{o\epsilon})z_{o\epsilon}) = F_R(P, H')$ provides the truth value 1 in (17). Consequently, $\mathcal{M}(Q^{\forall}_{Bi\ Ve}(\mathrm{Men}_{o\epsilon}, \mathrm{Hum}_{o\epsilon})) = 1$. Since the syllogism is strongly valid, it means in our model that $\mathcal{M}(P_1) \otimes \mathcal{M}(P_2) \leq \mathcal{M}(C)$.

For example, if $|P| = 100$ then the quantifier "most" means at least 91 people. By the discussed syllogism, if we know that "All men are humans" and "Most politicians are men", we conclude that at least 91 politicians are humans.

Analogously we can demonstrate validity of the following syllogisms:

$$P_1 : \text{No homework is fun.}$$
EPD-I: P_2: Almost all writting is homework.
$$\overline{C : \text{Most writting is not fun.}}$$

$$P_1 : \text{All women are well dressed.}$$
ATT-I: P_2 : Most people in the party are women.
$$\overline{C : \text{Most people in the party are well dressed.}}$$

4.3   Examples of Strongly Valid Syllogisms of Figure II

Below we introduce only examples of strongly valid syllogisms in T^{IQ}. The validity can be verified by the same construction as in the previous subsection.

$$P_1 : \text{All birds have wings.}$$
AEO-II: $P_2 : \text{No humans have wings.}$
$$\overline{C : \text{Some humans have not wings.}}$$

$$P_1 : \text{All informative thinks are useful.}$$
AGG-II: $P_2 : \text{Many website are not useful.}$
$$\overline{C: \text{Many website are not informative.}}$$

4.4   Examples of Strongly Valid Syllogisms of Figure-III

Let us consider the syllogism of Figure-III as follows:

$$P_1: \text{Almost all small children go to the basic school.}$$
PTI-III: $P_2: \text{Most small children drive bicycle.}$
$$\overline{C: \text{Some children driving bicycle go to the basic school.}}$$

Suppose the same frame as above. Let M_ϵ be a set of "children". Let $\text{Child}_{o\epsilon}$ be a formula "small children" with the interpretation $\mathcal{M}(\text{Child}_{o\epsilon}) = C \subseteq M_\epsilon$ where C is a classical set. Furthermore, let $\text{Basic}_{o\epsilon}$ be a formula representing "children go to the basic school" with the interpretation $\mathcal{M}(\text{Basic}_{o\epsilon}) = B \subseteq M_\epsilon$ where B is a classical set. Finally, let $\text{Bicycle}_{o\epsilon}$ be a formula representing "children driving bicycle" with the interpretation $\mathcal{M}(\text{Bi}_{o\epsilon}) = I \subseteq M_\epsilon$ where I is a classical set.

Major premise "Almost all small children go to the basic school." This is nonclassical intermediate quantifier. By the assumption, its interpretation in the above model

$$\mathcal{M}((\exists z_{o\epsilon})((\boldsymbol{\Delta}(z_{o\epsilon} \subseteq \text{Child}_{o\epsilon}) \,\&\, (\forall x_\epsilon)(z_{o\epsilon} x_\epsilon \Rightarrow \text{Basic}_{o\epsilon} x_\epsilon))$$
$$\wedge \, (Bi\ Ex)((\mu \text{Child}_{o\epsilon}) z_{o\epsilon}))) = 1 \quad (18)$$

means to find the biggest subset $\mathcal{M}(z_{o\epsilon}) = B' \subseteq C$ such that:

$$\mathcal{M}(\boldsymbol{\Delta}(z_{o\epsilon} \subseteq \text{Child}_{o\epsilon})) = 1, \quad (19)$$
$$\mathcal{M}((\forall x_\epsilon)(z_{o\epsilon} x_\epsilon \Rightarrow \text{Basic}_{o\epsilon})) = 1, \quad (20)$$
$$\mathcal{M}((Bi\ Ex)((\mu \text{Child}_{o\epsilon}) z_{o\epsilon})) = 1. \quad (21)$$

It can be verified that this holds if $B' = B$. From (21) and Figure 1 it follows that $\mathcal{M}((\mu \text{Child}_{o\epsilon}) z_{o\epsilon}) = F_R(C, B) \geq 0.97$. This means that if $|C| = 100$, then $|B| \geq 97$.

Minor premise "*Most small children drive bicycle.*" This is non-classical intermediate quantifier. By the assumption, its interpretation in the above model

$$\mathcal{M}((\exists z_{o\epsilon})((\Delta(z_{o\epsilon} \subseteq \text{Child}_{o\epsilon}) \,\&\, (\forall x_\epsilon)(z_{o\epsilon} x_\epsilon \Rightarrow \text{Bicycle}_{o\epsilon} x_\epsilon))$$
$$\wedge (Bi\ VR)((\mu \text{Child}_{o\epsilon}) z_{o\epsilon}))) = 1 \quad (22)$$

means to find the biggest subset $\mathcal{M}(z_{o\epsilon}) = I' \subseteq C$ such that:

$$\mathcal{M}(\Delta(z_{o\epsilon} \subseteq \text{Child}_{o\epsilon})) = 1, \quad (23)$$
$$\mathcal{M}((\forall x_\epsilon)(z_{o\epsilon} x_\epsilon \Rightarrow \text{Bicycle}_{o\epsilon})) = 1, \quad (24)$$
$$\mathcal{M}((Bi\ Ve)((\mu \text{Child}_{o\epsilon}) z_{o\epsilon})) = 1. \quad (25)$$

It can be verified that this holds if $I' = I$.

From (25) and Figure 1 it follows that $\mathcal{M}((\mu \text{Child}_{o\epsilon}) z_{o\epsilon}) = F_R(C, I) \geq 0.91$. This means that if $|C| = 100$, then $|I| \geq 91$.

Conclusion "*Some children driving bicycle go to the basic school.*" This is the classical general quantifier and so, its interpretation in the above model is

$$Q_{Bi\Delta}^{\exists}(\text{Bicycle}_{o\epsilon}, \text{Basic}_{o\epsilon}) := (\exists x_\epsilon)(\text{Bicycle}_{o\epsilon}(x_\epsilon) \wedge \text{Basic}_{o\epsilon}(x_\epsilon)) \quad (26)$$

which is interpreted by

$$\mathcal{M}(Q_{Bi\Delta}^{\exists}(\text{Bicycle}_{o\epsilon}, \text{Basic}_{o\epsilon})) = \bigvee_{m \in M_\epsilon} (\mathcal{M}(\text{Bicycle}_{o\epsilon}(m)) \wedge \mathcal{M}(\text{Basic}_{o\epsilon}(m))).$$
$$(27)$$

The interpretation $\mathcal{M}(\text{Bicycle}_{o\epsilon}(x_\epsilon) \wedge \text{Basic}_{o\epsilon}(x_\epsilon)) = I \cap B$. From both premises we have that $I' \subseteq C$ and $B' \subseteq C$ thus $I' \cap B' \subseteq C$. From validity of the syllogism we know that $\mathcal{M}(Q_{Bi\Delta}^{\exists}(\text{Bicycle}_{o\epsilon}, \text{Basic}_{o\epsilon})) = 1$ and so, we conclude that $I' \cap B' \neq \emptyset$.

Analogously we may verify the validity of the following syllogisms:

$$\textbf{PPI-III:} \quad \begin{array}{l} P_1 : \text{Almost all parliamentarian are men.} \\ P_2 : \text{Almost all parliamentarian have driver.} \\ \hline C : \text{Some drivers are men.} \end{array}$$

$$\textbf{TTI-III:} \quad \begin{array}{l} P_1 : \text{Most people on earth eat meat.} \\ P_2 : \text{Most people on earth are women.} \\ \hline C : \text{Some women eat meat.} \end{array}$$

4.5 Example of Invalid Syllogism

Peterson in his book gave also an example of invalid syllogism of Figure-III. We will show that this syllogism is invalid also in our theory.

$$P_1 : \text{Most good jokes are old.}$$
TAT-III: $P_2 : \text{All good jokes are funny.}$
$$\overline{C : \text{Most funny jokes are old.}}$$

Suppose the same model as above. Let M_ϵ be a set of "jokes". Let $\text{GJoke}_{o\epsilon}$ be a formula "good jokes" with the interpretation $\mathcal{M}(\text{GJoke}_{o\epsilon}) = J \subseteq M_\epsilon$ where J is a classical set. Furthermore, let $\text{Old}_{o\epsilon}$ be a formula "old" with the interpretation $\mathcal{M}(\text{Old}_{o\epsilon}) = O \subseteq M_\epsilon$ where O is a classical set. Finally, let $\text{Funn}_{o\epsilon}$ be a formula "funny" with the interpretation $\mathcal{M}(\text{Fun}_{o\epsilon}) = F \subseteq M_\epsilon$ where F is a classical set.

Major premise "Most good jokes are old." The assumption

$$\mathcal{M}((\exists z_{o\epsilon})((\Delta(z_{o\epsilon} \subseteq \text{GJoke}_{o\epsilon}) \,\&\, (\forall x_\epsilon)(z_{o\epsilon} x_\epsilon \Rightarrow \text{Old}_{o\epsilon} x_\epsilon))$$
$$\wedge (Bi\ Ve)((\mu\text{GJoke}_{o\epsilon})z_{o\epsilon}))) = 1 \quad (28)$$

means to find the biggest subset $\mathcal{M}(z_{o\epsilon}) = O' \subseteq J$ such that:

$$\mathcal{M}(\Delta(z_{o\epsilon} \subseteq \text{GJoke}_{o\epsilon})) = 1, \quad (29)$$
$$\mathcal{M}((\forall x_\epsilon)(z_{o\epsilon} x_\epsilon \Rightarrow \text{Old}_{o\epsilon})) = 1, \quad (30)$$
$$\mathcal{M}((Bi\ Ve)((\mu\text{GJoke}_{o\epsilon})z_{o\epsilon})) = 1. \quad (31)$$

It can be verified that this holds if $O' = O$.

From (31) and Figure 1 it follows that $\mathcal{M}((\mu\text{GJoke}_{o\epsilon})z_{o\epsilon}) = F_R(J, O) \geq 0.91$. This means that if $|J| = 100$, then $|O| \geq 91$.

Minor premise "All good jokes are funny." The assumption

$$\mathcal{M}((\forall x_\epsilon)(\text{GJoke}_{o\epsilon}(x_\epsilon) \Rightarrow \text{Funn}_{o\epsilon}(x_\epsilon))) =$$
$$\bigwedge_{m \in M_\epsilon} (\mathcal{M}(\text{GJoke}_{o\epsilon}(m)) \to \mathcal{M}(\text{Funn}_{o\epsilon}(m))) = 1 \quad (32)$$

means that $J \subseteq F$ and hence $O' \subseteq F$.

Conclusion "Most funny jokes are old." The conclusion is the following formula:

$$Q^\vee_{Bi\ Ve}(\text{Funn}_{o\epsilon}, \text{Old}_{o\epsilon}) :=$$
$$(\exists z_{o\epsilon})((\Delta(z_{o\epsilon} \subseteq \text{Funn}_{o\epsilon}) \,\&\, (\forall x_\epsilon)(z_{o\epsilon} x_\epsilon \Rightarrow \text{Old}_{o\epsilon} x_\epsilon)) \wedge (Bi\ Ve)((\mu\text{Funn}_{o\epsilon})z_{o\epsilon})).$$
$$(33)$$

From the first premise for $\mathcal{M}(z_{o\epsilon}) = O'$ we have that $\mathcal{M}((\forall x_\epsilon)(z_{o\epsilon} x_\epsilon \Rightarrow \text{Old}_{o\epsilon})) = 1$. From the second one we obtain that

$$\mathcal{M}(\Delta(z_{o\epsilon} \subseteq \text{Funn}_{o\epsilon})) = 1.$$

From Figure 1 and from $\mathcal{M}((\mu\text{Funn}_{o\epsilon})z_{o\epsilon}) = F_R(F, O') = 0.83$ we obtain that $\mathcal{M}((Bi\ Ve)((\mu\text{Funn}_{o\epsilon})z_{o\epsilon})) < 1$. Consequently, we conclude that

$$\mathcal{M}(Q^\vee_{Bi\ Ve}(\text{Funn}_{o\epsilon}, \text{Old}_{o\epsilon})) < 1$$

which means that $\mathcal{M}(P_1) \otimes \mathcal{M}(P_2) > \mathcal{M}(C)$ and so, the syllogism **TAT**-III is invalid.

5 Conclusions

In this paper, we briefly introduced a formal theory of intermediate quantifiers and constructed an intended model of it. In this model we demonstrated on several examples how validity of generalized syllogisms works.

Let us emphasize that our theory is very general because validity of all the syllogisms was proved syntactically. This means that the following holds *in every model* $\mathcal{M}$: a truth value of the conclusion C is greater or equal to a truth value of the strong conjunction of the premises $P_1 \,\&\, P_2$, i.e. $\mathcal{M}(P_1) \otimes \mathcal{M}(P_2) \leq \mathcal{M}(C)$. The considered measure and the concept of context included in the theory of evaluative expressions makes it possible also to estimate sizes of the (fuzzy) sets of elements characterized using intermediate quantifiers with respect to the concrete context, i.e., what does it mean, for example "most" when speaking about a set of 100 elements.

References

1. Andrews, P.: An Introduction to Mathematical Logic and Type Theory: To Truth Through Proof. Kluwer, Dordrecht (2002)
2. Cignoli, R.L.O., D'Ottaviano, I.M.L., Mundici, D.: Algebraic Foundations of Many-valued Reasoning. Kluwer, Dordrecht (2000)
3. Hájek, P.: Metamathematics of Fuzzy Logic. Kluwer, Dordrecht (1998)
4. Holčapek, M.: Monadic L-fuzzy quantifiers of the type $\langle 1^n, 1 \rangle$. Fuzzy Sets and Systems 159, 1811–1835 (2008)
5. Dvořák, A., Holčapek, M.: L-fuzzy Quantifiers of the Type $\langle 1 \rangle$ Determined by Measures. Fuzzy Sets and Systems 160, 3425–3452 (2009)
6. Keenan, E.L.: Quantifiers in formal and natural languages. In: Handbook of Logic and Language, pp. 837–893. Elsevier, Amsterdam (1997)
7. Murinová, P.: A Formal Theory of Generalized Intermediate Syllogisms. Fuzzy Sets and Systems 186, 47–80 (2012)
8. Novák, V.: On fuzzy type theory. Fuzzy Sets and Systems 149, 235–273 (2005)
9. Novák, V.: A comprehensive theory of trichotomous evaluative linguistic expressions. Fuzzy Sets and Systems 159(22), 2939–2969 (2008)
10. Novák, V.: A formal theory of intermediate quantifiers. Fuzzy Sets and Systems 159(10), 1229–1256 (2008)
11. Novák, V.: EQ-algebra-based fuzzy type theory and its extension. Fuzzy Sets and Systems 159(22), 2939–2969 (2008)
12. Novák, V.: Elements of Model Theory in Higher Order Fuzzy Logic. Fuzzy Sets and Systems 205, 101–115 (2012)
13. Peterson, P.: Intermediate quantifiers, Logic, linguistics, Aristotelian semantics. Ahgate, Aldershot (2000)
14. Peters, S., Westerståhl, D.: Quantifiers in Language and Logic. Claredon Press, Oxford (2006)
15. Keenan, E.L., Westerståhl, D.: Quantifiers in formal and natural languages. In: Handbook of Logic and Language, pp. 837–893. Elsevier, Amsterdam (1997)

Ranking Images Using Customized Fuzzy Dominant Color Descriptors

J.M. Soto-Hidalgo[1], J. Chamorro-Martínez[2], P. Martínez-Jiménez[2],
and Daniel Sánchez[1,3]

[1] Department of Computer Architecture, Electronics and Electronic Technology,
University of Córdoba
jmsoto@uco.es
[2] Department of Computer Science and Artificial Intelligence, University of Granada
{jesus,pedromartinez,daniel}@decsai.ugr.es
[3] European Centre for Soft Computing, Mieres, Asturias, Spain
daniel.sanchezf@softcomputing.es

Abstract. In this paper we describe an approach for defining customized color descriptors for image retrieval. In particular, a customized fuzzy dominant color descriptor is proposed on the basis of a finite collection of fuzzy colors designed specifically for a certain user. Fuzzy colors modeling the semantics of a color name are defined as fuzzy subsets of colors on an ordinary color space, filling the semantic gap between the color representation in computers and the subjective human perception. The design of fuzzy colors is based on a collection of color names and corresponding crisp representatives provided by the user. The descriptor is defined as a fuzzy set over the customized fuzzy colors (i.e. a level-2 fuzzy set), taking into account the imprecise concept that is modelled, in which membership degrees represent the dominance of each color. The dominance of each fuzzy color is calculated on the basis of a fuzzy quantifier representing the notion of dominance, and a fuzzy histogram representing as a fuzzy quantity the percentage of pixels that match each fuzzy color. The obtained descriptor can be employed in a large amount of applications. We illustrate the usefulness of the descriptor by a particular application in image retrieval.

Keywords: Customized Fuzzy Color, Dominant color descriptor, Fuzzy Quantification, Image retrieval.

1 Introduction

The use of visual descriptors, which describe the visual features of the contents in images, has been suggested for image retrieval purposes [9,13], among many other applications. These descriptors describe elementary characteristics such as color, texture and shape, which are automatically extracted from images [1,7]. Among these characteristics, color plays an important role because of its robustness and independence from image size and orientation [4,12,5].

In this paper we focus on the dominant color descriptor, one of the most important descriptors in the well-known MPEG-7 standard [15]. A dominant color descriptor must provide an effective, compact, and intuitive representation of the most representative colors presented in an image. In this context, many approaches to dominant color

H.L. Larsen et al. (Eds.): FQAS 2013, LNAI 8132, pp. 198–208, 2013.

extraction have been proposed in the literature [16,6]. Our objective in this paper is to develop a dominant color descriptor able to cope with the following aspects:

1. to fill the semantic gap between the color representation in computers and its human perception, i.e., the color descriptors must be based on colors whose name and semantics, provided by (and understandable for) the users, is far from the set of colors represented by color spaces employed by computers, like RGB for instance.
2. the fuzziness of colors. Colors as perceived by humans correspond to subsets of those colors represented in computers. However, these perceived colors don't have clear, crisp, boundaries. Hence, perceived colors can be better modeled as fuzzy subsets of crisp colors.
3. the subjectivity in the set of colors. The collection of colors to be employed in the indexing of images is different for every user and/or application. We may also have a different number of colors, the same color may be given different names, and the same color name may have a different semantics, depending on the user and/or application. Hence, we need a *customized* set of colors, both in name and modeling, for each case.
4. the imprecision in the human's perception of dominance. It is usual to consider degrees of dominance, that is, colors can be clearly dominant, clearly not dominant, or can be dominant to a certain degree.

The first two requirements can be fulfilled by using the notions of *fuzzy color* and *fuzzy color space* [11]. In order to fulfil the third one, and to obtain customized fuzzy color spaces, we shall employ a methodology for developing a fuzzy color space on the basis of a collection of color names and corresponding crisp representatives, to be provided by the user for the specific application. Finally, in order to fulfil the fourth requirement, we introduce in this paper a fuzzy descriptor for dominant colors as a level-two fuzzy set on the fuzzy colors comprising the customized fuzzy color space modelling the imprecise concept of dominance.

In order to obtain this descriptor we use a histogram of fuzzy colors, as the dominance is related with the frequency of the colors in the image. The frequency will be represented by a fuzzy cardinality measure. A fuzzy quantifier will be employed in order to represent the semantics of *dominant* on the basis of the amount of pixels having a certain color, in a natural way. The final degree of dominance for a certain fuzzy color will be obtained as the accomplishment degree of a quantified sentence, calculated as the compatibility between the quantifier and the fuzzy cardinality of the set of pixels with the fuzzy color.

The rest of the paper is organized as follows. In section 2 the fuzzy color modelling is presented. The dominance-based color fuzzy descriptor is defined in section 3, and an inclusion fuzzy operator for image retrieval is described in section 4. Results are shown in section 5, and the main conclusions and future work are summarized in section 6.

2   Fuzzy Modelling of Colors

In this section, the notions of fuzzy color (section 2.1) and fuzzy color space (section 2.2) we presented in a previous work [11] are summarized.

2.1 Fuzzy Color

For representing colors, several color spaces can be used. In essence, a color space is a specification of a coordinate system and a subspace within that system where each color is represented by a single point. The most commonly used color space in practice is RGB because is the one employed in hardware devices (like monitors and digital cameras). It is based on a cartesian coordinate system, where each color consists of three components corresponding to the primary colors red, green, and blue. Other color spaces are also popular in the image processing field: linear combination of RGB (like CMY, YCbCr, or YUV), color spaces based on human color term properties like hue or saturation (HSI, HSV or HSL), or perceptually uniform color spaces (like CIELa*b*, CIELuv, etc.).

In order to manage the imprecision in color description, we introduce the following definition of fuzzy color:

Definition 1. *A fuzzy color $\widetilde{C}$ is a normalized fuzzy subset of colors.*

As previously explained, colors can be represented as a triplet of real numbers corresponding to coordinates in a color space. Hence, a fuzzy color can be defined as a normalized fuzzy subset of points of a color space. From now on, we shall note XYZ a generic color space with components X, Y and Z[1], and we shall assume that a color space XYZ, with domains D_X, D_Y and D_Z of the corresponding color components is employed. This leads to the following more specific definition:

Definition 2. *A fuzzy color $\widetilde{C}$ is a linguistic label whose semantics is represented in a color space XYZ by a normalized fuzzy subset of $D_X \times D_Y \times D_Z$.*

Notice that the above definition implies that for each fuzzy color $\widetilde{C}$ there is at least one crisp color $\mathbf{r}$ such that $\widetilde{C}(\mathbf{r}) = 1$.

In this paper, and following [11], we will define the membership function of $\widetilde{C}$ as

$$\widetilde{C}(\mathbf{c}; \mathbf{r}, S, \Omega) = f\left(\left|\overrightarrow{\mathbf{rc}}\right|; t_1^c, \ldots, t_n^c\right) \tag{1}$$

depending on three parameters: $S = \{S_1, \ldots, S_n\}$ a set of bounded surfaces in XYZ verifying $S_i \cap S_j = \emptyset \ \forall i, j$ (i.e., pairwise disjoint) and such that $Volume(S_i) \subset Volume(S_{i+1})$; $\Omega = \{\alpha_1, \ldots, \alpha_n\} \subseteq (0, 1]$, with $1 = \alpha_1 > \alpha_2 > \cdots > \alpha_n = 0$, the membership degrees associated to S verifying $\widetilde{C}(\mathbf{s}; \mathbf{r}, S, \Omega) = \alpha_i \ \forall \mathbf{s} \in S_i$; and $\mathbf{r}$ a point inside $Volume(S_1)$ that is assumed to be a crisp color representative of $\widetilde{C}$.

In Eq.1, $f : \mathbb{R} \to [0, 1]$ is a piecewise function with knots $\{t_1^c, \ldots, t_n^c\}$ verifying $f(t_i^c) = \alpha_i \in \Omega$, where these knots are calculated from the parameters $\mathbf{r}$, S and Ω as follows: $t_i^c = \left|\overrightarrow{\mathbf{rp}_i}\right|$ with $\mathbf{p}_i = S_i \cap \overline{\mathbf{rc}}$ being the intersection between the line $\overline{\mathbf{rc}}$ (straight line containing the points $\mathbf{r}$ and $\mathbf{c}$) and the surface S_i, and $\left|\overrightarrow{\mathbf{rp}_i}\right|$ the length of the vector $\overrightarrow{\mathbf{rp}_i}$.

---

[1] Although we are assuming a three dimensional color space, the proposal can be easily extended to color spaces with more components.

2.2   Fuzzy Color Space

For extending the concept of color space to the case of fuzzy colors, and assuming a fixed color space XYZ, with D_X, D_Y and D_Z being the domains of the corresponding color components, the following definition is introduced:

Definition 3. *A fuzzy color space* $\widetilde{XYZ}$ *is a set of fuzzy colors that define a partition of* $D_X \times D_Y \times D_Z$.

As we introduced in the previous section (see Eq.1), each fuzzy color $\widetilde{C}_i \in \widetilde{XYZ}$ will have associated a representative crisp color $\mathbf{r}_i$. Therefore, for defining our fuzzy color space, a set of representative crisp colors $R = \{\mathbf{r}_1, \ldots, \mathbf{r}_n\}$ is needed.

For defining each fuzzy color $\widetilde{C}_i \in \widetilde{XYZ}$, we also need to fix the set of surfaces S_i and the associated memberships degrees Ω_i (see Eq.1). In this paper, we have focused on the case of convex surfaces defined as a polyhedra (i.e, a set of faces). Concretely, three surfaces $S_i = \{S_1^i, S_2^i, S_3^i\}$ have been used for each fuzzy color $\widetilde{C}_i$ with $\Omega_i = \{1, 0.5, 0\} \forall i$.

To obtain $S_2^i \in S_i \ \forall i$, a Voronoi diagram has been calculated [10] with R as centroid points . As results, a crisp partition of the color domain given by convex volumes is obtained (each volume will define a Voronoi cell). The surfaces of the Voronoi cells will define the surfaces $S_2^i \in S_i \ \forall i$. Once S_2^i is obtained, the surface S_1^i (resp. S_3^i) is calculated as a scaled surface of S_2^i with scale factor of 0.5 (resp. 1.5). For more details about the parameter values which define each polyhedra, see [11].

3   Dominance-Based Fuzzy Color Descriptor

For describing semantically an image, the dominant colors will be used. In this section, a Fuzzy Descriptor for dominant colors is proposed (section 3.2) on the basis of the dominance degree of a given color (section 3.1).

3.1   Dominant Fuzzy Colors

Intuitively, a color is dominant to the extent it appears frequently in a given image. As it is well known in the computer vision field, the histogram is a powerful tool for measuring the frequency in which a property appears in an image. The histogram is a function $h(x) = n_x$ where x is a pixel property (grey level, color, texture value, etc.) and n_x is the number of pixels in the image having the property x. It is common to normalize a histogram by dividing each of its values by the total number of pixels, obtaining an estimate of the probability of occurrence of x.

Working with fuzzy properties suggests to extend the notion of histogram to "fuzzy histogram". In this sense, a fuzzy histogram will give us information about the frequency of each fuzzy color. In this paper, the cardinality $h(x)$ of the fuzzy subset of pixels having property x, P_x, will be measured by means of the fuzzy cardinality ED [2] divided by the number of pixels in the image, which is a fuzzy subset of $[0, 1]$ calculated as follows:

$$h(x) = \sum_{\alpha_i \in \Lambda(P_x)} \frac{(\alpha_i - \alpha_{i+1})}{|(P_x)_{\alpha_i}|/N} \tag{2}$$

with $\Lambda(P_x) = \{1 = \alpha_1 > \alpha_2 > \cdots > \alpha_n > 0\}$ the level set of P_x union $\{1\}$ in case P_x is not normalized, and considering $\alpha_{n+1} = 0$.

Using the information given by the histogram, we will measure the "dominance" of a color fuzzy set. Dominance is an imprecise concept, i.e., it is possible in general to find colors that are clearly dominant, colors that are clearly not dominant, and colors that are dominant to a certain degree. On the other hand, the degree of dominance depends on the percentage of pixels where the color appears. Hence, it seems natural to model the idea of dominance by means of a fuzzy set over the percentages, i.e., a fuzzy quantifier defined by a non-decreasing subset of the real interval $[0, 1]$. More specifically, we define the fuzzy subset "Dominant", noted as *Dom*, as follows:

$$Dom(\alpha) = \begin{cases} 0 & \alpha \leq u_1 \\ \frac{\alpha - u_1}{u_2 - u_1} & u_1 \leq \alpha \leq u_2 \\ 1 & \alpha \geq u_2 \end{cases} \tag{3}$$

for each $\alpha \in [0, 1]$, where u_1 and u_2 are two parameters such that $0 \leq u_1 < u_2 \leq 1$.

Finally, in order to know whether a certain color is dominant in the image, we have to calculate the compatibility between its frequency and the quantifier defining the notion of "Dominant". For instance, in the case of a crisp color, its frequency is a crisp number x and its dominance is $Dom(x)$. In the case of a fuzzy color $\widetilde{C}$, its frequency is a fuzzy subset of the rational numbers calculated as in Eq. (2) and noted $h(\widetilde{C})$. In order to obtain the dominance of a fuzzy color $\widetilde{C}$ in an image the compatibility between the fuzzy cardinality of the fuzzy set of pixels $P_{\widetilde{C}}$ with color $\widetilde{C}$ in the image, and the quantifier *Dom*, corresponds to the evaluation of the quantified sentence "*Dom* of pixels in the image are $P_{\widetilde{C}}$". We shall use the method *GD* introduced in [3] as follows:

$$GD_{Dom}(P_{\widetilde{C}}) = \sum_{\alpha \in Supp(h(\widetilde{C}))} h(\widetilde{C})(\alpha) \times Dom(\alpha) \tag{4}$$

3.2 Dominance-Based Fuzzy Descriptors

On the basis of the dominance of colors, a new dominance image descriptor is proposed.

Definition 4. *Let $\mathscr{C}$ a finite reference universe of color fuzzy sets. We define the* Fuzzy Dominant Color Descriptor *as the fuzzy set*

$$FDCD = \sum_{\widetilde{C} \in \mathscr{C}} GD_{Dom}(\widetilde{C})/\widetilde{C} \tag{5}$$

with $GD_{Dom}(\widetilde{C})$ being the dominance degree of $\widetilde{C}$ given by Eq. 4.

4 Matching Operators

Fuzzy operators over fuzzy descriptors are needed in many practical applications. In this section, a "Fuzzy inclusion operator" (section 4.1) is proposed.

4.1   Fuzzy Inclusion Operator

Given two Fuzzy Dominant Color Descriptors, $FDCD^i$ and $FDCD^j$, the operator presented in this section calculates the inclusion degree of $FDCD^i$ in $FDCD^j$. The calculus is done using the *Resemblance Driven Inclusion Degree* introduced in [8], which computes the inclusion degree of two fuzzy sets whose elements are imprecise.

Definition 5. *Let $FDCD^i$ and $FDCD^j$ be two Fuzzy Dominant Color Descriptors defined over a finite reference universe of fuzzy sets $\mathscr{P}$, $FDCD^i(x)$ and $FDCD^j(x)$ the membership functions of these fuzzy sets, S the resemblance relation defined over the elements of $\mathscr{P}$, $\otimes$ be a t-norm, and I an implication operator. The inclusion degree of $FDCD^i$ in $FDCD^j$ driven by the resemblance relation S is calculated as follows:*

$$\Theta_S(FDCD^j | FDCD^i) = \min_{x \in \mathscr{P}} \max_{y \in \mathscr{P}} \theta_{i,j,S}(x,y) \tag{6}$$

where

$$\theta_{i,j,S}(x,y) = \otimes(I(FDCD^i(x), FDCD^j(y)), S(x,y)) \tag{7}$$

In this paper we use the minimum as t-norm, the compatibility as the resemblance relation S, and as implication operator the one defined in equation 8.

$$I(x,y) = \begin{cases} 1 & \text{if } x \leq y \\ y/x \text{ otherwise} \end{cases} \tag{8}$$

5   Results

In this section, we will use the proposed Fuzzy Dominant Color Descriptor (FDCD) and the inclusion operator (section 4.1) in order to define image retrieval queries based on dominant colors. Specifically, query refinement on results provided by the *Flickr* system using FDCD over an user customized fuzzy color space.

5.1   Customized Color Space: An Example

We have used an user customized fuzzy color space for fruit colors, designed on the basis of a collection of crisp colors and color names provided by the user, following the methodology described in section 2. A user outlined regions in several images and labeled each regions with the color name of a fruit. The crisp color corresponding to each region has been obtained as the centroid of all the colors in the corresponding regions in the RGB color space. For each fruit, 10 images have been used and one or more regions in each image considered as representative of each fruit have been outlined by the user. The results of this experiment are shown in table 1. (only two images for each fruit, not regions chosen by the user, are shown).

Table 1. Collection of images and corresponding representative colors. Selected regions for every image are not shown.

Color Name	Images	Representative color	RGB
banana			[254.0, 213.0, 0.0]
blackberry			[38.0, 42.0, 41.0]
green apple			[188.0, 227.0, 60.0]
lemon			[254.0, 234.0, 101.0]
orange			[255.0, 115.0, 1.0]
plum			[145.0, 145.0, 197.0]
raspberry			[253.0, 108.0, 128.0]
red apple			[162.0, 29.0, 34.0]
strawberry			[204.0, 12.0, 11.0]

(a)

(b)

Fig. 1. Retrieval results on *Flickr* using keyword "banana": (a) results from Flickr only with keyword and (b) query refinement using $FDCD^{query} = 1/banana$

(a)

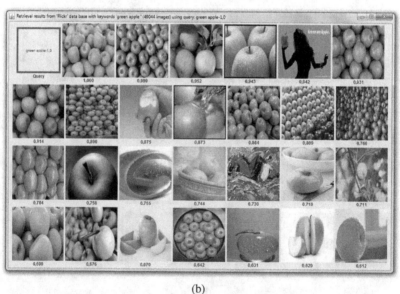

(b)

Fig. 2. Retrieval results on *Flickr* using keyword "green apple": (a) results from Flickr only with keyword and (b) query refinement using $FDCD^{query} = 1/greenapple$

5.2 Query Refinement Examples

In this illustrative application we have used the *Flickr* online image database. We have performed queries using color names as keywords. Then, we have refined the result of these queries using the Fuzzy Dominant Color Descriptor (FDCD) for each image and the fuzzy inclusion operator on the basis of the customized fuzzy color space just mentioned. Nowadays, there are more than 6 billion images hosted in *Flickr*, and our query has been applied to a collection of around 100000 images provided by the *Flickr* API [14] as the most interesting photos for a certain month.

Figure 1a shows the retrieval results for the query *banana*, ranked by relevance by *Flickr*. Figure 1b shows the result of the query refinement by ranking the previous result on the basis of our inclusion operator taking as criterion the dominance of the customized fuzzy color *banana* in the image. More specifically, we calculate the inclusion degree of the descriptor $1/banana$. This is equivalent in this particular case to calculate the degree of dominance of the color *banana* in the whole image. We have employed a definition of dominance based on a linguistic quantifier with parameters $u_1 = 0.1$ and $u_2 = 0.25$. Using this approach, the subjectivity of the color *banana* and the imprecision of the dominance in an image are considered and in our opinion, the refinement provides a better ranking of images.

A similar experiment has been performed using the color name *green apple*. Figure 2a shows the retrieval results for the query *green apple*, ranked by relevance by *Flickr*. Figure 2b shows the result of the query refinement using the dominance of the customized fuzzy color *green apple* in the image, using the same procedure employed for the previous case. Again, we consider that the refinement provides a better ranking of images than that of *Flickr*.

Please note that we do not claim that dominance of the previous customized fuzzy colors is enough on its own in order to recognize the presence of certain fruits in images. Other objects than bananas and/or green apples, but having the same colors, may be recognized when dominance is consider alone. However, they are very good for refining queries based on color labels, as those performed by *Flickr* and many other image retrieval systems.

6 Conclusions

In this paper, a new Fuzzy Dominant Color Descriptor has been proposed. This descriptor has been defined as a fuzzy set over a finite universe of fuzzy colors, in which membership degrees represent the dominance of each color. The color fuzzy sets have been defined taking into account the relationship between the color representation in computers and its human perception. In addition, fuzzy operators over the new descriptor have been proposed. We have illustrated the usefulness of our proposals with an application in image retrieval, specifically query refinement on results provided by the *Flickr* system.

Several future work related to this will be to apply the descriptor in the linguistic description of images, and the combination with other customized fuzzy concepts related to color as well as other basic features of images.

References

1. Chang, S.F., Sikora, T., Puri, A.: Overview of the mpeg-7 standard. IEEE Transaction on Circuits and Systems for Video Technology 11, 688–695 (2001)
2. Delgado, M., Martín-Bautista, M.J., Sánchez, D., Vila, M.A.: A probabilistic definition of a nonconvex fuzzy cardinality. Fuzzy Sets and Systems 126(2), 41–54 (2002)
3. Delgado, M., Sánchez, D., Vila, M.A.: Fuzzy cardinality based evaluation of quantified sentences. International Journal of Approximate Reasoning 23, 23–66 (2000)
4. Gabbouj, M., Birinci, M., Kiranyaz, S.: Perceptual color descriptor based on a spatial distribution model: Proximity histograms. In: International Conference on Multimedia Computing and Systems, ICMCS 2009, pp. 144–149 (2009)
5. Huang, Z., Chan, P.P.K., Ng, W.W.Y., Yeung, D.S.: Content-based image retrieval using color moment and gabor texture feature. In: 2010 International Conference on Machine Learning and Cybernetics (ICMLC), vol. 2, pp. 719–724 (2010)
6. Li, A., Bao, X.: Extracting image dominant color features based on region growing. In: 2010 International Conference on Web Information Systems and Mining, WISM 2012, vol. 2, pp. 120–123 (2010)
7. Islam, M.M., Zhang, D., Lu, G.: Automatic categorization of image regions using dominant color based vector quantization. In: Computing: Techniques and Applications, DICTA 2008, Digital Image, pp. 191–198 (2008)
8. Marín, N., Medina, J.M., Pons, O., Sánchez, D., Vila, M.A.: Complex object comparison in a fuzzy context. Information and Software Technology 45(7), 431–444 (2003)
9. Negrel, R., Picard, D., Gosselin, P.: Web scale image retrieval using compact tensor aggregation of visual descriptors. IEEE MultiMedia (99), 1 (2013)
10. Preparata, F.P., Shamos, M.I.: Computational geometry: algorithms and applications, 2nd edn. Springer, New York (1988)
11. Soto-Hidalgo, J.M., Chamorro-Martinez, J., Sanchez, D.: A new approach for defining a fuzzy color space. In: IEEE World Congress on Computational Intelligence (WCCI 2010), pp. 292–297 (July 2010)
12. van de Sande, K.E.A., Gevers, T., Snoek, C.G.M.: Evaluating color descriptors for object and scene recognition. IEEE Transactions on Pattern Analysis and Machine Intelligence 32(9), 1582–1596 (2010)
13. Wu, J., Rehg, J.M.: Centrist: A visual descriptor for scene categorization. IEEE Transactions on Pattern Analysis and Machine Intelligence 33(8), 1489–1501 (2011)
14. Yahoo! Flickr api. a programmers place to create applications @ONLINE (2013)
15. Yamada, A., Pickering, M., Jeannin, S., Jens, L.C.: Mpeg-7: Visual part of experimentation model version 9.0. ISO/IEC JTC1/SC29/WG11/N3914 (2001)
16. Yang, N., Chang, W., Kuo, C., Li, T.: A fast mpeg-7 dominant color extraction with new similarity measure for image retrieval. Journal of Visual Communication and Image Representation 19(2), 92–105 (2008)

Linguistic Descriptions: Their Structure and Applications

Vilém Novák, Martin Štěpnička, and Jiří Kupka

Centre of Excellence IT4Innovations,
Division of the University of Ostrava,
Institute for Research and Applications of Fuzzy Modeling,
30. dubna 22, 701 03 Ostrava 1, Czech Republic
{Vilem.Novak,Martin.Stepnicka,Jiri.Kupka}@osu.cz
http://irafm.osu.cz/

Abstract. In this paper, we provide a brief survey of the main theoretical and conceptual principles of methods that use the, so called, linguistic descriptions and thus, belong to the broad area of methods encapsulated under the term *modeling with words*. The theoretical frame is fuzzy natural logic — an extension of mathematical fuzzy logic consisting of several constituents. In this paper, we will deal with formal logical theory of evaluative linguistic expressions and the related concepts of linguistic description and perception-based logical deduction. Furthermore, we mention some applications and highlight two of them: forecasting and linguistic analysis of time series and linguistic associations mining.

1 Introduction — Necessity to Learn from Linguists and Logicians

This is an overview paper focusing on linguistic descriptions and expressions forming them, discussion of inference method on the basis of linguistic description, learning linguistic descriptions from data and demonstration of few applications in time series analysis and mining linguistic associations.

There are more meanings of the term "linguistic description". In linguistics, it is the task to analyze and describe objectively how language is spoken by a group of people. In this paper, we understand by it a piece of text taken as a set of sentences of natural language that characterizes some situation, strategy of behavior, control of some process, etc., and that provides rules (instructions) on how to decide or what to do. People are able to understand such text and follow practically the rules contained in it. Of course, such wide understanding assumes detailed analysis of natural language as a whole. For technical purposes, however, we do not need to consider arbitrary sentences but only a special part that includes linguistic expressions used for description of a certain kind of data.

To be able to work and apply linguistic descriptions in technical applications, we need a mathematical model of their meaning. We argue that this requires a careful analysis of the class of used expressions and their syntactical structure and also a logical analysis of their semantical properties. What are the linguistic

H.L. Larsen et al. (Eds.): FQAS 2013, LNAI 8132, pp. 209–220, 2013.

expressions used in linguistic descriptions? In the literature, one can meet the term "linguistic label" (c.f. [1,2,3]). A careful reading reveals that the class of linguistic expressions in concern is covered by a quite wide and very interesting class of *evaluative linguistic expressions*. This class, on one hand, does not have too complicated structure to be formalized but, on the other hand, it is a powerful constituent of natural language that enables us to express many kinds of evaluations of phenomena occurring in the surrounding reality.

Below, we will briefly characterize evaluative expressions from a linguistic point of view and outline a mathematical model of their meaning that is based on logical analysis and that proved to have many successful applications. We focus especially on the results obtained on the basis of the theory of the *fuzzy natural logic* that is extension of mathematical fuzzy logic in narrow sense (FLn) (cf. [4,5]).

2   Theory of Evaluative Linguistic Expressions

2.1   Grammatical Structure of Evaluative Expressions

Simple *evaluative linguistic expressions* (possibly with signs) have the general form

$$\langle\text{linguistic hedge}\rangle\langle\text{TE-adjective}\rangle \tag{1}$$

where ⟨TE-adjective⟩ represents a class of special adjectives (TE stands for "trichotomous evaluative") that includes *gradable adjectives* (big, cold, deep, fast, friendly, happy, high, hot, important, long, popular, rich, strong, tall, warm, weak, young), *evaluative adjectives* (good, bad, clever, stupid, ugly, etc.), but also adjectives such as *left, middle, medium*, etc. For the TE-adjectives it is characteristic that they can be grouped to form a *fundamental evaluative trichotomy* consisting of two antonyms and a middle member, for example *low, medium, high*; *clever, average, stupid*; *good, normal, bad*, etc. The triple of adjectives *small, medium, big* will further be taken as canonical.

The ⟨linguistic hedge⟩ represents a class of adverbial modifications that includes a class of *intensifying adverbs* such as "very, roughly, approximately, significantly", etc. Two basic kinds of linguistic hedges can be distinguished in (1): *narrowing* hedges, for example, "extremely, significantly, very" and *widening* ones, for example "more or less, roughly, quite roughly, very roughly". Note that narrowing hedges make the meaning of the whole expression more precise while widening ones have the opposite effect. Thus, "very small" is more precise than "small", which, on the other hand, is more precise (more specific) than "roughly small". The situation when ⟨linguistic hedge⟩ is not present at the surface level is dealt with as a presence of *empty linguistic hedge*. In other words, all simple evaluative expressions have the form (1), including examples such as "small, long, deep", etc.

Simple evaluative expressions of the form (1) can be combined using logical connectives (usually "and" and "or") to obtain *compound* ones. Let us emphasize, however, that they cannot be formed according to rules of boolean logic, i.e. we cannot form compound evaluative expressions as special boolean normal forms since the resulting expressions might loose their meaning. Combination of syntactic and semantic rules makes the class of evaluative expressions narrower. This concerns also the use of linguistic hedges since, e.g. "very medium" has no meaning. Even more difficult is the use of negation *not* where one faces a special linguistic phenomenon called *topic-focus articulation* (c.f. [6]). For example, *not very small* has (at least) two different meanings: either "(not very) small" or "not (very small)".

Evaluative linguistic expressions occur in the position of adjectival phrases characterizing features of some objects and are used either in predicative or attributive role. Example of the first use is " The man is very stupid". Example of the second one is " The very stupid man climbed a tree". The purpose of the first sentence is simply to communicate a particular quality of the sentence subject. However, the purpose of the second sentence is primarily to tell us what the subject did, i.e., climbed a tree; that the subject is very stupid is a secondary consideration.

In this paper, we will consider only predicative use and, therefore, expressions of the form "⟨noun⟩ is $\mathcal{A}$" where $\mathcal{A}$ is an evaluative expression will be called *evaluative (linguistic) predications*; for example "temperature is high, speed is very low", etc.. The ⟨noun⟩ is in technical applications often replaced by some variable to form a linguistic predication of the form "X is $\mathcal{A}$".

2.2   Semantics of Simple Evaluative Expressions

In any model of the semantics of linguistic expressions, we must distinguish their intension and extensions in various possible worlds (cf. [7]). A *possible world* is a state of our world at a given time moment and place. A possible world can also be understood as a *particular context* in which the linguistic expression is used. In this paper, we will prefer the term *linguistic context* (or simply *context*).

Intension of a linguistic expression is an abstract construction which conveys a property denoted by the linguistic expression. Thus, linguistic expressions are names of intensions. Intension is invariant with respect to various contexts.

Extension of a linguistic expression is a class of objects determined by its intension in a given context (possible world). Thus, it depends on a particular context of use and it is changed whenever the context (time, place) is changed. For example, the expression "deep" is the name of an intension being a certain property of depth, which in a concrete context may mean 1 cm when a beetle needs to cross a puddle, 3 m in a small lake, but 3 km or more in the ocean.

To construct the meaning of evaluative linguistic expressions, we must follow several principles. First, we must realize that their extensions are classes of elements delineated on *nonempty, ordered and bounded scales*. The scale is in each context vaguely partitioned to form the fundamental evaluative trichotomy. Thus, any element from the scale is contained in the extension of at most two neighboring expressions from this trichotomy. Each scale is determined by three distinguished points: *leftmost, central* and *rightmost*.

Evaluative expressions form pairs of antonyms characterizing opposite sides of scales. Sets of simple evaluative expressions differing in hedges are linearly ordered in the following sense: if an element of the scale falls in the extension of the "smaller" evaluative expression then it falls in the extension of all "larger" ones (provided that they exist). For example, each *very small value* is at the same time *small*, and at the same time *roughly small*, etc.

Finally, for each evaluative expression and in each context there exists a limit typical element. Extension of the expression falls inside a horizon running from it in the sense of the ordering of the scale. The horizon is determined in analogy with *sorites paradox* (c.f. [5,8]).

The set of contexts (possible worlds) for evaluative expressions can be mathematically modeled as

$$W = \{\langle v_L, v_C, v_R \rangle \mid v_L, v_C, v_R \in [0, \infty) \text{ and } v_L < v_C < v_R\}. \quad (2)$$

If $w \in W$ is a context, $w = \langle v_L, v_C, v_R \rangle$, then $u \in w$ means that $u \in [v_L, v_R]$ and we say that u *belongs to the context* w. Let us mention that extensions of the evaluative expressions characterizing small values lay between the points v_L, v_C and those characterizing big values lay between v_C, v_R. The point v_C is the "most typical" medium value and it need not lay in the middle of the interval $[v_L, v_R]$.

Let $\mathcal{A}$ be an evaluative expression (i.e. abstract expression or evaluative predication). Then its *intension* is a function

$$\text{Int}(\mathcal{A}) : W \to \mathcal{F}(\mathbb{R}) \quad (3)$$

where $\mathcal{F}(\mathbb{R})$ denotes a set of all fuzzy sets over $\mathbb{R}$. Given a context $w \in W$, an *extension* of $\mathcal{A}$ is a *fuzzy set* $A(w)$ on $\mathbb{R}$, i.e. $A(w) \subseteq \mathbb{R}$. Thus, the extension of an evaluative expression $\mathcal{A}$ in a context w is a fuzzy set (cf. Figure 1)

$$\text{Ext}_w(\mathcal{A}) = \text{Int}(\mathcal{A})(w) \subseteq [v_L, v_R] \subseteq \mathbb{R}. \quad (4)$$

Extensions of evaluative expressions $\mathcal{A}$ are fuzzy sets the shapes of which are determined on the basis of logical analysis of their meaning. The consequence is the following: given a context $w = \langle v_L, v_S, v_R \rangle \in W$. Then $\text{Ext}_w(\mathcal{A}) = \nu \circ H(w)$ where H is one of the linear functions $L, R, M : W \times \mathbb{R} \to [0,1]$ representing a left (right, middle) horizon in each context $w \in W$, and $\nu : [0,1] \to [0,1]$ represents a hedge (for the details, see [8]). Hence, we obtain three classes of intensions:

(i) S-intensions: $Sm_\nu : W \to \mathcal{F}(\mathbb{R})$ where $Sm_\nu(w) = L(w) \circ \nu \subseteq [v_L, v_S]$ for each $w \in W$,

(ii) M-intensions: $Me_\nu : W \to \mathcal{F}(\mathbb{R})$ where $Me_\nu(w) = M(w) \circ \nu \subseteq (v_L, v_R)$ for each $w \in W$,

(iii) B-intensions: $Bi_\nu(w) = R(w) \circ \nu \subseteq [v_S, v_R]$ for each $w \in W$. For the detailed justification of this model of the meaning of evaluative expressions — see [8].

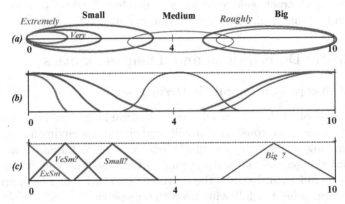

Fig. 1. Extensions of evaluative predications "X is *extremely small*", "X is *very small*", "X is *small*", "X is *medium*", "X is *big*" and "X is *roughly big*" in the context $\langle 0, 4, 10 \rangle$. Part (a) shows intuitive understanding of these expressions; (b) shows mathematical model of their extensions using fuzzy sets; (c) demonstrates incorrect model of their extensions using triangular fuzzy sets. Indeed, for example the value 0.1 is "extremely small" in a degree close to 1, but "very small" in a degree close to 0 and for sure not small at all. But this is apparently wrong.

Example 1. Let us consider "age" which is a typical feature of people. Then its context in Europe can be $w = \langle 0, 45, 100 \rangle$ where people below 45 years are "young" in some degree, those around 45 are "middle aged", and those over 100 are surely "old". Ages below 45 are "young" in various degrees, those around 45 are "middle aged" in various degrees, and those over 45 are "old" in various degrees. Of course, there are different contexts in different areas and also periods of time (e.g., age of people in Europe in medieval times was different from the above one).

Shapes of few possible extensions are drawn in Figure 1. One can see in part (a) that, for example, *small* are values on the left part of the scale (of course, relative to the given context). Then *very small* is proper subclass of small values and *extremely small* a proper subclass of the latter. But surely, every extremely small value is both very small as well as small and similarly for very small values. For precise definitions and explanation see [8].

We argue that extensions of all evaluative expressions can be constructed on the basis of the same principle (though, of course, they may differ for various expressions). In other words, we argue that for each kind of evaluative expression and in each context there exists a linearly ordered scale, inside which extensions of the characteristic shapes as in Figure 1(b) can be constructed. In case that the TE-expression is an abstract evaluative adjective, such as "good, smart, ugly", etc., the scale is formed by certain abstract degrees that can in mathematical model be formed, e.g., by the interval [0, 1]. Derivation of values in the scale can be in specific cases quite complicated, determined by many factors. For example, "good car" can be result of several factors, such as *speed, comfort, size, reliability,* etc. But still, an abstract scale of "degrees of goodness" exists; we can obtain it mathematically, using, e.g., some aggregation operator (cf. [9]).

3   Linguistic Descriptions and Their Semantics

3.1   The Concept of a Linguistic Description

People usually express their knowledge using natural language. They form a linguistic description which consists of conditional clauses specifying a certain kind of relation among various phenomena. When we want to formalize we realize, however, that the genuine natural language is unnecessarily complicated to be employed in its full form. Luckily, the contained knowledge can be expressed in a simplified form using the following fuzzy/linguistic rules:

$$\text{IF } X \text{ is } \mathcal{A} \text{ THEN } Y \text{ is } \mathcal{B} \tag{5}$$

where $\mathcal{A}, \mathcal{B}$ are evaluative linguistic expressions. The antecedent variable X attains values from some universe U and the consequent variable Y values from some universe V. Of course, the antecedent of (5) may consist of more evaluative predications joined by "AND". Let us emphasize that (5) is a *conditional clause of natural language* and, therefore, we will call it fuzzy/linguistic IF-THEN rule.

We will make the concept of linguistic description even narrower. Namely, a *linguistic description* is defined as a finite set of fuzzy/linguistic IF-THEN rules with common X and Y:

$$\text{IF } X \text{ is } \mathcal{A}_1 \text{ THEN } Y \text{ is } \mathcal{B}_1$$
$$\dots\dots\dots\dots\dots\dots\dots\dots\dots \tag{6}$$
$$\text{IF } X \text{ is } \mathcal{A}_m \text{ THEN } Y \text{ is } \mathcal{B}_m$$

Note that the linguistic description (6) is, in fact, a *special piece of text*. Then one or more linguistic descriptions may form a *knowledge base* which contains a complex knowledge about a situation such as decision one, characterization of behavior of some system, etc.

3.2   Semantics of Linguistic Description

Let $\mathcal{R}$ be a fuzzy IF-THEN rule (5) and $\text{Int}(X \text{ is } \mathcal{A})$, $\text{Int}(Y \text{ is } \mathcal{B})$ be intensions (cf. (3)). Then the intension of $\mathcal{R}$ is a function $\text{Int}(\mathcal{R}) : W \times W \to \mathcal{F}(\mathbb{R} \times \mathbb{R})$ assigning to each couple of contexts $w, w' \in W$ a fuzzy relation

$$\text{Int}(\mathcal{R})(w, w') = \text{Int}(X \text{ is } \mathcal{A})(w) \Rightarrow \text{Int}(Y \text{ is } \mathcal{B})(w'), \qquad w, w' \in W, \tag{7}$$

where the formula on the right-hand side of (7) represents a fuzzy relation composed pointwise from the extensions $\text{Int}(X \text{ is } \mathcal{A})(w)$, $\text{Int}(Y \text{ is } \mathcal{B})(w')$ using some implication operator $\Rightarrow$. Extensive practical experiences and various theoretical considerations lead to the assumption that $\Rightarrow$ is the Łukasiewicz implication. Formula (7) provides rules for computation of the *extension of* $\mathcal{R}$ in the contexts w, w' which is a fuzzy relation defined by $\text{Ext}_{\langle w,w' \rangle}(\mathcal{R})(x, y') = \text{Int}(X \text{ is } \mathcal{A})(w, x) \Rightarrow \text{Int}(Y \text{ is } \mathcal{B})(w', y)$ for all $x \in w, y \in w'$.

An important linguistic phenomenon that cannot be ignored is that of topic-focus articulation (c.f. [6]). In the case of linguistic descriptions, the *topic* is formed by a set of intensions $\{\text{Int}(X \text{ is } \mathcal{A}_j) \mid j = 1, \ldots, m\}$ and *focus* by $\{\text{Int}(Y \text{ is } \mathcal{B}_j) \mid j = 1, \ldots, m\}$. The phenomenon of topic-focus articulation plays, besides others, an important role in the inference method called perception-based logical deduction described below.

3.3   Learning of Linguistic Description from Data

The idea of learning linguistic description is based on the concept of perception as considered by L. A. Zadeh. It must be emphasized, however, that we do not consider a psychological term but rather a simple technical term: by *perception* we understand an evaluative expression assigned to the given value in the given context. It can be understood as a linguistic characterization of certain kind of "measurement" done by people in a concrete situation. Mathematically, we define a special function of *local perception* assigning to each value $u \in w$ for $w \in W$ an intension

$$\text{LPerc}^{\text{LD}}(u, w) = \text{Int}(X \text{ is } \mathcal{A}) \tag{8}$$

of the *sharpest* evaluative predication where by sharper we mean that the meaning of a sharper evaluative expression is more specific than that of less sharp expression. For example, the meaning of *very small* is sharper than that of *small*. Details can be found in [8]. Moreover, the value $u \in w$ must be the *most typical* element for the extension $\text{Ext}_w(X \text{ is } Ev_{\nu,j}^X)$. This principle enables us to generate linguistic descriptions (6) from data. Possible ways, how to apply it, are described below.

3.4   Perception-Based Logical Deduction

Both perception and linguistic description provide us with enough information and so, we can derive a conclusion on the basis of them. The procedure is called *perception-based logical deduction* (PbLD). It was in detail described in [10,11] and elsewhere and so, we will recall here only the very basic principles.

Let us consider a simple example of linguistic description:

$$\mathcal{R}_1 := \text{IF } X \text{ is } small \text{ THEN } Y \text{ is } big$$
$$\mathcal{R}_2 := \text{IF } X \text{ is } medium \text{ THEN } Y \text{ is } very \ big$$
$$\mathcal{R}_3 := \text{IF } X \text{ is } big \text{ THEN } Y \text{ is } small.$$

Furthermore, let contexts of the respective variables X, Y be $w = w' = \langle 0, 0.4, 1 \rangle$. Then small values are some values around 0.2 (and smaller) and big ones some values around 0.7 (and bigger). We know from the linguistic description that small input values correspond to big output ones and vice-versa. Therefore, given an input, e.g. $X = 0.2$, we expect the result $Y \approx 0.7$ due to the rule $\mathcal{R}_1$. The reason is that with respect to the above linguistic description, our perception of 0.2 (in the given context) is "small", and thus, in this case the output value of Y should be "big". Similarly, for $X = 0.75$ we expect the result $Y \approx 0.15$ due to the rule $\mathcal{R}_3$. For the value of X around 0.4 (a typical medium), the value of Y should be close to 1 because we expect "very big" output. Let us emphasize that conclusion is independent on the chosen context, i.e. when changing it, the conclusion in general will be the same. Of course, the output values will be different but again corresponding to perceptions of *big, roughly big, medium,* and *very small* in the new context. It is important to emphasize that PbLD works locally. This means that though vague, the rules are distinguished one from another one because they bring different local information about the phenomena which must be complied with. PbLD works very well in many kinds of applications.

4   Linguistic Descriptions in the Analysis and Forecasting of Time Series

There are plenty of applications of linguistic descriptions in various areas, such as automatic control [12], managerial decision making [13] and elsewhere. In this section, we will briefly describe a recent application to time series analysis and forecasting. The fundamental method is the *fuzzy transform* (F-transform) introduced in [14]. It is a technique that transforms a given real bounded continuous function $f : [a, b] \to [c, d]$ into a space of finite n-dimensional vectors $\mathbf{F}[f] = (F_1[f], \ldots, F_n[f])$ of real values called components. The vector $\mathbf{F}[f]$ represents the original function and may replace it in complex computations. The *inverse transform* transforms $\mathbf{F}[f]$ back into a continuous function $\hat{f}$ that approximates the original function f.

The given time series $X(t), t \in Q$ where Q is a finite set of natural numbers, is decomposed into a trend-cycle TC, a seasonal component $S(t)$ and an error component. The F-transform enables us to extract the trend-cycle TC with high fidelity (see [14,15,16]).

As noted in [17], the most important contribution of fuzzy systems is interpretability and transparency. A natural step to this goal are automatically generated linguistic descriptions from the data applied to the vector of components $(F_1[X], \ldots, F_T[X])$. On the basis of it and using the perception-based logical deduction, we can predict future components $F_{T+1}[X], \ldots, F_{T+h}[X]$. Then, applying the inverse F-transform to the latter, we obtain forecast of the trend-cycle.

Let us remark that the learned linguistic description contains autoregressive rules consisting of the components as well as of their first- and/or second-order differences: $\Delta F_i[X] = F_i[X] - F_{i-1}[X]$ and $\Delta^2 F_i[X] = \Delta F_i[X] - \Delta F_{i-1}[X]$. For example, a linguistic description generated from the well-known *pigs* series related (numbers of pigs slaughtered in state Victoria — R.J. Hyndman's Time Series Data Library) contained the following fuzzy rule:

IF Y_i is *big* AND ΔY_i is *quite roughly small* AND ΔY_{i-1} is *extremelly small*
THEN ΔY_{i+1} is *very roughly small*

It may be read as follows: "If the number of pigs slaughtered in the current year is big and the biannual increment is quite roughly small and the previous biannual increment was also positive with extremely small strength then the up coming biannual increment will be very roughly small." We argue, that such an understandable description may be very beneficial for further decision-making processes.

Furthermore, using the perception-based logical deduction and a generated linguistic description, we may forecast future F-transform components. The seasonal components may be forecasted either statistically [15,16] or with help of advanced computational intelligence methods [18].

5   Mining Linguistic Associations

Evaluative linguistic expressions can be effectively applied also in mining of associations among data characterized linguistically. Therefore, we will call them *linguistic associations*. Our method (originally published in [19]) allows us to search for linguistic associations from (usually two-dimensional) numerical data set.

5.1   The Original Method

Consider a data-set $\mathcal{D}$ of the following form

	X_1	X_2	$\ldots$	X_k
o_1	a_{11}	a_{12}	$\ldots$	a_{1k}
o_2	a_{21}	a_{22}	$\ldots$	a_{2k}
$\vdots$	$\vdots$	$\vdots$	$\ddots$	$\vdots$
o_m	a_{m1}	a_{m2}	$\ldots$	a_{mk}

where any real number $a_{ij} \in \mathbb{R}$ is a value of jth *attribute* (property) X_j on ith *object* (observation, transaction) o_i. To apply our method it is necessary to specify a meaningful *context* w_j of every attribute X_j. Then, according to Section 2, suitable evaluative linguistic expressions can be considered for all attributes.

Example 2. Let us consider a data set of medical records with an attribute X_j representing "age". Then as in Example 1 we can take $w = \langle 0, 45, 100 \rangle$ as its context. This choice seems to be suitable for patients o_i currently living in Europe.

When contexts of all attributes are specified, we are able to work with linguistic predications of the form $(X_j$ is $A_j)$, where A_j is any simple evaluative expression, and to combine them among attributes by using conjunctions. Finally, for two disjoint sets of attributes $\{Y_1, \ldots, Y_p\}$, $\{Z_1, \ldots, Z_q\}$, our goal is to search for associations of the form $\mathcal{C} \sim \mathcal{D}$, where

$$\mathcal{C} = \mathsf{AND}_{j=1}^{p} \, (Y_j \text{ is } A_j), \quad \mathcal{D} = \mathsf{AND}_{j=1}^{q} \, (Z_j \text{ is } B_j),$$

and $\sim$ expresses a relationship between the *antecedent* $\mathcal{C}$ and *consequent* $\mathcal{D}$ (for details and references see the text below).

To find linguistic associations in $\mathcal{D}$, two steps were proposed in the original method. In the first step, the function LPerc in (8) assigning the most typical linguistic evaluative predication $A_{i,j}$ to a pair (a_{ij}, w_j) is applied. By using this function, the numerical data set $\mathcal{D}$ is transformed into a set containing (evaluative) linguistic expressions instead of numbers:

$$
\begin{array}{c|cccc}
 & X_1 & X_2 & \ldots & X_k \\
\hline
o_1 & \mathcal{A}_{1,1} & \mathcal{A}_{1,2} & \ldots & \mathcal{A}_{1,k} \\
o_2 & \mathcal{A}_{2,1} & \mathcal{A}_{2,2} & \ldots & \mathcal{A}_{2,k} \\
\vdots & \vdots & \vdots & \ddots & \vdots \\
o_m & \mathcal{A}_{m,1} & \mathcal{A}_{m,2} & \ldots & \mathcal{A}_{m,k}
\end{array}
$$

As the second step, any associations mining method dealing with categorical values can be used. In [19], the GUHA method was used. Recall that GUHA is the first data-mining method [20][1]. As a result we obtain a set of linguistic associations that can be further studied. It should be emphasized that every found linguistic association should be considered just as a hypothesis about possible validity between sets of attributes.

The set of found linguistic associations can be further studied both from semantic as well as syntactic point of view. Among other results, we can reduce the number of found associations without loosing the discovered information. The most important feature of this approach is that the discovered knowledge is easily interpretable and hence understandable to experts from various fields.

When analyzing the meaning of evaluative expressions in the given context, we realize that what we obtained is a certain crisp decomposition of intervals carrying contexts of attributes X_j. It is necessary to have in mind the correct interpretation of linguistic expressions. The basic meaning of, e.g., the linguistic expression *more or less small* includes also all small values. In special cases, we want distinguish the values more subtly and so, we may consider the expression *more or less small but not small* (cf. [22]).

The use of fuzzy quantifiers instead of crisp (non-fuzzy) ones may produce better results (e.g. [23] and references therein) although the model of evaluative linguistic expressions is not used. This motivated us to work with slightly different model of linguistic expressions ([22]) (extending the model of evaluative linguistic expressions) and to study their properties ([24]). For completeness, we would like to point out that the model of evaluative linguistic expressions can also be used together with the F-transform ([25]) to detect (functional) dependencies among attributes.

---

[1] In that time, the term "data-mining" was not used and the authors spoke about "exploratory data analysis". The term "data-mining" was introduced in ninetieths in connection with works of R. Agrawal and R. Srikant; cf. [21].

6 Other Applications and Conclusion

In this paper, we gave overview of an important class of linguistic expressions called *evaluative linguistic expressions* and discussed a well working mathematical model of their semantics. We mentioned applications in time series analysis and prediction and mining linguistic associations. There are many more applications of linguistic descriptions, for example in control, where the linguistic description is taken as expert characterization of the control strategy of a given system [12,26,27]. This opens interesting possibilities, for example, learning linguistic description during monitoring successful control [28], modification of the context and there are many real applications of this method (c.f., e.g., [29]). There are also applications in decision making [13] and elsewhere.

Acknowledgments. The research was supported by the European Regional Development Fund in the IT4Innovations Centre of Excellence project (CZ.1.05/1.1.00/02.0070). Furthermore, we gratefully acknowledge partial support of the project KONTAKT II-LH12229 of MŠMT ČR.

References

1. Martínez, L., Ruan, D., Herrera, F.: Computing with words in decision support systems: An overview on models and applications. Int. J. of Comp. Intelligence Systems 3, 382–395 (2010)
2. Zadeh, L., Kacprzyk, J.E. (eds.): Computing with Words in Information/Intelligent Systems 1. STUDFUZZ, vol. 33. Springer, Heidelberg (1999)
3. Zadeh, L.A.: The concept of a linguistic variable and its application to approximate reasoning I, II, III. Information Sciences 8(9), 199–257, 301–357, 43–80 (1975)
4. Hájek, P.: Metamathematics of Fuzzy Logic. Kluwer, Dordrecht (1998)
5. Novák, V., Perfilieva, I., Močkoř, J.: Mathematical Principles of Fuzzy Logic. Kluwer, Boston (1999)
6. Hajičová, E., Partee, B.H., Sgall, P.: Topic-Focus Articulation, Tripartite Structures, and Semantics Content. Kluwer, Dordrecht (1998)
7. Duží, M., Jespersen, B., Materna, P.: Procedural Semantics for Hyperintensional Logic. Springer, Dordrecht (2010)
8. Novák, V.: A comprehensive theory of trichotomous evaluative linguistic expressions. Fuzzy Sets and Systems 159(22), 2939–2969 (2008)
9. Calvo, T., Mayor, G., Mesiar, R. (eds.): Aggregation operators. New trends and applications. Physica-Verlag, Heidelberg (2002)
10. Novák, V., Perfilieva, I.: On the semantics of perception-based fuzzy logic deduction. International Journal of Intelligent Systems 19, 1007–1031 (2004)
11. Novák, V.: Perception-based logical deduction. In: Reusch, B. (ed.) Computational Intelligence, Theory and Applications, pp. 237–250. Springer, Berlin (2005)
12. Novák, V.: Linguistically oriented fuzzy logic controller and its design. Int. J. of Approximate Reasoning 12, 263–277 (1995)
13. Novák, V., Perfilieva, I., Jarushkina, N.G.: A general methodology for managerial decision making using intelligent techniques. In: Rakus-Andersson, E., Yager, R.R., Ichalkaranje, N., Jain, L.C. (eds.) Recent Advances in Decision Making. SCI, vol. 222, pp. 103–120. Springer, Heidelberg (2009)

14. Perfilieva, I.: Fuzzy transforms: theory and applications. Fuzzy Sets and Systems 157, 993–1023 (2006)
15. Novák, V., Štěpnička, M., Dvořák, A., Perfilieva, I., Pavliska, V., Vavříčková, L.: Analysis of seasonal time series using fuzzy approach. International Journal of General Systems 39, 305–328 (2010)
16. Štěpnička, M., Dvořák, A., Pavliska, V., Vavříčková, L.: A linguistic approach to time series modeling with the help of the F-transform. Fuzzy Sets and Systems 180, 164–184 (2011)
17. Bodenhofer, U., Bauer, P.: Interpretability of linguistic variables: a formal account. Kybernetika 2, 227–248 (2005)
18. Štěpnička, M., Donate, J., Cortez, P., Vavříčková, L., Gutierrez, G.: Forecasting seasonal time series with computational intelligence: contribution of a combination of distinct methods. In: Proc. EUSFLAT 2011, pp. 464–471 (2011)
19. Novák, V., Perfilieva, I., Dvořák, A., Chen, Q., Wei, Q., Yan, P.: Mining pure linguistic associations from numerical data. International Journal of Approximate Reasoning 48, 4–22 (2008)
20. Hájek, P., Havránek, T.: Mechanizing Hypothesis formation, Mathematical Foundations for a General Theory. Springer, New York (1978)
21. Agrawal, R., Srikant, R.: Fast algorithms for mining association rules in large databases. In: 20th International Conference on Very Large Databases, Santiago, Chile (1994)
22. Kupka, J., Tomanová, I.: Some extensions of mining of linguistic associations. Neural Network World 20, 27–44 (2010)
23. Hüllermeier, E., Yi, Y.: In defense of fuzzy association analysis. IEEE Transactions on Systems, Man, and Cybernetics B 37(4) (2007)
24. Kupka, J., Tomanová, I.: Some dependencies among attributes given by fuzzy confirmation measures. In: Proc. EUSFLAT - LFA 2011, pp. 498–505 (2011)
25. Perfilieva, I., Novák, V., Dvořák, A.: Fuzzy transform in the analysis of data. International Journal of Approximate Reasoning 48(1) (2008)
26. Novák, V.: Linguistically oriented fuzzy logic controller. In: Proc. of the 2nd Int. Conf. On Fuzzy Logic and Neural Networks IIZUKA 1992, pp. 579–582. Fuzzy Logic Systems Institute, Iizuka (1992)
27. Novák, V.: Genuine linguistic fuzzy logic control: Powerful and successful control method. In: Hüllermeier, E., Kruse, R., Hoffmann, F. (eds.) IPMU 2010. LNCS (LNAI), vol. 6178, pp. 634–644. Springer, Heidelberg (2010)
28. Bělohlávek, R., Novák, V.: Learning rule base of the linguistic expert systems. Soft Computing 7, 79–88 (2002)
29. Novák, V., Kovář, J.: Linguistic IF-THEN rules in large scale application of fuzzy control. In: Da, R., Kerre, E. (eds.) Fuzzy If-Then Rules in Computational Intelligence: Theory and Applications, pp. 223–241. Kluwer Academic Publishers, Boston (2000)

Landscapes Description Using Linguistic Summaries and a Two-Dimensional Cellular Automaton

Francisco P. Romero and Juan Moreno-García

School of Industrial Engineering,
University of Castilla La Mancha,
Toledo, Spain
{FranciscoP.Romero,Juan.Moreno}@uclm.es

Abstract. Cellular automata models are used in ecology since they permit integrate space, ecological process and stochasticity in a single predictive framework. The complex nature of modeling (spatial) ecological processes has made linguistic summaries difficult to use within the traditional cellular automata models. This paper deals with the development of a computational system capable to generate linguistic summaries from the data provided by a cellular automaton. This paper shows two proposals that can be used for this purpose. We build our system by combining techniques from Zadeh's Computational Theory of Perceptions with ideas from the State Machine Theory. This paper discusses how linguistic descriptions may be integrated into cellular automata models and then demonstrates the use of our approach in the development of a prototype capable to provide a linguistic description of ecological phenomena.

Keywords: cellular automata, linguistic descriptions, machine state theory, computing with words.

1 Introduction

A landscape results from a succession of states evolving over a period of time [1]. Consequently, its constant evolution can lead to remarkable changes that may produce enormous ecological impacts. Hence, to better understand the landscape changes, landscape ecologists have focused on the development of dynamic simulation models, which attempt to replicate the possible paths of a landscape evolution and thereby evaluate future ecological implications [2].

Cellular automata models are increasingly used in ecology due to their easiness of implementation, ability to replicate spatial forms, and capacity to be quickly readapted to reproduce several types of dynamic spatial phenomena. Key factors to the explanation of biological interactions, stochasticity and the explicit consideration of space are easily incorporated in this models, increasing their ecological realism.

H.L. Larsen et al. (Eds.): FQAS 2013, LNAI 8132, pp. 221–232, 2013.

Cellular automata model consists of a regular n-dimensional array of cells that interact within a certain vicinity, according to a set of transition rules. Thus, in a cellular automata model, the state of each cell in an array depends on its previous state and the state of cells within a defined cartographic neighborhood, all cells being updated simultaneously at discrete time steps. The algorithm used to make the cells interact locally is known as the cellular automata local rule [3].

In cellular automata models information from different sources can be translated into a set of transition rules which defines the behavior of the system [4]. But the description of the landscape is not as simple as might be expected. A number of assumption and decisions regarding the data to be analyzed are required and the transition rules are not enough to provide a global view of the landscape evolution. For the correct definition and analysis of landscape many other aspects have jointly to be considered and at the community level of ecology any phenomena are complex and open to multiple interpretation. Consequently, the indicators-based description of the landscape can be unintelligible.

The work presented here is an example of a soft computing-based application capable to generate linguistic summaries of the evolution of a landscape modeling by a cellular automaton. There are several approaches to address the linguistic summarization of data, with a particular focus on fuzzy logic [5] and protoforms [6]. Linguistic summaries of time series [7] is another field increasingly explored for generating natural language-based reports.

Linguistic summaries are natural language-based description of data sets, which capture the core trends of the data[8]. These summaries are not meant to be a replacement for classical statistical analysis but rather an alternative mean of analyzing and describing complex systems in linguistic terms instead of numerical values

For this purpose, we extract certain relevant attributes from the cellular automaton and we used these attributes as input variables of two models: a State Machine-based model used for describing the behavior of each individual in the landscape and a linguistic model that represents the landscape (similar to [9]. We also used a fuzzy logic algorithm to obtain a set of sentences that is presented to the user as a linguistic description.

The main contributions of this paper are the two proposed models to create the linguistic description of a landscape phenomenon. The first model allows the designer to interpret the landscape and allows to make meaningful sentences for a final user, and the second model is a method to subsume the past, being more specific the past actions of an individual by presenting results in a very human consistent way, using natural language sentences.

The rest of the work is developed in the following way: Section 2 includes a concise description of cellular automata models and the linguistic summaries of data. Section 3 describes our method to obtain linguistic descriptions of the local events and in section 4 our method to obtain a global linguistic description of the landscape is explained. Finally, our conclusions and future work are outlined in Section 5.

2 Background

In this section, the fundamental concepts supporting our approach are explained beginning with the notion of cellular automata. Later, the main features of linguistic summaries and its applications are included.

2.1 Cellular Automata

According to [10] conventional cellular automata (CA) model consists of:

1. a Euclidean space divided into an array of identical cells (regular lattice);
2. a cell neighborhood of a defined size and shape;
3. a set of discrete cell states;
4. a set of transition rules, which determine the state of a cell as a function of the states of cells in a neighborhood;
5. discrete time steps with all cell states updated simultaneously.

A CA can be also described as a finite-state machine on a regular lattice. The input to the machine is the states of all cells in its neighborhood, the change of its state is based on the rules or transition functions. The states of all cells in the lattice are updated synchronously in discrete time steps. Each cell evolves according to the same rules which depend only on the state of the cell and a finite number of neighboring cells, and the neighborhood relation is local and uniform.

The simple cellular automata as defined above are capable of surprisingly complex behavior-complex both in the formal and in the intuitive sense. Thus, CA as a computational methodology has been applied to various science fields, such as numerical analysis, computational fluid dynamics, traffic analysis, growth phenomena modeling, etc.

Some features of CA are very useful for simulating biological and ecological systems [11]:

- the number of variables is huge but
- the number of states of each of the variables is typically small (thus allowing for a qualitative rather than a quantitative description),
- the interactions are often a set of rules expressed on a non algebraic form
- the locality of variables and of the interactions.
- the multilevel modeling ("*the variables in which the model is formulated are different from those which are to be observed in the model*"') [11].

There are several examples of CA used to simulate the behavior of a landscape with multi-species (ants, foxes, rabbits, etc.) interactions in a food chain hierarchy, for example the work presented in [12]. Another example is the work [13] which use CA to analyze the regeneration of endemic species of long-lived trees. Complex models of CA are also developed, for example, in [14] the CA presents multi-scale vicinity-based transitional functions, incorporation of spatial feedback approach to a stochastic multi-step simulation engine, and the application of logistic regression to calculate the spatial dynamic transition probabilities. Finally, the work of Mantelas et al. [15] presents a combination between cellular automata and fuzzy logic. The proposed model is based on a fuzzy system that incorporates cellular automata techniques and is used for urban modeling which accesses the multi-level urban growth dynamics and expresses them in linguistic terms.

2.2   Linguistic Summaries of Data

Linguistic summaries are meant to be a general, human consistent description of data sets, which capture the core trends of the data. These summaries are not meant to be a replacement for classical statistical analysis but rather an alternative means of representing the data focused on quick human understandability and interpretability; the summaries are brief descriptions of trends in the data stated in natural language.

The basic Yager's [16] approach to the linguistic summarization of sets of data set consist of : a summarizer, S (e.g. strong); a quantity in agreement, Q (e.g. most) and a validity degree (e.g. 0.8); as, e.g., T("most of the ants are strong") $= 0.8$

Given a set of data D, it is possible to fine a summarizer S and a quantity in agreement (Q) and the assumed measure of truth (validity) will indicate the truth (validity) of the statement that Q data items satisfy the statement (summarizer) S.

In summary, we consider a fuzzy linguistic summary [9] as a set of sentences which express knowledge about a situation in the world through the use of linguistic variables, fuzzy linguistic summarizers and fuzzy linguistic quantifiers.

3   Linguistic Summary of Individual Behavior

The process of obtaining a linguistic summary consists in two steps: (1) Firstly it is inferred what happened in order to describe an evolution (Sec. 3.1). To do this, a variation of the Finite State Machine (FSM) proposed by Chen [17] is used. (2) The linguistic summary is derived from the inferred FSM (Sec. 3.2).

3.1   Inferring an Instance of a State Machine Instance

Due to some limitations of FSM, we will use Finite State Machine with Parameters (FSMwP) [17] to model the behavior of the system. Modeling in FSM has the problem of state explosion. For example, to model a buffer of n capacity using FSM would require at least n states [17], while by using an integer parameter to describe the content of the buffer the number of states required can be reduced. Furthermore, if the capacity of the buffer changes only the range of the parameter must be changed, without remodeling the system.

Chen and Lin [17] proposed to employ both FSM and sets of parameters in modeling discrete event systems (FSMwP). FSMwP is similar to the Extended

Finite State Machines [18], but FSMwP is more general and was designed for modeling general discrete event systems.

Formally, an FSM is described using Equation 1.

$$FSM = \left(\sum, Q, \delta, q_0, Q_m\right) \tag{1}$$

where $\sum$ is the event set, Q is the state set, $\delta : \sum \times Q \to Q$ is the transition function, q_0 is the initial state, and Q_m is the final state.

Formally, an FSMwP is described using Equation 2.

$$FSMwP = \left(\sum, Q, \delta, P, G, (q_0, p_0), Q_m\right) \tag{2}$$

where $\sum$ is the event set, Q is the state set, $\delta : \sum \times Q \times G \times P \to Q \times P$ is the transition function, P is the vector of parameters, G is the set of guards, (q_0, p_0) are the initial state and parameters respectively, and Q_m is the final state.

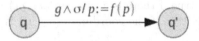

Fig. 1. A transition in FSMwP: q and q' are states, σ is an event, g is a guard and p a parameter

Now the way that the parameters are introduced into an FSM is explained [18]. Let $p \in P$ be a vector of parameters. Chen and Lin also introduced *guards* $g \in G$ that are predicates on the parameters p. δ was defined as a function from $\sum \times Q \times G \times P$ to $Q \times P$ as illustrated Figure 1. The transition shown can be interpreted as follows:

If at state q the guard g is true and the event σ occurs, then the next state is q' and the parameters will be updated to $f(p)$.

As you can see, FSMwP is appropriated for describing the individual behavior since our model works with two parameters that describe the position of the individual: x and y coordinates. The event notation notation must be expanded with two parameters to indicate the coordinates. For example, Chen represents an event like E, while our notation is $E(x_1, y_1)$, that is, two parameters are added.

The FSMwP of our problem (Figure 2) is formally described using the notation in [17] as follows:

- $\sum = \{B(x_1, y_1), E(x_1, y_1), M(x_1, y_1), WT(x_1, y_1), F(x_1, y_1), R(x_1, y_1), D(x_1, y_1)\}$.
- $Q = \{W, FE, D\}$.
- δ is described in Figure 2.

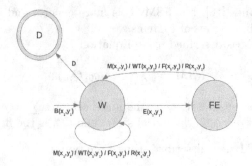

Fig. 2. The State Machine for Ants

- $P = \{x, y\}$.
- $G = \{\ [Neighbor(x_1, y_1)], [Plant(x_1, y_1)], [Empty(x_1, y_1)]\ \}$, where:
  - $Neighbor(x_1, y_1)$ is $[[x_1 = x - 1] \vee [x_1 = x] \vee [x_1 = x + 1]] \wedge [[y_1 = y - 1] \vee [y_1 = y] \vee [y_1 = y + 1]]$. This predicate means that the coordinates (x_1, y_1) and (x, y) are neighbors coordinates.
  - $Plant(x_1, y_1)$ is true if there is a plant in (x_1, y_1).
  - $Empty(x_1, y_1)$ is true if the coordinate (x_1, y_1) is free, that is, without plant or ant.
- $(q_0, p_0) = (W, (x, y))$.
- $Q_m = D$.

Figure 2 shows the graphical representation of the FSMwP of our problem where:

- $B(x_1, y_1)$ is $B(x_1, y_1)\ /\ x := x_1;\ y := y_1$. It means that the ant is born in the coordinate (x_1, y_1).
- $E(x_1, y_1)$ is $[Plant(x_1, y_1)] \wedge [Neighbor(x_1, y_1)] \wedge E(x_1, y_1)\ /\ x := x_1;\ y := y_1$. If there is a plant in a neighbor position and the event E occurs, the ant eats this plant and moves to the position of the plant.
- $M(x_1, y_1)$ is $[Empty(x_1, y_1)] \wedge [Neighbor(x_1, y_1)] \wedge M(x_1, y_1)\ /\ x := x_1;\ y := y_1$. If there is a free neighbor position and the event M occurs, the ant moves to this free position.
- $WT(x_1, y_1)$ is $[Plant(x_1, y_1)] \wedge [Neighbor(x_1, y_1)] \wedge WT(x_1, y_1)\ /\ x := x;\ y := y$. If there is a plant in a neighbor position and the event WT occurs, the ant waters this plant and maintains its position.
- $F(x_1, y_1)$ is $[Ant(x_1, y_1)] \wedge [Neighbor(x_1, y_1)] \wedge F(x_1, y_1)\ /\ x := x;\ y := y$. If exist a neighbor ant and the event F occurs, the ant fights with this ant.
- $R(x_1, y_1)$ is $[Ant(x_1, y_1)] \wedge [Neighbor(x_1, y_1)] \wedge R(x_1, y_1)\ /\ x := x;\ y := y$. If exist a neighbor ant and the event R occurs, the ant reproduces and maintains its position.
- D is the dead event, when it occurs the final state D is reached.

Table 1. Initial Situation of the Lattice

4			
3		$P2$	
2	$H1$	$P1$	$H2$
1			
	1	2	3	4

Table 2. Situation of the Lattice after one evolution

4			
3		$P2+$	
2		$H1$	$H2$
1			
	1	2	3	4

Figure 2 represents the birth of an ant (event $B(x_1, y_1)$). After that, the ant is waiting the occurrence of a new event (state W). Then, an event of the type $E(x_1, y_1)$, $M(x_1, y_1)$, $WT(x_1, y_1)$, $F(x_1, y_1)$ or $R(x_1, y_1)$ occurs. If the event $E(x_1, y_1)$ occurs the state FE is activated, and the ant is waiting for the occurrence of one of these events, $M(x_1, y_1)$, $WT(x_1, y_1)$, $F(x_1, y_1)$ or $R(x_1, y_1)$. Each event behaves as was indicated above. Finally, state W leads to death when the ant loses all its energy (event D).

To describe the FSMwP behavior a run of an FSMwP (a FSMwP instance) is defined as a sequence as follows [17]:

$$r = (q_0, p_0) \xrightarrow{l_1} (q_1, p_1) \xrightarrow{l_2} (q_2, p_2) \xrightarrow{l_3} \ldots \tag{3}$$

where l_i is the label of the ith transition.

3.2   Obtaining a Linguistic Summary from a State Machine Instance

A FSMwP instance (Equation 3) will be used to obtain the linguistic summary of an evolution of the system. To do this, our method needs to know the initial and the final situation of a local zone of the lattice. An example will be used to explain the proposed method. Table 1 reflects the initial situation of the lattice, where $H1$ and $H2$ are two ants situated in the cells $(2, 2)$ and $(2, 4)$ respectively, and $P1$ and $P2$ are two plants in the cells $(2, 3)$ and $(3, 3)$ respectively.

Suppose that after an evolution the lattice situation changes to the situation shown in Table 2, where $P2+$ represents that plant $P2$ has more energy because it has been watered. Equation 4 shows the FSMwP instance suffered by $H1$ in this evolution that provokes these changes.

$$(W, (2, 2)) \xrightarrow{E(2,3)} (FE, (2, 3)) \xrightarrow{WT(3,2)} (W, (2, 3)) \tag{4}$$

Table 3. Situation of the Lattice after two evolutions

4				
3		H2		
2				
1				
	1 2	3	4	

Equation 4 can be expressed with the following natural language expression:

H1 in coordinate $(2,2)$ *eats plant P1 in coordinate* $(2,3)$ *and moves to coordinate* $(2,3)$. *After, H1 waters plant P2 in coordinate* $(3,2)$.

Another possibility to express this more reductively is as follows:

H1$(2,2)$ *eats plant P1*$(2,3)$ *and moves to* $(2,3)$. *Then, H1*$(2,3)$ *waters plant P2*$(3,2)$.

Equation 5 shows the FSMwP instance suffered by $H2$ due to a new evolution (from Table 2 to 3).

$$(W,(2,4)) \xrightarrow{E(3,3)} (FE,(3,3)) \xrightarrow{F(2,3)} (W,(3,3)) \tag{5}$$

It natural language expression is as follows:

H2$(2,4)$ *eats plant P2*$(3,3)$ *and moves to* $(3,3)$. *Then, H2*$(3,3)$ *fights with ant H1*$(2,3)$ *winning the fight.*

The formal method to obtain the natural language expression will be done in a future work. As you can see, all necessary information is contained in the FSMwP instance.

4   Global Description of the Landscape Situation

In order to obtain a linguistic description of the landscape situation we used the granular linguistic model of a phenomenon introduced by Gracian et al. in [9]. According to this methodology it is necessary to define the "top-order perception", i.e., a definition of the general state of the phenomenon using natural language sentences. For this purpose, the template proposed in [19] is used as follows:

Q the landscape is R

where

- Q is a fuzzy quantifier [20] applied on the cardinality of the perception "*the landscape is R*".

- R is a summarizer. It is a constraint applied to the set of elements in the situation of the landscape.

The concept of *ecological balance* is used to describe the landscape. This concept could be defined as "*a state of dynamic equilibrium within a community of organisms in which genetic, species and ecosystem diversity remain relatively stable, subject to gradual changes through natural succession.*" or "*a stable balance in the numbers of each species in an ecosystem*" [21] . Therefore, our purpose is to answer the question "*Is the landscape in a stable equilibrium?*" using non-expert friendly natural language sentences.

Following the mentioned methodology the process to obtain these sentences consists of the following steps or phases:

1. Landscape Division: (crisp) areas and (fuzzy) regions
2. Definition and Computation of the basic measures: number of species, density, number of individuals, etc.
3. Definition of the first order perceptions: level of population and biodiversity
4. Definition of the top-order perceptions: Is the landscape in balance?

Figure 3 shows how the Granular linguistic model proposed in [9] could represents this process of summarization.

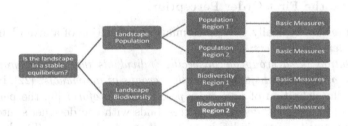

Fig. 3. Granular linguistic model of the landscape situation

4.1   Landscape Division

A landscape, modeled by a two-dimensional cellular automaton, can be divided into areas and regions defined as follows.

- An area is a block of contiguous cells which are strict boundaries (crisp) within the grid itself. These areas are disjoint but maintain a neighborhood relationship. An example of areas can be seen in Figure 4 where the white cells are empty and the colored cells are occupied by individuals.
- A region is defined by the union of the neighborhood of the different individuals within a particular area. If the static areas are disjoint, regions are dynamic and have fuzzy boundaries.

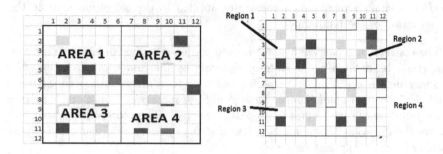

Fig. 4. Example of Areas and Regions

4.2   Definition of the Basic Measures

The data of interest for each region of the landscape are the following: Number of individuals of each specie within a region (E_i), number of species within a region $|E|$, total number of individuals within a region ($E = \sum_i E_i$), density ($D = E/|C|$ where $|C|$ is the region size, i.e the number of cells including in the region). At each time step, the computational system performs the previously defined measurements over each region of the landscape.

4.3   Define the First Order Perceptions

Two measures are typically used to estimate the situation of a given landscape, *population level* and *diversity level*.

A *population* is "*a group of conspecific individuals that is demographically, genetically, or spatially disjunct from other groups of individuals*" [22]. In agreement with the definition of *first-order perception protoform* [9], the perception protoform of "*Level of Population*" corresponds with the designer's interpretation of the quantity of the individuals, i.e., it is a tuple (U, y, g, T) where: U is the empty set, y is a numerical value of the input variable, g is a function based on membership functions , and T the linguistic template using to build the result ("*The region is {overpopulated / normally populated / depopulated}*").

In this preliminary work, we have used density as input variable to obtain the level of population, in future works, the use of more complex values like $\frac{r}{k}$ selection index [23] will be considered.

Species diversity is defined as "*the effective number of different species that are represented in a collection of individuals*". In our case, this concept is equivalent to *biodiversity*. The effective number of species refers to the number of equally-abundant species needed to obtain the same mean proportional species abundance as that observed in the dataset of interest (where all species may not be equally abundant). *Species diversity* consists of two components, species richness and species evenness. Species richness is a simple count of species, whereas species evenness quantifies how equal the abundances of the species are[24]. The perception protoform of "*Species Diversity*" is a tuple (U, Ox, g, T) where: U is the empty set, O_x is a linguistic label expressing the degree of species diversity of the region, g is a function based on membership functions , and T the linguistic template using to build the result ("*The region is {megadiverse / diverse / depopulated}*").

4.4 Definition of the Top-Order Perceptions

We needed to set the top-order perception to complete the model. The perception protoform of the *Balance of the landscape* is a tuple (U, S_x, g, T) where: U are the linguistic variables P_x population and D_x diversity, S_x is a linguistic variable expressing the balance of the region x, g is the aggregation function $W_s = g(P_x, D_x)$ were W_s is a vector of weights, T is the template "*the landscape is in a {high / medium / low} ecological balance*".

We implement the aggregation function g using a set of fuzzy rules (Mamdani style) with linguistic hedges.

5   Conclusions and Future Work

In this paper, we have used cellular automata to model a multi-species landscape and showed two methods to obtain linguistic summaries from the data provided by the evolution of the landscape. One for describing the individual behavior and another for describing the landscape situation.

There are a lot of work to do for further development of this proposal such as improving top-order perception using a more complex structure or multi-order perceptions, carrying out experiments to verify the feasibility of this methods in huge cellular automata, etc.

References

1. Forman, R., Godron, M.: Landscape Ecology. Wiley, New York (1986)
2. Baker, W.: A review of models of landscape change. Landscape Ecology 2(2), 111–133 (1989)
3. Sirakoulis, G., Karafyllidis, I., Thanailakis, A.: A cellular automaton model for the effects of population movement and vaccination on epidemic propagation. Ecological Modelling 133(3), 209–223 (2000)
4. Ermentrout, G.B., Edelstein-Keshet, L.: Cellular automata approaches to biological modeling. J. Theor. Biol. 160(1), 97–133 (1993)
5. Kacprzyk, J., Zadrozny, S.: Linguistic database summaries and their protoforms: towards natural language based knowledge discovery tools. Inf. Sci. 173(4), 281–304 (2005)
6. Yager, R.R.: Knowledge trees and protoforms in question-answering systems. JASIST 57(4), 550–563 (2006)
7. Kacprzyk, J., Wilbik, A., Zadrozny, S.: Linguistic summarization of time series using a fuzzy quantifier driven aggregation. Fuzzy Sets and Systems 159(12), 1485–1499 (2008)

8. van der Heide, A., Triviño, G.: Automatically generated linguistic summaries of energy consumption data. In: ISDA, pp. 553–559 (2009)

9. Triviño, G., Sanchez, A., Montemayor, A.S., Pantrigo, J.J., Cabido, R., Pardo, E.G.: Linguistic description of traffic in a roundabout. In: FUZZ-IEEE, pp. 1–8 (2010)

10. Langton, C.G.: Studying artificial life with cellular automata. Physica D 22D(1-3), 120–149 (1986)

11. Hogeweg, P.: Cellular automata as a paradigm for ecological modeling. Appl. Math. Comput. 27(1), 81–100 (1988)

12. Yang, X.S.: Characterization of multispecies living ecosystems with cellular automata. In: Proceedings of the Eighth International Conference on Artificial Life, ICAL 2003, pp. 138–141. MIT Press, Cambridge (2003)

13. Cannas, S.A., Páez, S.A., Marco, D.E.: Modeling plant spread in forest ecology using cellular automata. Computer Physics Communications 121/122, 131–135 (1999)

14. Soares-Filho, B.S., Cerqueira, G.C., Pennachin, C.L.: Dinamica – a stochastic cellular automata model designed to simulate the landscape dynamics in an amazonian colonization frontier. Ecological Modelling 154(3), 217–235 (2002)

15. Mantelas, L.A., Hatzichristos, T., Prastacos, P.: A fuzzy cellular automata modeling approach – accessing urban growth dynamics in linguistic terms. In: Taniar, D., Gervasi, O., Murgante, B., Pardede, E., Apduhan, B.O. (eds.) ICCSA 2010, Part I. LNCS, vol. 6016, pp. 140–151. Springer, Heidelberg (2010)

16. Yager, R.R.: A new approach to the summarization of data. Inf. Sci. 28(1), 69–86 (1982)

17. Chen, Y.L., Lin, F.: Modeling of discrete event systems using finite state machines with parameters. In: Proceedings of the 2000 IEEE International Conference on Control Applications, pp. 941–946 (2000)

18. Cheng, K.T., Krishnakumar, A.S.: Automatic generation of functional vectors using the extended finite state machine model. ACM Trans. Des. Autom. Electron. Syst. 1(1), 57–79 (1996)

19. Yager, R.R.: Fuzzy summaries in database mining. In: Proceedings of the 11th Conference on Artificial Intelligence for Applications, CAIA 1995, p. 265. IEEE Computer Society, Washington, DC (1995)

20. Zadeh, L.A.: A computational approach to fuzzy quantifiers in natural languages. Computers & Mathematics with Applications 9(1) (1983)

21. Perttu, K.: Ecological, biological balances and conservation. Biomass and Bioenergy 9(1-5), 107–116 (1995)

22. Silvertown, J.: Introduction to plant population ecology. Longman (1982)

23. Reznick, D., Bryant, M.J., Bashey, F.: r-and-K-Selection Revisited: The Role of Population Regulation in Life-History Evolution. Ecology 83(6), 1509–1520 (2002)

24. Hill, M.O.: Diversity and evenness: a unifying notation and its consequences. Ecology 54(2), 427–432 (1973)

Comparing f_β-Optimal with Distance Based Merge Functions

Daan Van Britsom*, Antoon Bronselaer, and Guy De Tré

Department of Telecommunications and Information Processing, Ghent University,
Sint-Pietersnieuwstraat 41, B-9000 Ghent, Belgium
{Daan.VanBritsom,Antoon.Bronselaer,Guy.DeTre}@UGent.be

Abstract. Merge functions informally combine information from a certain universe into a solution over that same universe. This typically results in a, preferably optimal, summarization. In previous research, merge functions over sets have been looked into extensively. A specific case concerns sets that allow elements to appear more than once, multisets. In this paper we compare two types of merge functions over multisets against each other. We examine both general properties as practical usability in a real world application.

Keywords: Merge functions, multisets, content selection.

1 Introduction

In an ever growing digitalised world the amount of data available to the end user has very quickly become extremely cluttered. When one selects different data inputs regarding a single topic, frequently referred to as coreferent information, there is always duplicate, conflicting and missing information out and about. Therefore, when working with coreferent information there are several techniques that allow one to merge this information in order to get a briefer and correct overview. One of these possible techniques concerns the use of the f-value, a measurement balancing correctness and completeness, of the proposed solution with respect to the sources. These so-called f-optimal merge functions have been discussed extensively in [1] and expanded to f_β-optimal merge functions that allow for a preference to be given to either correctness or completeness by means of a parameter β. This type of merge function is typically applied to sets that allow elements to occurs multiple times, multisets. A second possible group of techniques concerns the use of distance measurements in order to determine which possible solution is closest related to all the sources. In order to illustrate how both types of merge functions can be useful in a real world application we demonstrate how their respective solutions can be used to generate multi-document summarizations (MDSs).

The remainder of this paper is structured as follows. In Section 2 we describe a few preliminary definitions required to understand the comparison we wish to

---

* Corresponding author.

H.L. Larsen et al. (Eds.): FQAS 2013, LNAI 8132, pp. 233–244, 2013.
© Springer-Verlag Berlin Heidelberg 2013

establish in this document. Section 3 details how one is able to influence the outcome of an f_β-optimal merge function. A simple example of a distance based merge function is provided in Section 4, whilst the comparison of both types of merge functions is made in Section 5. In part one of the latter we examine a few general properties and in part two we illustrate how both merge techniques can lie at the basis of Multi-Document Summarizations. Finally, we conclude in Section 6 with some final remarks on how we will further test the possibilities and advantages of both types of merge function in the creation of Multi-Document Summarizations.

2   Preliminaries

As a first type of merge function we would like to use in this paper's comparison, we iterate the definition of f_β-optimal merge functions. As stated in the intro-duction this type of merge functions is typically applied to sets, more specifically sets that allow elements to occur multiple times, multisets. We briefly recall some important definitions regarding multisets [2].

2.1   Multisets

Informally, a multiset is an unordered collection in which elements can occur multiple times. Many definitions have been proposed, but within the scope of this paper, we adopt the following functional definition of multisets [2].

Definition 1 (Multiset). *A multiset M over a universe U is defined by a function:*

$$M : U \to \mathbb{N}. \tag{1}$$

For each $u \in U$, $M(u)$ denotes the multiplicity of an element u in M. The set of all multisets drawn from a universe U is denoted $\mathcal{M}(U)$.

The j-cut of a multiset M is a regular set, denoted as M_j and given as:

$$M_j = \{u | u \in U \wedge M(u) \geq j\}. \tag{2}$$

Whenever we wish to assign an index $i \in \mathbb{N}$ to a multiset M, we use the notation $M_{(i)}$, while the notation M_j is preserved for the j-cut of M. We adopt the definitions of Yager [2] for the following operators: $\cup$, $\cap$, $\subseteq$ and $\in$.

2.2   Merge Functions

The general framework of merge functions provides the following definition [3].

Definition 2 (Merge function). *A merge function over a universe U is de-fined by a function:*

$$\varpi : \mathcal{M}(U) \to U. \tag{3}$$

As explained in the introduction of this paper, we are interested in merge functions for (multi)sets rather than atomic elements. Therefore, we consider merge functions over a universe $\mathcal{M}(U)$ rather than a universe U. This provides us with the following function:

$$\varpi : \mathcal{M}\big(\mathcal{M}(U)\big) \to \mathcal{M}(U). \tag{4}$$

In order to avoid confusion, we shall denote S (a source) as a multiset over U and we shall denote M as a multiset over $\mathcal{M}(U)$ (a collection of sources). Thus, in general, M can be written as:

$$M = \{S_{(1)}, ..., S_{(n)}\}. \tag{5}$$

Finally, we shall denote $\mathscr{S} \in \mathcal{M}(U)$ as a general solution for a merge problem, i.e. $\varpi(M) = \mathscr{S}$. The most simple merge functions for multisets are of course the source intersection and the source union. That is, for any M:

$$\varpi_1(M) = \bigcap_{S \in M} S \tag{6}$$

$$\varpi_2(M) = \bigcup_{S \in M} S. \tag{7}$$

Within this paper, we consider a solution relevant if it is a superset of the source intersection or a subset of the source union. Therefore, we call the source intersection the lower solution (denoted $\underline{\mathscr{S}}$) and the source union the upper solution (denoted $\overline{\mathscr{S}}$). To conclude this section, we introduce the family of f-optimal merge functions, which are merge functions that maximize the harmonic mean of a measure of solution correctness (i.e. precision) and a measure of solution completeness (i.e. recall). This objective is better known as the f-value [4]. To adapt the notion of precision and recall to the setting of multiset merging, we define two *local* (i.e. element-based) measures [1].

Definition 3 (Local precision and recall). *Consider a multiset of sources* $M = \{S_{(1)}, ..., S_{(n)}\}$. *Local precision and recall are defined by functions p^* and r^* such that:*

$$\forall u \in U : \forall j \in \mathbb{N} : p^*(u, j | M) = \frac{1}{|M|} \sum_{S \in M \wedge S(u) \geq j} M(S) \tag{8}$$

$$\forall u \in U : \forall j \in \mathbb{N} : r^*(u, j | M) = \frac{1}{|M|} \sum_{S \in M \wedge S(u) \leq j} M(S). \tag{9}$$

One can see that p^* depicts the percentage of sources in which u occurs at least j times and r^* the percentage of sources in which u occurs a maximum of j times.

Definition 4 (f-optimal merge function). *A merge function ϖ is f-optimal if it satisfies for any $M \in \mathcal{M}(\mathcal{M}(U))$:*

$$\varpi(M) = \underset{\mathscr{S} \in \mathcal{M}(U)}{\arg\max} \, f(\mathscr{S} | M) = \underset{\mathscr{S} \in \mathcal{M}(U)}{\arg\max} \left(\frac{2 \cdot p(\mathscr{S}|M) \cdot r(\mathscr{S}|M)}{p(\mathscr{S}|M) + r(\mathscr{S}|M)} \right) \tag{10}$$

constrained by:

$$\left(\max_{\mathscr{S} \in \mathcal{M}(U)} f(\mathscr{S}|M) = 0 \right) \Rightarrow \varpi(M) = \emptyset \tag{11}$$

and where, with T *a triangular norm, we have that:*

$$p(\mathscr{S}|M) = \mathop{\mathrm{T}}_{u \in \mathscr{S}} \left(p^*(u, \mathscr{S}(u)|M) \right) \tag{12}$$

$$r(\mathscr{S}|M) = \mathop{\mathrm{T}}_{u \in \mathscr{S}} \left(r^*(u, \mathscr{S}(u)|M) \right). \tag{13}$$

3   Influencing the Content Selection

The f-optimal merge function as defined in Definition 4 doesn't allow one to influence the outcome $\mathscr{S} \in \mathcal{M}(U)$ of the merge function. Suppose one would want to select fewer elements in order to show a preference to precision rather than recall. In order to do so one could take a subset of $\mathscr{S}$ but then one would no longer have a solution with an optimal f-value. The merge function becomes even more restricting if one would want more elements as a solution, thus giving preference to recall rather than precision, for there is no option to gain more concepts. In order to influence the outcome of the f-optimal merge function we have chosen to use the weighted harmonic mean [5], and the merge function thus changes as follows.

Definition 5 (Weighted f_β-optimal merge func.). *A merge function* ϖ *is* f_β-*optimal if it satisfies for any* $M \in \mathcal{M}(\mathcal{M}(U))$:

$$\varpi(M) = \mathop{\arg\max}_{\mathscr{S} \in \mathcal{M}(U)} f_\beta(\mathscr{S}|M) = \mathop{\arg\max}_{\mathscr{S} \in \mathcal{M}(U)} \left(\frac{(1+\beta^2) \cdot p(\mathscr{S}|M) \cdot r(\mathscr{S}|M)}{\beta^2 \cdot p(\mathscr{S}|M) + r(\mathscr{S}|M)} \right) \tag{14}$$

still constrained by (11), $\beta \in [0, \infty]$ *and where, with* T *a triangular norm, (12) and (13) still apply.*

The parameter β expresses how much more weight is given to recall as opposed to precision, more specifically, recall has a weight of β times precision. Thus, when $\beta = 1$ precision and recall are weighted the same and this results in the non-weighted f-optimal merge function as defined in Definition 4. When $\beta < 1$, a preference is given to precision, for example when $\beta = 0.5$, recall is given half the weight of precision. When $\beta > 1$, a preference is given to recall, for example when $\beta = 2$, recall is given twice the weight of precision. When $\beta = 0$, f_β returns the precision and when β approaches infinity f_β results in the recall.

In previous research it has been shown that the specific case where $T = T_M$, the minimum t-norm as proposed by Zadeh, has interesting properties and therefore, for the remainder of this paper, we will restrict ourselves to this case [1].

4   Distance Based Merge Functions

Another approach to generate a result for a set of coreferent items one wishes
to merge consists of using distance based merge functions. There are quite a few
techniques to measure a distance between two sets, including Cosine similarity
and Minkowski distances such as the Manhattan and Euclidean distance. The
example we will be using throughout this paper is based on the Minkowski
distance, an effective and frequently used distance measurement.

Definition 6 (Simple distance function). *Consider two sources $S_{(1)}, S_{(2)}$
and a universe U consisting of u elements. The distance between these sources
according to the simple distance function δ is*

$$\delta(S_{(1)}, S_{(2)}) = \sum_{u \in U} |S_{(1)}(u) - S_{(2)}(u)| \tag{15}$$

with, as stated earlier, $S_{(i)}(u)$ the multiplicity of element u in source $S_{(i)}$

This distance function results in calculating the number of adjustments required
to get from one source to another, whilst only allowing additive and subtractive
operations.

One can now use the distance function δ to calculate the distance from a
single set with respect to several different sets.

Definition 7 (Simple distance based merge function). *Consider a multiset of sources $M = \{S_{(1)}, ..., S_{(n)}\}$ in a universe U. A distance based merge
function ϖ_δ returns the solution that has a minimal total distance to all provided
sources. For each element $m \in M$ for each $M \in \mathcal{M}(\mathcal{M}(U))$:*

$$\varpi_\delta(M) = \arg\min \sum_{i=1}^{n} \delta(\mathscr{S}, S_{(i)}) \tag{16}$$

Informally, the function ϖ_δ calculates the solution $\mathscr{S}$ that requires the least
total additive and subtractive operations to go from $\mathscr{S}$ to all the possible sources
S in M.

Due to the distributivity of the minimum over the summation we can formulate this distance function as follows.

Definition 8 (Element based merge func.). *Consider a multiset of sources
$M = \{S_{(1)}, ..., S_{(n)}\}$ in a universe U. A distance based merge function $\varpi_{\delta\epsilon}$
returns the solution that has a minimal total distance to all provided sources.
For each element $m \in M$ for each $M \in \mathcal{M}(\mathcal{M}(U))$:*

$$\varpi_{\delta\epsilon}(M) = \forall u \in U : \mathscr{S}(u) = \arg\min_{k \in \mathbb{N}} \sum_{i=1}^{n} |S_{(i)}(u) - k| \tag{17}$$

Informally, the function $\varpi_{\delta\epsilon}$ calculates the optimal multiplicity (range of multiplicities) $\mathscr{S}(u)$ for each element u so that requires the least total additive and subtractive operations to go from $\mathscr{S}(u)$ to the multiplicity of that element in every source $S_{(i)}$ in M.

Obviously, the complexity of the latter function is a lot smaller than the complexity of ϖ_{δ}. However, it quickly becomes apparent that if we were to apply this function on a realistic dataset of documents we would have an exponential amount of possible solutions to compare. If the dataset only consists of a universe of 100 words with a average multiplicity range of only five possibilities, we would have to generate and evaluate 5^{100}, roughly $7.8 * 10^{69}$ solutions. The solution space is however uniquely defined by the multiplicityset generated by $\varpi_{\delta\epsilon}$.

5   Making the Comparison

Now that both types of merge functions have been recapitulated we want to compare them to one another. In subsection 5.1 we go over a few useful properties concerning merge functions and see which ones apply on either one of the types of function. In subsection 5.2 we apply both functions to a real world application, the summarizing of multiple documents, more specifically the content selection step, and see which advantages or disadvantages the merge functions have.

5.1   Properties

Property 1 (Idempotence). A merge function ϖ for multisets over a ground universe U is idempotent if and only if, for any $M = \{S, ..., S\}$ we have that:

$$\varpi(M) = S. \tag{18}$$

As has been proven in [1], the f-optimal merge function is idempotent, the proof that the weighted f_β-optimal merge function is idempotent as well is trivial. It is obvious that the proposed distance based merge function is idempotent as well, considering that the solution S is the only one not requiring any additive or subtractive operations relative to all the sources.

Property 2 (Monotonicity). A merge function ϖ for multisets over a ground universe U is monotone if and only if, for any $M = \{S_{(1)}, ..., S_{(n)}\}$ and for any $M^* = \{S^*_{(1)}, ..., S^*_{(n)}\}$ such that:

$$\forall i \in \{1, ..., n\} : S_{(i)} \subseteq S^*_{(i)} \wedge M\left(S_{(i)}\right) = M^*\left(S^*_{(i)}\right) \tag{19}$$

we have that:

$$\varpi(M) \subseteq \varpi(M^*). \tag{20}$$

Where the defined global precision and recall functions are monotone as proven in [1], the f-optimal merge function is not and thus the weighted f_β-optimal merge function is neither. Due to the nature of the Minkowski distance, the proposed distance based merge function however, is monotone.

Property 3 (Quasi Robustness). A merge function ϖ over $\mathcal{M}(U)$ is quasi-robust if and only if, for any error-free $M \in \mathcal{M}(\mathcal{M}(U))$ (with $|M| > 1$) and for any erroneous source E, we have that:

$$\varpi(M \cup \{E\}) \cap E = \emptyset. \tag{21}$$

With E an erroneous source, as defined in [6], a source that has no element in common with any of the sources in M.

It has been proven in [6] that the f-optimal merge function is quasi robust. The f_β-optimal merge function however is not. When β approaches infinity the f_β-optimal merge functions approaches the union for which quasi robustness clearly doesn't hold. The proposed distance based merge function however is quasi robust as well from the moment that $|M| > 2$. The proof for this is trivial because the moment you have two sources not containing a certain element, including this element to the solution will always result in at least one more additive or subtractive operation relative to the sources as opposed to not including it into the solution.

5.2   Multi-Document Summarization

In order to illustrate other possible differences between distance based and f_β-optimal merge function we apply both algorithms to the Multi-Document Summarization problem (MDS problem) using the Document Understand Conference dataset of 2002 (DUC2002) and try to evaluate how we can influence both algorithms. Suppose we therefore define a cluster of sources from the DUC2002 set as a multiset M and every document of the n documents of that cluster as a source S so the equation $M = \{S_{(1)}, ..., S_{(n)}\}$ clearly still holds up. The solution $\mathscr{S}$ of the merge function can only contain elements from the sources, therefore the universe U does not consist of the entire English language but instead contains all the words from all the different sources $\{S_{(1)}, ..., S_{(n)}\}$ that are part of the cluster cluster combined.

It has been shown in previous research that once a set of key concepts has been identified for a cluster of coreferent documents, a summarization can be generated [7]. In this paper we will therefore focus on how both types of merge functions can generate a set of concepts that represent the key elements of the cluster automatically and as usable as possible. We will focus on two separate issues. First, we will try to establish how easy it is to find a single optimal set of key concepts defining the cluster. Secondly, we will examine to which extend it is possible to objectively influence this selection process.

f_β-Optimal Merge Function. If we were to illustrate the type of solution generated by the f_β-optimal merge function by using the first cluster of documents of the DUC2002 set we would get, for a value of β of 1, thus resulting in the non-weighted f-optimal merge function, the following result.

$\varpi_{\beta=1}(M) = \{\{weather=1, winds=5, rico=3 \dots$

$\dots, director=1, inches=1, service=1\} =1, \{caribbean=2, like=1, residents=1 \dots$

$\dots, civil=1, expected=1, only=1\}=1\}$

As one can see above, for the first cluster we get a multiset containing two other multisets with multiplicity one as a solution. When we calculate the solution for each cluster of the DUC2002 set we get a small multiset as a result each time, as one can see in Table 1. The distance based merge function however, as one can read further down in the paper, does not. This makes it a lot more difficult to choose one of the suggested multisets and later on influence this multiset.

Table 1. Number of solutions per clusterID for the DUC2002 dataset for the f_β-optimal function with $\beta = 1$

ID	$\sharp\mathscr{S}$	ID	$\sharp\mathscr{S}$	ID	$\sharp\mathscr{S}$	ID	$\sharp\mathscr{S}$	ID	$\sharp\mathscr{S}$	ID	$\sharp\mathscr{S}$
1	2	11	2	21	2	31	2	41	2	51	2
2	1	12	1	22	2	32	1	42	2	52	1
3	2	13	1	23	2	33	1	43	1	53	1
4	1	14	2	24	1	34	2	44	2	54	1
5	2	15	2	25	1	35	2	45	2	55	2
6	1	16	2	26	1	36	2	46	1	56	2
7	2	17	2	27	2	37	2	47	2	57	2
8	1	18	2	28	2	38	1	48	2	58	1
9	2	19	1	29	2	39	1	49	1	59	2
10	1	20	1	30	2	40	2	50	2		

The next evaluation step concerns testing the amenability of the f_β-optimal merge function. As has been recollected in Section 3 this can be done by the usage of the parameter β. We illustrate again by using the first cluster of the DUC2002 dataset.

$\varpi_{\beta=0.25}(M) = \{\{to=3, gilbert=2, storm=3, caribbean=1, mph=1, were=1, west=1, national=1, in=6, said=3, was=2, the=23, on=2, winds=1, s=3, hurricane=6, at=1, they=1, of=10, from=1, moving=1, for=1, center=1, a=4, coast=1, and=10\}=1\}$

$\varpi_{\beta=0.50}(M) = \{\{ national=1, center=2, puerto=1, gilbert=5, flooding=1, we=1, this=1, at=3, sustained=1, as=1, caribbean=1, would=1, moving=1, one=1, an=1, residents=1, 000=1, islands=1, weather=1, from=3, hurricane=6, they=1, into=1, was=4, miami=1, republic=1, west=1, about=2, people=1, dominican=1, coast=1, inches=1, it=3, is=1, the=30, in=10, on=2, said=5, of=12, mph=2, with=2, by=1, for=3, s=3, their=1, off=2, and=10, were=1, night=1, storm=4, reported=1, winds=4, to=6, a=5, sunday=1, heavy=1, there=1\}=1\}$

For values of $\beta < 1$ one obtains a subset of the original solution obtained from the non-weighted f-optimal merge function, as proven in [8]. As one can see above this may also result in the fact that the solution $\mathscr{S}$ no longer contains several multisets. The reason for this can be found in the fact that a preference is given to precision, to correctness, and therefore the likelihood of multiple multisets providing an equally optimal solution, drops. The same conclusion can be made as when $\beta > 1$, due to the fact that a preference is given to recall, the likelihood of multiple multisets being part of the optimal solution drops, as can be seen in the example.

Simple Distance Based Merge Function. As we have shown in Section 4 it might prove to be difficult to generate and display all possible results. But, as previously stated, the solution space is uniquely defined by the multiplicityset generated as described in Definition 8. We once more illustrate the results of this type of merge function by generating a solution for the first cluster of the DUC2002 set. The multiplicityset defining all the possible solutions for the first cluster can be found in Appendix A. Suffice to say it contains over 100 words, some of which with over 5 possible optimal multiplicities, which makes it very impractical to use due to the large amount of possible solutions.

Why there are so many possible solutions lies in the fact the more documents we have in which a word occurs, the higher the chance that there is not a single multiplicity defining the optimal balance. For instance, if a word u were to occur one time in the first source $S_{(1)}(u) = 1$, three times in the second $S_{(2)}(u) = 3$, $S_{(3)}(u) = 5$ and $S_{(4)}(u) = 7$, then the solution $\mathscr{S}(u)$ exists out of three possible multiplicities $\mathscr{S}(u) = [3, 4, 5]$ because from each multiplicity it only requires a total of eight additions or subtractions relative to the occurrences in the sources.

The reason why we still care about this difference is due to the fact that the semantic difference between a word w having multiplicity one or zero makes a huge difference in the interpretation by the user but for the distance function it makes virtually no difference at all. That is why there are so many possible optimal solutions. This of course only occurs when the sources are rather well balanced. It is also perfectly possible that there is only a single correct multiplicity for every word. However as one can clearly see in Table 2 depicting the amount of possible solutions for each cluster, as soon as there is not a single optimal solution the amount of possible solutions runs extremely high. This of course makes it very difficult to choose an optimal solution and afterwards influence the content selection.

Table 2. Number of solutions per clusterID for the DUC2002 dataset for the distance based merge function

ID	$\sharp\mathscr{S}$	ID	$\sharp\mathscr{S}$	ID	$\sharp\mathscr{S}$	ID	$\sharp\mathscr{S}$	ID	$\sharp\mathscr{S}$	ID	$\sharp\mathscr{S}$
1	3.377E44	11	5.107E18	21	1.297E18	31	5.629E16	41	3.486E21	51	3.799E15
2	1	12	1	22	6.333E15	32	1	42	2.111E14	52	1
3	1.480E31	13	1	23	3.239E54	33	1	43	1	53	1
4	1	14	6.984E40	24	1	34	1.367E17	44	1.159E11	54	1
5	1.776E36	15	6.648E19	25	1	35	1.669E13	45	3.298E13	55	1.489E31
6	1	16	1.290E25	26	1	36	1.202E16	46	1	56	1.513E18
7	8.881E35	17	2.988E22	27	4.669E19	37	2.350E27	47	1.056E14	57	2.757E21
8	1	18	7.124E15	28	4.178E30	38	1	48	8.977E16	58	1
9	9.277E11	19	1	29	3.804E15	39	1	49	1	59	6.274E26
10	1	20	1	30	1.197E36	40	1.284E32	50	1.811E9		

In order to illustrate how the proposed distance based merge function would generate a multiset of keywords κ of a set of documents, we select a few of the possible multisets of keywords with a minimal total distance to all the sources.

- κ_{min} generated by using the smallest multiplicity per element
- κ_{max} generated by using the largest multiplicity
- κ_{med} generated by using the median multiplicity of each element

One can find κ_{min} and κ_{max} completely in Appendix B. As one would suspect κ_{med} generated by using the median multiplicity of each element is analogue to κ_{max} with maximum multiplicity however it might introduce certain difference for elements that are on the cusp, for instance multiplicity range one to zero. It is therefore not present in Appendix B.

Practically speaking, besides the issue that there is an enormous amount of possible sets of key concepts, it is also quite difficult to objectively influence this selection process. One of the great advantages of the f_β-optimal merge function lies in the fact that through changing the parameter β one can influence the outcome of the function. When applying the merge function $\varpi_{\delta\epsilon}$ one frequently has an extreme amount of possible optimal solutions to select a set of concepts from. One might see the choice herein as possibly influencing the outcome, but one might lose valuable information just because other words appear in the same average frequency and get lost in the selection process. An objective way to influence the selection process would be to use another distance function but unless we find a more efficient technique to calculate the merge function the performance and usability of this merge function will be extremely poor.

6   Conclusion and Future Work

In this paper we have made a comparison between a weighted f_β-optimal merge function and a simple distance based merge function. We compared a few general properties concerning merge functions that showed that both functions have their merit, but when it came down to usability in a real life problem the f_β-optimal merge function proved to be performing better. The f_β-optimal merge function however has been developed more and is more advanced than the proposed distance based merge function. As previously stated there are several other distance functions we could apply in order to calculate the distance between two sources. Other possibilities include, but one is not restricted to, the Cosine similarity, Hamming distance and other variances on the Minkowksi distance. We are planning to investigate these further but the initial research concerning these measurements falls outside of the scope of this paper.

References

1. Bronselaer, A., Van Britsom, D., De Tré, G.: A framework for multiset merging. Fuzzy Sets and Systems 191, 1–20 (2012)

2. Yager, R.: On the theory of bags. International Journal of General Systems 13(1), 23–27 (1986)
3. Bronselaer, A., De Tré, G., Van Britsom, D.: Multiset merging: The majority rule. In: Melo-Pinto, P., Couto, P., Serôdio, C., Fodor, J., De Baets, B. (eds.) Eurofuse 2011. AISC, vol. 107, pp. 279–292. Springer, Heidelberg (2011)
4. Baeza-Yates, R., Ribeiro-Neto, B.: Modern information retrieval. ACM Press (1999)
5. van Rijsbergen, C.J.: Information Retrieval. Butterworths, London (1979)
6. Bronselaer, A., Van Britsom, D., De Tré, G.: Robustness of multiset merge functions. In: Greco, S., Bouchon-Meunier, B., Coletti, G., Fedrizzi, M., Matarazzo, B., Yager, R.R. (eds.) IPMU 2012, Part I. CCIS, vol. 297, pp. 481–490. Springer, Heidelberg (2012)
7. Van Britsom, D., Bronselaer, A., De Tré, G.: Automatically generating multi-document summarizations. In: Proceedings of the 11th International Conference on Intelligent Systems Design and Applications, pp. 142–147. IEEE (2011)
8. Van Britsom, D., Bronselaer, A., De Tré, G.: Concept identification in constructing multi-document summarizations. In: Greco, S., Bouchon-Meunier, B., Coletti, G., Fedrizzi, M., Matarazzo, B., Yager, R.R. (eds.) IPMU 2012, Part II. CCIS, vol. 298, pp. 276–284. Springer, Heidelberg (2012)

Appendices

A   MultiplicitySet Generated by $\varpi_{\delta\epsilon}$

MultiplicitySet = {*time:[1]*, *right:[0, 1]*, *3:[1]*, *2:[0, 1]*, *5:[0, 1]*, *lines:[0, 1]*, *a:[6]*, *m:[1]*, *s:[5, 6]*, *p:[0, 1]*, *zone:[0, 1]*, *bob:[0, 1]*, *hal:[0, 1]*, *strengthened:[0, 1]*, *had:[1, 2]*, *watch:[1]*, *areas:[0, 1]*, *reached:[0, 1]*, *000:[1, 2]*, *moved:[1]*, *expected:[1]*, *which:[0, 1]*, *there:[1, 2]*, *reported:[1, 2]*, *puerto:[3]*, *western:[0, 1]*, *hurricanes:[0, 1]*, *home:[0, 1]*, *television:[0, 1]*, *tropical:[1]*, *officials:[1]*, *gerrish:[0, 1]*, *cut:[0, 1]*, *jamaica:[3, 4, 5]*, *where:[0, 1]*, *hit:[1]*, *eye:[0, 1, 2]*, *damage:[0, 1]*, *strong:[0, 1]*, *streets:[0, 1]*, *gilbert:[5]*, *while:[0, 1]*, *east:[1]*, *into:[1]*, *night:[2]*, *along:[1]*, *miami:[1]*, *sunday:[1, 2]*, *caribbean:[1, 2]*, *seen:[0, 1]*, *south:[2, 3]*, *down:[0, 1]*, *province:[0, 1]*, *islands:[1, 2]*, *hurricane:[10, 11, 12, 13]*, *strength:[0, 1]*, *ripped:[0, 1]*, *high:[1]*, *people:[1, 2]*, *arrived:[0, 1]*, *slammed:[0, 1]*, *like:[0, 1]*, *coastal:[1]*, *now:[0, 1]*, *residents:[1]*, *radio:[0, 1]*, *but:[1]*, *saturday:[1]*, *north:[1, 2]*, *southeast:[0, 1, 2]*, *haiti:[0, 1, 2]*, *around:[1]*, *sheets:[1]*, *their:[1]*, *first:[0, 1]*, *said:[10]*, *higher:[0, 1]*, *storm:[4, 5, 6]*, *over:[1, 2]*, *government:[0, 1]*, *moving:[1, 2]*, *he:[0, 1]*, *miles:[3]*, *before:[1]*, *ocean:[0, 1]*, *sustained:[1]*, *warnings:[1, 2]*, *by:[1, 2]*, *long:[1]*, *kingston:[0, 1]*, *would:[2]*, *be:[0, 1]*, *get:[0, 1]*, *and:[18, 19, 20]*, *maximum:[0, 1]*, *island:[2, 3]*, *area:[0, 1]*, *edt:[0, 1]*, *formed:[1]*, *all:[1]*, *at:[4]*, *dominican:[2, 3]*, *as:[4, 5]*, *an:[1, 2]*, *off:[2, 3]*, *forecaster:[1]*, *they:[2]*, *no:[1, 2]*, *of:[19, 20, 21, 22, 23]*, *on:[4, 5]*, *only:[1]*, *or:[0, 1]*, *winds:[5, 6]*, *most:[1]*, *flights:[0, 1, 2]*, *larger:[0, 1]*, *second:[0, 1]*, *gulf:[1]*, *when:[0, 1]*, *certainly:[0, 1]*, *republic:[2, 3]*, *issued:[1]*, *heavy:[2, 3]*, *eastern:[0, 1]*, *this:[1, 2]*, *from:[3, 4, 5]*, *was:[4, 5]*, *is:[1, 2]*, *it:[5, 6, 7]*, *know:[0, 1]*, *in:[12]*, *hotel:[0, 1]*, *mph:[3, 4]*, *passed:[0, 1]*, *westward:[0, 1]*, *forecasters:[0, 1, 2]*, *cayman:[0, 1, 2]*, *windows:[0, 1]*, *25:[0, 1]*, *we:[1, 2, 3]*, *next:[0, 1]*, *15:[0, 1]*, *northwest:[1]*, *ve:[0, 1]*, *civil:[0, 1]*, *up:[0, 1]*, *10:[1]*, *to:[10, 11, 12]*, *reports:[0, 1]*, *mexico:[1]*, *that:[2, 3, 4, 5, 6]*, *about:[2]*, *re:[0, 1]*, *rain:[1, 2]*, *defense:[0, 1]*, *track:[0, 1]*, *inches:[1]*, *service:[1]*, *our:[0, 1]*, *out:[0, 1]*, *50:[0, 1]*, *flooding:[1]*, *flash:[0, 1]*, *for:[4, 5]*, *city:[1, 2]*, *center:[3]*, *weather:[1], national:[2, 3], director:[1], trees:[1], cuba:[0, 1, 2], evacuated:[0, 1], south-ern:[0, 1], 100:[1, 2], should:[1], canceled:[0, 1], little:[0, 1], were:[4, 5, 6, 7], three:[0, 1], power:[1], systems:[1], west:[2], other:[0, 1], one:[1, 2], coast:[3, 4], rico:[3], with:[3, 4], the:[40, 41, 42, 43, 44, 45, 46, 47, 48, 49, 50, 51, 52, 53, 54, 55, 56], roofs:[1], continue:[0, 1]}*

B   Complete Mergesets Generated by $\varpi_{\delta\epsilon}$

$\kappa_{min} = \{$*weather=1, winds=5, rico=3, their=1, power=1, puerto=3, hit=1, most=1, island=2, hurricane=10, issued=1, this=1, one=1, northwest=1, sustained=1, expected=1, islands=1, we=1, high=1, mexico=1, dominican=2, for=4, south=2, reported=1, about=2, systems=1, heavy=2, over=1, north=1, warnings=1, repub-lic=2, sunday=1, only=1, night=2, jamaica=3, rain=1, but=1, east=1, it=5, is=1, tropical=1, caribbean=1, in=12, sheets=1, before=1, residents=1, s=5, said=10, on=4, coastal=1, that=2, 100=1, off=2, m=1, with=3, 000=1, of=19, by=1, had=1, moving=1, around=1, a=6, from=3, time=1, should=1, national=2, no=1, and=18, to=10, formed=1, center=3, at=4, there=1, as=4, along=1, west=2, an=1, flooding=1, forecaster=1, 3=1, moved=1, they=2, would=2, people=1, officials=1, roofs=1, 10=1, storm=4, saturday=1, miles=3, city=1, mph=3, watch=1, all=1, gilbert=5, into=1, were=4, miami=1, was=4, coast=3, the=40, long=1, trees=1, gulf=1, director=1, inches=1, service=1*$\}$

$\kappa_{max} = \{$*caribbean=2, like=1, residents=1, that=6, seen=1, puerto=3, 100=2, damage=1, officials=1, warnings=2, inches=1, where=1, into=1, get=1, higher=1, sheets=1, trees=1, we=3, watch=1, western=1, jamaica=5, coast=4, national=3, hurricane=13, southern=1, service=1, around=1, mph=4, radio=1, reached=1, edt=1, ve=1, maximum=1, it=7, reports=1, is=2, hotel=1, in=12, up=1, which=1, evacuated=1, down=1, hit=1, the=56, was=5, gerrish=1, larger=1, certainly=1, city=2, arrived=1, little=1, heavy=3, track=1, he=1, one=2, to=12, center=3, but=1, north=2, first=1, defense=1, three=1, along=1, when=1, this=2, westward=1, south=3, next=1, sunday=2, republic=3, people=2, power=1, other=1, passed=1, right=1, and=20, eastern=1, high=1, islands=2, island=3, most=1, over=2, re=1, while=1, eye=2, gilbert=5, canceled=1, slammed=1, rain=2, miami=1, issued=1, 000=2, area=1, miles=3, haiti=2, night=2, ripped=1, 50=1, tropical=1, all=1, windows=1, time=1, ocean=1, about=2, television=1, their=1, flights=2, flooding=1, strength=1, strengthened=1, southeast=2, with=4, flash=1, storm=6, director=1, they=2, now=1, cuba=2, s=6, p=1, out=1, m=1, weather=1, long=1, our=1, or=1, systems=1, moving=2, on=5, kingston=1, cayman=2, coastal=1, gulf=1, a=6, of=23, formed=1, by=2, west=2, zone=1, dominican=3, said=10, areas=1, for=5, from=5, should=1, winds=6, moved=1, be=1, no=2, hurricanes=1, reported=2, 25=1, lines=1, cut=1, roofs=1, at=4, as=5, mexico=1, 5=1, an=2, before=1, bob=1, 3=1, 2=1, were=7, know=1, saturday=1, forecaster=1, east=1, streets=1, 15=1, sustained=1, 10=1, there=2, hal=1, province=1, would=2, government=1, second=1, home=1, had=2, rico=3, strong=1, northwest=1, continue=1, off=3, civil=1, forecasters=2, expected=1, only=1*$\}$

Flexible Querying
with Linguistic F-Cube Factory[*]

R. Castillo-Ortega[1], Nicolás Marín[1], Daniel Sánchez[1,2], and Carlos Molina[3]

[1] Department of Computer Science and A.I.
University of Granada, 18071, Granada, Spain
{rita,nicm,daniel}@decsai.ugr.es
[2] European Centre for Soft Computing,
33600, Mieres, Asturias, Spain
daniel.sanchezf@softcomputing.es
[3] Department of Languages and Computer Systems,
University of Jaén, 23071, Jaén, Spain
carlosmo@ujaen.es

Abstract. In this paper a new tool which allows flexible querying on multidimensional data bases is presented. *Linguistic F-Cube Factory* is based on the use of natural language when querying multidimensional data cubes to obtain linguistic results. Natural language is one of the best ways of presenting results to human users as it is their inherent way of communication. Data warehouses take advantage of the multidimensional data model in order to store big amounts of data that users can manage and query by means of OLAP operations. They are a context where the development of a linguistic querying tool is of special interest.

Keywords: Linguistic summarization, Time series, Multidimensional data model, OLAP, Business Intelligence, Fuzzy Logic.

1 Introduction

Companies and organizations have to deal with huge amounts of data in their daily operation. The necessity to handle these data has motivated the development of different data management tools. One such tool is data warehousing, based on the multidimensional data model.

The multidimensional data model is based on the use of data cubes. Data cubes are sets of data related to a given fact whose context is described by several dimensions. This way, each cell of the cube contains aggregated data related to elements along each dimension. The use of the multidimensional data

---

[*] Part of this research was supported by the Andalusian Government (Junta de Andalucía, Consejería de Innovación, Ciencia y Empresa) under project P07-TIC-03175 *Representación y Manipulación de Objetos Imperfectos en Problemas de Integración de Datos: Una Aplicación a los Almacenes de Objetos de Aprendizaje.* Part of this research was supported by the Spanish Government (Science and Innovation Department) under project TIN2009-08296.

H.L. Larsen et al. (Eds.): FQAS 2013, LNAI 8132, pp. 245–256, 2013.
© Springer-Verlag Berlin Heidelberg 2013

model and OLAP (OnLine Analytical Processing) operations to query them is of crucial importance in Business Intelligence [20], in which flexibility and understandability when showing the results of queries are key points.

In this context, we are interested in providing data warehousing tools with the possibility to perform flexible queries yielding results expressed in natural language, using a vocabulary adapted to the user. These linguistic results are highly understandable for the users, and are a very good approach for aggregating/summarizing information when performing certain OLAP operations. Historically, fuzzy set theory and fuzzy logic have played an essential role in the attempts to transform data into words to obtain linguistic descriptions understandable by humans [18,7]. Seminal work in this area must be credited to Ronald R. Yager [26,27,28]. Many other approaches have been developed since then [12,22,29,25,30,15,23,21,11]. The obtention of fuzzy summaries from multidimensional databases has been studied by A. Laurent [17].

More specifically, we are interested in providing linguistic results comparing time series obtained from data cubes with time dimension. The temporal dimension is among the most popular dimensions within a data cube structure. This is due to the importance of time in all the activities carried out by humans. These time series are easily obtained by using OLAP operations on datacubes with time dimension. Techniques providing linguistic descriptions of time series are called *time series summarization techniques* in the literature [8,1,13,16,14,5,4,6].

In this paper we describe the extension of an existing flexible data warehousing tool, F-Cube Factory [9], with a flexible querying interface of multidimensional data cubes that takes advantage of linguistic capabilities to produce outcomes expressed by means of natural language patterns. *Linguistic F-Cube Factory* is a friendly tool that allows the users the creation and management of data cubes with linguistic features. We will show how linguistic comparison of time series can be obtained when using *Linguistic F-Cube Factory* querying capabilities on data cubes with temporal dimension. As we will see, the platform includes implemented wizards that offer to the user valuable information to successfully accomplish the querying process.

2   Preliminary Research

This section is devoted to present some concepts and tools developed in previous works of the authors, in which the basis of the present work can be found. We will start with some ideas about the multidimensional data model implemented in F-Cube Factory. Then, we will mention a specific method to carry out linguistic comparison of time series data.

2.1   Multidimensional Data Model and F-Cube Factory

F-Cube Factory [9] is a Business Intelligence system to manage data. It implements different models of data storage as ROLAP, MOLAP and fuzzy MOLAP [19]. The main characteristics of the fuzzy multidimensional model implemented in F-Cube Factory are the following:

Definition 1. *A **dimension** is a tuple $d = (l, \leq_d, l_\perp, l_\top)$ where $l = \{l_1, \ldots, l_m\}$ so that each l_i is a set of values $l_i = \{c_{i1}, \ldots, c_{im_i}\}$ and $l_i \cap l_j = \emptyset$ if $i \neq j$, and $\leq_d$ is a partial order relation between the elements of l so that $l_i \leq_d l_k$ if $\forall c_{ij} \in l_i \Rightarrow \exists c_{kp} \in l_k / c_{ij} \subseteq c_{kp}$. $l_\perp$ and $l_\top$ are two elements of l so that $\forall l_i \in l,\ l_\perp \leq_d l_i \leq_d l_\top$.*

We denote level to each element l_i. To identify the level l of the dimension d we will use $d.l$. The two special levels $l_\perp$ and $l_\top$ will be called *base level* and *top level* respectively. The partial order relation in a dimension is what gives the hierarchical relation between levels.

Definition 2. *For each pair of levels l_i and l_j such that $l_j \in H_i$, we have the relation $\mu_{ij} : l_i \times l_j \to [0,1]$ and we call this the **kinship relation**. H_i is the set of children of l_i, $H_i = \{l_j | l_j \neq l_i \wedge l_j \leq_d l_i \wedge \neg \exists l_k, l_j \leq_d l_k \leq_d l_i\}$*

This relation represents a crisp hierarchy when we use only the values 0 and 1 for kinship and we only allow an element to be included with degree 1 by an unique element of its parent levels. The relaxation of these conditions using values in the interval [0,1] without any other limitation produces a fuzzy hierarchical relation.

Definition 3. *We say that any pair (h, α) is a **fact** when h is an m-tuple on the attributes domain we want to analyze, and $\alpha \in [0,1]$.*

The value α controls the influence of the fact in the analysis. The imprecision of the data is managed by assigning an α value representing this imprecision. Now we can define the structure of a fuzzy DataCube.

Definition 4. *A **DataCube** is a tuple $C = (D, l_b, F, A, H)$ such that $D = (d_1, \ldots, d_n)$ is a set of dimensions, $l_b = (l_{1b}, \ldots, l_{nb})$ is a set of levels such that l_{ib} belongs to d_i, $F = R \cup \emptyset$ where R is the set of facts and $\emptyset$ is a special symbol, A is an application defined as $A : l_{1b} \times \ldots \times l_{nb} \to F$, giving the relation between the dimensions and the facts defined, and H is an object of type history with information regarding the obtention of the current data cube.*

In order to provide support for the definition and management of data within this model, the system F-Cube Factory has been developed [9]. The system is built using client/server architecture. The server implements the main functionality over the data cubes (definition, management, queries, aggregation operators, user views operators, API for data cube access, etc.). The client is web based and is thought to be light enough to be used in a personal computer and to give an intuitive access to the functionality of the server (hiding the complexity of using a DML or DDL to the user). For a more detailed explication of the structure and the operations, see [19,10].

2.2 Linguistic Comparison of Time Series Data

In previous papers [5,4] we have presented a new general model to obtain linguistic summaries from time series data. This model can work with time series data obtained by means of OLAP operations on data cubes with time dimension. Suppose that the time dimension is described in its finest grained level of granularity by members $T = \{t_1, ..., t_m\}$. Then, through the use of appropriate queries we can associate fact values to each of the time members: this way, we have $TS = \{< t_1, v_1 >, ..., < t_m, v_m >\}$, where every v_i is a value of the basic domain of the variable V.

This model has been extended to the case of providing linguistic comparison of time series in terms of values [3]. For that purpose we assume that the time dimension is hierarchically organized in n levels, namely, $L=\{L_1, ..., L_n\}$. Each level L_i has associated a fuzzy partition $\{D_{i,1}, ..., D_{i,p_i}\}$ of the basic time domain.

Let TS_1 and TS_2 be two time series defined over the same variable V at a given period of time. Then,

$$\Delta TS_{local}(t_i) = \begin{cases} 0, \text{if } TS_1(t_i) - TS_2(t_i) = 0 \\ \dfrac{TS_1(t_i) - TS_2(t_i)}{max(TS_1(t_i), TS_2(t_i)) - gm}, \text{otherwise} \end{cases} \quad (1)$$

defines a time series comparing values of the original series, where t_i is a specific point in the time domain and gm is the global minimum of TS_1 and TS_2. Our approach to the linguistic comparison is to obtain a linguistic summary of this time series using a linguistic variable defined on $[-1, 1]$. In this paper, we use a partition with the labels *much lower*=(-1,-1,-0.8,-0.6), *lower*=(-0.8,-0.6,-0.3,-0.1), *similar*=(-0.3,0,0,0.3), *higher*=(0.1,0.3,0.6,0.8), and *much higher*=(0.6,0.8,1,1).

A linguistic comparison of two time series is a set of type II quantified sentences on the form "Q of D are A", where Q is a linguistic quantifier, and A, D are fuzzy subsets of a finite crisp set X. In our case, D is related to the time dimension and A is related to the comparison of values or trends of both series using the labels defined before.

In order to obtain sentences like the one showed before, apart from the fuzzy subsets, we need a family of linguistic quantifiers. In our case, we are going to use a totally ordered subset $\{Q_1, ..., Q_{qmax}\}$ of a coherent family of quantifiers $\mathcal{Q}$ defined in [24].

We have presented Greedy approaches that produce a single optimal linguistic comparison of time series [3,5] based on the algorithms for summarizing series described in [4,2].

3   Linguistic F-Cube Factory

The techniques for comparing series explained in the previous section have been implemented as linguistic OLAP operations within *Linguistic F-Cube Factory*. In this section, we show how to use these operations to perform flexible queries that involve series comparison in the system. We also show the interface that allows the user to interact with the results of queries.

For the sake of illustration, let us consider a data warehouse with information related to different medical centres in a particular area. We count on an existing data cube containing information about the patient inflow according to several dimensions, in this case *location, gender,* and *time.* Each dimension is described using hierarchies of partitions of linguistic labels. As we have already mentioned, if we apply a set of OLAP operations on this data cube, we can obtain time series data representing *"the patient inflow of a certain gender in a certain centre along a certain time period".*

Figure 1 helps to explain the process followed by the new *Linguistic F-Cube Factory* platform in a schematic way. In order to perform the linguistic summarization of the comparison of time series we need to obtain a new data cube containing time series data describing the comparison of time series (among them, the mentioned ΔTS_{local}). In this case, we want to compare the patient inflow in *centre n* with respect to the other centres. This cube is shown in 2) and, as we can see, it has the same number of dimensions but not the same number of elements (in this data cube *centre n* dissapears as is the one being compared with the rest of centres). The final step is the summarization of this data cube to obtain the data cube in 3) containing the linguistic summaries of the comparative time series.

Figure 2 represents the *Linguistic F-Cube Factory* screen in which general and particular information of this data cube are shown. For example, the user can navigate through the hierarchy of a certain dimension by selecting the corresponding *View / Edit* link.

Once the desired data cube has been selected the user has the option of obtaining comparative linguistic summaries in the sense of those commented in Section 2.2. The user has to click on the corresponding link and fill the required fields in the comparison wizard. The wizard is divided to fulfill two main tasks. The first task is to help the user to obtain a new data cube containing a set of different comparative time series data; in this step the user needs to provide the name of the new data cube, the dimension and the dimension member he/she wants to compare with the rest, the temporal dimension, and the dimension representing the comparison in terms of the variable under study. The second task is to build the appropriate linguistic framework to describe the comparative series; here the user has to introduce the number of labels he/she desires to use. At the end of the process the user will have a new data cube as in Figure (1.2).

Figure 3 shows the information of this new data cube. As we can see the data cube maintains the same dimensions plus the automatically generated series that contain the desired comparison framework.

To obtain a data cube as in Figure (1.3) the user has to click in the appropriate link: *expert mode* or *non-expert wizard* like the one in Figure 4, where a column shows the linguistic comparative summaries describing the patient inflow of a certain gender treated in a certain centre along a given time period with respect to the patient inflow in the selected centre. In this case, *centre A* is the one being compared with the rest of centres.

If the user wishes to get more information regarding a certain summary, he/she only needs to select it using the corresponding box. As a result of this action a new screen appears (Figure 5). The screen has three well differentiated areas.

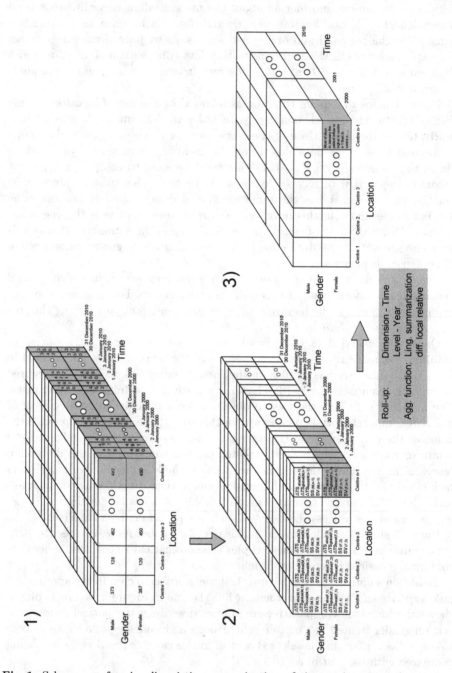

Fig. 1. Schema: performing linguistic summarization of time series comparison on a temporal data cube

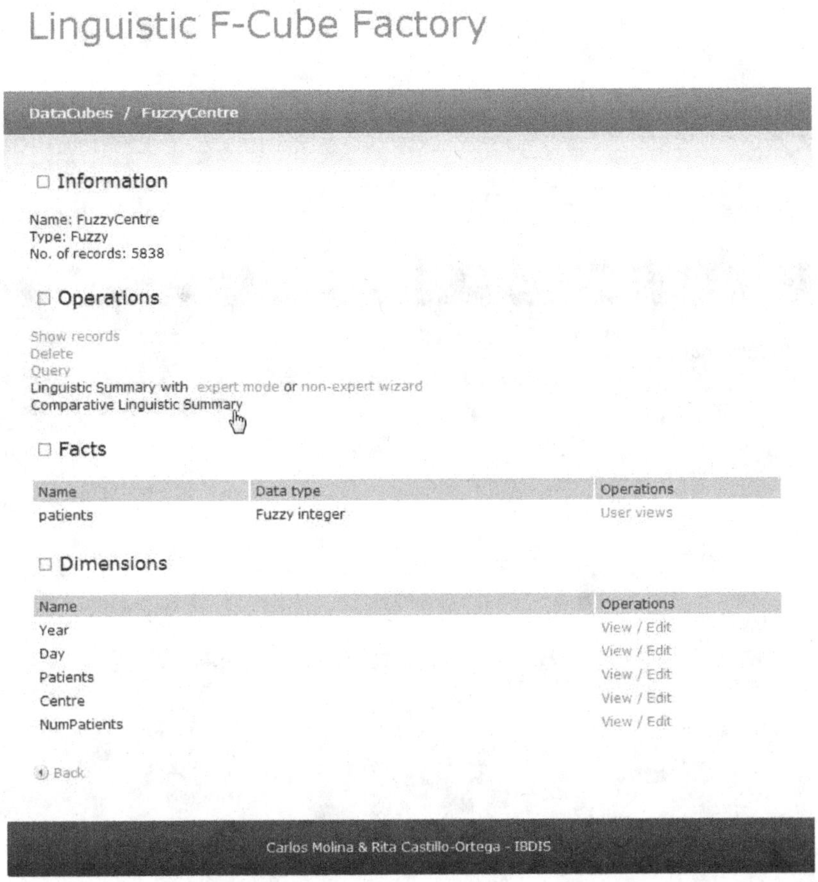

Linguistic F-Cube Factory

DataCubes / FuzzyCentre

□ Information

Name: FuzzyCentre
Type: Fuzzy
No. of records: 5838

□ Operations

Show records
Delete
Query
Linguistic Summary with expert mode or non-expert wizard
Comparative Linguistic Summary

□ Facts

Name	Data type	Operations
patients	Fuzzy integer	User views

□ Dimensions

Name	Operations
Year	View / Edit
Day	View / Edit
Patients	View / Edit
Centre	View / Edit
NumPatients	View / Edit

◁ Back

Carlos Molina & Rita Castillo-Ortega - IBDIS

Fig. 2. FuzzyCentre data cube information in *Linguistic F-Cube Factory*

The upper part shows the graphical representation of the selected comparative time series data. The X axis represents the temporal dimension while the Y axis represents the described dimension, in this case the local relative comparison of centre A and centre B male patient inflow during year 2009.

Then, the central zone represents the complete linguistic summary built by post-processing the original quantified sentences. Finally, the last zone is dedicated to show the user the list of quantified sentences compounding the result. This is called the *raw result*. The accomplishment degree of each sentence is also shown. There is the possibility of obtaining finest information about each sentence by clicking them. The user can listen the sentence by means of a dedicated media player and can see in the graphical representation the points that support the sentences by means of highlighted areas.

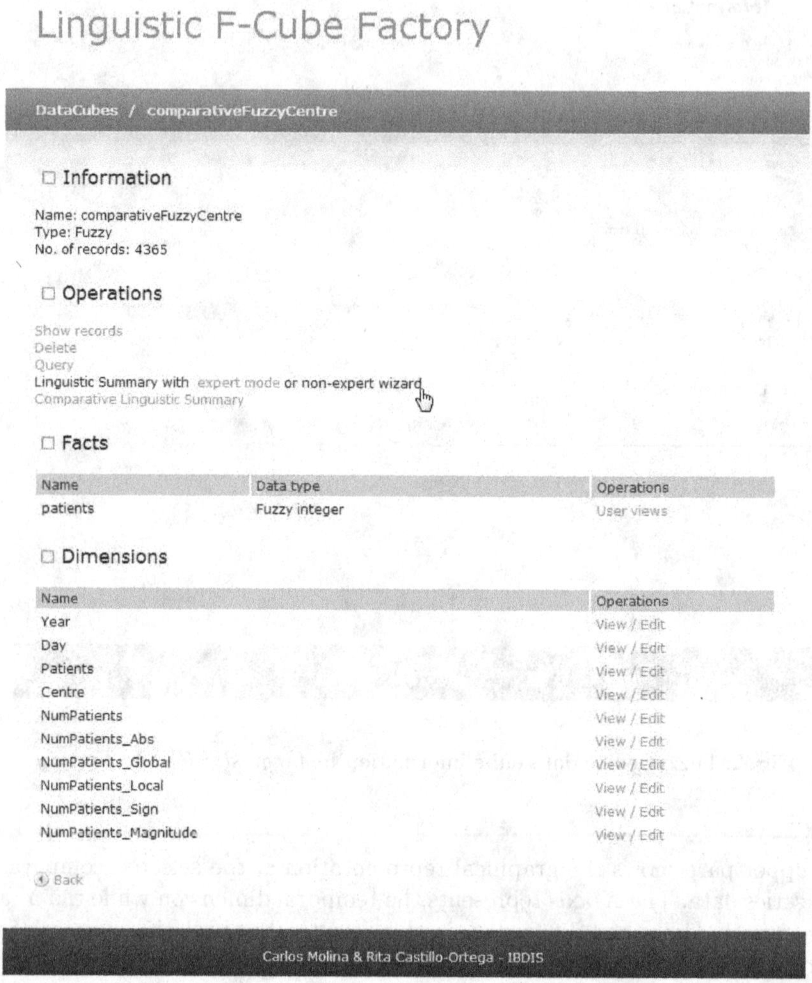

Linguistic F-Cube Factory

DataCubes / comparativeFuzzyCentre

☐ **Information**

Name: comparativeFuzzyCentre
Type: Fuzzy
No. of records: 4365

☐ **Operations**

Show records
Delete
Query
Linguistic Summary with expert mode or non-expert wizard
Comparative Linguistic Summary

☐ **Facts**

Name	Data type	Operations
patients	Fuzzy integer	User views

☐ **Dimensions**

Name	Operations
Year	View / Edit
Day	View / Edit
Patients	View / Edit
Centre	View / Edit
NumPatients	View / Edit
NumPatients_Abs	View / Edit
NumPatients_Global	View / Edit
NumPatients_Local	View / Edit
NumPatients_Sign	View / Edit
NumPatients_Magnitude	View / Edit

◁ Back

Carlos Molina & Rita Castillo-Ortega - IBDIS

Fig. 3. comparativeFuzzyCentre data cube information in *Linguistic F-Cube Factory*

Linguistic F-Cube Factory

DataCubes / comparativeFuzzyCentre / LinguisticSummary / comparativeSummaryN1

☐ **Result**

Year	Patients	Centre	Summary	
2009	Female	B	At least 70% of days with mild weather, the patient inflow is higher or much higher in centre B than in centre A. At least 80% of days with cold weather, the patient inflow is lower or higher in centre B than in centre A; in September is higher or much higher. Most of days in June, the patient inflow is higher or much higher in centre B than in centre A; in May is higher. In August, and in July the comparison results are highly variable.	☐
2009	Female	C	At least 70% of days with mild weather, the patient inflow is higher or much higher in centre C than in centre A; with cold weather is lower or higher; with hot weather is much lower or higher.	☐
2009	Male	B	Most of days with cold weather, and in November, the patient inflow is higher or much higher in centre B than in centre A. At least 70% of days with hot weather, and in September, the patient inflow is much lower or lower in centre B than in centre A. In April, in March, in May, and in October the comparison results are highly variable.	☑
2009	Male	C	Most of days with cold weather, the patient inflow is higher or much higher in centre C than in centre A. At least 80% of days with hot weather, the patient inflow is much lower or lower in centre C than in centre A. At least 70% of days with hot to cold weather, the patient inflow is lower or higher in centre C than in centre A. In April, in March, and in May the comparison results are highly variable.	☐
2010	Female	B	At least 70% of days with mild weather, the patient inflow is higher or much higher in centre B than in centre A. At least 80% of days with cold weather, the patient inflow is lower or higher in centre B than in centre A; in September are higher or much higher. Most of days in June, the patient inflow is higher or much higher in centre B than in centre A; in May are higher. In August, and in July the comparison results are highly variable.	☐
2010	Female	C	At least 70% of days with mild weather, the patient inflow is higher or much higher in centre C than in centre A; with cold weather is lower or higher; with hot weather is much lower or higher.	☐
2010	Male	B	Most of days with cold weather, and in November, the patient inflow is higher or much higher in centre B than in centre A. At least 80% of days with hot weather, the patient inflow is much lower or lower in centre B than in centre A. At least 70% of days in September, the patient inflow is much lower or lower in centre B than in centre A. In April, in March, in May, and in October the comparison results are highly variable.	☐
2010	Male	C	Most of days with cold weather, the patient inflow is higher or much higher in centre C than in centre A. At least 80% of days with hot weather, the patient inflow is much lower or lower in centre C than in centre A. At least 70% of days with hot to cold weather, the patient inflow is lower or higher in centre C than in centre A. In April, in March, and in May the comparison results are highly variable.	☐

Fig. 4. Resulting data cube with comparative linguistic summaries as facts

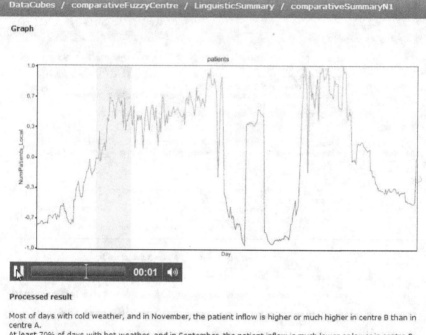

Linguistic F-Cube Factory

DataCubes / comparativeFuzzyCentre / LinguisticSummary / comparativeSummaryN1

Graph

Processed result

Most of days with cold weather, and in November, the patient inflow is higher or much higher in centre B than in centre A.
At least 70% of days with hot weather, and in September, the patient inflow is much lower or lower in centre B than in centre A.
In April, in March, in May, and in October the comparison results are highly variable.

Raw result

Select the sentence you want to see with more detail.

○ 1.0 / Most of days with cold weather, the patient inflow is higher or much higher in centre B than in centre A
○ 1.0 / At least 70% of days with hot weather, the patient inflow is much lower or lower in centre B than in centre A
○ 1.0 / in April the comparison results are highly variable
◉ 1.0 / in March the comparison results are highly variable
○ 1.0 / in May the comparison results are highly variable
○ 0.9655698 / Most of days in November, the patient inflow is higher or much higher in centre B than in centre A
○ 1.0 / in October the comparison results are highly variable
○ 1.0 / At least 70% of days in September, the patient inflow is much lower or lower in centre B than in centre A

◁ Back

Carlos Molina & Rita Castillo-Ortega - IBDIS

Fig. 5. Detailed information regarding to a specific linguistic comparative summary

4   Conclusions and Future Work

In this work we have presented how the *Linguistic F-Cube Factory* platform can be used for the creation and management of data cubes containing linguistic comparisons of time series.

This allows the users to carry out flexible queries in which the processes outcomes are linguistic results. The incorporation of linguistic capabilities to the platform by means of the use of Fuzzy Logic makes the tool nearer to human users that have at their disposal friendly results that are more useful to develop their activities.

We are working on extending the functionality of the linguistic platform through the use of different methods for the communication of the results to users.

References

1. Batyrshin, I.Z., Sheremetov, L.: Perception-based approach to time series data mining. Appl. Soft Comput. 8(3), 1211–1221 (2008)
2. Castillo-Ortega, R., Marín, N., Sánchez, D.: Linguistic summary-based query answering on data cubes with time dimension. In: Andreasen, T., Yager, R.R., Bulskov, H., Christiansen, H., Larsen, H.L. (eds.) FQAS 2009. LNCS, vol. 5822, pp. 560–571. Springer, Heidelberg (2009)
3. Castillo-Ortega, R., Marín, N., Sánchez, D.: Time series comparison using linguistic fuzzy techniques. In: Hüllermeier, E., Kruse, R., Hoffmann, F. (eds.) IPMU 2010. LNCS, vol. 6178, pp. 330–339. Springer, Heidelberg (2010)
4. Castillo-Ortega, R., Marín, N., Sánchez, D.: A fuzzy approach to the lingusitic summarization of time series. Journal of Multiple-Valued Logic and Soft Computing (JMVLSC) 17(2-3), 157–182 (2011)
5. Castillo-Ortega, R., Marín, N., Sánchez, D.: Linguistic query answering on data cubes with time dimension. International Journal of Intelligent Systems (IJIS) 26(10), 1002–1021 (2011)
6. Castillo-Ortega, R., Marín, N., Sánchez, D., Tettamanzi, A.G.B.: A multi-objective memetic algorithm for the linguistic summarization of time series. In: GECCO, Genetic and Evolutionary Computation Conference 2011, pp. 171–172 (2011)
7. Chen, G., Wei, Q., Kerre, E.E.: Fuzzy Logic in Discovering Association Rules: An Overview. In: Data Mining and Knowledge Discovery Approaches Based on Rule Induction Techniques, Massive Computing Series. Massive Computing Series, ch. 14, pp. 459–493. Springer, Heidelberg (2006)
8. Chiang, D., Chow, L.R., Wang, Y.: Mining time series data by a fuzzy linguistic summary system. Fuzzy Sets Syst. 112, 419–432 (2000)
9. Delgado, M., Molina, C., Rodríguez Ariza, L., Sánchez, D., Vila Miranda, M.A.: F-cube factory: a fuzzy OLAP system for supporting imprecision. International Journal of Uncertainty, Fuzziness and Knowledge-Based Systems 15(suppl.1), 59–81 (2007)
10. Delgado, M., Molina, C., Sánchez, D., Vila, M.A., Rodriguez-Ariza, L.: A linguistic hierarchy for datacube dimensions modelling. In: Current Issues in Data and Knowledge Engineering, Varsovia, pp. 167–176 (September 2004)
11. Díaz-Hermida, F., Bugarín, A.: Linguistic summarization of data with probabilistic fuzzy quantifiers. In: ESTYLF 2010, Proceedings of XV Congreso Español Sobre Tecnologías y Lógica Fuzzy, Huelva, Spain, Frebruary 3-5, pp. 255–260 (2010)

12. Kacprzyk, J.: Fuzzy logic for linguistic summarization of databases. In: IEEE International Fuzzy Systems Conference, pp. 813–818 (1999)
13. Kacprzyk, J., Wilbik, A., Zadrozny, S.: Linguistic summarization of time series using a fuzzy quantifier driven aggregation. Fuzzy Sets and Systems 159(12), 1485–1499 (2008)
14. Kacprzyk, J., Wilbik, A., Zadrozny, S.: An approach to the linguistic summarization of time series using a fuzzy quantifier driven aggregation. Int. J. Intell. Syst. 25(5), 411–439 (2010)
15. Kacprzyk, J., Zadrozny, S.: Data mining via protoform based linguistic summaries: Some possible relations to natural language generation. In: Proceedings of the IEEE Symposium on Computational Intelligence and Data Mining, CIDM 2009, part of the IEEE Symposium Series on Computational Intelligence 2009, Nashville, TN, USA, March 30- April 2, pp. 217–224 (2009)
16. Kobayashi, I., Okumura, N.: Verbalizing time-series data: With an example of stock price trends. In: IFSA/EUSFLAT Conf., pp. 234–239 (2009)
17. Laurent, A.: A new approach for the generation of fuzzy summaries based on fuzzy multidimensional databases. Intell. Data Anal. 7, 155–177 (2003)
18. Mitra, S., Pal, S.K., Mitra, P.: Data mining in soft computing framework: A survey. IEEE Transactions on Neural Networks 13, 3–14 (2001)
19. Molina, C., Rodríguez Ariza, L., Sánchez, D., Vila, M.A.: A new fuzzy multidimensional model. IEEE T. Fuzzy Systems 14(6), 897–912 (2006)
20. O'Brien, J.A., Marakas, G.M.: Management information systems, 8th edn. McGraw-Hill (2008)
21. Pilarski, D.: Linguistic summarization of databases with quantirius: a reduction algorithm for generated summaries. International Journal of Uncertainty, Fuzziness and Knowledge-Based Systems 18(3), 305–331 (2010)
22. Raschia, G., Mouaddib, N.: SAINTETIQ: a fuzzy set-based approach to database summarization. Fuzzy Sets Syst. 129(2), 137–162 (2002)
23. van der Heide, A., Triviño, G.: Automatically generated linguistic summaries of energy consumption data. In: Ninth International Conference on Intelligent Systems Design and Applications, ISDA 2009, Pisa, Italy, November 30-December 2, pp. 553–559 (2009)
24. Vila, M.A., Cubero, J.C., Medina, J.M., Pons, O.: The generalized selection: an alternative way for the quotient operations in fuzzy relational databases. In: Bouchon-Meunier, B., Yager, R., Zadeh, L. (eds.) Fuzzy Logic and Soft Computing. World Scientific Press (1995)
25. Voglozin, W.A., Raschia, G., Ughetto, L., Mouaddib, N.: Querying a summary of database. J. Intell. Inf. Syst. 26(1), 59–73 (2006)
26. Yager, R.R.: A new approach to the summarization of data. Information Sciences (28), 69–86 (1982)
27. Yager, R.R.: Toward a language for specifying summarizing statistics. IEEE Transactions on Systems, Man, and Cybernetics, Part B 33(2), 177–187 (2003)
28. Yager, R.R.: A human directed approach for data summarization. In: IEEE International Conference on Fuzzy Systems, pp. 707–712 (2006)
29. Yager, R.R., Petry, F.E.: A multicriteria approach to data summarization using concept ontologies. IEEE T. Fuzzy Systems 14(6), 767–780 (2006)
30. Zhang, L., Pei, Z., Chen, H.: Extracting fuzzy linguistic summaries based on including degree theory and FCA. In: Melin, P., Castillo, O., Aguilar, L.T., Kacprzyk, J., Pedrycz, W. (eds.) IFSA 2007. LNCS (LNAI), vol. 4529, pp. 273–283. Springer, Heidelberg (2007)

Mathematical Morphology Tools to Evaluate Periodic Linguistic Summaries

Gilles Moyse, Marie-Jeanne Lesot, and Bernadette Bouchon-Meunier

UPMC Univ. Paris 06, CNRS UMR 7606, LIP6, F-75005, Paris, France
surname.name@lip6.fr

Abstract. This paper considers the task of establishing periodic linguistic summaries of the form "Regularly, the data take high values", enriched with an estimation of the period and a linguistic formulation. Within the framework of methods that address this task testing whether the dataset contains regularly spaced groups of high and low values with approximately constant size, it proposes a mathematical morphology (MM) approach based on watershed. It compares the proposed approach to other MM methods in an experimental study based on artificial data with different forms and noise types.

Keywords: Fuzzy linguistic summaries, Periodicity computing, Mathematical Morphology, Temporal data mining, Watershed.

1 Introduction

Linguistic summaries aim at building human understandable representations of datasets, thanks to natural language sentences. They take different forms representing different kinds of patterns [27,28,12]. In this paper we consider this task in the case of time series for which regularity is looked for, more precisely summaries of the form "Regularly, the data take high values". If the data are membership degrees to a fuzzy modality A, the sentence can be interpreted as "regularly, the data are A". Moreover, if the sentence holds, a candidate period is computed and an appropriate linguistic formulation is generated, based on the choice of a relevant time unit, approximation and adverb. The final sentence can for instance be "Approximately every 20 hours, the data take high values".

The Detection of Periodic Events (DPE) methodology [21] defines a framework to address this task relying on the assumption that if a dataset contains regularly spaced high and low value groups of approximately constant size, then it is periodic. It consists in 3 steps: clustering, cluster size regularity and linguistic rendering. In [21], the first step is based on the calculation of an erosion score based on Mathematical Morphology [24]. In this paper, we propose to apply the DPE methodology using a new clustering method depending on a watershed approach [4] and to compare it in an enriched experimental protocol.

Section 2 presents an overview of related works. A reminder about the evaluation of periodic protoforms is given in Section 3. The proposed watershed method is described in Section 4. Lastly, Section 5 presents experimental results on artificial data comparing the two approaches as well as a baseline method.

H.L. Larsen et al. (Eds.): FQAS 2013, LNAI 8132, pp. 257–268, 2013.

2   Related Works

This section briefly describes the principles of linguistic summaries, temporal data mining and period detection in signal processing, at the crossroads of which the considered DPE methodology lies. To the best of our knowledge, DPE is the first approach combining these fields.

2.1   Linguistic Summaries

Linguistic summaries aim at building compact representations of datasets, in the form of natural language sentences describing their main characteristics. Besides approaches based on natural language generation techniques, they can be produced using fuzzy logic, in which case they are called fuzzy linguistic summaries (see [5,13] for a comparison between these two areas).

Introduced in the seminal papers [11,27,28], they are built on sentences called "protoforms", such as "QX are A" where Q is a quantifier (e.g. "most" or "around 10"), A a linguistic modality associated with one of the attributes (e.g. "young" for the attribute "age") and X the data to summarise. The relevance of a candidate protoform, measured by the truth degree of its instantiation for the considered data, depends on the Σ-count of the dataset according to the chosen fuzzy modality. Extensions have been defined to handle the temporal nature of data, using a "Trend" attribute [10] or considering fuzzy temporal propositions [6] to restrict the truth value of a summary to a certain period of time, but they do not cope with periodicity.

2.2   Temporal Data Mining

Temporal data mining is a domain that groups various issues related to data mining taking into account the temporal aspect of the data (see [8,15] for exhaustive states of the art). Some methods aim at discovering frequent patterns, using extensions of the Apriori algorithm [1], possibly dedicated to long sequences [19] or with time-window or duration constraints [16,20]. Although mining recurring events, these approaches are not concerned with periodicity. Cyclic association rules [22] are satisfied on a fixed periodic basis: the time axis is split into constant length segments against which association rules are tested. So as to automatically compute a candidate period, extensions based on a Fourier transform [3] or a statistical test over the average interval between events [17] can be used.

2.3   Signal Processing for Period Detection

Period detection is a well known problem in signal processing and several methods have been proposed to address it.

The most straightforward one is based on an analysis in the time domain, and computes the period by measuring the distance between two successive zero-crossings [14]. It is very sensitive to noise.

The two most common methods are autocorrelation [9] in the time domain and spectral analysis with Fourier transform [23] in the frequency domain. They are efficient on specific data, namely sinusoidal and stationary signals, in which the period remains constant. The short-time Fourier [2] and wavelet [18] transforms are more sophisticated methods in the time-frequency domain able to deal with non stationary signals. However, the former needs a window size parameter to be efficient, and the latter is non parametric but very sensitive to time shifts, which is a bias to be avoided in the context of a high-level linguistic interpretation of the data.

Statistical methods have also been proposed, applying to the specific case where the data is sinusoidal with a Gaussian noise [7].

Lastly, cross-domain approaches have been developed, adding further complexity: in [3], as already mentioned, a fast Fourier transform is used on top of cyclic association rule extraction to build a list of candidate periods. In [26], both autocorrelation and a periodogram are used.

3   Evaluation of Periodic Protoforms

The principle of the Detection of Periodic Events methodology (DPE) [21] relies on the assumption that if a dataset contains regularly spaced high and low value groups of approximately constant size, then it is periodic. This assumption guides the truth evaluation of sentences of the form "Every p, values are high". DPE is a modular methodology which can be seen as a general framework to evaluate periodic protoforms. This section describes its general architecture of the DPE methodology as well as its instanciation proposed in [21]. Section 4 presents a new watershed based method for its clustering step.

3.1   Input and Output

The input dataset, denoted X, is temporal and contains N normalised values (x_i), i.e. $X = \{x_i, i = 1, ..., N\}$ such that $\forall i$, $x_i \in [0, 1]$. The data are considered to be regularly sampled, i.e. at date $t_i = t_1 + (i - 1) \times \Delta t$ where t_1 is the initial measurement time and Δt is the sampling rate.

The outputs of the DPE methodology are a periodicity degree π, a candidate period p_c and a natural language sentence. The periodicity degree π indicates the extent to which the dataset is periodic: 1 means it is absolutely periodic and the value decreases as the dataset is less periodic. The sentence is a linguistic description of the period found in the dataset, designed for human understanding. It has the form "M every p $unit$, the data take high values", where M is an adverb as "roughly", "exactly", "approximately", p is the approximate value based on the candidate period p_c, and $unit$ is a unit considered the most appropriate to express the period [21].

3.2   Architecture

The DPE methodology works in four steps : first, it clusters the data into groups of successive high or low values, second it computes the regularity of the group sizes and the periodicity degree, third it computes a candidate period and finally it returns a natural language sentence. In the following, more details are given regarding these 4 steps.

High and Low Value Detection. The first step of DPE aims at detecting groups of high/low consecutive values. To this aim, a prediction function g returning the group type (H or L) of x_i is defined. Successive values classified H are gathered in high value groups, and conversely for low values.

A baseline function g_{BL} relies on a user-defined threshold t_{value} to distinguish high and low values :

$$g_{BL}(x_i) = \begin{cases} H & \text{if } x_i > t_{value} \\ L & \text{otherwise} \end{cases} \tag{1}$$

A function g_{ES} exploiting mathematical morphology tools is proposed in [21], based on the erosion score es defined as:

$$x_i^0 = x_i \qquad x_i^j = \min\left(x_{i-1}^{j-1}, x_i^{j-1}, x_{i+1}^{j-1}\right) \qquad es_i = \sum_{j=1}^{z} x_i^j$$

where z is the smallest integer such that $\forall i = 1...N$, $x_i^z = 0$. This erosion score transform, classically used to identify the skeleton of a shape, has the following characteristics: high x_i in high regions have high es, low x_i in high regions have quite a high es, isolated high x_i in low regions have low es. Thus, erosion scores provide an automatic adaptation to the data level.

Computing the erosion score on the data complement $\overline{X}$ where $\overline{x}_i = 1 - x_i$ allow to symmetrically identify low regions. We propose the prediction function:

$$g_{ES}(x_i) = \begin{cases} H & \text{if } es_i > \overline{es}_i \\ L & \text{otherwise} \end{cases} \tag{2}$$

Groups are defined as successive values of the same type as returned by g.

Periodicity Computing. The second step of DPE consists in evaluating the regularity of the sizes of the high and low value groups. If these sizes are regular, then the dataset is considered periodic according to the assumption defined at the beginning of this section.

First, the size of each group is computed, setting $s_j^H = \left|G_j^H\right|$ for the jth high value group and $s_j^L = \left|G_j^L\right|$ for the jth low value group. Experiments with fuzzy cardinalities have showed no significant difference [21].

The regularity ρ is then determined for high and low value groups based on the average value μ and the deviation d of their size (see [21] for justification):

$$\rho = 1 - \min\left(\frac{d}{\mu}, 1\right) \qquad \mu = \frac{1}{n}\sum_{j=1}^{n} s_j \qquad d = \frac{1}{n}\sum_{j=1}^{n} |s_j - \mu| \qquad (3)$$

both for high and low value groups n denotes the number of groups. The size dispersion is thus measured using the coefficient of variation $CV = d/\mu$: d is more robust to noise than standard deviation and the quotient with μ makes it relative and allows to adapt to the value level.

Finally, with the regularities of high value groups ρ^H and low value groups ρ^L, the periodicity degree π is returned as their average, i.e. $\pi = \left(\rho^H + \rho^L\right)/2$.

Candidate Period Computation. For a perfectly regular phenomenon, the period is defined as the time elapsed between two occurrences of an event, in this paper, "high value". Therefore the candidate period p_c is approximated as the sum of the average size of high and low value groups, i.e. $p_c = \mu^H + \mu^L$. p_c is relevant only if π is high enough, i.e. if the dataset is considered as periodic.

Linguistic Rendering. The last step yields a linguistic periodic summary of the form "M every p unit, the data take high values", significant only if π is high. As described in [21], the unit used to describe the data is calculated first on a set of units entered as prior knowledge. Then the period p_c is rounded in order to make it more natural for a human being. Lastly, an adverb is chosen based on the approximation error between the computed and the rounded value.

4 Watershed Based Method

We propose a new method to identify high and low value groups based on another Mathematical Morphology tool, namely watershed. It can be seen as a variant of g_{BL} where the threshold is automatically derived from the data: it reduces the required expert knowledge and automatically adapts to the data.

4.1 Principle

Watershed in Mathematical Morphology has been introduced in [4] to perform 2D image segmentation. Its underlying intuition comes from geography: the image greyscale levels are seen as a topographic relief which is flooded by water. Watersheds are the divide lines of the domains of attraction of rain falling over the region. An efficient implementation has been proposed in [25] based on an immersion process analogy: as illustrated in Fig. 1, when the level of water rises, basins appear. When it rises more, new ones are created while others merge. At the end, all basins merge into a single one.

We propose to apply watershed to detect groups, defining them from the identified basins for a given water line: low value groups are defined as the

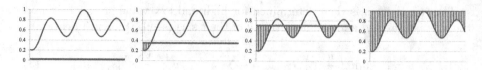

Fig. 1. Illustration of the immersion of a dataset for the watershed calculation

basins, i.e. consecutive values below the divide line, and high value groups as consecutive values above the divide line.

Furthermore, we propose to base the identification of the relevant water line, i.e. the threshold to separate high and low value groups, on the evolution of the basin structure. Indeed, the desired threshold should lead to a group identification that is robust to small local noise, i.e. making it a little greater or lower should not modify the number of identified groups. As water rises, basins appear (resp. disappear), when a gap (resp. a peak) crosses the water line: the threshold should be located at a level where no peak and gap, that represent local noise, are present. As formalised below, we propose to set the water line at the middle of the largest interval separating two consecutive basin structure changes.

4.2 Implementation

The changes of the basin structure are easily identified as they correspond to local peaks and gaps, i.e. values resp. greater or lower than their direct neighbours (previous and next values): when a peak or gap is identified, its level is recorded as a level where a basin structure change occurs. This principle is formalised below after the description of the pre-processing steps applied to the data.

Preprocessing. Before finding the peaks and gaps, 2 preprocessing steps are applied: first, a moving average on a window whose size w is chosen by expert knowledge is calculated to smooth the curve and to avoid oversegmentation.

Second the consecutive equal values are removed. Indeed, basin structure changes could occur in configurations named "plateaux" which are different from gaps and peaks. A plateau is a collection of consecutive equal points surrounded with points of lesser (convex plateaux) or greater (concave plateaux) values. So as to ease the structure change detection and once the data are smoothed, consecutive equal points are removed from the dataset, so that convex plateaux become peaks, and concave plateaux become gaps. Thus, all basin structure changes can be detected with a simple peak/gap analysis.

Processing. With the preprocessed data W, the determination of the levels where the basin structure changes is done with a single scan to detect local peaks and gaps. The levels at which the changes occur are stored in L:

$$L = \{w_i \in W / (w_i > w_{i+1} \wedge w_i > w_{i-1}) \vee (w_i < w_{i+1} \wedge w_i < w_{i-1})\}$$

Then the adaptable watershed-based threshold t_W is computed as:

$$t_W = \tfrac{1}{2}\left(L_m + L_{m+1}\right) \qquad\qquad m = \underset{i \in \{1...|L|-1\}}{\arg\max}\ L_{i+1} - L_i \qquad (4)$$

where L_j are the elements of L sorted in ascending order. Finally, the clustering function is defined as:

$$g_W\left(x_i\right) = \begin{cases} H & \text{if } x_i > t_W \\ L & \text{otherwise} \end{cases} \qquad (5)$$

5 Experimental Results

This section presents results obtained with artificial data, to compare the baseline, erosion score and watershed methods defined by (1), (2) and (5).

5.1 Data Generation

The datasets are generated as noisy series of periodic shapes, either rectangles or sines. They are created as a succession of high and low value groups, of size p^H and p^L respectively, on which two types of noise are applied: the group size noise ν_s randomly modifies the size values p^H and p^L, the value noise ν_y changes the values taken by the data within the groups. In the first step, the sizes of the high and low value groups are randomly drawn, adding some noise to the ideal values p^H and p^L: p^* generally denoting one of these two values, the size of each group is defined as:

$$s_j = \lceil 1 + \text{sgn}\left(0.5 - \epsilon_1\right) \times \nu_s \times \epsilon_2 \rceil p^*$$

where ϵ_1 and ϵ_2 are uniform random variables $\mathcal{U}(0,1)$. This distribution randomly increases or decreases the reference group size, through the $\text{sgn}(0.5 - \epsilon_1)$ coefficient, in a proportion defined as $\nu_s \times \epsilon_2$. The size of a group thus varies between $(1 - \nu_s)p^*$ and $(1 + \nu_s)p^*$. Group sizes are generated until their cumulative sum reaches the total desired number of points N.

After the group sizes have been determined for the rectangle shape, if the jth group spans from index a to b, X^* is set as:

$$\forall k \in \{a,\dots,b\} \quad x_k^* = \begin{cases} 1 & \text{if group } j \text{ is high} \\ 0 & \text{otherwise} \end{cases}$$

For the sine shape, if the jth group spans from index a to b, X^* is set as:

$$\forall k \in \{a,...,b\} \quad x_k^* = \frac{1}{2} + \frac{\lambda}{2}\sin\left(\pi\frac{k-a}{b-a}\right) \qquad \lambda = \begin{cases} 1 \text{ if group } j \text{ is high} \\ -1 \text{ otherwise} \end{cases}$$

This calculation for the sine shape creates a discontinuous break around 0.5. It does not seem to introduce biases in the results though.

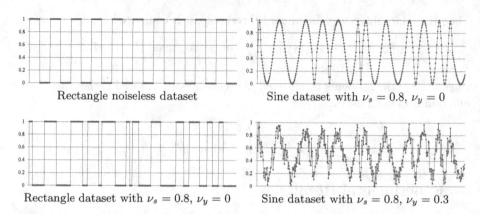

Fig. 2. Four examples of generated datasets

In the third step the value noise ν_y is added to X^* leading to $\widehat{X}$. The noise is applied downward for high value groups and upward for low value groups:

$$\widehat{x}_i = \begin{cases} x_i^* - \nu_y\epsilon & \text{if } \widehat{x}_i \text{ is in a high value group} \\ x_i^* + \nu_y\epsilon & \text{otherwise} \end{cases}$$

where ϵ is a uniform random variable $\mathcal{U}(0, 1)$.

Finally, the dataset X results from the normalisation of $\widehat{X}$ to $[0, 1]$. Fig. 2 illustrates 4 examples of generated datasets.

5.2 Experimental Protocol

As shown in Table 1, 16 test scenarios are implemented where the data series are generated with an increasing value noise ν_y and group size noise ν_s from 0 to 1 at a 0.05 pace (21 values) with a different combination of high / low value groups size, shape, group size noise and value noise.

The periodicity degree π, the candidate period p_c, the error in period evaluation Δp and the clustering accuracy Acc are computed with the 3 methods, baseline (BL), watershed (W) and erosion score (ES): Δp is defined as $\Delta p = |p_c - p| / p$. The period p to compare the candidate period with is computed as the sum of the average sizes of the two types of generated groups.

Table 1. The 16 test scenarios. The noise specified in the header is the constant one in the scenario, the not mentioned one is the increasing one.

Size (p^H/p^L)	Shape	$\nu_s = 0$	$\nu_y = 0$	$\nu_s = 0.5$	$\nu_y = 0.3$
Balanced (25/25)	Square	S1	S2	S9	S10
	Sine	S3	S4	S11	S12
Thin (10/40)	Square	S5	S6	S13	S14
	Sine	S7	S8	S15	S16

To compute *Acc*, the accuracy in the classification into high and low value groups, the labels are the group membership defined in the generation step. *Acc* is weighted so as to take into account the bias in the group size, for the "Thin (10/40)" scenarios.

5.3   Result Interpretation

General Results. From the results obtained with the 16 scenarios and not detailed here, it appears that the noise type (group size or value) is the most important parameter: all methods exhibit similar behaviours for the 3 measures π, Δp and *Acc* for a given noise type. The shape is the second most important parameter, since differences appear between squares and sines, especially with increasing ν_y. Other parameters, as the size of the groups (balanced or thin) or the combination of noises, do not seem to bear an important influence.

The results also show that all 3 methods return a periodicity of 1 when the data has no noise and are decreasing functions of the noise parameters.

In the following, we focus on scenarios 5 to 8. Figure 3 illustrates the outcomes of the experiments. Nevertheless, the results mentioned below are valid for all considered scenarios.

Baseline Approach. The comparison between methods shows that the baseline curve is very sensitive to noise. Indeed, with squares (Fig. 3a, b), π falls sharply and Δp rises sharply as soon as ν_y reaches 0.5. This is due to the fact that from this level of noise, some points labelled as high have a value smaller than 0.5 and are classified as low. Since very few points are misclassified, accuracy is still high but small groups are created within larger ones, generating a high deviation in the group size, yielding poor periodicity degree and period evaluation precision.

This behaviour appears for lower values of ν_y with sines (Fig. 3g, h) since this kind of misclassification is possible as soon as $\nu_y > 0$.

Interestingly, the baseline *Acc* is always comparable to the one obtained with the other methods (Fig. 3c, f, i, l). Indeed, the phenomenon just described slightly affects the clustering accuracy. Moreover, as ν_y increases, the accuracy decreases in approximately the same amount for all methods, so BL remains comparable with the others. This is why the *Acc* measure is not very relevant here to choose a method, whereas Δp is much more discriminant.

Erosion Score vs. Watershed. Generally speaking, erosion score is smoother than watershed in the sense that it varies less abruptly, ensuring steadiness in the evaluation. Moreover, it is generally more or equally precise in calculation of the period (11 scenarios out of 16) and in clustering accuracy (12 scenarios out of 16). Furthermore, the erosion score is also more precise over several experiences since it has a lower standard deviation than the watershed.

Regarding group size noise, watershed gives wrong period estimation when $\nu_g > 0.7$ (Fig. 3e). This is due to the fact that the moving average makes small peaks disappear. On the other hand, ES keeps these small peaks especially with

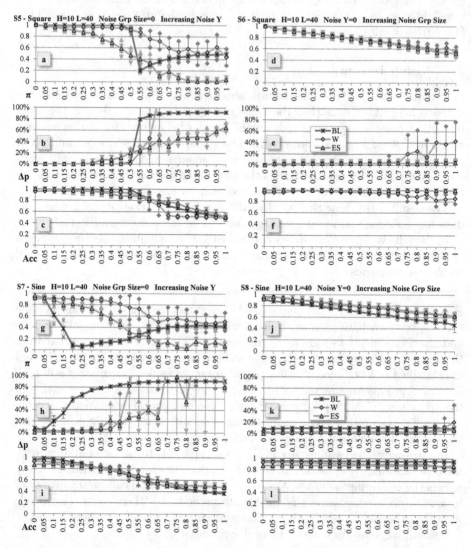

Fig. 3. Mean and standard deviation for π, Δp and Acc for scenarios 5 to 8

3 graphs per scenario: top is periodicity degree, middle candidate period estimation error Δp, bottom clustering accuracy Acc.

3 curves per graph: the cross/red is the baseline BL, triangle/green is the erosion score ES and diamond/blue is the watershed W.

rectangles, which is a context where data are highly contrasted (0 or 1). With less contrasted data like sines, the difference is attenuated and all methods perform very well (Fig. 3k).

As for ν_y, the period computation becomes wrong as it increases, especially with sines (Fig. 3h). However, ES seems to be more robust to ν_y than watershed since the latter increases sharply from $\nu_y = 0.5$. This can be linked with the erosion score not using a constant threshold to cluster the data as opposed to the watershed method. Since the groups are processed individually with ES, a misclassification for one group does not necessarily propagate to the others, whereas a threshold not chosen appropriately with the watershed leads to misclassification throughout the dataset, resulting in a bad evaluation of the period.

6   Conclusion

A new watershed clustering method is proposed as an alternative to assess the relevance of linguistic expression of the form "M every p unit, the data take high values" and tested within the framework of the DPE methodology [21]. Experimental results obtained with different shapes (rectangle and sine) and noises (value and group size) prove to be relevant. The DPE methodology is a good approach to classify the data in high and low value groups and to estimate the period of the dataset. The erosion score method seems more precise than the watershed one, due to its adaptable threshold for classification.

Future works aim at developing new fuzzy quantifiers as "from time to time", "often", "rarely", and detect periodicity in sub-parts of the dataset, which both can be developed with the high and low values clustering in DPE. Another direction is the definition of a quality measure to compare the different methods, among themselves as well as to existing approaches.

References

1. Agrawal, R., Srikant, R.: Mining sequential patterns. In: Proc. of the 11th Int. Conf. on Data Engineering, pp. 3–14 (1995)
2. Allen, J., Rabiner, L.: A unified approach to short-time Fourier analysis and synthesis. Proc. of the IEEE 65(11), 1558–1564 (1977)
3. Berberidis, C., Vlahavas, I.P., Aref, W.G., Atallah, M.J., Elmagarmid, A.K.: On the Discovery of Weak Periodicities in Large Time Series. In: Elomaa, T., Mannila, H., Toivonen, H. (eds.) PKDD 2002. LNCS (LNAI), vol. 2431, pp. 51–61. Springer, Heidelberg (2002)
4. Beucher, S., Lantuejoul, C.: Use of watersheds in contour detection. In: Proc. of the Int. Workshop on Image Processing, Real-time Edge and Motion Detection/estimation (1979)
5. Bouchon-Meunier, B., Moyse, G.: Fuzzy Linguistic Summaries: Where Are We, Where Can We Go? In: CIFEr 2012, pp. 317–324 (2012)
6. Cariñena, P., Bugarín, A., Mucientes, M., Barro Ameneiro, S.: A language for expressing fuzzy temporal rules. Mathware & Soft Computing 7(2), 213–227 (2000)
7. Castillo, I., Lévy-Leduc, C., Matias, C.: Exact adaptive estimation of the shape of a periodic function with unknown period corrupted by white noise. Mathematical Methods of Statistics 15(2), 146–175 (2006)

8. Fu, T.C.: A review on time series data mining. Engineering Applications of Artificial Intelligence 24(1), 164–181 (2011)
9. Gerhard, D.: Pitch Extraction and Fundamental Frequency: History and Current Techniques. Tech. rep., University of Regina (2003)
10. Kacprzyk, J., Wilbik, A., Zadrozny, S.: Linguistic summarization of time series using a fuzzy quantifier driven aggregation. Fuzzy Sets and Systems 159(12), 1485–1499 (2008)
11. Kacprzyk, J., Yager, R.R.: "Softer" optimization and control models via fuzzy linguistic quantifiers. Information Sciences 34(2), 157–178 (1984)
12. Kacprzyk, J., Zadrozny, S.: Protoforms of Linguistic Data Summaries: Towards More General Natural-Language-Based Data Mining Tools. In: Abraham, A., Ruiz-del Solar, J., Koeppen, M. (eds.) Soft Computing Systems, pp. 417–425 (2002)
13. Kacprzyk, J., Zadrozny, S.: Computing With Words Is an Implementable Paradigm: Fuzzy Queries, Linguistic Data Summaries, and Natural-Language Generation. IEEE Transactions on Fuzzy Systems 18(3), 461–472 (2010)
14. Kedem, B.: Spectral analysis and discrimination by zero-crossings. Proc. of the IEEE 74(11), 1477–1493 (1986)
15. Laxman, S., Sastry, P.S.: A survey of temporal data mining. Sadhana 31(2), 173–198 (2006)
16. Lee, C.H., Chen, M.S., Lin, C.R.: Progressive partition miner: An efficient algorithm for mining general temporal association rules. IEEE Transactions on Knowledge and Data Engineering 15(4), 1004–1017 (2003)
17. Ma, S., Hellerstein, J.L.: Mining partially periodic event patterns with unknown periods. In: Proc. of the 17th Int. Conf. on Data Engineering, pp. 205–214. IEEE Comput. Soc. (2001)
18. Mallat, S.: A theory for multiresolution signal decomposition: the wavelet representation. IEEE Trans. on PAMI 11(7), 674–693 (1989)
19. Mannila, H., Toivonen, H., Inkeri Verkamo, A.: Discovery of Frequent Episodes in Event Sequences. Data Mining and Knowledge Discovery 1(3), 259–289 (1997)
20. Méger, N., Rigotti, C.: Constraint-based mining of episode rules and optimal window sizes. In: Boulicaut, J.-F., Esposito, F., Giannotti, F., Pedreschi, D. (eds.) PKDD 2004. LNCS (LNAI), vol. 3202, pp. 313–324. Springer, Heidelberg (2004)
21. Moyse, G., Lesot, M.J., Bouchon-Meunier, B.: Linguistic summaries for periodicity detection based on mathematical morphology. In: IEEE SSCI 2013, pp. 106–113 (2013)
22. Ozden, B., Ramaswamy, S., Silberschatz, A.: Cyclic association rules. In: Proc. of the 14th Int. Conf. on Data Engineering, pp. 412–421. IEEE Comput. Soc. (1998)
23. Palmer, L.: Coarse frequency estimation using the discrete Fourier transform. IEEE Transactions on Information Theory 20(1), 104–109 (1974)
24. Serra, J.: Introduction to mathematical morphology. Computer Vision, Graphics, and Image Processing 35(3), 283–305 (1986)
25. Vincent, L., Soille, P.: Watersheds in digital spaces: an efficient algorithm based on immersion simulations. IEEE Trans. on PAMI 13(6), 583–598 (1991)
26. Vlachos, M., Yu, P.S., Castelli, V.: On periodicity detection and structural periodic similarity. In: Proc. of the 5th SIAM Int. Conf. on Data Mining, vol. 119, p. 449 (2005)
27. Yager, R.R.: A new approach to the summarization of data. Information Sciences 28(1), 69–86 (1982)
28. Zadeh, L.A.: A computational approach to fuzzy quantifiers in natural languages. Computers & Mathematics with Applications 9(1), 149–184 (1983)

Automatic Generation of Textual Short-Term Weather Forecasts on Real Prediction Data

A. Ramos-Soto[1], A. Bugarin[1], S. Barro[1], and J. Taboada[2]

[1] Centro Singular de Investigacion en Tecnoloxias da Informacion (CITIUS),
University of Santiago de Compostela, Spain
{alejandro.ramos,alberto.bugarin.diz,senen.barro}@usc.es
[2] MeteoGalicia, Spain
juan.taboada@meteogalicia.es

Abstract. In this paper we present a computational method which obtains textual short-term weather forecasts for every municipality in Galicia (NW Spain), using the real data provided by the Galician Meteorology Agency (MeteoGalicia). This approach is based on Soft-Computing based methods and strategies for linguistic description of data and for Natural Language Generation. The obtained results have been thoroughly validated by expert meteorologists, which ensures that in the near future it can be improved and released as a real service offering custom forecasts for a wide public.

1 Introduction

1.1 The Need for Linguistic Descriptions of Data

Nowadays, the massive availability of huge quantities of data demands the proposal of methods to extract descriptions that make them understandable by users, especially by those who are not experts. One interesting task in this context is the building of linguistic descriptions of data, understood as brief and precise general textual descriptions of (usually numeric) datasets [1]. A key issue for the expressions which compose the descriptions is them to be completely adapted in their syntax and semantics to the users' informative demands and/or styles.

Several approaches have been described in the literature for building linguistic descriptions of data. For instance, using fuzzy quantified propositions, in different realms of application, such as description of the patient inflow in health centers [1] or domestic electric consumption reports [2]. Other fuzzy approaches use more complex expressions involving relationships among different attributes ([3] in economic data). Out of the fuzzy field and within the Natural Language Generation (NLG) domain, other approaches use pre-defined grammatical rules and structures composed of several sentences ([4], [5] in meteorological prediction field) or consider expressions with fusion or translation of events ([6] in the description of physiological signals).

H.L. Larsen et al. (Eds.): FQAS 2013, LNAI 8132, pp. 269–280, 2013.

Some examples of linguistic descriptions of data from the Soft-Computing field are "Most of the days the consumption in the mornings is lower than the consumption in the evenings" [2], "Towards the end of the session prices lowered" [7] or "Several of the temperatures are warm and a few are cold" [8], [9]. In general, all of these approaches generate descriptions from numerical data sets, especially time series data. Most of them have been tested experimentally and there is not an established methodology for validation [10], [11] mostly due to their lack of application in real environments.

In this paper we present a linguistic description solution which generates automatic short-term weather forecasts in the form of natural language texts. For this, in the next subsection we introduce the context in which this method has been developed and justify the need for it. In Section 2 we describe the types of data and information which is necessary in order to generate the natural language texts. In Section 3 we describe the approach we have followed in detail and, finally, in Sections 4 and 5 we present the validation results for the method and the most relevant conclusions.

1.2   Short-Term Weather Forecasts in Galicia

The operative weather forecasting offered by the Galician Meteorology Agency (MeteoGalicia)[12] consisted until now of a general description of the short-term (four days) meteorological trend (Fig. 1). This service has been recently improved in order to provide visitors with symbolic forecasts for each of the 315 municipalities in Galicia, thus improving its quality and allowing users to obtain more precise information about specific locations of the Galician geography. However, this increase in the quantity of available numerical-symbolic data has a main downside, which lies in the lack of natural language forecasts which describe this set of data. This issue makes the forecast harder to understand, since a typical user needs to look at every symbol and detect which phenomena are relevant, whereas a natural language description could directly provide all this information.

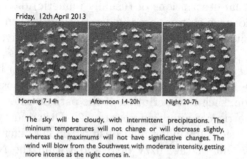

Friday, 12th April 2013

Morning 7-14h   Afternoon 14-20h   Night 20-7h

The sky will be cloudy, with intermittent precipitations. The mininum temperatures will not change or will decrease slightly, whereas the maximums will not have significative changes. The wind will blow from the Southwest with moderate intensity, getting more intense as the night comes in.

Fig. 1. Example of a real weather forecast for 12th April, 2013 for Galicia, published on [12]

In order to solve this lack, we present in this paper a computational method which, from short-term data for a given location, generates a linguistic description which highlights those meteorological phenomena considered important by an expert meteorologist, similar to the one presented in Fig. 1. This method is inspired by several research fields in Artificial Intelligence. On one hand, we have adopted techniques from the linguistic description generation field in Soft-Computing by using operators which extract linguistic information from numerical data. Also, we use techniques inspired by the Natural Language Generation field, by using grammars and templates which convert the extracted information into texts which are ready for human consumption.

2 Operative Weather Forecast Data

The required information for building the textual forecasts falls within three main categories, as in Fig. 2:

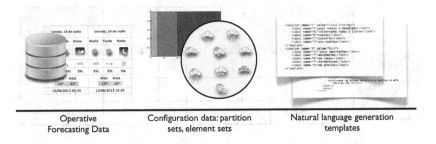

| Operative | Configuration data: partition | Natural language generation |
| Forecasting Data | sets, element sets | templates |

Fig. 2. Data and information sources used in the generation of the automatic weather forecasts

- **Operative forecasting data.** This set of data, available from MeteoGalicia's database, covers all the 315 Galician municipalities and includes data associated to the following forecast items:
  - Meteorological phenomena: They are stored as numerical codes and displayed by using graphical symbols which provide information about the two meteorological variables of interest, namely the sky state (cloud coverage) and precipitation.
  - Wind: Just as with the phenomena, each numerical code for the wind variable is associated to a specific wind intensity and direction, which is displayed graphically by using its corresponding symbol.
  - Temperature: Temperature data is stored as numerical values in degrees Celsius.

  Figure 3 shows that, for a given location, 12 values are stored for both meteorological phenomena and wind variables. Each value is associated to a time period within a given day (morning, afternoon, night), thus these 12 values translate into 4 forecast days which contain 3 values for each day. In the case of the temperature, for each forecast day maximum and minimum values are stored.

- **Configuration data.** It includes sets of partitions and elements that are used by the operators which extract the basic information from the forecast data and convert them into an intermediate language, independent from the final output language and its lexical-syntactical variants. In the case of the temperature crisp partitions are used (the numerical domain of the temperature variable is divided into intervals which are assigned to a linguistic label), while for the rest of the data we employ categories which classify the sets of meteor and wind symbols.

- **Natural language generation templates.** They include the sentences and generic descriptions for a given language, which are filled with the information contained in the intermediate language generated by the initial operators. Both templates and configuration data are encoded and stored in XML text structured files.

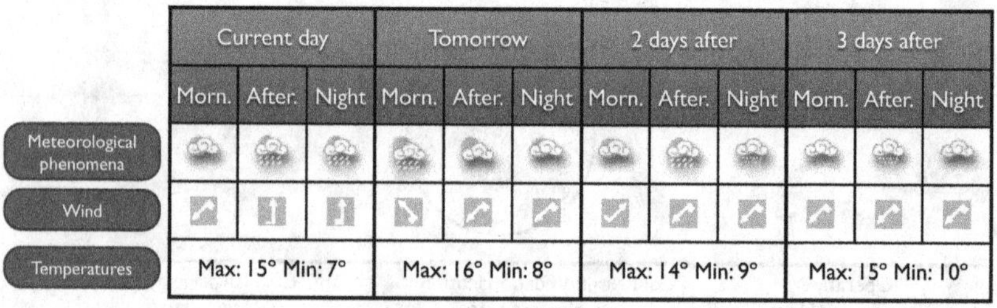

	Current day			Tomorrow			2 days after			3 days after		
	Morn.	After.	Night	Morn.	After.	Night	Morn.	After.	Night	Morn.	After.	Night
Meteorological phenomena												
Wind												
Temperatures	Max: 15° Min: 7°			Max: 16° Min: 8°			Max: 14° Min: 9°			Max: 15° Min: 10°		

Fig. 3. An example of short-term weather forecast data

3 Operative Forecasting Texts Generation Method

The method we have developed employs all of the data and information described in the previous section to generate the final output textual weather forecasts. It does it by processing the initial weather forecast data in two separate steps. During the first one, which is more related to linguistic description approaches, it converts the numerical-symbolic raw data into an intermediate code which is created by referencing values and linguistic labels. The second phase, which can be defined as a natural language generation step, translates this intermediate code into a text for one of the available final output human languages, which is ready for human consumption.

3.1 From Numerical Raw Data to an Intermediate Language

The method's first phase obtains an intermediate language which contains, for every variable, the information which is considered relevant to be included in the textual forecasts which are generated in the second phase of the processing. The process, as it can be seen in Fig. 4, is direct, and essentially consists of providing each extractor operator with its corresponding data – besides partition information – in order to generate the intermediate language. Each operator and the kind of data it receives is described in more detail in what follows:

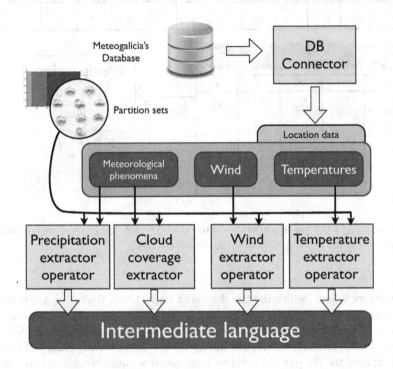

Fig. 4. Schema of the first phase of the textual forecasts generation method

- **Precipitation operator.** It extracts information from the meteorological phenomena which consists of those short-term periods when precipitations are expected to occur. These periods are classified according to the importance and kind of precipitations detected and encoded into the output intermediate code. For instance, if among the 12 values two rain periods are detected, codes like "1-3 i i i" and "7-9 i ni ni" could be obtained (Fig. 5).
- **Cloud coverage/sky state operator.** This operator also employs the information contained in the meteorological phenomena, but it classifies them according to the level of cloud coverage associated to each symbol. The current approach arranges the twelve values into three subintervals, corresponding to the beginning, the middle and the end of the whole short-term period. For each subinterval, the operator assigns one label according to the predominant sky state in that subinterval. Labels are calculated by evaluating the most predominant sky state for each subinterval by using trapezoidal

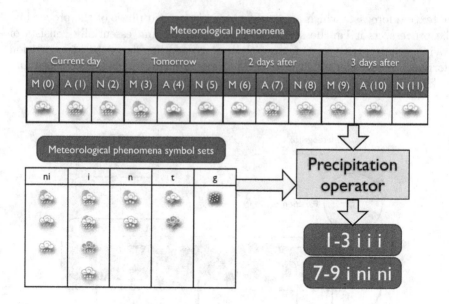

Fig. 5. Schema of the precipitation operator method with the current meteorological phenomena sets for precipitation and its associated labels

fuzzy sets which determine the degree of importance that each value has in relation to each subinterval (for instance, the first, second or third values, which correspond with the first day, are more important than the following ones when referring to the beginning of the short-term period). Thus, an example of the output obtained by this operator could consist of three labels "Clear Cloudy Clear".

- **Wind operator.** It follows a similar strategy to the precipitation operator, although in this case it only groups periods where the wind intensity is strong or very strong. This condition has been included as indicated by meteorologists from the beginning, since according to them this is the most relevant phenomenon that must be stated for this variable. Consequently, if there are two periods of strong wind we could obtain two intermediate codes such as "2-4 322,322" and "7-9 321,322,321". In this case, the numeric codes are associated to given intensity and directions, as it is defined in MeteoGalicia's guidelines for forecast interpretation.

- **Temperature operator.** This operator generates an intermediate code which reflects the temperature trend for the 4-day period, but it also obtains information about the climatic behaviour of the forecasted temperatures (whether the predicted temperatures are normal for the current season or not). This information is provided by the comparison between the forecasted temperature and the municipality climatic mean (the mean for the previous 30 years for the current season).

The obtained intermediate code can vary, depending on whether the trend of the maximum and minimum temperatures is the same and whether their difference against the climatic mean is similar or not. For this, two crisp partitions for the temperature variable are used, which measure the period trend on one side and the climatic behavior on the other, assigning several linguistic labels. For instance, if the maximum temperatures keep constant, the minimums increase moderately and both are normal for the current time of the year, the corresponding intermediate code would be "SC, AM Normal" (Fig. 6). In other cases up until four labels could be obtained, if all the trends were different from each other.

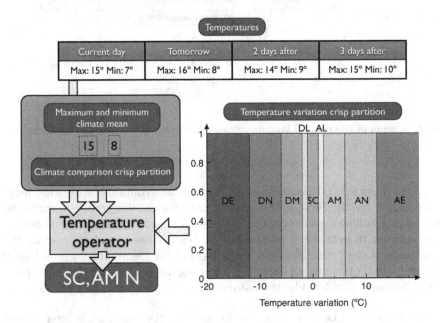

Fig. 6. Schema of the temperature operator, with the current definition of the temperature variation partition and its associated labels, which are the same for the climate comparison crisp partition (with different labels)

3.2 From Intermediate Language to Natural Language

Like the previous phase, the natural language generation step has been divided into different modules for each variable, so that changes in one of them do not affect the rest of the system. Each of these modules receives the intermediate code generated by their corresponding extractor operator and, using the information contained in the natural language generation templates for a given output language, generates the final textual forecast for the corresponding variable (Fig. 7).

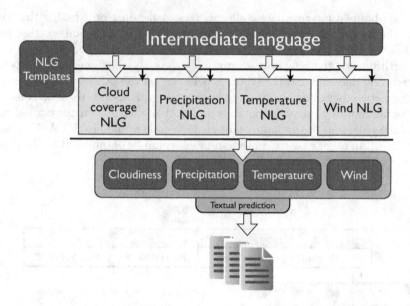

Fig. 7. Schema of the natural language generation phase of the textual forecasts generation method

Each of these natural language generation modules contains expert knowledge incorporated into the code, so that according to certain detectable events in the intermediate language, different alternatives can be selected in the variable template in order to generate a more complete text. These templates contain the most common cases, which serve to cover most of the identifiable meteorological phenomena from this kind of data. The following list enumerates the main meteorological cases considered by the natural language generators for each variable:

– Precipitation: The main cases distinguish whether the rain is intermittent or persistent, besides special phenomena like storm, snow or hail.
– Cloud coverage: Using the intermediate language provided by the cloud coverage operator, which consists of three labels (beginning, half and end of the short-term period), we distinguish three cases:
  • The three subintervals share the same label, thus the short-term period can be described as a whole.
  • Two subintervals share the same label, but the remaining one is different, which means that, although there is a certain predominance, the remaining subinterval must also be described.
  • Each subinterval has a different label associated to it. In this last case, the period is described as an evolution between the first and the end subinterval.
– Temperature: Four cases have been taken into account, which depend on whether the maximum and minimum values follow the same trend or not,

for both the temperature variation for the period and the difference against
the climatic mean.

- Wind: Strong or very strong wind periods are detected, including wind
changes of intensity or direction within them.

3.3 Implementation Details

This method is being developed in the multiplatform coding language Python,
with the use of libraries for mathematical and fuzzy calculations (*numpy, py-
fuzzy*) or text pattern recognition by grammars (*pyparsing*). The current implementation supports both Linux and Windows systems by employing different
database access modules. For accessing databases in Linux we employ *unixODBC*
and the driver *Microsoft SQL Server ODBC Driver 1.0*. For Windows, *pymssql*
is used. The initially supported languages include Spanish and Galician.

4   Results and Validation

MeteoGalicia's meteorologists have provided support for expert validation of the
results, which allows us to refine the solution in a way that will ensure it works
under realistic conditions and cases. The application was installed in MeteoGalicia so that the experts could test it and choose appropriate validation examples.
As a result, 40 textual test forecasts have been generated, which include real operative forecast data but also extreme cases where the forecast has been modified
in order to test the robustness of the application.

4.1   Validation

The validation methodology used for this application focuses in verifying that the
content of the textual forecasts reflects the important phenomena that appear in
the original data, aside from the text style and other issues related to linguistics
[10]. For this, the experts were asked to assess the contents and quality of the
automatically obtained forecasts using a numerical 1-5 scale. Table 1 shows the
global mean score for the validation dataset.

Table 1. Score by meteorologists for the validation test

Number of cases	Average score	Standard deviation
40	4.35	0.62

Although some minor issues caused by wrong definitions in the NLG templates were found, in general, the meteorologists' assessment about the content

of the forecasts is very positive. In this sense, the experts have directly identified the content and language of the generated results with the ones they would provide. The fact that both content and language from the automatic forecasts are almost indistinguishable from those that an expert would produce are the most important among the several quality aspects which can be measured [10], [11].

4.2   An Illustrative Example, Pontevedra, 21st March 2013

The following example has been obtained by executing the method for the forecast data of Pontevedra, on the 21st of March. As Fig. 8 shows, it is a rather rainy period, for which strong winds are forecasted during the initial values. As for temperatures, their trend keeps stable without changes. Figure 8 also includes the generated textual forecast, which has been obtained with the current version of the proposed method. As it can be appreciated, the most important phenomena, such as precipitations and strong winds, are highlighted.

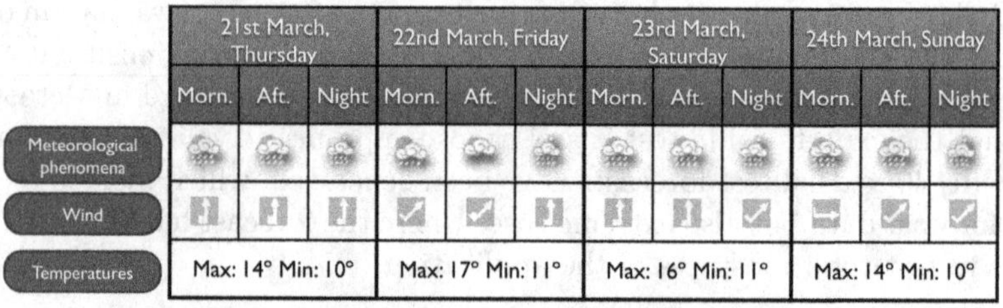

	21st March, Thursday			22nd March, Friday			23rd March, Saturday			24th March, Sunday		
	Morn.	Aft.	Night	Morn.	Aft.	Night	Morn.	Aft.	Night	Morn.	Aft.	Night
Meteorological phenomena												
Wind												
Temperatures	Max: 14° Min: 10°			Max: 17° Min: 11°			Max: 16° Min: 11°			Max: 14° Min: 10°		

Sky state: Cloudy sky
Precipitations: We expect persistent precipitations from Thursday morning to Friday morning. We expect persistent precipitations from Friday night to Sunday night.
Temperatures: Temperatures will be normal for the maximums and notably high for the minimums for this time of the year, without changes in general.
Wind: Strong wind from the South from Thursday morning to Thursday evening.

Fig. 8. Short-term forecast data and automatically generated weather forecast for Pontevedra, 21st March

4.3   An Illustrative Example: Pedrafita, 13th March 2013

The following example (Fig. 9), generated from data for Pedrafita do Cebreiro (a montainous region in Galicia), differs from the previous one in the kind of precipitations which are forecasted, which can also be snowy. This fact is reflected in the automatically generated textual forecast.

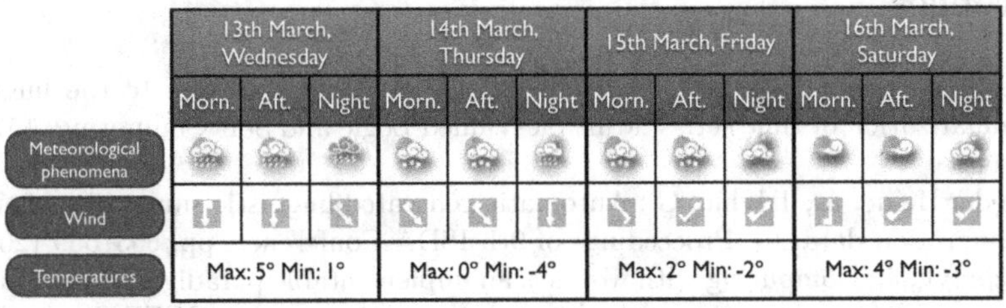

	13th March, Wednesday			14th March, Thursday			15th March, Friday			16th March, Saturday		
	Morn.	Aft.	Night	Morn.	Aft.	Night	Morn.	Aft.	Night	Morn.	Aft.	Night
Meteorological phenomena												
Wind												
Temperatures	Max: 5° Min: 1°			Max: 0° Min: -4°			Max: 2° Min: -2°			Max: 4° Min: -3°		

> **Sky state:** Cloudy sky
> **Precipitations:** We expect persistent precipitations from Wednesday morning to Friday afternoon, which can happen as snow.
> **Temperatures:** Temperatures will be extremely low for the maximums and notably low for the minimums for this time of the year, with maximums without changes and minimums decreasing moderately.
> **Wind:** Strong wind from the North from Wednesday morning to Wednesday afternoon.

Fig. 9. Short-term forecast data and associated automatic weather forecast for Pedrafita, 13th March

5 Conclusions and Future Work

In this paper we have presented a computational method which obtains textual short-term weather forecasts for the 315 municipalities in Galicia, using the real data provided by MeteoGalicia. As opposed to other linguistic descriptions approaches, this method is based on an applied development in a realistic application, whose definition and structure is inspired by the linguistic descriptions research field by using operators which extract relevant information, and also by the natural language generation field.

The main value of this application resides in its ability to cover and support a service of high interest for a wide number of users, which can be only provided by generating descriptions of data automatically, due to the high number of forecasts which must be obtained.

Furthermore, the automatic textual forecasts were evaluated by the meteorologists in order to assess the quality of their contents and to check whether their expert knowledge was included correctly. The obtained results show that the textual forecasts fulfill the experts' requirements in a high degree. This, at the same time, makes us optimistic about releasing the method as a real service in a near future, once the method is extended and improved in order to fully meet the meteorologists' requirements.

Acknowledgements. This work was supported by the Spanish Ministry for Economy and Competitiveness under grant TIN2011-29827-C02-02. It was also supported in part by the European Regional Development Fund (ERDF/FEDER) under the project CN2012/151 of the Galician Ministry of Education. A. Ramos-Soto is supported by the Spanish Ministry for Economy and Competitiveness (FPI Fellowship Program).

References

1. Castillo-Ortega, R., Marín, N., Sánchez, D.: A fuzzy approach to the linguistic summarization of time series. Multiple-Valued Logic and Soft Computing, 157–182 (2011)
2. van der Heide, A., Triviño, G.: Automatic generated linguistic summaries of energy consumption data. In: Proceedings of 9th ISDA Conference, pp. 553–559 (2009)
3. Kacprzyk, J.: Computing with words is an implementable paradigm: Fuzzy queries, linguistic data summaries, and natural-language generation. IEEE Trans. Fuzzy Systems, 451–472 (2010)
4. Goldberg, E., Driedger, N., Kittredge, R.I.: Using natural-language processing to produce weather forecasts. IEEE Expert 9, 45–53 (1994)
5. Reiter, E., Sripada, S., Hunter, J., Davy, I.: Choosing words in computer-generated weather forecasts. Artificial Intelligence 167, 137–169 (2005)
6. Portet, F., Reiter, E., Gatt, A., Hunter, J., Sripada, S., Freer, Y., Sykes, C.: Automatic generation of textual summaries from neonatal intensive care data. Artif. Intell. 173(7-8), 789–816 (2009)
7. Kobayashi, I., Okumura, N.: Verbalizing time-series data: With an example of stock price trends. In: Proceedings IFSA/EUSFLAT Conf., pp. 234–239 (2009)
8. Diaz-Hermida, F., Bugarin, A.: Linguistic summarization of data with probabilistic fuzzy quantifiers. In: Actas XV Congreso Español sobre Tecnologías y Lógica Fuzzy (ESTYLF), pp. 255–260 (February 2010)
9. Ramos-Soto, A., Díaz-Hermida, F., Bugarín, A.: Construcción de resúmenes lingüísticos informativos sobre series de datos meteorológicos: informes climáticos de temperatura. In: Actas XVI Congreso Español sobre Tecnologías y Lógica Fuzzy (ESTYLF), pp. 644–649 (February 2012)
10. Eciolaza, L., Pereira-Fariña, M., Trivino, G.: Automatic linguistic reporting in driving simulation environments. Applied Soft Computing (2012)
11. Ramos-Soto, A., Díaz-Hermida, F., Barro, S., Bugarín, A.: Validation of a linguistic summarization approach for time series meteorological data. In: 5th ERCIM International Conference, 133 (December 2012)
12. MeteoGalicia: Meteogalicia's website, www.meteogalicia.es

Increasing the Granularity Degree in Linguistic Descriptions of Quasi-periodic Phenomena

Daniel Sanchez-Valdes and Gracian Trivino

European Centre for Soft Computing (ECSC)
Mieres, Asturias, Spain
{daniel.sanchezv,gracian.trivino}@softcomputing.es

Abstract. In previous works, we have developed some computational models of quasi-periodic phenomena based on Fuzzy Finite State Machines. Here, we extend this work to allow designers to obtain detailed linguistic descriptions of relevant amplitude and temporal changes. We include several examples that will help to understand and use this new resource for linguistic description of complex phenomena.

Keywords: Linguistic description of data, Computing with Perceptions, Fuzzy Finite State Machine, Quasi-periodic phenomena.

1 Introduction

Computational systems allow obtaining and storing huge amounts of data about phenomena in our environment. Currently, there is a strong demand for computational systems that can interpret and linguistically describe the large amount of information that is being generated.

Some physical phenomena provide signals with a similar repetitive temporal pattern. These phenomena are called quasi-periodic due to their variations in period and amplitude. Examples of this type of signals are electrocardiograms, the breathing, accelerations produced during the human gait, vibrations of musical instruments, etc. Popular approaches to deal with quasi-periodic phenomena vary from Wavelets transform [1] to Hidden Markov Models [2] and Neural Networks [3]. Fuzzy Finite State Machines (FFSM) are specially useful tools to model dynamical processes that change in time, becoming an extension of classical Finite State Machines [4][5].

Our research line is based on the Computational Theory of Perceptions introduced by Zadeh [6][7]. In previous works, we have developed computational systems able to generate linguistic descriptions of different types of phenomena, e.g., gait analysis [8], activity recognition [9] and traffic evolution [10][11].

We have used FFSMs to model and linguistically describe the temporal evolution of quasi-periodic signals during a period of time. In this work, our goal consists of exploring the possibility of generating linguistic descriptions of those instants in which phenomena are significantly deviated from the model. Here, we identify each state and linguistically report the evolution of the phenomenon.

H.L. Larsen et al. (Eds.): FQAS 2013, LNAI 8132, pp. 281–292, 2013.

When the input signal does not completely suit with the available model, the computational system describes the reasons that cause this event, providing a finer monitoring of the input signal. This analysis is appropriate to overcome signal processing, predictive control and monitoring tasks.

This paper is organized as follows: Section 2 describes the architecture of computational systems able to create linguistic descriptions of complex phenomena. Section 3 explains how to apply it to the study of quasi-periodic signals. Section 4 shows a set of simple examples that demonstrate the potential of our approach. Finally, Section 5 provides some concluding remarks and future works.

2   Linguistic Description of Complex Phenomena

This section briefly describes the main concepts of our approach to linguistic description of complex phenomena.

2.1   Computational Perception (CP)

A CP is a tuple (A, W) described as follows:

$A = (a_0, a_1, a_2, \ldots, a_n)$ is a vector of linguistic expressions (words or sentences in Natural Language) that represents the whole linguistic domain of CP. The components of A are defined by the designer by extracting the most suitable sentences from the typically used ones in the application domain of language. In the application context, each component a_i describes the linguistic value of CP in each situation of the phenomenon with specific granularity degree. These sentences can be either simple, e.g., $a_i = $ "The temperature is quite high" or more complex, $a_i = $ "Today the weather is better than the last days". If the perception does not match with any of the available possibilities the model uses the linguistic expression a_0, e.g, "The available model cannot explain completely the current situation".

$W = (w_0, w_1, w_2, \ldots, w_n)$ is a vector of validity degrees $w_i \in [0, 1]$ assigned to each a_i. In the application context, w_i represents the suitability of a_i to describe the perception. After including a_0, the components of A form a strong fuzzy partition of the domain of existence of CP, i.e., $\sum_{i=0}^{n} w_i = 1$.

2.2   Perception Mapping (PM)

We use PMs to combine or aggregate CPs. A PM is a tuple (U, y, g, T) where:

U is a vector of input CPs, $U = (u_1, u_2, \ldots, u_m)$, where u_i are tuples (A_i, W_i) and m the number of input CPs. We call *first order perception mappings* (1PMs) when U are numerical values obtained , e.g., from a database.

y is the output CP, $y = (A_y, W_y)$.

g is an aggregation function $W_y = g(W_1, W_2, \ldots, W_m)$, where W_i are the vectors of validity degrees of the m input CPs. In Fuzzy Logic, many different types of aggregation functions have been developed. For example, g can be implemented using a set of fuzzy rules.

T is a text generation algorithm that allows generating the sentences in A_y. In our current approach, T is typically a basic linguistic template, e.g., *"Likely the temperature sensor is wrong"* and/or *"The temperature of this room is {high | medium | low}"*.

2.3   Granular Linguistic Model of Phenomena (GLMP)

GLMP consists of networks of PMs (see Fig. 2). We say that output CPs are explained by PMs using a set of input CPs. In the network, each CP covers an specific aspect of the phenomenon with certain granularity degree.

We call *first order computational perceptions* (1CPs) to those ones obtained from the system input and we call *second order computational perception* (2CPs) to those ones explained by previous CPs. By means of using different aggregation functions and different linguistic expressions, the GLMP paradigm allows the designer to model computationally her/his perceptions. Note that, after being instantiated with a set of input data, the GLMP provides a structure that, in medium size applications, could include hundreds of valid sentences. We will see in the following sections how to merge these sentences in a template to generate linguistic reports.

3   Linguistic Modeling of Quasi-periodic Phenomena

In order to illustrate the method for creating linguistic descriptions of quasi-periodic phenomena, we use a simple example built around a sinusoidal signal $u(t) = sin(\omega t)$. Fig. 1 shows the modeled signal $u(t)$, which is divided into four states q_1, q_2, q_3 and q_4. Each state q_i is repeated along the time and has a set of special characteristics that distinguish it from the other states.

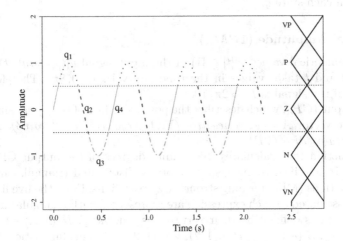

Fig. 1. Sinusoidal signal with related fuzzy labels

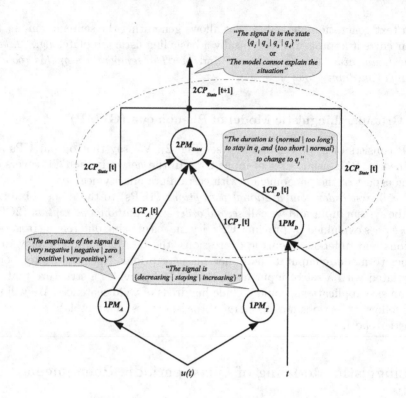

Fig. 2. GLMP that models the signal

Fig. 2 shows a GLMP designed to model this signal. $1CP_A$ and $1CP_T$ describe the amplitude and trend of the input signal respectively. $2CP_{FFSM}$ shows the state of the signal at every time instant t and $1CP_D$ presents the duration of the signal in each state q_i.

3.1 Signal Amplitude ($1PM_A$)

U are the numerical values ($u[t] \in \mathbb{R}$) of the input signal $u[t] = sin(\omega t)$, where $\omega = 10$ and t takes values in the time interval $t \in [0, 8\pi]$. Therefore, the period of the signal is $K = 2\pi/10$.

y is the output $1CP_A$, which describes the possible values of the signal amplitude with the vector $A =$(*Very negative (VN), Negative (N), Zero (Z), Positive (P), Very positive (VP)*).

g is the function that calculates the validity degrees of the output CP. These values are obtained by means of uniformly distributed triangular membership functions (MFs) forming strong fuzzy partitions. Here, the five linguistic labels associated to each expression are represented with triangular membership functions defined by their vertexes as follows: {VN $(-\infty, -2, -1)$, N $(-2, -1, 0)$, Z $(-1, 0, 1)$, P $(0, 1, 2)$, VP $(1, 2, \infty)$}. This linguistic labels are represented in Fig. 1, together with the input signal $u(t)$.

T is the text generator that produces linguistic expressions as follows: *"The amplitude of the signal is {very negative | negative | zero | positive | very positive}"*.

3.2   Signal Trend ($1PM_T$)

U are, as in $1PM_A$, the numerical values ($u[t] \in \mathbb{R}$) of the input signal.

y is the output $1CP_T$, which describes the possible values of the signal trend with the vector $A =$(*Decreasing (D), Staying (S), Increasing (I)*).

g is the output function that calculates the signal trend as $u[t] - u[t-1]$. It is defined by the vertexes of triangular MFs as follows: $\{D \ (-\infty, -0.01, 0), S \ (-0.01, 0, 0.01), I \ (0, 0.01, \infty)\}$.

T produces linguistic expressions defined as follows: *"The signal is {decreasing | staying | increasing}"*.

3.3   State Duration ($1PM_D$)

U are the input $2CP_{State}$ and the time instant t. Each state q_i has an associated duration d_i, which is numerically calculated with the time t.

y is the output $1CP_D$ which describes how long the phenomenon is in each state q_i. The possible linguistic expressions to describe this perception are contained in the vector: $A =$(*too short to change to q_j, normal to change to q_j, normal to stay in q_i, too long to stay in q_i*).

g is the output function that calculates the validity degrees of $1CP_D$. When the state q_i takes the value zero, its duration d_i takes value zero. However, when q_i takes values bigger than zero, its duration d_i increase its value according to the time, measuring the duration of the state.

Based on the duration d_i of each state we define two temporal conditions described as follows:

The **time to stay** is the maximum time that the signal is allowed to stay in state i. We associate two linguistic labels that describe this perception: *{normal to stay in q_i, too long to stay in q_i}*.

The **time to change** is the minimum time that the signal must be in state i before changing to state j. We associate other two linguistic labels that describe this perception: *{too short to change to q_j, normal to change to q_j}*.

Fig. 3 shows the linguistic labels used to define these temporal constraints. In this application, states q_1 and q_3 have the same duration and the linguistic labels are trapezoidal MFs that can be defined by their vertexes as follows:
Time to stay: *{normal* $(-\infty, -\infty, 5/12, 1/2)$, *too long* $(5/12, 1/2, \infty, \infty)\}$.
Time to change: *{too short* $(-\infty, -\infty, 1/4, 5/12)$, *normal* $(1/4, 5/12, \infty, \infty)\}$.
On the other hand, states q_2 and q_4 share the same duration and the linguistic labels can be defined by their vertexes as follows:
Time to stay: *{normal* $(-\infty, -\infty, 1/3, 1/2)$, *too long* $(1/3, 1/2, \infty, \infty)\}$.
Time to change: *{too short* $(-\infty, -\infty, 1/6, 1/4)$, *normal* $(1/6, 1/4, \infty, \infty)\}$.

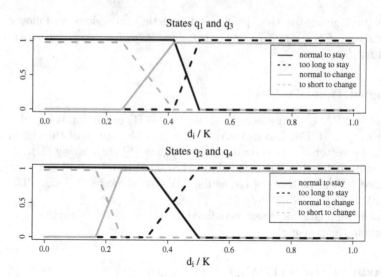

Fig. 3. Times to stay and times to change of states q_1, q_2, q_3 and q_4

These vertexes are expressed in terms of K as a percentage of the signal period. They have been carefully calculated attending to the signal structure. For example, if the signal is in the state q_1 and, at specific time t, d_1 is equal to 0.2 ($d_1/K = 0.318$), the aggregation function g indicates that this duration is *normal to stay in q_1* (1), *too long to stay in q_1* (0), *normal to change to q_2* (0.4) and *too short to change to q_2* (0.6).

T produces linguistic expressions as follows: *"The duration is {normal, too long} to stay in q_i and {too short, normal} to change to q_j"*, with $i \in \{1, 2, 3, 4\}$.

3.4 Signal State ($2PM_{State}$)

With the information about the amplitude, trend, duration of each state and the signal state in t, this $2PM_{State}$ calculates the next state of the signal in $t + 1$.

U are the input CPs: $\{1CP_A, 1CP_T, 1CP_D, 2CP_{State}\}$

y is the output $2CP_{State}$, that describes the possible states of the signal as the states of a FFSM. The vector of linguistic expressions is $A = (q_0, q_1, q_2, q_3, q_4)$, where q_0 means that the input signal does not fit with any of the other fuzzy states (see Fig. 4). The definition of the number of states and allowed transitions have to be designed according to the modeled signal, i.e., the higher the complexity of the monitored signal the bigger the number of states to monitor its evolution fine.

g is the aggregation function implemented using a set of fuzzy rules. We have used twelve fuzzy rules, namely, R_{ii} to remain in the state i and R_{ij} to change from state i to state j. The whole rules base is listed as follows:

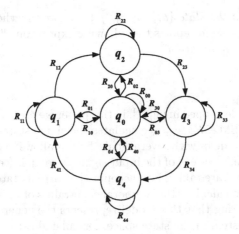

Fig. 4. State diagram of the FFSM for the sinusoidal signal

R_{11}: IF ($State[t]$ is q_1) AND ($amplitude$ is P) AND ($duration$ is $normal\ to\ stay$ $in\ q_1$) THEN ($State[t+1]$ is q_1)

R_{22}: IF ($State[t]$ is q_2) AND ($amplitude$ is Z) AND ($trend$ is D) AND ($duration$ is $normal\ to\ stay\ in\ q_2$) THEN ($State[t+1]$ is q_2)

R_{33}: IF ($State[t]$ is q_3) AND ($amplitude$ is N) AND ($duration$ is $normal\ to\ stay$ $in\ q_3$) THEN ($State[t+1]$ is q_3)

R_{44}: IF ($State[t]$ is q_4) AND ($amplitude$ is Z) AND ($trend$ is I) AND ($duration$ is $normal\ to\ stay\ in\ q_4$) THEN ($State[t+1]$ is q_4)

R_{12}: IF ($State[t]$ is q_1) AND ($amplitude$ is Z) AND ($trend$ is D) AND ($duration$ is $normal\ to\ change\ to\ q_2$) THEN ($State[t+1]$ is q_2)

R_{23}: IF ($State[t]$ is q_2) AND ($amplitude$ is N) AND ($trend$ is D) AND ($duration$ is $normal\ to\ change\ to\ q_3$) THEN ($State[t+1]$ is q_3)

R_{34}: IF ($State[t]$ is q_3) AND ($amplitude$ is Z) AND ($trend$ is I) AND ($duration$ is $normal\ to\ change\ to\ q_4$) THEN ($State[t+1]$ is q_4)

R_{41}: IF ($State[t]$ is q_4) AND ($amplitude$ is P) AND ($trend$ is I) AND ($duration$ is $normal\ to\ change\ to\ q_1$) THEN ($State[t+1]$ is q_1)

R_{01}: IF ($State[t]$ is q_0) AND ($amplitude$ is P) AND ($trend$ is I) THEN ($State[t+1]$ is q_2)

R_{02}: IF ($State[t]$ is q_0) AND ($amplitude$ is Z) AND ($trend$ is D) THEN ($State[t+1]$ is q_3)

R_{03}: IF ($State[t]$ is q_0) AND ($amplitude$ is N) AND ($trend$ is D) THEN ($State[t+1]$ is q_4)

R_{04}: IF ($State[t]$ is q_0) AND ($amplitude$ is Z) AND ($trend$ is I) THEN ($State[t+1]$ is q_1)

R_{i0}: ELSE ($State[t+1]$ is q_0)

T produces linguistic expressions that be adapted depending on the situation. When the signal is in the states q_1, q_2, q_3 or q_4 the template is the following:

"*The signal is in the state* $\{q_1, q_2, q_3, q_4\}$". However, when the signal is in the state q_0 the system reports the following expression: "*The model cannot explain the current situation*".

3.5   Remarks

Following the works developed in [8][11], in this paper we present in detail the CPs existing into the FFSM. $2PM_{States}$ calculates the signal state at each instant t. The output $2CP_{States}$ deals with several goals: first, it allows to generate linguistic descriptions about each state q_i of the input signal; second, it feeds the $2PM_{States}$ to calculate the next state taking into account the current state at t; and finally, it feeds the $1PM_D$ to calculate the temporal constraints of the FFSM.

It is worth remarking that this model represents the general form of a time-invariant discrete system in the state space, formulated as:

$$\begin{cases} X[t+1] = f(X[t], U[t]) \\ Y[t] = g(X[t], U[t]) \end{cases}$$

where:

- U is the input vector of the system: $(u_1, u_2, ..., u_{n_u})$, with n_u being the number of input variables.
- X is the state vector: $(x_1, x_2, ..., x_n)$, with n being the number of states.
- Y is the output vector: $(y_1, y_2, ..., y_{n_y})$, with n_y being the number of output variables.
- f is the function which calculates the state vector at time step $t+1$.
- g is the function which calculates the output vector at time step t.

In our approach, the model could be formulated as:

$$2CP_{State}[t+1] = f(2CP_{State}[t], U[t])$$

where, $2CP_{State}$ is the state vector at time step t, U is the input vector composed by the input variables $1CP_A$, $1CP_T$ and $1CP_D$, and f is the function that calculates the state vector at time step $t+1$. In this application, we have used the state vector as output vector and we have not used g.

4   Examples

In order to illustrate how the model works, Figs. 6, 7, 8 and 9 show different behaviors of the text generator depending on the input signal values. Each figure represents the input signal. Below of the signal is illustrated the evolution of each state q_i (w_{q_i}). The vertical line indicates the time instant where the description is generated. As mentioned before, the modeled FFSM was designed to recognize the states of an input signal of the form $u(t) = sin(10t)$.

The main objective is to linguistically describe the evolution of the signal, i.e., to indicate the state q_i in which the signal is at every time instant and, in case of deviation from model, to specify the details. Note that the way to analyze the reasons why the model cannot explain the signal in a specific period consists of going over the GLMP in the reverse direction the model was built.

The linguistic reports obtained can be used by experts to understand changes in the signal and foreseeing its future behavior. We have applied the report template shown in Fig. 5. This template changes the report depending on the validity degrees of the sentences. The sentences with highest validity degree are chosen at each time instant for each CP.

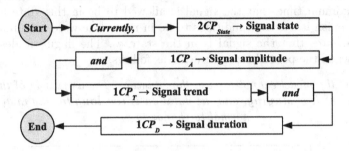

Fig. 5. Template for the linguistic report

4.1 Example 1

Fig. 6 shows the fuzzy states evolution when the input signal is exactly the expected one, i.e., when the set of rules represent perfectly its evolution. Therefore, w_{q_0} will be equal to zero along the time. A example of instant report when the signal matches completely the model is as follows

"Currently, the signal is in the state q_1. The amplitude of the signal is positive, it is increasing, and the duration is normal to stay in q_1 and too short to change to q_2 ".

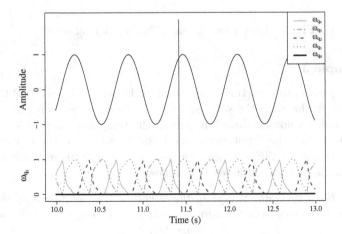

Fig. 6. States when the signal is exactly the expected one

4.2   Example 2

When the input signal has a perturbation or an unforeseen situation, the set of rules does not recognize its behavior and the model indicates that the signal is in an uninterpretable state (q_0). The activation of the state q_0 informs that the input signal does not match with the typical sinusoidal signal. Fig. 7 shows a first example of this type of situations. In this case, the input signal has a perturbation that maintains its value constant during a period of time. At the time of failure, the signal was in the state q_1 and the temporal constraint related to the maximum time that the signal is allowed to be in this state force the system to finish it. As the conditions to be in next state were not fulfilled, the system indicates that the signal is in the state q_0. The linguistic description obtained when the perturbation occurs is as follows:

*"Currently, the model cannot explain the situation. The amplitude of the signal is positive, it is **staying**, and the duration is **too long** to stay in q_1 and normal to change to q_2".*

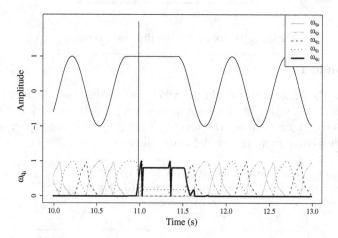

Fig. 7. States when the signal has a temporal error

4.3   Example 3

In this example, another perturbation modifies the normal behavior of the input data (Fig. 8). In this case, the state q_1 decreases its duration, being in this state less than a half its normal duration. Again, the temporal conditions of the set of rules detect that the input signal does not fit well with the model in this instant. The reason is different with respect to the previous example and now the linguistic description obtained is as follows:

*"Currently, the model cannot explain the situation. The amplitude of the signal is zero, it is decreasing, and the duration is normal to stay in q_1 and **too short** to change to q_2".*

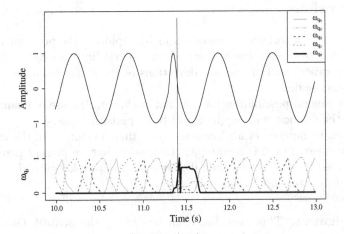

Fig. 8. States when the signal has another type of temporal error

4.4   Example 4

This last example shows the modeling of a different type of perturbation (Fig. 9). In this case the error is produced by an amplitude perturbation that forces the input signal to be almost twice bigger than the expected one during the state q_3. The conditions that model the signal amplitude detect a strange behavior. This is reflected in the evolution of q_0, which takes high values in these moments. The linguistic description obtained is as follows:

*"Currently, the model cannot explain the situation. The amplitude of the signal is **very negative**, it is decreasing, and the duration is normal to stay in q_3 and too short to change to q_1"*

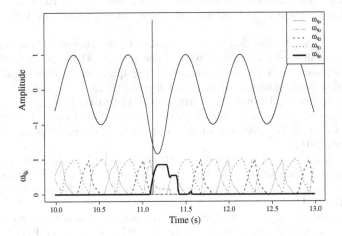

Fig. 9. States when the signal has an amplitude error

5 Conclusions

We have developed previous research works by exploring the possibility of providing detailed linguistic descriptions of quasi-periodic phenomena. Here, we focus on the description of relevant deviations of the signal from the available computational model.

There are several possibilities for pending work, e.g., the use of quantifiers to summarize the frequency of happening of these perturbations.

An important number of applications could take advantage of the contribution of this research line. For example, the results that we present here can be directly applied to analyze and describe anomalies in physiological signals, such as electrocardiogram and human gait.

Acknowledgment. This work has been funded by the Spanish Government (MICINN) under project TIN2011-29827-C02-01.

References

1. Burrus, C.S., Gopinath, R.A., Guo, H.: Introduction to wavelets and wavelet transforms: a primer. Prentice Hall, pper Saddle River (1998)
2. Rabiner, L.R.: A tutorial on hidden Markov models and selected applications in speech recognition. Proceedings of the IEEE 77(2), 257–286 (1989)
3. Ripley, B.D.: Pattern recognition and neural networks. Cambridge university press (2008)
4. Lawson, M.V.: Finite Automata. Chapman and Hall/CRC (2003)
5. Davis, J.: Finite State Machine Datapath Design, Optimization, and Implementation (Synthesis Lectures on Digital Circuits and Systems). Morgan and Claypool Publishers (2008)
6. Zadeh, L.A.: From computing with numbers to computing with words - from manipulation of measurements to manipulation of perceptions. IEEE Transactions on Circuits and Systems - I: Fundamental Theory and Applications 45(1), 105–119 (1999)
7. Zadeh, L.A.: A new direction in AI. Towards a Computational Theory of Perceptions of measurements to manipulation of perceptions. AI Magazine 22 (2001)
8. Alvarez-Alvarez, A., Trivino, G.: Linguistic description of the human gait quality. Engineering Applications of Artificial Intelligence (2012)
9. Sanchez-Valdes, D., Eciolaza, L., Trivino, G.: Linguistic Description of Human Activity Based on Mobile Phone's Accelerometers. In: Bravo, J., Hervás, R., Rodríguez, M. (eds.) IWAAL 2012. LNCS, vol. 7657, pp. 346–353. Springer, Heidelberg (2012)
10. Eciolaza, L., Trivino, G., Delgado, B., Rojas, J., Sevillano, M.: Fuzzy linguistic reporting in driving simulators. In: Proceedings of the 2011 IEEE Symposium on Computational Intelligence in Vehicles and Transportation Systems, Paris, France, pp. 30–37 (2011)
11. Alvarez-Alvarez, A., Sanchez-Valdes, D., Trivino, G., Sánchez, Á., Suárez, P.D.: Automatic linguistic report about the traffic evolution in roads. Expert Systems with Applications 39(12), 11293–11302 (2012)

A Model-Based Multilingual Natural Language Parser — Implementing Chomsky's X-bar Theory in ModelCC

Luis Quesada, Fernando Berzal, and Juan-Carlos Cubero

Department of Computer Science and Artificial Intelligence, CITIC,
University of Granada, Granada 18071, Spain
{lquesada,fberzal,jc.cubero}@decsai.ugr.es

Abstract. Natural language support is a powerful feature that enhances user interaction with query systems. NLP requires dealing with ambiguities. Traditional probabilistic parsers provide a convenient means for disambiguation. However, they incorrigibly return wrong sequences of tokens, they impose hard constraints on the way lexical and syntactic ambiguities can be resolved, and they are limited in the mechanisms they allow for taking context into account. In comparison, model-based parser generators allow for flexible constraint specification and reference resolution, which facilitates the context consideration. In this paper, we explain how the ModelCC model-based parser generator supports statistical language models and arbitrary probability estimators. Then, we present the ModelCC implementation of a natural language parser based on the syntax of most Romance and Germanic languages. This natural language parser can be instantiated for a specific language by connecting it with a thesaurus (for lexical analysis), a linguistic corpus (for syntax-driven disambiguation), and an ontology or semantic database (for semantics-driven disambiguation).

Keywords: Natural languages, disambiguation, query parsing.

1 Introduction

Lexical ambiguities occur when an input string simultaneously corresponds to several token sequences [10], which may also overlap. Syntactic ambiguities occur when a token sequence can be parsed into several parse trees [7]. A common approach to disambiguation consists of performing probabilistic scanning (i.e. probabilistic lexical analysis) and probabilistic parsing (i.e. probabilistic syntactic analysis), which assign a probability to each possible parse tree. However, existing techniques for probabilistic scanning and parsing present several drawbacks: probabilistic scanners may produce incorrect sequences of tokens due to wrong guesses or to occurrences of words that are not in the lexicon, and probabilistic parsers cannot consider relevant context information such as resolved references between language elements, and both probabilistic scanners and parsers

H.L. Larsen et al. (Eds.): FQAS 2013, LNAI 8132, pp. 293–304, 2013.

impose hard constraints on the way lexical and syntactic ambiguities can be resolved.

Model-based language specification techniques [8] decouple language design from language processing. ModelCC [13,12] is a model-based parser generator that includes support for dealing with references between language elements and, thus, instead of returning mere abstract syntax trees, ModelCC is able to obtain abstract syntax graphs and consider lexical and syntactic ambiguities.

In this paper, we explain how ModelCC supports probabilistic language models and we present the implementation of a natural language parser. Section 2 provides an introduction to probabilistic parsing techniques and to the model-based language specification techniques employed by the ModelCC parser generator. Section 3 explains the probabilistic model support in ModelCC. Section 4 describes the implementation of a natural language parser. Finally, Section 5 presents our conclusions and pointers for future work.

2   Background

In this section, we provide an analysis of the state of the art on probabilistic parsing and on model-based language specification.

2.1   Probabilistic Parsing

There are many approaches to part-of-speech tagging using probabilistic scanners and for language disambiguation using probabilistic parsers.

Probabilistic scanners based on Markov-like models [9] consider the existence of implicit relationships between words, symbols or characters found close in sequences, and irrevocably guess the type of a lexeme based on the preceding ones. When using such techniques, a single wrong guess renders the whole parsing procedure irremediably erroneous, as no correct parse tree that uses a wrong token can be found.

Probabilistic scanners based on lexicons [7] assign probabilities to a lexeme belonging to different word classes from the statistical analysis of lexicons. Scanning a lexeme that belongs to a particular word class but never belonged to that class in the training lexicon provides wrong scanning results, which, in turn, render the whole parsing procedure useless.

Probabilistic parsers [11] compute the probability of different parse trees by considering token probabilities and grammar production probabilities, which are empirically obtained from the analysis of linguistic corpora. The probability of a symbol is defined as the product of the probability of the grammar rule that produced the symbol and the probabilities of all the symbols involved in the application of that rule. The probability of a parse tree is that of its root symbol. These techniques do not take context into account.

Probabilistic lexicalized parsers [4,2] associate lexical heads and head tags to the grammar symbols. Grammar rules are then decomposed and rewritten to include the different combinations of symbols, lexical heads, and head tags.

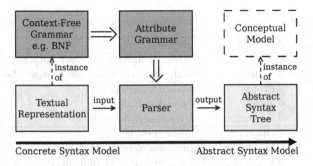

Fig. 1. Traditional language processing

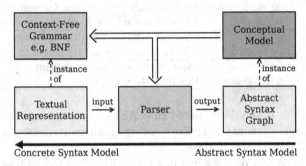

Fig. 2. Model-based language processing

Different probabilities can be associated to each of the new rules. When using this technique, the grammar significantly expands and a more extensive analysis of linguistic corpora is needed to produce accurate results. It should be noted that this technique is not able to consider relevant context information such as resolved references between language elements.

Conventional probabilistic scanners and parsers do not allow the use of arbitrary probability estimators or statistical models that take advantage of more context information.

2.2 Model-Based Language Specification

In its most general sense, a model is anything used in any way to represent something else. In such sense, a grammar is a model of the language it defines. The idea behind model-based language specification is that, starting from a single abstract syntax model (ASM) that represents the core concepts in a language, language designers can develop one or several concrete syntax models (CSMs). These CSMs can suit the specific needs of the desired textual or graphical representation for language sentences. The ASM-CSM mapping can be performed, for instance, by annotating the abstract syntax model with the constraints needed to transform the elements in the abstract syntax into their concrete representation.

A diagram summarizing the traditional language design process is shown in Figure 1, whereas the corresponding diagram for the model-based approach is

Table 1. Summary of the metadata annotations supported by ModelCC

Constraints on...	Annotation	Function
Patterns	@Pattern	Pattern definition for basic language elements.
	@Value	Field where the input element will be stored.
Delimiters	@Prefix	Element prefix(es).
	@Suffix	Element suffix(es).
	@Separator	Element separator(s).
Cardinality	@Optional	Optional elements.
	@Minimum	Minimum element multiplicity.
	@Maximum	Maximum element multiplicity.
Evaluation order	@Associativity	Element associativity (e.g. left-to-right).
	@Composition	Eager or lazy composition for nested composites.
	@Priority	Element precedence level/relationships.
Composition order	@Float	Element member position may vary.
	@FreeOrder	All the element members positions may vary.
References	@ID	Identifier of a language element.
	@Reference	Reference to a language element.

shown in Figure 2. It should be noted that ASMs represent non-tree structures whenever language elements can refer to other language elements, hence the use of the 'abstract syntax graph' term.

ModelCC [13,12] is a parser generator that supports a model-based approach to the design of language processing systems. Its starting ASM is created by defining classes that represent language elements and establishing relationships among those elements. Once the ASM is created, constraints can be imposed over language elements and their relationships as the metadata annotations [6] shown in Table 1 in order to produce the desired ASM-CSM mappings.

Although probabilistic language processing techniques and model-based language specification have been extensively studied, to the best of our knowledge, there are no techniques that allow model-driven probabilistic parsing. In the next section, we explain ModelCC's support for probabilistic language models.

3   Probabilistic Parsing in ModelCC

ModelCC combines model-based language specification with probabilistic parsing by allowing the specification of arbitrary probabilistic language models.

Subsection 3.1 introduces ModelCC support for probabilistic language models and presents ModelCC's *@Probability* annotation. Subsection 3.2 discusses the use of contextual information in ModelCC probabilistic parsers. Subsection 3.3 explains how symbol probabilities are computed.

3.1   Probabilistic Language Models

ModelCC's *@Probability* annotation allows the specification of probability values for language elements and language element members. Probability values can be specified for syntactic elements of the languages and for lexical components, in which case it should be noted that the lexical analyzer behaves as a part-of-speech tagger in natural language processing.

Such probability values can be specified using three alternatives: a probability value as a real number between 0 and 1, a frequency as an integer number, or a custom probability evaluator that computes the probability value from the analysis of the language element and its context.

Since ModelCC supports lexical and syntactic ambiguities and the combination of language models, one of the main novelties of ModelCC with respect to existing techniques is that it allows the modular specification of probabilistic languages, that is, it is able to produce parsers from composite language specifications even when some of the language elements overlap or conflict.

ModelCC also supports alternative models for the representation of uncertainty (e.g. possibilistic models, models based on Dempster-Shafer theory, or any other soft computing models), provided that an evaluation operator for language element instances is provided and an evaluation operator for the application of grammar rules is provided. Optionally, a casting operator that translates the estimated value in one model into a value valid for a different kind of model allows the specification of modular languages even when different mechanisms for representing natural language ambiguities are employed for different parts of the language model.

3.2   Context Information

ModelCC provides context information that custom probability evaluators and constraints can take into account when processing a language element.

The context information includes the current syntax graph and the parse graph symbol corresponding to the language element being evaluated. If the language element instance is a reference, the context information also includes the referenced language element instance, its corresponding parse graph symbol, and the context graph, which is the smallest graph that contains both the reference and the referenced object.

It should be noted that, from this information, it is possible to compute traditional metrics such as the distance between the reference and the referenced object in the input or in the syntax graph and whether the reference is anaphoric, cataphoric, or recursive.

However, in contrast to existing probabilistic parsing techniques, ModelCC also allows the specification of complex syntactic constraints, semantic constraints, and probability evaluators that use extensive context information such as resolved references between language elements.

3.3   Probability Evaluation

The probability of a particular parse graph G for a sentence $w_{1:m}$ of length m is defined as the product of the probabilities associated to the n instances of language elements E_i in the parse graph G:

$$P(G|w_{1:m}) = \prod_{i=1}^{n} P(E_i|w_{s_i:e_i}) \tag{1}$$

Given a language element E that represents a part-of-speech tag and a word w, the lexical analyzer acts as a POS tagger and provides $P(E|w)$.

Given a language element E with $M_1..M_n$ members in its definition, some of which are optional, the probability $P(E|M_{1:n})$ is computed as follows. Let $OPT(E)$ be the set of optional elements for E. Assuming that their appearance is statistically independent, we can estimate the probability of E given its observed elements O:

$$P(E|O_{1:k}) = P(E) \prod_{\substack{M_i \in OPT(E), \\ M_i \in O_{1:k}}} P(M_i|E) \prod_{\substack{M_j \in OPT(E), \\ M_j \notin O_{1:k}}} (1 - P(M_j|E)) \tag{2}$$

Given an ambiguous sentence $w_{1:n}$, its disambiguation is done by picking the parse graph $\hat{G}$ with the highest probability for that sentence:

$$\hat{G}(w_{1:m}) = \arg \max_G \{P(G|w_{1:m})\} \tag{3}$$

We now present the ModelCC implementation of a natural language parser.

4   A Natural Language Parser

In this section, we present a model-based specification for a probabilistic natural language parser. Subsection 4.1 outlines the natural language features. Subsection 4.2 provides the ModelCC ASM specification of the natural language. Subsection 4.3 explains how the natural language can be instantiated. Subsection 4.4 presents a sample English language parser.

4.1   Natural Languages in Terms of Chomsky's X-bar Theory

Chomsky's X-bar theory [3] claims that certain human languages share structural similarities. Our language supports this theory by comprehending these common structures, which we proceed to explain.

In our model, a sentence consists of a clause (i.e. a complete proposition), a clause can be either a simple clause or the coordinate clause composite that creates a compound sentence, a simple clause consists of an optional nominal phrase and a verbal phrase, and a coordinate clause composite consists of a set of clauses and an optional floating coordinating conjunction.

In our natural language model, a complement is a phrase used to complete a predicate construction, and a head is a complement that plays the same grammatical role as the whole predicate construction.

Our natural language model supports nominal, verbal, adverbial, adjectival, and prepositional complements.

Nominal complements comprise nominal phrases, nominal composites, and nominal clauses. A nominal phrase consists of an optional determiner, a noun, and an optional set of complements. A nominal composite consists of an optional determiner, a set of nominal complements and an optional floating conjunction. A nominal clause consists of an optional determiner, an optional subordinating conjunction and a subordinate clause. Nouns comprise common nouns, proper nouns, and pronouns. Pronouns, in turn, reference nouns and proper nouns.

Verbal complements comprise verbal phrases and verbal composites. A verbal phrase consists of a set of floating verbs and an optional floating preposition. A verbal composite consists of a set of verbal complements and an optional floating conjunction.

Adverbial complements comprise adverbial phrases, adverbial composites, and adverbial clauses. An adverbial phrase consists of an adverb. An adverbial composite consists of a set of adverbial complements and an optional floating conjunction. An adverbial clause consists of an optional subordinating conjunction and a subordinate clause.

Adjectival complements comprise adjectival composites and adjectival clauses. An adjectival composite consists of a set of adjectival complements and an optional floating conjunction. An adjectival clause consists of an optional subordinating conjunction and a subordinate clause.

Prepositional complements comprise prepositional phrases and prepositional composites. A prepositional phrase consists of a floating preposition and a head. A prepositional composite consists of a set of prepositional complements and an optional floating conjunction.

It should be noted that this natural language embraces Romance languages such as Spanish, Portuguese, French, and Italian, as well as Germanic languages such as English and German.

4.2   Specification of the Natural Language ASM

In order to implement our natural language parser using ModelCC, we first provide a specification of the language ASM as a set of UML class diagrams shown in Figures 3, 4, 5, 6, and 7. Adjectival complements and adverbial complements can be specified in a manner similar to nominal complements. As it can be observed from the figures, the model-based specification of the natural language matches the language description.

As the specified model is an abstract syntax model, it does not correspond to any particular language. The ASM is more like the Mentalese language postulated by the Language Of Thought Hypothesis [5]. In the next subsection, we explain how different fully-functional natural language parsers can be instantiated from this model by defining additional language-specific constraints.

4.3   Specification of the Natural Language CSM

In order to implement a parser for a particular natural language, the ASM-CSM mapping has to be specified. A pattern matcher is assigned to each lexical

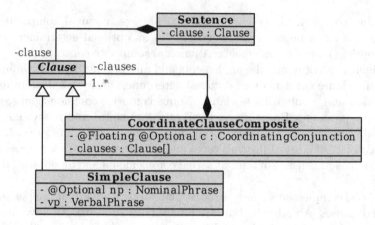

Fig. 3. ModelCC specification of the sentence and clause elements of our natural language

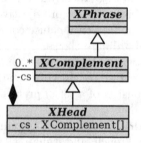

Fig. 4. ModelCC specification of the phrase, head, and complement elements of our natural language

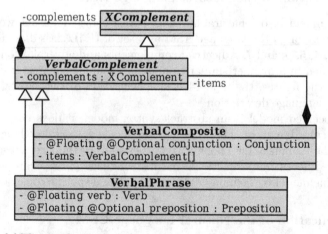

Fig. 5. ModelCC specification of the verbal complement language elements in our natural language

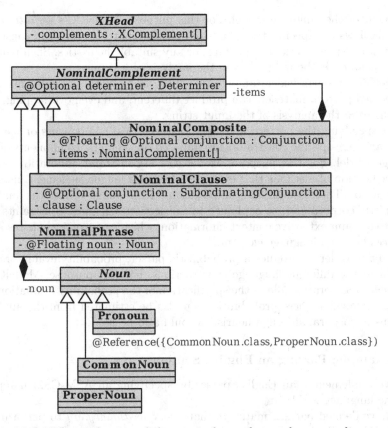

Fig. 6. ModelCC specification of the nominal complement language elements in our natural language

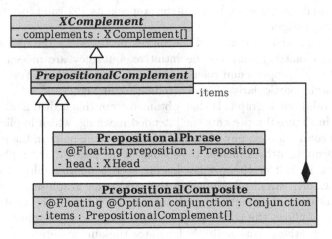

Fig. 7. ModelCC specification of prepositional complement language elements in our natural language

component of the language model. For this purpose, ModelCC's *@Pattern* annotation allows the specification of custom pattern matchers that can consist of regular expressions, dictionary lookups, or any suitable heuristics. Such pattern matchers can easily be induced from the analysis of lexicons.

ModelCC supports lexical ambiguities apart from syntactic ambiguities, so the specified pattern matchers can produce different, and even overlapping, sets of tokens from the analysis of the input string.

After specifying the pattern matchers for the lexical components of the language, language-specific constraints are imposed on syntactic components of the language model. For this purpose, ModelCC's *@Constraint* annotation allows the specification of methods that evaluate whether a language element instance is valid or not. These constraints can be automatically induced from the analysis of linguistic corpora and, as explained in Subsection 3.2, these constraints can take into account extensive context information, which can even include resolved references between language elements.

Finally, in order to produce a probabilistic parser, probability evaluators are assigned to the different language constructions. For this purpose, ModelCC's *@Probability* annotation allows the specification of the probability evaluation for language elements. These probabilities can also be estimated from the analysis of linguistic corpora, although heuristics could also be used.

4.4   Example: Parsing an English Sentence

We have implemented an English parser by specifying an ASM-CSM mapping from the language ASM.

We have defined pattern matchers that query *wiktionary.org* to perform the lexical analysis. We have approximated probability values derived from the analysis of the Google n-gram datasets to different lexemes and constructions.

As an example, we have parsed the sentence "I saw a picture of New York". The lexical analysis graph for this sentence represents 192 valid token sequences and is shown in Figure 8.

It should be noted that some of the tokens produced by the *wiktionary.org*-based pattern matcher may not be intuitive, but they are indeed valid. For example, the "I" proper noun refers to the ego, and the "New York" adjective refers to a style, particularly of food, originating in New York.

A set of valid parse graphs is then obtained from this lexical analysis graph. The graph in Figure 9 represents the intended meaning, which implies that the speaker did see a photography of New York city at some point in the past. Apart from this meaning, others such as the speaker using a saw to cut a photography of New York in half or the speaker having seen a photography of a recently founded York city can also be found in the valid parse graph set.

The application of probability evaluators deducted from the analysis of linguistic corpora allows the parser pinpointing the intended meaning.

Furthermore, there exist multiple language thesauri annotated with specific semantic knowledge such as the 20 question game [1] database, which have information on whether a word refers to an animal, vegetable, mineral or other

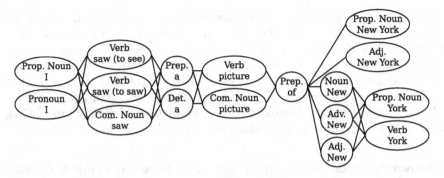

Fig. 8. Lexical analysis graph for the sentence "I saw a picture of New York"

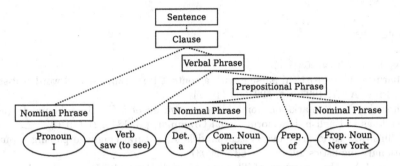

Fig. 9. Correct parse graph for the sentence "I saw a picture of New York"

(such as specific actions, processes, or abstract concepts), on whether if a word refers to real or imaginary concepts, and on the look and feel or effect of most concepts. Such a database can be combined with our language parser by means of probability evaluators to figure out semantic constraints that greatly narrow the valid parse graph set by discarding absurd cases.

A specific ontology designed for a query answering system can also be used to fine tune the natural language parser for a particular application. For example, let us consider the transitive verb "to rub" and the direct object "table". Since the verb implies a physical effect on the direct object and the direct object has several meanings, such as a piece of furniture or as a data matrix, the latter interpretation is discarded as it cannot undertake a physical effect.

5   Conclusions and Future Work

Lexical and syntactic ambiguities are always present in natural languages. NLP-based query answering systems require dealing with ambiguities. A common approach to disambiguation consists of performing probabilistic scanning and probabilistic parsing. Such techniques present several drawbacks: they may produce wrong sequences of tokens, they impose hard constraints on the way ambiguities can be resolved, and they only take advantage of small amounts of context information.

ModelCC is a model-based parser generator that supports lexical ambiguities, syntactic ambiguities, the specification of constraints, and reference resolution. ModelCC solves the aforementioned drawbacks of existing techniques.

In this paper, we have described ModelCC support for probabilistic language models. We have also described the ModelCC implementation of a natural language parser, and we have provided an English-language instantiation of it.

In the future, we plan to research on the automatic induction of probabilistic language models, syntactic constraints, and semantic constraints.

Acknowledgments. Work partially supported by research project TIN2012-36951, "NOESIS: Network-Oriented Exploration, Simulation, and Induction System".

References

1. 20q, http://www.20q.net
2. Charniak, E.: Statistical parsing with a context-free grammar and word statistics. In: Proc. AAAI 1997, pp. 598–603 (1997)
3. Chomsky, N.: Remarks on nominalization. In: Jacobs, R., Rosenbaum, P. (eds.) Readings in English Transformational Grammar, pp. 184–221 (1970)
4. Collins, M.: Head-driven statistical models for natural language parsing. Computational Linguistics 29(4), 589–637 (2003)
5. Fodor, J.A.: The Language of Thought. Crowell Press (1975)
6. Fowler, M.: Using metadata. IEEE Software 19(6), 13–17 (2002)
7. Jurafsky, D., Martin, J.H.: Speech and Language Processing: An Introduction to Natural Language Processing, Computational Linguistics and Speech Recognition, 2nd edn. Prentice Hall (2009)
8. Kleppe, A.: Towards the generation of a text-based IDE from a language metamodel. In: Akehurst, D.H., Vogel, R., Paige, R.F. (eds.) ECMDA-FA. LNCS, vol. 4530, pp. 114–129. Springer, Heidelberg (2007)
9. Markov, A.A.: Dynamic Probabilistic Systems (Volume I: Markov Models). In: Howard, R. (ed.) Extension of the Limit Theorems of Probability Theory to a Sum of Variables Connected in a Chain, pp. 552–577. John Wiley & Sons (1971)
10. Nawrocki, J.R.: Conflict detection and resolution in a lexical analyzer generator. Information Processing Letters 38(6), 323–328 (1991)
11. Ney, H.: Dynamic programming parsing for context-free grammars in continuous speech recognition. IEEE Transactions on Signal Processing 39(2), 336–340 (1991)
12. Quesada, L.: A model-driven parser generator with reference resolution support. In: Proceedings of the 27th IEEE/ACM International Conference on Automated Software Engineering, pp. 394–397 (2012)
13. Quesada, L., Berzal, F., Cubero, J.-C.: A language specification tool for model-based parsing. In: Yin, H., Wang, W., Rayward-Smith, V. (eds.) IDEAL 2011. LNCS, vol. 6936, pp. 50–57. Springer, Heidelberg (2011)

Correlated Trends: A New Representation for Imperfect and Large Dataseries

Miguel Delgado, Waldo Fajardo, and Miguel Molina-Solana*

Department Computer Science and Artificial Intelligence,
Universidad de Granada, 18071 Granada, Spain
{mdelgado,aragorn,miguelmolina}@ugr.es

Abstract. The computational representation of dataseries is a task of growing interest in our days. However, as these data are often imperfect, new representation models are required to effectively handle them. This work presents *Frequent Correlated Trends*, our proposal for representing uncertain and imprecise multivariate dataseries. Such a model can be applied to any domain where dataseries contain patterns that recur in similar —but not identical— shape. We describe here the model representation and an associated learning algorithm.

1 Introduction

Databases from most of industrial and biological areas often contain timestamped or ordered records. Therefore, dataseries are gaining weight as a suitable source of information, and working with them has become an important machine learning task. The following are a few illustrative examples of phenomena that can be described and represented by series of observations along different dimensions:

- The weather in a given location, represented as a series of observations at different time instants, including information such as temperature, precipitations or wind speed [1].
- The way of playing an instrument. A particular performance of a piece of music can be represented as a series of notes with its respective duration and volume, among others attributes [2].
- The way a human being behaves within Ambient Assisted Living, with the aim of identifying strange actions and situations of potential danger [3].
- The interactions between currency exchanges. Several works have studied how some currencies behave against each other at different financial situations [4].

Two main goals of dataseries analysis are found in literature: forecasting and modeling. The aim of forecasting is to accurately predict the next values of the series, whereas modeling aims to describe the whole series. Even though they can be sometimes related, they usually differ as a proper model for the long-term evolution might not predict well the short-term evolution, and viceversa.

H.L. Larsen et al. (Eds.): FQAS 2013, LNAI 8132, pp. 305–316, 2013.

In either case, and whatever the goal of a particular dataseries analysis is, data representation is a crucial task anyway. It is hence required a formal representation capable of modeling the complexity of the particular data. This representation must be more reduced than representing all the observations of the phenomenon, but still describe it accurately enough.

An additional problem is that real-world information is hardly certain, complete and precise; more on the contrary, it is usually incomplete, imprecise, vague, fragmentary, not fully reliable, contradictory, or imperfect in some other way. Historically, two ways of addressing imperfection have been employed for representing information in a computer [5]:

- Restricting the model to only that part of the information that is accurate and reliable. This approach avoids further complications of representation, but lacks the capacity of capturing the whole rich notion of information in human cognition.
- Developing models capable of representing imperfect information. This is the preferred approach as it allows a greater number of applications. However, a data simplification stage is often needed, to allow traditional tools to treat the data.

Referring to dataseries, some schemes have been proposed for directly handling imperfect information [6], but most of the research has focused on similarity measures to deal with imperfection. Hence there is still a need for further research and new practical systems capable of accurately modeling (not only computing distance between) imperfect dataseries.

This paper addresses this necessity by describing our ongoing work towards a novel approach for representing imperfect multivariate dataseries —concretely uncertain and imprecise. We show how underlying local trends in data can be represented in an easy and effective way, without a complicated formalism. Some other works have identified the necessity of focusing in frequent local cues for behavior modeling [7,8].

Specifically, our proposal summarizes dataseries through capturing their general footprint by means of discovering repetitive patterns in one dimension and their interdependence with patterns in other dimensions. That is, by computing the frequency distribution of the co-occurrence of patterns in different dimensions. The process can be divided in the following three stages:

1. high-level abstraction of the observations within each dimension;
2. tagging according to the patterns identified in one of the dimensions;
3. characterization of the dataseries as sets of frequency distributions.

Two advantages of our proposal are of special relevance. In first place, the representation of the dataseries is finite and constant in size for a given problem, regardless of the number of observations. This fact contrasts with other works [8,9] in which the size of the representation completely depends on the amount of data available.

The second advantage is an incremental representation which can be calculated on-line very easily: when a new value is observed, this information is included in the representation, which is immediately updated. That is not the case in other proposals [9,4] which need to recalculate the whole representation when new observations are available.

Our method is therefore able to automatically offer a representation of the dataseries until a given observation, and allow it to deal with dataseries of infinite length. The proposal is hence specially aimed to those phenomena with a large number of observations.

The rest of the paper is organized as follows. Section 2 introduces the concepts of dataseries and imperfect data. In Section 3, we describe our proposed model, *Frequent Correlated Trends* for representing imperfect dataseries, and the developed system, including data gathering, representation and distance measurement. This section also introduces the formal notation and an illustrative example. The paper concludes with a summary, some final considerations, and pointing out future work in Section 4.

2   Definitions

2.1   Data Series

A dataseries A can be intuitively defined as an ordered sequence (finite or not) of values obtained at successive intervals of an indexing variable —often *time*. Each one of these observations A_i takes values from a domain $\mathcal{U}^n$.

According to the value of n, we have several kinds of series. If $n = 1$, each element is a single (scalar) value and we have an *univariable* series. On the other hand, if $n > 1$, each element is a vector with n components and the dataseries is a *multivariable* one. We can distinguish two kinds of multivariable series. A *multidimensional* series is a series in which the majority of variables are independent, whereas a *multivariate* series has many dependent variables that are correlated to each other to varying degrees.

In many dataseries, data follow recurring seasonal patterns. A dataseries has serial dependence if the value at some point A_i is statistically dependent on the value at another point A_j. However, in many situations, values in series A^r (the series with the r-th component of each value of A) could be dependent of those in series A^s. That is the case in multivariate series. Our hypothesis is that the correlation among patterns of different dimensions of the same series is an interesting source of information should we want to summarize the series.

Whilst dataseries can be infinite in length (i.e. a continuous variable, or a non-ending series of observations), computers can only deal with a finite number of pieces of information. Therefore, in order to represent an arbitrary dataseries, it is mandatory to employ an approximate representation, reducing the size of data, and implying a trade-off between accurately capturing the data and representing it in the finite memory of a computer.

2.2 Imperfect Data

We experience that information in most domains is usually incomplete, imprecise, vague, not fully reliable, contradictory, or imperfect in some other way. Imperfect information might result from using unreliable information sources, it can be the unavoidable result of information gathering methods that require estimation or judgment, or be produced by a restricted representation model.

In general, these various information deficiencies may result in different types of imperfection [5], being imprecision and uncertainty the most common types found in data.

Imprecision and *vagueness*, in particular, are related with the impossibility to give a concrete value to an element. Although the correct value is within a range of values, it is not possible to know which one. For example, 'between 100 and 120 kilograms' and 'very heavy' for John's weight are imprecise and vague values respectively. Vague information is usually represented by linguistic terms.

Uncertainty, on the other hand, indicates the degree of certainty of a value. It expresses how sure one can be about a statement. 'It is almost sure that John is his brother' is an example of information uncertainty.

Although related, imprecision and uncertainty state for different things and hence, they are not interchangeable. Zadeh [10] enunciated the principle of incompatibility between certainty and precision, and several data models have been proposed to handle uncertainty and imprecision, and most of them are based on the same paradigms. Imprecision is generally modeled with fuzzy sets, and uncertainty with fuzzy measures.

The theory of fuzzy sets, introduced by Lotfi A. Zadeh, is an extension of the notion of classical sets, allowing elements to have degrees of membership, not just the binary terms of belonging and not-belonging. As said, it has been (and it is still) widely used to represent imprecision.

Uncertainty, on the other hand, is generally addressed by the theory of fuzzy measure, which indicates the degree of evidence or certainty that a particular element belongs in the set. Three special cases of fuzzy measure theory have been widely studied: evidence theory, possibility theory and probability theory. Historically, the last one has been by far the most employed in literature.

Our approach lies within this general framework —dealing with imprecision by means of fuzzy sets, and with uncertainty by employing probability. This representation is general enough to allow its application to several problems and domains.

We can also find in the literature the concept of the probability of a fuzzy event. Zadeh [11] proposed it and Yager [12] addressed the issue by proposing a new definition of this probability. However, the problem of establishing a theory of probability for fuzzy events still remains an open one [13].

Because of this lack of well-definition we do not follow this approach in the present paper. We aim for a well-defined framework of representation that fuzzy probabilities cannot yet offer.

3   Our Proposal: Frequent Correlated Trends

So far in this paper we have introduced the concepts of dataseries and imperfect information, and described the problems their representation pose. In the current section, we propose and describe a new model for representing imperfect multivariate dataseries —concretely, uncertain and imprecise— which are widely found in real-life problems. The model aims to represent underlying local trends in the data in an easy and effective way, without a complicated formalism. We also describe the corresponding learning algorithm.

Our proposal computes the frequency distribution for the co-occurrence of patterns in different dimensions of the dataseries. In order to apply our model and learning algorithm, the following three hypothesis with respect to the data are assumed:

- The series of data contain one or several recurrent patterns. Those patterns will be the ones we will try to identify.
- If a certain pattern was observed in the past, it will eventually happen again in the future.
- Dataseries are long enough for the information to be representative. The longer the dataseries is, the more accurate the representation.

3.1   Representation Model

The mathematical model of our proposal is as follows. First, let T be an ordered set. We define X as a multivariate dataseries with observations at points of T. Each observation $x_i \in X$ (at a point $i \in T$) is a vector of n scalars that takes values from a domain $\mathcal{U}^n$. Each component $d \in D$, with $|D| = n$, is considered a dimension of the dataseries X; X^d is the univariate series[1] of component d, and its domain is defined by the d-th component of $\mathcal{U}^n$.

In an ideal situation, we would have very precise measurements for each observation x_i, and any repetition X' of the phenomenon described by X would be identical. In practice, it is hardly the case, as the accuracy of the observations is not fully reliable. In fact, what we usually have are approximated values derived from imprecise measurements. Nevertheless, there are some commonalities between several instances of the same phenomenon, as we are able to identify repetitions of them.

As most dataseries describing a phenomenon are inherently imprecise and changing, it is mandatory to employ a representation model capable of handling these imperfections. This fact does not imply any constraints; more on the contrary, it appraises the general case which is actually the most common one.

Our approach for dealing with the representation of dataseries is based on the acquisition of what we called *frequent correlated trends*, that characterize each particular series. Specifically, these trends model the relationships between structural patterns in a reference dimension, and patterns in other dimensions,

---

[1] Note that a multivariate dataseries can be seen as a set of several univariate dataseries with values at the same points of an indexing variable.

that is, their co-occurrences. Notice that correlated trends are not trying to characterize the whole dataseries with respect to an expected global value, but only finding relevant relations between patterns in different dimensions.

It is important to note here that we focus on the computational representation of the observations that came from several sensors, regardless of which ones they are. Therefore, our approach is independent of the technologies used to collect the information.

For illustrative purposes we will present a concrete example about weather, along with the formal description, in order to clarify concepts and procedures regarding both the method and model. In particular, we will focus on the weather in a given location, which can be understood as a behavior. One of the most common ways of representing these data is by means of a series of observations at different time instants $(T = time)$. The model we propose in this paper is indeed appropriate for modeling this phenomenon because the exact values are not really important, the repetitions are not exactly the same, and nonetheless weather has an inherent seasonal nature.

Therefore, the behavior X will be the weather in a given location, and it is described as a set of observations $x_i, i \in \{1 \cdots 15\}$. Each observation x_i has three components corresponding to three $(n = 3)$ aspects of the weather at that point i. The set of those aspects is $D = \{temperature, precipitations, wind\ speed\}$, and the concrete values of the observations x_i^d for each dimension $d \in D$ are the following[2]:

$$X^{temperature} = \{19, 21, 21, 25, 23, 22, 19, 17, 21, 22, 22, 18, 19, 22, 24\}$$
$$X^{precipitations} = \{98, 102, 110, 93, 95, 71, 67, 52, 63, 75, 80, 95, 72, 104, 98\}$$
$$X^{windspeed} = \{7, 6, 1, 3, 4, 7, 11, 15, 14, 7, 9, 11, 9, 9, 7\}$$

3.2   Learning Algorithm

In order to transform real data to the frequent correlated trends formalism, a learning algorithm is necessary. In the rest of this section, we describe the process and discuss how we do such transformation following the example. The procedure is summarized in Algorithm 1.

As previously said, our representation aims to describe a behavior X in terms of the relationships between patterns from a reference dimension $I \in D$ and patterns from the rest of dimensions of X. Patterns in X^I will be used to index repeating patterns in other dimensions. In other words, patterns in series $X^d, d \neq I$ will be related with those in X^I.

In most behaviors, the selection of I is clear by the semantics of the domain and does not represent a problem. In the case of not knowing which series is appropriate to use as reference, the correlations between dimensions should be

---

[2] As our method is aimed for behaviors with a large number of observations, it will not be able to get any meaningful information from these data. However, as an illustrative example of the process, we will employ this reduced number.

studied and the dimension with higher correlations with the rest should be selected. This way, we tend to maximize the amount of correlated patterns. In fact, this might very well be the way to proceed in all cases; however, using the semantics can clarify the representation in many domains, should we later want to interpret the results.

Algorithm 1. Learning algorithm for Frequent Correlated Trends

Definitions:
  $T \subset \mathbb{R}$ is an ordered set
  X is a multivariate series indexed by T
  D is the set of dimensions of X
  X^d is the univariate series with only observations from dimension d
  P is the set of possible directions of an interval: *ascending, descending, null*
  s is the number of observations for each segment
  $\mathcal{P}_s$ is the set of indexing patterns of size s
  $\mathcal{Q}_s$ is the set of qualitative shapes of size s
Inputs:
  X, D, s, $\mathcal{P}_s$, $\mathcal{Q}_s$
Outputs:
  Frequency distributions of co-occurrences of patterns in different dimensions
Algorithm:
  Select a reference dimension $I \in D$
  Tag each interval in X^I with a value from P, obtaining X'^I
  Segment X'^I in groups of size s
  Tag each one of those segments with an element from $\mathcal{P}_s$, obtaining $\mathcal{X}^I$
  for all dimensions $d \in D, d \neq I$ **do**
      Calculate $\bar{X}^d$ as the average value of observations of X^d
      Transform each observation $x_i^d \in X^d$ as follows:
      if $x_i^d \geq \bar{X}^d$ **then**
          $x_i'^d \leftarrow +$
      else
          $x_i'^d \leftarrow -$
      end if
      Segment X'^d in groups of size s
      Tag each one of those segments with an element from $\mathcal{Q}_s$, obtaining $\mathcal{X}^d$
      for all $p \in \mathcal{P}_s$ **do**
          Count the number of simultaneous occurrences of p in $\mathcal{X}^I$ and $q \in \mathcal{Q}_s$ in $\mathcal{X}^d$
          Build a frequency distribution for $\mathcal{Q}_s$ given p from those values
      end for
  end for

It should be noted here that, although we have defined X as being indexed by T, this dimension is not a choice for I, because T is an ordered set and thus no patterns can arise from its consecutive values, apart from the monotone growth. Also, $T \notin D$ according to how we have defined D.

The next step comprises transforming the reference series X^I using a simple contour criteria. Two consecutive (imprecise) observations form an interval. If we consider just the directions of the intervals (slopes), three options are available: (a)scending, (d)escending and (n)ull. Each interval in the reference series X^I is then tagged with one of these three options.

At this point, we have to decide the size s of patterns. Small patterns are preferred because they are more easily found in the series. The longer the pattern is, the more difficult it is to find occurrences. We define the set of indexing patterns of size s, $\mathcal{P}_s$, as the set of all permutations with repetitions of size $s-1$ of elements from $P = \{a, d, n\}$. If groups of three values (i.e. two intervals) are considered, nine different patterns arise, $\mathcal{P}_3 = \{aa, ad, an, da, dd, dn, na, nd, nn\}$.

We then segment the indexing series X^I in groups of size s and tag each group with the corresponding indexing pattern from $\mathcal{P}_s$. We will obtain a new series $\mathcal{X}^I$. To do so, we propose to use a sliding window of s consecutive observations that moves in steps of one.

Regarding our example, we will select *temperature* as the indexing dimension I, because we would like to relate patterns in dimensions *precipitations* and *wind speed* with those in *temperature*. We aim to obtain relations such as: "65% of the times, when the temperature rises in two consecutive days, the second day it rains more than the first". Once selected, the segmentation of that series will take place as formerly described, obtaining $\mathcal{X}^I$. In summary, $X^{temperature}$ is transformed into X'^I (transforming each pair of observations to a value from P) and after into $\mathcal{X}^I$:

$$X^{temperature} = \{19, 21, 21, 25, 23, 22, 19, 17, 21, 22, 22, 18, 19, 22, 24\}$$
$$X'^I = \{a, n, a, d, d, d, d, a, a, n, d, a, a, a\}$$
$$\mathcal{X}^I = \{an, na, ad, dd, dd, dd, da, aa, an, nd, da, aa, aa\}$$

Similarly, the rest of dataseries $X^d (d \in D, d \neq I)$ conforming X are transformed by employing a qualitative binary transformation in the following way: each value x_i^d is compared to the average value $\bar{X}^d$ from the dataseries, and it is transformed into a qualitative value where $+$ means 'the value is higher than or equal[3] to the average', and $-$ means 'the value is lower than the average'.

Note here that if the dataseries is finite, the average value (a vector with the average values for each component) can be easily computed with the usual formula:

$$\bar{X} = 1/N \cdot \sum_{1 \leq i \leq N} x_i$$

Otherwise, the average can be calculated incrementally. However, an initial estimation of it (by any means) would be required. In this situation the average value is different depending on the position of the observation been calculated, but tends to become stable as more values are computed.

---

[3] In our current model, we have arbitrarily decided to include equal values in the $+$ set. However, another qualitative value, namely $=$, could be used to distinguish between those values that are greater than the average and those which are equal (or almost equal).

In our example, we will proceed to transform series *precipitations* and *wind speed*. To do so, we first calculate the average value for those series using all available observations. Those average values are 85 mm and 8 mph, for *precipitations* and *wind speed* respectively. We will then compare all observations $x \in X^{precipitations}$ with 85 and transform them to $+$ (if the value is greater than or equal to 85), or to $-$ (if lower). We do the same process with observations in $X^{windspeed}$, obtaining the following series:

$$X^{precipitations} = \{98, 102, 110, 93, 95, 71, 67, 52, 63, 75, 80, 95, 72, 104, 98\}$$
$$X'^{precipitations} = \{+, +, +, +, +, -, -, -, -, -, -, +, -, +, +\}$$
$$X^{windspeed} = \{7, 6, 1, 3, 4, 7, 11, 15, 14, 7, 9, 11, 9, 9, 7\}$$
$$X'^{windspeed} = \{-, -, -, -, -, -, +, +, +, -, +, +, +, +, -\}$$

Being s the size of the segment and r the number of different qualitative values, there are r^s possible resulting shapes. In the current example, since we are segmenting the series in groups of three observations and using two qualitative values, eight (2^3) different patterns may arise. We note these possibilities as $\mathcal{Q}_3 = \{---, --+, -+-, -++, +--, +-+, ++-, +++\}$. We then segment $X'^{precipitations}$ and $X'^{windspeed}$ in groups of three observations (employing an sliding window of size 1) and tag these groups with patterns from $\mathcal{Q}_3$, obtaining $\mathcal{X}^{precipitations}$ and $\mathcal{X}^{windspeed}$ respectively:

$$\mathcal{X}^{precipitations} = \{+++, +++, +++, ++-, +--, ---, ---,$$
$$---, ---, --+, -+-, +-+, -++\}$$
$$\mathcal{X}^{windspeed} = \{---, ---, ---, ---, --+, -++, +++,$$
$$++-, +-+, -++, +++, +++, ++-\}$$

Once we have transformed and tagged all the series, we can represent them with the frequent correlated trend formalism. A frequent correlated trend for a given dimension $d \in D$ of a behavior X is represented by a set of discrete frequency distributions for that given dimension d. Each of these frequency distributions represents the way patterns in $\mathcal{X}^d$ behaves with respect to indexing patterns from $\mathcal{P}_s$ in $\mathcal{X}^I$. In other words, we have a frequency distribution for each element in $\mathcal{X}^d$.

To populate the representation of a particular dimension d of a dataseries X, we take advantage of the tagging already performed over the observations. We construct a histogram for each indexing pattern $p \in \mathcal{P}_s$. Histograms have a bin for each element $q \in \mathcal{Q}_s$, and are constructed by calculating the percentage of points $i \in T$ where $\mathcal{X}_i^I = p$ and $\mathcal{X}_i^d = q$. These histograms can be understood as discrete frequency distributions of $\mathcal{Q}^s$ given p. Thus, frequent correlated trends capture statistical information of how a certain phenomenon behaves. Combining frequent correlated trends from different dimensions of the phenomenon, we improve its representation.

Our representation of the dataseries might be output to users as frequency distributions. However, this is not the more readable of the representations, being our method not specially aimed to provide explanation to users. Further processing should be done in order to provide users with a more friendly and approachable representation.

3.3   Computing Distance

In many occasions, the final goal of representing dataseries is their posterior identification and classification. If that is the case, a dataseries —which we describe as a collection of frequent correlated trends— can be use as an instance to compare with new instances. Each dataseries can then be viewed as a point in a space of dimensionality $n \cdot |\mathcal{P}_s| \cdot |\mathcal{Q}_s|$, where n is the number of dimensions in the dataseries; $|\mathcal{P}_s|$ is the cardinality of the set of indexing patterns; and $|\mathcal{Q}_s|$ is the cardinality of the set of qualitative shapes for representing the rest of series. In that case, the usual hypothesis is that similar instances tend to be close in the representation space.

This representation tends to have a high dimensionality, a source of problems if the number of samples is not large enough. Due to the so-called *curse dimensionality* (that is, the exponential growth of the space as a function of dimensionality), multivariate spaces of increasing dimensionality tend to be sparse. For that reason, if frequent correlated trends are to be used for learning and classification, it is mandatory to study the dimensionality of each concrete problem and the number of available samples. If these samples are not enough, it might be necessary to reduce the dimensionality of the model by means of a suitable feature extraction mechanism.

In any case, to compare different dataseries, a distance measure is required. As an illustrative example, we describe in the following lines a simple one based on Manhattan distance.

The distance d_{XY} between two multivariate dataseries X and Y (represented with the frequent correlated trend formalism), is defined as the weighted sum of distances between the frequency distributions of each structural pattern:

$$d_{XY} = \sum_{p \in \mathcal{P}} w^p_{XY} \cdot dist(p_X, p_Y) \qquad (1)$$

where $\mathcal{P}$ is the set of the different structural patterns considered; $dist(p_X, p_Y)$ is the distance between two frequency distributions (see (3) below); and w^p_{XY} are the weights assigned to each pattern. Weights have been introduced for balancing the importance of patterns with respect to the number of times they appear. Weights are defined as the mean of cardinalities of respective histograms for a given pattern p:

$$w^p_{XY} = (N^p_X + N^p_Y)/2 \qquad (2)$$

Finally, distance between two frequency distributions is calculated by measuring the absolute distances between respective patterns:

$$dist(u, v) = \sum_{q \in \mathcal{Q}} |u_q - v_q| \qquad (3)$$

where u and v are two frequency distributions for the same pattern; and $\mathcal{Q}$ is the set of all possible values they can take.

When several dimensions are considered, we propose to simply aggregate the individual corresponding distances.

4   Conclusions

The computational representation of dataseries is a task of great interest for researchers as more and more sensors are available continuously collecting information.

However, the recurring patterns within dataseries that describe phenomena are everything but perfect. Although the problem of representing imperfect data has been addressed many times in the past, the lack of a general solutions obligates to build ad-hoc solutions for different problems. For that reason, there is still a need to find new solutions and models to represent such information.

Our proposal, which assumes that some kind of commonality exists among patterns in a dataseries, represents imperfect dataseries as a set of frequency distributions. To do so, it first transforms the imperfect observations into qualitative values. Then, it selects a dimension of the dataseries and uses it to look for correlations with the rest of dimensions. These correlations are expressed as discrete frequency distributions of co-occurrences of patterns.

The main advantage of our method is that it employs a finite and constant representation in size for dataseries, regardless of their length. It also allows for an incremental representation of the observations until a particular moment.

On the other hand, the main limitation of our model is that it is incapable of explaining the predictions it makes. Additionally, as the method is aimed for long series, its best performance is achieved when a large number of observations is available.

4.1   Further Work

Our next step is applying the *Frequent Correlated Trend* model to different domains to test their feasibility, and compare the performance of our proposal with respect to others. Some of these domains and problems have been described here as illustrative examples, but many others might be suggested.

We also plan to study the behavior of different distance measures. We have defined in this paper a very simple one, but many others could be proposed to better account for the particular semantics of different applications.

Finally, we are interested in investigate how the outputs of this representation might be presented to users in effective ways. Our current approach lacks this feature as it is more aimed to computationally representing them rather than obtaining human-friendly interpretations.

Acknowledgments. The authors would like to thank the Spanish Ministry of Education for their funding under the project TIN2009-14538-C02-01. Miguel Molina-Solana is also funded by the Research Office at the University of Granada.

References

1. Kriegler, E., Held, H.: Utilizing belief functions for the estimation of future climate change. Int. Journal of Approximate Reasoning 39(2-3), 185–209 (2005)
2. Molina-Solana, M., Arcos, J.L., Gómez, E.: Identifying Violin Performers by their Expressive Trends. Intelligent Data Analysis 14(5), 555–571 (2010)
3. Delgado, M., Ros, M., Vila, M.A.: Correct behavior identification system in a Tagged World. Expert Systems with Applications 36(6), 9899–9906 (2009)
4. Zhang, Y.Q., Wan, X.: Statistical fuzzy interval neural networks for currency exchange rate time series prediction. Applied Soft Computing 7(4), 1149–1156 (2007)
5. Motro, A.: Sources of Uncertainty, Imprecision, and Inconsistency in Information Systems. In: Motro, A., Smets, P. (eds.) Uncertainty Management in Information Systems: From Needs to Solutions, pp. 9–34. Kluwer Academic Publishers (1996)
6. Liao, S.S., Tang, T.H., Liu, W.Y.: Finding relevant sequences in time series containing crisp, interval, and fuzzy interval data. IEEE Transactions on Systems, Man, and Cybernetics 34(5), 2071–2079 (2004)
7. Herbst, G., Bocklisch, S.F.: Short-Time Prediction Based on Recognition of Fuzzy Time Series Patterns. In: Hüllermeier, E., Kruse, R., Hoffmann, F. (eds.) IPMU 2010. LNCS, vol. 6178, pp. 320–329. Springer, Heidelberg (2010)
8. Saleh, B., Masseglia, F.: Discovering frequent behaviors: time is an essential element of the context. Knowledge and Information Systems 28(2), 311–331 (2010)
9. Xu, W., Kuhnert, L., Foster, K., Bronlund, J., Potgieter, J., Diegel, O.: Object-oriented knowledge representation and discovery of human chewing behaviours. Engineering Applications of Artificial Intelligence 20(7), 1000–1012 (2007)
10. Zadeh, L.A.: Outline of a new approach to the analysis of complex systems and decision processes. IEEE Trans. on Systems, Man and Cybernetics 3(1), 28–44 (1973)
11. Zadeh, L.A.: Probability measures of fuzzy events. Journal of Mathematical Analysis and Applications 23(2), 421–427 (1968)
12. Yager, R.: A note on probabilities of fuzzy events. Information Sciences 18(2), 113–129 (1979)
13. Trillas, E., Nakama, T., García-Honrado, I.: Fuzzy Probabilities: Tentative Discussions on the Mathematical Concepts. In: Hüllermeier, E., Kruse, R., Hoffmann, F. (eds.) IPMU 2010. LNCS, vol. 6178, pp. 139–148. Springer, Heidelberg (2010)

Arc-Based Soft XML Pattern Matching

Mohammedsharaf Alzebdi[1], Panagiotis Chountas[1], and Krassimir Atanassov[2]

[1] University of Westminster, ECSE Department, 115 New Cavendish Street
W1W 6UW London, UK
{alzebdm,chountp}@wmin.ac.uk
[2] Bulgarian Academy of Sciences, CLBME Department Sofia, Bulgaria
krat@argo.bas.bg

Abstract. The internet is undoubtedly the biggest data source ever with tons of data from different sources following different formats. One of the main challenges in computer science is how to make data sharing and exchange between these sources possible; or in other words, how to develop a system that can deal with all these differences in data representation and extract useful knowledge from there. And since XML is the de facto standard for representing data on the internet, XML query matching has gained so much popularity recently. In this paper we present new types of fuzzy arc matching that can match a pattern arc to a schema arc as long as the correspondent parent and child nodes are there and have reachability between them. Experimental results shown that the proposed approach provided better results than previous works.

Keywords: Pattern Tree, Intuitionistic Fuzzy Trees, Soft Arc Matching.

1 Introduction

Due to the heterogeneity in XML schemas, traditional crisp querying techniques are not efficient for analyzing XML data, because they require exact query matching. Therefore, there is a need for new approaches that can achieve approximate query matching instead. This means that not just the exact answer of a query, but also approximate answers will be retrieved.

Even though many studies have addressed approximate query matching in the literature; we believe that their approaches have some limitations and that an Intuitionistic Fuzzy approach is very useful to achieve approximate XML query matching by considering matching a pattern tree with multiple data sources and then joining sub-results together in order to construct a complete answer to a query. The focus of this paper is on matching Pattern Trees (Pt) based on Soft Node Matching as well as Soft Arc Matching. Matching is mainly based on the primitive tree structure, the arc, meaning that an answer of a query can be constructed from different arcs or twigs. Following that, new methods of soft arc matching are presented in this paper.

2 XML Schema Heterogeneity

XML (eXtensible Markup Language) is W3C Recommendation considered as the standard format for structured documents and data on the Web. Because of being the

H.L. Larsen et al. (Eds.): FQAS 2013, LNAI 8132, pp. 317–327, 2013.

most common standard for data transmissions between heterogeneous systems, XML has gained so much popularity recently, especially in web applications. As the amounts of data transmitted and stored in XML are rapidly growing, the ability to efficiently query XML is becoming increasingly important. However, current XML Query languages have some performance-related and structural heterogeneity challenges that need to be addressed [1].

For example, suppose that we are interested in querying a group of XML documents to retrieve information about university departments with research groups along with any projects and/or publications for these groups. According to our understanding of that domain, we might form a query that looks like Pt in Figure 1 below. Nodes with single circle shape indicate structural nodes that are not part of the output, whereas double-circled ones refer to output nodes. Node labels that are underlined, e.g. dname and @id, signify ID nodes acting as Primary Keys.

Looking at how subtrees (twigs) of Pt are matched to the data schemas s1, s2 and s3, we can see that twig 1, can be matched to s1. However, the element node location in Pt needs to be matched to the attribute node @location in S1. For twig 2, it can be noticed that the arc (group, project) in Pt is structured as (project, group) in s2. Lastly, twig 3 can be fully matched to the correspondent twig in s3; however, we cannot determine which publication belongs to which group because the arc (group, publication) does not have a match. Nevertheless, there is an indirect connection between the group and publication using the ID/IDREF directives.

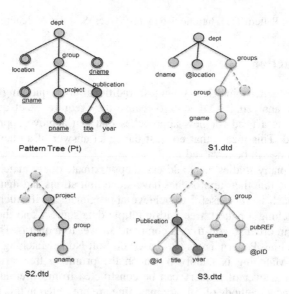

Fig. 1. Matching Pt with several data schemas

To put it in a formal way, the reasons behind diversity in data schemas can be:

- Representation of a certain domain can be scattered in multiple schemas instead of one single schema.
- A node, such as '*location*', can be modelled either as an element node or attribute node, and this is mainly due to flexibility of XML. However, if the node is planned to have child(ren) then it has to be an element node.
- Many-to-many relationships between two nodes, such as *group* and *project*, can be modelled as an arc (group, project) or (project, group). Even in case of one-to-many relationships, two nodes can still be modelled differently.
- Sometimes separating node(s) can be found between a parent and a child node e.g. the arc (dept, group) in Pt can be matched with the arc (dept, group) in S1 even though there is a separating node (*groups*) between the parent and the child.
- Some XML documents are *normalised* i.e. ID/IDREF are used to connect "entities" together, just like primary and foreign key connections in relational databases.

The above forms of heterogeneity in XML schemas can often be found in reality, especially when schemas belong to different sources. Everyone has his own perception of a certain domain, and s/he models it in a different way.

3   XML Pattern Tree Matching

Since XML Queries are modelled as Query Patterns or Pattern Trees (Pt) in computer science literature, XML Query matching research cannot be separated from XML schema matching. In this section we present the most common approaches of XML Query/Schema matching in both database and IR communities. There are plenty of surveys on previous approaches [2-5] where they were classified into different groups according to different criteria.

In this research we focus on studies addressing the issue of matching the structure of a query pattern against the structures of a set of data sources in order to get over the structural heterogeneity. In this case, the pattern query is written based on one's common knowledge about the domain in hand without necessarily knowing the structure of the underlying sources. This is one of the biggest challenges for data interoperability between different systems.

While huge amount of research was dedicated to twig matching for query processing and optimisation purposes, less effort was focused on structural pattern matching. Tree Edit Distance based approaches were considered for calculating XML tree similarity and inclusion; however, there were few attempts to calculate tree similarity by considering the number of common nodes and/or arcs. In this section we present few IR-style approaches on approximate structural matching of pattern trees.

Polyzotis [6] proposed a study that focused on approximate answers for twig queries considering the structural part of the problem. A new XML similarity metric, termed Element Simulation Distance (ESD), was proposed which, according to the author, outperforms previous syntax-based metrics by capturing regions of approximate similarity between XML trees. It considers both overall path structure and the distribution of document edges when computing the distance between two XML trees. According to that metric, two elements are considered to be more similar if they have more matching children, and less similar if they have fewer children in common. This can be reasonable when tree nodes have close semantics to each other. However, this is not always the case; it is common to find nodes with weak semantic connections such as group and department in figure 1-S1.

In other studies [7, 8], Sanz et al. proposed tree similarity measures between a pattern tree and sub-trees within an XML document as a solution for approximate XML query matching. Similarity is calculated based on the number of matching nodes without considering the semantics of parent-child connections. The process consists of two steps: first, identifying the portions of documents that are similar to the pattern (fragments and regions identification). Second, the structural similarity between each of these portions and the pattern is calculated. The proposed node similarity metric does not only depend on the label, but it also depends on the depth of the node "distance-based similarity" such that the similarity linearly decreases as the number of levels of difference increase. However, when matching a pattern tree to an XML data tree, the hierarchical organization of the pattern and the region is not taken into account [8] i.e. matching is only on the node level but not on the arc/edge level.

Another significant study addressed element similarity metrics in structural pattern matching [9]. Authors introduced two types of element similarity measures, internal and external. The internal similarity measure depends on feature similarity which includes i)Node name similarity: this in turn can be classified into syntactic (label) and semantic (meaning) similarity. Node names are first normalised into tokens that are compared using different string similarity approaches such as string edit distance. Semantic similarity, on the other hand, is calculated by using measures from WordNet ii) Data type similarity measure, iii) Constraint similarity measure, mainly cardinality constraint, and iv) Annotation similarity measure, which considers the provided annotations for tree nodes.

External similarity measure, however, considers the position (context) of the element in the schema tree i.e. it considers the element relationship with the surrounding elements, descendants, ancestors, and siblings. A function was used to combine internal and external measures and give the overall similarity measure between two nodes.

Following a totally different approach, Agarwal et al. [10], proposes XFinder, a system for top K XML subtree search that works on exact and approximate pattern matching with focus on approximate structural matching between ordered XML trees. XML query trees as well as document trees were transformed into sequences which are then compared against each other. This technique was adopted in other studies as well [11-13].

4   Soft XML Query Matching

The proposed approach is called IF (Intuitionistic Fuzzy) Tree matching. Detailed definitions of Intuitionistic Fuzzy Trees were presented in previous works [14-18]. The benefit of using IF matching is that it gives more information about how much a pattern tree Pt matches an underlying schema tree St. The "source" tree does not need to be completely included in the "destination" one; it can be partially included. We propose two ways of softening matching rules, soft node matching and soft arc matching.

4.1   Soft Node Matching

An algorithm is developed to softly match nodes of a Pt with nodes of St's [15]. Two nodes do not have to have the same label in order to be considered matching. A linguistic (lexical) ontology is utilised to add semantics to node labels and then a function is invoked to compare these semantics and calculates the similarity (Semantic closeness) according to the distance between them. This is defined as:

semantics (n) ≈ semantics(n′) where the symbol '≈' reads "close to"

Additionally, any two nodes can be matched together even if they do not have the same node type. Stated differently, an element node in a Pt can be matched to an attribute node in St provided that the element node is a leaf node (See AttNode arc matching in Figure 2).

Pt nodes can be classified into different types according to their role in the query tree or the schema tree. In addition to element nodes (e.g. dept) and attribute nodes (e.g. @dname) in figure 2 (d), we preset a set of definitions.

Definition 1. A node can be an **ID node** if it can uniquely identify any instance of its parent node e.g. @dname, in figure 2 (d), can identify its parent node (dept) a. For optimum query matching results, each parent node in Pt has to have an ID node. The reason is that it enables joining sub-trees from different schemas based on the ID of the common node. The labels of ID nodes are underlined in pattern trees to signify their role as a 'Primary key' of their parents.

Definition 2. An **Output node** is a node which value is to be returned in the query. It is distinguished by having a shape of double circles in pattern trees. An output node is either a leaf element node or an attribute node.

Definition 3. A node is called **Intermediate Node** if it is neither a root node nor a leaf node i.e. it is a node that has a parent node and one or more child nodes. Those can be either element nodes or attribute nodes.

4.2   Soft Arc Matching

As arc is the fundamental unit of structure in data schemas, we propose different ways of approximate matching of a pattern arc with a schema arc. The main idea is to adapt to the different ways of modeling a parent-child relationship in different data sources. Figure 2 shows six different ways of arc matching.

Before these types are discussed, we present few definitions on types of arcs: Leaf arc, Non-leaf arc, Pattern arc and Schema arc.

Definition 4. A **leaf arc** is an arc whose child node is a leaf.

Definition 5. A **non-leaf arc** is an arc whose child node is not a leaf.

Depending on whether the arc is a leaf or non-leaf arc, different arc matching techniques apply. In the following sections, we present different types of arc matching along with formal definitions and we explain to which types of arcs they apply.

Definition 6. A **pattern arc** A(m, n) is any arc within a pattern tree Pt.

Definition 7. A **schema arc** A(m', n') is any arc within a schema tree.

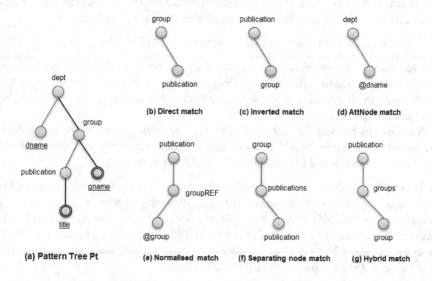

Fig. 2. Types of soft arc matching

Definition 8. The degree of membership (or matching) of a pattern arc A(m, n) and a schema arc A(m', n') is defined by μ_A such that:

Definition 9. An **ID arc** is an arc which child node is an ID node.

$$0 \leq \mu_A \geq 1$$

Definition 10. desc(n)= the descendant nodes of n.

Definition 11. anc(n): the ancestor nodes of n

The following is an explanation of each type of soft arc matching.

Direct Match: In this type of match, a pattern arc A(m, n), where m is the parent of n, is matched to an identical schema arc A(m`, n`) where m→m` and n→n`. This type of matching applies to both leaf arcs and non-leaf arcs which means than n can be either an element or an attribute node. Figure 2 (b) shows an arc matched in that way. Formally, direct arc match is defined as:

$$A(m, n) \rightarrow_D A(m`, n`) \text{ iff } m \rightarrow m' \text{ and } n \rightarrow n' \text{ and } n'.parent()=m'$$

Since this is an exact match, μ_A will be equivalent to 1.0.

Inverted Match: Unlike direct match, inverted match occurs when a pattern arc A(m, n) maps to a schema arc A(m`, n`) where m→n` and n→m`. This mismatch of modeling a relationship between two nodes m & n is common in XML documents depending on the modeler's perception, or point of interest. Arcs matching in this way should be non-leaf arcs because a leaf node cannot be modeled as child in one tree and as parent in another e.g. the arc A(dept, dname) in figure 2 (a) cannot be found as A(dname, dept) because the node dname is a leaf node that is correspondent to a text XML element, which cannot have children. However, if the arc is a non-leaf arc, such as A(group, publication) in figure 2 (a), it can be modelled as A(publication, group) in other schemas such as in figure 2 (c). Formally, inverted arc match is defined as:

$$A(m, n) \rightarrow I \ A(m`, n`) \text{ iff } m \rightarrow n' \text{ and } n \rightarrow m'$$

Since this is not an exact match, μ_A will be less than 1.0. For the proposed approach, 10% of belief is deducted as a result i.e. $\mu_A=0.9$ to distinguish this match from Direct match.

AttNode Match: In relational databases, an entity type such as dept (department) has attributes such as name, location etc. In XML, however, a department name can be modelled either as an attribute node (@dname) or an element node (dname). However, since attribute nodes cannot have children, this type of arc match only applies to leaf arcs such as the one shown in figure 2 (d). Formally, AttNode arc match is defined as:

$$A(m, n) \rightarrow_A A(m`, @n`) \text{ iff } m \rightarrow m' \text{ and } n \rightarrow @n'$$

Again, 10% of belief is deducted as a result non exact match. Thus, $\mu_A=0.9$.

Normalized Match: This match can be found in cases where an XML document is normalized i.e. each entity is modelled as a sub-tree (twig) within the document which can be referenced by using IDREF instructions. This can be thought of as analogy of the primary and foreign keys in relational modelling. To return data from more than one twig, an XML query joins the correspondent twigs based on a common node while having an attribute or element that acts as the ID of the common node. In Figure 3 the node pubREF is a reference node that refers to publication node within the same document. Consequently, we can say that the pattern arc (group, publication) is matching with (group, pubREF) using normalized arc match. Obviously, arcs matched in this way should be non-leaf arcs because the child node (e.g. publication) should have a child ID node on which the join will take place.

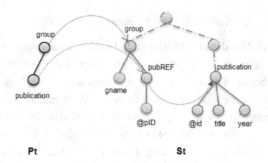

Fig. 3. Normalized arc match

This type of match is complicated and it requires more processing resources i.e. time and memory. This is mainly because it is not enough to achieve this type of match based on the data schema only; the actual XML data document is also required because it is not possible to know to which element an IDREF attribute is referring without traversing the actual XML document. For our approach, we use a function that picks an instance of an IDREF element, such as pubREF, from the XML document, and scans it to find the parent of that ID e.g. publication node in our example. Formally, normalised arc match is defined as:

$$A(m, n) \rightarrow N\ A(m`, n`)\ \text{iff}\ m \rightarrow m'\ \text{and}\ n \rightarrow \text{de-ref}(n')$$

Where de-ref(n') is a function that returns the node referenced by n'.

Same as previous types of matching, 10% of belief is deducted making $\mu_A = 0.9$.

Separating Node (SepNode) Match: Sometimes there are separating nodes between the parent and the child. In this case the arc still can be matched if the number of separating nodes does not exceed a predefined threshold. Formally, A(m, n) can be matched with A(m`, n`) using this method if m is an ancestor of n.

This has been proposed by many previous studies [19, 20]. However, this can result in getting the wrong result especially that this type applies to both leaf and non-leaf arcs. To explain this, suppose that we have a leaf arc A(dept, location), that is to be matched with a schema as in figure 4. Clearly, there is no matching arc in St because the node dept does not have a child location. However, using the separating node arc match, dept has a descendant node called location, which means that the arc can be approximately matched. But the problem is that the node location does not refer to the location of department, it refers to the location of the project. Thus, blind approximate matching using this technique can return wrong matchings.

To solve this dilemma, we consider that intermediate nodes are strong nodes as they usually represent independent entities (concepts) in relational models. Leaf nodes, however, usually represent attributes of their parents, and therefore they are weak nodes. If an arc A (m, n) is to be matched, m is a strong node and n is a weak node, any separating nodes between them can induce weak semantics e.g. the node location in figure 4 refers to its parent (project) but not to it ascendant (dept). On the other hand if both nodes (m and n) are strong nodes e.g. (dept, publication), then an intermediate node, such as group in our example, is unlikely to affect the semantics. In the same example, the arc A(dept, publication) in Pt can be matched against St even if there is a separating node (group). This can be explained because a department consists of research groups, and these have publications. Thus, we can say that those publications belong to the department.

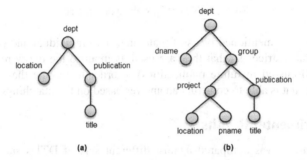

Fig. 4. (a) A pattern Pt, (b) a schema tree St

Formally, separating node match is defined as:

A(m, n)$\rightarrow$S A(m`, n`) iff m$\rightarrow$ m' and n $\rightarrow$ n' and m' $\in$ desc(n') and m' is not the parent of n'

Unlike previous types of approximate matching, belief of this type depends on the number of separating nodes. Therefore, it is not correct to just deduct 10% of the belief; the amount deducted has to be proportional to the number of separating nodes. Belief of this type is defined by the following equation:

$$\mu_A = 1/(1+\partial/2)$$ where ∂ is the number of separating nodes.

The reason why ∂ was divided by 2 is to reduce the effect of increasing number of nodes and make the belief reasonable.

Hybrid Arc Match: Not only can an arc be matched using the aforementioned approaches individually; but it also can be matched using combinations of them.

For example, a combination of both AttNode and SepNode (AS) match (figure 5) where the pattern arc A(dept, location) is matched to the schema arc A(dept, @location) while having a separating node (info) that separates the parent and child nodes. Formally, AttNode-SepNode hybrid match is defined as:

$$A(m, n) \rightarrow_{AS} A(m`, n`) \text{ iff } m \rightarrow m', n \rightarrow n', m' \in anc(n') \text{ and } n' \text{ is an attribute node}$$

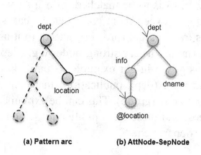

(a) Pattern arc   (b) AttNode-SepNode

Fig. 5. AttNode-SepNode hybrid match

Overall, the aforementioned types of matching, both individual and combined, are saved in several matrices so that they are used as input to the query rewriting algorithm (to be discussed in future publications) in order to rewrite the original query into a new one that is able to construct an answer based on the matchings.

5   Experimental Results

The Pt in figure 1 was compared against different sets of DTDs starting from 10 DTDs and adding 10 each time up to 50 DTDs. For each DTD, arc matching was performed and the results (mappings) were kept in the correspondent mapping matrices which will be passed to next phase, query rewriting (to be addressed in future publications). All of the proposed types of soft arc matching were detected by the matching algorithm. Only 7 DTDs had matching arcs with the Pt whereas the rest were dummy DTDs (unrelated).

6   Conclusion and Further Work

The proposed solution presented a new approach to approximate XML query matching based on new types of soft arc matching which proved to be more efficient than previous works. This was achieved by a number of novel algorithms for soft matching of XML pattern trees and schema trees. A prototype was developed for experimental result and it was able to identify the proposed soft arc matching techniques.

For future work, there is still a lot of work to be done before the heterogeneity of XML schemas is resolved efficiently. Further work can be done on querying XML documents with different schemas where data and meta-data are mixed e.g. the tag <Africa> refers to a name of area i.e. it is data; however, it is treated as meta-data in this example. Additionally, the IF query matching approach can be extended to matching contents in addition to structures. This can be useful in identifying and removing duplications. Furthermore, there is potential to improve the current approach by using semantic web technologies such as using reasoning to improve soft arc matching techniques.

References

1. Harold, E., Means, W.: XML in a Nutshell. O'Reilly Media, Incorporated (2004)
2. Tekli, J., Chbeir, R., Yetongnon, K.: An overview on XML similarity: Background, current trends and future directions. Computer Science Review 3, 151–173 (2009)
3. Gang, G.: Efficiently Querying Large XML Data Repositories: A Survey. IEEE Transactions on Knowledge and Data Engineering 19, 1381–1403 (2007)
4. Sukomal Pal, M.M.: XML Retrieval: A Survey. Internet Policies and Issues 8, 229–272 (2006)
5. Marouane, H.: A Survey of XML Tree Patterns. IEEE Transactions on Knowledge and Data Engineering 99 (2011)
6. Polyzotis, N., Garofalakis, M., Ioannidis, Y.: Approximate XML query answers. Presented at the Proceedings of the 2004 ACM SIGMOD International Conference on Management of Data, Paris, France (2004)
7. Sanz, I., Mesiti, M., Guerrini, G., Berlanga, R.: Fragment-based approximate retrieval in highly heterogeneous XML collections. Data Knowl. Eng. 64, 266–293 (2008)
8. Sanz, I., Mesiti, M., Guerrini, G., Llavori, R.B.: Approximate subtree identification in heterogeneous XML documents collections. Presented at the Proceedings of the Third International Conference on Database and XML Technologies, Trondheim, Norway (2005)
9. Algergawy, A., Nayak, R., Saake, G.: Element similarity measures in XML schema matching. Information Sciences 180, 4975–4998 (2010)
10. Agarwal, N., Oliveras, M.G., Chen, Y.: Approximate Structural Matching over Ordered XML Documents. Presented at the Proceedings of the 11th International Database Engineering and Applications Symposium (2007)
11. Rao, P., Moon, B.: PRIX: Indexing And Querying XML Using Sequences. Presented at the Proceedings of the 20th International Conference on Data Engineering (2004)
12. Wang, H., Meng, X.: On the Sequencing of Tree Structures for XML Indexing. Presented at the Proceedings of the 21st International Conference on Data Engineering (2005)
13. Wang, H., Park, S., Fan, W., Yu, P.S.: ViST: a dynamic index method for querying XML data by tree structures. Presented at the Proceedings of the 2003 ACM SIGMOD international conference on Management of data, San Diego, California (2003)
14. Alzebdi, M., Chountas, P., Atanassov, K.: Approximate XML Query Matching and Rewriting Using Intuitionistic Fuzzy Trees. Presented at the IEEE IS Sofia (2012)
15. Alzebdi, M., Chountas, P., Atanassov, K.: Intuitionistic Fuzzy XML Query Matching. In: Christiansen, H., De Tré, G., Yazici, A., Zadrozny, S., Andreasen, T., Larsen, H.L. (eds.) FQAS 2011. LNCS, vol. 7022, pp. 306–317. Springer, Heidelberg (2011)
16. Alzebdi, M., Chountas, P., Atanassov, K.T.: An IFTr approach to approximate XML query matching. Presented at the SMC (2011)
17. Alzebdi, M., Chountas, P., Atanassov, K.: Enhancing DWH Models with the Utilisation of Multiple Hierarchical Schemata. In: IEEE SMC 2010, pp. 488–492 (2010)
18. Chountas, P., Alzebdi, M., Shannon, A., Atanassov, K.: On Intuitionistic Fuzzy Trees. Notes on Intuitionistic Fuzzy Sets 15(2), 30–32 (2009)
19. Wiwatwattana, N., Jagadish, H.V., et al.: X^ 3: A Cube Operator for XML OLAP. In: IEEE 23rd International Conference on Data Engineering, ICDE 2007, pp. 916–925 (2007)
20. Chen, Y., Chen, Y.: A new tree inclusion algorithm. Inf. Process. Lett. 98, 253–262 (2006)

Discrimination of the Micro Electrode Recordings for STN Localization during DBS Surgery in Parkinson's Patients

Konrad Ciecierski[1], Zbigniew W. Raś[2,1], and Andrzej W. Przybyszewski[3]

[1] Warsaw Univ. of Technology, Institute of Comp. Science, 00-655 Warsaw, Poland
[2] Univ. of North Carolina, Dept. of Comp. Science, Charlotte, NC 28223, USA
[3] UMass Medical School, Dept. of Neurology, Worcester, MA 01655, USA
K.Ciecierski@ii.pw.edu.pl, ras@uncc.edu,
Andrzej.Przybyszewski@umassmed.edu

Abstract. During deep brain stimulation (DBS) treatment of Parkinson disease, the target of the surgery is a small (9 x 7 x 4 mm) deep within brain placed structure called *SubthalamicNucleus* (*STN*). It is similar morphologically to the surrounding tissue and as such poorly visible in CT[1] or MRI[2]. The goal of the surgery is the permanent precise placement of the stimulating electrode within target nucleus. Precision is extremely important as wrong placement of the stimulating electrode may lead to serious mood disturbances. To obtain exact location of the *STN* nucleus an intraoperative stereotactic supportive navigation is being used. A set of 3 to 5 parallel micro electrodes is inserted into brain and in measured steps advanced towards expected location of the nucleus. At each step electrodes record activity of the surrounding neural tissue. Because *STN* has a distinct physiology, the signals recorded within it also display specific features. It is therefore possible to provide analytical methods targeted for detection of those *STN* specific characteristics. Basing on such methods this paper presents clustering and classification approaches for discrimination of the micro electrode recordings coming from the *STN* nucleus. Application of those methods during the neurosurgical procedure might lessen the risks of medical complications and might also shorten the – out of necessity awake – part of the surgery.

Keywords: Parkinson's disease, DBS, STN, DWT, RMS, LFB, HFB, K-Means, EM Clustering, C4.5, Random Forest.

Introduction

Parkinson Disease (PD) is chronic and advancing movement disorder. The risk factor of the disease increases with the age. As the average human life span elongates also the number of people affected with PD steadily increases. PD is primary related to lack of the dopamine, that after several years causes not only

---

[1] Computer Tomography.
[2] Magnetic Resonance Imaging.

H.L. Larsen et al. (Eds.): FQAS 2013, LNAI 8132, pp. 328–339, 2013.

movement impairments but also often non-motor symptoms like depressions, mood disorder or cognitive decline and therefore it has a very high social cost. People as early as in their 40s, otherwise fully functional are seriously disabled and require continuous additional external support. The main treatment for the disease is pharmacological one. Unfortunately, in many cases the effectiveness of the treatment decreases with time and some other patients do not tolerate anti PD drugs well. In such cases, patients can be qualified for the surgical treatment of the PD disease. This kind of surgery is called DBS[3]. Goal of the surgery is the placement of the permanent stimulating electrode into the STN nucleus. This nucleus is a small – deep in brain placed – structure that does not show well in CT or MRI scans.

Having only an approximate location of the STN, during DBS surgery a set of parallel micro electrodes are inserted into patient's brain. As they advance, the activity of surrounding neural tissue is recorded. Localization of the STN is possible because it has distinct physiology and yields specific micro electrode recordings. It still however requires an experienced neurologist / neurosurgeon to tell whether recorded signal comes from the STN or not [3].

That is why it is so important to provide some objective and human independent way to group and classify recorded signals. Analytical methods, clusterings and classifiers described in this paper have been devised for that purpose. Taking as input recordings made by set of electrodes at subsequent depths they are to provide information as to which of the recordings have been made inside the STN.

1   Recording's Attributes

From the micro electrode recorded signal two kinds of information can be obtained [7]. First one comes from spikes – neuronal action potentials – i.e. electrical activity of neurons being near the electrode recording point. Second information is derived from the background noise present in the signal. This background noise is a cumulative electrical activity coming from the more distant neurons. Both approaches require prior detection of the spikes [3] [4] [8] [9]. First one uses information about spikes in a direct way. The background noise analysis requires prior spike removal. Removal of course implies prior spike detection.

For presented analysis a set of 11 characteristic attributes is defined for each recording:

- three features derived directly from the spike occurrence
- four features derived from the signal background
- four meta–features derived indirectly from the signal background

1.1   Spike Derived Attributes

Characteristics derived from the spike occurrences focus on the observed spiking frequency. Assuming that for recording rec, the total number of spikes detected

---

[3] Deep Brain Stimulation.

in it is denoted as $SpkCnt(rec)$ and its length in seconds is $RecLen(rec)$ then the first attribute is given by equation 1.

$$AvgSpkRate(rec) = \frac{SpkCnt(rec)}{RecLen(rec)} \tag{1}$$

Assuming that for recording rec spikes occurred chronologically at times $t_0, \ldots, t_{n-1}$ and that $s_j = t_{j+1} - t_j$ then a set of intra–spike intervals S is defined by equation 2.

$$S(rec) = \{s_0, \ldots, s_{n-2}\} \tag{2}$$

Let us define now the sets of *bursting* and *frequent* intra–spike intervals as follows:

$$S_{brst}(rec) = \{s_j \in S(rec) : s_j < 33.3ms\} \tag{3}$$

$$S_{frq}(rec) = \{s_j \in S(rec) : 33.3 \leq s_j < 66.6ms\} \tag{4}$$

Such intervals are also used in construction of intra–spike interval histograms [3]. Histograms show [3] that majority of bursting activity occurs for intervals below 33.3 ms and very little of it can be observed for intervals longer than 66.6 ms. Having the above defined, the final two spike characteristic attributes can be given:

$$BurstRatio(rec) = \frac{\overline{\overline{S_{brst}(rec)}}}{\overline{\overline{S(rec)}}} \tag{5}$$

$$FrquentRatio(rec) = \frac{\overline{\overline{S_{frq}(rec)}}}{\overline{\overline{S(rec)}}} \tag{6}$$

For recordings having no spikes, $BurstRatio$ and $FrquentRatio$ are defined with zero value.

Fig. 1 shows value of spike derived attributes on consecutive depths (μm) for the micro electrode that has entered the STN at depth -4000 μm.

1.2   Background Activity Derived Attributes

Background characteristics do not depend so heavily on probability of micro electrode being next to the active neuron cell and as such are more reliable [3] [9]. They concentrate on the amplitude of the background noise and on the aspects related to the signal's power.

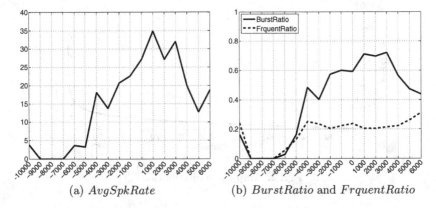

(a) *AvgSpkRate*   (b) *BurstRatio* and *FrquentRatio*

Fig. 1. Spike derived attributes

As electrodes used in deep brain recordings may slightly differ in their electrical impedance, results from methods relying on background noise have to be normalized.

Assuming that a specific method mth takes as input vector of recordings from subsequent depths, it produces a vector of characteristic coefficients C.

$$mth(rec_{-10000}, ..., rec_{+5000}) = (c_{-10000}, ..., c_{+6000}). \tag{7}$$

The base value of C is then defined as

$$C_{base} = \frac{c_{-10000} + ... + c_{-6000}}{5} \tag{8}$$

and, finally the normalized C has the form

$$C_{NR} = (\frac{c_{-10000}}{C_{base}}, ..., \frac{c_{+6000}}{C_{base}}). \tag{9}$$

Reason why average of the first five values is used has anatomical and physiological background and it is explained in [3] and [9]. All attributes described in this chapter have their values normalized in described way.

Percentile Based Attribute

Spikes amplitude is by far greater than the background noise. Because of that, one can find an amplitude value below which no spikes are present or above which spike must rise. This feature is commonly used in many neurological appliances, for example the 50th percentile - *median* of amplitude's module is used for both spike detection and artifact removal process [6] [8] [9]. Fig. 2 shows amplitude's module distribution for a signal recorded within the STN.

Already the 95th (or even pinpointed 99th) percentile shows background activity and discards almost all samples coming from the spikes. To be however safely independent from any spike activity, even lower percentile can be used. In

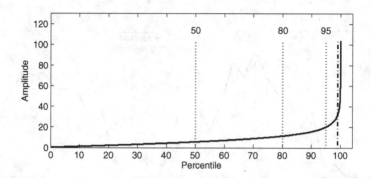

Fig. 2. Amplitude distribution

this paper the 80th percentile is used. For signal rec this attribute is denoted as $Prc_{80}(rec)$. Using higher percentile increases the risk of including high amplitude of the spikes in the attribute value, this in turn might lead to falsely elevated attribute value in non STN areas. On the other hand, using low percentile causes the difference between attribute values for recordings coming from the STN and from outside of it to be less evident.

Fig. 3a shows the value of the 80th percentile of amplitude module. Percentile has been calculated for the same data set as with Figures 1a and 1b.

RMS Based Attribute
This attribute bases on a fact that STN is known to produce lots of spikes with high amplitude and also has loud background noise. So it can be reasonably expected that signals recorded from it would present elevated Root Mean Square Value. This parameter is, among others, also used for Bayesian calculations in [5] [9]. If we assume that recording rec has n samples $\{x_0, ..., x_{n-1}\}$ then attribute $RMS(rec)$ is given by equation (10).

$$RMS(rec) = \sqrt{\frac{\sum_{i=0}^{n-1} x_i^2}{n}} \qquad (10)$$

Fig. 3b shows the value of the RMS attribute. RMS has been calculated for the same data set as with Figures 1a, 1b, and 3a.

Attributes Based on Signal's Power
It has been postulated in [9], [10] and [11] that background neural activity can be divided into two frequency areas. First, power of an activity in range below 500 Hz is called Low Frequency Background(LFB). Second contains frequencies in range 500 Hz to 3000 Hz and is thus called High Frequency Background (HFB). In mentioned paper authors use properties of HFB to pinpoint STN location. Here also, the increased LFB and HFB values are used as hallmark of recordings coming from the STN. Power of the signal in both bands is calculated using DWT coefficients [1]. As described in greater detail in [9] both methods

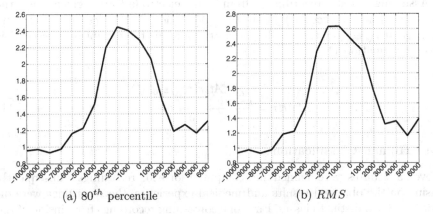

(a) 80th percentile (b) RMS

Fig. 3. Prc_{80} and RMS attributes

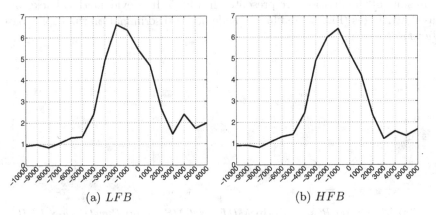

(a) LFB (b) HFB

Fig. 4. LFB and HFB attributes

require that the signal has to be firstly filtered from any contamination artifacts and naturally occurring spikes.

Fig. 4 shows the value of the $LFB(rec)$ and $HFB(rec)$ attributes for recordings coming from depths ranging from -10000 μm to 6000μm. Both attributes have been calculated for the same data set as with Figures 1a,1b, 3a and 3b.

1.3 Meta Attributes

In previous section for any recording rec four its background attributes were given: $Prc_{80}(rec)$, $RMS(rec)$, $LFB(rec)$ and $HFB(rec)$. To obtain attributes that are less affected by local variations at certain depths, four meta attributes were introduced. Meta attributes are calculated as 1–padded five element wide moving average of the proper background attribute.

Assuming that recordings from an electrode are given as the list $(rec_{-10000}, \ldots, rec_{6000})$ and that $\forall_{j<-10000}\ Attr(rec_j) = 1$ and $\forall_{j>6000}\ Attr(rec_j) = 1$, then the meta attribute for the attribute $Attr(rec_j)$ is given by equation 11.

$$MAttr = \frac{\sum\limits_{k=-2}^{2} Attr(rec_{j+1000k})}{5} \tag{11}$$

1.4   Human Classification

Eleven attributes were defined and computed for 9502 recordings. Subsequently basing on the obtained results and medical experience, the recordings were divided into two distinct classes. First one containing recordings from inside of the STN and second from its outside. First class labeled STN contained 2161 (22.74 %) recordings. Second, labeled $\neg STN$ contained 7341 (77.26 %) recordings.

Results of the division are presented in Fig. 5. It is evident that background based attributes (Fig. 5a and 5b) provide better discrimination than spike based $BurstRatio$ (Fig. 5c).

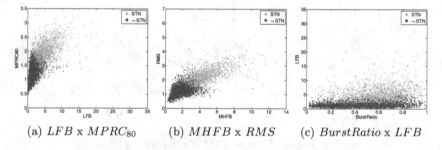

(a) LFB x $MPRC_{80}$ (b) $MHFB$ x RMS (c) $BurstRatio$ x LFB

Fig. 5. Division of recordings on selected planes

2   Clustering of the Recordings

In the following section two clustering methods were used to check if it is feasible to discriminate recordings into two classes in an automatic way. By automatic we mean that it is using only 11 attributes mentioned above without any human feedback.

2.1   K-means Clustering

Following results were obtained by running $K - means$ clustering on all eleven attributes from 9502 recordings. Clustering was performed with target number of clusters set to two.

	Human classification		
	STN	$\neg STN$	Total
$K-means$ clustering STN	2016	2296	4312
$\neg STN$	145	5045	5190
Total	2161	7341	9502

$$sensitivity = \frac{2016}{2016+145} \approx 0.933 \qquad specificity = \frac{5045}{5045+2296} \approx 0.687$$

Only 145 out of 2161 recordings assigned by human to the STN category have been mislabeled. Unfortunately as much as 2296 out of 7341 recordings assigned by human to the $\neg STN$ category have been mislabeled. Results yield good sensitivity and poor specificity. $K-means$ can be used with good certainty for predicting recordings originating in the STN. It is however too prone to assign STN label to recordings made outside of the STN. This can be observed comparing Figures 5c and 6c. Specificity lower then 0.8 according to [12] means that $K-means$ method cannot be reliably used for medical discrimination of the recordings.

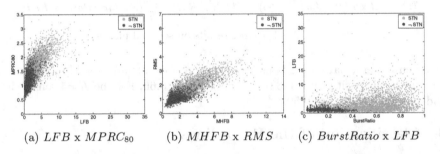

(a) LFB x $MPRC_{80}$ (b) $MHFB$ x RMS (c) $BurstRatio$ x LFB

Fig. 6. $K-means$ clustering results on selected planes

When larger k parameter is used the obtained results are steadily worsening. For the $K=3$ sensitivity is 0.70 and specificity is 0.62. For the $K=4$ sensitivity is 0.66 and specificity is 0.59.

2.2   EM Clustering

EM is a distribution based clustering, it finds a set of Gaussian distributions that matches given data. In this way for each attribute, its mean and σ are obtained for each resulting cluster. Clustering has been done using Weka provided EM clustering method. The target number of clusters was this time also set to two.

	Human classification		
	STN	$\neg STN$	Total
EM clustering STN	2155	1737	3892
$\neg STN$	6	5604	5610
Total	2161	7341	9502

$$sensitivity = \frac{2155}{2155+6} \approx 0.997 \qquad specificity = \frac{5604}{5604+1737} \approx 0.763$$

Only 6 out of 2161 recordings assigned by human to the STN category have been mislabeled. Less then with $K-means$, but still 1737 out of 7341 recordings assigned by human to the $\neg STN$ category have been mislabeled. Results yield very good sensitivity and better specificity. EM clustering results can be used with very good certainty for predicting recordings originating in the STN. Still the specificity of 0.75 being below the 0.8 threshold means that too many $\neg STN$ recordings might be assigned the STN label.

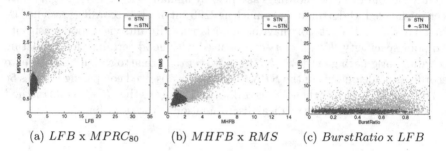

(a) LFB x $MPRC_{80}$ (b) $MHFB$ x RMS (c) $BurstRatio$ x LFB

Fig. 7. EM clustering results on selected planes

When larger k parameter is used the obtained results are steadily worsening. For the $K=3$ sensitivity is 0.88 and specificity is 0.66. For the $K=4$ sensitivity and specificity both are 0.59.

3   Classification of the Recordings

Clustering methods – having sensitivity of 0.93 and 1.00 respectively – proved to be good for correctly identifying recordings made inside the STN. Both clusterings however have relatively low specificity meaning that basing on their results the stimulating electrode might be placed in a wrong region incorrectly identified as a good one. That of course in not acceptable. Looking for the significantly better results two classification attempts have been made using all computed attributes and human provided class attribute.

3.1   C4.5 Classification

First classification has been done using C4.5 methods (Weka J48). Classifier has been build with pruning confidence factor 0.25 and minimum number of instances per leaf set to 2. Resulting tree has 30 leaves and its size is 59.

		Human classification		
		STN	$\neg STN$	Total
$C4.5$ classifier	STN	2021	147	2168
	$\neg STN$	140	7194	7334
	Total	2161	7341	9502

$$sensitivity = \frac{2021}{2021+140} \approx 0.935 \qquad specificity = \frac{7194}{7194+147} \approx 0.980$$

Only 140 out of 2161 recordings made within the STN have been misclassified.

In contrast to clusterings only 147 out of 7341 recordings assigned by human to the $\neg STN$ category have been mislabeled as STN. Results yield nearly perfect sensitivity and specificity.

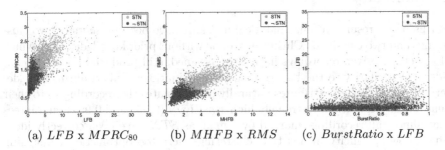

(a) $LFB \times MPRC_{80}$ (b) $MHFB \times RMS$ (c) $BurstRatio \times LFB$

Fig. 8. $C4.5$ classification results on selected planes

3.2   Random Forest Classification

Second classification has been done using Random Forest classifier. In this approach a set of decision trees is constructed, individual trees are constructed basing on randomly selected subsets of the selected attributes. Later, new object is checked against each tree and class returned by the majority of the trees is assigned to it. For this paper, classifier has been build with 10 trees, each tree is considering 4 randomly selected attributes. Resulting trees have their sizes between 421 and 469.

		Human classification		
		STN	$\neg STN$	Total
Random Forest classifier	STN	2138	6	2144
	$\neg STN$	23	7335	7358
	Total	2161	7341	9502

$$sensitivity = \frac{2138}{2138+23} \approx 0.989 \qquad specificity = \frac{7335}{7335+6} \approx 0.999$$

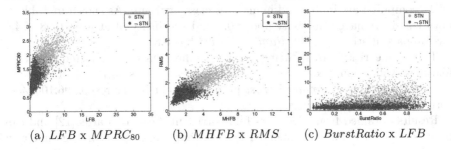

(a) $LFB \times MPRC_{80}$ (b) $MHFB \times RMS$ (c) $BurstRatio \times LFB$

Fig. 9. *RandomForest* classification results on selected planes

Only 23 out of 2161 recordings made within the STN have been misclassified.

Even less than in $C4.5$ case – only 6 out of 7341 – recordings assigned to the $\neg STN$ category by human have been mislabeled as STN. This time also, results yield nearly perfect sensitivity and specificity.

4   Summary

In this paper results of four analytical methods have been presented. Two clusterings and two classifiers. Clusterings, done without prior knowledge as to which of the classes given recordings have been assigned, still produced results of considerable similarity to the human made discriminations. This confirms that described set of defined attributes naturally discriminates the recording as to their anatomical origination. Extremely high sensitivity (0.933 and 0.997) guarantees that almost all recordings marked by human as STN have been correctly identified. Low specificity means that clusterings were too optimistic in assigning recordings to the STN group. While according to [12] specificity below 0.8 is not acceptable some other sources consider results with specificity of at least 0.75 valid provided that sensitivity is over 0.95. In such assumption generally the results provided by the EM clustering might be worth considering (see Table 1). Specific problem of electrode placement described in this paper however requires as high specificity as possible. Due to the closeness to the limbic part of the brain [2] placement of the electrode in a place that has been wrongly considered as a proper one might endanger the mental health of the patient.

Table 1. Sensitivity and specificity summary

	sensitivity	specificity
$K - menas$ clustering	0.933	0.687
EM clustering	0.997	0.763
$C4.5$ classifier	0.935	0.980
$RandomForest$ classifier	0.989	0.999

Much better results as shown in Sections 3.1 and 3.2 can be obtained using the classifiers. Both $C4.5$ and $RandomForest$ classifiers have been build using all 11 computed attributes and human set class attribute. As shown in Table 1, both classifiers have sensitivity above 0.9 and specificity of at least 0.98. $C4.5$ has kappa statistics of 0.879, $RandomForest$ has 0.888.

To exclude risk of over fitting, both classifiers have been tested with 10–fold cross-validation. Score of cross–validation for $C4.5$ is 95.769 % of correctly classified recordings. $RandomForest$ produced event better result: 96.106 % of correctly classified recordings.

Results obtained from classifiers fulfill strict medical standards [12] and can be used as a base for recommender system that identifies recordings made in the STN.

References

1. Jensen, A.: A Ia Cour-Harbo. Ripples in Mathematics. Springer (2001)
2. Nolte, J.: The Human Brain, Introduction to Functional Anatomy. Elsevier (2009)
3. Israel, Z., et al.: Microelectrode Recording in Movement Disorder Surgery. Thieme Medical Publishers (2004)
4. Alexander, B., et al.: Wavelet Filtering before Spike Detection Preserves Waveform Shape and Enhances Single-Unit Discr. J. Neuroscience Methods, 34–40 (2008)
5. Moran, A., et al.: Real-Time Refinement of STN Targeting Using Bayesian Decision-Making on the RMS Measure. J. Mvmt. Disorders 21(9), 1425–1431 (2006)
6. Quian Quiroga, R., Nadasdy, Z., Ben-Shaul, Y.: Unsupervised Spike Detection and Sorting with Wavelets and Superparamagnetic Clustering. MIT Press (2004)
7. Gemmar, P., et al.: MER Classification for DBS, 6th Heidelberg Innov. Forum (2008)
8. Ciecierski, K., Raś, Z.W., Przybyszewski, A.W.: Selection of the Optimal Microelectrode during DBS Surgery in Parkinson's Patients. In: Kryszkiewicz, M., Rybinski, H., Skowron, A., Raś, Z.W. (eds.) ISMIS 2011. LNCS, vol. 6804, pp. 554–564. Springer, Heidelberg (2011)
9. Ciecierski, K., Raś, Z.W., Przybyszewski, A.W.: Foundations of Recommender System for STN Localization during DBS Surgery in Parkinson's Patients. In: Chen, L., Felfernig, A., Liu, J., Raś, Z.W. (eds.) ISMIS 2012. LNCS, vol. 7661, pp. 234–243. Springer, Heidelberg (2012)
10. Novak, P., Przybyszewski, A.W., et al.: Localization of the subthalamic nucleus in Parkinson disease using multiunit activity. J. Neur. Sciences 310, 44–49 (2011)
11. Novak, P., et al.: Detection of the subthalamic nucleus in microelectrographic recordings in Parkinson disease using the high-frequency (> 500 Hz) neuronal background. J. Neurosurgery 106, 175–179 (2007)
12. Walker, H.K., Hall, W.D., Hurst, J.W. (eds.): Clinical Methods: The History, Physical, and Laboratory Examinations, 3rd edn. Butterworths, Boston (1990)

Image Classification Based on 2D Feature Motifs

Angelo Furfaro[1], Maria Carmela Groccia[1], and Simona E. Rombo[2]

[1] Dipartimento di Ingegneria Informatica, Modellistica, Elettronica e Sistemistica
Università della Calabria
Via P. Bucci 41C, 87036 Rende (CS), Italy
{a.furfaro@deis.,mariacarmela.groccia@}unical.it
[2] Dipartimento di Matematica e Informatica
Università degli Studi di Palermo
Via Archirafi n.34, Palermo, Italy
simona.rombo@math.unipa.it

Abstract. The classification of raw data often involves the problem of selecting the appropriate set of features to represent the input data. In general, various features can be extracted from the input dataset, but only some of them are actually relevant for the classification process. Since relevant features are often unknown in real-world problems, many candidate features are usually introduced. This degrades both the speed and the predictive accuracy of the classifier due to the presence of redundancy in the candidate feature set.

In this paper, we study the capability of a special class of motifs previously introduced in the literature, i.e. 2D irredundant motifs, when they are exploited as features for image classification. In particular, such a class of motifs showed to be powerful in capturing the relevant information of digital images, also achieving good performances for image compression. We embed such 2D feature motifs in a bag-of-words model, and then exploit K-nearest neighbour for the classification step. Preliminary results obtained on both a benchmark image dataset and a video frames dataset are promising.

1 Introduction

In the last few years, several image classification methods, based on supervised and on unsupervised techniques or on a mix of them, have been proposed in the literature (see [6,18,22] for suitable surveys on the topic). Classic approaches use low-level image features, such as color or texture histograms, while more recent techniques rely on intermediate representations made of local information extracted from *interesting* image patches referred to as *keypoints* [6]. Image keypoints are automatically detected using various techniques, and then represented by means of suitable descriptors. Keypoints are usually clustered based on their *similarity*, and each cluster is interpreted as a *"visual word"*, which summarizes the local information pattern shared by the belonging keypoints [32]. The set of all the visual words constitutes the *visual vocabulary* or *codebook*. For classification purposes, an image is then represented as a histogram of its local features, which is analogous to the *bag-of-words* model for text documents. Examples of commonly exploited keypoint detectors are: Difference of Gaussian (DoG) [17], Sample Edge Operator [5], Kadir-Brady [15]. Feature descriptors are often based on SIFT (Scale Invariant Feature Transform) [16].

H.L. Larsen et al. (Eds.): FQAS 2013, LNAI 8132, pp. 340–351, 2013.

As in many other real-world classification problems, relevant features to exploit for image classification are in general unknown a priori. Therefore, various candidate features can be introduced, many of which are either partially or completely irrelevant/redundant to the target concept [8]. A relevant feature is neither irrelevant nor redundant to the target concept; an irrelevant feature does not affect the target concept in any way, and a redundant feature does not add anything new to the target concept [14]. In this work, we analyze a special kind of features for image classification, based on the concept of *2D irredundant motifs* (also known as *2D basis*) introduced in [3].

2D motifs are rectangular portions of an input image I, that are frequently repeated in I (Figure 1 illustrates a simple kind of 2D motifs). When all the possible repeated portions of I are considered, also taking into account sub-regions that are *similar* but not identical, the number of such features can grow exponentially with the size of I, and many of them are irrelevant and/or redundant.

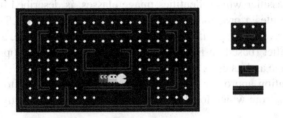

Fig. 1. Some simple 2D motifs in Pac-man

2D basis, successfully exploited for image compression in [1], showed to be useful in encoding the relevant information from an input image, by suitably reducing its representation. The main idea is that of eliminating all the 2D motifs that are *redundant* with respect to the other ones in the set of candidates. The notion of redundancy here is related to the coverage of motifs, since if several motifs occur on the same portions of I, then only the most informative ones will be kept in the basis.

Given an input training set of images already classified, we propose a classification approach based on the extraction of 2D feature motifs, that are exploited to build a codebook containing the motifs in the 2D basis of the input images. After feature selection, image classification is then performed based on the K-Nearest Neighbour approach.

The paper is organized as follows. In Section 2, we briefly summarize some of the main techniques presented in the literature for image classification, that are related with our work. Section 3 illustrates some preliminary notions, while in Section 5 the proposed approach is described in detail. In Section 5 we show some preliminary results we obtained on real datasets, also comparing them with those returned by other methods proposed in the literature. Finally, conclusive remarks are drawn in Section 6.

2   Related Work

One of the most challenging issues for image classification is the appropriate choice of the features needed to build the codebook. In [20], a classification algorithm is presented that uses as features square subwindows of random sizes, which are sampled at random positions from training images and then suitably embedded with decision trees. A variant of the method is proposed in [19] for biomedical images classification. More recently, SURF (Speeded Up Robust Features) have been introduced in [4], relying on integral images for image convolutions.

Clustering algorithms are often employed for building the codebook from the images of the training set. Two variants of the k-means algorithm, namely Approximate k-means (AKM) and Hierarchical k-means (HKM), have been used in [24] to improve the vocabulary of visual words obtained from large datasets. The reported experiments show that both AKM and HKM scale well with respect to the dataset size. More in particular, AKM achieves very similar precision performance to exact k-means but with a lower computational cost.

A hybrid classification scheme, based on a combination of an unsupervised generative step, used to discover relevant *object categories* as image *topics*, followed by a discriminative classifier which identifies image classes, is described in [7]. The generative stage executes a probabilistic Latent Semantic Analysis (pLSA)[12] over the training set to derive topic specific distributions. In the second stage, a multiclass discriminative classifier (based on KNN or SVM) is trained given the topic ditribution of each training image and its class label.

A multi-resolution approach, where each image is sub-sampled to obtain a series of scaled versions from which local features are extracted, has recently been proposed in [33].

Some dissimilarity measures between histograms representing images, which are able to deal with *partial matching* issues, i.e. with the presence of irrelevant clutters, and to take into account *co-occurence* information between visual words, are discussed in [31].

Approaches based on *local invariant descriptors* and on two different indexing techniques for efficient localization of the nearest neighbour in feature space, have been proposed in [28] and [30].

In this paper, we propose an approach that, differently than [7,12,24], is supervisioned. Furthermore, we introduce features of different type w.r.t. those used by the approaches summarized above.

3   Preliminary Notions

A digitized image can be represented as a rectangular array I of $N = m \times n$ *pixels*, where each pixel i_{ij} is a *character* (typically, encoding an integer) over an alphabet Σ, corresponding to the set of colours occurring into I (see Figure 2). In the following, we will refer to I as *encoded image*.

We are interested in finding a compact descriptor for an encoded image I, able to keep only the essential information in I, in such a way that, given a set S of related images, the corresponding set $\mathcal{D}$ of descriptors associated to the images in S is able to suitably characterize S. To this aim, we search for the repetitive content of I, under the assumption that if a small block of I is sufficently repeated in I, then it represents a larger portion of I that is characteristic for the original digitized image (e.g., sky,

(a)

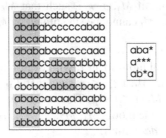

(b)

Fig. 2. (a) A digitized image (Lena). (b) The corresponding encoded image I over the alphabet of colours $\Sigma = \{c_1, c_2, ..., c_k\}$ (each element i_{ij} represents a pixel of Lena with a specific colour $c_l \in \Sigma$).

```
ababccabbabbbac
abababccccabab
abcaababaccaaaa
abababacccccaaa        aba*
ababccabaaabbbb        a***
abaaababcbcbabb        ab*a
cbcbcbabbacbacb
abaccaaaaaaabb
abbbbbbbbacacac
abbabbbaaaaaccc
```

Fig. 3. An encoded image defined on $\Sigma = \{a, b, c\}$ (on the left) and an extended image defined on $\{a, b, c, *\}$ (on the right). The image on the right represents all the four portions highlighted in the image on the left.

see, a forest, etc.). Such repeated blocks can be not necessarily identical, but somewhat identical unless some pixel, due to different shades or lightness in the original picture. Thus, in addition to the *solid* characters from Σ, we also deal with a special *don't care* character, denoted by '*', that is a wildcard matching any character in $\Sigma \cup \{*\}$. Don't cares are useful in order to take into account *approximate* repetitions occurring in I, as illustrated in Figure 3, where both an encoded image defined on $\Sigma = \{a, b, c\}$ and a small image defined on $\{a, b, c, *\}$ are shown. In particular, the smaller image represents a block that is repeated in the encoded image, in the four highlighted regions. We refer to an array A defined on an alphabet $\Sigma \cup \{*\}$ as *extended image*.

A *2D motif* of an encoded image I of size $m \times n$ (see also [3]) is an extended image M of size $m' \times n'$ such that:

1. $m' \leq m$ and $n' \leq n$;
2. there is at least one solid character adjacent to each edge of M;
3. there are at least 2 *approximate occurrences* of M in I, where an approximate occurrence is a position $[k, l]$ ($k \leq m - m' + 1$ and $l \leq n - n' + 1$) in I such that $M[i, j] = I[k + i - 1, l + j - 1]$, or $M[i, j] = *$, for $1 \leq i \leq m'$ and $1 \leq j \leq n'$.

The set of all the approximate occurrences of M in I is the *occurrences list* $\mathcal{L}_M^I$ of M in I.

With reference to the concept of approximate occurrence, we also say that a 2D motif M_1 (or simply motif) is a *sub-motif* of another motif M_2, if M_1 has at least an approximate occurrence in M_2.

When approximate occurrences are taken into account, and motifs with don't cares are thus considered, the number of all the possible motifs that one can extract from an input encoded image can grow drastically, often becoming exponential in the size of the input image. In order to limit such a growth, suitable notions of *maximality* and *irredundancy* have been proposed in both one and two dimensions [2,3,11,23,25,26], leading to the concept of *motif basis* that we recall below.

Maximal Motif. Let M_1, M_2, ..., M_f be the motifs in an encoded image I, and let $\mathcal{L}_{M_1}$, $\mathcal{L}_{M_2}$, ..., $\mathcal{L}_{M_f}$ be their occurrence lists, respectively. A motif M_i is *maximal* if and only if there exists no motif M_j, $j \neq i$, such that M_i is a sub-motif of M_j and $|\mathcal{L}_{M_i}| = |\mathcal{L}_{M_j}|$. In other words, M_i cannot be substituted by M_j without loosing some of the M_i occurrences.

Irredundant Motif. A maximal motif M in I, with occurrence list $\mathcal{L}_M$, is *redundant* if there exist maximal sub-motifs M_i, $1 \leq i \leq p$, such that $\mathcal{L}_M = \mathcal{L}_{M_1} \cup \mathcal{L}_{M_2} \cup \ldots \cup \mathcal{L}_{M_p}$, up to some offsets, and $M \preceq M_i$, $1 \leq i \leq p$. This means that every occurrence of M on I is already covered by one of the motifs $M_1, M_2, \ldots, M_p$. A maximal motif that is not redundant is called an *irredundant* motif.

Motif Basis. Consider now the set $\mathcal{M}$ of all maximal motifs in I. A subset $\mathcal{B}$ of $\mathcal{M}$ is called a *basis* of $\mathcal{M}$ if the following hold:

1. for each $M \in \mathcal{B}$, M is irredundant with respect to $\mathcal{B} - \{M\}$;
2. let $G(X)$ be the set of all the redundant maximal motifs implied by the set of motifs X: then $\mathcal{M} = G(\mathcal{B})$.

In [3] it has been proved that the basis $\mathcal{B}$ of k-motifs for an image I on an alphabet Σ is unique for any k, and that it is made of $O(N)$ motifs if N is the size of I. This helps in obtaining a compact descriptor for I, almost due to the fact that, in practice, the size of the basis is usually much smaller than N.

We conclude this section by illustrating an example showing the basis on the encoded image I defined on $\Sigma = \{a, b\}$. In particular, Figure 4 depicts I and its maximal motifs. Note that $N = 12$ and the number of maximal motifs is 8. Only 5 out of them are worth to belong to the basis of I, that are, M_3, M_4, M_6, M_7 and M_8 (as an example, M_1 is made redundant by M_4 and M_6).

4   Proposed Approach

We now describe the approach proposed in this paper for the classification of digital images. In particular, we consider an input image dataset $T = \{I_1, I_2, \ldots I_n\}$, where images are subdivided in h classes $C_1, C_2, \ldots, C_h$, and a test image I (we refer in the following to the encoded images corresponding to the input digital images). The output of our approach will be the label C_i ($i = 1, \ldots, h$) of the class which I is predicted to belong to.

$$I_{[4,3]} = \begin{Vmatrix} a & b & b \\ b & b & a \\ a & b & a \\ b & a & b \end{Vmatrix} \quad M_1 = \begin{Vmatrix} a & b \end{Vmatrix} \quad \mathcal{L}_{M_1} = \{[1,1],[3,1],[4,2]\}$$

$$M_2 = \begin{Vmatrix} b \end{Vmatrix} \quad \mathcal{L}_{M_2} = \{[1,2],[1,3],[2,1],[2,2],[3,2],[4,1],[4,3]\}$$

$$M_3 = \begin{Vmatrix} a \\ b \end{Vmatrix} \quad \mathcal{L}_{M_3} = \{[1,1],[3,1],[3,3]\}$$

$$M_4 = \begin{Vmatrix} a & b \\ b & * \end{Vmatrix} \quad \mathcal{L}_{M_4} = \{[1,1],[3,1]\} \qquad M_5 = \begin{Vmatrix} a \end{Vmatrix} \quad \mathcal{L}_{M_5} = \{[1,1],[2,3],[3,1],[3,3],[4,2]\}$$

$$M_6 = \begin{Vmatrix} b & * \\ a & b \end{Vmatrix} \quad \mathcal{L}_{M_6} = \{[2,1],[3,2]\} \qquad M_7 = \begin{Vmatrix} b & * \\ b & a \end{Vmatrix} \quad \mathcal{L}_{M_7} = \{[1,2],[2,2]\}$$

$$M_8 = \begin{Vmatrix} * & b \\ b & * \end{Vmatrix} \quad \mathcal{L}_{M_8} = \{[1,1],[1,2],[3,1]\}$$

Fig. 4. An input encoded image I and its maximal motifs

For each image $I_j \in T$, we extract the basis $\mathcal{B}_j$ of 2D motifs. Let $\mathcal{D} = \cup_{j=1}^n \mathcal{B}_j$ be the set of all the motifs in the basis extracted this way, i.e. the codebook. We exploit motifs in such basis as features in order to characterize the different classes as follows. For each motif $m_i \in \mathcal{D}$, and for each image $I_j \in T$ ($j = 1, \ldots, n$), an array (histogram) W_j is built such that $W_j[i]$ is the number of occurrences of m_k in I.

Let now I be the test image. Let W be an array such that $W[i]$ is the number of occurrences of the motif $m_i \in \mathcal{D}$ in I. The K-Nearest Neighbour technique is then applied by computing the euclidean distances d_j between each W_j and W, and by choosing as the output class that one having the larger consensus, among those k images scoring the minimum values for d_j. Figure 5 shows the pseudocode of the proposed classification algorithm.

The BASIS EXTRACTION procedure called at line 3 implements the technique proposed in [3], that at first extracts the set $\mathcal{M}$ of all the maximal motifs occurring in I, then computes their occurrence lists, and finally eliminates the residue redundancy in $\mathcal{M}$ and returns in output both the motifs of the basis of I and their occurrence lists.

The COMPUTE OCCURRENCES procedure receives in input a motif m and an image I, and returns in output the number of occurrences of m in I.

The following example clarifies how the proposed approach works. Let us consider the training set $T = \{I_1, I_2, I_3, I_4\}$ defined on $\Sigma = \{a, b, c, d, e, f, g, h\}$, where:

$$I_1 = \begin{Vmatrix} b & c & e \\ b & c & e \\ c & h & h \end{Vmatrix}, \quad I_2 = \begin{Vmatrix} b & c & e & c \\ f & a & h & h \\ b & h & h & a \end{Vmatrix}, \quad I_3 = \begin{Vmatrix} d & f & a \\ c & f & a \end{Vmatrix} \quad \text{and} \quad I_4 = \begin{Vmatrix} g & c & a \\ d & f & a \\ g & c & d \end{Vmatrix}.$$

Input:	an image dataset T of n images grouped in h classes $C_1, C_2, \ldots, C_h$;
	a test image I;
	an integer k;
Output:	a label $x \in \{1, 2, \ldots, h\}$ corresponding to the class predicted for I;

1.	$\mathcal{D} = \emptyset$;
2.	**for each** image $I_j \in T$ **begin**
3.	**call** BASIS EXTRACTION and obtain $\mathcal{B}_j$;
4.	$\mathcal{D} = \mathcal{D} \cup \mathcal{B}_j$;
5.	**end**
6.	**for each** image $I_j \in T$ **begin**
7.	let W_j be an empty array of $|\mathcal{D}|$ integer values;
8.	**for each** motif $m_i \in \mathcal{D}$ **begin**
9.	**call** COMPUTE OCCURRENCES on m_i and I_j and obtain $nocc_{j,i}$;
10.	$W_j[i] = nocc_{j,i}$;
11.	**end**
12.	**end**
13.	let W be an empty array of $|\mathcal{D}|$ integer values;
14.	**for each** motif $m_k \in \mathcal{D}$ **begin**
15.	**call** COMPUTE OCCURRENCES on m_k and I and obtain $nocc_i$;
16.	$W[i] = nocc_i$;
17.	**end**
18.	let E be an empty array of n real values;
19.	**for each** image $I_j \in T$ **begin**
20.	$E[j]$ is set as the Euclidean distance between W_j and W;
21.	**end**
22.	sort E in increasing order;
23.	let S be the set of images in T corresponding to the first k elements in E;
24.	**return** the label of the most popular class in S;

Fig. 5. The Classification Algorithm

T is partitioned in two classes $C_1 = \{I_1, I_2\}$ and $C_2 = \{I_3, I_4\}$. Suppose we want to classify the following test image defined on Σ:

$$I_t = \begin{Vmatrix} b & c & e & b \\ f & h & h & h \\ b & h & h & h \\ c & a & h & a \end{Vmatrix}$$

In order to build the codebook, the algorithm first extracts the basis $\mathcal{B}_i$ from each image I_i of the training set. The obtained bases are as follows (we do not include trivial motifs made of only a character):

$\mathcal{B}_1 = \{M_1\}$ where $M_1 = \begin{Vmatrix} b & c & e \end{Vmatrix}$

$\mathcal{B}_2 = \{M_2, M_3\}$ where $M_2 = \begin{Vmatrix} * & h \\ h & * \end{Vmatrix}$ and $M_3 = \begin{Vmatrix} h & h \end{Vmatrix}$

$\mathcal{B}_3 = \{M_4\}$ where $M_4 = \begin{Vmatrix} f & a \end{Vmatrix}$

$\mathcal{B}_4 = \{M_5\}$ where $M_5 = \begin{Vmatrix} g & c \end{Vmatrix}$

The obtained codebook is then: $\mathcal{D} = \{M_1, M_2, M_3, M_4, M_5\}$.

For each image of the training set, the corresponding histogram is determined by counting the occurrences of the motifs that build up the codebook. The resulting vectors are: $W_1 = [2, 0, 1, 0, 0]$, $W_2 = [1, 2, 2, 1, 0]$, $W_3 = [0, 0, 0, 2, 1]$ and $W_4 = [0, 0, 0, 1, 2]$.

The histogram of the test image, which is similarly computed, is: $W_t = [1, 3, 4, 0, 0]$. The Euclidean distances between each W_i and W_t are then determined and are: $d(W_1, W_t) = 4.35$, $d(W_2, W_t) = 2.5$, $d(W_3, W_t) = 5.57$ and $d(W_4, W_t) = 5.57$.

By using $k = 1$, the smaller distance is $d(W_2, W_t)$ and therefore I_t is assigned to the class C_1.

5 Experimental Validation

We now describe an experimental validation campaign performed to test the image classification approach presented in this paper. In particular, we exploited at first a benchmark image dataset, the ZuBuD dataset, that is described below, and then also a dataset of images obtained by storing frames from digital videos of some Italian television news.

5.1 Tests on the ZuBuD Dataset

The ZuBuD dataset is a publicly available collection of images that depict 201 buildings in Zurich [29]. The training set contains five pictures per building, each of which has been taken from a different viewpoint. The test set contains 115 pictures of a subset of the 201 buildings. The pictures were acquired by two different cameras in different seasons and under different weather conditions. Training images have a resolution of 640×480 pixel, whereas that of test images is of 320×240 pixel.

Because of the high number of classes, we first tested our classifier by using a subset made of only five classes. In a second experiment, we considered ten classes and finally the full training set. In the choice of the classes to be included for the experiments, attention has been paid to select both clearly distinct classes and subsets of similar classes. Test images are classified by using $k = 1$ for the K-Nearest Neighbour algorithm. This way, the test image is assigned to the class of the nearest (most similar) image of the training set.

In order to evaluate the performances of the considered classification techniques, we refer to the *error rate* of an experiment as the percentage of misclassified test images.

The subset of the training images used for the first experiment is shown in Fig. 6, where each row corresponds to a different class. It can be noted that in this case, there is a clear distinction among the classes. The classification algorithm performed well by achieving a 0% error rate, i.e. all the test images were correctly assigned to the belonging class.

For the second experiment, five more classes were added to the training set of the previous case. This time, classes are not not sharply separated as before: there are pictures belonging to different classes which are similar, especially for the color of some of their portions. Also in this case, the performance of the classifier were good in that it achieved an error rate of 20% despite the presence of similarities among images from different classes.

For the last experiment we used the full training set and the classifier obtained an error rate of 27.8%. Table 1 summarizes the results of this series of experiments.

Fig. 6. Images from the exploited dataset (ZuBuD)

Table 1. Results on the dataset Zubud with different numbers of classes

	Error rate
5 classes	0%
10 classes	20%
201 classes	27.8%

In order to improve the accuracy of 2D feature motifs based classification approach, we also performed two further experiments. At first, we applied a preprocessing step by quantizing the input images of the training set, in order to reduce noice and to increase the density of the extracted motifs. The second experiment was that of reducing the size of the codebook as follows. For each set of motifs $\{m_1, m_2, \ldots, m_k\}$ where there exists m_j such that each m_i is a submotif of m_j ($i \neq j$, $i, j = 1, \ldots, k$), then the entire set is condensed and only m_j is kept in the codebook (i.e., each m_i is substituted by m_j). Note that, for images belonging to the same class, finding repeated submotifs across them is not unusual.

The ZuBuD datased has been employed in the literature to evaluate the performance of other classifiers [21,30,28]. Table 2 summarizes the results of our algorithm, including also the last two variants we considered, and compares them with those achieved by other approaches. The best variant for our approach is that exploiting condensed codebook, and the results we obtained are comparable with those of the other approaches. Note that the features we exploited rely only on pixel-level information. The accuracy of the proposed approach could be improved by combining them by other higher level information.

5.2   Tests on Video Frames

In order to further analyze the behaviour of 2D motif features, we applied the approach described in Section 4 to a dataset made of 60 images extracted as frames from the digital video of three Italian television news (i.e., TG1, TG2 and TGLa7). Some of the images in such a dataset, that we refer to as VideoTG in the following, are shown in Figure 7.

Table 2. Comparison on the dataset Zubud

Method	Error rate
Approach presented in [20]	4.35%
Approach presented in [30]	13.9%
2D Feature Motifs with condensed codebook	19.13%
2D Feature Motifs with preprocessing	23.48%
2D Feature Motifs	27.8%
Approach presented in [28]	59%

(a)   (b)   (c)

Fig. 7. (a) Frames from Italian TG1. (b) Frames from Italian TG2. (c) Frames from Italian TGLa7.

We performed two different tests on VideoTG, and applied in both cases cross-validation by dividing the input dataset in three disjoint sets, then performing three different tests, with a different training/test set for each of them. Furthermore, we fixed $k = 1$. In particular, in the first test we considered only the frames corresponding to the two classes TG1 and TG2, while in the second one we exploited all the VideoTG dataset, thus considering three classes. The results we obtained are shown in Table 3.

Table 3. Tests performed exploiting only the two TG1 and TG2 classes (2 classes), and the three classes TG1, TG2 and TGLa7 (3 classes)

	Error rate (2 classes)	Error rate (3 classes)
Experiment 1	0%	13.33%
Experiment 2	20%	20%
Experiment 3	0%	20%
AVG	6.7%	17.78%

These tests confirm that the performances of the method become worse when the number of classes increases. However, since the classes are well separated, the approach seems to be successfull in distinguishing frames coming from different videos.

6   Conclusion

We proposed an image classification approach based on 2D feature motifs suitably embedded in a bag-of-words model and used in cascade with K-Nearest Neighbour. We performed an experimental validation campaign on both a benchmark image dataset and a video frames dataset, obtaining promising results. Indeed, the error rate achieved by our technique is comparable with those of other existing approaches, and improves when suitable refinements are provided either in preprocessing or in codebook building steps.

We plan to explore several directions as our future work. Among them, we will study how combining 2D feature motifs with classification approaches alternative to K-Nearest Neighbour, such as for example Support Vector Machines. We also will analyze suitable invariants, such as scaling and/or rotations [10,13], that could improve the ability of 2D basis in capturing the essential information of digital images, and suitable similarity measures and approaches coming from time series pattern discovery [9,27].

Acknowledgements. This research was partially supported by the Italian Ministry for Education, University and Research under the grants PON 01_02477 2007-2013 - FRAME and FIT (IDEAS: "Un Ambiente Integrato per lo Sviluppo di Applicazioni e Soluzioni").

References

1. Amelio, A., Apostolico, A., Rombo, S.E.: Image compression by 2D motif basis. In: Data Compression Conference (DCC 2011), pp. 153–162 (2011)
2. Apostolico, A., Parida, L.: Incremental paradigms of motif discovery. J. of Comp. Biol. 11(1), 15–25 (2004)
3. Apostolico, A., Parida, L., Rombo, S.E.: Motif patterns in 2D. Theoretical Computer Science 390(1), 40–55 (2008)
4. Bay, H., Ess, A., Tuytelaars, T., Van Gool, L.: Speeded-Up Robust Features (SURF). Computer Vision and Image Understanding 110(3), 346–359 (2008)
5. Berg, A.C., Berg, T.L., Malik, J.: Shape matching and object recognition using low distortion correspondences. In: Proc. of IEEE Computer Society Conference on Computer Vision and Pattern Recognition (CVPR 2005), vol. 1, pp. 26–33 (2005)
6. Bosch, A., Muñoz, X., Martí, R.: Review: Which is the best way to organize/classify images by content? Image Vision Comput. 25(6), 778–791 (2007)
7. Bosch, A., Zisserman, A., Muñoz, X.: Scene classification using a hybrid generative/discriminative approach. IEEE Trans. Pattern Anal. Mach. Intell. 30(4), 712–727 (2008)
8. Dash, M., Liu, H.: Feature selection for classification. Intelligent Data Analysis 1, 131–156 (1997)
9. Keogh, E.J., et al.: Supporting exact indexing of arbitrarily rotated shapes and periodic time series under euclidean and warping distance measures. VLDB J. 18(3), 611–630 (2009)
10. Fredriksson, K., Mäkinen, V., Navarro, G.: Rotation and lighting invariant template matching. Information and Computation 205(7), 1096–1113 (2007)
11. Grossi, R., Pisanti, N., Crochemore, M., Sagot, M.-F.: Bases of motifs for generating repeated patterns with wild cards. IEEE/ACM Trans. Comp. Biol. Bioinf. 2(3), 159–177 (2000)
12. Hofmann, T.: Unsupervised learning by probabilistic latent semantic analysis. Machine Learning 42(1-2), 177–196 (2001)
13. Hundt, C., Liskiewicz, M., Nevries, R.: A combinatorial geometrical approach to two-dimensional robust pattern matching with scaling and rotation. Theoretical Computer Science 410(51), 5317–5333 (2009)
14. John, G.H., Kohavi, R., Pfleger, K.: Irrelevant features and the subset selection problem. In: Machine Learning: Proceedings of the Eleventh International, pp. 121–129 (1994)

15. Kadir, T., Brady, M.: Saliency, scale and image description. Int. J. Comput. Vision 45(2), 83–105 (2001)
16. Lowe, D.G.: Object recognition from local scale-invariant features. In: Proc. of the 7th IEEE International Conference on Computer Vision, vol. 2, pp. 1150–1157 (1999)
17. Lowe, D.G.: Local feature view clustering for 3D object recognition. In: Proc. of the IEEE Computer Society Conference on Computer Vision and Pattern Recognition (CVPR 2001), pp. 682–688 (2001)
18. Lu, D., Weng, Q.: A survey of image classification methods and techniques for improving classification performance. International Journal of Remote Sensing 28(5), 823–870 (2007)
19. Marée, R., Geurts, P., Piater, J.H., Wehenkel, L.: Biomedical image classification with random subwindows and decision trees. In: Liu, Y., Jiang, T.-Z., Zhang, C. (eds.) CVBIA 2005. LNCS, vol. 3765, pp. 220–229. Springer, Heidelberg (2005)
20. Marée, R., Geurts, P., Piater, J., Wehenkel, L.: Random subwindows for robust image classification. In: Proc. of International Conference on Computer Vision and Pattern Recognition (CVPR), pp. 34–40 (2005)
21. Matas, J., Obdržálek, S.: Object recognition methods based on transformation covariant features. In: 12th European Signal Processing Conference (2004)
22. Nanni, L., Lumini, A., Brahnam, S.: Survey on LBP based texture descriptors for image classification. Expert Syst. Appl. 39(3), 3634–3641 (2012)
23. Parida, L., Pizzi, C., Rombo, S.E.: Characterization and extraction of irredundant tandem motifs. In: Calderón-Benavides, L., González-Caro, C., Chávez, E., Ziviani, N. (eds.) SPIRE 2012. LNCS, vol. 7608, pp. 385–397. Springer, Heidelberg (2012)
24. Philbin, J., Chum, O., Isard, M., Sivic, J., Zisserman, A.: Object retrieval with large vocabularies and fast spatial matching. In: IEEE Conference on Computer Vision and Pattern Recognition (CVPR 2007), pp. 1–8 (2007)
25. Rombo, S.E.: Optimal extraction of motif patterns in 2D. Information Processing Letters 109(17), 1015–1020 (2009)
26. Rombo, S.E.: Extracting string motif bases for quorum higher than two. Theor. Comput. Sci. 460, 94–103 (2012)
27. Rombo, S.E., Terracina, G.: Discovering representative models in large time series databases. In: Christiansen, H., Hacid, M.-S., Andreasen, T., Larsen, H.L. (eds.) FQAS 2004. LNCS (LNAI), vol. 3055, pp. 84–97. Springer, Heidelberg (2004)
28. Shao, H., Svoboda, T., Ferrari, V., Tuytelaars, T., Van Gool, L.: Fast indexing for image retrieval based on local appearance with re-ranking. In: Proc. of International Conference on Image Processing (ICIP 2003), vol. 2, pp. III-737–III740 (2003)
29. Shao, H., Svoboda, T., Van Gool, L.: Zubud - Zurich building database for image based recognition. Technical Report TR-260, Computer Vision Lab, Swiss Federal Institute of Technology, Switzerland (2003)
30. Shao, H., Svoboda, T., Tuytelaars, T., Van Gool, L.: HPAT indexing for fast object/scene recognition based on local appearance. In: Bakker, E.M., Lew, M., Huang, T.S., Sebe, N., Zhou, X.S. (eds.) CIVR 2003. LNCS, vol. 2728, pp. 71–80. Springer, Heidelberg (2003)
31. Xie, N., Ling, H., Hu, W., Zhang, X.: Use bin-ratio information for category and scene classification. In: IEEE Conference on Computer Vision and Pattern Recognition (CVPR 2010), pp. 2313–2319 (2010)
32. Yang, J., Jiang, Y.-G., Hauptmann, A.G., Ngo, C.-W.: Evaluating bag-of-visual-words representations in scene classification. In: Proceedings of the International Workshop on Workshop on Multimedia Information Retrieval, MIR 2007, pp. 197–206 (2007)
33. Zhou, L., Zhou, Z., Hu, D.: Scene classification using a multi-resolution bag-of-features model. Pattern Recognition 46(1), 424–433 (2013)

Wildfire Susceptibility Maps Flexible Querying and Answering*

Paolo Arcaini, Gloria Bordogna, and Simone Sterlacchini

CNR – IDPA – National Research Council of Italy – Institute for the Study of the
Dynamics of Environmental Processes
{paolo.arcaini,gloria.bordogna,simone.sterlacchini}@idpa.cnr.it

Abstract. Forecasting natural disasters, as wildfires or floods, is a mandatory activity to reduce the level of risk and damage to people, properties and infrastructures. Since estimating real-time the susceptibility to a given phenomenon is computationally onerous, susceptibility maps are usually pre-computed. So, techniques are needed to efficiently query such maps, in order to retrieve the most plausible scenario for the current situation. We propose a flexible querying and answering framework by which the operator, in charge of managing an ongoing disaster, can retrieve the list of susceptibility maps in decreasing order of satisfaction with respect to the query conditions. The operator can also describe trends of the conditions that are related with environmental parameters, assessing what happens if a dynamic parameter is increasing or decreasing in value.

1 Introduction

The European Civil Protection, by its *Disaster Prevention, Preparedness and Intervention* programs[1], states that the improvement of the methods both for data sharing and communication, and the enhancement of the public information, education and awareness are two of the major strategic modes of actions proposed to prevent the risks and damage to people, properties and infrastructures and, in so doing, to reduce vulnerabilities and risks to hazards. In the same way, the World Conference on Disaster Reduction (2005) promoted actions to identify, assess and monitor disaster risks and enhancing early warning.

By relying on these guidelines, the *SISTEMATI* project has the main aim of developing a system to improve the capabilities of both planners and managers (from now on named *operators*) in preparedness and response to wildfire disaster-related activities by providing the following functionality, as depicted in Fig. 1:

- *Hazard and risk maps generation*: at the current stage of the project development, we started generating multiple wildfire *susceptibility* maps of the territory by using a statistic modeling approach that classifies the spatial units

---

* This work has been carried out within the project SISTEMATI "Geomatics Supporting Environmental, Technological and Infrastructural Disaster Management", funded by the Italian Ministry of Research jointly with Regione Lombardia.
[1] http://ec.europa.eu/echo/civil_protection/civil/prote/cp14_en.htm

H.L. Larsen et al. (Eds.): FQAS 2013, LNAI 8132, pp. 352–363, 2013.

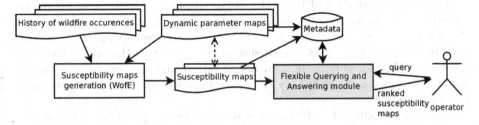

Fig. 1. Wildfire prevention framework

of the territory with respect to their predisposition to wildfires occurrence. The approach uses historical records of wildfires and knowledge of both static and dynamic parameters surveyed on the territory under analysis. In fact, some information needed to obtain reliable predictions of wildfires is highly dynamic, such as the wind conditions when the historic wildfires events occurred. The availability of susceptibility maps and the parameters associated with them will increase the knowledge of both the potential occurrence of destructive events and the meteorological conditions that determined such events. Based on this piece of information together with economic value and vulnerability of infrastructure, risk maps will be generated in a further phase of the *SISTEMATI* project, in order to provide information on the direct/indirect effects of the destructive events at a regional scale.

- *Hazard and risk maps flexible querying and answering*: the system will allow the real-time retrieval and visualization of hazard and risk maps by providing the operators in charge of the territorial planning with the ability to flexibly query the metadata associated with the stored maps (presently, wildfire susceptibility maps): such queries specify the current or forecast values of the static and dynamic parameters in the form of (soft) constraints on the metadata, and the system retrieves, from the repository, the maps generated with the values of the parameters that best satisfy the (soft) constraints. This way the system performs a case-based reasoning to retrieve the most plausible scenarios by comparing the current conditions to analogous environmental situations: the closer the actual circumstances are to the model assumptions, the better the prediction will be.

This paper focuses on the querying and answering model of the *SISTEMATI* framework that is defined based on a fuzzy database approach [4,3]. The operator can formulate *flexible* queries by providing either approximate values of the environmental parameters, or the trends of the dynamic parameters that characterize the current situation (status). The system will retrieve the susceptibility maps depicting the most plausible scenarios whose metadata match the query at least to a given degree. This fuzzy database approach is motivated by the fact that very often the highly dynamic parameters characterizing current or future conditions are ill-known and thus precise values cannot be stated. Moreover, the maps are generated with values of the parameters that are invariably affected by imprecision and errors. Additionally, very often the dynamic parameter maps have a coarser resolution of the static parameter maps. Given these premises, adopting a classic database approach would be inadequate.

Fusion strategies are applied to provide the two maps corresponding to the most plausible optimistic/pessimistic susceptibility scenarios that match the flexible query. This is also important for risk managers and planners to better evaluate the expected risks, since they know the optimistic and pessimistic scenarios for any given decision.

Section 2 introduces the definition of susceptibility maps and of dynamic parameter maps; moreover, it presents a technique to fuse susceptibility maps. Section 3 describes our proposal of flexible querying susceptibility maps, introducing two techniques: query on current parameters and query on trend. Section 4 presents some related work and Section 5 concludes the paper.

2   Wildfire Susceptibility Maps Generation

The main aim is to generate some *susceptibility maps* representing the scenarios in which a particular natural/environmental disaster can happen in a specific area. The aim is to provide planners with information concerning the potential of a territory to be affected by future damaging natural events, i.e., any occurrence (of wildfires, floods, landslides and avalanches) "that causes damage, ecological disruption, loss of human life, or deterioration of health and health services on a scale sufficient to warrant an extraordinary response from outside the affected community or area" [1]. At this stage of the prototypal system implementation, we generate wildfire *susceptibility* maps, which is the first step necessary to successively compute *hazard* and *risk* maps, that will be dealt with in a successive phase. By definition, a hazard map provides much more information with respect to a susceptibility map, given that it also supplies the magnitude and the return period of the event under analysis. In line with this, risk refers to the combined effects of hazard and vulnerability (in terms of physical, social, economic and environmental vulnerability) that elements of the territory could potentially experience.

Some wildfire susceptibility maps have been modelled at a regional scale, in order to investigate different spatially geo-referenced scenarios in which the natural events under analysis may be expected. The concept of susceptibility refers to the likelihood of a wildfire occurring in an area on the basis of local static and dynamic conditions. A susceptibility map is normally a raster-based map that can represent the spatial domain under study by either continuous values (probability values in [0,1]) or ordinal values representing susceptibility classes obtained by applying distinct classification methods. A susceptibility value is associated to each pixel or aggregation of pixels, representing the level of susceptibility to wildfires, graphically represented by different colours, that can be either a susceptibility class identifier or a range of probability values of occurrence. Each susceptibility map is also associated with *parameter maps* in which the values of the parameters specify under which static and dynamic conditions each distinct position in the map has been modelled. Such parameter maps are synthesized by metadata, organized into a knowledge base available for querying. We distinguish between *dynamic* parameters that change over time, such as those denoting meteorological conditions, and *static* parameters that, given a geographic area, can be considered invariable over the time of analysis, such as land use, vegetation, altitude, slope aspect and gradient, among the others [12]. In the specific case of wildfire susceptibility maps, modelled within the *SISTEMATI* project, the dynamic parameters we deal with are represented by temperature and rainfall with reference to the day before the occurrence, one week before, and one month before; for wind direction and speed, instead, we only use data related to the day before the occurrence.

Each susceptibility map provides the stakeholders with different scenarios and it can be generated by applying either physical or statistical models concerning the natural event under analysis. In this paper we use the latter approach, applying the Weights of Evidence modelling technique (WofE) [2]. WofE is a log-linear form of the data-driven Bayesian probability model that exploits known occurrences as training points to derive a predictive output (in our context the susceptibility maps). This latter is generated from multiple, weighted evidences (evidential themes representing explanatory variables), influencing the spatial distribution of the occurrences in the study area [9]. Training points are represented by historical occurrences of wildfires[2]. Notice that the WofE technique does not compute "fuzzy" susceptibility maps, i.e., each pixel is associated with a single value.

Since modelling susceptibility maps is a computationally time consuming activity, it is really difficult or even impossible to calculate them real-time when needed in order to face possible emergencies in terms of preparedness and response activities. The availability of a model able to describe the state-of-the-nature is as important as the assessment of the location of future wildfires. For this reason, in order to provide real-time information to the operators in charge of managing the pre-alarm/alarm phase, we decided to store many susceptibility maps, each one generated off-line with different values and combinations of the static and dynamic parameters above mentioned. This may allow the operator to query the knowledge base associated with the stored maps and retrieve the most "similar" scenario(s) to the current situation in relation to the environmental setting and the meteorological conditions. Each map is associated with some ranges of values for the dynamic parameters.

In order to implement a querying/answering mechanism, it is necessary to represent the contents of the parameter maps (used in the WofE technique) to optimize access and query evaluation. To this aim a knowledge base is generated which contains, for each susceptibility map, metadata that synthesize the contents of the associated parameter maps.

The querying must be flexible in order to allow the specification of *soft* conditions on the parameters values, and must be tolerant to uncertainty, thus

---

[2] We considered five different temporal subsets chosen homogeneous with respect to their meteorological conditions, while evidences are the static and dynamic data collected within the study area.

ranking the susceptibility maps in decreasing order of satisfaction to the query conditions. To model this flexible querying, we rely on the framework of fuzzy databases [3,4,6].

Furthermore, the result of a flexible query must be an informative answer. Besides retrieving the most likely susceptibility maps, two *virtual* maps, representing the most likely optimistic and pessimistic scenarios that may occur, together with the variability of susceptibility levels in each pixel, are generated. To provide this additional answer, one must perform some map fusion operations.

2.1 Definition of Susceptibility Map and Its Metadata

In the following we introduce the representations of both the data and the metadata that will be used in the following sections.

Susceptibility Map. At the current stage of the project development, we define a *susceptibility map* sm_k as an image where the value $sm_k(r, c)$ of pixel (r, c) represents the probability that an hazardous event can happen in the corresponding area (obviously, the area represented by a pixel depends on the scale of the map). A susceptibility map is usually discretised in a small number of *susceptibility levels* (or *classes*). The map partition into susceptibility classes is crisp and not fuzzy; this is because each class is associated with a given emergency or mitigation process that must be executed by the civil protection, and thus we need to identify with precise boundaries the regions having a given susceptibility class. In order to obtain h levels, we identify $h + 1$ values $v_0, \ldots, v_h$ in the interval $[0, 1]$: we set $v_0 = 0$, $v_h = 1$ and we set values $v_1, \ldots, v_{h-1}$ such that $\forall l \in [1, h]$: $v_{l-1} < v_l$. In order to visualize the discretized maps, we assign a colour cl_l to each susceptibility level $l = 1, \ldots, h$, corresponding to the interval $[v_0, v_1]$ if $l = 1$, and to the interval $(v_{l-1}, v_l]$ if $l = 2, \ldots, h$.

In our context, the wildfire susceptibility maps have been discretized using five levels; each level corresponds to a specific mitigation or emergency procedure that civil protection must apply in the region.

Generally, for generating a susceptibility map we need some static and dynamic parameters. While the static parameter maps portray values that can be assumed invariant in time and, for this reason, constant for all the susceptibility maps, the dynamic parameter maps are dependent on the dates of the wildfire training points used by WofE, and thus they are specific for each generated susceptibility map. So it does make sense to only query the latter ones, to retrieve the scenarios most similar to the current situation. In the following we only define dynamic parameter maps.

Dynamic Parameter Maps. The susceptibility maps are generated using p dynamic parameters, each of which refers to a different time period. For each dynamic parameter j (with $j = 1, \ldots, p$), we have t_j parameter maps $dp^{j,t}$ (with $t = 1, \ldots, t_j$). In a map $dp^{j,t}$, we identify the value of a pixel (r, c) with $dp^{j,t}(r, c)$. In the following, when it is not necessary to identify the time periods in which the dynamic parameters have been recorded, we use only one index to iterate over the dynamic parameters, i.e., we identify a dynamic parameter as dp^j, with $j = 1, \ldots, N$ where $N = \sum_{i=1}^{p} t_i$.

For the wildfire susceptibility maps, we have used four dynamic parameters: *rainfall, temperature, wind speed, wind direction*. The first two parameters have been measured in three different time periods: *previous month, previous week* and *previous day*. The latter two parameters, instead, have been measured only in the *previous day*. So, associated with each susceptibility map, we have eight parameter maps. For example, the temperatures maps of the previous month, week and day are identified with $dp^{2,1}$, $dp^{2,2}$ and $dp^{2,3}$ if we consider the time period, or with dp^4, dp^5 and dp^6 if we do not consider the time period.

Let us consider in the following to have M susceptibility maps sm_k (with $k = 1, \ldots, M$), each one associated with N parameter maps dp_k^j (with $j = 1, \ldots, N$).

Metadata. For each dynamic parameter map dp^j associated with a susceptibility map sm_k, we generate a metadata that synthesizes its content. This metadata is structured into the following fields:

$$mdp^j = \{HazardName, k, ParameterName, TimePeriod, D, H\}$$

where $HazardName$ is the name of the type of hazard we are considering (in the prototypal system it is always equal to *wildfire*), k is the index of the susceptibility map sm_k, $ParameterName$ and $TimePeriod$ are the names of the parameter and of the time period to which it is referred to, D is a primitive data type identifying the domain of values of the parameter (for example integer), and $H = \{< range, freq >, \}$ is the histogram of frequencies. $range = [v_a, v_b]$ identifies a range of values within the extremes v_a and v_b on D, and $freq$ is a value in $[0, 1]$, that is the frequency of the $range$ values which belong to the parameter map with the name $ParameterName$. More clearly, H is the normalized histogram of frequencies of the values in dp^j discretized by assuming a given bin. There exists a function $hist^j : mdp^j.D \rightarrow [0, 1]$ such that, given $x \in mdp^j.D$, it returns the value $mdp^j.H.freq$ associated with $x \in mdp^j.H.range$. Finally, we define $\widehat{hist}^j = \max_{x \in mdp^j.D} hist^j(x)$ as the maximum value of histogram $mdp^j.H$.

2.2 Fusing Susceptibility Maps

Given m susceptibility maps (with $m \leq M$), we can merge them in order to obtain the most *pessimistic/optimistic* scenario.

For the optimistic scenario sm_{opt}, the value of a pixel (r, c) is computed as follows:

$$sm_{opt}(r, c) = \min_{k=1}^{m} sm_k(r, c) \qquad (1)$$

In the optimistic map, we expect that the best case happens, i.e., we select the lowest susceptibility level within each pixel.

For the pessimistic scenario sm_{pes}, the value of a pixel (r, c) is computed as follows:

$$sm_{pes}(r, c) = \max_{k=1}^{m} sm_k(r, c) \qquad (2)$$

In the pessimistic map, we expect that the worst case happens, i.e., we select the greatest susceptibility level within each pixel.

Of course, intermediate maps can be obtained by taking intermediate values between the best and the worst case; we could use OWA operators [13]. As future work, we plan to investigate which are the best OWA operators that allow to model distinct decision attributes.

2.3   Variability of Susceptibility Maps

We are also interested in measuring the variability of the expected wildfire susceptibility among the maps. Given m susceptibility maps, we build the variability map Var_{sm}, where the value of each pixel is defined as the variance of the pixel within the susceptibility maps.

Pixels with low variability identify areas in which the susceptibility level is independent of the values of the dynamic parameters: the susceptibility could be either low or high. In case it is high, the territorial planners should provide mitigation interventions or continuous resources to face emergencies in such areas of high susceptibility. Pixels with high variability, instead, identify areas in which the susceptibility level highly depends on the dynamic parameters. In such cases it would be necessary to provide means to know which are the ranges of the parameters that determine high levels, in order to plan the allocation of the resources when the meteorological conditions are becoming close to those values.

3   Flexible Querying Susceptibility Maps

An operator should be able to perform three different kinds of flexible queries over the maps:

- *direct query on current parameters*: the operator specifies some (soft) conditions on the current parameters, possibly in a (sub)region of the spatial domain, and obtains as a result the list of maps in decreasing order of satisfaction of the query conditions. An example of the specified conditions could be: "the temperature has been *hot* in the last week, the rainfall has been *low* in all the previous month and the wind yesterday was blowing in *NS direction* with a *high* speed" where *hot, low* and *high* are defined as soft constraints on the parameters domains.
- *direct query on trend*: the operator specifies the current parameters relative to some time periods (month, week and day) in a possible (sub)region, and the system determines increasing/decreasing trend of the parameters, thus defines soft constraints that represent the parameters trend. For example, the operator could specify a constraint as: "the temperature has been *mildly hot* in the last month, *hot* in the last week and *very hot* yesterday".
- *inverse query*: the operator specifies a susceptibility level (or an interval of consecutive susceptibility levels) in a (sub)region with the aim of retrieving the values of the parameters that more likely produce the specified susceptibility level in the specified (sub)region. An example of query could be: "what

are the values of the dynamic parameters that more likely produce a susceptibility level greater than medium?". This kind of query is similar to the *reverse search* in [12].

Direct queries (both on current parameters and on trend) are useful for online risk management, when an operator observes the current status of meteorological conditions and checks if there are areas with a high degree of susceptibility. Inverse queries, instead, are useful for risk management planning, when a risk management team, responsible of a given territorial area, determines the lower/upper bounds of some environmental parameters which cause a given susceptibility level. This information can be used to define pre-alert emergency plans in case of meteorological conditions within the lower and upper bounds. In this work we only analyse direct queries.

3.1 Direct Query on Current Parameters

In the direct query on current parameters, the operator can specify:

- the bounding box of the area of interest (if omitted it is assumed to be the whole spatial domain); since the maps are generated at regional level and the dynamic parameters are generally known with a coarser resolution, we think that in most of the queries the bounding box will be omitted, since one is interested in the scenarios of the whole territory;
- for each of the parameters that s(he) wants to constraint (e.g., the temperature of last week), a *soft* constraint approximating the observed values of the parameter. The membership function of a soft constraint sc, such as *hot*, can be defined with a trapezoidal shape μ_{sc} through four values (a, b, c, d) with $a \leq b \leq c \leq d$ belonging to a domain D (e.g., temperature). The definition of the function is as follows:

$$\mu_{sc}(x) = \begin{cases} 0 & if \ x \leq a \lor x \geq d \\ \frac{x-a}{b-a} & if \ a < x < b \\ 1 & if \ b \leq x \leq c \\ \frac{d-x}{d-c} & if \ c < x < d \end{cases} \tag{3}$$

In case the operator specifies the average value of the parameter avg and the standard deviation σ, we set $b = c = avg$, $a = avg - \sigma$ and $d = avg + \sigma$.

Similarity Index of a Single Parameter. Once the operator has specified a soft constraint sc over a dynamic parameter dp^j, we must compare sc with the values of the M parameter maps dp_k^j (with $k = 1, \ldots, M$) associated with the susceptibility maps $\{sm_1, \ldots, sm_M\}$. For each parameter dp^j, we must compute a *similarity index* $si_{sc,k}^j$ specifying the similarity of the current situation sc with the situation dp_k^j that generated the susceptibility map sm_k.

Let us see how a single similarity index $si_{sc,k}^j$ is computed, by considering a soft constraint on the temperature parameter in the previous week, as shown in Fig. 2. For each susceptibility map sm_k, we retrieve the corresponding parameter map (in this case dp_k^5, i.e., the map of the temperatures in the previous

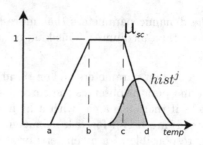

Fig. 2. Computation of the similarity index for the temperature parameter in the previous week

week). From this map we compute the normalized histogram of frequencies of the temperatures in the previous week $hist^j_{BB}$ in the selected bounding box BB. Notice that, in the case in which the bounding box is omitted, we do not need to compute the histogram, since it is already precomputed and available in the correspondent metadata field $mdp^5_k.H$. The similarity index can be computed based on a measure of the overlapping of the two curves, as shown in Fig. 2. One can use the Jaccard coefficient [8]:

$$si^j_{sc,k} = Jaccard(\mu_{sc}, mdp^j_k) = \frac{\sum\limits_{x \in mdp^j_k.D} \min(\mu_{sc}(x), hist^j(x))}{\sum\limits_{x \in mdp^j_k.D} \max(\mu_{sc}(x), hist^j(x))}$$

A simpler definition of the similarity index could be:

$$si^j_{sc,k} = \max_{x \in mdp^j_k.D} \min(\mu_{sc}(x), hist^j(x))$$

Note Note that the soft constraints describe the parameter conditions of the whole spatial domain (or the subregion delimited by a bounding box): we do not support the specification of different constraints for different sub-areas and this could be a limitation in some cases. Let's consider the temperature parameter. For limited areas, the assumption that the temperature is uniform is acceptable. For wide areas, instead, such assumption can not be done. Actually, also for limited areas, such assumption may be not acceptable: for example, in an area that spans between a valley and the peak of a mountain, the difference of temperature could be significant.

Global Similarity Indexes. In order to define the similarity between a susceptibility map sm_k and the situation described by the operator who formulates queries, we must define a *global* similarity index $\tilde{si}$ obtained by the aggregation of the single similarity indexes $\{si_{1,k}, \ldots, si_{N,k}\}$ defined for the single parameters. We could obtain the aggregation as the minimum of the single indexes; such approach has the advantage that the minimum operator is idempotent. Another solution is to make the product of the single indexes; this solution has the advantage that it better ranks the susceptibility maps, and the disadvantage that the product operator is not idempotent.

Result. The result of the *direct query on the current parameters* is a list $\widehat{sm}_1, \ldots, \widehat{sm}_m$ of the susceptibility maps with the m highest similarity indexes, ranked in decreasing order of their similarity indexes $\{\widetilde{si}_1 \geq \ldots \geq \widetilde{si}_m\}$.

The similarity indexes of the retrieved maps can be used to assign *plausibility* values to the optimistic and pessimistic maps defined in formulae 1 and 2 in Section 2.2. We could use optimistic/pessimistic maps generated using all the M susceptibility maps or using only the first $m < M$ maps retrieved with the query. Given an optimistic/pessimistic *fused* susceptibility map fsm, function Om_{fsm} retrieves, for each pixel, the index k of map sm_k from which the value of the map fsm has been derived. In case of optimistic and pessimistic maps obtained by formulae 1 and 2 (i.e., $fsm = opt$ or $fsm = pes$), functions Om_{opt} and Om_{pes}, given a pixel (r, c), are respectively defined as follows:

$$k = Om_{opt}(r, c) = \arg \min_{i=1}^{m} \widehat{sm}_i(r, c) \qquad k = Om_{pes}(r, c) = \arg \max_{i=1}^{m} \widehat{sm}_i(r, c)$$

So, given the similarity indexes $\{\widetilde{si}_1, \ldots, \widetilde{si}_m\}$ computed with respect to the current query, we build the plausibility map pl_{fsm} of map fsm by computing the value of each pixel in the following way:

$$pl_{fsm}(r, c) = \widetilde{si}_k \qquad \text{where} \quad Om_{fsm}(r, c) = k$$

So, the plausibility of an optimistic/pessimistic map fsm in a given pixel (r, c) is determined by the similarity index $\widetilde{si}_k$ of the susceptibility map $\widehat{sm}_k$ from which the value $fsm(r, c)$ has been derived. The plausibility index, associated with a pixel classified into a given susceptibility class, allows to estimate the uncertainty of the classification. One could define minimal acceptable levels on the plausibility degree to identify regions with a minimal certainty of classification.

3.2 Direct Query on Trend

In this case the operator provides the possibly approximate values of a dynamic parameter in the previous month, week and day (c^m, c^w, c^d). Such kind of query is possible only over the parameters whose values are given for all the three time periods (e.g., rainfall and temperature for the wildfire alert).

We suppose that the operator provides three values that follow a trend among the values of the previous month, week and day; the trend is increasing ($\uparrow$) when $c^m \leq c^w \leq c^d$, and decreasing ($\downarrow$) when $c^m \geq c^w \geq c^d$.

The query is executed in three steps:
1) definition of the three (soft) constraints based on the values specified by the operator for the parameter;
2) filtering of the susceptibility maps whose parameters values have the same *trend* of the values specified by the operator;
3) among the maps identified in the previous step, evaluation of the soft constraints as done in the *direct query on current parameters* (Section 3.1).

Given a dynamic parameter dp, in order to answer to a query, we retrieve, for each susceptibility map sm_k, the metadata histograms $mdp.H^m$, $mdp.H^w$ and

$mdp.H^d$ (related to the measurements of the previous month, week and day), and get their maximum values, identifying them with $\widehat{hist}^{\,m}$, $\widehat{hist}^{\,w}$ and $\widehat{hist}^{\,d}$. Then, we select only the susceptibility maps that follow the trend specified by the operator. For example, if the user specified an increasing trend, we select only those maps in which $\widehat{hist}^{\,m} \leq \widehat{hist}^{\,w} \leq \widehat{hist}^{\,d}$. Then, for these maps, we compute the similarity indexes and the global similarity indexes as it has been described in Section 3.1.

4   Related Work

Several projects use susceptibility maps for forecasting events: in [7], for example, susceptibility maps for several types of hazards have been generated. However these platforms usually only permit to display the susceptibility maps and to do standard geospatial data manipulation operations (e.g., zooming and panning), but they do not support flexible querying as the one we propose in this paper.

Others applications have modelled environmental risks based on dynamic variables such as in [5] where malaria disease is modelled by considering the climatic changes, or in [11] where soil erosion is modelled by considering the rainfalls so that all storm rainfall above a critical threshold (whose value depends on soil properties and land cover) is assumed to contribute to runoff, and erosion is assumed to be proportional to runoff.

A work similar to ours is described in [12]. However, their work differs from ours since they use fire simulators to predict fires, and they use the history of fires only to calibrate the input variables of the simulator (using genetic algorithms). They use a nearest neighbour search to find, in the fire history, the fire event that occurred with a configuration of the dynamic parameters most similar to the current configuration.

5   Conclusions and Future Work

Our proposal is original as far as three aspects. The first aspect, related to the application, is the storage and retrieval of susceptibility maps which make sense in a real settings due to the fact that these maps are generated based on dynamic parameters. The management of pre-alarm and emergency phases may benefit by the retrieval of plausible wildfire scenarios given the current values of the dynamic parameters, that would be impossible to generate real-time.

The other two original aspects are the application of soft computing methods to model both the flexible querying and the soft fusion of susceptibility maps. Flexible queries can provide a user friendly means to the operator for expressing the values of the current dynamic parameters, that are often ill-known (generally one has the mean and the standard deviation). Thus, the retrieval must be able to tolerate imprecision in the definition of the selection conditions of susceptibility maps and must provide ranked answers. Further novelty, as far as the retrieval of spatial data, is generating "virtual" susceptibility maps representing the most pessimistic and optimistic susceptibility scenarios matching the current values of the dynamic parameters. These "virtual" scenarios provide a richer information than the single plausible susceptibility maps, since they account for the level of maximum and minimum susceptibility of an area of the territory given the current dynamic parameters values.

As future work we plan to experiment different operators for computing the similarity indexes of the single parameters and the global similarity index, and determine those that provide the better forecasting results. Moreover we plan to generate the susceptibility maps applying a fuzzy partition [10], instead of a crisp one as currently done. This would allow to describe more precisely the *smooth* boundaries that actually exist between the sub-regions classified with the five susceptibility levels. To this purpose, we should adapt the WofE method for producing maps with fuzzy regions, and the querying and answering framework proposed in this paper for handling the new type of maps.

References

1. Coping with major emergencies: WHO strategy and approach to humanitarian action. Tr, World Health Organization (1995)
2. Agterberg, F.P., Bonham-Carter, G.F., Wright, D.: Weights of Evidence modelling: a new approach to mapping mineral potential. In: Statistical Applications in the Earth Sciences. Geological Survey of, Canada, pp. 171–183 (1989)
3. Bordogna, G., Psaila, G.: Fuzzy-spatial SQL. In: Christiansen, H., Hacid, M.-S., Andreasen, T., Larsen, H.L. (eds.) FQAS 2004. LNCS (LNAI), vol. 3055, pp. 307–319. Springer, Heidelberg (2004)
4. Bosc, P., Kaeprzyk, J.: Fuzziness in database management systems. Physica-Verlag (1996)
5. Craig, M., Snow, R., le Sueur, D.: A Climate-based Distribution Model of Malaria Transmission in Sub-Saharan Africa. Parasitology Today 15(3), 105–111 (1999)
6. Galindo, J. (ed.): Handbook of Research on Fuzzy Information Processing in Databases. IGI Global (2008)
7. Institutional building for natural disaster risk reduction (DRR) in Georgia, http://drm.cenn.org
8. Miyamoto, S., Nakayama, K.: Similarity measures based on a fuzzy set model and application to hierarchical clustering. IEEE Transactions on Systems, Man and Cybernetics 16(3), 479–482 (1986)
9. Raines, G.: Evaluation of Weights of Evidence to Predict Epithermal-Gold Deposits in the Great Basin of the Western United States. Natural Resources Research 8(4), 257–276 (1999)
10. Schneider, M.: Fuzzy spatial data types for spatial uncertainty management in databases. In: Handbook of Research on Fuzzy Information Processing in Databases, pp. 490–515 (2008)
11. van der Knijff, J., Jones, R., Montanarella, L.: Soil Erosion Risk Assessment in Europe. Tr, European Commission. European Soil Bureau (2000)
12. Wendt, K.: Efficient knowledge retrieval to calibrate input variables in forest fire prediction. Master's thesis, Escola Tècnica Superior d'Enginyeria. Universitat Autònoma de Barcelona (2008)
13. Yager, R.R.: On ordered weighted averaging aggregation operators in multicriteria decisionmaking. IEEE Transactions on Systems, Man and Cybernetics 18(1), 183–190 (1988)

Enhancing Flexible Querying Using Criterion Trees

Guy De Tré[1], Jozo Dujmović[2], Joachim Nielandt[1], and Antoon Bronselaer[1]

[1] Dept. of Telecommunications and Information Processing, Ghent University,
Sint-Pietersnieuwstraat 41, B-9000 Ghent, Belgium
`{Guy.DeTre,Joachim.Nielandt,Antoon.Bronselaer}@UGent.be`
[2] Dept. of Computer Science, San Francisco State University,
1600 Holloway Ave, San Francisco, CA 94132, U.S.A.
`jozo@sfsu.edu`

Abstract. Traditional query languages like SQL and OQL use a so-called WHERE clause to extract only those database records that fulfil a specified condition. Conditions can be simple or be composed of conditions that are connected through logical operators. Flexible querying approaches, among others, generalized this concept by allowing more flexible user preferences as well in the specification of the simple conditions (through the use of fuzzy sets), as in the specification of the logical aggregation (through the use of weights). In this paper, we study and propose a new technique to further enhance the use of weights by working with so-called criterion trees. Next to better facilities for specifying flexible queries, criterion trees also allow for a more general aggregation approach. In the paper we illustrate and discuss how LSP basic aggregation operators can be used in criterion trees.

Keywords: Fuzzy querying, criterion trees, LSP, GCD.

1 Introduction

1.1 Background

Traditionally, WHERE-clauses have been used in query languages to extract those database records that fulfil a specified condition. This condition should then reflect the user's preferences with respect to the records that should be retrieved in the query result. Most traditional query languages like SQL [10] and OQL [2] only allow WHERE-conditions which can be expressed by Boolean expressions. Such Boolean expression can be composed of simple expressions that are connected by logical conjunction ($\wedge$), disjunction ($\vee$) and negation ($\neg$) operators. Parentheses can be used to alter the sequence of evaluation.

Adequately translating the user's needs and preferences into a representative Boolean expression is often considered to be a difficult and challenging task. This is especially the case when user requirements are complex and expressed in natural language. Soft computing techniques help developing fuzzy approaches

H.L. Larsen et al. (Eds.): FQAS 2013, LNAI 8132, pp. 364–375, 2013.

for flexible querying that help to solve these difficulties. An overview of 'fuzzy' querying techniques can, among others, be found in [18].

In this paper, 'fuzzy' querying of a regular relational database is considered. Such a database consists of a collection of relations, represented by tables [3], comprising of attributes (columns) and tuples (rows). Each relation R is defined by a relation schema

$$R(A_1 : T_1, \ldots, A_n : T_n)$$

where the $A_i : T_i$'s are the attributes of R, each consisting of a name A_i and an associated data type T_i. This data type, among others, determines the domain dom_{T_i} consisting of the allowed values for the attribute. Each tuple

$$t_i(A_1 : v_1, \ldots, A_n : v_n)$$

with $v_i \in dom_{T_i}$, $1 \leq i \leq n$ represents a particular entity of the (real) world modelled by the given relation.

The essence of 'fuzzy' querying techniques is that they allow to express user preferences with respect to query conditions using linguistic terms which are modelled by fuzzy sets. The basic kind of preferences, considered in 'fuzzy' database querying, are those which are expressed *inside* an elementary query condition that is defined on a single attribute $A : T$. Hereby, fuzzy sets are used to express in a gradual way that some values of the domain dom_T are more desirable to the user than others. For example, if a user is looking for 'cheap houses', a fuzzy set with membership function μ_{cheap} on the domain of prices, as depicted in Fig. 1 can be used to reflect what the user understands by its linguistically expressed preference 'cheap house'.

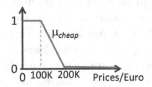

Fig. 1. The modelling of 'Cheap house prices'

During query processing, basically all relevant database tuples t are evaluated to determine whether they satisfy the user's preferences (to a certain extent) or not. Hereby, each elementary query criterion c_i, $i = 1, \ldots, m$ of the query is evaluated, resulting in an elementary matching degree $\gamma_{c_i}(t)$ which is usually modelled by a real number of the unit interval $[0, 1]$ (where $\gamma_{c_i}(t) = 1$ represents that the tuple t fully satisfies the criterion and $\gamma_{c_i}(t) = 0$ denotes no satisfaction). For example, evaluating the elementary criterion 'cheap price' for a tuple t with price attribute value $t[Price] = 110K$ results in an elementary satisfaction degree $\gamma_{cheap\_price}(t) = \mu_{cheap}(110K) = 0.9$, which expresses that a house of 110K satisfies the criterion 'cheap house' to an extent 0.9.

Next, the elementary degrees are aggregated to compute the overall matching degree $\gamma(t)$ of the tuple. In its simplest form, the aggregation of elementary matching degrees is determined by the fuzzy logical connectives conjunction, disjunction and negation which are respectively defined as follows:

$$\gamma_{c_1 \wedge c_2}(t) = i(\gamma_{c_1}(t), \gamma_{c_2}(t)) \tag{1}$$

$$\gamma_{c_1 \vee c_2}(t) = u(\gamma_{c_1}(t), \gamma_{c_2}(t)) \tag{2}$$

$$\gamma_{\neg c}(t) = 1 - \gamma_c(t) \tag{3}$$

where i and u resp. denote a t-norm and its corresponding t-conorm [12].

In a more complex approach, users are allowed to express their preferences related to the relative importance of the elementary conditions in a query, hereby indicating that the satisfaction of some query conditions is more desirable than the satisfaction of others. Such preferences are usually denoted by associating a relative weight w_i ($\in [0,1]$) to each elementary criterion c_i, $i = 1, \ldots, m$ of the query. Hereby, as extreme cases, $w_i = 0$ models 'not important at all' (i.e., should be omitted), whereas $w_i = 1$ represents 'fully important'. Assume that the matching degree of a condition c_i with an importance weight w_i is denoted by $\gamma_{c_i^*}(t)$. In order to be meaningful, weights are assumed to satisfy the following requirements [4]:

- In order to have an appropriate scaling, at least one of the associated weights has to be 1, i.e., $\max_i w_i = 1$.
- If $w_i = 1$ and the associated elementary matching degree for c_i equals 0, i.e., $\gamma_{c_i}(t) = 0$, then the weight's impact should be 0, i.e., $\gamma_{c_i^*}(t) = 0$.
- If $w_i = 1$ and $\gamma_{c_i}(t) = 1$, then $\gamma_{c_i^*}(t) = 1$.
- If $w_i = 0$, then the weight's impact should be such as if c_i does not exist.

The impact of a weight can be computed by first matching the condition as if there is no weight and then second modifying the resulting matching degree in accordance with the weight. A modification function that strengthens the match of more important conditions and weakens the match of less important conditions is used for this purpose. From a conceptual point of view, a distinction has been made between static weights and dynamic weights.

Static weights are fixed, known in advance and can be directly derived from the formulation of the query. These weights are independent of the values of the tuple(s) on which the query criteria act and are not allowed to change during query processing. As described in [4], some of the most practical interpretations of static weights can be formalised in a universal scheme. Namely, let us assume that query condition c is a conjunction of weighted elementary query conditions c_i (for a disjunction a similar scheme has been offered). Then the matching degree $\gamma_{c_i^*}(t)$ of an elementary condition c_i with associated implicative importance weight w_i is computed by

$$\gamma_{c_i^*}(t) = (w_i \Rightarrow \gamma_{c_i}(t)) \tag{4}$$

where $\Rightarrow$ denotes a fuzzy implication connective. The overall matching degree of the whole query composed of the conjunction of conditions c_i is calculated using a standard t-norm operator. Implicative weighting schemes in the context of information retrieval and weights that have the maximum value 1 were mostly investigated by Larsen in [13,14].

The approach for static weights, has been refined to deal with a dynamic, variable importance $w_i \in [0,1]$ depending on the matching degree of the associated elementary condition. Extreme (low or high) matching degrees could then for example result in an automatic adaptation of the weight.

A further, orthogonal distinction has been made between static weight assignments, where it is also known in advance with which condition a weight is associated (e.g., in a situation where the user explicitly states his/her preferences) and dynamic weight assignments, where the associations between weights and conditions depend on the actual attribute values of the record(s) on which the query conditions act (e.g., in a situation where most criteria have to be satisfied, but it is not important which ones). OWA operators [16] are an example of a technique with dynamic weight assignments.

Another aspect of 'fuzzy' querying concerns the aggregation of (partial) query conditions to be guided by a linguistic quantifier (see, e.g., [11,9]). In such approaches conditions of the following form are considered:

$$c = \Psi \text{ out of } \{c_1, \ldots, c_k\} \tag{5}$$

where Ψ is a linguistic (fuzzy) quantifier and c_i are elementary conditions to be aggregated. The overall matching degree $\gamma_c(t)$ of c can be computed in different ways. Commonly used techniques are for example based on liguistic quantifiers in the sense of Zadeh [17], OWA operators [16] and the Sugeno integral [1].

1.2   Problem Description

In many real-life situations users tend to group and structure their preferences when specifying selection criteria. Quite often, criteria are generalised or further specialised to obtain a better insight in what one is looking for. Such generalisations and specialisations then result in a hierarchically structured criteria specification, which will further on be called a *criterion tree*.

For example, for somebody who is searching for a house in a real estate database it is quite natural to require affordability (acceptable price and maintenance costs) and suitability (good comfort, good condition and a good location). Good comfort might be further specified by living comfort and basic facilities, where living comfort refers to at least two bathrooms, three bedrooms, garage etc. and basic facilities refer to gas, electricity, sewage, etc. Good condition might be specified by recent building and high quality building material. Finally, good location might be subdivided by accessibility, healthy environment, nearby facilities etc. The criterion tree corresponding to these user requirements is given in Fig. 2.

Query languages have no specific facilities for efficiently handling criterion trees. Indeed, criterion trees have to be translated to logical expressions, but for large criterion trees containing many criteria, this translation becomes difficult

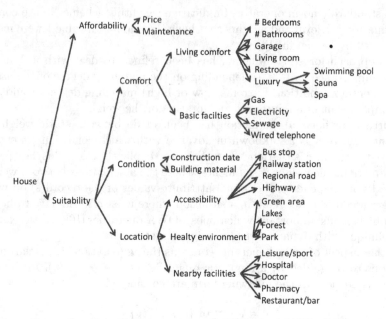

Fig. 2. Criterion tree for house selection

to interpret. Moreover, when working with weighted criteria, users often like to express preferences over subgroups of criteria. For example, a user might want to specify that the condition of a house is more important than its location. Such kinds of preferences require weight assignments to the internal nodes of a criterion tree. Translating such weights to weights for individual criteria is artificial and requires a significant effort as it becomes almost impossible to obtain a weight set which correctly reflects the preferences of the user. The latter even holds more generally: the more criteria we have to deal with, the more difficult it is to assign meaningful weights to the criteria [8]. Weight assignment is especially difficult for queries of high complexity which can contain hundreds of criteria and of which the weights should be easily adjustable. These kind of difficulties can be avoided by only considering manageable subsets of semantically related criteria to which weights are assigned in accordance with the user's preferences. Remark that humans use a similar approach when specifying complex requirements.

1.3 Objectives

In this paper we propose a novel flexible query specification and handling technique which is based on LSP (Logic Scoring of Preference) [5], a methodology which originates from decision support. The presented technique supports working with criterion trees and moreover allows for more flexibility in aggregating elementary degrees of satisfaction. The latter being obtained by providing the user with a selected number of generalized conjunction/disjunction (GCD) operators. Furthermore we illustrate that the use of criterion trees and GCD aggregation more adequately reflects the way how users reason while specifying their preferences related to their database search.

The remainder of the paper is organised as follows. In the next Section 2, criteria specification in criterion trees is discussed. The issues respectively dealt with are hierarchic query specification, weight specification and GCD selection. Next, the evaluation of a criterion tree is presented in Section 3. This evaluation is an important component of query processing and results in an associated overall satisfaction degree for each tuple that is relevant to the query result. In Section 4 we give an illustrative example based on house selection in order to justify the use of criterion trees with soft computing aggregation like GCD. Finally, in Section 5 the main contributions of the paper are summarised, conclusions are stated and some directions for future research are given.

2   Specification of Criterion Trees

A *criterion tree* is a hierarchical structure that is recursively defined as a collection of nodes starting at a root node. Each node can be seen as a container for information and can on its turn be connected with zero or more other nodes, called the child nodes of the node, which are one level lower in the tree hierarchy. A node that has a child is called the child's parent node. A node has at most one parent. A node that has no child nodes is called a leaf.

The leaf nodes of a criterion tree contain an elementary query condition c_A that is defined on a single database attribute $A : T$ as described in the introduction Section 1. This condition expresses the user's preferences related to the acceptable values for attribute $A : T$ in the answer set of the query.

All non-leaf nodes, i.e., the internal nodes, of a criterion tree contain a symbol representing an aggregation operator. Each child node n_i of a non-leaf node n has an associated weight w_i reflecting its relative importance within the subset of all child nodes of the non-leaf node. Hereby, for a non-leaf node with k child nodes it must hold that $\sum_{i=1}^{k} w_i = 1$. With this choice, we follow the semantics of the LSP methodology [5], which are different form those presented in [4].

Using Extended BNF (EBNF) notation [15], a criterion tree can be described by:

```
aggregator = "C" | "HPC" | "SPC" | "A" | "SPD" | "HPD" | "D"
criterion tree = elementary criterion | composed criterion
composed criterion = aggregator "(" criterion tree":"weight","
        criterion tree":"weight {"," criterion tree":"weight}")"
elementary criterion = attribute "IS {("min value"," suitability")"
    {",(" value"," suitability")" } ",("max value"," suitability")}"
```

where { } means 'repeat 0 or more times'. The values in elementary criterion must form a strictly increasing sequence.

The supported aggregators are denoted by 'C' (conjunction), 'HPC' (hard partial conjunction), 'SPC' (soft partial conjunction), 'A' (neutrality), 'SPD' (soft partial disjunction), 'HPD' (hard partial disjunction) and 'D' (disjunction).

This set is in fact a selection of seven special cases from the infinite range of generalized conjunction/disjunction (GCD) functions and can be easily extended when required.

The seven aggregators can be combined yielding nine combined aggregators as presented in Fig. 3.

Symbol	Aggregator
∨	Disjunction (D)
$\bar{\triangledown}$	Hard partial disjunction (HPD)
$\underline{\triangledown}$	Soft partial disjunction (SPD)
$\ominus$	Neutrality (A)
$\underline{\triangle}$	Soft partial conjunction (SPC)
$\bar{\triangle}$	Hard partial conjunction (HPC)
∧	Conjunction (C)
$\triangledown \in \{\bar{\triangledown}, \underline{\triangledown}\}$	Partial disjunction (PD)
$\triangle \in \{\bar{\triangle}, \underline{\triangle}\}$	Partial conjunction (PC)
$\bar{\bar{\triangledown}} \in \{\bar{\triangledown}, \vee\}$	Extended HPD (EHPD)
$\underline{\underline{\triangledown}} \in \{\underline{\triangledown}, \ominus\}$	Extended SPD (ESPD)
$\underline{\underline{\triangle}} \in \{\underline{\triangle}, \ominus\}$	Extended SPC (ESPC)
$\bar{\bar{\triangle}} \in \{\bar{\triangle}, \wedge\}$	Extended HPC (EHPC)
$\tilde{\triangledown} \in \{\underline{\triangledown}, \bar{\bar{\triangledown}}\}$	Total disjunction (TD)
$\tilde{\triangle} \in \{\underline{\triangle}, \bar{\bar{\triangle}}\}$	Total conjunction (TC)
$\Diamond \in \{\tilde{\triangledown}, \tilde{\triangle}\}$	Generalized conjunction/ disjunction (GCD)

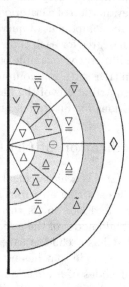

Fig. 3. Basic and combined simultaneity and replaceability operators

Two basic special cases of GCD are the partial conjunction (PC) and the partial disjunction (PD). Partial conjunction is a model of simultaneity, whereas partial disjunction is a model of replaceability. If we want to use GCD as an aggregator in a criterion tree, we have to select one of the supported aggregators based on the desired degree of simultaneity or replaceability.

Both 'C' and 'HPC' are models of high simultaneity and mandatory requirements. All inputs must be (partially) satisfied, and therefore they reflect mandatory requirements. If any input in an aggregated group of preferences is 0, the output is going to be 0. 'SPC' is also a model of simultaneity, but its (adjustable) level of simultaneity is lower than in the case of HPC. No input is mandatory. A single nonzero input is sufficient to produce a (small) nonzero output.

'D', 'HPD', and 'SPD' are models of replaceability symmetrical to 'C', 'HPC', and 'SPC'. 'D' and 'HPD' are models of high replaceability and sufficient requirements. If only one input is completely satisfied, that is sufficient to completely satisfy the whole group and the values of other inputs are insignificant. Each input can fully compensate (replace) all remaining inputs. 'SPD' is also a model of replaceability, but its (adjustable) level of replaceability is lower than in the case of HPD. No input is sufficient to completely satisfy the whole group, but any nonzero input is sufficient to produce a nonzero output.

The neutrality aggregator A (arithmetic mean) provides a perfect logic balance between simultaneity and replaceability. Thus, the logic interpretation of the arithmetic mean is that it represents a 50-50 mix of conjunctive and disjunctive properties; that is explicitly visible in the case of two inputs:

$$x_1 \theta x_2 = \frac{x_1 + x_2}{2} = \frac{(x_1 \wedge x_2) + (x_1 \vee x_2)}{2}. \tag{6}$$

For any number of inputs, all inputs are desired and each of them can partially compensate the insufficient quality of any other of them. No input is mandatory and no input is able to fully compensate the absence of all other inputs. In other words, the arithmetic mean simultaneously, with medium intensity, satisfies two contradictory requests: (1) to simultaneously have all good inputs, and (2) that each input has a moderate ability to replace any other input.

The arithmetic mean is located right in the middle of GCD aggregators but we cannot use it as a single best representative of all of them. The central location of the arithmetic mean is not sufficient to give credibility to additive scoring methods. Indeed, it is difficult to find an evaluation problem without mandatory requirements, or without the need to model various levels of simultaneity and/or replaceability. These features are ubiquitous and indispensable components of human evaluation reasoning. Unfortunately, these features are not supported by the arithmetic mean. Therefore, in the majority of evaluation problems the additive scoring represents a dangerous oversimplification because it is inconsistent with observable properties of human evaluation reasoning.

Once specified, criterion trees can be used in the specification of the WHERE-clause of a query. Their evaluation for a relevant database tuple t results in a criterion satisfaction specification, which can then be used in the further evaluation and processing of the query. In the next section, it is presented how criterion trees are evaluated.

3 Evaluation of Criterion Trees

Criterion trees are evaluated in a bottom-up way. This means that, when considering a relevant database tuple t, firstly, the elementary criteria c_i of the leaf nodes are evaluated. Any elementary criterion specification used in 'fuzzy' querying can be used. In its simplest form, c_i is specified by a fuzzy set F denoting the user's preferences related to an attribute $A : T$, as illustrated in Fig. 1. Criterion evaluation then boils down to determining the membership value of the actual value $t[A]$ of A for t, i.e.,

$$\gamma_{c_i}(t) = \mu_F(t[A]). \tag{7}$$

Next, all internal nodes (if any) are evaluated, bottom-up. An internal node n can be evaluated as soon as all its child nodes n_i, $i = 1, \ldots, k$ have been evaluated. For evaluation purposes, an implementation of GCD is required [7]. We can use the following implementation based on weighted power means (WPM):

$$M(x_1, \ldots, x_n; r) = \begin{cases} \left(\sum_{i=1}^n w_i x_i^r\right)^{1/r} & \text{, if } 0 < |r| < +\infty \\ \prod_{i=1}^n x_i^{w_i} & \text{, if } r = 0 \\ \min(x_1, \ldots, x_n) & \text{, if } r = -\infty \\ \max(x_1, \ldots, x_n) & \text{, if } r = +\infty \end{cases} \tag{8}$$

where $x_i \in [0,1]$, $1 \le i \le n$ are the input values which in the context of flexible querying represent satisfaction degrees (hereby, 0 and 1 respectively denote 'not satisfied at all' and 'fully satisfied'); the normalised weights $0 < w_i \le 1$, $1 \le i \le n$, $\sum_{i=1}^n w_i = 1$ specify the desired relative importance of the inputs and the computed exponent $r \in [-\infty, +\infty]$ determines the logic properties of the aggregator. Special cases of exponent values are: $+\infty$ corresponding to full disjunction 'D', $-\infty$ corresponding to full conjunction 'C', and 1 corresponding to weighted average 'A'. The other exponent values allow to model other aggregators, ranging continuously from full conjunction to full disjunction and can be computed from a desired value of orness (ω), and for this form of GCD function we can use the following numeric approximation [5]:

$$r(\omega) = \frac{0.25 + 1.89425x + 1.7044x^2 + 1.47532x^3 - 1.42532x^4}{\omega(1 - \omega)} \tag{9}$$

where

$$x = \omega - 1/2.$$

Suitable orness-values are the following: $\omega = 1/6$ for 'HPC', $\omega = 5/12$ for 'SPC', $\omega = 4/6$ for 'SPD' and $\omega = 5/6$ for 'HPD'.

Implication $w \Rightarrow x = \overline{w} \vee x = \overline{w} \wedge \overline{x}$ means that 'it is not acceptable that w is high and x is low', or 'important things must be satisfied'. The product wx used in Eq. (8) is also a form of implication because the effect is similar, i.e., again 'important things must be satisfied'.

Considering tuple t, the query satisfaction degree $\gamma_n(t)$ corresponding to n, computed using Eq. (8) with arguments $\gamma_{n_i}(t)$, $i = 1, \ldots, k$, w_i being the weight that has been associated with n_i, $i = 1, \ldots, k$, and r being the value that models the aggregator that is associated with n.

The overall satisfaction degree for tuple t using a criterion tree is obtained when the root node n_{root} of the tree is evaluated, i.e., this satisfaction degree yields

$$\gamma_{n_{root}}(t). \tag{10}$$

4   An Illustrative Example

As an example we reconsider the search for a house in a real estate database as presented in Fig. 2. Assuming that a user is looking for an affordable house with good comfort, good condition and a good location and wants to specify each of these subcriteria in more detail. Using GCD aggregators, the criterion tree given in Fig. 2 can be further detailed as shown in Fig. 4.

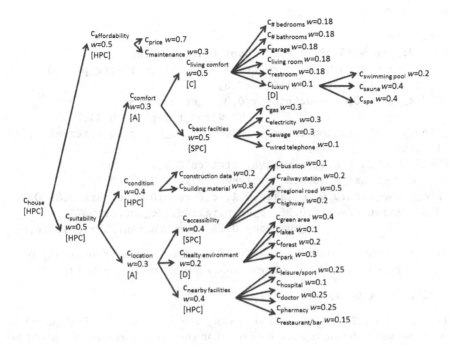

Fig. 4. Detailed criterion tree for house selection

Such a criterion tree can then be specified in an SQL statement that is used to query a regular relational database. Our approach is to use a predefined function $TREE$ which takes a criterion tree as argument and computes the overall satisfaction degree of the tuples being processed by the query. This is illustrated with the following query which includes a regular join condition and a condition tree *c_house*.

```
SELECT id, address, price, TREE(c_house) AS satisfaction
FROM real_estates r, location l
WHERE (r.location_id=l.id) AND satisfaction>0.5
ORDER BY satisfaction
```

The tree condition tree *c_house* is further specified by

```
c_house=HPC(c_affordability:0.5,c_suitability:0.5)
```

where

```
c_affordability=HPC(c_price:0.7, c_maintenance:0.3)
c_suitability=HPC(c_comfort:0.3, c_condition:0.4, c_location:0.3)
```

Furthermore,

```
c_comfort=
A(C(c_#bedrooms:0.18, c_#bathrooms:0.18, c_garage:0.18,
     c_living_room:0.18, c_restroom:0.18, D(c_swimming_pool:0.2,
     sauna:0.4, c_spa:0.4):0.1):0.5,
   SPC(c_gas:0.3, c_electricity:0.3, c_sewage:0.3,
                             c_wired_telephone:0.1):0.5)
c_condition=HPC(c_construction_date:0.2, c_building_material:0.8)
c_location=
A(SPC(c_bus_stop:0.1, c_railway_station:0.2,
        c_regional_road:0.5, c_highway:0.2):0.4,
   D(c_green_area:0.4, c_lakes:0.1, c_forest:0.2, c_park:0.3):0.2,
   HPC(c_leisure/sport:0.25, c_hospital:0.1, c_doctor:0.25,
                      c_pharmacy:0.25, c_restaurant/bar:0.15):0.4)
```

The elementary criteria can generally be handled using soft computing techniques as presented in Fig. 1. For example, c_price can be specified by

```
r.price IS {(100,1), (200,0)}
```

where $\{(100, 1), (200, 0)\}$ is used to specify the fuzzy set that is depicted in Fig. 1. So the criterion c_price denotes that the price of the house should be compatible with the linguistic term 'cheap'. Compatibility is then determined by the membership grade of the stored *price* value of the house under consideration. For the other elementary conditions, similar preference specifications can be provided. Once all elementary conditions are evaluated, the criterion tree for the house under consideration can be evaluated as described in Section 3 and the resulting value will be returned by the function $TREE$ (in the example labelled as $satisfaction$). If preferred, a basic condition acting as a threshold condition on the satisfaction degrees can be added in the WHERE-clause of the query ($satisfaction > 0.5$ in the example). The satisfaction degrees can also be used to rank the tuples in the query result.

5   Conclusions and Future Work

In this paper, we proposed the concept of a criterion tree. Criterion trees offer flexible facilities for specifying complex query conditions. More specifically, they provide adjustable, generalized aggregators and weights denoting relative preferences among (sub)criteria can be assigned to all non-root criteria of the tree. These are requirements for adequately reflecting human decision making, to the best of our knowledge not being considered in flexible querying up to now. The proposed work is currently being implemented within the framework of the open source PostgreSQL object-relational database system.

In the presented work, only basic GCD aggregators have been considered. However, it is clear that not all criterion specifications necessary to reflect human reasoning can be modelled using the current approach. Therefore, the current work has to be extended with other aggregators, what will be subject to future work. One such extension concerns the handling of bipolarity and the ability to deal with mandatory, desired and optional conditions.

References

1. Bosc, P., Lietard, L., Pivert, O.: Sugeno fuzzy integral as a basis for the interpretation of flexible queries involving monotonic aggregates. Information Processing and Management 39(2), 287–306 (2003)
2. Cattell, R.G.G., Barry, D.K. (eds.): The Object Data Standard: ODMG 3.0. Morgan Kaufmann, San Francisco (2000)
3. Codd, E.F.: A Relational Model of Data for Large Shared Data Banks. Communications of the ACM 13(6), 377–387 (1970)
4. Dubois, D., Prade, H.: Using fuzzy sets in flexible querying: why and how? In: Andreasen, T., Christiansen, H., Larsen, H.L. (eds.) Flexible Query Answering Systems. Kluwer Academic Publishers, Dordrecht (1997)
5. Dujmović, J.J.: Preference Logic for System Evaluation. IEEE Transactions on Fuzzy Systems 15(6), 1082–1099 (2007)
6. Dujmović, J.J., Larsen, H.L.: Generalized conjunction/disjunction. Int. Journal of Approximate Reasoning 46, 423–446 (2007)
7. Dujmović, J.J.: Characteristic Forms of Generalized Conjunction/Disjunction. In: Proc. IEEE World Congress on Computational Intelligence, Hong Kong (2008)
8. Dujmović, J.J., De Tré, G.: Multicriteria Methods and Logic Aggregation in Suitability Maps. Int. Journal of Intelligent Systems 26(10), 971–1001 (2011)
9. Galindo, J., Medina, J.M., Cubero, J.C., Garcia, M.T.: Relaxing the Universal Quantifier of the Division in Fuzzy Relational Databases. Int. Journal of Intelligent Systems 16(6), 713–742 (2001)
10. ISO/IEC 9075-1:2011: Information technology – Database languages – SQL – Part 1: Framework (SQL/Framework) (2011)
11. Kacprzyk, J.: Ziółkowski, A.: Database queries with fuzzy linguistic quantifiers. IEEE Transactions on Systems, Man and Cybernetics 16, 474–479 (1986)
12. Klement, E.P., Mesiar, R., Pap, E. (eds.): Triangular Norms. Kluwer Academic Publishers, Boston (2000)
13. Larsen, H.L.: Efficient Andness-directed Importance Weighted Averaging Operators. Int. Journal of Uncertainty, Fuzziness and Knowledge-Based Systems 12(suppl.), 67–82 (2003)
14. Larsen, H.L.: Importance weighting and andness control in De Morgan dual power means and OWA operators. Fuzzy Sets and Systems 196(1), 17–32 (2012)
15. Wirth, N.: What Can We Do About the Unnecessary Diversity of Notation for Syntactic Definitions. Communications of the ACM 20(11), 822–823 (1977)
16. Yager, R.R., Kacprzyk, J.: The Ordered Weighted Averaging Operators: Theory and Applications. Kluwer Academic Publishers, Norwell (1997)
17. Zadeh, L.A.: A computational approach to fuzzy quantifiers in natural languages. Computational Mathematics Applications 9, 149–184 (1983)
18. Zadrozny, S., De Tré, G., De Caluwe, R., Kacprzyk, J.: An Overview of Fuzzy Approaches to Flexible Database Querying. In: Galindo, J. (ed.) Handbook of Research on Fuzzy Information Processing in Databases, pp. 34–54. IGI Global, Hershey (2008)

A Possibilistic Logic Approach
to Conditional Preference Queries

Didier Dubois, Henri Prade, and Fayçal Touazi

IRIT, CNRS & University of Toulouse, France
{dubois,prade,faycal.touazi}@irit.fr

Abstract. The paper presents a new approach to deal with database preference queries, where preferences are represented in the style of possibilistic logic, using symbolic weights. The symbolic weights may be processed without the need of a numerical assignment of priority. Still, it is possible to introduce a partial ordering among the symbolic weights if necessary. On this basis, four methods that have an increasing discriminating power for ranking the answers to conjunctive queries, are proposed. The approach is compared to different lines of research in preference queries including skyline-based methods and fuzzy set-based queries. With the four proposed ranking methods the first group of best answers is made of non dominated items. The purely qualitative nature of the approach avoids the commensurability requirement of elementary evaluations underlying the fuzzy logic methods.

1 Introduction

One may consider that there are two main research trends in the preference queries literature, namely the fuzzy set-based approach [1,2] on the one hand, and skyline methods [3,4,5] on the other hand. Besides, in artificial intelligence, CP-nets [6] for conditional preference statements developed in the last decade have become a popular setting. It is based on a graphical representation, and obeys the ceteris paribus principle. Its potential use for dealing with preference queries has even been stressed [7]. Besides, the use of possibilistic logic for the modeling of preferences queries has been advocated more recently [8,9].

Fuzzy sets have been often proposed for the modeling of flexible queries as it provides a basis for rank-ordering the retrieved items. However, this requires the specification of membership functions, possibly of priority weights, and more importantly it is based on the implicit assumption of the commensurability of the elementary evaluations. Skyline methods single out non dominated elements, but do not rank-order them (if the query is not iterated on the remaining items), up to a few exceptions [10]. In this paper, we investigate the use of a possibilistic logic approach to the handling of data base (conditional) preference queries, which remains as symbolic as possible, but preserves a capability for rank-ordering the answers.

The paper is organized as follows. First, a short background on possibilistic logic, and the use of symbolic weights is provided in Section 2. Then, a running example is

H.L. Larsen et al. (Eds.): FQAS 2013, LNAI 8132, pp. 376–388, 2013.

proposed, that will be used for comparing the different methods discussed in the paper. Section 3 presents four methods for rank-ordering query answers, with an increasing discriminating power. The first method handles preferences as conditional possibility constraints. The second method, which proves to be more refined, relies on an inclusion-based ordering. In the third and fourth methods, additional constraints are added between the symbolic weights of the possibilistic logic encoding, in the spirit of CP-nets and CP-theories [11] (a generalization of CP-nets) respectively. Then Section 4 briefly compares our proposal to related work on skyline and fuzzy set approaches.

2 Technical Prerequisites and Running Example

We consider a propositional language where formulas are denoted by $p_1, ..., p_n$, and Ω denotes its set of interpretations. The logical conjunctions, disjunctions and negations are denoted by $\land$, $\lor$ and $\neg$, respectively.

2.1 Possibilistic Logic

Let $B^N = \{(p_j, \alpha_j) \mid j = 1, \ldots, m\}$ be a possibilistic logic base where $\alpha_j \in \mathcal{L} \subseteq [0, 1]$ is a priority level attached to formula p_i [12]. Each formula (p_j, α_j) means that $N(p_j) \geq \alpha_j$, where N is a necessity measure, i.e., a set function satisfying the property $N(p \land q) = \min(N(p), N(q))$. A necessity measure is associated to a possibility distribution π on the set of interpretations, as follows:

$$N(p) = \min_{\omega \notin M(p)} (1 - \pi(\omega)) = 1 - \Pi(\neg p),$$

where Π is the possibility measure associated to N and $M(p)$ is the set of models induced by the underlying propositional language for which p is true.

The base B^N is associated to the possibility distribution on interpretations:

$$\pi_B^N(\omega) = \min_{j=1,\ldots,m} \pi_{(p_j, \alpha_j)}(\omega) \tag{1}$$

where $\pi_{(p_j, \alpha_j)}(\omega) = 1$ if $\omega \in M(p_j)$, and $\pi_{(p_j, \alpha_j)}(\omega) = 1 - \alpha_j$ if $\omega \notin M(p_j)$. An interpretation ω is all the more possible as it does not violate any formula p_j having a higher priority level α_j. So, if $\omega \notin M(p_j)$, $\pi_B^N(\omega) \leq 1 - \alpha_j$, and if $\omega \in \bigcap_{j \in J} M(\neg p_j)$, then $\pi_B^N(\omega) \leq \min_{j \in J}(1 - \alpha_j)$. It is a description "from above" of π_B^N, which is the least specific possibility distribution in agreement with the knowledge base B^N.

2.2 Symbolic Weights

The weights associated to possibilistic logic formulas, which can be understood as priority or certainty levels, may be processed symbolically. By that, we mean that we are not assigning a value to the weights. So doing, we in general lose the benefit of the total ordering existing between values in a scale. Still, a partial ordering remains between symbolic expressions, e.g., we do know that $\min(\alpha, \beta) \leq \alpha$ whatever the values of α and β. Moreover, one may introduce some further constraints between symbolic weights, when available, e.g., $\alpha > \beta$, and exploit them. This includes the particular case where one knows the complete ordering between all the symbolic weights introduced. Generally speaking, when several constraints are introduced, it is important to make sure that they are consistent.

Since one may not know precisely how imperative preferences are in general, it is convenient to handle weights in a symbolic manner, without having to assign precise values. Having symbolic weights still allows us to construct a vector for each outcome that will serve in their rank-ordering. Indeed, a query will be viewed as a (weighted) conjunction of logical formulas, and items in the database are then rank-ordered according to the level to which they satisfy this conjunction. Then, the vector components are nothing but the arguments of the min in equation (1) defining the semantics of a possibilistic base made of formulas (p_j, α_j) expressing goals and their importance. In this paper we explain how these vectors are obtained, and discuss how vectors can be ordered on this basis.

2.3   Running Example

Throughout the paper, we will use the following running example in order to illustrate the proposed approach to preference queries. This example is implemented on an experimental platform in information processing developed at IRIT in Toulouse (*http:/www.irit.fr/PRETI*) (see [13]). The data base stores pieces of information about houses to let that are described in terms of 25 attributes.

Example 1. *We want to express the following preferences:*

- *The number of persons accommodated should be more than 10, imperatively;*
- *It is preferred to have a house where animals are allowed,*
- *It is preferred to be close to the sea by a distance between 1 and 20 km;*
- *If the house is far from the sea by more than 20 km, it is preferred to have a tennis court at less than 4 km*
- *If moreover the distance of the house to the tennis court is more than 4 km, it is desirable to have a swimming pool be at a distance less than 6 km*

These preference constraints can be encoded by the following possibilistic logic formulas. Indeed, in our approach any query is represented by a possibilistic logic base. Here there is one imperative constraint, the other constraints being soft.

 - Hard preference constraint
    - $\phi_0 = (Accomod. \geq 10, 1)$
 - Soft preference constraints
    - $\phi_1 = (Animal, \alpha_1)$
    - $\phi_2 = (1 \leq Sea \leq 20, \alpha_2)$
    - $\phi_3 = (\neg(Sea > 20) \vee Tennis \leq 4, \alpha_3)$
    - $\phi_4 = (\neg(Sea > 20) \vee \neg(Tennis \leq 4) \vee Pool \leq 5, \alpha_4)$

3   Handling Preference Queries

What makes the possibilistic logic setting particularly appealing for the representation of preferences is not only the fact that the language incorporates priority levels explicitly, but the existence of different representation formats [14,15], equally expressive

[16,17], even if more or less natural or suitable for expressing preferences. Namely, preferences can be represented as prioritized goals, i.e. possibilistic formulas, or in terms of conditionals (i.e. statements of the form if p is true then having q true is preferred to having it false), or even as a Bayesian-like networks, since a possibilistic logic base can be encoded either as a qualitative or a quantitative possibilistic network and vice-versa [18]. In the next subsection, we recall how conditional preferences are represented in possibilistic logic with symbolic weights. Then in the three next subsections, different ways of handling symbolic priorities for processing the queries are discussed.

3.1 Preference Encoding in Possibilistic Logic

The unconditional preference of the form "q is preferred to $\neg q$" may be understood in the possibility theory setting as the constraint $\Pi(q) > \Pi(\neg q)$, which expresses that at least one model of q is preferred to any interpretation that makes q false. More generally, the possibilistic encoding of conditional preferences of the form "in context p, q is preferred to $\neg q$" is a constraint of the form $\Pi(p \wedge q) > \Pi(p \wedge \neg q)$. This includes the previous case where p is a tautology. Using conditioning, this constraint is still equivalent to $\exists \alpha$ s. t. $N(q|p) \geq \alpha > 0$, where $N(q|p) = 1 - \Pi(\neg q|p)$, such that $\Pi(r|p) = 1$ if $\Pi(p \wedge r) \geq \Pi(p \wedge \neg r)$ and $\Pi(r|p) = \Pi(p \wedge r)$ otherwise.

This constraint can be encoded by the possibilistic formula $(\neg p \vee q, \alpha)$, which expresses the requirement $N(\neg p \vee q) \geq \alpha$, which is itself equivalent here to the above constraint on the conditional necessity measure $N(q|p) \geq \alpha$ (see, e.g., [12]).

More generally, if we need to consider more than two mutually exclusive alternatives, this can be encoded by means of several possibilistic formulas. For instance, the two weighted formulas $\{(\neg p \vee q \vee r, 1), (\neg p \vee q, \alpha)\}$ state that if p is true, it is imperative to have $q \vee r$, and that q is preferred to $r \wedge \neg q$ since $\alpha > 0$. This extends to n alternatives. For instance, $\{(\neg p \vee q \vee r \vee s, 1), (\neg p \vee q \vee r, \alpha), (\neg p \vee q, \beta)\}$ with $\beta < \alpha < 1$ says that in context p, one wants to have q true, or at least r true, or at least s true; see [9] for further discussions.

In the next subsections, we shall exhibit different methods to rank-order outcomes based on a possibilistic logic base encoding preferences, but we first discuss the direct exploitation of constraints of the form $\Pi(p \wedge q) > \Pi(p \wedge \neg q)$.

3.2 Weak Comparative Preferences

The handling of a set of possibilistic constraints ϕ_i of the form $\Pi(p_i \wedge q_j) > \Pi(p_i \wedge \neg q_j)$ amounts here to looking for the largest possibility distribution π compatible with these constraints applying the minimum specificity principle, see, e.g., [14]. The largest solution π, which always exists if the set of constraints is consistent, can be computed using Algorithm 1 below, and represents a preference profile that rank-orders interpretations, in agreement with the preference requirements. The minimal specificity principle ensures that all the constraints are satisfied, but only these constraints are (in other words, no extra preferences are introduced). In the algorithm, the possibility distribution π is represented as a well-ordered partition $(E_1, ..., E_m)$ of Ω, associated with the ordering $\succ_{WCP}$ such that: $\forall \omega, \omega' \in \Omega, \omega \succ_{WCP} \omega'$ iff $\omega \in E_i, \omega' \in E_j$ and $i < j$.

The well-ordered partition given in Algorithm 1 satisfies the minimum specificity principle. The most satisfactory set E_1 is made of the interpretations that satisfy some $L(\phi_i)$, and do not satisfy any $R(\phi_j)$. Then the set of constraints whose left part is satisfied by an interpretation of E_1 are deleted, and the procedure is iterated on the remaining constraints as long as there are some. This procedure yields a possibility distribution

---

Algorithm 1. Minimal specificity ranking algorithm

Require: Φ a set of constraints ϕ_i of the form: $\Pi(L(\phi_i)) > \Pi(R(\phi_i))$
    E_i a set of interpretations classified at the ith rank
    Ω the set of all interpretations
    m=1;
    while $\Phi \neq \emptyset$ **do**
        Put in E_m any interpretation ω_l that satisfies some $L(\phi_i)$, and does not satisfy any $R(\phi_j)$
        if $E_m = \emptyset$ **then**
            The preference base is inconsistent
        else
            Delete E_m from Ω
            Delete all pairs $(L(\phi_i), R(\phi_i))$ from Φ such that $L(\phi_i)$ is satisfied by at least one element of E_m
            m=m+1;
        end if
    end while
    return $\{E_1, \cdots, E_m\}$

---

whose number of values is at most $n + 1$ where n is the number of constraints. Indeed, it is clear that at least one constraint is deleted at each iteration step.

Proposition 1. *Let a query Q composed of n preference constraints. The number of elements of the well-ordered partition $\{E_1, \cdots, E_m\}$ produced by Algorithm 1 is at most $m = n + 1$.*

Example 2. *Let 4 preference constraints be given as follows ($L(\phi_i), R(\phi_i)$ are replaced by sets of interpretations): $\phi_1 = (\{t_1, t_2, t_3\}, \{t_4, t_5, t_6, t_7, t_8\})$; $\phi_2 = (\{t_4, t_5\}, \{t_6, t_7, t_8\})$; $\phi_3 = (\{t_6\}, \{t_7\})$; $\phi_4 = (\{t_7\}, \{t_8\})$. Applying Algorithm 1 gives 5 preference levels: $E_1 = \{t_1, t_2, t_3\}$, $E_2 = \{t_4, t_5\}$, $E_3 = \{t_6\}, E_4 = \{t_7\}$ and $E_5 = \{t_8\}$.*

Example 1 (continued). The imperative preference constraint (ϕ_1) restricts the list to 15 houses. Considering the preference constraints in the running example the result of applying Algorithm 1 to constraints of the form $\Pi(\phi_i) > \Pi(\neg\phi_i), i = 1, \ldots, 4$ is given in Table 1 where $m = 2$ and $E_1 = \{539\}$. As there is no conflict detected by the algorithm, there are only two classes of outcomes.

3.3   Lexicographic Comparaison

We now consider a possibilistic logic encoding of the preference requirements, i.e., a possibilistic logic base Σ. For each interpretation ω, we can build a vector $\omega(\Sigma)$ in the following way, for each preference constraint ϕ_i for $i = 1, \cdots, n$:

Table 1. Weak comparison preference ranking

Id	Sea	Animal	Tennis	Rank
539	4.00	1	3.00	1
119	100	0	0.50	2
191	100	1	0.60	2
261	83.00	1	4.50	2
320	23.00	1	1.50	2
339	100	1	8.50	2
366	100	0	0.50	2
434	100	0	3.50	2
435	89.00	1	6.50	2
507	83.00	1	4.00	2
519	58.00	0	1.50	2
530	100	1	0.50	2
536	83.00	0	1.50	2

Table 2. Lexicographic ranking

Id	ϕ_0	ϕ_1	ϕ_2	ϕ_3	ϕ_4	Rank
539	1	1	1	1	1	1
191	1	1	$1-\alpha_2$	1	1	2
320	1	1	$1-\alpha_2$	1	1	2
530	1	1	$1-\alpha_2$	1	1	2
119	1	$1-\alpha_1$	$1-\alpha_2$	1	1	3
261	1	1	$1-\alpha_2$	$1-\alpha_3$	1	3
339	1	1	$1-\alpha_2$	1	$1-\alpha_4$	3
366	1	$1-\alpha_1$	$1-\alpha_2$	1	1	3
434	1	$1-\alpha_1$	$1-\alpha_2$	1	1	3
435	1	1	$1-\alpha_2$	$1-\alpha_3$	1	3
507	1	1	$1-\alpha_2$	$1-\alpha_3$	1	3
519	1	$1-\alpha_1$	$1-\alpha_2$	1	1	3
536	1	$1-\alpha_1$	$1-\alpha_2$	1	1	3

- if ω satisfies ϕ_i we put '1' in the ith component of the vector;
- otherwise, we put $1 - \alpha_i$ (α_i is the weight associated to preference constraint ϕ_i).

in agreement with the minimally specific possibility distribution π associated with Σ (see Section 2). Indeed, since we are dealing with symbolic weights, we cannot compute the result of the min operator aggregation of the vector components. So, we keep the vectors as they are, and order them using the classical lexicographic ordering, see, e.g., [12], thus defining an order denoted $\succ_{leximin}$ between vectors.

In the standard case of a totally ordered scale, the *leximin* order is defined by first reordering the vectors in an increasing way, and then applying the min order to the subparts of the reordered vectors without identical components. Since we deal with a partial order over the priority weights (at least, we know that $1 > 1 - \alpha_i, \forall i$, and $1 - \alpha_i \geq \min(1 - \alpha_i, 1 - \alpha_j)$ and so on), the reordering of vectors is no longer unique, and we have to generalize the definition in the following way:

Definition 1 (leximin). *Let v and v' be two vectors having the same number of components. First, delete all pairs (v_i, v'_j) such that $v_i = v'_j$ in v and v' (each deleted component can be used only one time in the deletion process). Thus, we get two non overlapping sets $r(v)$ and $r(v')$ of remaining components, namely $r(v) \cap r(v') = \emptyset$. Then, $v \succ_{leximin} v'$ iff $\min(r(v) \cup r(v')) \subseteq r(v')$ (where $\min$ here returns the set of minimal elements of the partial order between the priority weights).*

Example 1 (continued). When applying the possibilistic logic semantics to query evaluation, we deal not only with interpretations, but also with items (several items may correspond to the same interpretation of the requirement). Thus, considering the house with id 339, its associated vector is $v(339) = (1, 1, 1 - \alpha_2, 1, 1 - \alpha_4)$ (see Table 2). The house satisfies the two first preference constraints (number of people accommodated and animals allowance) and also satisfies the 5th preference concerning the distance to a swimming pool. But it falsifies the preference about distance to the sea (it is 10 km far), and it falsifies the preference about the distance to a tennis court. Now let us compare this house with the house with id 292 and vector $v(292) = (1, 1, 1 - \alpha_2, 1, 1)$,

applying the *leximin* order. Then, the reduced associated vectors have one compo-
nent here: $r(292) = (1)$ and $r(339) = (1 - \alpha_4)$. Then, we have $\min(r(292) \cup r(339)) = \{1 - \alpha_4\} \subseteq r(339)$. So, $v(292) \succ_{leximin} v(339)$, and by extension, we
write $house_{292} \succ_{leximin} house_{339}$.

Considering all the items in the running example, the result of the lexicographic
comparison over the 15 houses is given in Table 2.

One can observe that $\succ_{leximin}$ may induce up to $n + 1$ layers, since an item may
violate $0, 1, \cdots,$ or n preference constraints. Indeed, items are ranked according to the
number of preferences violated.

Proposition 2. *If a query Q is composed of n preference constraints, then the maximal
number of levels generated by $\succ_{leximin}$ is $n + 1$.*

Contrary to what Tables 1 and 2 suggest, $\succ_{leximin}$ does not refine $\succ_{WCP}$ as the latter generally introduces constraints between weights that are not present in the method
of this section. However in the running example, the WCP is equivalent to applying
classical logic, ending up in two classes of interpretations only. In that special case,
$\succ_{leximin}$ trivially refines $\succ_{WCP}$, since then the latter separates outcomes ω that satisfy all constraints from those that violate at least one of them, while $\succ_{leximin}$ always
classifies outcomes in terms of the number of violated constraints. However, $\succ_{leximin}$
does not use priorities induced by the WCP approach (Algorithm 1). Of course it is
also possible to refine the ordering of outcomes induced by WCP using $\succ_{leximin}$, or
equivalently to refine the $\succ_{leximin}$ with symbolic weights, by exploiting the priorities
found by Algorithm 1.

4   Adding Constraints between Symbolic Weights

In the previous subsection, the partial order between priority weights, underlying the
use of the lexicographic comparison, was not requiring any information on the relative
values of the symbolic weights associated with the preference requirements. It should be
clear that the lexicographic ordering between vectors (and thus between interpretations,
and between items) will be refined by the knowledge of some additional information on
the relative importance of requirements. For instance, if being not too far from a tennis
court is less important than being somewhat close to the sea, then we can enforce in
comparisons that $\alpha_i > \alpha_j$, where α_i, α_j are the respective weights associated to the
tennis and sea requirements. However, it is important to keep in mind that when we
consider two possibilistic logic formulas (φ, α) and (ψ, β) such that $\varphi \vDash \psi$ then we
should have $\beta \geq \alpha$. This is in agreement with the fact that if one requires $\varphi = \neg p \vee q$
and $\psi = \neg p \vee q \vee r$ (i.e. in context p, q must be true, or at least r), satisfying φ cannot be
more important than satisfying ψ if we do not want to trivialize the latter requirement
ψ (since satisfying φ entails satisfying ψ). To ensure this kind of coherence property,
one may compute the degree to which each requirement is entailed by the other ones
(which may result in attaching to formulas symbolic expressions involving max and
min of other symbolic weights).

If no extra information is available between priorities, one may apply some general
principle for introducing inequalities between symbolic weights. In the following we

discuss two options that enable us to obtain a more refined leximin-based ordering. The first option is inspired from the CP-net representation of preferences [6], and the second one from its refinement in terms of CP-theories [11].

4.1 Constraints between Weights in CP-net Style

This method is inspired from CP-nets, a well-known framework for representing preferences in AI [6]. It is a graphical representation that exploits conditional preferential independence in structuring the preferences provided by a user. These preferences take the form $u : x_i > \neg x_i$, i.e., x is preferred to $\neg x$ in context u, (u can be tautological). CP-nets are underlain by a ceteris paribus principle that amounts to giving priority to preferences attached to parent nodes over preferences attached to children nodes in the CP-net structure. Besides, it has been noticed that a CP-net ordering can be approximated by a possibilistic logic representation with symbolic weights [19,8]. The priority in favor of father nodes carries over to the possibilistic setting in the following way. For each pair of formulas of the form $(\neg u \vee x_i, \alpha_i)$ and $(\neg u \vee \neg x_i \vee x_j, \alpha_j)$, x_i plays the role of the father of x_j in a CP-net. Indeed, the first formula expresses a preference in favor of having x_i true (in context u), while in the second formula the context is refined from u to $u \wedge x_i$, which establishes a particular type of links between the two formulas where the second formula is in some sense a descendant of the first one. Then, the following constraint between the corresponding weights is applied $\alpha_i > \alpha_j$, in a CP-net spirit. These constraints between symbolic weights can be obtained systematically by Algorithm 2, which computes the partial order between symbolic weights from a possibilistic logic base. Applying this procedure allows us to add constraints among symbolic weights and to get a more refined ranking of items, as we notice in the following example.

Algorithm 2. Relative importance between possibilistic formulas in a CP-net spirit

---

Require: C a set of constraints of the form (p_i, α_i)
    CBW=$\emptyset$: the set of constraints between weights
    for c_j in C **do**
        if c_j is of the form (u_i, α_j) **then**
            for c_k in C **do**
                if c_k is of the form $(\neg u_i \vee x_i, \alpha_k)$ **then**
                    CBW $\leftarrow$ CBW $\cup \{\alpha_j > \alpha_k\}$
                end if
            end for
        end if
    end for
    return CBW

Example 1 (continued). Considering the preference constraints in the running example, the result of the lexicographic comparison of vectors adding the CP-nets-like constraints between weights, namely here $\alpha_2 > \alpha_3$ and $\alpha_3 > \alpha_4$, is given in Table 3, where a more refined ranking is obtained. In particular, house 339 is preferred to house 261,

Table 3. Lexicographic ranking with additional **Table 4.** Lexicographic ranking with additional CP-net constraints tional CP-theory constraints

Id	Sea	Animal	Tennis	Weights	Rank
539	4.00	1	3.00	1	1
191	100	1	0.60	$1\text{-}\alpha_2$	2
320	23.00	1	1.50	$1\text{-}\alpha_2$	2
530	100	1	0.50	$1\text{-}\alpha_2$	2
119	100	0	0.50	$1\text{-}\alpha_1,1\text{-}\alpha_2$	3
339	100	1	8.50	$1\text{-}\alpha_2,1\text{-}\alpha_4$	3
366	100	0	0.50	$1\text{-}\alpha_1,1\text{-}\alpha_2$	3
434	100	0	3.50	$1\text{-}\alpha_1,1\text{-}\alpha_2$	3
519	58.00	0	1.50	$1\text{-}\alpha_1,1\text{-}\alpha_2$	3
536	83.00	0	1.50	$1\text{-}\alpha_1,1\text{-}\alpha_2$	3
261	83.00	1	4.50	$1\text{-}\alpha_2,1\text{-}\alpha_3$	4
435	89.00	1	6.50	$1\text{-}\alpha_2,1\text{-}\alpha_3$	4
507	83.00	1	4.00	$1\text{-}\alpha_2,1\text{-}\alpha_3$	4

Id	Sea	Animal	Tennis	Weights	Rank
539	4.00	1	3.00	1	1
191	100	1	0.60	$1\text{-}\alpha_2$	2
320	23.00	1	1.50	$1\text{-}\alpha_2$	2
530	100	1	0.50	$1\text{-}\alpha_2$	2
339	100	1	8.50	$1\text{-}\alpha_2,1\text{-}\alpha_4$	3
261	83.00	1	4.50	$1\text{-}\alpha_2,1\text{-}\alpha_3$	4
435	89.00	1	6.50	$1\text{-}\alpha_2,1\text{-}\alpha_3$	4
507	83.00	1	4.00	$1\text{-}\alpha_2,1\text{-}\alpha_3$	4
119	100	0	0.50	$1\text{-}\alpha_1,1\text{-}\alpha_2$	4
366	100	0	0.50	$1\text{-}\alpha_1,1\text{-}\alpha_2$	5
434	100	0	3.50	$1\text{-}\alpha_1,1\text{-}\alpha_2$	5
519	58.00	0	1.50	$1\text{-}\alpha_1,1\text{-}\alpha_2$	5
536	83.00	0	1.50	$1\text{-}\alpha_1,1\text{-}\alpha_2$	5

435, and 507 since $1 - \alpha_4 > 1 - \alpha_3$. Houses 119, 366, 434, 519, 536 are clearly not as good as houses 539, 191, 320, and 530; moreover they can be compared with neither house 339, nor with houses 261, 435, and 507 (since α_1 cannot be compared with α_4 or α_3); this is why houses 119, 366, 434, 519, 536 are put in the highest possible layer: i.e., below the lowest one where houses are preferred to them and in the highest one where there is an incomparable item.

4.2 Constraints between Weights in CP-Theories Style

CP-theories as introduced in [11], are a generalization of CP-nets. Also based on a graphical representation, CP-theories offer a more expressive language where preference priority can be made explicit between the preference constraints. Thus, such constraints have the same form as in CP-nets $u : x > \neg x$ [W]; in addition we have the set of variables (attributes) W for which it is known that the preference associated to x does not depend on any value assignment of an attribute in W (i.e., the preference attached to the concerned attribute holds *irrespective* of values of attributes in W). It has been suggested that possibilistic logic is able to approximate this representation by adding more priority constraints over the symbolic weights [20]. Formally, a possibilistic preference constraint of the form $u : x > \neg x$ [W], with an *irrespective* requirement w. r. t. variables in W is encoded by a possibilistic preference statement $(\neg u \vee x, \alpha_i)$, to which we shall add the constraint $\alpha_i > \alpha_j$ for any α_j over symbolic weights, such that $(\neg u \vee w, \alpha_j)$ is a possibilistic preference statement, with the same context u, over one variable (or more) $w \in W$. These constraints over weights can be obtained by Algorithm 3.

Example 1 (continued). We consider the preference constraints in the running example. In addition, it is natural to assume the preference for animals allowance holds irrespectively of the preference concerning the distance to the sea ($\top : Animals > \neg Animals$ [Sea]). Then, Algorithm 3 yields $\alpha_1 > \alpha_2 > \alpha_3 > \alpha_4$ and the result of applying lexicographic comparison over the 15 houses, is given in Table 4, which leads here to an even more refined ranking. We can establish that the different ranking procedures discussed so far agree on the best selected items. Besides, we can show that the maximal number of layers induced by the lexicographic ordering may be greatly increased by the presence of additional constraints:

Proposition 3. *Let E^1 denote the set of non dominated models of a consistent possibilistic base. This set remains unchanged under the weak comparative preference ordering and the lexicographic ordering (in the presence of additional constraints or not).*

$$E^1_{\succ WCP} = E^1_{\succ leximin}$$

Proof of Proposition 3: Let $E^1_{\succ WCP}$ be the set of non dominated interpretations obtained by the weak comparative preferences method. We know that $\omega \in E^1_{\succ WCP}$ if only if ω satisfies all preference constraints. Let $\omega(\Sigma)(1, \cdots, 1)$ be the vector associated to ω. It is clear that any item that has an associated vector only made of '1' components is preferred to any other vector containing at least one component $\neq$'1' according to *leximin* order. So $\omega \in E^1_{\succ leximin}$. $\qquad\qquad\square$

Algorithm 3. Relative importance between possibilistic formulas in a CP-theories spirit

---

Require: C a set of constraints of the form (P_i, α_i)
    for $c_i \in C$ a preference constraint associated with an irrespective requirement **do**
        if $W_i = \emptyset$ **then**
            CBW $\leftarrow$ CBW $\cup$ **Algorithm 2**
        else
            for c_j in C **do**
                if c_j is of the form $(\neg u_i \lor \neg x_i \lor v, \alpha_j)$ or $(\neg u_i \lor z, \alpha_j)/z \in W_i, v \in \{V - U\}$
    then
                CBW $\leftarrow$ CBW $\cup (\alpha_i > \alpha_j)$
            end if
        end for
        end if
    end for
    return CBW

4.3 Hybridizing Weak Comparative Preferences and Lexicographic Methods

As shown in the previous subsections, the above three leximin-based methods lead to different, but compatible rankings of items, with increasing discrimination. The respective complexities of the comparative preferences and lexicographic methods are *Polynomial* and $\Pi^P_2 - complete$ [21,22]. Indeed, it can be observed in practice that the lexicographic method is more costly from a computational point of view:

Proposition 4. *Let Q be a query made of n preference constraints, then the maximal number of levels generated by $\succ_{leximin}$ with additional constraints over weights is 2^n.*

Indeed, this number of layers is obtained by refining the ordering of items violating the same number of preferences in case of a total ordering of the weights. Since the

Table 5. Comparative Table of different approaches dealing with preference queries

	Formulation		Context		Ranking	
	Qualitat.	Quantitat.	Uncond. req.	Cond. req.	Skyli.	Top-k
Lacroix Lavency [24]	✓		✓			✓
Chomicki 2002 [4]	✓		✓	✓	✓	
Kießling 2002 [5]	✓	✓	✓		✓	
Fagin et al 2001 [2]		✓	✓			✓
Fuzzy logic [13]		✓	✓	✓		✓
Symbolic weight possibilistic logic	✓		✓	✓	✓	✓

lexicographic method leads to a more refined rank-ordering, one may think of first using the weak comparative preferences method to stratify items, and then, each layer (except the top one because of Proposition 3) may be refined by one of the lexicographic methods by considering each level as a new intermediate database. This hybrid method may be of interest for computing a rank-ordering for *top k* items (when k is larger than $|E^1|$!). However, for refining the ranking of items inside a layer, we need to process it as a whole, even if it leads to considering more than k items in the ranking.

5 Related Work

Different types of approaches for handling preference queries have been proposed in the literature. Table 5 (whose evaluation criteria are taken from [23]) provides a comparative assessment of the possibilistic approach along with a representative subset of other approaches.

As for other qualitative methods, the approach presented is capable of expressing preference between attribute values or between tuples of attribute values, since we use general logical formulas (e.g., considering only the two attributes 'Price' and 'Distance to sea' with values *low, medium* and *high* for a house to let, a query may express that one prefers *low price and low distance*, or at least *low price and medium distance* or at least *medium price and medium distance*). As can be noticed, only the Kießling approach [5] can express both qualitative and quantitative preferences. Besides, only the possibilistic and the Chomicki [4] approaches can deal with conditional preferences. It can be seen that the advantages of this approach and of the fuzzy logic one [25] are complementary, which suggests to try and hybridize them in the future. To this end, it would be necessary to compare vectors including both symbolic and numerical weights.

6 Conclusion

The interest for preference representation in the possibilistic logic framework first stems from the logical nature of the representation. Moreover, the possibilistic representation can express preferences of the form "or at least", or "and if possible"(see [20] for an introductory survey), and can handle partial orders thanks to the use of symbolic weights, without enforcing implicit preferences (as it is the case for father node preferences in CP nets). We have proposed three types of methods in order to rank-order items, which are characterized by an increasing refinement power with manageable complexity, especially using the hybrid method explained in Subsection 4.3. Still, much remains to be done. First, the use of symbolic weights is really advantageous but we still miss some properties of numerical weights. One may think of combining these two formats to be as much expressive as possible. Moreover, this approach should be able to deal with null values, which create specific difficulties in preference queries.

References

1. Bosc, P., Pivert, O.: Some approaches for relational databases flexible querying. Journal of Intelligent Information Systems 1, 323–354 (1992)
2. Fagin, R., Lotem, A., Naor, M.: Optimal aggregation algorithms for middleware. In: Proc. 20th ACM SIGACT-SIGMOD-SIGART Symp. on Principles of Database Syst. (2001)
3. Börzsönyi, S., Kossmann, D., Stocker, K.: The skyline operator. In: Proc. 17th IEEE International Conference on Data Engineering, pp. 421–430 (2001)
4. Chomicki, J.: Preference formulas in relational queries. ACM Transactions on Database Systems 28, 1–40 (2003)
5. Kiessling, W.: Foundations of preferences in database systems. In: Proc. of the 28th International Conference on Very Large Data Bases (VLDB 2002), pp. 311–322 (2002)
6. Boutilier, C., Brafman, R.I., Domshlak, C., Hoos, H., Poole, D.: CP-nets: A tool for representing and reasoning with conditional ceteris paribus preference statements. J. Artificial Intelligence Research (JAIR) 21, 135–191 (2004)
7. Brafman, R.I., Domshlak, C.: Database preference queries revisited. Technical Report TR2004-1934, Cornell University, Computing and Information Science (2004)
8. HadjAli, A., Kaci, S., Prade, H.: Database preference queries - A possibilistic logic approach with symbolic priorities. Ann. Math. Artif. Intell. 63, 357–383 (2011)
9. Bosc, P., Pivert, O., Prade, H.: A possibilistic logic view of preference queries to an uncertain database. In: Proc. IEEE Inter. Conf. on Fuzzy Systems (FUZZ-IEEE 2010), Barcelona, Spain, July 18-23, pp. 1–6 (2010)
10. Hadjali, A., Pivert, O., Prade, H.: On different types of fuzzy skylines. In: Kryszkiewicz, M., Rybinski, H., Skowron, A., Raś, Z.W. (eds.) ISMIS 2011. LNCS, vol. 6804, pp. 581–591. Springer, Heidelberg (2011)
11. Wilson, N.: Computational techniques for a simple theory of conditional preferences. Artif. Intell. 175, 1053–1091 (2011)
12. Dubois, D., Prade, H.: Possibilistic logic: a retrospective and prospective view. Fuzzy Sets and Systems 144, 3–23 (2004)
13. de Calmès, M., Dubois, D., Hüllermeier, E., Prade, H., Sedes, F.: Flexibility and fuzzy case-based evaluation in querying: An illustration in an experimental setting. Int. Journal of Uncertainty, Fuzziness and Knowledge-Based Systems 11, 43–66 (2003)
14. Benferhat, S., Dubois, D., Prade, H.: Towards a possibilistic logic handling of preferences. Applied Intelligence 14, 303–317 (2001)
15. Dubois, D., Kaci, S., Prade, H.: Representing preferences in the possibilistic setting. In: Bosi, G., Brafman, R.I., Chomicki, J., Kießling, W. (eds.) Preferences: Specification, Inference, Applications. Number 04271 in Dagstuhl Seminar Proceedings (2006)
16. Benferhat, S., Dubois, D., Kaci, S., Prade, H.: Bridging logical, comparative, and graphical possibilistic representation frameworks. In: Benferhat, S., Besnard, P. (eds.) ECSQARU 2001. LNCS (LNAI), vol. 2143, pp. 422–431. Springer, Heidelberg (2001)
17. Benferhat, S., Dubois, D., Kaci, S., Prade, H.: Graphical readings of possibilisitc logic bases. In: 17th Conf. Uncertainty in AI (UAI 2001), Seattle, August 2-5, pp. 24–31 (2001)

18. Benferhat, S., Dubois, D., Garcia, L., Prade, H.: On the transformation between possibilistic logic bases and possibilistic causal networks. Inter. J. of Approx. Reas. 29, 135–173 (2002)
19. Dubois, D., Prade, H., Touazi, F.: Conditional preference nets and possibilistic logic. In: van der Gaag, L.C. (ed.) ECSQARU 2013. LNCS, vol. 7958, pp. 181–193. Springer, Heidelberg (2013)
20. Dubois, D., Prade, H., Touazi, F.: Handling partially ordered preferences in possibilistic logic. In: ECAI 2012 Workshop on Weighted Logics for Artificial Intelligence, pp. 91–98 (2012)
21. Benferhat, S., Yahi, S.: Complexity and cautiousness results for reasoning from partially pre-ordered belief bases. In: Sossai, C., Chemello, G. (eds.) ECSQARU 2009. LNCS, vol. 5590, pp. 817–828. Springer, Heidelberg (2009)
22. Yahi-Mechouche, S.: Raisonnement en présence d'incohérence: de la compilation de bases de croyances stratifiées à l'inférence à partir de bases de croyances partiellement préordon-nées. Université d'Artois Faculté des Sciences Jean Perrin, Lens (2009)
23. Stefanidis, K., Koutrika, G., Pitoura, E.: A survey on representation, composition and application of preferences in database systems. ACM Trans. Database Syst. 36, 19:1–19:45 (2011)
24. Lacroix, M., Lavency, P.: Preferences: Putting more knowledge into queries. In: Proc. of the 13th Inter. Conference on Very Large Databases (VLDB 1987), pp. 217–225 (1987)
25. Bosc, P., Pivert, O.: Fuzzy Preference Queries to Relational Databases. Imperial College Press (2012)

Bipolar Conjunctive Query Evaluation for Ontology Based Database Querying

Nouredine Tamani[1], Ludovic Liétard[2], and Daniel Rocacher[3]

[1] IATE/INRA-Supagro, France
nouredine.tamani@supagro.inra.fr, ntamani@gmail.com
[2] IRISA/IUT/Univ. Rennes 1, France
ludovic.lietard@univ-rennes1.fr
[3] IRISA/ENSSAT/Univ. Rennes 1, France
rocacher@enssat.fr

Abstract. In the wake of the flexible querying system, designed in [21], allowing the expression of user preferences as bipolar conditions of type "and if possible" over relational databases and ontologies, we detail in this paper the user query evaluation process, under the extension of the logical framework to bipolarity of type "or else" [15,14]. Queries addressed to our system are bipolar conjunctive queries made of bipolar atoms, and their evaluation relies on three-step algorithm: (i) *atom substitution process* that details how bipolar subsumption axioms defined in the bipolar ontology are used, (ii) *query derivation process* which delivers from each atom substitution a complementary query, and (iii) *translation process* that translates the obtained set of queries into bipolar SQLf statements, subsequently evaluated over a bipolar relational database.

1 Introduction

Flexible querying systems allow users to express preferences in their queries. In this context, several preference operators and languages relying on various logical frameworks have been proposed, such as Preference SQL [5], SQLf language [2], winnow operator [11], among many other works. We consider in this paper fuzzy set theory [24] as a general model to express flexibility through fuzzy conditions. In this line, many works have been carried out to allow the expression and evaluation of preferences as fuzzy bipolar conditions [1,7,8,10,23,25,22].

In this bipolar context, a personalized flexible querying system based on fuzzy bipolar conditions of type *"and if possible"*, fuzzy bipolar DLR-Lite [21] and Bipolar SQLf language [19] has been designed in [21]. More recently, the theoretical framework has been extended in [14] to generalized fuzzy bipolar conditions (*"and if possible"*, *"or else"*), to build a more expressive system. We then detail in this paper the process of query evaluation of the generalized flexible querying system. The considered queries are bipolar conjunctive queries made of bipolar atoms corresponding to bipolar concepts defined with bipolar conditions of both types *"and if possible"* and *"or else"*. The proposed evaluation process relies on an algorithm based on three steps: *(i) substitution* step which details an elementary atom substitution based on a bipolar subsumption defined in the

H.L. Larsen et al. (Eds.): FQAS 2013, LNAI 8132, pp. 389–400, 2013.

ontology, *(ii) query derivation* step which derives from the original query its set
of complementary conjunctive queries, and *(iii) translation* step which translates
conjunctive queries issued from the query derivation process into bipolar SQLf
statements, which are then evaluated upon the relational database attached to
the system, and hosted by PostgreSQLf DBMS extended to bipolar datatypes
and functions. The resulted tuples are finally merged and ranked from the most
to the least satisfactory. The query evaluation process is implemented to show
the feasibility of the approach. We have pointed out that substitutions based
on subsumptions simplify a complex conjunctive query, and reduce the amount
of data to consider in the evaluation so that only the most relevant ones are
targeted to resolve the queries.

In section 2, we recall the considered bipolar approach and Bipolar SQLf language. In section 3, we introduce the fuzzy bipolar DLR-Lite extended to bipolarity (*"and if possible"* and *"or else"*). Section 4 details the proposed algorithms
for bipolar conjunctive query evaluation and sums up preliminary obtained results. Section 5 recalls our contributions and introduces some perspectives.

2   Bipolarity in Flexible Querying of Relational Databases

We recall in this section the considered bipolar framework, its related logical operators, and the definition of fuzzy bipolar relations and Bipolar SQLf language.

2.1   Fuzzy Bipolar Conditions

A bipolar condition is made of two poles: a negative pole and a positive pole.
The former expresses a constraint that every accepted elements must satisfy. The
latter expresses a wish that distinguishes among accepted elements those which
are optimal. In this framework, we consider two kinds of bipolar conditions:
(i) conjunctive bipolar conditions of type *"and if possible"*, and (ii) disjunctive
bipolar conditions of type *"or else"*.

The former has been widely studied ([1,7,23,22,25]). In this paper, we consider
Dubois and Prade interpretation [10,8,9,7] in which such conditions, denoted
(c, w), mean "satisfy c and if possible satisfy w" [15]. Formally, in a bipolar
condition of type *"and if possible"* [15]: *(i) the negation of c ($\neg c$) refers to
the rejection, and c expresses the acceptability, (ii) w corresponds to the optimal
elements, (iii) the acceptability condition c is more important than the optimality
condition w, (iv) the coherence property: the set of optimal elements is included
in the set of accepted elements ($w \subseteq c$).* The latter type (*"or else"*) has been
introduced formally in [15], in which an *"or else"* condition, denoted $[e, f]$, means
"satisfy e or else satisfy f". Formally, in such conditions we have [15]: *(i) the
negation of f ($\neg f$) refers to the rejection, and f expresses the acceptability, (ii)
e corresponds to the optimal elements, (iii) the optimality condition e is more
important than the acceptability condition f, (iv) the coherence property: the set
of optimal elements is included in the set of accepted elements ($e \subseteq f$).*

Bipolar conditions generalize fuzzy conditions ϕ in the same way: "ϕ and if
possible ϕ" or "ϕ or else ϕ" and means that accepted elements are also optimal.

2.2   Evaluation of Fuzzy Bipolar Conditions

In bipolar conditions of type *"and if possible"*, if c and w are boolean, the satisfaction w.r.t. (c, w) is an ordered pair from $\{(1,1), (1,0), (0,0)\}$ ($(0,1)$ does not satisfy the coherence property). Tuples are returned in decreasing order of degrees (tuples scored with $(1,1)$ at the top of the delivered list, then those attached with $(1,0)$). In the case of fuzzy conditions (defined on universe U), the satisfaction to (c, w) is an ordered pair in $[0,1]^2$, denoted $(\mu_c(u), \mu_w(u))$ s.t. $u \in U$ and $\mu_c(u)$ (resp. $\mu_w(u)$) is the grade of satisfaction of u w.r.t. c (resp. w). The coherence property becomes then: $\forall u \in U, \mu_w(u) \leq \mu_c(u)$, corresponding to the Zadeh inclusion [24]. To satisfy this property, a bipolar condition (c, w) is evaluated as $(c, c \wedge w)$. Similarly, in bipolar conditions of type *"or else"*, if e and f are Boolean, the satisfaction w.r.t. $[e, f]$ is an ordered pair from $\{[1,1], [0,1], [0,0]\}$ ($[1,0]$ does not satisfy the coherence property). Tuples are returned in decreasing order of degrees (tuples scored with $[1,1]$ at the top of the delivered list, then those attached with $[0,1]$). In the case of fuzzy conditions (defined on universe U), the satisfaction to $[e, f]$ is an ordered pair in $[0,1]^2$, denoted $[\mu_e(u), \mu_f(u)]$ The coherence property is also a Zadeh inclusion: $\forall u \in U, \mu_e(u) \leq \mu_f(u)$. To satisfy to this property, a bipolar condition $[e, f]$ is evaluated as $[e \wedge f, f]$.

To rank the delivered tuples, previous bipolar approaches [7,23,25] consider the aggregation of the constraint and the wish. In our case, based on Dubois and Prade conclusions [7,8], about the limitations of aggregation methods, which lead to the lost of semantics attached to each pole (acceptability and optimality and their level of importance), we consider the lexicographical order. Indeed, in bipolar conditions of type *"and if possible"* (resp. *"or else"*), we rank tuples according to the acceptability (resp. optimality) condition, followed by the optimality (resp. acceptability) one; a total order is then obtained with $(1,1)$ as the greatest element and $(0,0)$ as the smallest one. It has been shown in [15] that both kind of bipolar conditions are compatible, since they are defined on the same priority principle over poles, and they can be expressed in a single query such as in *"find flights from Paris to Berlin which are (fast and if possible not expensive) and (having an early departure or else a morning departure)"*.

In the remainder of the paper, we denote by $\langle a, b \rangle$ a generalized bipolar condition, and we identify by its attached ordered pairs of grades $\langle \alpha, \beta \rangle \in [0,1]^2$ its nature such that: if $\alpha \leq \beta$ then it is of type *"or else"*, if $\alpha = \beta$ then it is a generalized fuzzy condition, and if $\alpha \geq \beta$ it is of type *"and if possible"*.

Based on the lexicographical order, the *lmin* and *lmax* operators [12,3] have been introduced as extended t-norm and t-conorm, to define the conjunction and the disjunction of bipolar conditions of type *"and if possible"*. They have been extended to bipolar conditions of type *"or else"* in [18] and are defined as:

$$([0,1] \times [0,1])^2 \rightarrow [0,1] \times [0,1]$$

$$(\langle x,y \rangle, \langle x',y' \rangle) \mapsto lmin(\langle x,y \rangle, \langle x',y' \rangle) = \begin{cases} \langle x,y \rangle & \text{if } x < x' \vee (x = x' \wedge y < y') \\ \langle x',y' \rangle & \text{otherwise.} \end{cases}$$

$$(\langle x,y \rangle, \langle x',y' \rangle) \mapsto lmax(\langle x,y \rangle, \langle x',y' \rangle) = \begin{cases} \langle x,y \rangle & \text{if } x > x' \vee (x = x' \wedge y > y') \\ \langle x',y' \rangle & \text{otherwise.} \end{cases}$$

As shown in [18], the $lmin$ (resp. $lmax$) operator is commutative, associative, idempotent and monotonic. The ordered pair of grades $\langle 1, 1 \rangle$ is the neutral (resp. absorbing) element of the operator $lmin$ (resp. $lmax$) and the ordered pair $\langle 0, 0 \rangle$ is the absorbing (resp. neutral) element of the operator $lmin$ (resp. $lmax$).

To complete our set of logical operators, we consider the negation operator introduced in [15] such that $\neg\langle a, b \rangle = \langle \neg a, \neg b \rangle$.

2.3   Fuzzy Bipolar Relations and Bipolar SQLf Language

When a query involving a fuzzy bipolar condition $\langle a, b \rangle$ is evaluated on a relation R, each tuple $t \in R$ is attached with an ordered pair of grades $\langle \mu_a(t), \mu_b(t) \rangle$ that expresses its satisfaction, and a so-called fuzzy bipolar relation is obtained. A tuple t is denoted $\langle \mu_a, \mu_b \rangle / t$. Tuples in R are ranked lexicographically using the operator $\geq_{lex}$ s.t. $t_1 \geq_{lex} t_2$ (or $\langle \mu_a(t_1), \mu_b(t_1) \rangle \geq_{lex} \langle \mu_a(t_1), \mu_b(t_2) \rangle$), iff $\mu_a(t_1) \geq \mu_a(t_2)$ or $(\mu_a(t_1) = \mu_a(t_2) \wedge \mu_b(t_1) \geq \mu_b(t_2))$. Its strict counterpart, denoted $>_{lex}$ is defined as: $\langle \mu_a(t_1), \mu_b(t_1) \rangle >_{lex} \langle \mu_a(t_1), \mu_b(t_2) \rangle$ iff $\mu_a(t_1) > \mu_a(t_2)$ or $(\mu_a(t_1) = \mu_a(t_2) \wedge \mu_b(t_1) > \mu_b(t_2))$. Operators $\leq_{lex}$ and $<_{lex}$ are defined similarly.

It can be noticed that a fuzzy relation is a fuzzy bipolar relation such that $\forall x, \mu_a(x) = \mu_b(x)$, and a regular relation is a fuzzy bipolar relation in which grades $\langle 1, 1 \rangle$ are associated to its tuples.

Based on [13,3], a bipolar relational algebra extended to both types of bipolar conditions and enriched with the negation operator has been introduced in [14,18]. We have also shown its backward compatibility with the fuzzy relational algebra [2]. Non-algebraic operators such as nesting and partitioning with linguistic quantifiers have also been extended in [14,18].

Bipolar SQLf Language. Bipolar SQLf language [19] allows the expression of simple and complex queries (based on projections, restrictions, nesting, partitioning with bipolar quantified propositions, divisions involving bipolar relations). It has been extended to both types of bipolar conditions in [18]. Its basic statement is a combination of a bipolar projection and a bipolar selection. It is expressed as:

Select *[**Distinct**]* $[n|t|(t_1, t_2)|n, t|n, (t_1, t_2)]$ *attributes* **From**
    *relations [**As** alias] **Where** $\langle a, b \rangle$;*

The bipolar condition $\langle a, b \rangle$ can refer to either *"and if possible"* (denoted (a, b)) or *"or else"* (denoted then $[a, b]$) fuzzy bipolar condition. It can also refer to an aggregation (and, or, not) of several fuzzy bipolar conditions.

With the calibration parameters, we can express bipolar top-k queries (parameter $n \in \mathbb{N}$), which returns the n best tuple w.r.t. their ordered pair of grades. A qualitative filtering can also be expressed by either parameter $t \in]0, 1]$ or $\langle t_1, t_2 \rangle$. In the former case, t can be written $(t, 0)$ and delivers tuples u such that $\langle \mu_a(u), \mu_b(u) \rangle \geq_{lex} (t, 0)$, and in the latter case, the delivered tuples u are those satisfying $\langle \mu_a(u), \mu_b(u) \rangle \geq_{lex} \langle t_1, t_2 \rangle$.

Example 1. Let *journeys* be a relational table of journeys from *Paris* to *Berlin*. The query: *"Find journeys from Paris to Berlin which are (fast, and if possible not expensive) or having (an early departure, or else a morning departure)"*, can be expressed in the bipolar SQLf as:

Select *#Journey* **From** *Journey* **As** *J* **Where** *J.from =* *'Paris' and*
J.to = 'Berlin' and (fast(J.duration), not expensive(J.rate)) or
[early(J.departure), morning(J.departure)];

3   Fuzzy Bipolar DLR-Lite

The Fuzzy DLR-Lite [6,17] relies on DLR-Lite [4]; its knowledge base $\mathcal{K}$ is made of facts component $\mathcal{F}$, ontology component $\mathcal{O}$, and abstraction component $\mathcal{A}$. Based on its extension to bipolarity of type *"and if possible"* [21], we extend it here to bipolarity of both types *"and if possible"* and *"or else"*.

Bipolar Facts Component $\mathcal{F}$. A bipolar fact $R(v_1, ..., v_n)\langle s_{c_1}, s_{c_2}\rangle$, is a line in a fuzzy bipolar relation R defined by the bipolar condition $\langle c_1, c_2\rangle$, $v_{i,i=1,...,n}$ are constants which form tuple $u = [v_1, ..., v_n]$, $s_{c_1} = \mu_{c_1}(u)$ and $s_{c_2} = \mu_{c_2}(u)$.

Bipolar Abstraction Component $\mathcal{A}$. Simple and complex abstraction rules are considered. *Simple abstraction rule* is of form $R_1 \longmapsto R_2(v_1, ..., v_n)\langle s_{c_1}, s_{c_2}\rangle$. It links an ontology concept R_1 with a projection of n columns $v_1, ..., v_n$, s.t. $n \leq m$ of an m-ary fuzzy bipolar relation R_2 defined by $\langle c_1, c_2\rangle$, and $\langle s_{c_1}, s_{c_2}\rangle$ is the ordered pair of grades of satisfaction attached to tuple $[v_1, ..., v_n]$. *Complex abstraction rule* has the form $R_1 \longmapsto (v_1, ..., v_n).sqlfb$, and links the concept R_1 to the bipolar relation, made of n-ary tuples $[v_1, ..., v_n]$, delivered by the bipolar SQLf query *sqlfb*.

Bipolar Ontology Component $\mathcal{O}$. This component defines intersection axioms ($\sqcap$), simple and conditional projection axioms, and subsumption axioms ($\sqsubseteq$).

• *Intersection axiom* is denoted $R_1 \sqcap R_2$, where R_1 and R_2 are compatible fuzzy bipolar relations, of same arity, defined resp. by fuzzy bipolar conditions $\langle c_1, c_1'\rangle$ and $\langle c_2, c_2'\rangle$. It is interpreted by the relational intersection operator:

$$(R_1 \sqcap R_2)^{\mathcal{I}} \Leftrightarrow \forall u, \langle s_c(u), s_{c'}(u)\rangle = lmin(\langle \mu_{c_1}(u), \mu_{c_1'}(u)\rangle, \langle \mu_{c_2}(u), \mu_{c_2'}(u)\rangle) \quad (1)$$

• *Simple projection axiom* is denoted $\exists[i_1, ..., i_k]R$, and corresponds to the projection on columns $i_1, ..., i_k$ of the fuzzy bipolar relation R of arity n such that $n \geq k$. It is interpreted as a bipolar projection of distinct elements on fields $i_1, ..., i_k$ of R defined by the fuzzy bipolar condition $\langle c, c'\rangle$:

$$\forall u, (\exists[i_1, ..., i_k]R)(u)^{\mathcal{I}} = \langle s_c(u), s_{c'}(u)\rangle = \underset{t \in R \wedge t[i_1,...,i_k]=u}{lmax} (\langle \mu_c(t), \mu_{c'}(t)\rangle) \quad (2)$$

• *Restricted projection axiom* is a bipolar projection and selection of tuples from R defined by the bipolar condition $\langle c, c'\rangle$. It is denoted $\exists[i_1, ..., i_k]R.(C_1 \sqcap ... \sqcap C_h)$, s.t. $C_{i,i=1...h} = ([i]\theta v)$ are conditions of selection s.t. $\theta \in \{<, \leq, =, \neq, >, \geq\}$, $[i]$ is a column in R, v is a constant and $\sqcap$ is the operator *and*. It is interpreted as:

$$\forall u, (\exists[i_1, ..., i_k]R.(C_1 \sqcap ... \sqcap C_h))(u)^{\mathcal{I}} = \langle s_c(u), s_w(u)\rangle =$$
$$\underset{t \in R \wedge t[i_1,...,i_k]=u \wedge C_1^{\mathcal{I}}(u)=1 \wedge C_2^{\mathcal{I}}(u)=1 \wedge ... \wedge C_h^{\mathcal{I}}(u)=1}{lmax} (\langle \mu_c(t), \mu_{c'}(t)\rangle) \quad (3)$$

• *Subsumption axiom* is denoted $(R_l \sqsubseteq R_r)\langle n_1, n_2 \rangle$, s.t. R_l and R_r are fuzzy bipolar relations of the same arity, defined resp. by fuzzy bipolar conditions $\langle c_l, c_l' \rangle$ and $\langle c_r, c_r' \rangle$. As introduced in [20], its interpretation is based on fuzzy subsumption defined as:

$$((R_l \sqsubseteq R_r)[n])^{\mathcal{I}} \Leftrightarrow min_x(R_l(x) \rightarrow_{G\ddot{o}} R_r(x)) = n, \tag{4}$$

where $\rightarrow_{G\ddot{o}}$ is the Gödel fuzzy implication.

The extension of formula (4) to bipolarity is obtained by the substitution of t-norm min for its extension to bipolarity $lmin$, and Gödel fuzzy implication for a new implication operator denoted $\rightarrow_{BG\ddot{o}}$, as expressed in formula (5):

$$((R_l \sqsubseteq R_r)\langle n_1, n_2 \rangle)^{\mathcal{I}} \Leftrightarrow lmin_x(R_l(x) \rightarrow_{BG\ddot{o}} R_r(x)) = \langle n_1, n_2 \rangle \tag{5}$$

where $\rightarrow_{BG\ddot{o}}$ is the Gödel fuzzy implication extended to bipolarity, and the ordered pair of degrees $\langle n_1, n_2 \rangle$ is interpreted as the minimal threshold of inclusion of R_l in R_r. To define the operator $\rightarrow_{BG\ddot{o}}$, we extend the operator defined in [21] to our framework of bipolarity as expressed in formula (6):

$$I_R(\langle a_{c_1}, a_{c_1'} \rangle, \langle b_{c_1'}, b_{c_2'} \rangle) = lmax\{\langle z, z' \rangle \in [0,1]^2 :$$
$$lmin(\langle a_{c_1}, a_{c_1'} \rangle, \langle z, z' \rangle) \leq_{lex} \langle b_{c_1'}, b_{c_2'} \rangle\} \tag{6}$$

Then, we obtain the following interpretation of the bipolar subsumption:

$$(R_l(x) \rightarrow_{BG\ddot{o}} R_r(x))^{\mathcal{I}} = \begin{cases} \langle 1, 1 \rangle & if \langle \mu_{c_l}(x), \mu_{c_l'}(x) \rangle \leq_{lex} \langle \mu_{c_r}(x), \mu_{c_r'}(x) \rangle, \\ \langle \mu_{c_r}(x), \mu_{c_r}(x) \rangle, & otherwise. \end{cases} \tag{7}$$

Formula (7) is compatible with formula (6), and it defines an extension of the Gödel fuzzy implication to bipolarity. The proof is similar to the one of [21]. It is also easy to prove that formula (7) verifies properties of R-implications extended to bipolarity (monotonicity, neutrality, exchange, identity and ordering).

This interpretation ensures the validity of the transitivity property of natural inclusions in the context of bipolarity. Indeed, in a series of bipolar inclusion axioms of form: $A_1 \sqsubseteq_{\langle \alpha_1, \alpha_1' \rangle} A_2 \sqsubseteq_{\langle \alpha_2, \alpha_2' \rangle} \cdots \sqsubseteq_{\langle \alpha_{n-1}, \alpha_{n-1}' \rangle} A_n$, transitive links as $\forall i, j, i < j, A_i \sqsubseteq_{\langle \alpha, \alpha' \rangle} A_j$, s.t. $\langle \alpha, \alpha' \rangle = lmin(\langle \alpha_i, \alpha_i' \rangle, ..., \langle \alpha_{j-1}, \alpha_{j-1}' \rangle)$ hold.

4 Fuzzy Bipolar Conjunctive Query Evaluation

Queries addressed to a fuzzy bipolar DLR-Lite knowledge base consist of extended conjunctive queries to bipolarity. The main form of such queries is:

$$\underbrace{q(x)\langle s_c, s_{c'} \rangle}_{head} \leftarrow \underbrace{\exists y R_1(z_1)\langle s_{c_1}, s_{c_1'} \rangle, ..., R_l(z_l)\langle s_{c_l}, s_{c_l'} \rangle,}_{body}$$

$$\underbrace{OrderBy(\langle s_c, s_{c'} \rangle = lmin(\langle s_{c_1}, s_{c_1'} \rangle, ..., \langle s_{c_l}, s_{c_l'} \rangle)),}_{scoring\ function} \tag{8}$$

s.t. q is a bipolar m-ary relation, x (resp. y) is a vector of bound (resp. unbound) variables, every $R_i(z_i)$ is either a bipolar m_i-ary relation or a concrete predicate of form $(z\theta v)$, with z is a variable in x or y, $\theta \in \{<, \leq, =, \neq, >, \geq\}$ and v is a constant, z_i is a tuple of variables in x or y. Ordered pairs of degrees $\langle s_{c_i}, s_{c_i'} \rangle_{i=1,...,l}$ expressed at each atom $(R_i(z_i))$ are aggregated by the $lmin$ operator, due to the conjunctive nature of the query, to compute pairs of degrees $\langle s_c, s_{c'} \rangle$ to attach to the resulting tuples. Syntactically, ordered pairs $\langle s_{c_i}, s_{c_i'} \rangle$ equal $\langle 1, 1 \rangle$ can be omitted. The query evaluation process requires a step of query rewriting [6,17], which we extend hereinafter to bipolarity, based on the operator $lmin$.

Bipolar Conjunctive Query Evaluation Process. As in [6,17], query rewriting process consists of deriving complementary queries q_i according to the applicability of the inclusion axioms on the query atoms. The importance of this process is shown through the following example in the fuzzy case.

Example 2. Let *SpeedJourney* and *FavoriteJourney*, denoted SJ and FJ resp., be fuzzy concepts in a fuzzy DLR-Lite knowledge base, made of (i) facts SJ $(t_1, 0.9)$, SJ $(t_2, 0.6)$, SJ $(t_3, 0.4)$, FJ $(t_3, 0.8)$, FJ $(t_4, 0.5)$, and (ii) inclusion axiom $SJ \sqsubseteq_{0.8} FJ$, in which a *speed* journey is a *favorite* one to grade at least 0.8. Let q be a query about *favorite* journeys: $q(t)[s] \leftarrow FJ(t), OrderBy(s)$.

If q is evaluated without considering the inclusion axioms, the system delivers a partial set of answers $S_q = \{(t_3, 0.8), (t_4, 0.5)\}$. To make it complete, the query rewriting process delivers the complementary query q', based on inclusion axiom $(SJ \sqsubseteq FJ)[0.8]$: $q'(t)[s'] \leftarrow SJ(t), OrderBy(s' = min(0.8, speed^{\mathcal{I}}(t)))$, and its evaluation over facts returns the set $S'_q = \{(t_1, 0.8), (t_2, 0.6)(t_3, 0.4)\}$ which is merged with S_q, s.t. we keep tuples that scored highest in the case of duplication, to obtain the complete set of answers $S = \{(t_1, 0.8), (t_2, 0.6), (t_3, 0.8), (t_4, 0.5)\}$.

The variables of a conjunctive query can be divided on bound variables referring to distinguished variables, variables appearing more than once, and valued variables, and on unbound variables, denoted by $-$. Atom of form $\exists[i_1, ..., i_k]R$ is denoted $R(-, ..., -, x_{i_1}, ..., x_{i_k}, -, ..., -)$.

Substitution Process. Let $\alpha'(\alpha, \tau)$ be the set of atoms resulted from the application of the inclusion axiom τ on atom α, and $\theta(\alpha, \tau)$ is a grade substitution function which substitutes $\langle s_1, s_2 \rangle$, the ordered pair of grades attached to atom α, for $lmin(\langle n_1, n_2 \rangle, \langle s_{c_1}, s_{c_1'} \rangle, ..., \langle s_{c_l}, s_{c_l'} \rangle)$ such that $\langle n_1, n_2 \rangle$ is the pair of grades of satisfaction of τ, and $\langle s_{c_j}, s_{c_j'} \rangle_{j=1,...,l}$ are ordered pairs of grades attached to fuzzy bipolar relations in the left-hand side of τ. Algorithm 1 details an elementary bipolar substitution of an atom α based on subsumption axiom τ involving α at its right-hand side.

Query Generation Process. Each atom substitution derives from q a complementary query q_i, added to the set of derived query $\mathcal{Q}$. This process is detailed in algorithm 2. Algorithm 2 has an exponential complexity $(O(2^k))$ depending on both number of substitutable atoms k and number of axioms applicable for each atom. In real application, queries do not exceed 4 substitutable atoms for which only one axiom is applicable; the number of generated queries is then 15.

Algorithm 1. Substitution process

Input: atom α, inclusion axiom τ; Output: $\alpha'(\alpha, \tau)$, $\theta(\alpha, \tau)$.

BEGIN

If $(\alpha = A(x)\langle s_1, s_2 \rangle$ and $\tau = R_1 \sqcap ... \sqcap R_m \sqsubseteq A)\langle n_1, n_2 \rangle)$ Then

    For $t = 1$ to m do If $Rl_t = B_t$ Then $C_t(x) = B_t(x)$;

                Else If $Rl_t = \exists[j]R$ Then $C_t(x) = \exists z_1, ..., z_d.R_t(z)$;

                Else If $Rl_t = \exists[j]R.(C_1, ..., C_h)$ Then

                    $z = \langle z_1, ..., z_{j-1}, x, z_{j+1}, ..., z_d \rangle$; /*$d$ is the arity of R*/

                    $C_t(x) = \exists z_1, ..., z_d.R_t(z) \wedge C_1(z) \wedge ... \wedge C_h(z)$;

        End If

    End For

    /*$\langle s_{c_t}, s_{c'_t} \rangle, t \in \{1,...,m\}$ is the satisfaction of x w.r.t. C_t*/

    $\alpha'(\alpha, \tau) = \{C_1(x)\langle s_{c_1}, s_{c'_1} \rangle, ..., C_m(x)\langle s_{c_m}, s_{c'_m} \rangle$;

    $\theta(\alpha; \tau) = lmin(\langle n_1, n_2 \rangle, \langle s_{c_1}, s_{c'_1} \rangle, ..., \langle s_{c_m}, s_{c'_m} \rangle)$;

Else If $(\alpha = R(-, ..., -, x_{i_1}, ..., x_{i_k}, -, ..., -)\langle s_1, s_2 \rangle$ and

                        $\tau = (Rl_1 \sqcap ... \sqcap Rl_m \sqsubseteq \exists[i_1, ..., i_k]R)\langle n_1, n_2 \rangle)$ Then

    For $t = 1$ to m do If $Rl_t = B_t \wedge k = 1$ Then $C_t(x_{i_1}, ..., x_{i_k}) = B_t(x)$;

        Else If $Rl_t = \exists[j_1, ..., j_k]R_t$ Then $C_t(x_{i_1}, ..., x_{i_k}) = \exists z_1, ..., z_d.R_t(z)$;

        Else If $Rl_t = \exists[j_1, ..., j_k]R_t.(C_1, ..., C_h)$ Then

                      $z = \langle z_1, ..., z_{j_1-1}, x_{j_1}, ..., x_{j_k}, z_{j_k+1}, ..., z_d \rangle$;

                    $C_t(x_{i_1}, ..., x_{i_k}) = \exists z_1, ..., z_d.R_t(z) \wedge C_1(z) \wedge ... \wedge C_h(z)$;

        End If

    End For

    $\alpha'(\alpha, \tau) = \{C_1(x_{i_1}, ..., x_{i_k})\langle s_{c_1}, s_{c'_1} \rangle, ..., C_m(x_{i_1}, ..., x_{i_k})\langle s_{c_m}, s_{c'_m} \rangle\}$;

    $\theta(\alpha; \tau) = lmin(\langle n_1, n_2 \rangle, \langle s_{c_1}, s_{c'_1} \rangle, ..., \langle s_{c_m}, s_{c'_m} \rangle)$;

End If;

END.

As future improvement, we can discard inclusion axioms having a weak degrees of truth, since it is meaningless to consider the substitution of concept A for B, when $A \sqsubseteq B$ at degrees $\approx \langle 0.01, 0.01 \rangle$, for instance.

Conjunctive Query Translation Process. This step consists of the translation of the delivered queries into bipolar SQLf statements to be evaluated over a relational database. To perform this translation, we need to rewrite a bipolar query in its dual form. Let $q = \{(\alpha(\mathcal{V}_q), sub\langle s_1, s_2 \rangle), \alpha \in \mathcal{O}\}$ be a conjunctive query, s.t. $\mathcal{V}_q$ is the set of variables expressed in q, sub is the concept to use in substitution of α, and attached to its grades of substitution $\langle s_1, s_2 \rangle$, and $\mathcal{O}$ is the ontology component of the related fuzzy bipolar DLR-Lite knowledge base. The dual form of q is q^d in which each variable is related to the atom on which it is expressed: $q^d = \{v([(\alpha_1, pos_1), ..., (\alpha_n, pos_n)], type, [c_1, ..., c_m]), [\langle s_1^1, s_2^1 \rangle, ..., \langle s_1^k, s_2^k \rangle], \alpha_i \in \mathcal{O}, v \in \mathcal{V}_q\}$, where pos_i is the position of variable v in concept α_i, $type$ is its type (distinguished or not), $c_{j,j=1,...,m}$ are the eventual conditions expressed on v, and $\langle s_1^r, s_2^r \rangle_{r=1,...,k}$, s.t. $k \leq n$ are the ordered pairs of grades of substitutions performed on q to obtain α_r. The dual form is easy to perform by scanning the atoms and saving for each variable a reference to the atom in which it appears, its position in q, its type and eventual conditions expressed on.

Algorithm 2. Query generation process

Input: Query $q = \{\alpha_i, i = 1, ..., n\}$, subsumption axioms $\mathcal{S} = \{\tau_j, j = 1, ..., m\}$
Output: $\mathcal{Q}$ /*set of generated queries*/
BEGIN
Add the original query to $\mathcal{Q}$: $\mathcal{Q} \leftarrow q$;
Build for each atom in q **a set of its applicable axioms:**
For each $\alpha_i \in q$ do
  For each $\tau_j \in \mathcal{S}$ do If (τ_j is applicable to α_i) Then $\mathcal{S}_i \leftarrow \tau_j$;
Form the initial query with substitutions: $q' \leftarrow (\alpha_i, \mathcal{S}_i)$;
Init the set $\mathcal{Q}'$: $\mathcal{Q}' \leftarrow q'$ **and apply the substitutions on** $\mathcal{Q}'$:
While ($\mathcal{Q}' \neq \emptyset$) do
  Select a query from $\mathcal{Q}'$: $q_j \leftarrow \mathcal{Q}'$
  For $i = 1$ to m_j do /*$m_j = |q_j| = $ number of couples $(\alpha_i, \mathcal{S}_i) \in q_j$ */
    For each $\tau_k \in \mathcal{S}_i$ such that $(\alpha_i, \mathcal{S}_i) \in q_j$ do
    $q_{jk}^{\theta} \leftarrow Substitute(\alpha_i, \tau_k, \alpha', \theta)$ /*algorithm 1*/
    Add the derived query q_{jk}^{θ} **in** $\mathcal{Q}$: $\mathcal{Q} \leftarrow q_{jk}^{\theta}$;
    Copy q_{jk}^{θ} **in** q'_j **from** $(\alpha_{i+1}, \mathcal{S}_{i+1})$ **to** $(\alpha_{m_j}, \mathcal{S}_{m_j})$: $copy(q'_j, q_{jk}^{\theta}, i+1, m_j)$;
    Add q'_j **in** $\mathcal{Q}'$: $\mathcal{Q}' \leftarrow q'_j$;
    End for
  End for
Delete q_j **from** $\mathcal{Q}'$: $remove(\mathcal{Q}', q_j)$;
End While
END.

Example 3. Let q be a conjunctive query about *preferred* journeys (denoted PJ) from *Paris* to *Brussels*, and $(FavoriteJourney \sqsubseteq PJ)\langle 0.8, 0.6 \rangle$ a subsumption axiom denoted $(FJ \sqsubseteq PJ)\langle 0.8, 0.6 \rangle$:

$q = \{$ *(PJ (num?, source, destination, duration?, rate?, comfort),*

        $FJ\langle 0.8, 0.6 \rangle$, *source = 'Paris', destination = 'Brussels'}*;

such that variables *num, duration* and *rate* are distinguished (labeled by symbol '?') and their values form the returned tuples, variables *source* and *destination* are bound (they appear more than once in q) and *comfort* is ignored since it is free. Algorithms 1 and 2 deliver the set $\mathcal{Q} = \{q, q_1\}$, such that:

$q_1 = \{$ *(FJ (num?, source, destination, duration?, rate?, comfort),*

    source = 'Paris', destination = 'Brussels', lmin(Favorite(num), $\langle 0.8, 0.6 \rangle$))};

The dual form of q_1 is $q_1^d = \{$ *num (FJ, 1, distinguished, void), source (FJ, 2, bound, = 'Paris'), destination (FJ, 3, bound, = 'Brussels'), duration (FJ, 4, distinguished, void), rate (FJ, 5, distinguished, void),* $[\langle 0.8, 0.6 \rangle]\}$.

Algorithm 3 details the translation of a conjunctive query into a bipolar SQLf statement. Obtained bipolar SQLf statements are then evaluated over the system database, hosted by PostgreSQLf[1] DBMS extended to bipolar datatype and functions [18] (pair of grades datatype, *lmin* and *lmax* operators). The delivered sets of tuples are finally merged such that we keep tuples that scored highest in the case of duplication.

---

[1] https://github.com/postgresqlf/PostgreSQL_f/blob/master/doc

---

Algorithm 3. Conjunctive query translation process

Input: A query $q = \{\alpha(\mathcal{V}_q), \alpha \in \mathcal{O}\}$, $\mathcal{O}$, Output: Bipolar SQLf statement $sqlf$;
BEGIN
Compute the dual form of q: $q^d\{v([(\alpha_1, pos_1), ..., (\alpha_n, pos_n)], type, Conds), v \in \mathcal{V}_q\}$;
Compute the projected attributes: For each distinguished variable $v_i \in \mathcal{V}_q$ do
• Retrieve the first atom in which v_i is involved: $(\alpha_1, pos_1) \leftarrow getAtoms(v_i, q^d)$;
• Retrieve from $\mathcal{O}$ attribute name of v_i in concept α_1: $getRealAttribute(\mathcal{O}, \alpha_1, pos_1)$;
• Form the list of projected attributes: proj;
End For
Form the join clause of the query based on non-distinguished variable and related concepts: JoinCond = α_1.attribute;
For $j = 2$ to n do JoinCond = JoinCond + " = " + $\alpha_j.getRealAttribute(\mathcal{O}, \alpha_j, pos_j)$;
Form the selection conditions (SCond) based on query concrete atoms:
If $(Conds \neq void)$ Then
For each condition $c_j \in Conds$ SCond = SCond + attribute + c_j;
Compute the scoring function based on substitutions:
        If $(grades \neq void)$ Then For each ordered pair of grades $\langle a, b \rangle \in grades$ do
           If (scores = "") Then $score \leftarrow lmin((getScore1(), getScore2()), \langle a, b \rangle)$
               Else $score \leftarrow lmin(score, lmin((getScore1(), getScore2()), \langle a, b \rangle))$
End For
Build the From clause of the Bipolar SQLf query:
For each $\alpha_k \in Atoms$ do From = From + " , " + α_j;
Build the whole sqlf statement:
$sqlf \leftarrow$ Proj + score + " From " + From
If (JoinCond $\neq$ "" or SCond $\neq$ "") Then $sqlf \leftarrow sqlf$ + "Where" + JoinCond + SCond;
END.

Example 4. The translation of q, q_1 (example 3) delivers the following queries:
$q = Select\ numJourney,\ durationJourney,\ rateJourney\ From\ PreferredJourney$
$Where\ sourceJourney = 'Paris'\ and\ destinationJourney = 'Brussels';$
$q_1 = Select\ numJourney,\ durationJourney,\ rateJourney,\ lmin\ ((getScore1(),$
$getScore2()),\ (0.8, 0.6))\ From\ FavoriteJourney\ Where\ sourceJourney = 'Paris'$
$and\ destinationJourney = 'Brussels';$

Prototyping and Preliminary Tests. To show the feasibility of our approach, we have extended the prototype developed in [21], limited to bipolar conditions of type *"and if possible"*, to bipolar conditions of type *"or else"*. It is developed in the field of multimodal transport system described by the Ontology Transportation Networks [16]. The data are collected manually from different mono-modal transport systems (train, TGV, plane, bus) and their related services (restaurant, hotel, etc.) of the Brittany Region, which link cities of *Lannion, Brest, Rennes, Plouaret Tregor,* and *Guingump* to *Paris*. We computed from the collected data all (437) feasible journeys stored in table *Journey*. We submitted to the system the following query: *"Find favorite journeys (not expensive and if possible having few changes), which departure is around 2h pm with a good hotel (close to transport station or else not expensive) at the destination"*, and expressed in conjunctive query as q:

$q = FavoriteJourney\ (num?,\ source?,\ dest?,\ -,\ hDep?,\ Harr?,\ -,\ -,\ -),\ source = "Paris",\ dest = "Paris"\ ,\ hDep \sim "14:00",\ GoodHotel\ (dest);$

From available data, we defined subsumption axiom *FavoriteStation* $\sqsubseteq_{\langle 0.73, 0.73 \rangle}$ *GoodHotel* on which algorithms 1 and 2 are applied to compute the set of derived queries $\mathcal{Q} = \{q, q'\}$. Then, algorithm 3 is applied to translate q and q' into bipolar SQLf statements, which are evaluated over the facts component.

Preliminary tests show that the inclusion axiom simplifies the query q before its evaluation and allows to target only *favorite* journeys having a destination close to good hotels without considering the table of hotels. Once journeys are computed, the system focuses only on this subset to solve the second part of query (about good hotels). Concerning the time consumption, the developed system builds once a local database in which it stores facts of the fuzzy bipolar DLR-Lite specification. Then, the conjunctive query evaluation overload consists in query derivation and translation (algorithms 1, 2 and 3). This overload is balanced by the reduction of the amount of targeted data. Indeed, the research space is reduced when fuzzy bipolar conditions and subsumptions are expressed and applied. Indeed, in our experiments, we noticed that the table of *favorite* journeys contains 41% of tuples of table *journey* (so, a reduction of 59%). For the hotels, the reduction of space is about 37% when the bipolar condition (*not expensive or else close to transport station*) is considered, and it is up to 66% when the subsumption rule *FavoriteStation* $\sqsubseteq_{\langle 0.73, 0.73 \rangle}$ *GoodHotel* is applied.

5   Conclusion

We have developed in this paper a three-step algorithm which is the basis of the proposed bipolar conjunctive query evaluation approach. The first step details the rules to apply to substitute atoms of a given query based on the applicable subsumption axioms. The second one shows the way in which substitutions are applied on a query to derive the set of its complementary queries. The third one is aimed at translating a conjunctive query into bipolar SQLf statements. This evaluation process is developed under a bipolar knowledge base relying on bipolar conditions of both types *"and if possible"* and *"or else"*, implemented as a standalone application and saved in a relational database managed by PostgreSQLf DBMS extended to the aforementioned framework of bipolarity.

As future works, we plan to reduce the complexity of query generation algorithm and to perform further tests to study the system performances. We aim also to extend the core of the bipolar PostgreSQLf DBMS (limited to bipolar projection, restriction and join operations) to encompass a larger set of bipolar SQLf statements, particularly, those based on nesting and grouping operators.

References

1. Bordogna, G., Pasi, G.: A fuzzy query language with a linguistic hierarchical aggregator. In: Proc. of the ACM SAC 1994, pp. 184–187. ACM, USA (1994)
2. Bosc, P., Pivert, O.: Sqlf: A relational database langage for fuzzy querying. IEEE Transactions on Fuzzy Systems 3(1), 1–17 (1995)
3. Bosc, P., Pivert, O., Liétard, L., Mokhtari, A.: Extending relational algebra to handle bipolarity. In: 25th ACM SAC, pp. 1717–1721 (2010)

4. Calvanese, D., Giacomo, G.D., Lembo, D., Lenzerini, M., Rosati, R.: Data complexity of query answering in description logics. In: KR, pp. 260–270 (2006)
5. Chomicki, J.: Querying with intrinsic preferences. In: Proceedings of the 8th International Conference on Extending Database Technology, pp. 34–51 (2002)
6. Colucci, S., Noia, T.D., Ragone, A., Ruta, M., Straccia, U., Tinelli, E.: Informative Top-k retrieval for advanced skill management. In: Semantic Web Information Management, ch. 19, pp. 449–476. Springer, Heidelberg (2010)
7. Dubois, D., Prade, H.: Bipolarity in flexible querying. In: Andreasen, T., Motro, A., Christiansen, H., Larsen, H.L. (eds.) FQAS 2002. LNCS (LNAI), vol. 2522, pp. 174–182. Springer, Heidelberg (2002)
8. Dubois, D., Prade, H.: Handling bipolar queries in fuzzy information processing. In: Handbook of Research on Fuzzy Information Processing in Databases, pp. 97–114. IGI Global (2008)
9. Dubois, D., Prade, H.: An introduction to bipolar representations of information and preference. Inter. Jour. of Intelligent Systems 23 (2008)
10. Dubois, D., Prade, H.: An overview of the asymmetric bipolar representation of positive and negative information in possibility theory. Fuzzy Sets and Systems 160(10), 1355–1366 (2009)
11. Kiesling, W.: Foundation of preferences in database systems. In: Proceedings of the 28th VLDB Conference, Hong Kong, China (2002)
12. Liétard, L., Rocacher, D.: On the definition of extended norms and co-norms to aggregate fuzzy bipolar conditions. In: IFSA/EUSFLAT, pp. 513–518 (2009)
13. Liétard, L., Rocacher, D., Bosc, P.: On the extension of sql to fuzzy bipolar conditions. In: 28th North American Information Processing Society Conference (2009)
14. Liétard, L., Rocacher, D., Tamani, N.: A relational algebra for generalized fuzzy bipolar conditions. In: Flexible Approaches in Data, Information and Knowledge Management. Springer (to appear, 2013)
15. Liétard, L., Tamani, N., Rocacher, D.: Fuzzy bipolar conditions of type "or else". In: FUZZ-IEEE, pp. 2546–2551 (2011)
16. Lorenz, B., Rosener, M.: Ontology of transportation networks. Tech. rep., REWERSE. Euro. Com. and Swiss Federal Office for Education and Science (2005)
17. Straccia, U.: Softfacts : a top-k retrieval engine for a tractable description logic accessing relational databases. Tech. rep., ISTI-CNR, Italy (Jan 2009)
18. Tamani, N.: Interrogation personnalisée des systémes d'information dédiés au transport: une approche bipolaire floue. Ph.D. thesis, Université de Rennes 1 (2012)
19. Tamani, N., Liétard, L., Rocacher, D.: Bipolar sQLf: A flexible querying language for relational databases. In: Christiansen, H., De Tré, G., Yazici, A., Zadrozny, S., Andreasen, T., Larsen, H.L. (eds.) FQAS 2011. LNCS, vol. 7022, pp. 472–484. Springer, Heidelberg (2011)
20. Tamani, N., Liétard, L., Rocacher, D.: Extension of an ontology for flexible querying. In: FUZZ-IEEE, pp. 2033–2038 (2011)
21. Tamani, N., Liétard, L., Rocacher, D.: A fuzzy ontology for database querying with bipolar preferences. Inter. Jour. of Intelligent Systems 28(1), 4–36 (2013)
22. de Tré, G., Zadrozny, S., Bronselaer, A.: Handling bipolarity in elementary queries to possibilistic databases. IEEE Trans. on Fuzzy Systems 18(3), 599–612 (2010)
23. De Tré, G., Zadrożny, S., Matthé, T., Kacprzyk, J., Bronselaer, A.: Dealing with positive and negative query criteria in fuzzy database querying. In: Andreasen, T., Yager, R.R., Bulskov, H., Christiansen, H., Larsen, H.L. (eds.) FQAS 2009. LNCS (LNAI), vol. 5822, pp. 593–604. Springer, Heidelberg (2009)
24. Zadeh, L.: Fuzzy sets. Information and Control 8(3), 338–353 (1965)
25. Zadrożny, S., Kacprzyk, J.: Bipolar queries: A way to enhance the flexibility of database queries. In: Ras, Z.W., Dardzinska, A. (eds.) Advances in Data Management. SCI, vol. 223, pp. 49–66. Springer, Heidelberg (2009)

Bipolar Querying of Valid-Time Intervals Subject to Uncertainty

Christophe Billiet[1], José Enrique Pons[2], Olga Pons[2], and Guy De Tré[1]

[1] Department of Telecommunications and Information Processing, Ghent University,
Sint-Pietersnieuwstraat 41, B-9000, Ghent, Belgium
`{Christophe.Billiet,Guy.DeTre}@UGent.be`
[2] Department of Computer Science and Artificial Intelligence, University of Granada,
C/Periodista Daniel Saucedo Aranda, S/N, E-18071, Granada, Spain
`{jpons,opc}@decsai.ugr.es`

Abstract. Databases model parts of reality by containing data representing properties of real-world objects or concepts. Often, some of these properties are time-related. Thus, databases often contain data representing time-related information. However, as they may be produced by humans, such data or information may contain imperfections like uncertainties. An important purpose of databases is to allow their data to be queried, to allow access to the information these data represent. Users may do this using queries, in which they describe their preferences concerning the data they are (not) interested in. Because users may have both positive and negative such preferences, they may want to query databases in a bipolar way. Such preferences may also have a temporal nature, but, traditionally, temporal query conditions are handled specifically. In this paper, a novel technique is presented to query a valid-time relation containing uncertain valid-time data in a bipolar way, which allows the query to have a single bipolar temporal query condition.

Keywords: Bipolar Querying, Valid-time Relation, Valid Time, Temporal Databases, Uncertainty, Possibility Theory, Ill-known Intervals.

1 Introduction

Generally, database systems model (parts of) reality. For this, their databases contain data representing properties of real-world objects or concepts. Some essential properties of real-world objects or concepts are time-related. Thus, databases often contain data representing temporal values [1], which are basically indications of time and describe such properties. These temporal values are usually either time intervals [1] or instants [1], which may informally be seen as infinitesimally short 'periods' or 'points' in time. Based on their interpretation and purpose, temporal values can be classified into several categories, but the presented work will only consider valid-time indications, which indicate when corresponding data is a valid or true representation of the reality modelled by its database [1–4].

H.L. Larsen et al. (Eds.): FQAS 2013, LNAI 8132, pp. 401–412, 2013.

A lot of database data are produced by humans, but human-made data are prone to imperfections, as some of these data may be vague or imprecise [5], contradictory, incomplete or uncertain [3], [4], [6]. Of course, data representing temporal values may contain such imperfections too [2–4], [7]. The work presented in this paper will consider databases containing data representing valid-time intervals subject to uncertainty and will assume all non-temporal data in these databases to contain no imperfections.

One of the most important purposes of a database is to allow its data to be queried, to allow the information or knowledge represented by this data to be retrieved. A user may query a database in a 'regular' way: the user describes the data which he or she finds desired or satisfactory and thus wants to retrieve, by perfectly describing the allowed values of these data. A user may also query a database in a 'fuzzy' way: the user describes the data which he or she finds desired or satisfactory by imperfectly describing the allowed values of these data [8]. These imperfect descriptions may contain vagueness or imprecision, often through the use of linguistic terms [9], [10]. A user may also query a database in a 'bipolar' way. Generally, two main approaches to this exist. One is for the user to describe the data which he or she finds acceptable and to describe the data among this acceptable data, which he or she finds really desired, both by describing the allowed values of these data [11]. The other is for the user to independently describe both the data which he or she finds desired or satisfactory and the data which he or she finds undesired or unsatisfactory, both by describing the allowed values of these data [12], [13]. The descriptions used in bipolar querying may contain imperfections. The presented work will only consider the latter approach to bipolar querying and will allow a simple form of imprecision in non-temporal elementary query conditions.

Compared to non-temporal user preferences, temporal user preferences usually have an uncommon nature and interpretation and thus expressing them relies on uncommon mechanics in querying: users usually prefer using specific temporal operators to express temporal preferences. Hence, several proposals have considered specific sets of temporal operators, often based on the possible temporal relationships between two time indications [2], [14], [15]. Such temporal relationships define semantically meaningful relationships with a temporal nature, between two time indications. In [16], a collection of temporal relationships between two time intervals (and as a special case instants) is introduced and this collection is considered groundbreaking. Of course, to query temporal data in a fuzzy way, fuzzy variants of such temporal operators are necessary and several proposals have thus introduced such variants [4], [14], often based on fuzzy variants of such temporal relationships [15], [17].

Techniques for the regular or fuzzy querying of valid-time databases containing valid-time data subject to uncertainty are considered by several existing proposals [2], [4], [17]. However, to the knowledge of the authors, only one proposal has considered the bipolar querying of valid-time databases (in [18], a technique is proposed to query a valid-time database containing temporal data subject to imprecision in a bipolar way) and none have considered the bipolar querying of valid-time databases containing temporal data subject to uncertainty. Thus, this paper presents a novel technique to query a valid-time relation containing valid-time data subject to uncertainty in a bipolar way, which allows the user to specify a single bipolar temporal query condition. This paper is structured as follows: in section 2, some preliminary concepts and techniques are described, in section 3, the novel technique which is the main contribution of the work presented in this paper, is explained and in section 4, the conclusions of this paper and some directions for future research are given.

2   Preliminaries

2.1   General Preliminaries, Notations and Nomenclature

Databases may contain data representing temporal values. Based on their purpose and interpretation, such time indications can be classified into different categories [1], [3]. The work presented in this paper only considers temporal values of the category *valid time*. Their purpose or interpretation is for every valid-time indication to correspond to a collection of data and to indicate a period of time during which this data is a valid or true representation of reality.

The work presented in this paper concerns time indications subject to uncertainty. This uncertainty is always assumed to be caused by a (partial) lack of knowledge: the exact, intended time indication is not known, eventhough there is only one time indication intended and as such no variability. Confidence about exactly which time indication is the intended one in the context of such uncertainty is modelled using possibility theory [17], [19]. In the presented work, 'possibility' and 'necessity' are always interpreted as measures of plausibility, respectively necessity, given all available knowledge. Time intervals not subject to any imperfection are called *crisp time intervals* (CTI) in this paper.

2.2   Valid-Time Relations

A *valid-time relation* (VTR) always has *valid-time attributes*. These are attributes describing a single valid time [1] for the objects or concepts represented by the VTR's tuples. A VTR may contain different tuples corresponding to the same real-world concept or object. The non-valid-time attribute values of such a tuple represent the capacities of the properties described by their corresponding attributes which were, are or will be true or valid for the object or concept corresponding to the tuple during the period in time indicated by the valid-time indication represented by the tuple's valid-time attribute values. Thus, such a tuple represents the 'version' of the object or concept corresponding to this tuple which was, is or will be real or valid or the 'state' this object or concept was, is or will be in, during the period in time indicated by the valid-time indication represented by the tuple's valid-time attribute values. In the presented work, such valid-time indications will always be time intervals [1] and will always be referred to as *valid-time intervals* (VTI).

2.3 Uncertainty in Valid-Time Intervals

The presented work allows VTI to be subject to uncertainty, by allowing them to be *ill-known valid-time intervals* (IKVTI). The concept of IKVTI is based on the concepts of *possibilistic variables* (PV) and *ill-known intervals* (IKI) [3], [17], [20].

Definition 1. *A possibilistic variable (PV) X on a universe U is a variable taking exactly one value in U, but for which this value is (partially) unknown. The possibility distribution π_X on U models the available knowledge about the value that X takes: for each $u \in U$, $\pi_X(u)$ represents the possibility that X takes the value u.*

Now consider a set U containing single values (and not collections of values). When a PV X_v is defined on such a set U, the unique value X_v takes, which is (partially) unknown, will be a single value in U and is called an *ill-known value* (IKV) in U [3], [17], [20]. In this paper, IKV will be denoted using lower-case letters. The work presented in this paper uses a specific kind of IKI, defined as follows, although other definitions exist [3], [4].

Definition 2. *Consider an ordered set U. An* ill-known interval *(IKI) in U is an interval in U of which both boundary values are IKV in U.*

Specifically concerning valid time, an IKI in a time domain represented by the domains of a VTR's valid-time attributes is called an *ill-known valid-time interval* (IKVTI). The work presented in this paper requires the possibility distributions defining an IKVTI's IKV to be convex [17]. In this paper, an IKVTI with boundary IKV s and e will be noted $[s, e]$.

2.4 Evaluation of Temporal Relationships

To express temporal elementary query conditions, operators based on temporal relationships are necessary. In the presented work, only Allen relationships [16] between a CTI and a IKVTI are considered. To evaluate such relationships, the *ill-known constraints* (IKC) framework presented in [17] is used. It relies on the concept of IKC.

Definition 3. *Given an ordered set U, an ill-known constraint (IKC) C = (R, v) on U is specified by means of a binary relation $R \subseteq U^2$ and a fixed IKV v in U. Any set $A \subseteq U$ now satisfies IKC C = (R, v) if and only if:*

$$\forall a \in A : (a, v) \in R$$

The satisfaction of an IKC C by a set A will be noted $C(A)$ in this paper. Consider an ordered set U, an IKC $C = (R, v)$ on U and a set $A \subseteq U$. Due to the uncertainty inherent to v, it is uncertain whether A satisfies C or not. The degree of possibility $\text{Pos}(C(A))$ that A satisfies C and the degree of necessity $\text{Nec}(C(A))$ that A satisfies C, can be calculated as follows [17]:

$$\text{Pos}(C(A)) = \min_{a \in A} \left(\sup_{(a,w) \in R} \pi_{X_v}(w) \right) \tag{1}$$

$$\text{Nec}(C(A)) = \min_{a \in A} \left(\inf_{(a,w) \notin R} 1 - \pi_{X_v}(w) \right) \tag{2}$$

Given an ordered set U, degrees of possibility and necessity that a set $A \subseteq U$ satisfies a boolean combination of IKC on U can be found by using the possibilistic extensions of boolean operators 'and' ($\wedge$), 'or' ($\vee$) and 'not' ($\neg$), as described in [3], [4], [17].

The IKC framework now allows evaluating a given Allen relationship AR between a given CTI I and a given IKVTI $J = [s, e]$ by allowing the calculation of the degrees of possibility and necessity that $I \ AR \ J$ holds. For this, the combination of AR and J is translated to a specific boolean combination of specific IKC. These translations are shown in table 1. Every row of this table corresponds to a given Allen relationship between I and J, indicated by the row's value in the 'Allen Relationship' column. The collections of specific IKC for given Allen relationships are shown in the 'Constraints' column (every $C_i, i \in \{1, 2, 3, 4\}$ denotes an IKC) and the specific combination of these IKC used for evaluation of the Allen relationships are shown in the 'Combination' column. Finally, the degrees of possibility and necessity that $I \ AR \ J$ holds are then the degrees of possibility, respectively necessity that I satisfies the specific aggregation of specific IKC found as translation of the combination of AR and J. Using the formulas shown above, the requested possibility and necessity degrees can be calculated from these.

2.5   Bipolar Querying

As mentioned before, humans may express their query preferences using both positive and negative query conditions [12], [13]. If the semantics of these

Table 1. The translations of Allen relationships to the IKC framework

Allen Relationship	Constraints	Combination
I before J	$C_1 \overset{\triangle}{=} (<, s)$	$C_1(I)$
I equal J	$C_1 \overset{\triangle}{=} (\geq, s)$, $C_2 \overset{\triangle}{=} (\neq, s)$ $C_3 \overset{\triangle}{=} (\leq, e)$, $C_4 \overset{\triangle}{=} (\neq, e)$	$C_1(I) \wedge \neg C_2(I) \wedge$ $C_3(I) \wedge \neg C_4(I)$
I meets J	$C_1 \overset{\triangle}{=} (\leq, s)$ $C_2 \overset{\triangle}{=} (\neq, s)$	$C_1(I) \wedge \neg C_2(I)$
I overlaps J	$C_1 \overset{\triangle}{=} (<, e)$, $C_2 \overset{\triangle}{=} (\leq, s)$, $C_3 \overset{\triangle}{=} (\geq, s)$	$C_1(I) \wedge \neg C_2(I) \wedge \neg C_3(I)$
I during J	$C_1 \overset{\triangle}{=} (>, s)$, $C_2 \overset{\triangle}{=} (\leq, e)$ $C_3 \overset{\triangle}{=} (\geq, s)$, $C_4 \overset{\triangle}{=} (<, e)$	$(C_1(I) \wedge C_2(I)) \vee$ $(C_3(I) \wedge C_4(I))$
I starts J	$C_1 \overset{\triangle}{=} (\geq, s)$, $C_2 \overset{\triangle}{=} (\neq, s)$	$C_1(I) \wedge, \neg C_2(I)$
I finishes J	$C_1 \overset{\triangle}{=} (\leq, e)$, $C_2 \overset{\triangle}{=} (\neq, e)$	$C_1(I) \wedge \neg C_2(I)$

conditions are non-symmetric, meaning that the positive preferences can not be derived from the negative or vice versa, the bipolarity in this query is called *heterogenous* [13]. The presented work will concern only such heterogenous query bipolarity.

A query usually takes the form of a boolean combination of elementary query conditions. Every elementary query condition then expresses the user's demands concerning a single attribute. Bipolarity in a query can either be specified between or inside the elementary query conditions. In [13], it is shown that combining both approaches makes no sense and that the approach where bipolarity is specified inside elementary query conditions, using intuitionistic fuzzy sets [21], is a more intuitive one. In the presented work, only the latter approach is used. In this approach, elementary query conditions express both what is accepted and what is not accepted by the query, at once, and are called *bipolar query conditions* (BQC) [13].

Consider a relation attribute A. Let dom_A be the domain of A's data type, let μ_{c_A} and ν_{c_A} be membership functions from dom_A to the unit interval $[0, 1]$, where $\mu_{c_A}(x)$ represents to what extent $x \in dom_A$ is satisfactory and $\nu_{c_A}(x)$ to what extent x is unsatisfactory to a user, then a BQC c_A expressing this user's preferences concerning A can be modelled by an Intuitionistic Fuzzy Set (IFS) [21] as [12], [13]:

$$c_A = \{(v, \mu_{c_A}(v), \nu_{c_A}(v)) : v \in dom_A\} \tag{3}$$

Note that to allow overspecification of the user's preferences, the IFS's consistency condition can be relaxed, which means that there may exist values $v \in dom_A$ for which $\mu_{c_A}(v) + \nu_{c_A}(v) > 1$ [12], [13].

If the user explicitly defines μ_{c_A}, but doesn't define ν_{c_A}, then ν_{c_A} will be assumed to be the inverse of μ_{c_A} [13]. If the user explicitly defines ν_{c_A}, but doesn't define μ_{c_A}, then μ_{c_A} will be assumed to be the inverse of ν_{c_A} [13]. Thus, in the absence of clear heterogenousness of the bipolarity in a query condition, the bipolarity will be assumed homogenous [13].

The evaluation of a BQC results in a so-called *bipolar satisfaction degree* (BSD) [13], which is a pair

$$(s, d), \ s, d \in [0, 1]$$

where s is called the *satisfaction degree* and d is called the *dissatisfaction degree* [13]. Here, s and d are independent from each other and express to which extent the BSD respectively represents 'satisfied' and 'dissatisfied' [13]. Extreme values for s and d are 0 ('not at all') and 1 ('fully'). For example: the BSD $(1, 0)$ represents 'fully satisfied, not dissatisfied at all' [13].

As explained in [13], there is no consistency condition for BSD's and for a BSD (s, d), $s + d > 1$ is allowed. The motivation is that BSD's try to reflect heterogenous bipolarity in human reasoning, which can sometimes be inconsistent.

In general, the evaluation of a BQC c_A on relation attribute A for a tuple r will result in a BSD $(s^r_{c_A}, d^r_{c_A})$, which is calculated as follows. Let $r[A]$ denote the value of tuple r for attribute A, then [13]:

$$(s^r_{c_A}, d^r_{c_A}) = (\mu_{c_A}(r[A]), \nu_{c_A}(r[A])) \tag{4}$$

Remark that the traditional approach to fuzzy querying using regular fuzzy sets can be obtained from this as a special case, where the bipolarity involved is homogenous. In that case, a user only specifies positive query preferences [13].

3 A Novel Querying Approach

3.1 Valid-Time Relations Subject to Uncertainty

The presented proposal will concern VTR where the VTI are IKVTI. Generally, such a VTR R can be seen as constructed in the following way. Let R have n non-temporal attributes $A_i, 1 \leq i \leq n, i \in \mathbb{N}$ and two valid-time attributes VST and VET. Every tuple T of R represents an object or concept version or state which is valid during the time period indicated by the tuple's IKVTI I_T. This IKVTI I_T is now defined by two IKV, which respectively describe the starting and ending instants of I_T and are represented by the tuple's VST, respectively VET values. The interpretation is that the version or state corresponding to a tuple was, is or will be valid during a period of time, but exactly which period this is intended to be, is unknown. Confidence about exactly which period is intended, is modelled by the tuple's IKVTI [2], [4].

Table 2. The example relation used in this paper

ID	Author	VST	VET
1	Aloïsius	[4/4/1208, 6/4/1208, 16/4/1208]	[10/12/1208, 1/1/1209, 26/1/1209]
2	Theofilus	[2/4/1209, 12/4/1209, 22/4/1209]	[21/12/1209, 1/1/1210, 21/1/1210]
3	Gerardus	[14/1/1209, 15/1/1209, 16/1/1209]	[21/12/1209, 15/1/1210, 25/1/1210]
4	Euforius	[21/12/1210, 1/1/1211, 11/1/1211]	[21/12/1211, 1/1/1212, 11/1/1212]
5	Ambrosius	[11/12/1213, 21/12/1213, 15/1/1214]	[9/10/1216, 10/10/1216, 15/10/1216]
6	Aloïsius	[21/12/1213, 1/1/1214, 11/1/1214]	[9/6/1217, 9/6/1217, 12/6/1217]
7	Gerardus	[29/12/1214, 1/1/1215, 8/1/1215]	[9/6/1217, 10/6/1217, 12/6/1217]

In this paper, the relation shown in table 2 will be used as example relation. This relation models the being in effect of medieval legal acts. Since the properties of a legal act cannot change once it has taken effect, every legal act has only one version or state. This was deliberately done to simplify the example. Thus, every tuple of the relation corresponds to a legal act. A tuple's value for attribute 'ID' is a number uniquely identifying the legal act corresponding to the tuple. A tuple's value for attribute 'Author' is a character string representation of the name of the author of the legal act corresponding to the tuple. A tuple's IKVTI represents the time period during which the act corresponding to the tuple was in effect. For this, every value for VST, respectively VET represents an IKV describing the day on which the legal act respectively took effect and stopped taking effect. For this, every value for VST or VET is a triple $[d_1, d_2, d_3]$, where d_1, d_2, d_3 are elements of the ordered set of days in history. Such a triple $[d_1, d_2, d_3]$ now defines a triangular (and thus convex) possibility distribution π

which defines the mentioned IKV and which is defined by (differences in dates in this function prescription are expressed in amounts of days):

$$\pi(x) = \begin{cases} \frac{x-d_1}{d_2-d_1}, & \text{if } d_1 \leq x < d_2 \\ \frac{d_3-x}{d_3-d_2}, & \text{if } d_2 \leq x \leq d_3 \\ 0, & \text{else} \end{cases} \tag{5}$$

3.2 Querying Using Bipolar Valid-Time Conditions

The presented work introduces a novel querying technique. The most interesting aspect of this technique is that it allows the user to specify a bipolar valid-time demand. According to this technique, a user query Q consists of two separate parts Q_n and Q_t: $Q = (Q_n, Q_t)$. Here, Q_n is a boolean combination of BQC on non-valid-time attributes, expressing the user's non-temporal demands. Q_t expresses the user's valid-time demands and is a single crisp temporal BQC $((AR_+, I_+), (AR_-, I_-))$, where both AR_+ and AR_- are Allen relationships and both I_+ and I_- are CTI. The interpretation is that the user requires an object or concept that has a version or state that complies with his or her non-temporal demands and was, is or will be valid during a time interval which is in Allen relationship AR_+ with I_+ and wasn't, isn't or won't be valid during a time interval which is in Allen relationship AR_- with I_-.

Consider the example relation shown in table 2. Now assume a user queries the relation to find all legal acts of which the author is preferably named Aloïsius, less preferably Euforius and perhaps Eugenius, and of which the author is preferably not named Ambrosius, rather not Theofilus and perhaps not Antonius and which took effect preferably before 2/1/1210 and preferably not after 1/1/1214. These demands can now be translated to a query Q_{ex} in the following way:

$$Q_{ex} = (Q_{n,ex}, Q_{t,ex}) = (Q_{n,ex}, ((AR_{+,ex}, I_{+,ex}), (AR_{-,ex}, I_{-,ex}))),$$
$$Q_{n,ex} = \{(x, \mu_{ex}(x), \nu_{ex}(x)), \forall x \in \mathbb{S}\}$$
$$AR_{+,ex} = AR_{-,ex} = DURING$$
$$I_{+,ex} = \;]-infinity, 1/1/1210], \quad I_{-,ex} = [1/1/1214, +infinity[$$

and μ_{ex} and ν_{ex} are the membership functions of the fuzzy sets:

$$\{(Aloisius, 1), (Euforius, 0.7), (Eugenius, 0.1)\}, \text{ respectively}$$
$$\{(Ambrosius, 1), (Theofilus, 0.7), (Antonius, 0.1)\}$$

and $\mathbb{S}$ is the set of all author names in the example relation's 'Author' attribute domain.

3.3 Elementary Query Condition Evaluation

Generally, a first step in determining which objects or concepts corresponding to tuples to present to a user as answer to his or her query, is evaluating the query's elementary conditions for every tuple. In the presented work, given a user query Q constructed as proposed in section 3.2 and using the same notations, the following is done separately for every tuple T of VTR R:

- every non-temporal BQC in Q_n is evaluated as described in [13], resulting in a BSD for each. The interpretation is as described in [13]: the BSD's satisfaction, respectively dissatisfaction degrees express to which extent T satisfies, respectively dissatisfies the user preferences expressed by the BQC.
- the query's temporal BQC Q_t is evaluated as follows. Let I_T be the tuple's IKVTI. Then, independently, the statements 'I_T AR_+ I_+', respectively 'I_T AR_- I_-' are evaluated using the IKC framework as described in [2], [4], [17]. These evaluations result in a possibility degree $Pos_+(I_T)$ and a necessity degree $Nec_+(I_T)$, respectively a possibility degree $Pos_-(I_T)$ and a necessity degree $Nec_-(I_T)$. The interpretation is that $Pos_+(I_T)$ and $Nec_+(I_T)$ express the possibility, respectively necessity, that the time interval during which the version or state represented by T is valid and which is intended by I_T, is in relationship AR_+ with I_+ and thus complies with the user's positive temporal demand. Furthermore, the interpretation is that $Pos_-(I_T)$ and $Nec_-(I_T)$ express the possibility, respectively necessity, that the time interval during which the version or state represented by T is valid and which is intended by I_T, is in relationship AR_- with I_- and thus complies with the user's negative temporal demand.

Table 3 shows the results of the evaluation of the example query's elementary query conditions for the tuples of the example relation.

Table 3. Elementary query condition evaluation results for the example relation

ID	$BSD_{Q_{N,ex}}(T)$	$(Pos_+(I_T), Nec_+(I_T))$	$(Pos_-(I_T), Nec_-(I_T))$
1	$(1,0)$	$(1,1)$	$(0,0)$
2	$(0,0.7)$	$(1,0)$	$(0,0)$
3	$(0,0)$	$(11/25,0)$	$(0,0)$
4	$(0.7,0)$	$(0,0)$	$(0,0)$
5	$(0,1)$	$(0,0)$	$(14/25,0)$
6	$(1,0)$	$(0,0)$	$(1,0)$
7	$(0,0)$	$(0,0)$	$(1,1)$

3.4   Aggregation and Ranking

Generally, a second step in determining which objects or concepts corresponding to tuples to present to a user as answer to his or her query, is aggregating, for every tuple, the tuple's evaluation results for the query's elementary conditions, in order to determine how well the tuple complies with the entire user request expressed by the combination of the elementary query conditions. Usually, these evaluation results are quantifications of (dis)satisfaction. However, in the presented proposal, two different types of evaluation results can be discerned:

- the BSD's (dis)satisfaction degrees constitute quantifications of (dis)satisfaction: they quantify to which extent a tuple's attribute values (dis)satisfy a user's non-temporal preferences and thus assess an answer to the question: 'To what extent does a version or state represented in the relation (dis)satisfy the user's request and could thus be a (un)wanted result?'

- the possibility and necessity degrees Pos_+, Nec_+, respectively Pos_-, Nec_- constitute quantifications of possibility and necessity: they quantify the possibility that a tuple's intended crisp VTI does (not) comply with the user's temporal demands by quantifying confidence about exactly which crisp VTI is the tuple's intended VTI. Thus, they assess an answer to the question: 'Given all available knowledge, how plausible is it that a version or state represented in the relation (that may or may not (dis)satisfy the user's non-temporal preferences) actually existed, exists or will exist during the time period indicated by the user?'

A fundamental question arises now: should one consider combining quantifications of these different categories? On one hand, such quantifications have clearly different semantics and it would not be clear what exactly the meaning would be of the result of such a combination or what the semantically most coherent ways would be to further process such combination results. Thus, it is important, for every query result tuple presented to the user, to certainly keep both different types of evaluation results as separate (meta)data. On the other hand, without an unambiguous and straightforward ranking of the query result tuples, the user cannot clearly discern the result tuples which comply well with his or her demands from those which don't. This would defeat the purpose of querying. Reasonably, such a ranking should be based on the elementary query condition evaluation results. As there cannot exist a ranking between quantifications of categories with different semantics, a combination of quantifications of satisfaction and possibility seems to be required. In most existing proposals requiring a combination of quantifications of satisfaction and possibility, both quantifications are combined as to restrict one another. The result is usually seen as a quantification of possibility. Below, this approach is translated to the specific situation encountered in the presented work.

For every tuple T, the BSD's which are the evaluation results of the non-temporal BQC are combined to a single BSD $(s_n(T), d_n(T))$ as described in [13]. In this combination method, satisfaction degrees are combined with each other and separately, dissatisfaction degrees are combined with each other. This reasoning is now extended to include $Pos_+(I_T)$, $Nec_+(I_T)$, $Pos_-(I_T)$ and $Nec_-(I_T)$: a couple $(Pos_+(T), Nec_+(T))$, respectively $(Pos_-(T), Nec_-(T))$ is calculated, expressing the possibility and necessity that the version or state corresponding to T complies with all of the user's positive, respectively negative demands. This calculation is done as follows:

$$Pos_+(T) = \min(s_n(T), Pos_+(I_T))$$

$$Nec_+(T) = \begin{cases} 0, & \text{if } Pos_+(T) < 1 \\ \min(s_n(T), Nec_+(I_T)), & \text{else} \end{cases}$$

$$Pos_-(T) = \min(d_n(T), Pos_-(I_T))$$

$$Nec_-(T) = \begin{cases} 0, & \text{if } Pos_-(T) < 1 \\ \min(d_n(T), Nec_-(I_T)), & \text{else} \end{cases}$$

Table 4. Aggregation and ranking results for the example relation and query

ID	$Pos_+(T)$	$Nec_+(T)$	$Pos_-(T)$	$Nec_-(T)$	$BSD_{Q_{n,ex}}$	$(Pos_+(I_T), Nec_+(I_T))$	$(Pos_-(I_T), Nec_-(I_T))$
1	1	1	0	0	$(1,0)$	$(1,1)$	$(0,0)$
2	0	0	0	0	$(0,0.7)$	$(1,0)$	$(0,0)$
3	0	0	0	0	$(0,0)$	$(11/25,0)$	$(0,0)$
4	0	0	0	0	$(0.7,0)$	$(0,0)$	$(0,0)$
6	0	0	0	0	$(1,0)$	$(0,0)$	$(1,0)$
7	0	0	0	0	$(0,0)$	$(0,0)$	$(1,1)$
5	0	0	14/25	0	$(0,1)$	$(0,0)$	$(14/25,0)$

Next, a tie-break approach is used to rank the versions or states represented in R: ranking is done based on the value of $Pos_+(T)$, with $Nec_+(T)$ as tie-breaker for $Pos_+(T)$, with $Pos_-(T)$ as tie-breaker for $(Pos_+(T), Nec_+(T))$ and finally $Nec_-(T)$ as tiebreaker for $(Pos_+(T), Nec_+(T), Pos_-(T))$. The results of this aggregation and ranking approach for the example relation and query are shown in table 4.

The introduced technique for aggregating elementary query condition evaluation results and determining a ranking does not require the combination of quantifications with different interpretations and still presents the query result tuples in a ordered manner, where this ordering is consistent with the expected extend to which each tuple is usefull to the user. However, the technique still requires the user to decide which objects or concepts constitute the best query answer, although this decision is heavily supported.

4   Conclusions and Future Work

In this paper, a novel technique to query a valid-time relation containing valid-time data subject to uncertainty in a bipolar way, is presented. This technique allows the user to specify a single valid-time bipolar query condition. A major issue concerning the need to combine quantifications of (dis)satisfaction with quantifications of possibility resulting from·this technique is presented and shortly discussed, along with a solution for this issue. In the near future, the interactions between these types of quantifications and between uncertainty in valid-time data and bipolar querying will be further studied. Also, an approach to allow the valid-time bipolar query condition to be fuzzy will be considered.

References

1. Jensen, C.S., et al.: The consensus glossary of temporal database concepts - february 1998 version. In: Etzion, O., Jajodia, S., Sripada, S. (eds.) Dagstuhl Seminar 1997. LNCS, vol. 1399, p. 367. Springer, Heidelberg (1998)
2. Pons, J.E., Marín, N., Pons, O., Billiet, C., De Tré, G.: A Relational Model for the Possibilistic Valid-time Approach. International Journal of Computational Intelligence Systems 5(6), 1068–1088 (2012)
3. Billiet, C., Pons, J.E., Pons Capote, O., De Tré, G.: Evaluating Possibilistic Valid-Time Queries. In: Greco, S., Bouchon-Meunier, B., Coletti, G., Fedrizzi, M., Matarazzo, B., Yager, R.R. (eds.) IPMU 2012, Part I. CCIS, vol. 297, pp. 410–419. Springer, Heidelberg (2012)

4. Pons, J.E., Billiet, C., Pons Capote, O., De Tré, G.: A possibilistic valid-time model. In: Greco, S., Bouchon-Meunier, B., Coletti, G., Fedrizzi, M., Matarazzo, B., Yager, R.R. (eds.) IPMU 2012, Part I. CCIS, vol. 297, pp. 420–429. Springer, Heidelberg (2012)
5. Medina, J.M., Pons, O., Amparo Vila, M.: Gefred: A Generalized Model of Fuzzy Relational Databases. Information Sciences 76(1-2), 87–109 (1994)
6. Bosc, P., Pivert, O.: Modeling and Querying Uncertain Relational Databases: A Survey of Approaches Based on the Possible Worlds Semantics. International Journal of Uncertainty, Fuzziness and Knowledge-Based Systems 18(5), 565–603 (2010)
7. Dyreson, C.E., Snodgrass, R.T.: Supporting Valid-Time Indeterminacy. ACM Transactions on Database Systems 23(1), 1–57 (1998)
8. Zadrozny, S., De Tré, G., De Caluwe, R., Kacprzyk, J.: An overview of fuzzy approaches to flexible database querying. In: Handbook of Research on Fuzzy Information Processing in Databases. IGI Global (2008)
9. Devos, F., Maesfranckx, P., De Tré, G.: Granularity in the Interpretation of Around in Approximative Lexical Time Indications. Journal of Quantitative Linguistics 5(3), 167–173 (1998)
10. Kacprzyk, J., Zadrozny, S.: Computing with Words in Intelligent Database Querying: Standalone and Internet-based Applications. Information Sciences 134(1-4), 71–109 (2001)
11. Dubois, D., Prade, H.: Bipolarity in Flexible Querying. In: Andreasen, T., Motro, A., Christiansen, H., Larsen, H.L. (eds.) FQAS 2002. LNCS (LNAI), vol. 2522, pp. 174–182. Springer, Heidelberg (2002)
12. De Tré, G., Zadrozny, S., Bronselaer, A.: Handling Bipolarity in Elementary Queries to Possibilistic Databases. IEEE Transactions on Fuzzy Systems 18(3), 599–612 (2010)
13. Matthé, T., De Tré, G., Zadrozny, S., Kacprzyk, J., Bronselaer, A.: Bipolar Database Querying Using Bipolar Satisfaction Degrees. International Journal of Intelligent Systems 26(10), 890–910 (2011)
14. Galindo, J., Medina, J.M.: FTSQL2: Fuzzy Time in Relational Databases*. In: Proceedings of the 2nd International Conference in Fuzzy Logic and Technology, pp. 47–50 (2001)
15. Schockaert, S., De Cock, M., Kerre, E.E.: Fuzzifying Allen's Temporal Interval Relations. IEEE Transactions on Fuzzy Systems 16(2), 517–533 (2008)
16. Allen, J.F.: Maintaining Knowledge about Temporal Intervals. Communications of the ACM 26(11), 832–843 (1983)
17. Pons, J.E., Bronselaer, A., De Tré, G., Pons, O.: Possibilistic evaluation of sets. International Journal of Uncertainty. Fuzziness and Knowledge-Based Systems (2012); Accepted for publication in the International Journal of Uncertainty, Fuzziness and Knowledge-Based Systems
18. Billiet, C., Pons, J.E., Matthé, T., De Tré, G., Pons Capote, O.: Bipolar fuzzy querying of temporal databases. In: Christiansen, H., De Tré, G., Yazici, A., Zadrozny, S., Andreasen, T., Larsen, H.L. (eds.) FQAS 2011. LNCS, vol. 7022, pp. 60–71. Springer, Heidelberg (2011)
19. Dubois, D., Prade, H.: Possibility Theory. Plenum Press (1988)
20. Dubois, D., Prade, H.: Incomplete Conjunctive Information. Computers & Mathematics with Applications 15(10), 797–810 (1988)
21. Atanassov, K.T.: Intuitionistic Fuzzy Sets. Fuzzy Sets and Systems 20(1), 87–96 (1986)

Declarative Fuzzy Linguistic Queries
on Relational Databases

Clemente Rubio-Manzano[1], Pascual Julián-Iranzo[2], Esteban Salazar-Santis[1],
and Eduardo San Martín-Villarroel[1]

[1] Dep. of Information Systems, Univ. of the Bío-Bío, Chile
clrubio@ubiobio.cl
[2] Dep. of Information Technologies and Systems, Univ. of Castilla-La Mancha, Spain
Pascual.Julian@uclm.es

Abstract. In this paper we propose a declarative method to formulate
fuzzy linguistic queries on Relational Database Management Systems.
That is, flexible queries containing linguistic terms associate to the attributes of a table of a relational database. To this end, we adapt techniques originate from a proximity-based Logic Programming Language
called Bousi~Prolog.

Keywords: Relational databases, Information retrieval, Fuzzy Linguistic Queries, Proximity/Similarity Relations, Fuzzy Logic Programming.

1 Motivation and Introduction

Most corporations have been storing a large amount of data into **Database Management Systems** (RDBMS, for short) for years. Data manipulation queries are
performed by using a specific and formal language supported by a particular
RDBMS, being the most commonly used the **Structed Query Language** (SQL,
for short). These query languages have been designed to retrieve information
from databases containing precise data and where the user needs are indicated
specifying a set of selection conditions which require strict commands. In many
situations, users of a database system are not completely familiar with the data
stored in it, hence it is customary that a query is formulated by an expert user
who translates a natural language question into precise constrains on rigid values, what implies the possibility to miss interesting answers.

On the other hand, a portion of knowledge is not incorporated in an appropriate way within present-day database systems [17], since the real world information is often permeated by vagueness and/or imprecision. Therefore, database
systems should deal with this kind of information if they want to be capable of
getting a flexible, expressive or user-friendly interface. Also, they should permit
to retrieve information in a flexible way.

In this respect the techniques based on fuzzy set theory [19] are very useful
for modeling the vagueness and imprecision. In fact, fuzzy query processing for
relational database systems is an important application of the fuzzy set theory
because it allows to retrieve data from relational database systems with queries
containing linguistic terms [4].

H.L. Larsen et al. (Eds.): FQAS 2013, LNAI 8132, pp. 413–424, 2013.

In this paper we propose a declarative method to formulate fuzzy linguistic queries on Relational Database Management Systems. This new method allows us to retrieve data from those systems by using the inference mechanism of a proximity-based Logic Programming Language called Bousi~Prolog. To this end, firstly, a set of rows from a relational table is transformed into a set of Bousi~Prolog facts and, subsequently, a fuzzy relation is established between the constant arguments of such facts and the linguistic labels defined for a linguistic variable. This relation provides a mapping between the data records in the database and the fuzzy concepts that may be used in queries. This fuzzy relation is determined by way of a standard matching algorithm and it feeds an efficient semantic unification mechanism which has been incorporated into Bousi~Prolog. After that, a simple query (introduced by the user thought a graphical interface) is translated to a Bousi~Prolog query and then launched to the system, that will provide the corresponding answers. In order to implement this method, the Bousi Gunfire and Bousi Exporta tools have been built. Bousi Gunfire is a reduced version of the Bousi~Prolog system i.e. it only contains the core of this language. Bousi Exporta is a tool that allows the connection with a particular Relational Database Management System, visualizes its structure (tables, attributes and values) and transforms the rows and the linguistic variables into a Bousi~Prolog program. The result is a system which is able to formulate fuzzy linguistic queries on a particular Relational Database Management System. Moreover, our system is multi-platform since the extraction of data employs the meta-information provided by a Relational Database Management System.

Ending this section, it is fair to say that, in the present work (which is only a first step) we have not considered performance or scalability issues.

2   Fuzziness in Databases and Related Work

Imprecision and vagueness are two major types of imperfect information. Intuitively, the imprecision and vagueness appear when we do not know what exact value to choose for a particular attribute. A fuzzy set [19] may be used to represent vague and/or imprecise information. At least, there are two lines of work using fuzzy sets in combination with databases. The one where the data model is leaving intact and query processing is enhanced to allow vague concepts, which are represented by fuzzy sets and/or linguistic variables, [2,18,17] and the one that applies the fuzzy set theory to relational data models modifying relational calculus and algebra [3,13,16].

The first line of work seems to be more practical and promising since the relational model remains the most widely used alternative. In fact, inside of this first line, many methods have been presented for fuzzy query processing in relational database systems. Most of existing approaches for handling fuzziness in queries require explicit definitions of membership functions. Other approaches adopt clustering techniques as a tool to generate the mapping between fuzzy terms, defined at a higher level of abstraction, and the database records [4,9]. While the first option requires an exact specification of the meaning of each fuzzy term known by the system, the second one makes the definition task much easier since instead of explicitly defining the range of concept values corresponding to each term, the users need only to define the relative order of linguistic terms and the system, thru the clustering algorithm, will match each linguistic term to the adequate cluster of records [9].

The work presented in this paper can be seen as a hybrid approach since it makes use of membership functions, which give meaning to the linguistic terms, and clustering techniques to perform the query processing[1]. Hence, we will focus on these issues. In this context, the development of a fuzzy querying system consists in building a user interface to get information from a crisp database, making references to fuzzy linguistic concepts and putting in relation those fuzzy linguistic concepts with the precise data from the relational database.

Additionally, it is important to determinate the type of fuzzy terms on which a method can act. Following [9], we consider that a fuzzy term can be classified as: a qualitative numeric descriptor, a qualitative non-numeric descriptor or a quantification description. A qualitative numeric descriptor is a word that describes some numeric value or a range of numeric values. A qualitative non-numeric descriptor is a word which describes some non-numeric concept. A quantification description is a word which describes the quantity of responses desired in a natural language query. In contrast to [9] which are concerned only with qualitative numeric descriptors, our approach can work both with qualitative non-numeric and numeric descriptors. Although in this paper we concentrate our attention in the treatment of qualitative numeric descriptors. This ability is because our approach is based on proximity-based Logic Programming, which is able to deal with both kinds of descriptors.

In order to treat a qualitative numeric descriptor, we are going to use techniques developed for the construction of a semantic unification algorithm[7] which will put in relation a numeric descriptor with (possibly) several linguistic concepts, i.e. our method acts like a fuzzy clustering algorithm.

3   Preliminary Definitions, Concepts and Notations

Formally, our approach relies on a combination of concepts coming from diverse areas such as relational databases [5], fuzzy logic [19], and logic programming [11] which are briefly presented in this section.

The Relational Database Model. In the relational database model, a table is the main structure used to represent a class of real world entities. A table is defined by a set of columns corresponding to the attributes of a modeled class of entities. Each entity is represented by a row (tuple) in the table. Formally, a table T is a set of tuples and a tuple t is a set of pairs $\langle attribute - domain, value \rangle$ of the form: $t = \{\langle A_1, v_1 \rangle, \langle A_2, v_2 \rangle, \ldots, \langle A_n, v_n \rangle\}$ such that $v_i \in D_i$, being D_i a domain.

---

[1] Note that our semantic unification mechanism can be seen as an efficient fuzzy clustering algorithm.

Fuzzy Relations and Syntactic Domains. The concept of fuzzy relation was introduced by Zadeh in [19]. A *binary fuzzy relation* on a set U is a fuzzy subset on $U \times U$ (that is, a mapping $U \times U \longrightarrow [0,1]$). A binary fuzzy relation $\mathcal{R}$ is said to be a *proximity relation* if it fulfills the *reflexive* property (i.e. $\mathcal{R}(x,x) = 1$ for any $x \in U$) and the *symmetric* property (i.e. $\mathcal{R}(x,y) = \mathcal{R}(y,x)$ for any $x,y \in U$). If in addition it has the *transitive* property (i.e., $\mathcal{R}(x,z) \geq \mathcal{R}(x,y)\triangle\mathcal{R}(y,z)$ for any $x,y,z \in U$; where the operator '$\triangle$' is an arbitrary t-norm), it is called a *similarity relation*. Proximity relations also are called *tolerances* and they provide a mathematical tool for studying the indiscernibility of phenomena and objects.

We are mainly concerned with proximity relations on a syntactic domain. A similarity relation $\mathcal{R}$ on the alphabet of a first order language can be extended to terms by structural induction in the usual way [15].

Fuzzy Logic Programming and Bousi~Prolog. Fuzzy Logic Programming [10] is a research area which investigates how to introduce fuzzy logic concepts into logic programming in order to deal with the vagueness in a declarative way. When the imprecision is modeled by using similarity relations then we must speak of Similarity-based Logic Programming [15]. Bousi~Prolog (BPL, for short) [6,8] is an extension of the standard Prolog language that materializes this line of work and leads to a proximity-based logic programming framework. Its operational semantics is an adaptation of the SLD resolution principle where classical unification has been replaced by a fuzzy unification algorithm based on proximity relations. Informally, the *weak unification* algorithm states that two terms $f(t_1,\ldots,t_n)$ and $g(s_1,\ldots,s_n)$ weakly unify if the root symbols f and g are close and each of their arguments t_i and s_i weakly unify. Therefore, the weak unification algorithm does not produce a failure if there is a clash of two syntactical distinct symbols, whenever they are approximate, but a success with a certain approximation degree. Hence, Bousi~Prolog computes answers as well as approximation degrees. Bousi~Prolog makes a clear distinction between precise and vague knowledge. In a BPL program Precise knowledge is specified by a set of Prolog facts and rules, Vague Knowledge is mainly specified by a set of (what we call) proximity equations[2], defining a fuzzy binary relation (which are expressing how close are two concepts), and Control is let automatic to the system, through an enhanced SLD resolution procedure with a fuzzy unification algorithm based on fuzzy binary relations. These features are evidenced by the following example.

Example 1. Suppose a fragment of a database which stores information about people. Suppose some fuzzy subsets over the domain age, from which it have been obtained the proximity degrees between the linguistic labels young, middle and old. This knowledge can be coded by a set of proximity equations, as the following fragment of a deductive database shown.

---

[2] A *proximity equation*, denoted $a \sim b = \alpha$, represents an entry of a fuzzy binary relation, its intuitive reading is that two constants, n-ary function symbols or n-ary predicate symbols, a and b, are approximate with a certain degree, α.

```
% FACTS AND RULES (precise knowledge)
age(mary, middle).        age(sam, young).        age(john, old).
friend(X,Y):-age(X,Z), age(Y,Z), X \= Y.

% PROXIMITY EQUATIONS (vague knowledge)
young ~ middle = 0.75.  old ~ young = 0.25.  middle ~ old = 0.75.
```

In a standard Prolog system, if we ask about whether mary is a friend of sam, "?- friend(mary,sam)", the system fails. However the BPL system allows us to obtain the answer "Yes with 0.75", thanks to its ability to manage proximity equations during the unification process.

Linguistic Variables. A Linguistic Variable [20] is a quintuple $\langle X, T(X), U, G, M \rangle$ where: X is the variable name, $T(X)$ is the set of linguistic terms of X (i.e., the set of names of linguistic values of X, also known as linguistic labels), U is the domain or universe of discourse, G is a grammar that allows to generate $T(X)$ and M is a semantic rule which assigns to each linguistic term x in $T(X)$ its meaning (i.e., a fuzzy subset of U —characterized by its membership function μ_x—). It is usual to make the distinction between *atomic terms* (also called, *primary terms*) and *composite terms* which are composed of primary terms. The meaning of primary terms is defined axiomatically, assigning to each atomic term a fuzzy subset on U. In other words, the fuzzy subsets that M applies to composite terms are calculated, while the ones applied to primary terms are defined (in a subjective and context-dependent way).

In order to implement the concept of a linguistic variable, Bousi~Prolog only pays attention to its semantic component. That is, for a given variable X, only the domain U and the fuzzy subsets which are associated to primary linguistic terms in $T(X)$ are considered for their definition; the rest of composite terms are calculated automatically. On the other hand, it does not make a lexical distinction between the syntactic and the semantic component of X. Hence, Bousi~Prolog makes use of two directives to define and declare the structure of a linguistic variable X.

The domain directive allows to declare and define the universe of discourse or domain associated to a linguistic variable. The concrete syntax of this directive is: ":-domain(Dom_Name(n,m,Magnitude)).", where, Dom_Name is the name of the domain, n and m (with $n < m$) are the lower and upper bounds of the real subinterval $[n, m]$, and Magnitude is the name of the unit wherein the domain elements are measured.

The fuzzy_set directive allows to declare and define a list of fuzzy subsets (which are associated to the primary terms of a linguistic variable) on a predefined domain. The concrete syntax of this directive is:

```
:-fuzzy_set(Dom_Name,[SubS_1(a1,b1,c1[,d1]),...,SubS_n(an,bn,cn[,dn]])).
```

Fuzzy subsets are defined by indicating their name, SubS_i, and membership function type. At this time, two types of membership functions are possible: either trapezoidal functions, if four arguments are given, or triangular functions, if three arguments are used.

Additionally, a BPL program may include what we call "domain points". A *domain point* is our practical artifice to represent a precise crisp value in the universe of discourse, aiming to compare it with other linguistic terms. In its simplest syntactic form, a domain point is denoted by "Dom_Name#Dom_Val", where Dom_Name is a domain name and Dom_Val is a crisp value of that domain.

It is noteworthy that the implementation of a linguistic variable makes possible the manipulation of its linguistic labels as standard identifiers of the BPL language. Therefore, they may be used as regular symbols of a first order alphabet, that is, as constants, functions or, even, predicate symbols (See [7] for an extensive discussion).

4   Fuzzy Linguistic Queries on Relational Databases

In this section we propose an efficient declarative generic method to formulate fuzzy linguistic queries on a particular RDBMS. For this propose, two algorithms are used. The first one takes as input a relational table and defines a set of linguistic variables (on some domain-attributes of the input table). Then some table attributes are linked to linguistic variables and their values transformed into domain points (relative to that linguistic variables). Afterwards, each tuple is converted into a BPL fact. Finally, it returns as output a BPL program which is formed by a set of facts plus a set of domain and fuzzy_set directives that represent the linguistic variables previously defined. This BPL program (facts and directives) will be the input of the second algorithm that puts in relation the precise information from the database (which have been transformed into BPL domain points) with the respective linguistic labels of the linguistic variables defined by the user. Then, we could formulate fuzzy linguistic queries on the relational database.

Both algorithms are explained with detail in the following sections.

4.1   From Crisp Data to Imprecise Knowledge

The aim of this phase is to transform a set of tuples from a relational table into a set of Bousi~Prolog facts. Additionally, we want to transform each linguistic variable which has been defined on some vague attribute-domain into a set of domain and fuzzy_set directives. For this task, we use an Algorithm which generates a fact for each tuple of a table and one constant argument for each ⟨*Attribute, Value*⟩ pair of the tuple. The generated facts have the following form: "*predicate_symbol*($c_1, \ldots, c_n$).", where the name used for the predicate symbol is the name of the own table. We have two cases for the generation of the constants: if the pair ⟨*attribute_name, attribute_value*⟩ analyzed has an attribute which is vague in nature, then (i) generate a domain point formed by the name of the attribute concatenated to the symbol "#" and its associated attribute_value (i.e. "*attribute_name#attribute_value*", which is the BPL syntax for a domain point), else (ii) take its associated *attribute_value* as a constant argument.

The following example illustrates the first phase of our method which is implemented by this algorithm that has just informally described.

Example 2. Assume a fragment of a database that stores information about people and their jobs. We want to know who is middle-aged and likes science. Suppose a relational table named "**person**". Here, a person is defined by a key, his name, age and job.

Key	Name	Age	Job
01	John	24	Programmer
02	Paul	30	Engineer
03	Mary	34	Teacher
04	Warren	45	Football player

In order to render the database querying more flexible, our method must define a linguistic variable on a qualitative non-numeric descriptor (vague attribute). In this example the attribute Age is associated to the linguistic variable **age**, which is defined by means of a graphical tool. This graphical tool undertakes the task of generating the appropriate **domain** and **fuzzy_set** directives. Then each row is converted into a BPL fact, formed by the name of the table, as the predicate symbol, and the values of the attributes, as the constant arguments. If an attribute is vague, a domain point is formed instead, because we want to associate a concrete value with a linguistic variable. For example, the constant "age#24" indicating that the value 24 will be linked to the linguistic variable **age** (and hence with its respective linguistic terms) in a later phase. In the end, the BPL program generated by the algorithm is the following:

```
%% Linguistic variable defined on attribute age
:-domain(age(0,100,years)).
:-fuzzy_set(age,
          [young(0,0,20,50), middle(20,40,60,80), old(50,80,100,100)]).
%% Facts
person(01,john,age#24,programmer).  person(03,mary,age#34,teacher).
person(02,paul,age#30,engineer).  person(04,warren,age#45,football_player).
```

4.2 Generation of Proximity Equations

This section describes the second phase of our method, where the BPL program Π produced by the first algorithm is analyzed to extract all the information related to the linguistic variables that it contains. We start with a syntactical analysis process where the directives "**domain**" and "**fuzzy_set**" are read and the domains and associated fuzzy subsets are built. Also the domain points in the facts are collected. Conceptually, this process may be understood as one that builds a table of linguistic terms along with their meanings, as Table 1 illustrates.

Once the table of linguistic terms has been built, the next phase is focused on generating fuzzy relations between the stored linguistic terms in order to compile all the semantic information associated with them. The idea is the following: Suppose that $T(X) = \{x_i \mid i \in I\}$, where I is a set of indexes. For each x_i and x_j, with $i, j \in I$, we generate the entry of a fuzzy relation on $T(X)$: $\mathcal{R}(x_i, x_j) = \alpha$. The relationship degree α can be calculated as the relation between the fuzzy

Table 1. Table of linguistic terms: Memory representation for a fragment concerning a linguistic variable X on a domain Dom. For domain points no meaning has been assigned, since they are not properly considered as fuzzy subsets in our approach. This is coded by a null entry ($\bot$) int the membership function field.

Term	Domain	Membership Function	Term	Domain	Membership Function
x_1	Dom	μ_{x_1}	$Dom\#u_1$	Dom	$\bot$
x_2	Dom	μ_{x_2}	$Dom\#u_2$	Dom	$\bot$
...			...		

subsets $M(x_i)$ and $M(x_j)$ associated to these terms as meaning. The case where $M(x_j) = \bot$, that is x_j is a domain point, should be addressed specifically. More precisely, the generation phase is implemented by means of the following algorithm, which is an adaptation of one developed in [7] for the special case where the distinction between general and specific knowledge [14] is not taken into account when dealing with domain points.

Algorithm 1

Input: $S = \{\langle x_i, \mu_{x_i} \rangle \mid 1 \leq i \leq n\}$, *Subset of terms/meanings of a linguistic variable X.*
Output: *A set $\mathcal{R}$ of entries which defines a fuzzy relation on S.*
Initialization: $\mathcal{R} := \emptyset$
For each $\langle x_i, \mu_{x_i} \rangle$ **and** $\langle x_j, \mu_{x_j} \rangle$, **with** $i, j \in I$, **do**

    Case of
        1. $\mu_{x_i} \neq \bot$ and $\mu_{x_j} \neq \bot$: $\mathcal{R} := \mathcal{R} \cup \{\mathcal{R}(x_i, x_j) = match(\mu_{x_j}, \mu_{x_i})\}$;
        2. $\mu_{x_i} \neq \bot$ and $\mu_{x_j} = \bot$: **Let** $x_j = dom\#u_j$ **in** $\mathcal{R} := \mathcal{R} \cup \{\mathcal{R}(x_i, x_j) = \mu_{x_i}(u_j)\}$;
    endCase

endFor
Return $\mathcal{R}$

It is noteworthy that, in last algorithm, the subset S only contains the primary terms in $T(X)$ an those composite terms occurring in the program Π. The matching function calculates the degree of relation between two fuzzy subsets and it has been implemented using a technique successfully tested in the system *FuzzyClips* [12].

Once the relation, $\mathcal{R}$, has been generated, the operational mechanism of the BPL language manipulates the linguistic variable X and, more precisely, the terms in $T(X)$ (domain points included) in a totally standard way. That is, as symbols of a first order language which are capable of participating in a weak unification process at the same level as the rest of symbols of the language alphabet. Therefore, we are able to manipulate a semantic process, which involves a fuzzy linguistic query, by pure syntactical means.

Example 3. Continuing with Example 2. The set of proximity equations generated from the BPL program, by using the second phase of our method and Algorithm 1 is the following[3]:

---

[3] For the sake of simplicity, only the relations between linguistic terms and domain points are shown. The relations of linguistic terms with themselves are omitted.

```
young~age#24 = 0.86.    middle~age#24 = 0.2.    old~age#24 = 0.0.
young~age#30 = 0.66.    middle~age#30 = 0.5.    old~age#30 = 0.0.
young~age#34 = 0.53.    middle~age#34 = 0.7.    old~age#34 = 0.0.
young~age#45 = 0.16.    middle~age#45 = 1.0.    old~age#45 = 0.0.
```

Now, it is possible to ask about ``who is middle aged'' by lauching the query "?.-person(_,X,middle,_)." to the Bousi Gun fire tool (See next section). Then, it is able to find the answers: X=john with 0.2, X=paul with 0.5, X=mary with 0.7 and X=warren with 1.0; since it has been internally created a set of proximity equations between the crisp data 24, 30, 34 y 45, and the linguistic term middle and it is possible to establish a link between them by using weak unification.

5 Bousi Exporta and Bousi Gun Fire

In this section, the architecture of the software prototype implementing our method for formulating fuzzy linguistic queries on relational databases is detailed. This is a software composed of three layers: a user interface, Bousi Gunfire and Bousi Exporta.

The user interface layer allows users to interact with the Bousi Exporta and Bousi Gunfire tools. It implements the Bousi Exporta window and the input/output window:

• The Bousi Exporta window provides a graphical list of relational tables. Once a relational table has been selected, its columns and data can be visualized. Then, a linguistic variable can be defined by choosing a column of the selected table and establishing a magnitude and minimum/maximum values. After that, a set of linguistic labels may be defined by using a graphical editor of linguistic variables.

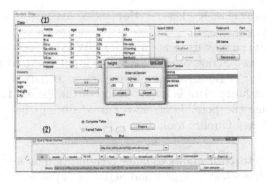

Fig. 1. Bousi Exporta (1) and input/output (2) window

• The input/output window provides the functionality to formulate fuzzy linguistic queries and to visualize the answers returned by the Bousi Gunfire tool.

In our system, a query is created by means of a graphic interface in which is shown a list of tables (see Figure 1). Once a table has been selected, one or several linguistic terms can be indicated in order to complete the query. For example, in order to formulate the query "find tall person with light eyes", the user must select the table "person" and the linguistic terms "tall" and "light eyes". Note that, this way of formulating a query is temporal since a natural language interface should be created for the future (it is comment in the Section 6).

The Bousi Gunfire tool implements the core of the BPL system which has been adapted to be capable of operating autonomously. Specifically, it receives as input a BPL program generated by the Bousi Exporta tool and a query formulated by the user. Then the inference mechanism of BPL returns a (possibly empty) set of answers which will be send to user interface. This layer mainly implements the functionality of a weak unification algorithm. Bousi Gunfire is a software tool which has been implemented in Java. It consists of over 35 classes divided in 4 java packages:

• The Compiler: this package is compound of 7 classes. It compiles the program generated by the Bousi Exporta tool and the query formulated with the user interface. Also, it returns the machine code which will be executed by the SWAM machine. The compiler is detailed in [6].

• The SWAM (Similarity-based Warren Abstract Machine): this package is composed of 9 classes. It executes the machine code generated in the compilation phase. It differs from the one presented and detailed in [6] in which, now, the input/output is performed by means of an especial communication package that is summarized in a following item.

• Utilities: this package is compound of 17 classes. It implements the data structures used for the rest of packages.

• The Communication package: this is the input/output package. It is composed of 3 classes. These classes are used for the communication with others applications (e.g. Bousi Exporta).

- The BousiAnswer class is a list of "$\langle VariableSubstitution, Degree \rangle$" objects from the utilities package.
- The BousiGunfire class is compound of a SWAM object that executes the BPL program, two String objects acting as input (path of the archive and query), and a list of BousiAnswer objects acting as output.
- The IncorrectQuery class is an extension of the Exception class. This is used to manage the possible errors occurred during the formulation of a query to the Bousi Gunfire tool.

The Bousi Exporta tool allows to make a connection on a particular DBMS (Currently MySQL, Oracle, Postgres and SQL Server) in order to give the user access to its structure (tables attributes and values). This connection is made thru a component of the user interface layer (See Figure 1) that, as has been explained, allows the user to define and associate linguistic variables to vague attributes of a table. Finally, both data and linguistic variables are transformed into a BPL program by means of the algorithms of our method.

Bousi Exporta is a software tool which has been mainly implemented with java.sql and javax.sql packages, using Netbeans as development environment. It consists of over 11 classes divided into 2 packages:

• MetaDataExporta package: this is compound of 8 classes. The main class is the DatabaseMetaData which allows getting the meta-information about a particular DBMS. To this end, it implements 3 important methods: `getTables()`, `getColumns()` and `getTypeColumns()`. Additionally, the `Connection` class is used to make a connection to the selected DBMS and the Statement and ResultSet are employed to recover the data.

• Query Exporta package: this is composed of 3 classes. It allows the communication with the Bousi Gunfire tool by using the `BousiAnswer`, `BousiGunfire` and `IncorrectQuery` classes referenced in previous items.

6 Conclusions and Future Work

In this work, a new method to formulate fuzzy linguistic queries on a database management system has been presented. To this end, a set of rows from a relational table is transformed into a set of Bousi~Prolog facts and a fuzzy relation is established between the constant arguments of such facts and the linguistic labels defined for a linguistic variable what provides a mapping between the data records in the database and the fuzzy concepts that may be used by queries. We have presented a software prototype that implements this method. We can signal out some possible limitations: i) the user interface and the queries are very simple; ii) the transformation from crisp data to imprecise knowledge is performed only on a single table.

Although this paper is not centered on applications, we think that our approach can be very useful in the following cases: (a) When the user of a relational database is not an expert and he/she does not know its structure (tables, columns, domains), our method helps such kind of inexperienced users to formulate fuzzy linguistic queries without the necessity of knowing the database structure or a specific language; (b) To adapt crisp information stored in a database in order to develop soft computing applications, since this method allows us to put in relation vague concepts with crisp data stored in databases and to transform crisp data into vague data; (c) Because database information is transformed into sentences (facts) of a fuzzy logic programming language, and therefore into a declarative program, it can be used to infer new knowledge; (d) The combination of the last two abilities (connexion of crisp data with vague concepts and deductive capabilities) allows us to employ crisp data in several unexpected applications such as: flexible deductive databases, fuzzy experts systems or approximate reasoning.

Regarding future work lines, we want to improve the user interface allowing more complex queries expressed in natural language. On the other hand, the transformation process has to be performed on multiple relational tables i.e. the relationship between relational tables must be taken into account. Finally, we want to study how this method can be used on XML databases in order to formulate fuzzy linguistic queries on them.

References

1. Bosc, P., Pivert, O.: SQLf: a relational database language for fuzzy querying. IEEE T. Fuzzy Syst. 3(1), 1–17 (1995)
2. Balamurugan, V., Kannan, K.S.: A Framework for Computing Linguistic Hedges in Fuzzy Queries. The Int. J. of Database Management Systems 2(1) (2010)
3. Buckles, B., Petry, F.: A fuzzy model for relational databases. Fuzzy Sets and Syst. 7, 213–226 (1985)
4. Chen, S.M., Hsiao, H.R.: A New Approach for Fuzzy Query Processing Based on Automatic Clustering Techniques. Information and Management Sciences 18(3), 223–240 (2007)
5. Codd, E.F.: A Relational Model of Data for Large Shared Data Banks. Communications of the ACM 13(6), 377–387 (1970)
6. Julián-Iranzo, P., Rubio-Manzano, C.: A similarity-based WAM for bousi˜Prolog. In: Cabestany, J., Sandoval, F., Prieto, A., Corchado, J.M. (eds.) IWANN 2009, Part I. LNCS, vol. 5517, pp. 245–252. Springer, Heidelberg (2009)
7. Julián, P., Rubio, C.: An Efficient Fuzzy Unification Method and its Implementation into the Bousi∼Prolog System. In: FUZZ-IEEE 2010, pp. 658–665 (2010)
8. Julián, P., Rubio, C., Gallardo, J.: Bousi∼Prolog: a Prolog Extension Language for Flexible Query Answering. Electronic Notes in Theoretical Computer Science 248, 131–147 (2009)
9. Kamel, M., et al.: Fuzzy Query using Clustering techniques. Information Processing and Management 26(2), 279–293 (1990)
10. Lee, R.C.T.: Fuzzy Logic and the Resolution Principle. Journal of the ACM 19(1), 119–129 (1972)
11. Lloyd, J.W.: Foundations of Logic Programming. Springer, Berlin (1987)
12. Orchard, R.A.: FuzzyClips Version 6.04A. User's Guide. Integrated Reasoning. Institute for Information Technology. Canada (1998)
13. Prade, H., Testemale, C.: Generalizing database relational algebra for the treatment of incomplete/uncertain information and vague queries. Information Science 34, 115–143 (1984)
14. Rios-Filho, L.G., Sandri, S.A.: Contextual Fuzzy Unification. In: Proc. of IFSA 1995, pp. 81–84 (1995)
15. Sessa, M.I.: Approximate reasoning by similarity-based SLD resolution. Theoretical Computer Science 275(1-2), 389–426 (2002)
16. Shenoi, S., Melton, A.: Proximity relations in the fuzzy relational database model. Fuzzy Sets and Systems 100, 51–62 (1999)
17. Tahami, V.: A conceptual framework for fuzzy query processing - a step toward very intelligent databases systems. Information Processing and Management 13, 289–303 (1977)
18. Takahashi, Y.: A fuzzy query language for relational databases. In: Kacprzyk, J., Bosc, P. (eds.) Fuzziness in Database Management Systems. Physica-Verlag, Berlin (1995)
19. Zadeh, L.A.: Fuzzy Sets. Information and Control 8(3), 338–353 (1965)
20. Zadeh, L.A.: The Concept of a Linguistic Variable and its Applications to Approximate Reasoning I, II and III. J. of Information Sciences 8 & 9 (1975)

Finding Similar Objects in Relational Databases
— An Association-Based Fuzzy Approach

Olivier Pivert[1], Grégory Smits[2], and Hélène Jaudoin[1]

[1] Irisa – Enssat, University of Rennes 1
Technopole Anticipa 22305 Lannion Cedex France
[2] Irisa – Enssat, IUT Lannion
Technopole Anticipa 22305 Lannion Cedex France
{pivert,jaudoin}@enssat.fr, gregory.smits@univ-rennes1.fr

Abstract. This paper deals with the issue of extending the scope of a
user query in order to retrieve objects which are similar to its "strict answers". The approach proposed exploits associations between database
items, corresponding, e.g., to the presence of foreign keys in the database
schema. Fuzzy concepts such as typicality, similarity and linguistic quantifiers are at the heart of the approach and make it possible to obtain a
ranked list of similar answers.

1 Introduction

The practical need for endowing information systems with the ability to exhibit
cooperative behavior has been recognized since the early '90s. As pointed out in
[9], the main intent of cooperative systems is to provide correct, non-misleading
and useful answers, rather than literal answers to user queries. Cooperative answers also aim at better serving the user's needs and expectations. The idea
developed in this paper, inspired notably by Stefanidis *et al.* [13], consists in
providing the user with answers which are not only "strict answers" to his/her
query, but also objects that he/she might like ("You May Also Like") — as
in recommender systems. In this paper, one investigates a fuzzy-set-based approach, which can also be seen as an extension of nearest neighbor queries where
the notion of neighborhood considered is based on associations between entities
(modeled for instance by foreign keys in a relational database context). As an
introductory example, let us consider the bibliographic database composed of
the relations:

- (A) *Author*(*a*, *name*) of key *a*;
- (P) *Publi*(*t*, *title*, *journal*) of key *t*;
- (K) *Keyword*(*w*, *word*) of key *w*;
- (W) *Written_by*(*t*, *a*) of key (*t*, *a*), with foreign keys *t* and *a*;
- (D) *Deals_with*(*t*, *w*) of key (*t*, *w*), with foreign keys *t* and *w*.

Let us consider a query retrieving names of authors and let us assume that the
user, after scanning the result, is interested in finding authors similar to one of

H.L. Larsen et al. (Eds.): FQAS 2013, LNAI 8132, pp. 425–436, 2013.
© Springer-Verlag Berlin Heidelberg 2013

the answers, say Codd. A possible meaning of "similar" in this context may be that the authors to be retrieved must publish in a set of journals similar to the set of journals where Codd publishes, must publish about a similar set of topics, and have a set of co-authors that is similar to Codd's. The approach we propose is based on a fuzzy comparison between the set of typical objects (journals, topics, co-authors) associated with a given target object (Codd in this example) and the set of typical objects associated with every other researcher present in the database. By doing so, a degree of matching can be measured for every researcher, which makes it possible to produce a top-k list of authors somewhat similar to Codd. Let us assume for instance that the fuzzy set of typical journals associated with Codd is $T_{Codd} = \{0.8/\text{PVLDB}, 0.6/\text{TKDE}, 0.3/\text{DKE}\}$. Then, taking into account the similarity based on journals, a researcher X will be considered all the more similar to Codd as his own set of typical journals T_X is close to T_{Codd} in the sense of an appropriate matching measure.

The remainder of the paper is structured as follows. Section 2 presents diverse approaches that may be used to compute the typical values of a multiset, i.e., that make it possible to convert a multiset E into a fuzzy set T describing the values that are the most typical in E. Section 3 discusses a sample of measures aimed at assessing the extent to which two fuzzy sets (of typical values, here) are similar. Such measures may be used to interpret the matching operator mentioned above. Section 4 discusses implementation aspects. Section 5 describes a preliminary experimentation that was carried out on the IMDb movie database. Section 6 discusses related work and situates our approach with respect to other proposals. Finally, Section 7 recalls the main contributions of the paper and outlines a few perspectives for future work.

2   Computing the Typical Values of a Multiset

Let us denote by f_i the relative frequency of a value x_i in a multiset E:

$$f_i = \frac{n_i}{n} \tag{1}$$

where n_i is the number of copies of x_i in E and n is the cardinality of E.

In order to assess the extent to which x_i is a typical value in E, two cases have to be taken into account: that where a metric — on which a similarity relation can be based — over the considered domain is available, and that where such a metric is not available and strict equality must be used. In any case, starting from a multiset E, the objective is to obtain a fuzzy set T such that $\forall x_i$, $\mu_T(x_i)$ expresses the extent to which x_i is typical in E.

2.1   Typicality Based on Strict Equality

In the absence of any similarity measure, an obvious solution is to take

$$\mu_T(x_i) = f_i. \tag{2}$$

Let us notice however that with this frequency-based approach, every element is considered somewhat typical. Those which have a low frequency get a low degree of typicality, but the elements which have a rather high frequency may also get a typicality degree significantly smaller than 1, since there are often several representative elements in a collection. Let us consider for instance a collection (multiset) of hundred animals including thirty dogs, thirty cats, and various other animals with only one occurrence each. The element "dog" has the frequency value 0.3, as well as the element "cat". Now, it could appear desirable to express that "dog" and "cat" are the two typical elements of the collection, to a high degree. One may then use:

$$\mu_T(x_i) = \mu_{most}(f_i) \tag{3}$$

where *most* is a fuzzy quantifier [17] whose general form is given in Figure 1. In order to get the desired behavior, one may use low values for δ and γ, for instance $\delta = 0.1$ and $\gamma = 0.5$ (which corresponds of course to a rather lax vision of *most*).

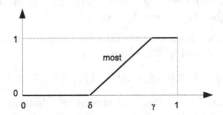

Fig. 1. A representation of *most*

2.2 Typicality Based on Similarity

When a similarity relation S over the considered domain is available, one may use the definition proposed by Dubois and Prade [8], which says, following Zadeh's interpretation [18], that an element x_i is all the more typical in a multiset E as it is both frequent in E and similar to most of the values of E:

$$\mu_T(x_i) = \frac{1}{n} \sum_{j=1}^{n} \mu_S(x_i, x_j) \tag{4}$$

with

$$\mu_S(x_i, x_j) = \max(0, \min(1, \frac{\alpha + \beta - d_{ij}}{\beta})), \tag{5}$$

where d_{ij} denotes the distance between x_i and x_j with respect to S, and the values α and β with $\alpha \leq \beta$ are positive real numbers which define a threshold of "indistinguishability" around each value x.

Example 1. Let us consider the multi-set (of cardinality $n = 30$):

$$E = \langle 1/0,\ 1/3,\ 1/4,\ 4/5,\ 7/6,\ 5/7,\ 3/8,\ 5/9,\ 2/12,\ 1/23 \rangle$$

where k/x_i means that element x_i has k copies in E. With $\alpha = 2$ and $\beta = 2$, and $d_{ij} = |x_i - x_j|$, one gets the fuzzy set of typical values:

$$T = \{0.05/0,\ 0.33/3,\ 0.52/4,\ 0.65/5,\ 0.77/6,\ 0.82/7,\ 0.73/8,\ 0.58/9,$$
$$0.15/12,\ 0.03/23\}$$

where μ/x_i means that element x_i belongs to T (i.e., is typical in E) to the degree μ.◊

Again, Formula (4) can be softened by applying a linguistic (fuzzy) quantifier *most* (meaning here "a significant proportion"):

$$\mu_T(x_i) = \mu_{most}\left(\frac{1}{n}\sum_{j=1}^{n}\mu_S(x_i,\ x_j)\right). \tag{6}$$

Here, the fuzzy quantifier *most* can be more drastic (for instance $\delta = 0.4$ and $\gamma = 0.8$) that in the strict equality case, since taking similarity into account generally leads to higher typicality degrees.

Example 2. Let us come back to the data of Example 1. Using the quantifier *most* defined by $\delta = 0.4$ and $\gamma = 0.8$, one gets:

$$T = \{0.3/4,\ 0.62/5,\ 0.92/6,\ 1/7,\ 0.82/8,\ 0.45/9\}.$$

Remark 1. In the case of nonnumerical attributes, defining the similarity measure (function μ_S) is not an easy task. A solution can be to use a domain ontology when it is available. See e.g. [2] where diverse similarity measures based on ontologies are discussed.

3   Fuzzy Matching Operator

Several interpretations of the condition E_1 *matches* E_2 — where E_1 and E_2 are two regular multisets of attribute values associated respectively with the target object and a candidate answer — can be thought of. The problem comes down to assessing the equality of two fuzzy sets, and many measures have been proposed for doing so, see, e.g., [12,5]. One may for instance:

- test the equality of the two fuzzy sets T_1 and T_2 of (more or less) typical elements in E_1 and E_2 respectively, for example by means of the Jaccard indice:

$$\mu_{matches}(E_1,\ E_2) = \frac{\sum_{x \in U}\min(\mu_{T_1}(x),\ \mu_{T_2}(x))}{\sum_{x \in U}\max(\mu_{T_1}(x),\ \mu_{T_2}(x))} \tag{7}$$

where U denotes the underlying domain of E_1 and E_2 — but this is rather drastic —, or by means of a measure such as:

$$\mu_{matches}(E_1,\ E_2) = \inf_{x \in U}\ 1 - |\mu_{T_1}(x) - \mu_{T_2}(x)|. \tag{8}$$

– check whether there exists at least one element which is typical both in E_1 and in E_2 (which corresponds to a rather lax view):

$$\mu_{matches}(E_1,\ E_2) = \sup_{x \in U}\ \min(\mu_{T_1}(x),\ \mu_{T_2}(x)). \tag{9}$$

– assess the extent to which most of the elements which are typical in E_1 are also typical in E_2 and reciprocally:

$$\mu_{matches}(E_1,\ E_2) = \min(\mu_{most \in T_2}(T_1),\ \mu_{most \in T_1}(T_2)). \tag{10}$$

The evaluation of Formula (10) is based on (one of) the interpretation(s) of fuzzy quantified statements of the form $Q\ X\ A\ are\ B$ where A and B are fuzzy predicates and Q is a fuzzy quantifier. See [17,15,16]. The most simple interpretation was proposed by Zadeh [17] and is based on the ratio of elements which are A and B among those which are A:

$$\mu(Q\ X\ A\ are\ B) = \mu_Q \left(\frac{\sum_{x \in X} \top(\mu_A(x),\ \mu_B(x))}{\sum_{x \in X} \mu_A(x)} \right) \tag{11}$$

where $\top$ denotes a triangular norm, for instance the minimum. Then, Equation (10) rewrites (taking $\top = \min$):

$$\mu_{matches}(E_1,\ E_2) = \min(\mu_{most} \left(\frac{\sum_{x \in X} \min(\mu_{T_1}(x),\ \mu_{T_2}(x))}{\sum_{x \in X} \mu_{T_2}(x)} \right),$$
$$\mu_{most} \left(\frac{\sum_{x \in X} \min(\mu_{T_1}(x),\ \mu_{T_2}(x))}{\sum_{x \in X} \mu_{T_1}(x)} \right)). \tag{12}$$

Example 3. Let us consider the bibliographic database introduced in Section 1 and assume that two authors are considered similar if the typical sets of journals in which they publish are similar. Let us consider the typical sets:

$$T_1 = \{0.2/PVLDB,\ 0.3/TKDE,\ 0.6/JIIS,\ 0.6/DKE\}$$

and

$$T_2 = \{0.3/DKE,\ 0.4/TODS,\ 0.6/PVLDB,\ 0.8/IJIS\}.$$

The similarity degrees obtained using the previous measures are:

– with Formula (7): $\mu_{matches}(E_1,\ E_2) = \frac{0.5}{3.3} \approx 0.15$
– with Formula (8): $\mu_{matches}(E_1,\ E_2) = inf(0.6, 0.7, 0.4, 0.7, 0.6, 0.2) = 0.2$
– with Formula (9): $\mu_{matches}(E_1,\ E_2) = sup(0.2, 0, 0, 0.3, 0, 0) = 0.3$
– with Formula (12) using a quantifier "most" defined by $\delta = 0.1$ and $\gamma = 0.5$:
$\mu_{matches}(E_1,\ E_2) = min(\mu_{most}(\frac{0.5}{2.1}),\ \mu_{most}(\frac{0.5}{1.7}))$
$= min(\mu_{most}(0.24),\ \mu_{most}(0.29))$
$= min(0.34, 0.48) = 0.48.\diamond$

Semantic proximity between values (if available) can also be taken into account during the computation of the similarity of two fuzzy sets. Such a matching measure, called *interchangeability*, is proposed in [4].

Remark 2. In the case where the query aimed at retrieving similar objects involves a conjunction of similarity conditions, the different degrees of matching may be aggregated by means of a triangular norm (e.g., the minimum operator), according to fuzzy set theory. However, some associations may be considered more important than others, and an aggregation operator such as the weighted average or the weighted minimum [7] can then be used instead of the minimum.

Remark 3. When the initial user query retrieves *several* objects (targets), one must look for similar items for each of these targets. Then, it seems reasonable to combine the different sets of similar items in a disjunctive manner: an object is in the extended result if it is similar to one strict answer (one target) at least. The degrees coming from the different sets of similar items are then combined by means of a triangular co-norm (e.g., the maximum operator).

4   Implementation Aspects

Let us first show how the different tools described in Sections 2 and 3 can be applied to the computation of a extended set of answers. First, let us emphasize the difficulty of defining a fully automated process in this case. Indeed, it does not seem possible, in general, to guess in what sense the end-user considers that an object is similar to another object (in the introductory example, for instance, is it because the authors have written papers on the same topics, and/or in the same journals, etc). The system needs some hints in order to derive the appropriate query. The outline of an interactive strategy could be as follows. After the strict answers are returned, the system asks the user "do you wish to get similar objects?". If his/her answer is "yes", the user is asked to check some boxes (predefined by a domain expert on the basis of primary key/foreign key constraints, in particular, or the discovery of so-called metapaths as defined in [14]) in order to specify his vision of "similar". For instance, in the context of the introductory example, the options could be: i) publications in similar journals, ii) publications on similar topics, iii) publications with similar co-authors.

Algorithm 1 describes the basic strategy for retrieving the objects similar to a target object c. In this algorithm, the $E_i(x)$'s are the multisets obtained by processing the n subqueries referring to x, and $T_i(x)$ is the fuzzy set of typical elements in $E_i(x)$.

Let m denote the cardinality of the relation containing the target objet c (and the objects one wants to retrieve). The previous algorithm implies to process $n \times m$ subqueries. An optimization consists in prefiltering the relation by means of a selection condition based on the typical sets associated with c, in order to avoid an exhaustive scan of the relation.

Example 4. Let us consider again the query from Section 1. Let us assume that the subquery:

select *journal* **from** P, W, A
where $P.t = W.t$ **and** $P.a = A.a$ **and** *name* = 'Codd'

returns a multiset whose associated fuzzy set of typical elements is

$$T_1 = \{0.1/\text{JODS}, 0.2/\text{PVLDB}, 0.3/\text{TKDE}, 0.6/\text{JIIS}, 0.6/\text{DKE}\}.$$

Then, only the authors a belonging to the result of the following query should be considered:

select a **from** P, W **where** $P.a = W.a$
and *journal* **in** ('JODS', 'PVLDB', 'TKDE', 'JIIS', 'DKE')

since the other authors have no chance to be somewhat satisfactory (they do not share any journal with Codd).$\diamond$

Input: a target object c; n specifications of multisets (i.e., n subqueries) ;
a threshold $\alpha \in (0, 1]$
Output: a fuzzy set $S(c)$ of objects similar to c
begin
    $S(c) \leftarrow \emptyset$;
    for $i \leftarrow 1$ **to** n **do**
     | compute $E_i(c)$; compute $T_i(c)$ from $E_i(c)$;
    end
    foreach *item x in the relation concerned* **do**
        for $i \leftarrow 1$ **to** n **do**
         | compute $E_i(x)$; compute $T_i(x)$ from $E_i(x)$;
         | compute the degree of matching μ_i between $T_i(c)$ and $T_i(\text{x})$;
        end
        $\mu \leftarrow \min_{i=1..n} \mu_i$;
        if $\mu \geq \alpha$ **then**
         | $S(c) \leftarrow S(c) \cup \{\mu/x\}$
        end
    end
end

Algorithm 1: Base algorithm

In the previous example, the selection condition is based on the support of the fuzzy set T_1. Notice that if one uses Equation (2) to compute the typicality degree, the support of a given $T_i(c)$ may be rather large. On the other hand, with Equation (3), one applies the measure of matching (cf. Section 3) to sets of objects that are *sufficiently typical* (i.e., whose frequency is sufficiently high). This makes it possible to use a more restrictive selection condition for filtering the relation. For instance, in the previous example, using a quantifier *most* defined by $\delta = \gamma = 0.25$, one would get the more selective prefiltering query:

select a **from** P, W **where** $P.a = W.a$
and *journal* **in** ('TKDE', 'JIIS', 'DKE')

since both JODS and PVLDB are not sufficiently typical with respect to the target objet to be taken into account while searching for similar authors.

5   Preliminary Experimentation

In order to check the efficiency and effectiveness of the approach, we performed a preliminary experimentation using the IMDb[1] movie database illustrated in Figure 2, where the cardinality of each table is given in brackets. In this context, queries may involve conditions on attributes such as actor's name, movie title, production date, etc.

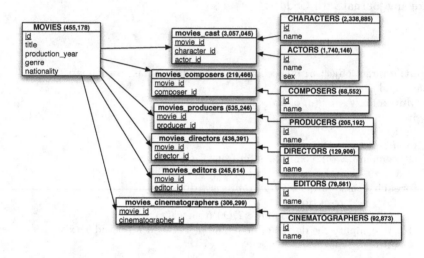

Fig. 2. Schema of IMDb

We use typicality based on strict equality (Formula (2)) since no similarity relations on the domains are available. A conjunction of two criteria is used to define similarity between actors:

1. Director criterion: two actors are considered similar if they have often worked with the same directors;
2. Actor criterion: two actors are considered similar if they have often played with the same other actors.

We have carried out this experimentation using the RDBMS PostgreSQL running on a PC with Intel Core^{TM} Duo CPU T7700 @ 2.40GHz, 2024MB of RAM, a processor cache of 4096 KB, and a hard disk with 16 MB of cache.

5.1   Response Time

Algorithm 1, even modified as explained above (before Example 4), is still rather inefficient inasmuch as it computes the typical sets associated with too many candidate objects (actors, here). In order to improve its performances, we introduced two additional filters:

---

[1] www.imdb.com

1. Director criterion: one only considers the actors who have worked with at least one of the five most typical directors associated with the target object;
2. Actor criterion: one only considers the actors with which the target actor has played more than once.

We considered twenty target actors and computed the average gain brought by the use of these filters. The results are as follows:

- without any filter, the average response time is 16.18 seconds (2500 objects are evaluated on average), which is obviously unrealistic;
- using both filters, the average response time is 1.51 seconds (138 objects are evaluated), which corresponds to a gain of 90.7%.

5.2 Comparison of the Answer Sets

A second objective was to compare the sets of answers produced by the different matching measures presented in Section 3. We considered Formulas (7), (8), (9), and (12) with $\mu_{most}(x) = x$. Again, we used twenty target actors and considered the top-20 results in each set of answers obtained. We compared every pair of sets of answers (L_i, L_j) using the Jaccard indice: $jac(L_i, L_j) = \frac{|L_i \cap L_j|}{|L_i \cup L_j|}$. The results are given in Table 1 (each number corresponds to an average computed over the twenty target actors). In this table, L_1, L_2, L_3, and L_4 correspond to the results obtained using Formulas (7), (8), (9), and (12) respectively.

Table 1. Comparison of the sets of top-20 answers (Jaccard indice)

	L_1	L_2	L_3	L_4
L_1	1	0.23	0.29	0.99
L_2		1	0.15	0.23
L_3			1	0.29
L_4				1

It appears that the matching measures produce significantly different results, except for measures (7) and (12) that produce answer sets that are almost always the same: $jac(L_1, L_4) = 0.99$. Indeed, we observed that for these two measures, the matching degrees obtained are different, but the sets of top-20 answers are the same, and the elements are even ordered in the same way. Consequently, Formula (12) will not be considered in the subsequent tests.

5.3 Relevance of the Answers

In order to assess the relevance of the answers obtained using the matching measures corresponding to Formulas (7), (8), and (9), we conducted a study involving 16 users who were asked to assess the top-20 results obtained using each of these measures for three target actors (namely Tom Cruise, Julia Roberts, and Robin Williams) corresponding to three individual results that one tries to expand. Again, a conjunction of two criteria ("director" and "actor", cf. above) was used to define the similarity between two actors (but the users were not aware of the similarity criteria used). Each answer could be assessed using one of the three choices "relevant", "not relevant", or "I don't know". For every set of answers L_i and every user u_j, we computed the precision attached to each of the three measures the following way:

$$prec(L_i, u_j) = \frac{|answers\ judged\ ``relevant"\ in\ L_i\ by\ u_j|}{|answers\ judged\ ``relevant"\ or\ ``not\ relevant"\ in\ L_i\ by\ u_j|}.$$

Finally, for each matching measure and each target actor, we computed the average precision over the 16 users. The results appear in Table 2.

Table 2. Average relevance

	L_1	L_2	L_3
Tom Cruise	**0.51**	0.45	0.43
Julia Roberts	**0.47**	0.39	0.13
Robin Williams	0.74	0.78	**0.81**
Average	**0.58**	0.54	0.46

It appears that the matching measure that yields the best precision on average is that based on the Jaccard indice (Formula (7)). It remains to be studied whether a combination of the lists of answers produced by different matching measures could improve the precision of the global result. Let us also mention that the rate of "I don't know" assessments is rather high (between 60 and 75% on average for each of the three measures considered). This also deserves a complementary analysis, in order to evaluate the proportion of such answers that would finally be considered relevant once the user gets familiar with them. Notice that this high rate of "I don't know" assessments does not constitute a weakness of the approach since it is indeed desirable to make the user discover new objects, as in any recommender system.

6 Related Work

Some recent work on keyword search over databases proposes to return "joining networks" of related tuples that together contain a given set of keywords where the tuples are related by foreign key-primary key links [1,3,10]. However, the goal of these approaches is not to retrieve the objects most similar to a given target, but to better cover an initial keyword query.

In [6], the authors consider a class of queries called the "object finder" queries, and their goal is to return the top-k objects that best match a given set of keywords by exploiting the relationships between documents and objects. Contrary to us, the authors consider an information retrieval context where the objects to

be retrieved are documents and where the weight of a link between a document
and an entity has to be computed by means of a full text search process.

In [11], the authors provide a framework and an engine for the declarative
specification of a recommendation process over structured data. The recommendation process in [11] is specified through a series of interconnected operators,
which apart from the traditional relational operators, includes also a number of
operators specific to the recommendation process, such as the *recommend* operator, that recommends a set of tuples of a specific relation with regards to their
relationship with the tuples of another relation. In this approach, the recommendation strategy is rather classical since it is based on similarity between values,
not on association between entities.

In [14], the authors study similarity search that is defined among the same
type of objects in heterogeneous networks. Intuitively, two objects are similar if
they are linked by many paths in the network. Again, this definition is different
from ours inasmuch as we do not focus on the number of links between two
objects but on the number of entities that are connected to both objects.

In [13], the authors focus on the specific recommendation process of computing
YMAL ("You May Also Like") results related to a specific user query. The
approach we proposed in this paper clearly belongs to the category of methods
that the authors of [13] call "current-state approaches" (where there is no other
information available other than a query q posed by a user u and its result $R(q)$),
but its originality lies in the exploitation of the notion of association between
entities from different tables, which is not explicitly mentioned by the authors
of [13].

7   Conclusion

In this paper, we have proposed an approach aimed at retrieving the items similar
to a target object from a database, on the basis of the associations that exist
between the different entities of the considered database. Intuitively, two objects
are similar if they are related with similar sets of entities. Since in general there
may exist several links between two given objets, multisets have to be considered,
and one makes use of fuzzy set theory in order to i) compute the fuzzy sets of
typical entities associated with a given one, ii) compare two fuzzy sets of typical
entities. This approach can be seen as the basis of a recommendation mechanism
in a structured database context. Of course, it can be used jointly with a value-
based similarity approach (objects may be considered similar if they have close
values on some attributes *and* are linked to similar sets of entities), which would
certainly improve the precision of the result. Preliminary experimental results
show that the approach described here is both tractable and promising in terms
of relevance/interest of the answers produced.

Among perspectives for future work, we intend to carry out a more complete
user study using different databases in diverse applicative contexts (bibliographic
database, classified ads database, etc). It is also worth studying how the notion
of similarity which makes the most sense for a given type of object could be

learned or defined *a priori*. Finally, a perspective concerns the definition of a mechanism aimed at providing the user with *explanations* related to the results produced.

References

1. Agrawal, S., Chaudhuri, S., Das, G.: DBXplorer: A system for keyword-based search over relational databases. In: Proc. of ICDE 2002, pp. 5–16 (2002)
2. Andreasen, T., Bulskov, H.: Query expansion by taxonomy. In: Galindo, J. (ed.) Handbook of Research on Fuzzy Information Processing in Databases, pp. 325–349. Information Science Reference, Hershey (2008)
3. Bhalotia, G., Hulgeri, A., Nakhe, C., Chakrabarti, S., Sudarshan, S.: Keyword searching and browsing in databases using BANKS. In: Proc. of ICDE 2002, pp. 431–440 (2002)
4. Bosc, P., Pivert, O.: On the comparison of imprecise values in fuzzy databases. In: Proc. of the 6th IEEE International Conference on Fuzzy Systems (FUZZ-IEEE 1997), Barcelona, Spain, pp. 707–712 (1997)
5. Bouchon-Meunier, B., Coletti, G., Lesot, M.-J., Rifqi, M.: Towards a conscious choice of a fuzzy similarity measure: A qualitative point of view. In: Hüllermeier, E., Kruse, R., Hoffmann, F. (eds.) IPMU 2010. LNCS, vol. 6178, pp. 1–10. Springer, Heidelberg (2010)
6. Chakrabarti, K., Ganti, V., Han, J., Xin, D.: Ranking objects based on relationships. In: Proc. of SIGMOD 2006, pp. 371–382 (2006)
7. Dubois, D., Prade, H.: Weighted minimum and maximum operations in fuzzy set theory. Information Sciences 39, 205–210 (1986)
8. Dubois, D., Prade, H.: On data summarization with fuzzy sets. In: Proc. of IFSA 1993, pp. 465–468 (1993)
9. Gaasterland, T.: Relaxation as a platform for cooperative answering. Journal of Intelligent Information Systems 1(3-4), 296–321 (1992)
10. Hristidis, V., Papakonstantinou, Y.: DISCOVER: Keyword search in relational databases. In: Proc. of VLDB 2002, pp. 670–681 (2002)
11. Koutrika, G., Bercovitz, B., Garcia-Molina, H.: Flexrecs: expressing and combining flexible recommendations. In: Proc. of SIGMOD 2009, pp. 745–758 (2009)
12. Pappis, C., Karacapilidis, N.: A comparative assessment of measures of similarity of fuzzy values. Fuzzy Sets and Systems (1993)
13. Stefanidis, K., Drosou, M., Pitoura, E.: You may also like" results in relational databases. In: Proc. of PersDB 2009 (2009)
14. Sun, Y., Han, J., Yan, X., Yu, P.S., Wu, T.: Pathsim: Meta path-based top-k similarity search in heterogeneous information networks. PVLDB 4(11), 992–1003 (2011)
15. Yager, R.: General multiple-objective decision functions and linguistically quantified statements. International Journal of Man-Machine Studies 21(5), 389–400 (1984)
16. Yager, R.: Interpreting linguistically quantified propositions. International Journal of Intelligent Systems 9(6), 541–569 (1994)
17. Zadeh, L.: A computational approach to fuzzy quantifiers in natural languages. Computing and Mathematics with Applications 9, 149–183 (1983)
18. Zadeh, L.: A computational theory of dispositions. International Journal of Intelligent Systems 2, 39–63 (1987)

M2LFGP: Mining Gradual Patterns over Fuzzy Multiple Levels

Yogi S. Aryadinata, Arnaud Castelltort, Anne Laurent, and Michel Sala

University Montpellier 2
LIRMM - CNRS UMR 5506,
161, Rue Ada, 34392 Montpellier Cedex 5, France
{yogi.aryadinata,arnaud.castelltort,
laurent,michel.sala}@lirmm.fr
http://www.lirmm.fr

Abstract. Data are often described at several levels of granularity. For instance, data concerning fruits that are purchased can be categorized regarding some criteria (such as size, weight, color, etc.). When dealing with data from the real world, such categories can hardly be defined in a crisp manner. For instance, some fruits may belong both to the *small* and *medium*-sized fruits. Data mining methods have been proposed to deal with such data, in order to take benefit from the several levels when extracting relevant patterns. The challenge is to discover patterns that are not too general (as they would not contain relevant novel information) while remaining typical (as detailed data do not embed general and representative information). In this paper, we focus on the extraction of gradual patterns in the context of hierarchical data. Gradual patterns describe covariation of attributes such as *the bigger, the more expensive*. As our proposal increases the number of combinations to be considered since all levels must be explored, we propose to implement the parallel computation in order to decrease the execution time.

1 Introduction

Databases are often considered in order to extract relevant information that describe the patterns occurring. For instance, gradual patterns such as *the bigger, the more expensive* can be extracted from data describing purchases of fruits.

Example 1. As presented in [1], we consider the database containing sales from a shop selling fruits as shown in Table 1. For the sake of simplicity, we consider here natural numbers (e.g., number of kg) but our approach also works on other domains, provided the fact that they are provided with a partial order Each tuple from the database corresponds to a cashier ticket.

Gradual patterns are extracted from databases where data can be ordered (e.g., numeric databases).

Several algorithms have been proposed for discovering such gradual patterns, based on the ones that have been proposed in the literature. These algorithms

H.L. Larsen et al. (Eds.): FQAS 2013, LNAI 8132, pp. 437–446, 2013.

Table 1. Fruit Sales Database

Id	Pineapple	RedApples	Cherries	Durian
T1	0	3	0	0
T2	2	1	1	2
T3	4	4	2	3
T4	2	1	1	1
T5	7	0	3	0

have been studied either to optimise performances (in terms of memory and/or runtimes) [2–4], or to discuss various manners of computing to which extent a pattern is present in a database or which patterns and rules can be discovered [5–8].

Fuzzy extensions have been defined in order to deal with real life applications where data and knowledge are often not crisp. [9] studies the possibility that the graduality is not over all the attribute but may be hidden somewhere in the domain of values. For instance, when mining gene expression, it may be the case that there is no pattern such as "The more the expression of gene G_i, the less the expression of gene G_j" over the whole interval but that it is rather the case that the pattern correlates values within the interval of values. Such a pattern may be "The more the expression of gene G_i is *almost* 0.2, the less the expression of gene G_j is *almost* 0.8". The approach proposes a definition of such fuzzy patterns and algorithms based on genetic programming in order to discover the most relevant parts of the universe (e.g. *almost* 0.2 and *almost* 0.8 *in the above example*).

[10] proposes to consider fuzzy orders instead of crisp orders so as to tackle the problem of data where differences between values may not always convey a crisp decision. For instance, it may be the case that an expression of gene of 0.1887 may not be that lower than an expression of gene of 0.1888.

More recently, we have studied how hierarchies can be managed in order to discover patterns at several levels of granularity, based on the work from the literature addressing multiple level data mining [11, 12].

Hierarchies can be either horizontal or vertical.

- horizontal hierarchies allow to merge several columns, for instance to put together small fruits,
- vertical hierarchies allow to merge several lines, for instance to merge purchases made within the same day.

Example 2. In Table 2, we consider the database from Example 1, where T1, T2 and T3 are from Monday and T4 and T5 are from Tuesday.

Table 2. Vertical Aggregation

Id	Pineapple	RedApples	Cherries	Durian
Monday (T1-3)	6	8	3	5
Tuesday (T4-5)	9	1	4	1

Example 3. In Table 3, we consider that the first two columns may be merged.

Table 3. Horizontal Aggregation

Id	WithoutKernel (Pineapple + RedApples)	WithKernel (Cherries)
T1	3	0
T2	3	1
T3	8	2
T4	3	1
T5	7	3

Several hierarchies can be defined.

Table 4. RedFruit Aggregation

	NotRed (Pineapple)	Red (RedApples + Cherries)
T1	0	3
T2	2	2
T3	4	6
T4	2	2
T5	7	3

However, real world data are often described by fuzzy hierarchies (e.g., a people can hardly be always crisply categorized into "young" or "old", a city can hardly be always crisply categorised into "south" and "north").

In this paper, we thus study how such fuzzy hierarchies can help for discovering relevant gradual patterns at multiple levels of granularity. As it introduces computation complexity, we consider the use of supercomputers in order to remain efficient and scalable over large and complex databases.

2   Gradual Patterns: Preliminary Definitions

Gradual patterns have been studied in the literature. Several notations have been proposed, we have chosen to consider the ones given below. Roughly speaking, gradual patterns are extracted by discovering the largest subsets of data that can be ordered when considering the corresponding attributes. For instance, from Example 1, we may consider the pattern "The higher the number of pineapples, the higher the number of durians" since the first four tuples can be ordered as $<T_1, T_4, T_2, T_3 >$ with $T_1.Pineapple \leq T_1.Durian$ and $T_4.Pineapple \leq T_4.Durian$ and $T_2.Pineapple \leq T_2.Durian$ and $T_3.Pineapple \leq T_3.Durian$.

Definition 1. *Gradual-Attribute. A gradual attribute I is defined over a domain $dom(I_j)$ on which an order $\leq_j$ (or simply $\leq$) is defined.*

Definition 2. *Gradual-DB. A gradual database is a set of tuples $\mathcal{T}$ defined over the schema $\mathcal{S} = \{Id, I_1, \ldots, I_n\}$ of n gradual attributes where Id is an identifier (primary key).*

Example 1 shows an example of a database which schema is $S=\{Pineapple, RedApples, Cherries, Durian\}$ containing 5 tuples defined over three attributes which domains are $\mathbb{N}$.

Definition 3. *Gradual item. A gradual item is a pair (i, v) where i is an item and v is variation $v \in \{\uparrow, \downarrow\}$. $\uparrow$ stands for an increasing variation while $\downarrow$ stands for a decreasing variation.*

For example, $(Pineapple, \uparrow)$ is a gradual item.

Definition 4. *Gradual Pattern (also known as Gradual Itemset). A gradual pattern is a set of gradual items, denoted by $GP = \{(i_1, v_1), \ldots, (i_n, v_n)\}$. The set of all gradual patterns that can be defined is denoted by GP.*

For example, $\{(Pineapple, \uparrow), (RedApples, \uparrow)\}$ is a gradual itemset.

Definition 5. *Tuple Ordering Over a Set of Attributes A. The tuples from a gradual database are ordered by defining an order $\prec$ with respect to a gradual pattern $GP\{(i_1, v_1), \ldots, (i_n, v_n)\}$. Two tuples t and t' can be ordered with respect to GP, denoted by $t \prec_{GP} t'$ if all the values of the corresponding items can be ordered with respect to the variations: for every $i_k(k \in [1, n])$, $t.i_k \leq t'.i_k$ if $v_l =\uparrow$ and $t'.i_k \leq t.i_k$ if $v_l =\downarrow$.*

When mining for gradual patterns, the goal is to extract frequent patterns. We thus have to determine what *frequent* means.

Definition 6. *Gradual Support. The support of a gradual pattern over a gradual database GDB is a function supp from $\mathcal{GP}$ to $[0, 1]$ that holds the following property (anti-monotonicity): for all GP_1, $GP_2 \in \mathcal{GP}$, $GP_1 \subseteq GP2 \Rightarrow supp(GP_1) \geq supp(GP_2)$. The support is $support(GP) = \frac{max_{L \in l}(|L|)}{(|R|)}$, where is the longest list of tuples that respects the gradual itemset of GP and $L = t_1, t_2, ..., t_m$ is a list of tuples from a set of tuples R.*

Definition 7. *Frequent Gradual Pattern. Given a "minimum support" threshold σ, a gradual pattern GP is said to be frequent if $supp(GP) \geq \sigma$.*

When dealing with hierarchies, we consider the case where columns or lines can be merged regarding their belonging to a common category in a taxonomy as shown in Table 4.

A multiple level attribute MLA is an attribute equipped with a hierarchy. This hierarchy is defined as a set of levels where every level is represented as a partition, all the partitions being embedded.

Definition 8. *ML-Attribute. An ML-Attribute MLA is defined by:*

- *a label,*
- *a domain $dom(MLA)$,*
- *a set of embedded levels of granularity $\mathcal{L} = L_0(MLA), \ldots, L_g(MLA)$ such as every domain $dom(L_i)$ is ordered with a relation $\leq_{L_i}$.*

Frequent gradual patterns are extracted from such databases by building, as described in [1], all possible hierarchies from the raw data. MLGP Operators are considered to aggregate values when merging columns and lines. For instance, the MLGP operator can be denoted by $\oplus$ and values are summed up as in the following example. In Table 3, results are built by summing up values for Pineapples and RedApples in the one hand and Cherries in the other hand.

3   M2LFGP: Dealing with Fuzzy Hierarchies

When dealing with fuzzy hierarchies, we consider that an item can be embedded within several upper categories at some extent, this extent being expressed by a degree ranging from 0 to 1 [13].

For instance, fruits belong to the category of small or big fruits at a certain extent, as described on Fig. 1. It should be noted that we do not require that all the degrees from a given category sum up to 1.

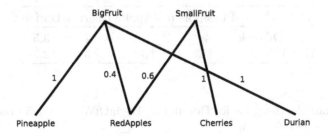

Fig. 1. Horizontal Hierarchy: Fruit Size

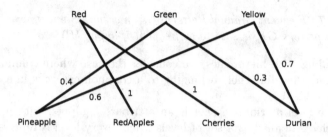

Fig. 2. Horizontal hierarchy: Fruit Color

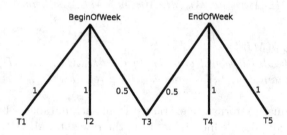

Fig. 3. Vertical Hierarchy: TimeStamps

When dealing with such cases, values must be merged differently from the crisp case. In order to take fuzzy hierarchies into account, we consider an MFLGP Operator which fuzzifies MLGP Operators. For this purpose, values are merged by integrating the degree from the hierarchy.

For this purpose, n-ary operations built from T-norms are considered.

For instance, we consider the product T-norm and the sum as the MLGP operator $\oplus$.

In the following table, we consider an example that merges lines for Begin of week and End of Week.

Table 5. Vertical Fuzzy Aggregation

Id	Pineapple	RedApples	Cherries	Durian
BeginOfWeek	4	6	2	3.5
EndOfWeek	11	3	3	2.5

In this example, the value for Durian for "BeginOfWeek" is 3.5, computed as: $0*1 + 2*1 + 3*0.5 = 3.5$.

When merging lines with an horizontal hierarchy, we consider the computation of the merged columns by considering the degree D of an item value v to be mapped with hierarchy value h as $D = v * degree_h$. These degrees are then merged using the $\oplus$ operator.

For instance, the following example shows how columns can be merged for small and big fruits with $\oplus = +$:

Table 6. Horizontal Fuzzy Aggregation

Id	BigFruit	SmallFruit
T1	1.2	1.8
T2	4.4	1.6
T3	8.6	5.4
T4	2.4	1.6
T5	7	3

In this example, the value for SmallFruit for tuple 1 is 1.8, computed as: $3 * 0.6 + 0 * 1 = 1.8$.

4   M2LFGP: Algorithms

In this part, we show how to extend our previous work to handle fuzziness. The algorithms being considered are given below. They allow to build databases from the source database provided with fuzzy hierarchies. Roughly speaking, the algorithms must combine fuzzy horizontal and vertical hierarchies.

The databases being generated from such transformations are then mined by classical algorithms for discovering relevant gradual patterns.

When generating the databases, the introduction of fuzzy hierarchies requires both to:

- represent fuzzy degrees;
- take the degrees into account when transforming databases for considering hierarchies.

Fuzzy hierarchies are represented using an XML description, as shown by Fig. 4.

Every hierarchy, should it be horizontal or vertical, is taken into account for transforming the database, in associated operations: Horizontal transformation, Vertical transformation and Horizontal-Vertical transformation which merge columns or lines regarding the definition of the hierarchy.

5   M2LFGP: Experimental Results

In our experimentation, the algorithm is implemented in Java for the dataset preparation / transformation and C++ for the parallel gradual pattern mining algorithm [3]. We study the efficiency of our method to get more valuable results.

```
<graph>
  <node id="n0">
    <name>BigFruit</name>
    <level>1</level>
  </node>
  <node id="n1">
    <name>SmallFruit</name>
    <level>1</level>
  </node>
  <node id="n2" >
    <name>PineApple</name>
    <level>0</level>
  </node>
  .
  .
  <edge id="e0" source="n0" target="n2">
    <weight>0.4</weight>
  </edge>
  <edge id="e1" source="n1" target="n2">
    <weight>0.6</weight>
  </edge>
  .
  .
</graph>
```

Fig. 4. XML Representation of a Fuzzy Hierarchy

In order to decrease runtimes, we consider parallel programming and we run our code on a server proposing up to 32 processing cores. This server provides a 8 AMD Opteron 852 (every processor being provided with 4 cores), 64 GB of RAM running Linux Centos 5.

We consider a synthetic database integrating hierarchies. The hierarchies are automatically generated from the databases. The dataset, C150A30, contains 150 tuples and 30 attributes, with 3 generated hierarchies (2 horizontal hierarchies and 1 vertical hierarchies). The horizontal hierarchies that have been considered contain 2 top level nodes and 3 top level nodes. The vertical hierarchy contains 2 top level nodes.

Figures 5 and 6 show the evolution of runtimes and speed up. Speed up displays how runtime can decrease when several processors and cores are considered. The more linear the speed up, the better, showing that runtimes decreases when the number of threads increases. For example, the comparison between sequential and parallel executions shows that sequential execution takes about 350 seconds while the runtime with 2 threads is reduced down to 168 seconds, as well the runtime with 3 threads that goes down to 80 seconds. As we can see, the execution time is greatly reduced until 4 threads. This happens because the processing of dataset has nearly reached the most efficient number of threads.

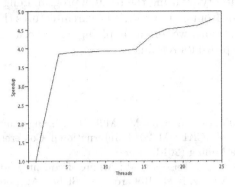

Fig. 5. Speed Up

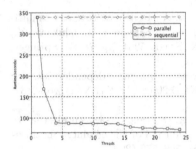

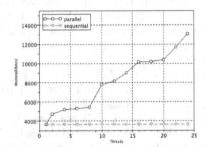

Fig. 6. Runtime over the number of threads

Fig. 7. Memory usage over the number of threads

The memory consumption is very low regarding the number of threads and the size of our dataset, as shown by Figure 7. In order to produce a better result, we need to improve our experiments. In particular, we plan to use the real-world datasets from environmental domain.

6 Conclusion and Future Work

In this paper, we propose the original method M2LFGP for mining relevant gradual patterns over fuzzy multiple levels (hierarchies). This work is very important as many real world databases contain fuzzy hierarchies in order to describe the data. We provide the necessary definitions together with algorithms that have been tested through experiments.

Future works include the study of the discovery of fuzzy gradual patterns (i.e., fuzzy items) over fuzzy hierarchies. Our work can also be improved by studying how the properties of the fuzzy hierarchy (e.g., degrees summing up to 1) can be exploited in order to design more efficient algorithms. Moreover, we aim at further studying scalability by using the parallel programming paradigms, especially for taking benefit from high performance architectures that are specialised in data distribution. Finally, we aim at applying our algorithms on several real databases in order to prove its relevance.

References

1. Laurent, A., Aryadinata, Y., Sala, M.: M2LGP: Mining multiple level gradual patterns. In: Proc. of ICKDDM 2013: International Conference on Knowledge Discovery and Data Mining (2013)
2. Di-Jorio, L., Laurent, A., Teisseire, M.: Mining frequent gradual itemsets from large databases. In: Adams, N.M., Robardet, C., Siebes, A., Boulicaut, J.-F. (eds.) IDA 2009. LNCS, vol. 5772, pp. 297–308. Springer, Heidelberg (2009)
3. Do, T.D.T., Laurent, A., Termier, A.: PGLCM: Efficient parallel mining of closed frequent gradual itemsets. In: Proc. of ICDM 2010, The 10th IEEE International Conference on Data Mining, pp. 138–147 (2010)
4. Laurent, A., Negrevergne, B., Sicard, N., Termier, A.: PGP-mc: Towards a multi-core parallel approach for mining gradual patterns. In: Kitagawa, H., Ishikawa, Y., Li, Q., Watanabe, C. (eds.) DASFAA 2010. LNCS, vol. 5981, pp. 78–84. Springer, Heidelberg (2010)
5. Hüllermeier, E.: Association rules for expressing gradual dependencies. In: Elomaa, T., Mannila, H., Toivonen, H. (eds.) PKDD 2002. LNCS (LNAI), vol. 2431, pp. 200–211. Springer, Heidelberg (2002)
6. Berzal, F., Cubero, J.C., Sanchez, D., Vila, M.A., Serrano, J.M.: An alternative approach to discover gradual dependencies. Int. Journal of Uncertainty, Fuzziness and Knowledge-Based Systems (IJUFKS) 15(5), 559–570 (2007)
7. Laurent, A., Lesot, M.-J., Rifqi, M.: GRAANK: Exploiting rank correlations for extracting gradual itemsets. In: Andreasen, T., Yager, R.R., Bulskov, H., Christiansen, H., Larsen, H.L. (eds.) FQAS 2009. LNCS, vol. 5822, pp. 382–393. Springer, Heidelberg (2009)
8. Bouchon-Meunier, B., Laurent, A., Lesot, M.J., Rifqi, M.: Strengthening fuzzy gradual rules through "all the more" clauses. In: FUZZ-IEEE (2010)
9. Ayouni, S., Yahia, S.B., Laurent, A., Poncelet, P.: Fuzzy gradual patterns: What fuzzy modality for what result? In: Proc. of the Second International Conference of Soft Computing and Pattern Recognition (SoCPaR), pp. 224–230. IEEE (2010)
10. Quintero, M., Laurent, A., Poncelet, P.: Fuzzy orderings for fuzzy gradual patterns. In: Christiansen, H., De Tré, G., Yazici, A., Zadrozny, S., Andreasen, T., Larsen, H.L. (eds.) FQAS 2011. LNCS, vol. 7022, pp. 330–341. Springer, Heidelberg (2011)
11. Han, J., Fu, Y.: Mining multiple-level association rules in large databases. IEEE Trans. Knowl. Data Engin. 11(5), 798–804 (1999)
12. Plantevit, M., Laurent, A., Laurent, D., Teisseire, M., Choong, Y.W.: Mining multidimensional and multilevel sequential patterns. ACM Transactions on Knowledge Discovery from Data (TKDD) 4(1) (2010)
13. Buckley, J.J., Feuring, T., Hayashi, Y.: Fuzzy hierarchical analysis revisited. European Journal of Operational Research 129(1), 48–64 (2001)

Building a Fuzzy Valid Time Support Module on a Fuzzy Object-Relational Database

Carlos D. Barranco[1], Juan Miguel Medina[2], José Enrique Pons[2], and Olga Pons[2]

[1] Division of Computer Science, School of Engineering, Pablo de Olavide University
Ctra. Utrera km. 1, 41013 Seville, Spain
cbarranco@upo.es
[2] Department of Computer Science and Artificial Intelligence, University of Granada
C/ Periodista Daniel Saucedo Aranda s/n, 18071 Granada, Spain
{medina,jpons,opc}@decsai.ugr.es

Abstract. In this work we present the implementation of a Fuzzy Valid Time Support Module on top of a Fuzzy Object-Relational Database System, based on a model to deal with imprecision in valid-time databases. The integration of these modules allows to perform queries that combines fuzzy valid time constraints with fuzzy predicates. Both modules can be deployed in Oracle Relational Database Management System 10.2 and higher. The module implements the mechanisms that overload the SQL sentences: Insert, Update, Delete and Select to allow fuzzy temporal handling. The implementation described supports the crisp valid time model as a particular case of its fuzzy valid time support provided.

Keywords: Fuzzy Databases, Fuzzy Object-Relational Data Base System, Fuzzy Temporal Databases.

1 Introduction

The relational model developed by Codd [3] has been widely used and extended to model reality in a closer way. There are two main research areas on this field. The first one aims to represent more complex data, and the second aims to query in a more human friendly way. This is also known as flexible querying.

The possibility theory [5] is the main mathematical framework that supports the proposed extensions to the relational model. The main proposals for fuzzy databases [2,20,22,23] deal with both representation and querying. Among them, the GEFRED model [12] is a synthesis including the main features of the above proposals. In GEFRED, both fuzzy relational algebra and relational calculus are defined. Flexible querying is provided by these two query languages.

Besides, the concept of time in databases has been studied in depth [7,21,14] in order to represent and handle time-variant or time related concepts. There are even some implementations that manage time in a database, for example the Oracle Workspace Manager [15] for crisp time intervals.

There are some theoretical models for dealing with uncertainty with respect to the time in a database [13,11,18], among them some able to represent uncertain

H.L. Larsen et al. (Eds.): FQAS 2013, LNAI 8132, pp. 447–458, 2013.
© Springer-Verlag Berlin Heidelberg 2013

time intervals [8,19] in the database, but there is not a complete implementation for the management of time-dependent objects in a database.

In this work we present an implementation of a Fuzzy Valid Time Support Module (FVTM) on top of a Fuzzy Object-Relational Database System [4,1] based on a theoretical model to deal with imprecision in valid-time databases described in [18].

The rest of the paper is organized as follows. In Section 2 some preliminary concepts in time modelling and temporal databases, as well as the representation of imperfect time intervals are introduced. Section 3 introduces the Fuzzy Object-Relational Management System that is the basis for the implementation. Next to that, in Section 4, the implementation of the Fuzzy Valid Time Support Module is described together with some illustrative examples of the system capabilities. Finally, Section 5 presents the main conclusions and lines for future work.

2   Preliminaries

In this section we are going to introduce some basic concepts on time modelling and temporal databases. Then, some theoretical concepts and a mathematical framework are provided to deal with imperfections in time intervals. Finally, a high level description of the data manipulation language (DML) is given.

2.1   Time Modelling and Temporal Databases

A *temporal database* is a database that manages some aspects of time in its schema [7]. The reality a temporal database tries to model, contains some temporal notions which have to be handled specifically in order to maintain a consistent modelling behavior. A *chronon* is the shortest duration of time supported by the database. Time can be represented either as points [6] or intervals [9] that may be subject to imperfection.

The temporal notions in temporal databases can be classified into four types based on their interpretation and modelling purpose.

- *User-defined time*: has no specific impact on the database consistency.
- *Transaction time*: time when the fact is stored in the database [10].
- *Valid time*: time when the fact is true in the modelled reality [21].
- *Decision time*: time when an event was decided to happen [14].

Database models can also named as *bi-temporal* (both valid and transaction-time) or *tri-temporal* (bi-temporal and decision time) models.

In the following, we will study the representation of imperfect time intervals.

Representation of Imperfect Time Intervals. In order to deal with uncertainty in time intervals, there are several proposals to consider. Here, two approaches are described: the first one, based on *Fuzzy Validity Periods* and the second one, based on *Possibilistic Valid-time Periods*.

Definition 1. *A **Possibilistic Valid-time Period** [17] (PVP) is a time interval in which one or both the starting or the ending points are not precisely known. In this representation, the starting and ending points of the interval are, respectively, two independent fuzzy sets. We note a PVP as $I = [S, E]$, where S, E are the starting and ending points respectively.*

Definition 2. *A **Fuzzy Validity Period** [9] (FVP) is defined as a fuzzy time interval specifying when the data regarding an object are valid.*

There are several proposals in the literature for transforming a PVP into a FVP. Figure 1 illustrates the use of the convex-hull transformation defined in [9]. An extended comparative between the properties of the FVP and the PVP is done in [17].

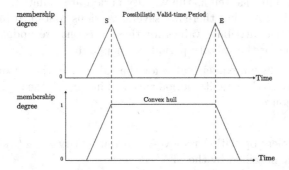

Fig. 1. Convex hull transformation to obtain a fuzzy validity period, FVP from a possibilistic valid-time period, PVP

Data Manipulation Language. The data manipulation language DML is implemented by using three operations: *insert, update* and *delete*. The user may set a temporal framework for the DML operations which consist on a valid-time interval. A context validity period (called session context FVP or SFVTP for short) may be defined by the user. This context validity period establishes a temporal framework which concerns all the data manipulation language sentences.

Insert. The user wants to store an entity which is valid during the time interval specified. The tuple is only inserted if there is no conflict with any other version.

Update. This operation adds new information about an existing entity (given by the tuple r). It is necessary to distinguish between two cases:

- The tuple r has no explicit valid-time value. In this case, the validity period for the update sentence is set by the SFVTP value. There are three main cases:

1. If there exists any other version of the tuple r in the database containing the time period specified by SFVTP, then the update is rejected.
2. If the context validity period contains the validity period of only one version (namely r'), then the validity period of that version is updated with the context validity period. The values for the attributes in r' are updated with the values for the attributes given in r.
3. The SFVTP contains several validity periods of other versions and may overlaps to the left and to the right with some other versions. In this case, the procedure is the following:
   (a) First, to solve the overlappings to the left and to the right of the SFVTP. This is done by modifying the starting and ending points of the overlapping intervals to fit the starting and ending points of the SFVTP.
   (b) Then, if there exist only one tuple contained by the SFVTP, the attributes as well as the validity period are updated.
   (c) In the case that there are several versions contained in the SFVTP, only the attribute values for these versions are updated, while the value for the validity period is not updated.

- If the tuple t has a explicit value for the valid-time, then if only one other version overlaps the validity period specified in $r[S, E]$, the tuple r is inserted in the relation v^T.

Delete. The delete operation removes a current entity $r \in v^T$ which overlaps the time interval specified by the SFVTP.

3 The Fuzzy Object-Relational Management System

The FVTM proposed in this work is based on top of Fuzzy Object-Relational Database Management System (FORDBMS) introduced in some of our previous work [4,1]. This system is developed as an extension of a market leader Database Management System (DBMS), Oracle®, by using its advanced object-relational features. This strategy let us take full advantage of the host Object-Relational DBMS (ORDBMS) features (high performance, scalability, availability, etc.) adding the ability of representing and handling of fuzzy data.

The FORDBMS extension includes a set of user-defined types (shown in Fig. 2) to allow the representation of a wide variety of fuzzy data, as the following:

- Atomic fuzzy types (AFT) to represent possibility distributions over ordered (OAFT) or non ordered (NOAFT) domains.
- Fuzzy collections (FC) able to represent fuzzy sets of objects, fuzzy or not, with conjunctive (CFC) or disjunctive (DFC) semantics.
- Fuzzy objects (FO) including attributes of crisp or fuzzy types, with an associated degree for each attribute to weigh its importance in object comparison.

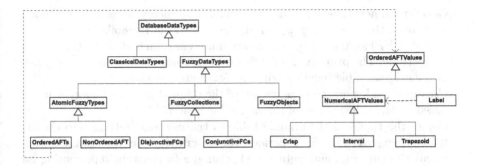

Fig. 2. Data type hierarchy for the FORDBMS

All fuzzy types define a Fuzzy Equal operator (FEQ) that computes the degree of fuzzy equality for each pair of instances. Particularly, for OAFT, the main data type used in this work, the system uses the possibility measure to implement FEQ. Additionally, OAFT type implements another version the equality operator using the necessity measure named NFEQ. Further details on this operator can be found in [1].

4   The Fuzzy Valid Time Support Module

This section describes the basic elements and use of the proposed implementation for the possibilistic valid time model. This implementation is called FVTM (Fuzzy Valid Time Support Module) and is developed on top of Oracle® ORDBMS 10.2 and higher. The FVTM is integrated into our FORDBMS, uses its OrderedAFT type and the operators defined on it, to perform some computations related with fuzzy valid time evaluations. As a result of this integration, we can perform queries and DML operations that combine fuzzy time constraints and fuzzy conditions. Another feature of the FVTM is that it supports the crisp time valid model as a particular case of its fuzzy valid time support.

To illustrate the elements and use of FVTM we will use an example table, deliberately oversimplified, that stores information about the employees of a company. As below DDL code shows, the table includes a salary attribute of type fsalary (orderedOAFT) that allows store fuzzy data on the salary domain and to perform queries asking for imprecise salary values:

```
EXECUTE orderedOAFT.extends('fsalary');
create table employee (name varchar2(16) primary key, salary fsalary);
```

To enable a table for fuzzy valid time support (for instance our employee table), it is necessary execute the sentence:

```
EXECUTE SDS_FTDB.enableFValidTime('employee');
```

As result of this execution, the FVTM, taking the argument table (employee in this example), performs the following tasks:

– Generates a new table with the _fvt suffix, that incorporates two additional attributes: the attribute pvp_valid, that stores the Possibilistic Valid-time Period (PVP) for the tuple, and the attribute version that stores the version of the tuple. The primary key for this table is comprised by the primary key of the original table together with the attribute version.
– Creates an index on the primary key of the original table and another index based on the primary key of the new table generated.
– Copies the tuples of the original table to the new one, setting into the attribute pvp_valid the FVT period: from current_time (a constant to represent the current time) until until_changed (a constant representing an undefined future time) and 1 into the attribute version.
– Generates a view named like the original table (employee in this example) that only shows the tuples that overlap the current PVP period with a value greater or equal than 0.5. We use the Strictly Consistent approach for FVT overlapping [18]. This approach establishes that the overlapping of two versions of the same tuple cannot be higher than 1. The choice of this threshold prevents the insertion or update of tuples (trough this view) that violate the Strictly Consistent restriction. Also generates a view with the _cp suffix that, further, shows the degree that each tuple overlaps this period with a value greater than 0.0.
– Builds an *insert* trigger and an *update* trigger on the first view, that intercept such DML operations, to adapt the PVP period for the affected tuples. For *delete* operations, it is not necessary to build a delete trigger, because the view it self determines the tuples affected by the deletion taking into account the PVP period.

From now on, the user can set the FVT period and perform queries and DML operations through the employee view. The Section 4.2 is devoted to illustrate by examples some of the possibilities that the FVTM provides for such kind of operations.

The next subsection describes the main components of the FVTM and is functioning.

4.1   Components and Functioning

The FVTM is composed by the following elements:

– The TSDS_PERIOD user defined type, that represents a Fuzzy Validity Period. It comprises four attributes of TIMESTAMP type: relaxedValidityFrom, validityFrom, validityTill and relaxedValidityTill, that represent, respectively, the trapezoidal possibility distribution [relaxedValidityFrom, validityFrom, validityTill, relaxedValidityTill].
  It includes several constructors to create its instances using diverse date formats. Also includes the method to_oaft to transform a TSDS_PERIOD instance into an OrderedAFT instance. To do this, for each attribute, it converts its timestamp value as a number value expressed in seconds since the

January 1, 4712 BC. This allows to transform several fuzzy valid time operations into fuzzy operations as `feq` or `nfeq`. The method `foverlaps` uses the method `to_oaft`, and computes the degree of overlapping (in $[0,1]$) of two `TSDS_PERIOD` instances. Moreover, this type includes the method `to_string(date_format)` to transform `TSDS_PERIOD` instances into strings taking into account the date format provided.

- The `TSDS_PVP` user defined type, that represents a Possibilistic Valid-time Period. It is composed by two attributes of type `TSDS_PERIOD`: `Tinitial`, `Tfinal`, that represent, respectively, the upper and lower of the PVP period. It includes several constructors to create its instances using diverse date formats. The included method `to_convexFVP` transforms a `TSDS_PVP` instance into a `TSDS_PERIOD` instance using the convex hull approach (see Fig. 1). With the help of this method, the method `foverlaps` computes the degree of overlapping (in $[0,1]$) of two `TSDS_PVP` instances. Another provided method is `closeR(J)`, that closes the self `TSDS_PVP` instance with the J instance (implements the case 3(a) for update operations as described in 2.1). Also, this type includes the method `to_string(date_format)` to transform `TSDS_PVP` instances into strings taking into account the date format provided.

- Several SQL user defined operators: the `overlaps()` operator that applies on two `TSDS_PERIOD` instances, using the implementation provided by the method `TSDS_PERIOD.foverlaps` or, on two `TSDS_PVP` instances, using the implementation provided by the method `TSDS_PVP.foverlaps`, the operator `to_string()` that applies on instances of `TSDS_PERIOD` or `TSDS_PVP` using the corresponding method implementation and, finally, the operator `closeR()` that applies on two instances of `TSDS_PVP` using the implementation provided by `TSDS_PVP.closeR()`.

- A session context called `TSDS_CTX`, managed by means of the `tsds_ctx_pkg` package. By means of this session context, the user can set, for the current session, the FVT period that will determine all its DML operations and queries on the database. If the FVT period is not established for the current user session then, by default, the FVT period is from `current_time` until `until_changed`.

- The package `SDS_FTDB` that includes constants, procedures and functions, to support the functioning of the FVTM, such as:
  - The `MIN_TIME`, and `MAX_TIME` `TIMESTAMP` constants that represent, respectively, the minimum and the maximum date that the system can represent. Another constant often used, is `UNTIL_CHANGED`, which takes the same value that the `MAX_TIME` constant.
  - The `enableFValidTime(table)` procedure (described in Section 4) generates the necessary functionality to provide FVT support for a table. The `disableFValidTime(table,keepPVPValid)` procedure disables the FVT support on the table passed as argument. If the second argument is `true` (default), then the system keeps the `pvp_valid` column and its

data, else the system removes the `pvp_valid` column and its data. In this latter case, only the current row for each primary key value is kept.
- The `setFValidTime` procedure, that has several overloadings, allows set the FVT period for the current user session context. If used without arguments, the FVT period from `current_time` until `until_changed` will be established in the current user session context. When this procedure is invoked with the string argument `'ANYTIME'` then the FVT period from the `MIN_TIME` value to `MAX_TIME` value will be established.
- The `getFValidTime` function returns, from the current user session context, a `TSDS_PVP` value that represents the FVT period for the current session.

4.2 Data Manipulation Operations

To illustrate the capabilities of the FVTM to handling DML operations affected by a FVT period, we will use the Table 1 that shows the tuples that result of the *insert*, *update* and *delete* statements described next. It is worth mentioning that all the examples included in this section have been directly executed on the FVTM prototype, using the proposed SQL operators to handle valid-time data. To manage datetime values, the FVTM uses the same datetime formats and functions as Oracle DBMS (see [16]). For the sake of simplicity, we use short datetime intervals in the examples shown in this section.

For each session, FVTM establishes by default an FVT period that comprises from `current\_time` until `until\_changed`. The user can change this FVT period using the `setFValidTime` procedure, for example:

```
-- Establishes the day 24th of the current month and year as the start of the fvt
-- and until_changed as the end.
execute SDS_FTDB.setFValidTime(tsds_pvp(to_date('24','dd'),SDS_FTDB.UNTIL_CHANGED));
```

Table 1. Employee_fvt table showing DML operations

Row	Name	Salary	PVP_valid	Ver.
1	Fred	40000	[[04,05,06,07];[08,14,17,21]]	1
1b	Fred	42000	[[03,04,04,05];[28,29,29,30]]	1
2	Ronald	35000	[[24,24,24,24]; UNTIL_CHANGED]	1
3	Frank	[32000,35000,40000,42000]	[[24,24,24,24]; UNTIL_CHANGED]	1
4	Tom	[38000,40000,42000,45000]	[[15,16,16,18]; UNTIL_CHANGED]	1
5	Michael	40000	[[04,05,06,07];[12,14,16,18]]	1
5b	Michael	40000	[[04,05,06,07];[6,7,15,15]]	1
6	Michael	45000	[[16,18,21,22]; UNTIL_CHANGED]	2
6b	Michael	50000	[[16,16,16,16]; UNTIL_CHANGED]	2

The next *insert* operations, performed through the `employee` view, taking into account the FVT period set above, add to the `employee_fvt` base table, the tuples shown in Table 1 with value 1, 2, 3, 4 and 5 for the `Row` column:

```
INSERT INTO EMPLOYEE VALUES ('Fred',fsalary(crisp(40000)),
    tsds_pvp(tsds_period('4','5','6','7','dd'),tsds_period('8','14','17','21','dd')));
INSERT INTO EMPLOYEE VALUES ('Ronald',fsalary(crisp(35000)),null);
INSERT INTO EMPLOYEE VALUES ('Frank',fsalary(trapezoid(32000,35000,40000,42000)),
    tsds_pvp(systimestamp,SDS_FTDB.UNTIL_CHANGED));
INSERT INTO EMPLOYEE VALUES ('Tom',fsalary(trapezoid(38000,40000,42000,45000)),
tsds_pvp(tsds_period('16',1,2,'dd'),sds_ftdb.until_changed));
INSERT INTO EMPLOYEE VALUES ('Michael',fsalary(crisp(40000)),
tsds_pvp(tsds_period('4','5','6','7','dd'),tsds_period('12','14','16','18','dd')));
INSERT INTO EMPLOYEE VALUES ('Michael',fsalary(crisp(45000)),
tsds_pvp(tsds_period('16','18','21','22','dd'),sds_ftdb.until_changed));
```

For the sake of simplicity of the the examples, dates are specified only by the month day, assuming the current month and year. As we can see, the statements show the use of several constructors for the `tsds_pvp` type. Also, several fuzzy values for the `salary` column are inserted to show that our system also can handle fuzzy data. The insert trigger created on `employee` view, intercepts and validates all insertions, rejecting these that are not compatible with the FVT constraints. For example, the next insert operation is rejected because the `PVP_valid` value for the existing tuple of Michael (row 5) overlaps with the `PVP_valid` value for the new tuple in a value greater than 0.5 (0.67 in this case):

```
INSERT INTO EMPLOYEE VALUES ('Michael',fsalary(crisp(50000)),
tsds_pvp(tsds_period('17',1,1,'dd'),sds_ftdb.until_changed));
```

FVTM provides two types of update operations: *sequential* updates and *non-sequential* updates. When the update statement does not establish a value for the `PVP_valid` column in the set clause, the FVTM handles the operation as a sequential update. If `PVP_valid` column is established in the update statement, then the FVTM handles it as a non-sequential update.

In the following example, we set the FVT period the 16th of current month and until *until_changed*. Then, we perform the salary update for Michael in this FVT period. This triggers the following sequential update: the `PVP_valid` value of row 5 is right closed, using the current context FVT period. The row 5 is replaced by the row 5b in this table. The row 6b is inserted with the new salary value for Michael but, as the `PVP_valid` value for row 6b includes the `PVP_valid` value for row 6, row 6b replaces the row 6 in the table `employee_fvt`.

```
execute SDS_FTDB.setFValidTime(tsds_pvp(to_date('16','dd'),SDS_FTDB.UNTIL_CHANGED));
update employee set salary=fsalary(crisp(50000)) where name ='Michael';
```

Next statement is a non sequential update because we have set a value for the `PVP_valid` column. If the new `PVP_valid` value does not overlaps with the currently value for Fred, the system shows the error message: 'Tuple to be updated does not exist'. If they overlap then the update is successful and the row 1 in the `employee_fvt` table is replaced by the row 1b, with the new `PVP_valid` and salary values set:

```
update employee set salary=fsalary(crisp(42000)), pvp_valid=tsds_pvp('4',1,1,'29',1,1,'dd')
where name='Fred';
```

After the previous sequence of insert and update sentences, finally, the table `employee_fvt` will contain the tuples with row: 1b,2,3,4,5b and 6b.

The *delete* statements are constrained by the context FVT period established. For example if, with the previously established context FVT period, we execute the statement: `delete from employee;`, the tuple 5 would be the only tuple that would remain.

Table 2. Results of the queries executed on the Employee_cd view

Row	Name	Salary	PVP_valid	FTDeg1	FTDeg2	Cd(Sal)
1b	Fred	42000	[[03,04,04,05];[28,29,29,30]]	1	0.67	0.6
2	Ronald	35000	[[24,24,24,24]; UC]	1	0	1
3	Frank	[32000,35000,40000,42000]	[[24,24,24,24]; UC]	1	0	1
4	Tom	[38000,40000,42000,45000]	[[15,16,16,18]; UC]	1	0	1
5b	Michael	40000	[[04,05,06,07];[6,7,15,15]]	0	0.33	1
6b	Michael	50000	[[16,16,16,16]; UC]	1	0	0

If the last *delete* statement is omitted then, finally, the table `employee_fvt` will contain the tuples with row: 1b,2,3,4,5b and 6b.

The following examples illustrates some additional capabilities of the FVTM. The FVTM system deals with both temporal and fuzzy predicates in the same sentence. Table 2 shows the rows of `employee` after the execution of the example queries. To illustrate in the same table the result set for all queries, we have added three columns that show the degree of temporal overlapping (`FTDeg1`, `FTDeg2`) and the degree of compliance of the fuzzy condition (`Cd(Sal)`) for each tuple with respect to each query.

Query 1. *""Retrieve all data from employees when the context fuzzy valid time period is from day 16th of the current month until changed"*

The following code performs this query and, as `FTDeg1` column shows, the tuple whith row 5b is the only one that it is not retrieved, because its `PVP_valid` value does not overlap the current FVT period:

```
execute SDS_FTDB.setFValidTime(tsds_pvp(to_date('16','dd'),SYSTIMESTAMP));
SELECT name,to_string(salary) "Salary", to_string(pvp_valid,'dd') "PVP_VALID",ftcdeg "FTDeg1"
FROM employee_cd
```

Query 2. *"Retrieve all data from employees with 'middle' salary (with degree greatest than 0.5) when the context fuzzy valid time period is from day 16th of the current month until changed"*

The following code shows the definition of the *'middle'* label for `salary` by means of the trapezoidal possibility distribution: [30000,35000,40000,45000], and then shows the query statement. The columns `FTDeg1` and `Cd(Sal)`. of the Table 2 shows that the rows retrieved by this statement are: 1b, 2, 3 and 4 because these have a value greater than 0.5 for `Cd(Sal)` and, its overlapping with the context FVT period, is greater or equal to 0.5.

```
execute orderedAFT.labelDef('Fsalary','middle',trapezoid(30000,35000,40000,45000));
SELECT name,to_string(salary) "Salary",to_string(pvp_valid,'dd') "PVP_VALID",ftcdeg "FTDeg1",
cdeg(1) "Cd(Sal)" FROM employee_cd WHERE fcond(feq(salary,fsalary('middle')),1) >0.5;
```

If, using the same fuzzy condition, we change the context FVT period, for example:

```
execute SDS_FTDB.setFValidTime(tsds_pvp('02',1,1,'03',1,2,'dd'));
```

The query only retrieves the tuple with row number 1b, because it is the only one that has a degree greater to 0.5 for the salary condition (Cd(Sal) column) and its overlapping with the context FVT period is greater or equal to 0.5 (FTDeg2 column).

5 Concluding Remarks and Future Works

In this work, we present an implementation on top of a FORDBMS of a complete valid-time model to represent and handle imprecise temporal intervals. The paper includes the formal definition of possibilistic valid-time periods in order to represent the time, to define suitable operators and to control integrity. This is the first implemented formal model in the literature for possibilistic valid-time intervals in object-relational databases. The semantics and the implementation of the DML operations are described within this work.

The main advantage of our proposal is that it allows to integrate valid-time handling (crisp and ill-defined), and even flexible querying handling, into a conventional ORDBMS.

Referential integrity is not included in the presented implementation yet. It will be subject of a upcoming version. Additionally, as future research, we plan to improve the implementation of the FVTM prototype presented here to increment its performance. For example, we will consider time indexing and alternative access methods to accelerate the retrieval response.

Acknowledgment. This work has been supported by the "Consejería de Innovación Ciencia y Empresa de Andalucía" (Spain) under research projects P07-TIC-02611, and the Spanish Ministry of Science and Innovation (MICINN) under grants TIN2009-08296, TIN2008-02066 and TIN2011-28956-C02-01.

References

1. Barranco, C.D., Campaña, J.R., Medina, J.M.: Towards a fuzzy object-relational database model. In: Galindo, J. (ed.) Handbook of Research on Fuzzy Information Processing in Databases, pp. 435–461. IGI Global (2008)
2. Buckles, B.P., Petry, F.E.: A fuzzy representation of data for relational databases. Fuzzy Sets and Systems 7(3), 213–226 (1982)
3. Codd, E.F.: Extending the database relational model to capture more meaning. ACM Transactions on Database Systems 4, 397–434 (1979)
4. Cubero, J.C., Marín, N., Medina, J.M., Pons, O., Vila, M.A.: Fuzzy object management in an object-relational framework. In: Proc. 10th Int. Conf. on Information Processing and Management of Uncertainty in Knowledge-Based Systems, IPMU 2004, pp. 1767–1774 (2004)
5. Dubois, D., Prade, H.: Possibility Theory: An Approach to Computerized Processing of Uncertainty. Plenum Press, New York (1988)

6. Dubois, D., Prade, H.: Processing fuzzy temporal knowledge. IEEE Transactions on Systems, Man, and Cybernetics 19, 729–744 (1989)

7. Dyreson, C., Grandi, F.E.A.: A consensus glossary of temporal database concepts. SIGMOD Rec. 23, 52–64 (1994)

8. Galindo, J., Medina, J.M.: Ftsql2: Fuzzy time in relational databases. In: EUSFLAT Conf. 2001, pp. 47–50 (2001)

9. Garrido, C., Marin, N., Pons, O.: Fuzzy intervals to represent fuzzy valid time in a temporal relational database. Int. J. Uncertainty Fuzziness Knowlege-Based Syst. 17, 173–192 (2009)

10. Jensen, C.S., Mark, L., Roussopoulos, N.: Incremental implementation model for relational databases with transaction time. IEEE Trans. Knowl. Data Eng. 3, 461–473 (1991)

11. Kurutach, W.: Modelling fuzzy interval-based temporal information: a temporal database perspective. Fuzzy Systems 2, 741–748 (1995)

12. Medina, J., Pons, O., Cubero, J.: GEFRED. a generalized model of fuzzy relational databases. Information Sciences 76, 87–109 (1994)

13. Mitra, D., Srinivasan, P., Gerard, M., Hands, A.: A possibilistic interval constraint problem: Fuzzy temporal reasoning. In: Proc. of the Third IEEE Conference on IEEE World Congress on Computational Intelligence Fuzzy Systems, vol. 2, pp. 1434–1439 (June 1994)

14. Nascimento, M.A., Eich, M.H.: Decision time in temporal databases. In: Proceedings of the 2nd Int. Workshop on Temporal Representation and Reasoning, pp. 157–162 (1995)

15. Oracle: Oracle database 11g. workspace manager overview (2012)

16. Oracle: Sql language reference. In: Oracle Database v. 11.2. Documentation, pp. 1–1522 (July 2012), http://docs.oracle.com/cd/E11882_01/server.112/e26088.pdf

17. Pons, J.E., Billiet, C., Pons Capote, O., De Tré, G.: A possibilistic valid-time model. In: Greco, S., Bouchon-Meunier, B., Coletti, G., Fedrizzi, M., Matarazzo, B., Yager, R.R. (eds.) IPMU 2012, Part I. CCIS, vol. 297, pp. 420–429. Springer, Heidelberg (2012)

18. Pons, J.E., Marín, N., Pons Capote, O., Billiet, C., de Tre, G.: A relational model for the possibilistic valid-time approach. International Journal of Computational Intelligence Systems 5(6), 1068–1088 (2012)

19. Pons, J.E., Pons Capote, O., Blanco Medina, I.: A fuzzy valid-time model for relational databases within the hibernate framework. In: Christiansen, H., De Tré, G., Yazici, A., Zadrozny, S., Andreasen, T., Larsen, H.L. (eds.) FQAS 2011. LNCS, vol. 7022, pp. 424–435. Springer, Heidelberg (2011)

20. Prade, H., Testemale, C.: Generalizing database relational algebra for the treatment of incomplete or uncertain information and vague queries. Information Sciences 34(2), 115–143 (1984)

21. Snodgrass, R.: The temporal query language tquel. In: Proceedings of the 3rd ACM SIGACT-SIGMOD, pp. 204–213. ACM, New York (1984)

22. Umano, M., Fukami, S.: Retrieval processing from fuzzy databases. Technical Reports of IECE of Japan 80(204), 45–54 (1980)

23. Zemankova-Leech, M., Kandel, A.: Fuzzy relational databases - a key to expert systems. Journal of the American Society for Information Science 37, 272–273 (1984)

Predictors of Users' Willingness to Personalize Web Search

Arjumand Younus[1,2], Colm O'Riordan[1], and Gabriella Pasi[2]

[1] Computational Intelligence Research Group, Information Technology,
National University of Ireland, Galway, Ireland
[2] Information Retrieval Lab, Informatics, Systems and Communication,
University of Milan Bicocca, Milan, Italy
{arjumand.younus,colm.oriordan}@nuigalway.ie, pasi@disco.unimib.it

Abstract. Personalized Web search offers a promising solution to the task of user-tailored information-seeking; however, one of the reasons why it is not widely adopted by users is due to privacy concerns. Over the past few years social networking services (SNS) have re-shaped the traditional paradigm of information-seeking. People now tend to simultaneously make use of both Web search engines and social networking services when faced with an information need. In this paper, using data gathered in a user survey, we present an analysis of the correlation between the users' willingness to personalize Web search and their social network usage patterns. The participants' responses to the survey questions enabled us to use a regression model for identifying the relationship between SNS variables and willingness to personalize Web search. We also performed a follow-up user survey for use in a support vector machine (SVM) based prediction framework. The prediction results lead to the observation that SNS features such as a user's demographic factors (such as age, gender, location), a user's presence or absence on Twitter and Google+, amount of activity on Twitter and Google+ along with the user's tendency to ask questions on social networks are significant predictors in characterising users who would be willing to opt for personalized Web search results.

1 Introduction

When using a Web search engine to tackle an information-seeking task, users typically oversimplify the complexity of their information needs by selecting a few keywords (query). Moreover, given the lack of information regarding the context of the query, it is hard to attain a reasonable user satisfaction when using Web search engines. In recent years personalized Web search has emerged as a promising way to improve the search quality through customization of search results for people with different information interests and goals. However, concerns about the privacy of users have introduced reluctance in the adoption of personalized Web search systems [8,17] such as iGoogle [13].

In the pre-digital age the most common way to find useful information was via interaction with friends, colleagues, or domain experts. With the advent of

H.L. Larsen et al. (Eds.): FQAS 2013, LNAI 8132, pp. 459–470, 2013.

social networking services (SNS) the information-seeking patterns of users have considerably changed [10] leading to an intersection between traditional and modern approaches of information-seeking [16]. Social networking services are now rekindling the pre-digital information-seeking pattern in the digital world.

Over the past few years, many research efforts have focused on both personalized Web search and social search [4,6]. In this paper, we investigate the correlation between social network usage patterns of users and their openness to opt for Web search personalization. This can open doors for understanding the user's willingness to adopt Web search personalization based on observable correlations. As a motivating example, let us consider a scenario involving two users: user A is highly active on various social networking services as he/she communicates his/her thoughts over a range of topics (politics, religion, economics etc.), while user B is much less active when compared to user A in sharing his/her thoughts on social networking services. In line with this scenario it would be interesting to investigate the correlations of the behaviors of both users A and B and their openness to Web search personalization. Some commercial systems have attempted a similar integration (as is evident in recent social search approaches by Google and Bing: e.g., Bing's Facebook integration[1] and Google's "Search Plus Your World" Google+ integration[2]). However, to the best of our knowledge, there is no literature that describes the correlations that we investigate in this paper.

Apparently the existence of a correlation between the willingness to personalize Web search and the inclination to use social networks seems intuitively obvious and hence, trivial. However, a thorough analysis of the various usage patterns within different types of popular social networking services is something that needs careful investigation so as to form a more coherent basis for the development of meaningful and well-accepted personalized search systems. With the aim of finding a characterisation of users who would prefer personalized search results more than non-personalized search results we conducted a user survey. We designed the user survey so that we can investigate the social network usage patterns of users along with their privacy concerns with respect to Web search personalization, and their preferences to opt for personalized Web search. We also looked into various SNS tools (more specifically Facebook, Twitter, LinkedIn etc.) as well as at the characteristics of SNS usage (such as frequency of SNS usage, frequency of posting SNS updates, number of friends on SNS, frequency of asking questions on SNS) which we hypothesise might relate more closely to users' willingness for Web search personalization. We conducted a large-scale user survey in two parts where the first part gathered responses from 380 people from various countries, and the second part gathered responses from 113 people from various countries. This data was then used in a regression model to analyse the correlation between SNS variables and willingness to personalize Web search. The data was also used in a support vector machine (SVM) model

---

[1] http://www.bing.com/community/site_blogs/b/search/archive/2011/05/16/news-announcement-may-17.aspx

[2] http://www.google.com/insidesearch/plus.html

to explore the potential to make predictions about users who would be willing to opt for personalized Web search results. We wish to explore if the prediction accuracy would be sufficient for a real personalised Web search system. In doing these analyses we discovered a number of useful patterns and behaviours about users' openness to Web search personalization; these outcomes can be of help in the future developments of personalized Web search and social search systems.

The remainder of this paper is organized as follows. Section 2 presents some related work relevant to our investigation along with an explanation of how we differ from past work. Section 3 describes our survey methodology in detail. Section 4 discusses the results of the survey while Section 5 presents the results of the prediction framework along with an examination of the features that better predict users who would opt for Web search personalization. Section 6 presents an overview of implications from this study along with some limitations. Finally, Section 7 concludes the paper.

2   Related Work

For over a decade Web search personalization has been viewed as an effective solution for user-tailored information-seeking. A huge amount of research proposes the use of implicitly-gathered user information such as browser history, query and clickthrough history and desktop history to improve search results ranking on a per-user basis [4,12,15]. The collection of such implicit user information often raises serious privacy concerns which is one of the major barriers in deploying personalized search applications [7,11]. The relationship between users' social network usage patterns and their willingness for Web search personalization is a promising direction that can help significantly towards the deployment of personalized search applications. The decision to investigate the correlation between social network usage patterns and willingness for Web search personalization was taken on account of many studies that have shown people to spend a considerable amount of time on social networks [3]. This high amount of social network usage by users in turn has also affected their information-seeking habits and specifically, the way they interact with search engines [9]. Hence, we argue in this contribution that it is necessary to revisit the notion of Web search personalization from this perspective.

The literature closest to our identified research questions addresses the use of social information in Web search [2,5]. However, these efforts do not aim towards using social network activities for inferring users' willingness towards Web search personalization. Research by Morris et al. pursues a comparison of information-seeking using search engines and social networks and proposes the design of future search systems that can help direct users to social resources in some circumstances [9]. We propose a somewhat similar notion through an analysis of the correlation between user's willingness for Web search personalization and his/her social network usage patterns.

3   Survey and Survey Results

In this section we describe the survey methodology along with the measures and variables that are analysed. The survey comprised 20 close-ended questions with five questions of a general nature (collecting basic information about the participants), five questions related to various aspects of Web search personalization (explained in section 3.2) and ten questions related to SNS usage (explained in section 3.2).

3.1   Participants and Survey Content

In order to understand how SNS usage patterns affect people's willingness to personalize Web search results, we designed a survey and dispatched it to a wide range of people in various countries (i.e. Ireland, Italy, Spain, France, United Kingdom, Finland, United States, Canada, Pakistan, India, South Korea). In the first phase, the survey was completed by 380 people. Demographic characteristics for the survey respondents in the first phase are shown in Table 1. Participants were recruited via university distribution lists (both online and offline) and social networking sites (chiefly, Facebook and Twitter) and we recruited a diverse range of people; distribution lists were employed to avoid recruiting only those participants who have a social network presence and thereby avoiding high skew in the results. In addition to collecting basic demographic information, we also collected information about participants' use of SNS tools such as which SNS accounts they have and which of them they use the most (shown in Table 2 and Table 3 respectively).

Table 1. Demographic Variables (n=380)

Demographics	N (%)
Male	235 (61.8%)
Female	145 (38.2%)
Europe	206 (54.2%)
America	21 (5.5%)
Asia	153 (40.3%)
10-20	0 (0%)
21-30	259 (68.2%)
31-40	87 (22.9%)
41-50	19 (5%)
Above 50	15 (3.9%)

Table 2. Statistics for SNS Accounts of Survey Respondents

SNS Tool Details	N (%)
Facebook Presence	356 (93.7%)
Twitter Presence	241 (63.4%)
Google+ Presence	239 (62.9%)
LinkedIn Presence	272 (71.6%)
Bookmarking Sites Presence	60 (15.8%)

Table 3. Statistics for Highly Used SNS Accounts by Survey Respondents

SNS Usage Details	N (%)
Facebook As Most Used	325 (85.5%)
Twitter As Most Used	106 (27.9%)
Google+ As Most Used	30 (7.9%)
LinkedIn As Most Used	17 (4.5%)

3.2   Measures and Variables

The following variables were included in the analyses:

Personalization Response: Respondents were asked whether or not they considered personalized search results to be of any benefit to them[3]. This was a binary variable with a "Yes" or "No" response. Additionally, a likert-scale variable was used corresponding to respondents' agreement with Web search personalization making the information-seeking process less painstaking; the scale ranged from "Strongly Disagree" (1) to "Strongly Agree" (5). Furthermore, three additional binary variables related to Web search personalization features of current search engines were also investigated. Respondents were asked about their awareness of the personalization feature in existing Web search engines in addition to their awareness about search engines making use of their search history data for the process of Web search personalization and finally, whether or not they were comfortable with such use. We report these statistics about Web search personalization in Table 4.

Table 4. Statistics on Users' Desirability for Web Search Personalization

Personalized Web Search Results Considered as Useful	N (%)
Yes	188 (49.5%)
No	192 (50.5%)
Awareness about Personalization Feature of Web Search Engines	N (%)
Yes	275 (72.4%)
No	105 (27.6%)
Awareness about Web Search Engines using Search History Data	N (%)
Yes	337 (88.7%)
No	43 (11.3%)
Comfortable with Web Search Engines using Search History Data	N (%)
Yes	196 (51.6%)
No	184 (48.4%)
Likert Scale for Agreement on Worth of Web Search Personalization	MEAN (SD)
	3.58 (0.98)

Facebook Usage: Considering the high usage of Facebook as reported in Table 3, some questions in the survey were particularly focused towards Facebook usage. In particular, respondents were asked about the frequency of their Facebook usage (with scale ranging from "several times a day" (7) to "never" (1)), frequency of posting something (status update, photo or link) on Facebook (with scale ranging from "frequently" (4) to "never" (1)), frequency of liking something on Facebook (with scale ranging from "frequently" (4) to "never" (1)) and the approximate number of Facebook friends. We report these statistics about Facebook usage in Table 5.

---

[3] For the ease of survey respondents, we included in the survey an explanation of what Web search personalization is along with an explanation of how implicit user data is used for this process. Further, we asked the survey respondents to contact us in case of any confusion with respect to Web search personalization and some of them asked us questions to understand the personalization process better.

Twitter Usage: We also included some Twitter-specific measures in our analysis. The survey did not ask for these measures explicitly. Instead, the survey respondents who used Twitter were asked to provide their Twitter handles which were then used to fetch all their tweets. From these tweets, we extracted for the survey respondents their number of mentions (the mention feature of Twitter enables its users to address a specific user within a tweet) and number of retweets (the retweet feature of Twitter enables its users to re-post a tweet posted by someone else). Additionally we also extracted the number of topics contained in survey respondents' tweets through the use of Twitter-LDA [18].

Table 5. Statistics on Facebook Usage

Frequency of Posting on Facebook	MEAN (SD)
	2.99 (0.95)
Frequency of Facebook Likes	MEAN (SD)
	3.22 (0.94)
No. of Facebook Friends	N (%)
Less than 100	62 (17.4%)
100-200	88 (24.7%)
200-300	85 (23.9%)
300-400	50 (14.0%)
400-500	28 (7.9%)
More than 500	43 (12.1%)

Q & A Activity on SNS Tools: Lastly, given the significance of Q & A activity on SNS [10], the survey also included questions about users' Q & A activities on SNS tools. Respondents were asked whether they had ever used SNS tools for information-seeking and whether they considered Q & A activity on SNS as useful. These were binary variables with a "Yes" or "No" response. If respondents preferred Q & A activity on SNS, we further asked them about their frequency of asking questions on SNS along with the frequency with which they considered answers coming from SNS as more reliable than answers obtained from search engines (with scale ranging from "most of the time" (4) to "never" (1)). We report these statistics about Q & A activity on SNS in Table 6.

Table 6. Statistics on Social Network Q & A Activity

Ever Used Social Networks for Information-Seeking	N (%)
Yes	272 (71.6%)
No	108 (28.4%)
Q & A Activity on Social Networks Considered as Useful	N (%)
Yes	187 (49.2%)
No	193 (50.8%)
Frequency of Asking Questions on SNS	MEAN (SD)
	1.99 (0.95)
Frequency of Considering Answers on SNS More Reliable than Search Engines	MEAN (SD)
	2.38 (1.03)

4   Analyses and Findings

We first examine the associations between variables representing Web search personalization willingness while controlling for demographic and other factors. Table 7 shows predictors of acts for Web search personalization willingness where the rows of the table represent these acts. Here, the acts are basically the variables corresponding to presence on various SNS tools, frequency of usage of various SNS tools, Facebook usage, Twitter usage, Q & A Activity on SNS as explained in section 3.2. Table 7 shows the results of logistic regressions with binary outcomes in the following dependent variables:

- User's trust in search personalization as a beneficial process (shown as *WP i.e., Willingness of Personalization* in Table 7)
- User's awareness of personalized search services such as iGoogle (shown as *AP i.e., Awareness of Personalization* in Table 7)
- User's awareness that search engines use their search history data for the process of Web search personalization (shown as *AH i.e., Awareness of History* in Table 7)
- User's acceptance (comfort level) of the fact that search engines use their search history data for the process of Web search personalization (shown as *WH i.e., Willingness of History* in Table 7)

Table 7. Logit regression showing the odds of users' willingness towards Web search personalization

1		WP	AP	AH	WH
2	Male	1.635**	2.643**	6.318***	1.480*
3	American	0.001	0.004	4.478	0.002
4	Asian	0.001	0.003	0.004	0.001
5	European	0.001	0.004	0.001	0.001
6	Age	1.069	0.997	0.981	0.841
7	Facebook Presence	0.982	0.561	0.860	1.844
8	Twitter Presence	1.544*	1.692	3.651***	1.519*
9	Google+ Presence	1.816***	1.427	1.364	1.626**
10	LinkedIn Presence	0.940	2.475*	2.031**	1.150
11	Bookmarking Sites Presence	1.289	2.535	1.153	1.771**
12	High Usage of Facebook	1.599	0.736	0.488	1.222
13	High Usage of Twitter	1.166*	1.574	3.986**	1.353
14	High Usage of Google+	3.042***	1.565	0.637	2.292**
15	High Usage of LinkedIn	1.193	2.069	1.127	0.971
16	Facebook Usage Frequency	0.898	1.204	1.051	1.231
17	Facebook Posting Frequency	1.637***	0.450*	0.893	1.246
18	Facebook Liking Frequency	0.920	1.776	0.922	0.924
19	No. of Facebook Friends	0.873*	1.031	1.181	0.899
20	Twitter Mentions	1.000	0.983	0.997	0.998*
21	Twitter Retweets	1.001	0.982	0.999	0.999
22	No. of Topics in Tweets	0.997	0.969	1.036	1.012
23	No. of Tweets	0.999	1.015*	1.001	1.001*
24	Prefers Q & A Activity on SNS	1.821***	1.474	0.972	1.492*
25	Considers Q & A Activity on SNS as Useful	1.771***	1.173	0.916	1.318
26	Frequency of Q & A Activity on SNS	1.366**	0.940	0.884	1.273**
27	Frequency of Considering Responses				
28	from SNS More Useful than Search Engines	1.374***	1.232	0.950	1.228**

Note *p<.05, **p<.01, ***p<.001.

Table 8. OLS regression showing the level of users' agreement towards Web search personalization

	Personalization Agreement
High Usage of Facebook	0.278*
High Usage of Twitter	0.143
High Usage of Google+	0.338*
High Usage of LinkedIn	0.036
Facebook Usage Frequency	0.052
Facebook Posting Frequency	0.139***
Facebook Liking Frequency	0.108*
No. of Facebook Friends	-0.026
Twitter Mentions	-0.171*
Twitter Retweets	-0.188
No. of Topics in Tweets	0.148
No. of Tweets	0.145*
Prefers Q & A Activity on SNS	0.051
Considers Q & A Activity on SNS as Useful	0.083
Frequency of Q & A Activity on SNS	0.165**
Frequency of Considering Responses from SNS More Useful than Search Engines	0.025

Note *p<.05, **p<.01, ***p<.001.

The results from Table 7 show that males are more likely to consider Web search personalization as beneficial (row 2 corresponding to WP) while location and age do not have much effect on willingness for Web search personalization (row 3-6 corresponding to WP). Furthermore, the presence of a user on Twitter and/or Google+ is a strong indicator that he/she will consider Web search personalization as beneficial and similar is the case for his/her high usage of Twitter and/or Google+ (row 8-9 corresponding to WP and row 13-14 corresponding to WP); the increase is more significant for user presence on Google+ and for high usage of Google+. Additionally, as expected an increase in posting frequency on Facebook increases the odds of willingness to personalize Web search and contrary to expectation, an increase in the number of users' friends on Facebook decreases the odds of willingness to personalize Web search (row 17 and row 19 corresponding to WP). Lastly, users who use SNS more frequently for Q & A activities are more likely to be willing to opt for Web search personalization in addition to users who frequently consider that responses coming from SNS as more reliable than responses from search engines (row 24-27 corresponding to WP).

For the dependent variable reflecting whether or not the user is aware of the personalization feature in current Web search engines (AP), the likelihood is increased if the user is a male, has a presence on LinkedIn, has a high posting frequency on Facebook, and has a high number of tweets on Twitter. Similarly for the dependent variable reflecting whether or not the user is aware of search engines making use of his/her search history data for the process of Web search personalization (AH), the likelihood is increased for males, for users with Twitter and/or LinkedIn presence, and for highly frequent Twitter users. Lastly, for the dependent variable reflecting whether or not the user is comfortable with search engines making use of his/her search history data for the process of Web search personalization WH, the likelihood is increased for males, for users of Twitter,

Google+ and/or bookmarking sites, for more frequent Google+ users, for users with less mentions and/or high number of tweets on Twitter, for users who frequently use SNS for Q & A activities and consider responses from SNS to be more reliable than responses coming from search engines.

Table 8 shows that users having a high usage frequency on Facebook and/or Google+ tend to agree more strongly with the notion of Web search personalization making the information-seeking process easier and less painstaking. This is also the case for users who have a high posting and liking frequency on Facebook along with a high tweeting frequency on Twitter. On the other hand, users' level of agreement with the notion that Web search personalization makes the information-seeking process easier decreases with their frequency of mentions on Twitter. Lastly, users with a high frequency of Q & A activity on SNS agree to a higher degree with the notion of Web search personalization making the information-seeking process easier.

5   Prediction Model for Web Search Personalization Willingness

In this section we explain the prediction model that we developed to predict whether or not a user would be interested in Web search personalization. We performed the predictions on a second set of user survey data (test data for our prediction model) and for this we collected responses of 113 people using the same survey designed for the correlation analysis of Section 3 and 4. The data collected in the first phase of the survey served as the training data (i.e., the 380 responses gathered for correlation analysis) for our model.

We utilized five types of information described in Section 3 that can help characterize a user's willingness for Web search personalization. Here, we refer to the demographic features (Table 1) as *demographics*, features denoting presence on various SNS tools (Table 2) as *sns_presence*, features describing high usage of various SNS tools (Table 3) as *sns_highusage*, features describing Facebook usage (Table 5) as *fb_usage* and features describing Q & A activity (Table 6) as *qa_activity*. The features we explored are used in conjunction with a supervised machine learning framework providing a model for predicting users who would (and wouldn't) opt for personalized Web search results. The baseline feature we used is setting of all Web search personalization predictions to "having no willingess for Web search personalization." As a learning algorithm, we used Support Vector Machines which is a state-of-the-art machine learning algorithm. We found the set of features that best predicted the variable *WP i.e., Willingness of Personalization* of Table 7. The goal of our prediction model is two-fold: first, to see if the prediction accuracy would be sufficient for a real personalized Web search system and second, to explore the value of various types of information in the process of automatically determining willingness for Web search personalization.

Figure 1 shows the classification performance of the personalization willingness prediction model, incrementally examining the role of each feature in the prediction accuracy[4]. In-line with intuition, we witnessed a consistent, gradual increase in performance as additional information is made available to the classifier. It is interesting to note the significant performance boost when we move from the baseline to *sns_presence* and *fb_usage* and also when the *qa_activity* features are added to the model. We first examined each feature in isolation (a maximum of 53.9% was obtained with *sns_presence*); after this we applied all possible permutations of features and Figure 1 shows the best possible order. Furthermore, within each category the following features led to the biggest performance gain:

- *demographics*: Gender, Location and Age
- *sns_presence*: Twitter Presence, Google+ Presence and Bookmarking Sites Presence
- *sns_highusage*: High Usage of Google+
- *fb_usage*: Facebook Usage Frequency, Facebook Posting Frequency and No. of Facebook Friends
- *qa_activity*: Prefers Q & A Activity on SNS, Considers Q & A Activity on SNS as Useful, Frequency of Q & A Activity on SNS and Frequency of Considering Responses from SNS More Useful than Search Engines

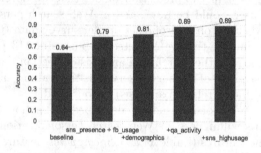

Fig. 1. Overall classification results for Various Feature Types

6   Discussion

We will now discuss the theoretical and practical implications of our study along with some limitations. The correlations (and their results) investigated here may seem intuitively obvious in terms of high engagement in social networks signifying a higher degree of readiness to accept the (at least partial) loss of privacy that is inevitably involved in search personalization. However, recent research indicates that this is not the case as users of SNS have shown growing privacy concerns [14]. Boyd and Hargittai [1] found that the majority of young adult users of Facebook have engaged with managing their privacy settings on the site at least to some extent and noted a rise in such privacy settings' engagement between

---

[4] We show the combinations that are statistically significant at the 0.95 level with respect to the baseline.

2009 and 2010, a year in which Facebook unveiled many controversial privacy changes that made more of information on the site public. In fact even as our data shows the percentage of survey respondents who do not consider Web search personalization as beneficial is higher than expected; similar is the case for survey respondents who are not comfortable with Web search engines using their search history data (refer to Table 4 for these statistics).

6.1 Implications

Users' privacy concerns have proven to be a significant challenge with respect to Web search personalization, and the issue of when to personalize and when not personalize represents an important challenge, which has not been deeply investigated yet. We argue through the investigations in this paper that social network usage patterns of users can serve as a significant predictor for determination when to personalize and when not to personalize. Understanding the target audience of personalized search systems is an important aspect for the development of meaningful and well-accepted systems, and hence, this work serves as a first step in that dimension. This could aid towards removing the need for the user to state privacy requirements and to infer these settings through social network usage patterns.

6.2 Limitations

Despite our maximum efforts to recruit a diverse range of participants for the study the results may be somewhat skewed towards considerable levels of use of SNS (Table 2) and this may not be representative of the general population. Another possible limitation may be users' lack of information about their answers to the survey questions due to not having sufficient insight regarding what personalized search actually comprises [8]. Nevertheless, there are reasons to be confident in our conclusions. First, the high skewness towards SNS is natural given the immense popularity of these services and evidence suggests that almost every Web user maintains SNS presence (the important difference lies in SNS activity which is a variable thoroughly investigated by us). Second, as explained previously we thoroughly explained the process of Web search personalization to the users so as to inform them of privacy considerations that arise from it.

7   Conclusion

In this paper, we utilized a survey methodology to gather relevant data and then investigated the correlations between users' social network usage patterns and their openness to opt for Web search personalization. We believe the findings of our study have significant implications for the design of future personalized search and social search applications. More significantly, it provides indicators as to where researchers could focus attention when attempting to deliver automatically adjustable privacy-enhanced solutions for Web search personalization.

By exploiting SNS data, predictions can be made about user preferences without requiring explicit user input. As future work, we aim to investigate long-term interaction of users with personalized Web search results obtained via SNS inferences as outlined in this paper.

References

1. Boyd, D.M., Hargittai, E.: Facebook privacy settings: Who cares? First Monday 15(8) (2010)
2. Carmel, D., Zwerdling, N., Guy, I., Ofek-Koifman, S., Har'el, N., Ronen, I., Uziel, E., Yogev, S., Chernov, S.: Personalized social search based on the user's social network. In: CIKM 2009, pp. 1227–1236. ACM, New York (2009)
3. DiSalvo, D.: Are social networks messing with your head? Scientific American Mind 20, 48–55 (2010)
4. Dou, Z., Song, R., Wen, J.-R.: A large-scale evaluation and analysis of personalized search strategies. In: WWW 2007, pp. 581–590. ACM, New York (2007)
5. Heymann, P., Koutrika, G., Garcia-Molina, H.: Can social bookmarking improve web search? In: WSDM 2008, pp. 195–206. ACM, New York (2008)
6. Horowitz, D., Kamvar, S.D.: The anatomy of a large-scale social search engine. In: Proceedings of the 19th International Conference on World Wide Web, WWW 2010, pp. 431–440. ACM, New York (2010)
7. Karat, C.-M., Brodie, C., Karat, J.: Usable privacy and security for personal information management. Commun. ACM 49(1), 56–57 (2006)
8. Kobsa, A.: Privacy-enhanced web personalization. In: Brusilovsky, P., Kobsa, A., Nejdl, W. (eds.) Adaptive Web 2007. LNCS, vol. 4321, pp. 628–670. Springer, Heidelberg (2007)
9. Morris, M.R., Jaime, T., Panovich, K.: A Comparison of Information Seeking Using Search Engines and Social Networks. In: ICWSM 2010, pp. 291–294 (2010)
10. Morris, M.R., Teevan, J., Panovich, K.: What do people ask their social networks, and why?: a survey study of status message q & a behavior. In: CHI 2010, pp. 1739–1748. ACM, New York (2010)
11. Sackmann, S., Strüker, J., Accorsi, R.: Personalization in privacy-aware highly dynamic systems. Commun. ACM 49(9), 32–38 (2006)
12. Shen, X., Tan, B., Zhai, C.: Implicit user modeling for personalized search. In: CIKM 2005, pp. 824–831. ACM, New York (2005)
13. Shen, X., Tan, B., Zhai, C.: Privacy protection in personalized search. SIGIR Forum 41(1), 4–17 (2007)
14. Stutzman, F.D.: Networked Information Behavior in Life Transition. PhD thesis, The University of North Carolina at Chapel Hill (2011)
15. Teevan, J., Dumais, S.T., Horvitz, E.: Personalizing search via automated analysis of interests and activities. In: SIGIR 2005, ACM, pp. 449–456. New York (2005)
16. Teevan, J., Morris, M.R.: Exploring the complementary roles of social networks and search engines. In: HCIC 2012 (2012)
17. Wang, Q., Jin, H.: Exploring online social activities for adaptive search personalization. In: CIKM 2010, pp. 999–1008. ACM, New York (2010)
18. Zhao, W.X., Jiang, J., Weng, J., He, J., Lim, E.-P., Yan, H., Li, X.: Comparing twitter and traditional media using topic models. In: Clough, P., Foley, C., Gurrin, C., Jones, G.J.F., Kraaij, W., Lee, H., Mudoch, V. (eds.) ECIR 2011. LNCS, vol. 6611, pp. 338–349. Springer, Heidelberg (2011)

Semantic-Based Recommendation of Nutrition Diets for the Elderly from Agroalimentary Thesauri

Vanesa Espín, María V. Hurtado, Manuel Noguera, and Kawtar Benghazi

Departamento de Lenguajes y Sistemas Informáticos, University of Granada
E.T.S.I.I.T., c/Daniel Saucedo Aranda s/n 18071, Granada, Spain
rebel@correo.ugr.es, {mhurtado,mnoguera,benghazi}@ugr.es

Abstract. The wealth of information on nutrition and healthy diets along the web, as health web magazines or forums, often leads to confuse users in several ways. Reliability and completeness of information, as well as extracting only the relevant one becomes a critical issue, especially for certain groups of people such as the elderly. Likewise, heterogeneity of information representation and without a clear semantics hinders knowledge sharing and enrichment. In this paper, it is introduced a method to compute the semantic similarity between foods used in NutElCare, an ontology-based recommender system capable of collecting and representing relevant nutritional information from expert sources in order to providing adequate nutrition tips for the elderly. The knowledge base of NutElCare is an OWL ontology built from AGROVOC FAO thesaurus.

Keywords: Recommender systems, semantic web, ontology-based representation, semantic similarity.

1 Introduction

Proper nutrition is a key factor in human health, but is particularly important for vulnerable sectors of the population such as children, the elderly or people with certain diseases or disabilities that require special attention. To meet these needs, nutrition experts work on designing suitable healthy diet plans [20].

The big volume of data is, on the one hand, what contributes to its great power, but on the other hand, makes its weaknesses more evident. It is becoming difficult for users to know what information they find on the web is reliable or complete. Also, handling the large volume of data and extracting the information correctly to suit their needs efficiently is not an easy task. Recommender systems provide a way to address these problems. Nonetheless, classic recommender systems suffer from some general drawbacks. One of the most important is the problem of heterogeneity of the representation of information, which leads to communication problems between agents that participate, even between agents and users, and prevents this information from being reused by other processes or applications [14].

H.L. Larsen et al. (Eds.): FQAS 2013, LNAI 8132, pp. 471–482, 2013.

To overcome some of the problems of traditional recommender systems, semantic recommender systems have emerged, which use Semantic Web techniques in their development, such as ontologies [12]. Ontologies are known to be wide sources of knowledge, this is largely due to its reasoning ability and the semantic they bring [13]. They provide formal, uniform and shareable representations about a domain [4] and have been applied to different recommendation systems to reduce heterogeneity and improve content retrieval [5], [23], [6].

In the nutritional context, several recommender systems have been proposed [1], [8], [19], [21]. Most of them are focused on obtaining recipes that appeal to users and there are very few oriented to health care. The fact that users who are intended to use such systems belong to more sensitive areas of the population, nutritionally speaking, is taken into account in an even smaller number of them.

The expected growth in the world aging population, the ubiquity and affordability of smart devices, and control over health care costs, lead to new scientific challenges to be addressed in order to provide new life opportunities for aging population [3]. In this paper it is introduced NutElCare (Nutrition for Elder Care), a recommender system that retrieves reliable and complete nutritional information from expert sources, either humans (e.g. nutritionists, gerontologists, bromatologists) or computerized (e.g. information systems, nutritional databases, World Health Organization -WHO- and Spanish Society of Parenteral and Enteral Nutrition -SENPE- recommendations). It assists older people to take advantage of these tips and make up their own diet plans in an easy way. One of the purposes of this system is that final users could use it on their own at home. In this way, elderly self-sufficiency is boosted, giving a major value to the system.

In NutElCare, the representation of information is improved by applying Semantic Web technologies. An OWL [22] ontology is used in order to make use of reasoning capabilities that description logics provides. This ontology contains a food taxonomy enriched with nutritional features and relations. AGROVOC FAO Thesaurus [7], in its OWL version, has been used as a starting point in the development of the NutElCare ontology for reusability and internationalization purposes. Furthermore, specific procedures to automate information retrieval and computation of the semantic similarity between concepts and terms have been also devised.

The remainder of the paper is organised as follows. Section 2 introduces some work related to nutritional recommender systems. In Section 3, it is explained the development of NutElCare: architecture, ontology and recommendation production. Finally, in Section 4, conclusions and future work are presented.

2   Related Work

Some work related to nutritional recommender systems exists in the literature. Many of the current systems aim to provide recipes that the user would like, a minor part aim to improve the user health and an even smaller part is intended for the vulnerable sectors of the population.

For example, members of CSIRO (Commonwealth Scientific and Industrial Research Organisation) [9] developed a recommendation system of recipes for all kinds of users. It uses a collaborative filtering method that takes into account ratings from the user community of the recipes and food that integrate it to generate recommendations and create new recipes. However, nutritional aspects or user profiles are not taken into account.

FOODS [19] is a semantic recommender system that uses the PIPS [15] food classification modelled as ontology in its knowledge base. It recommends recipes for all kinds of users but classified in profiles. While taking into account the user age and nutritional factors of the ingredients, it allows users to select those who they want and it does not take into account what other foods have been eaten during the week, so does not cover weekly needs if the user does not know them.

Aberg [1] proposes an intelligent food planning system that the elderly could use at home, to avoid the risk of malnutrition experienced by this sector of the population. The system considers different variables in recommendation such as: user taste preferences, cost of meals, preparation difficulty, availability of ingredients and nutritional needs. The system allows users to select values as interval ranges for different variables such as fat, energy or cholesterol. Thus, the author suggests that the use of the system must be guided by a carer, so it does not foster the self-sufficiency of the elderly.

Van Pinxteren et al [21] propose a similarity measure between recipes for nutritional recommender systems, that can be used to recommend alternatives to selected meals. It considers providing different choices so that users can ask for more vegetables or less fat, thus obtaining slight variations to selected recipes. A feature vector on foods is set up using weights according to the occurrence frequency of each feature and the Euclidean weighted distance between two meals is calculated to obtain their similarity. To provide healthy recipes, similar recipes to the selected healthy ones are provided but, in this case, similarity is not nutrient guided.

3   NutElCare

NutElCare is a semantic recommender system to providing healthy diet plans for the elderly. It makes use of an ontological description of the user profile from the body mass index (BMI), time of year and geographical environment, physical activity and the level of mastication and swallowing of the user. Once this first part of the profile is carried out, NutElCare provides a basic healthy diet plan based on it. The recommendation process starts, allowing the users to make variations to the diets according to their own preferences. The recommendations must be flexible and also take into account the personal tastes of the users, their allergenic contraindications and what food has been taken during the week, offering alternatives to the original diet plan based on these factors. The recommendations are always nutrient guided, providing alternative suggestions of similar conditions, to continue meeting the original healthy requirements of the diet.

The system allows the user the possibility of modifying this diet so that, without losing its nutritional and healthy value, it can better suit his individual interests and may become more to his taste. For this purpose, the recommender system must know both the physical/demographic characteristics of the user and their interests [18]. Hence, the first time that the user logs onto NutElCare, he completes a form with the relevant information about himself. This information will be the basis for selecting the diet model to recommend.

To offer a model diet, in addition to typical demographic data (such as name, gender, age, address), the following information is required:

− *Physical properties.*
  • The weight and height of the user in order to calculate the body mass index (BMI). Regarding the classification of the nutritional state in terms of the BMI of the elderly in the consensus document created by SENPE [17], NutElCare will classify the nutritional state of the users as: malnutrition, under-weight, normal-weight, over-weight or obesity.
  • Deglution and chewing level, to distinguish whether the user needs normal, soft, semi-liquid, or easily digested diets.
− *Environmental factors.* Season and geographical area in which recommendations are made.
− *Activity factors.* Amount of exercise a week (active, standard or sedentary life).

Next, the system collects the explicit interests of the user by means of questions about allergies and taste preferences. If one of these food elements or nutrients, considered as *no-interesting*, appears on the diet model to recommend, the system must reason which aliment from the knowledge base is the most similar and offer it instead. In the following sections, the architecture of the system, the proposed ontology and the process for semantic similarity calculation are explained.

3.1   NutElCare Architecture

NutElCare architecture is based on the following components of the semantic recommender systems:

− Knowledge base and items representation.
− User profiling and learning techniques of user interests.
− Obtaining and providing recommendations about items in the knowledge base through semantic similarity measures.

We can thus represent the architecture of the system as shown in figure 1. The following describes a summary of its components.

1. **User Interface.** One of the main success factors of the system is to be easy to use, almost as a game, while motivating the users to its correct and serious utilization, keeping in mind that it will contribute to improving their health and quality of life.

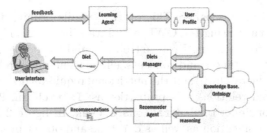

Fig. 1. NutElCare basic architecture

2. **User Profile.** It contains the model of the user that runs the system. The user modelling is composed by the static model, obtained at the first contact with the new user, and the dynamic model that is built with the help of the learning agent studying the behaviour of the user in the system.

3. **Learning Agent.** As the user interacts with the system and sends feedback about his behaviour, the agent uses learning techniques to study such behaviour, extracting patterns and communicating them to user profile to build the model.

4. **Diet Manager.** It is responsible for retrieving the suitable diet models for a user from the static part of his user profile. Once a diet model is retrieved, it builds a personalized diet for the current user and sends it to user interface.

5. **Recommender Agent.** It obtains the nutritional recommendations about food in order to provide it to the user or to the diet manager. It extracts information from the knowledge base through reasoning to calculate, with the help of a semantic similarity technique, a recommendation list of the selected food.

6. **Knowledge Base (Ontology).** It is the core of NutElCare. In it, all the nutritional information needed in the system operation is stored. It is modelled through the NutElCare ontology, where food items are represented with instances.

3.2 NutElCare Ontology

NutElCare contains a knowledge base represented as an OWL ontology, to which, the nutritional expert knowledge has been transferred from the expert sources. The ontology contains that information which the system reasons with, in order to obtain proper recommendations and it is mainly formed by a food taxonomy enriched with nutritional properties and the alimentary factors that combine in the recommendation process. As one of the major motivations in the use of ontologies is the reusability of a domain knowledge possibility, the initial work was to search for and select one ontology in the domain of nutrition and feeding. After some research, the AGROVOC FAO Thesaurus [7] of the United Nations was chosen for the sake of reusability and internationalisation.

For space limitations, the process followed to reduce AGROVOC and keep only the relevant information, as well as the development of the new ontology, are not explained in this paper. This process entails pruning the AGROVOC branch ''Products'' using the OWL API Java Library [2]. Then, it is carried out an iterative process, using Protégé [16], to build the ontology up consisting of: creation of new instances and concepts, elimination of non relevant concepts, restructure of the hierarchy to suit the nutritional requirements, and the addition of some nutritional relations and properties as: Is rich in, Protein level, Caloric level or Recommended rations a week. In figure 2 it is shown the new basic food classification as well as datatype and object properties created.

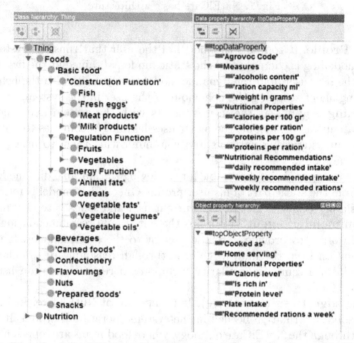

Fig. 2. Basic food classification. Datatype and object properties of NutElCare.

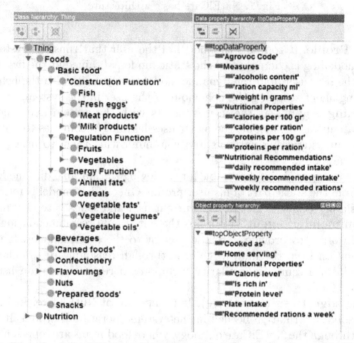

Fig. 3. Recommended rations a week uses

Finally, some SWRL[11] rules have been added to enrich the semantics and stimulate the inference of new knowledge. An example of SWRL rules to classify aliments by their caloric level is shown in figure 4.

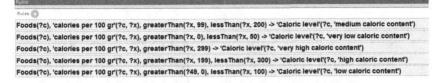

Fig. 4. NutElCare SWRL rules

3.3 Recommendation Production

The NutElCare recommendation process mainly consists of providing recommendations of alternatives to the different meals contained in the diets offered to the users. These recommendations are closely linked to the nutritional properties of food, represented both in their relations **is-a** and their semantic properties contained within type **Nutritional Properties**.

When the system explicitly obtains the characteristics and interests of the user (allergies and tastes), it becomes able to provide the user model diet. This initial model comes only from explicit user information. When the user obtains the diet that fits his physical profile complemented with his explicit interests, he may seek advice in order to vary those dishes that, for any reason (availability, palatability, preparation time, and so on), he does not want at this time.

To obtain the new food to recommend, replacing that specified by the user, the system calculates the semantic similarity of the selected food with other foods in the ontology, resulting in a list of those which are most nutritionally similar to that specified, and which complies with the user interests. The technique used in the calculation of the semantic similarity between foods is described below.

3.3.1 Distance

To define a semantic similarity measure between two individuals, we can apply an approach using the subsumption links in the domain ontology [10]. When calculating the semantic similarity between two foods, several factors are taken into account. First, the foods to offer must match in some point of the ontology hierarchy so that some important nutritional factors of both foods are the same. This point is the *stop node* defined with the property **Recommended weekly intake**, which gathers the aliments according to the number of rations to ingest of a type of food. For instance, the foods belonging to the class **fresh meat** can be ingested at least 3 times and at most 4 times a week. The ontology design allows all food instances to have some ancestor concept in the **is-a** hierarchy that matches one of these stop nodes. In the example of figure 5, all descendant classes of the stop node **fresh meat** are shown. In this case, all the instances of these classes need to go up to the superclass of the class they belong to. So, the comparisons would be between all the descendant foods of the class **fresh meat**.

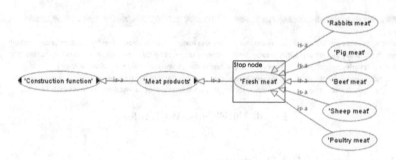

Fig. 5. Descendant classes of `fresh meat` stop node

The *distance* between two food instances i and i is defined by the equation (1), where ca is the common ancestor that subsumes both instances.

$$distance(i, j) = min\{ca(i, j)\} \ . \tag{1}$$

In the example of figure 5, the *distance* between two instances i and j will have value 1 if they belong to the same class and value 2 if the superclass of the class they belong to is common.

3.3.2 Nutritional Object Properties

The nutritional datatype properties are classified into object properties, i.e. those whose super-property is `Nutritional Properties`, with the goal of calculating the similarity between the different nutritional levels that the food belongs to.

For this reason, two different types of nutritional object properties are defined:

– *Level Properties* (*LP*): classify the nutritional level of some nutritional datatype property. They are inferred in the reasoning (through SWRL rules) and the values of their range are annotated with an annotation axiom of type `isDefinedBy` with integers from 1 to N, where N is the number of values that the property can take. For instance, `Caloric level`, can take 5 different values, so N for this *LP* is 5:

$$isDefinedBy(CaloricLevel, x) = \begin{cases} \text{very low caloric level} & x = 1 \\ \text{low caloric level} & x = 2 \\ \text{medium caloric level} & x = 3 \\ \text{high caloric level} & x = 4 \\ \text{very high caloric level} & x = 5 \end{cases}$$

As the *LP* are functional properties, it is possible to calculate the similarity of the property for two instances i and i with the formula (2).

$$sim_{lp}(i, j) = \frac{N_{lp} - abs(v_{lp_i} - v_{lp_j})}{N_{lp}} \tag{2}$$

where N_{lp} is the N value for the property level lp and v_{lp_i} and v_{lp_j} are the annotation values isDefinedBy of the property lp for i and j.

For example, in the case of a comparison between "Chicken meat" (cm) and "Duck meat" (dm):

- "Chicken meat": very high protein level $\rightarrow$ isDefinedBy $= 5$
- "Duck meat": high protein level $\rightarrow$ isDefinedBy $= 4$

$$sim_{lp}(cm, dm) = \frac{5 - abs(4 - 5)}{5} = 0.8$$

where the number of values that the lp Protein Level can take is 5 (N).

- *Typical Properties* (TP): are the remaining properties. The similarity between nutritional object properties TP is calculated using a *weight* for each instance being compared. For a property TP, two identical values (tp_{ij}) of such a property from among the N_{tp} possible values in the instance is calculated and the weight for this property in the instances i and j is obtained from the formula (3).

$$weight_{pt}(i) = w_{tp_i} = \frac{tp_{ij}}{N_{tp_i}} \qquad (3)$$

$$weight_{pt}(j) = w_{tp_j} = \frac{tp_{ij}}{N_{tp_j}}$$

where tp_{ij} is the number of coincidences of the TP "is rich in" between both instances and N_{tp_i} and N_{tp_j} the number of occurrences of the property in each instance i and j.

The object properties of the previous example are shown in figure 6.

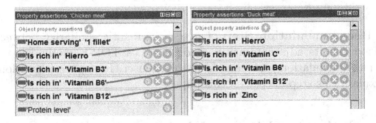

Fig. 6. Typical properties of Chicken meat and Duck meat in NutElCare

Considering "is rich in" as the nutritional object property P, the weights of the example may be calculated as:

$$w_{tp_{cm}} = \frac{tp_{cm,dm}}{N_{tp_{cm}}} = \frac{3}{4} = 0.75$$

$$w_{tm_{dm}} = \frac{tp_{cm,dm}}{N_{tp_{dm}}} = \frac{3}{5} = 0.60$$

From the obtained weights, the similarity is calculated as the arithmetic mean between both weights for the TP property, as shown in equation (4). $sim_{tp}(i,j)$ is in the interval $[0,1]$.

$$sim_{tp}(i,j) = \frac{w_{tp_i} + w_{tp_j}}{2} \ . \tag{4}$$

In the example:

$$sim_{tp}(cm, dm) = \frac{0.75 + 0.6}{2} = 0.675 \ .$$

3.3.3   Semantic Similarity

Finally, the semantic similarity between two instances is calculated as the arithmetic mean between all the similarities of their properties divided by the taxonomic distance between them. It is assumed that all the properties have the same importance. Thus, the semantic similarity between two individuals i and j may be calculated with the equation (5).

$$sim(i,j) = \frac{\frac{1}{T} \sum_{k=1}^{N} sim_{tp_k}(i,j) + \frac{1}{T} \sum_{l=1}^{M} sim_{lp_l}(i,j)}{distance(i,j)} \tag{5}$$

where $T = N + M$ is the total number of compared properties.

If we apply the equation (5) to the previous example, the semantic similarity between Chicken meat and Duck meat results in:

$$sim(cm, dm) = \frac{\frac{1}{3}(0.8 + 0.8) + \frac{1}{3}(0.675)}{1} = 0.7583 \ .$$

3.3.4   Recommendation Step.

Once the semantic similarity between the target food and its adjacents has been obtained, a TOP-K method is used to form a recommendation and provide it to the user. For space limitations, we won't explain this step deeper on this paper.

4   Conclusions and Future Work

Some general problems of the traditional recommender systems arise from heterogeneity in the data representation, leading to knowledge sharing problems. This deficiency also affects the reusability of the system or its components, by other agents, processes or applications. Several nutritional recommender systems have been proposed so far, but, they usually focus on user tastes only and do not consider nutritional properties of foods or user health status either.

In this paper we have presented NutElCare, a semantic recommender system of diets which makes use of a knowledge base represented in an OWL ontology and that provides nutritional advice for the elderly. Recommendations are motivated by the nutritional features of the foods and the particular interest of the users (explicit or implicit). An existent ontology, AGROVOC FAO in its OWL version, has been used to develop the NutElCare ontology, extracting the relevant concepts, and incorporating additional nutritional information. Moreover, NutElCare and AGROVOC ontologies are aligned so that roundtrip knowledge sharing is enabled.

A recommendation strategy for suggesting foods to the elderly has been designed taking into account healthy, nutrient guided and user interesting variations of diets based on the user profile. A method to calculate semantic similarity of foods based on their nutritional properties has been proposed. As future work, we plan to extend and improve the knowledge in the ontology by adding new nutritional properties and concepts, so that, with appropriate adaptations, other groups of people may be targeted and further inferences can be reasoned. Likewise, we intend to develop a graphical front-end so that interaction with NutElCare be more user-friendly. To end, the design and implementation of a learning agent for the study and comparison of different learning techniques in order to discover behaviour patterns, enhancing recommendations with the learned interests of the users, is also part of future work.

Acknowledgements. This research work has been partially funded by the Innovation Office from the Andalusian Government under project TIC-6600 and the Spanish Ministry of Economy and Competitiveness under project TIN2012-38600.

References

1. Aberg, J.: Dealing with Malnutrition: A Meal Planning System for Elderly. In: AAAI Spring Symposium: Argumentation for Consumers of Healthcare, pp. 1–7 (2006)
2. OWL API, Java OWL API reference implementation, http://owlapi.sourceforge.net/
3. Benghazi, K., Hurtado, M.V., Hornos, M.J., Rodríguez, M.L., Rodríguez-Domínguez, C., Pelegrina, A.B., Rodríguez-Fórtiz, M.J.: Enabling correct design and formal analysis of Ambient Assisted Living systems. Journal of Systems and Software 85(3), 498–510 (2012)
4. Bermudez, M., Noguera, M., Hurtado-Torres, N., Hurtado, M.V., Garrido, J.L.: Analyzing a firm's international portfolio of technological knowledge: A declarative ontology-based OWL approach for patent documents. Advanced Engineering Informatics (2013)
5. Blanco-Fernández, Y., Pazos-Arias, J.J., Gil-Solla, A., Ramos-Cabrer, M., López-Nores, M., García-Duque, J., Fernández-Vilas, A., Díaz-Redondo, R.P.: Exploiting synergies between semantic reasoning and personalization strategies in intelligent recommender systems: A case study. Journal of Systems and Software 81(12), 2371–2385 (2008)
6. Cobos, C., Rodriguez, O., Rivera, J., Betancourt, J., Mendoza, M., León, E., Herrera-Viedma, E.: A hybrid system of pedagogical pattern recommendations based on singular value decomposition and variable data attributes. Information Processing & Management 49(3), 607–625 (2013)

7. FAO: Agrovoc linked open data, http://aims.fao.org/aos/agrovoc/
8. Farsani, H.K., Nematbakhsh, M.: A semantic recommendation procedure for electronic product catalog 3, 86–91 (2006)
9. Freyne, J., Berkovsky, S.: Intelligent food planning: personalized recipe recommendation. In: Proceedings of the 15th International Conference on Intelligent User Interfaces, IUI 2010, pp. 321–324. ACM (2010)
10. Garrido, J.L., Hurtado, M.V., Noguera, M., Zurita, J.M.: Using a CBR approach based on ontologies for recommendation and reuse of knowledge sharing in decision making. In: Eighth International Conference on Hybrid Intelligent Systems, HIS 2008, pp. 837–842. IEEE (2008)
11. Horrocks, I., Patel-Schneider, P.F., Boley, H., Tabet, S., Grosof, B., Dean, M.: SWRL: A Semantic Web Rule Language Combining OWL and RuleML, http://www.w3.org/Submission/SWRL/
12. Loizou, A., Dasmahapatra, S.: Recommender systems for the semantic web (2006)
13. Noguera, M., Hurtado, M.V., Rodríguez, M.L., Chung, L., Garrido, J.L.: Ontology-driven analysis of uml-based collaborative processes using owl-dl and cpn. Science of Computer Programming 75(8), 726–760 (2010)
14. Peis, E., del Castillo, J.M., Delgado-López, J.: Semantic recommender systems. analysis of the state of the topic. Hipertext. Net 6, 1–5 (2008)
15. PIPS: Personalised information platform for health and life services, http://www.csc.liv.ac.uk/~floriana/PIPS/PIPSindex.html
16. PROTÉGÉ: Protégé ontology editor, http://protege.stanford.edu/
17. SENPE, SEGG: Valoración nutricional en el anciano. Documento de Consenso. Ed., Galenitas-Nigra Trea (2007)
18. Serrano-Guerrero, J., Herrera-Viedma, E., Olivas, J.A., Cerezo, A., Romero, F.P.: A google wave-based fuzzy recommender system to disseminate information in university digital libraries 2.0. Inf. Sci. 181(9), 1503–1516 (2011)
19. Snae, C., Bruecker, M.: FOODS: A food-oriented ontology-driven system, pp. 168–176. IEEE (2008)
20. Tufts University, S.o.N.S.: Keep fit for life meeting the nutritional needs of older persons. World Health Organization (2002)
21. Van Pinxteren, Y., Geleijnse, G., Kamsteeg, P.: Deriving a recipe similarity measure for recommending healthful meals. In: Proceedings of the 16th International Conference on Intelligent user Interfaces, IUI 2011, pp. 105–114. ACM, New York (2011)
22. W3C: OWL web ontology language overview, http://www.w3.org/TR/owl-features/
23. Wang, R.Q., Kong, F.S.: Semantic-enhanced personalized Recommender System. In: 2007 International Conference on Machine Learning and Cybernetics, vol. 7, pp. 4069–4074. IEEE (2007)

Enhancing Recommender System
with Linked Open Data

Ladislav Peska and Peter Vojtas

Faculty of Mathematics and Physics
Charles University in Prague
Malostranske Namesti 25, Prague, Czech Republic
{Peska,vojtas}@ksi.mff.cuni.cz

Abstract. In this paper, we present an innovative method to use Linked Open Data (LOD) to improve content based recommender systems. We have selected the domain of secondhand bookshops, where recommending is extraordinary difficult because of high ratio of objects/users, lack of significant attributes and small number of the same items in stock. Those difficulties prevents us from successfully apply both collaborative and common content based recommenders. We have queried Czech language mutation of DBPedia in order to receive additional attributes of objects (books) to reveal nontrivial connections between them. Our approach is general and can be applied on other domains as well. Experiments show that enhancing recommender system with LOD can significantly improve its results in terms of object similarity computation and top-k objects recommendation. The main drawback hindering widespread of such systems is probably missing data about considerable portion of objects, which can however vary across domains and improve over time.

Keywords: Recommender systems, Linked Open Data, implicit user preference, content based similarity.

1 Introduction

Recommending and estimating user preferences on the web are both important commerce application and interesting research topic. The amount of data on the web grows continuously and it is impossible to process it directly by a human. Keyword search engines were adopted to fight information overload but despite their undoubted successes, they have certain limitations. Recommender systems can complement onsite search engines especially when user does not know exactly what he/she wants.

Many recommender systems, algorithms or methods have been presented so far. Initially, the majority of research effort was spent on the collaborative systems and explicit user feedback. Collaborative recommender systems suffer from three well known problems: cold start, new object and new user problem. New user / object problem is a situation, where recommending algorithm is incapable of making relevant prediction because of insufficient feedback about current user / object. The cold start problem refers to a situation short after deployment of recommender system, where it cannot relevantly predict because it has insufficient data generally.

H.L. Larsen et al. (Eds.): FQAS 2013, LNAI 8132, pp. 483–494, 2013.

The new object problem became even more important when there is only limited amount of items per object type and thus object fluctuation is higher. Using attributes of the objects (and hence content based or hybrid recommender systems) can speed up learning curve and reduce both cold start and new object problems. Various domains however differ greatly in how many and how useful attributes can be provided in machine readable form.

The Linked Open Data (LOD) is a community project aiming to provide free and public data under the principles of linked data [3]. LOD datasets are available on the web under an open license in machine readable format (RDF triples) and linked together with unique URIs.

DBPedia [2] is one of the cornerstones of LOD cloud. It extracts data available on Wikipedia topic pages (primarily Wikipedia Infoboxes), publishes it as LOD and hence provides data about wide range of topics. Several language mutations of DBPedia are available varying in both in size of corresponding Wikipedia language mutation and completeness of transcription rules converting Wikipedia pages into the RDF.

1.1   Motivation, Problem Domain

Despite the widespread of recommending systems, there are still domains, where creating useful recommendations is very difficult.

- Auction servers or used goods shops have often only one object of given type available which prevents us from applying collaborative filtering directly.
- For some domains e.g. films, news articles or books it is difficult to define and maintain all important attributes hindering content based methods.
- Websites with relatively high ratio between *#objects* / *#users* will generally suffer more from cold start and new user problems.

For our study, we have selected the secondhand bookshops domain which includes all mentioned difficulties. The main problem of the book domain is that similarity between books is often hidden in vast number of attributes (characters appeared, location, art form or period, similarity of authors, writing form etc.). Although those attributes are difficult to design and fill, it is not impossible. But in the most cases, only one book is available in the secondhand bookshop, so creating a new record must be fast and simple enough that potential purchase could eventually cover the costs of work. For example average book price in our dataset is approximately 7EUR, so potential income per sold book is 2-3EUR. Collaborative recommender systems can be used, but their efficiency is hindered both by high ratio between *#objects* / *#users* and the fact that each object can be purchased only once.

On the other hand, Wikipedia covers the book domain quite well, so our main research question is whether we can effectively use information available on Linked Open Data cloud (e.g. DBPedia) to improve recommendation on difficult domains such as secondhand book shop.

1.2   Main Contribution

The main contribution of our work is:

- Proposing on-line method to automatically enrich object attributes with Linked Open Data and enhance content-based recommending.
- Experimental evaluation against both content-based and collaborative filtering on secondhand bookshops – a domain with high object fluctuation, high ratio *objects/users* and not much meaningful attributes.
- Identifying key problems hindering further development of similar systems.

The rest of the paper is organized as follows: we finish section 1 with review of some related work. In section 2 we describe in general how the LOD datasets can be used to enhance recommender systems and section 3 describes its implementation for Czech secondhand bookshop domain. Section 4 contains results of our experiments and finally section 5 concludes the paper and points out some future work.

1.3   Related Work

Although areas of recommender systems and Linked Open Data have been extensively studied recently, papers involving both topics are rather rare. The closest to our work is the research by Di Noia et al. [4], who developed content based recommender system for a movie domain based sole on several LOD datasets. Similarly A. Passant [12] developed *dbRec* – the music recommender system based on DBPedia dataset. The main difference between these and our approaches is that we start with real system and use LOD only as an instrument to improve recommendations on it. Objects of the system don't have to correspond to the data available in LOD and so we need to cope with problems like absence of unique mapping, missing records etc. Among other work connecting areas of recommender systems and LOD we would like to mention paper by Heitmann and Hayes [7] using Linked Data to cope with new-user, new-item and sparsity problems and Goldbeck and Hendler's work on FilmTrust [6]: movie recommendations enhanced by FOAF trust information from semantic social network.

As for the recommender systems, we suggest the papers by Adomavicius and Tuzhilin [1] or Konstan and Riedl [9] for overview. A lot of recommending algorithms aims to do decompose user's preference on the object into the preference of the objects attributes e.g. Eckhardt [5], which is a parallel to our content based similarity method. In our recent work [13], we present several methods how to combine attribute preferences together and analogically we plan to experiment with e.g. T-conorms to combine attribute similarities.

We suggest paper by Bizen, Heath and Berners-Lee [3] as a good introduction to the Linked Open Data. Our current experiment was based on Czech version of DBPedia[1], originally developed by Chris Bizer et al. [2].

---

[1] http://cs.dbpedia.org

2   Enhancing Content-Based Recommending with Linked Data

In this section, we would like to describe general principles how to enhance recommending systems with data from LOD cloud. We want to stress that one of our intentions was to keep the architecture simple and straightforward to enable its further widespread. Several extensions are possible. The top-level architecture is shown on Figure 1.

The system maintains one or more SPARQL endpoints to various LOD datasets. The connections are usually REST APIs or simple HTTP services. Whenever an object of the system is created or edited, the system will automatically query each SPARQL connection with unique object identifier or other best possible object specification, if no common unique ID is available (see Figure 2 for example of SPARQL query to Czech DBPedia).

The important step while using more than one LOD dataset is matching identical resources. The difficulty of this step is highly dependent on the datasets, but mostly involves (recursive) traversing of owl:sameAs links. Resources are then stored in the system triples store, which can be however quite simple (relational database table was sufficient during our experiment).

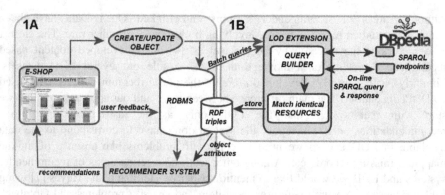

Fig. 1. Top-Level architecture of enhancing recommender system with LOD. Figure 1A represents original e-commerce system with a content-based recommender; 1B is its extension for querying and storing LOD datasets.

Each object of the system should be also queried periodically as the LOD datasets can provide more information over time. We have also considered using local copies of LOD datasets. This approach would however result in excessive burden to both data storage and system maintenance and also prevent us from using up-to-date dataset.

There are several possible approaches to map RDF triples *<RDF subject, RDF predicate, RDF object>* into the binary relation *<object, attribute>*. Each pair *<RDF predicate, RDF object>* (or *RDF object* only if we wish to omit *predicates*) can be identified as an attribute of *RDF subject* (object of the system) forming Boolean matrix of attributes $A_{|objects| \times |RDF\ objects|}$. This approach leads to various methods of matrix factorization [10]. Strength of such approach is that each resource is considered separately so more important resources may have more significant impact.

Also resource similarities e.g. SimRank [8] can be applied. However the number of resources can grow unlimited could make such solution computationally infeasible.

Our approach is to identify each *RDF predicate* as an attribute of *RDF subject* (object of the system). The value of the attributes is then a set of all relevant *RDF objects*. Two sets can be compared relatively fast with e.g. Jaccard similarity $(a_x \cap a_y)/(a_x \cup a_y)$. Total number of attributes will be limited and allows us to use e.g. linear combination as a similarity metrics. We lose the ability to weight each resource separately, but we still can weight each *RDF predicate*. The method can be further simplified by omitting *RDF predicate* and use set of all *RDF objects* as single attribute.

2.1 Content Based Similarity

The key part of any content based recommender system is defining similarity of its objects. We expect that each object of the system is represented by its attributes vector. Attributes can be one of following type:

- Numerical
- String
- Unordered (enumeration, category etc.)
- Set

Our content-based similarity is based on the idea of two-step preference model published by our research group [5]. The method works in two steps: for objects x and y, it first computes local similarity $\mu_a : D_a^2 \rightarrow [0,1]$ for each attribute $a \in A$. Global similarity of objects S is defined as linear combination of local similarities:

$$S = \sum_{i \in A} w_i * \mu_i \tag{1}$$

The choice of proper local similarity measure might depend on the domain and a nature of each attribute. As we aims to keep our method domain independent, we have chosen rather simpler, widely adopted metrics than ones closely specialized on particular domain. For objects x, y and values of their attribute a_x and a_y local similarity μ_a is:

- Normalized absolute value of the difference for numerical attributes.

$$\left| a_x - a_y \right| \Big/ \max_{\forall objects\ o} \left(|a_o| \right) \tag{2}$$

- Relative levenshtein distance for string attributes.

$$1 - (levenshtein(a_x, a_y)) \big/ \max(lenght(a_x), lenght(a_y)) \tag{3}$$

- Equality of unordered attributes.
- Jaccard similarity of set attributes.

$$(a_x \cap a_y)/(a_x \cup a_y) \tag{4}$$

Metrics (2), (3), (4) allows us to compute local similarity smoothly which can be beneficial as we expect to have rather high ratio *#objects/#attributes*. It is possible to compare all attributes in terms of equality, but it won't make much sense for many real-world attributes (e.g. small difference in price or issuing date implies high similarity). For string attributes, our aim was to both cope with minor typos and inconsistencies (e.g. Arthur Conan Doyle vs. A. C. Doyle) or to track series of books (usually containing common substring). There are analogies for this behavior in other domains as well.

Weight w_a of each local similarity μ_a is set by Linear Regression maximizing similarity of objects x, y co-occurring in the list of preferred objects P_u for some user u and minimizing otherwise. Optimizing formula is defined as follows:

$$\arg\min_{w}(RSS(w)) = \sum_{x,y \in O}\left(\pi(x,y) - w^T\mu\right)^2$$

$$where \ \pi(x,y) = \begin{cases} 1 \leftrightarrow \exists u : x, y \in P_u \\ 0 \ \text{OTHERWISE} \end{cases} \tag{5}$$

2.2 Content-Based and Collaborative Top-n Recommending

We are yet to describe the procedure of deriving recommendations. For arbitrary fixed user u, set of his/her preferred objects P_u and set of candidate objects C_u, the recommending algorithm computes similarity $S(o_i, o_j)$ for each $o_i \in P_u$ and $o_j \in C_u$.

Candidate objects are then ranked according to their average similarity:

$$Rank(o_i) = Rank\left(AVG(S(o_i, o_j); \forall o_i \in P_u)\right) \tag{6}$$

Similarity measure S can be one of:

- Content based similarity as described in section 2.1. In the experiment section, this method is further divided according to the usage of LOD.
- Item-to-item collaborative similarity as described in [11]. Similarity of objects x and y is measured as cosine similarity of vectors of users U_x and U_y preferring x and y respectively.

$$S_{Cosine} = \frac{U_x \bullet U_y}{\|U_x\|\|U_y\|} \tag{7}$$

In production recommender system, we should take into account also other metrics like diversity, novelty, serendipity, or pre-filter candidate objects, but for purpose of our study, we will focus on similarity only.

2.3 Hybrid Top-n Recommending

In order to benefit from both content-based and collaborative filtering, we have implemented also a hybrid approach enhancing content-based filtering with collaborative data: Collaborative similarity of objects (7) is used as an additional attribute of content-based similarity (we expand vectors w and μ from equation (5)).

3 Using Linked Open Data for Secondhand Bookshop

First domain dependent task is typically selecting proper LOD dataset. There are several datasets in the LOD cloud[2] containing information about books. We can

---

[2] http://datahub.io/group/lodcloud

mention i.e. DBPedia[3], Freebase[4], Europeana project[5] or British National Bibliography (BNB)[6]. One of our concerns was quality of the available data. From this point of view although BNB provides good coverage, it contains only a little more data (mostly about authors), than what is already available within the bookshop.

However the main drawback of the most datasets is language: without common identifier, it is difficult to connect data between various languages (e.g. Czech as the bookshop language and English as a foremost LOD language). ISBN does not work well in this case as it identifies each issue, publisher or language version separately and we also have to deal with books from the pre-ISBN era. The best way to identify data records is using textual comparison of book title and author name. So our key constraint for selecting dataset is availability of book titles (and person names) in Czech. Another constraint given by the architecture is availability of SPARQL endpoint in order to query effectively for demanded data.

Given the constraints we identified Czech DBPedia dataset as the best option. We plan to use automated translation services in future to be able to query also non-Czech datasets.

3.1   Querying Czech DBPedia

The example of SPARQL query to Czech DBPedia is shown on Figure 2a, portion of result on Figure 2b. Some restrictions on *RDF predicates* were applied thereafter to filter out useless data. We do not show them for the sake of clarity. In order to access more available data, we have conducted also separate authors search.

Czech DBPedia in its current state unfortunately doesn't support directly object categories (purl:subject in English DBPedia). It contains information about explicit links to other Wikipedia pages (dbpedia:wikiPageWikiLink), where categories can be extracted by "Category:" prefix. However the dataset doesn't contain its supercategories which make them far less useful, so we decided not to distinguish them from other links. We collect data also from preceding/following books in the series, source country, genre, translator of the book, author occupation, influences etc.

Our dataset contains information about 8802 objects (books). By querying Czech DBPedia, we were able to gather data about 87 books and authors of 577 books altogether recording additional information about 7.3% of books. This small coverage, as we investigated, was caused mainly by missing infoboxes on the vast majority of books in Czech Wikipedia, causing DBPedia to fail identifying them as books. The problem could be bypassed by traversing categories/supercategories which is however not possible in current version of Czech DBPedia. We are working with Czech DBPedia administrators to eliminate this problem.

---

[3] http://dbpedia.org
[4] http://www.freebase.com/
[5] http://datahub.io/dataset/europeana-lod
[6] http://bnb.data.bl.uk/

2A	2B book_name	p	o
PREFIX rdf: <http://www.w3.org/1999/02/22-rdf-syntax-ns#> PREFIX rdfs: <http://www.w3.org/2000/01/rdf-schema#> PREFIX dbPedia: <http://dbpedia.org/ontology/>	"Návrat krále"@cs	rdf:type	dbpedia:WrittenWork
	"Návrat krále"@cs	rdf:type	schema.org/Book
	"Návrat krále"@cs	rdfs:label	"Návrat krále"@cs
select ?book_name, ?p, ?o where {	"Návrat krále"@cs	dbpedia:numberOfPages	409
?book rdfs:label ?book_name . ?book_name bif:contains '("Návrat" and "krále")' . ?book rdf:type dbPedia:WrittenWork . ?book dbPedia:author ?author.	"Návrat krále"@cs	dbpedia:publisher	csdbpedia:Mlad%C3%A1_fronta
	"Návrat krále"@cs	dbpedia:sourceCountry	csdbpedia:Spojen%C3%A9_kr%C3%A
?author rdfs:label ?author_name . ?author_name bif:contains '("J. R. R. Tolkien")'.	"Návrat krále"@cs	dbpedia:author	csdbpedia:J._R._R._Tolkien
	"Návrat krále"@cs	dbpedia:previousWork	csdbpedia:Dv%C4%9B_v%C4%9B%C
?book ?p ?o. }	"Návrat krále"@cs	dbpedia:wikiPageWikiLink	csdbpedia:1955
	"Návrat krále"@cs	dbpedia:wikiPageWikiLink	csdbpedia:John_Ronald_Reuel_Tolkien
	"Návrat krále"@cs	dbpedia:wikiPageWikiLink	csdbpedia:Sauron
	"Návrat krále"@cs	dbpedia:wikiPageWikiLink	csdbpedia:Mordor

Fig. 2. Example of SPARQL query about The Return of the King (in Czech "Navrat krale") by J. R. R. Tolkien and a portion of returned data.

4   Experiments

In order to prove our theory as feasible, we performed a series of off-line experiments on real users of a Czech secondhand bookshop. The objects (books) have following content-based attributes: book name, author name, book category and book price. The experimental website does not provide an interface to collect explicit feedback (e.g. object rating), so we had to rely on implicit feedback only. For the purpose of experiments, we consider page-view action (user u visit object o) as an expression of positive user preference. The list of preferred objects P_u is then precisely list of visited objects. We have also restricted the domain of books to only those for which we were able to mine some data from DBPedia. Both similarity metrics and recommending algorithm can however predict for all objects. Finally we use all collected *RDF objects* as a value (set) of single attribute to show that even sole presence of additional data can improve recommendations.

The experiment had two phases. First, we compared differences between similarities of objects visited by the same user, which we believe should be more similar to those which were not visited. The first phase gives us rough picture about how can various similarity methods explain user behavior. Thereafter in the second phase we compared methods in more realistic recommending top-k objects scenario.

4.1   Similarity of Objects

In the first phase of our experiment, our aim was to demonstrate that adding semantic information from LOD can improve recommender systems capability to distinguish between similar and non-similar objects from the user perspective. For each pair of objects visited by the same user, we have computed content based similarity either including or not including DBPedia data. Similarities of mutually visited objects were then compared against average object similarity in general population as we want to test their resolving power.

Experiment results are shown in Table 1. The result seems to corroborate our theory that additional DBPedia data can help to distinguish between similar and dissimilar and thus also preferred and non-preferred objects.

Table 1. Results for similarity of objects: table shows average similarity of mutually visited objects $S_{visited}$, general population $S_{general}$ and their quotient for content based similarity with and without using DBPedia resources. Both Content only and Content with DBPedia methods have $S_{visited}$ significantly higher than $S_{general}$ (p-value $< 10^{-7}$). Content with DBPedia has also significantly higher $S_{visited} / S_{general}$ ratio (p-value $< 10^{-7}$).

Similarity Method	$S_{visited}$	$S_{general}$	$S_{visited} / S_{general}$
Content-based only	0.261701	0.101041	2.59005
Content-based with DBPedia	0.339476	0.084035	4.03970

4.2   Top-n Recommending

Although similarity can provide us with rough image about benefits of external Linked Data, the real usage scenario is different. Most typically, we have a user currently visiting an object or a category and we want to recommend him/her some other options. We expect that user already visited few objects.

Our dataset contains implicit feedback from the users. For arbitrary fixed user, his/her sequence of visited objects, let n be its size, was sorted according to the timestamp. Then we split the sequence and denote first k objects as a *train set* and nonempty remaining portion of the sequence as a *test set*. Note that only users with at least two visited objects qualify for the experiment. Train set sizes was set from 1 up to 5 because not enough users would qualify for larger train set sizes. For the purpose of the top-n recommending experiment, we define *train set* as preferred objects P_u and all objects expect those from *train set* as candidate objects: $C_u = O \backslash P_u$. We compute the list of recommended objects as described in (6). Following methods were tested:

- *Collaborative* filtering forming our baseline.
- *Content*-based filtering with no additional data.
- Content-based filtering with DBPedia data handled as a single attribute (referred as ***Content+DBPedia*** in results).
- Hybrid approach as described in 2.3 with DBPedia data (referred as ***Hybrid+DBPedia*** in results).

Only the feedback available at the time when current user visits the last object from *train set* is used. This simulates real situation (we cannot use data which will be created in the „future").

We consider as success if methods rank objects from the *test set* "good enough". In another words, recommending methods should be able to predict objects, which will user visit in the "future". In order to compare the recommending methods, we have set three success metrics:

- *Average position* of *test set* objects in the recommended list.
- Normalized Discounted Cumulative Gain (*nDCG*), where relevance is defined as 1 for objects from *test set* and 0 otherwise.

- ***Presence*** of *test set* object(s) ***at top-10*** as this is a typical size of list of recommended objects shown to the user.

Figure 3 shows results of the experiment: 3a, 3b and 3c show trend of the metrics while enlarging *train set* sizes, 3d contains average data for all users.

Both content based and hybrid methods outperform greatly the item-item collaborative filtering in all observed metrics (p-value $< 10^{-7}$ according to Tukey HSD test). The added value of *Content + DBPedia* method was observed mainly for the smaller train set sizes. It dominates over *Content* method consistently in both *nDCG* (p-value: 0.065) and *Presence at TOP-10* (p-value: 0.0399), however the absolute difference was rather small with growing train set sizes indicating, that additional data was most useful for new users of the system.

Hybrid with DBPedia method dominates significantly in average position (p-value $\leq 5.1*10^{-4}$) but has only small effect on other metrics suggesting that it only improves position of badly ranked objects. Even though it might be useful if the candidate set C_u is pre-filtered.

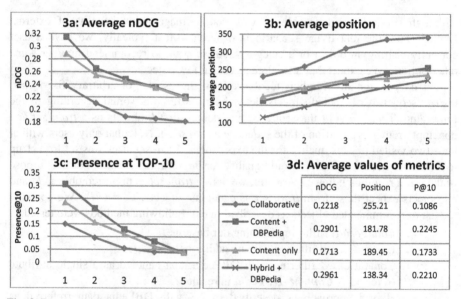

Fig. 3. Recommending experiment results: all metrics shown according to the train set size per user. Figure **3a** displays nDCG, **3b** average position and **3c** relative presence of preferred objects in top 10. Figure **3d** shows average values per all train set sizes. In nDCG *Content + DBPedia* outperformed *Content only* method, but with rather low significance (p-value: 0.065). Similar results, however a bit more conclusive provides also P@10 (p-value: 0.0399). In average position Hybrid + DBPedia dominates over all other methods (p-value $\leq 5.1*10^{-4}$).

Interesting phenomenon was decrease of all methods performance while increasing train set size. This might be partially caused by inappropriately chosen aggregation function (average) supported by the assumption, that objects visited further in the past has only limited impact on future visiting objects. Further result analysis also shown that majority of users who received best recommendations had only small number of visited object and so did not qualify for larger train sets. We will further investigate whether this is only coincidence and how to avoid such problem.

5   Conclusions and Future Work

In this paper, we aim to improve recommendations on problematical e-commerce systems by using additional data collected from Linked Open Data Cloud. We have shown several difficulties which current recommender systems may encounter and a domain (secondhand bookshops) where all those difficulties can be found. We have presented method for enhancing content based recommender system with data from LOD cloud. This on-line method is straightforward and quite general, so it can be applied on various domains.

The off-line experiments held on the real visitors of Czech secondhand bookshop corroborates our assumption, that enhancing recommender systems with LOD data can improve recommendation quality.

One of the main drawbacks of our approach is low coverage of objects within the Czech DBPedia. Therefore our future work involves improving Czech DBPedia mapping rules, refining SPARQL queries used during the experiment and exploring other available datasets. The possibility of automatic translation of book names should be also considered. Last but not least the experiments have shown unexpected behavior while increasing train set sizes which should be further investigated.

Acknowledgments. This work was supported by the grant SVV-2013-267312, P46 and GAUK-126313.

References

1. Adomavicius, G., Tuzhilin, A.: Toward the next generation of recommender systems: A survey of the state-of-the-art and possible extensions. IEEE Trans. Knowl. Data Eng. 17, 734–749 (2005)
2. Bizer, C., Cyganiak, R., Auer, S., Kobilarov, G.: DBpedia.org - Querying Wikipedia like a Database. In: WWW 2007, Banff, Canada (May 2007)
3. Bizer, C., Heath, T., Berners-Lee, T.: Linked Data - The Story So Far. Int. J. Semantic Web Inf. Syst. 5, 1–22 (2009)
4. Di Noia, T., Mirizzi, R., Ostuni, V.C., Romito, D., Zanker, M.: Linked open data to support content-based recommender systems. In: Proc. of I-SEMANTICS 2012, pp. 1–8. ACM (2012)
5. Eckhardt, A.: Similarity of users (content-based) preference models for Collaborative filtering in few ratings scenário. Expert Syst. Appl. 39, 11511–11516 (2012)
6. Golbeck, J., Hendler, J.: FilmTrust: movie recommendations using trust in web-based social network. In: Proc. of CCNC 2006, vol. 1, pp. 282–286. IEEE (2006)
7. Heitmann, B., Hayes, C.: Using Linked Data to Build Open, Collaborative Recommender Systems. In: AAAI Spring Symposium: Linked Data Meets Artificial Intelligence (2010)
8. Jeh, G., Widom, J.: SimRank: a measure of structural-context similarity. In: KDD 2002, pp. 538–543. ACM (2002)

9. Konstan, J., Riedl, J.: Recommender systems: from algorithms to user experience. In: User Modeling and User-Adapted Interaction, vol. 22, pp. 101–123 (2012)

10. Koren, Y., Bell, R., Volinsky, C.: Matrix Factorization Techniques for Recommender Systems. IEEE Computer 42, 30–37 (2009)

11. Linden, G., Smith, B., York, J.: Amazon.com recommendations: item-to-item collaborative filtering. IEEE Internet Computing 7, 76–80 (2003)

12. Passant, A.: dbrec - Music Recommendations Using DBpedia. International Semantic Web Conference (2) Springer, LNCS, 2010, 209-224

13. Peska, L., Vojtas, P.: Negative Implicit Feedback in E-commerce Recommender Systems. To appear on WIMS 2013 (2013),
    http://www.ksi.mff.cuni.cz/~peska/wims13.pdf

Making Structured Data Searchable via Natural Language Generation
with an Application to ESG Data

Jochen L. Leidner and Darya Kamkova

Thomson Reuters Global Resources
Catalyst Lab, Neuhofstrasse 1,
CH-6340 Baar, Switzerland
jochen.leidner@thomsonreuters.com

Abstract. Relational Databases are used to store structured data, which is typically accessed using report builders based on SQL queries. To search, forms need to be understood and filled out, which demands a high cognitive load. Due to the success of Web search engines, users have become acquainted with the easier mechanism of natural language search for accessing *un*structured data. However, such keyword-based search methods are not easily applicable to *structured* data, especially where structured records contain non-textual content such as numbers.

We present a method to make structured data, including numeric data, searchable with a Web search engine-like keyword search access mechanism. Our method is based on the creation of surrogate text documents using Natural Language Generation (NLG) methods that can then be retrieved by off-the-shelf search methods.

We demonstrate that this method is effective by applying it to two real-life sized databases, a proprietary database comprising corporate Environmental, Social and Governance (ESG) data and a public-domain environmental pollution database, respectively, in a federated scenario. Our evaluation includes speed and index size investigations, and indicates effectiveness ($P@1 = 84\%$, $P@5 = 92\%$) and practicality of the method.

1 Introduction

Keyword-based search offered by modern Web search engines (like Bing or Google) have become pervasive, and terabytes of public, unstructured Web pages available on the Internet are made available for searching that way. However, *structured* data, information stored in relational databases, is not as easily findable using the same mechanism, for various reasons. First, databases are often not exposed on the Web as static HTML pages, for example due to rights violations concerns. Instead, a Web form might give access selectively after filling in a set of fields that comprise a query. The totality of information available that way is known as the *Deep Web* [1]. Second, structured information often comprises

H.L. Larsen et al. (Eds.): FQAS 2013, LNAI 8132, pp. 495–506, 2013.

non-textual information such as numbers, which need to be interpreted in context: for instance, a number like *1970* might either be a date of birth of a person, the annual revenue (in million USD) of a company, or the amount (in tons) of a toxic substance released into the air in an industrial accident. To make sense of numbers in a database, the application using the database together with the database schema provide the appropriate context, and *database report builders* are typically used to search the data. Unfortunately, report builders are often complex and therefore difficult to use.

In this paper, we propose a novel solution to this findability problem for structured data: we address the problem of how to make structured databases searchable *by generating unstructured documents artificially*, which can then be indexed with off-the-shelf inverted index files like any unstructured data and retrieved using e.g. the vector space model. This permits the application of keyword search to a realm it was previously not applicable to. In some sense, we turn the information extraction problem upside down: in information extraction, structured data needs to be extracted from unstructured document collections. Imagine again a value like *1970*: the database schema tells us that it is a numeric entity, but it does not reveal whether it is a year or a monetary amount.[1] We describe a mechanism of generating documents based on simple natural language generation (NLG) rules that permits a non-programmer to write a small set of rules for a database that makes its content findable.

We apply our method to two databases from the environmental domain. The first data set comprises proprietary information about the reported Environmental, Sustainability and Governance (ESG) performance of companies, an area of growing interest in the context of sustainable and ethical investing and good governance [2].[2] ESG databases monitor corporate scandals, environmental issues, ethical concerns such as conflicts of interest, bribery, investment in education or well-being of its workforce (such as training and development program investments in dollars per employee per year), and similar aspects not covered by a company's financial fundamental analysis. In sustainable investment, the hope is that in the longer term, more ethical companies and those that focus not just on quarterly-measured shareholder value will fare better than companies that ignore ethical, social and environmental concerns. Our second data set is a database of toxic spill events in the USA that were reported to meet compliance obligations.

Paper Plan. The remainder of this paper is structured as follows: in Section 2, we discuss previous work in the fields of (structured) database search and text generation, respectively. In Section 3 we present our new method, and Section 4 describes our *E-Mesh* search engine, which implements it. Section 5 describes

---

[1] Note that even if the database schema contains a column named PERSON_BIRTHYEAR, this does not automatically facilitate the search *per se*, as the machine is oblivious to the meaning of these names even if they are mnemonic to humans.

[2] This data set is commercially available as a data feed, e.g. for research or investment analysis purposes.

our data. Section 6 reports on our evaluation. In Section 7, we summarize and conclude the paper with suggestions for future work.

2 Related Work

We present related work from database and Business Intelligence (BI) search, natural language text generation, and our application area, information systems for environmental and ESG (Environmental, Social and Governance) data.

BI Search. [3] is a general textbook on the state of the art in information retrieval, which form the basis of question answering systems over unstructured data as found in document collections (e.g. [4]). In contrast, [5] surveys traditional natural language database interface techniques. [6] present *SODA*, a system that can search stuctured data across a set of data warehouses in a company. *SODA* does not carry out a natural language analysis of the query, but is capable of recognizing concepts and operators in a free-form text query, from .which a query graph is constructed. To implement Soda for a new set of data warehouses, a domain ontology needs to be devised by humans for optimal results. [7] outline a tutorial on keyword search across structured, semistructured and graph data. [8] present a method for making structured data searchable that is based on adding additional tables to any database instance, and by adding a processing layer between Microsoft SQL Server and the ODBC API layer. They index all rows (direct or via foreign key join) such that each row contains all keywords. Unlike ours, their methods does not permit multi-lingual search, nor can it cope with finding numeric content. [9] apply smoothed relevance language models to ranked retrieval of structured data. Their experiments include Wikipedia, IMDB (both more unstructured than structured) and MONDIAL, a small structured data base comprising 9 MB and 17k relational tuples. Again, numeric information and multilinguality are beyond its scope. Traditionally, graph-based algorithms [10–12] have been tried to enable keyword search on relational and other data.

Environmental and ESG Information Systems. "Green" data, i.e. data from the realm of protecting the environment, has recently become available at scale (e.g. [13]). Not surprisingly, data management [14] and search are becoming increasingly important. The United Nations Environment Programme (UNEP) developed UNEP Explorer, a Web-based software for navigating environmental data sets in the public domain [15]. Unlike the method presented in this paper, UNEP Explorer does not have an easy-to-use keyword-based search function that delivers focused results. *ESG*, the Environmental Scenario Generator [16], is an advanced distributed tool to explore environmental data for scenario analysis, particularly around weather situations. It uses maps and a menu-based, i.e. non-textual GUI to solicit input and does not permit keyword search. [17] present *KOIOS*, a keyword search engine for environmental data built on top of the commercial EIS Cadenza (by Dizy GmbH). They take a very different route from the approach presented here: given a structured database, three indices are generated, namely a data index (a graph), a keyword index (which captures unstructured data parts) and a schema index, which represents classes and relationships. All three indices are represented in RDF, and queries are translated into SPARQL. No evaluation regarding retrieval effectiveness, query processing efficiency or index size overhead are described.

[18] make the case for so-called ESG data (short for "Environmental, Social and Governance" company data) to contain valuable evidence that can inform trading.

Natural Language Generation. We are not aware of any previous work that attempts to make structured data searchable by applying natural language *generation* techniques. [19] and [20] survey the state of the art in text generation methods, including systems. [21] is one of the few textbooks on NLG.

3   Method

Imagine a database like the toy example in Figure 1. We cannot query easily what is in the database with a Google-style search engine because we do not have an inverted file index – after all keyword search basically responds to a query with a set of ranked pages that contain the query keywords, and for our database no such (natural language) index exists.

Table 1. A Sample Toy Database Table T_PERSON

ID	Name	Weight	Age
1	John	80	35
2	Anna	60	33

To address this, we generate English prose documents, one per table row, using a set of template generation rules as shown in Table 2. Each rule comprises a `pattern` section containing a template with canned text that may contain slots that may be filled with values bound to variables (e.g. $P). We say the template gets *instantiated* and we call the resulting document *surrogate document*, as its main purpose is to make the data in the database row *findable*. There is no need to physically store surrogate documents in a file system, as their only purpose is to feed an indexing process, and they can be stored memory-efficiently together with the index itself in compressed form. The `keyword` section of the generation rule adds terms to the surrogate document to increase recall, but marks them as "not for rendering", as the keywords listed here are not part of any system response to the user. In particular, we aim for the pattern to be linguistically well-formed, whereas the `keyword` section is just a bag of terms and phrases that may be useful for retrieval. The `datapoint` section finally provides a binding between rows in the database row under consideration and the parameter variables in the `pattern` section. Database rows get processed one by one, so in the first iteration $P gets bound to John (i.e. T_PERSON.NAME), in the second iteration $P gets bound to Anna, and so on. Field names from the schema are added to the `keyword` section automatically, thus potentially supplementing any human-written generation rules. Table 2 shows a toy rule for explanation purposes; Appendix A lists a real generation rule in XML format as used by our system.

Table 2. A Natural Language Generation Rule That Generates Table 3 From the First Row in Table 1

pattern: $P weighs $W kilograms and is $A years old.
keywords: kg age
datapoint: `T_PERSON.NAME`|`T_PERSON.Weight`|`T_PERSON.AGE`

Table 3. Generated Two-Part Surrogate Document

John weighs 80 kilograms and is 35 years old.
kg age

Default rules control the output for non-existing data-points. All text generation rules are defined per language, as they have an ISO-639 language attribute; a separate index is automatically generated for each language that has at least one generation rule. Thus our method can generate a set of search engines in various languages from a single XML specification file.

The number of generation rules that need to be authored for a database table with $|T|$ tables and $|R|$ rows each is at most $|T| \times |R|$; in practice, one rule can generate surrogates covering multiple columns (Table 2 used a single rule for all four columns).

Ease of Customization. A key benefit of our method is that there are no software changes necessary to deal with a change of the database schema or with the addition of a new database: the only modification/extension required is the adaptation or extension of the text generation rules in a single XML file.

4   System

We implemented the method described in the previous section in Java. The architecture of our system *E-Mesh* is shown in Figure 1. A data reader component based on the Apache Hibernate framework [22] iterates over all database rows for each database tables; we will henceforth call each row *data points*. A text generator component (based on the Apache Velocity template engine) reads a set of NLG rules and instantiates them by binding data points mentioned in rules to variable parameters. Generation rules like the one in Table 2 were represented in XML. Variable values are inserted where they occur inside of generation templates that comprise the core part of rules, thus instantiating *surrogate documents*. An indexer component (based on the Apache Lucene search library) indexes these documents on the fly, creating one index per language for which generation rules exist. This completes the offline processing. The actual search engine provides online access to the indexed documents and makes them available to a GUI (again based on Lucene and the Glassfish servlet container). Figure 2 shows the *E-Mesh* search engine's user interface.

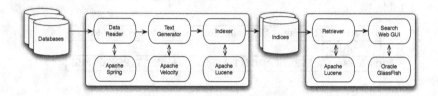

Fig. 1. *E-Mesh* System Architecture

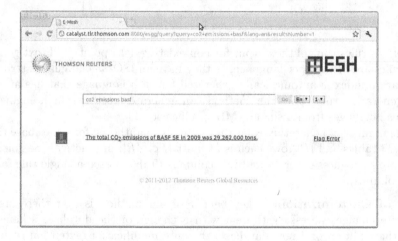

Fig. 2. Screen Capture of *E-Mesh*'s Web GUI

5   Data

We apply the methods described above to two data sets. The first data set is a proprietary database comprising Environmental, Social and Governance (ESG) information such as data points describing a company's environmental perfor-mance (e.g. carbon emissions), social/ethical performance (e.g. investments in employee training, ongoing harassment litigation) or governance performance (e.g. reports on insider trading, corruption charges against management). The data set is part of the Datastream product sold by Thomson Reuters, so the method is also applied to a second data set for easier replication of our study. We use a toxic spill database covering environmental pollutions in the U.S., data which was put into the public domain by the U.S. government [23].[3] This data set, henceforth TOXICSPILL for short, contains information about release events, namely time of the event, the name of the company responsible for the spill, the location, and the name of the chemical accidentally released. Table 4 summarizes the size and content of the two data sets and their corresponding generation rules.

Table 4. Data Setup Used in Our Experiments

Database	Size	Tables	Average Columns/Table	Generation Rules
ESG	21,000 MB	111	9.81	28 (EN) / 28 (DE)
TOXICSPILL	401 MB	1	87	2 (EN) / 2 (DE)

6   Evaluation

6.1   Experimental Design

Evaluating natural database interfaces or question answering systems over structured data poses the problem that the user does not know what he or she can ask, as the database schema is not known to them. Knowledge of the schema, on the other hand, would biases the user and hence any evaluation. There is also no standard evaluation database and query corpus available, which is partly due to the fact that many query systems are domain specific and/or database specific. To address these issues, we created two query corpora as follows: we generated a random list of companies, a random list of data point names, and a random list of years. We then asked human subjects not involved in this study to formulate one human natural language search query for each ⟨company; data point; year⟩ triple. With a probability 1/2, we included a the year from the interval [2008; 2012] into the query (years covered in our snapshop of the databases). The resulting two query corpora (one used as a development set and another one used as a test set) comprise 50 queries each, 25 asking for data from the ESG and the other 25 asking for data from the TOXICSPILL data set. We then ran these two query sub-corpora against the *E-Mesh* system, and evaluated the correctness of the results, measuring precision at ranks one and five, respectively. This methodology ensures independence while generating queries that are as natural as feasible and at the same time the resulting queries are not "unfair" to the system in the sense that the information needs encoded by the queries should in principle be answerable by the data in the databases at hand.

Baseline & Hypothesis. We compare the precision of the implemented method compared to a baseline which applies a TFIDF ranking to a "raw" index of data rows together with their names (taken from the schema), without applying our proposed generation templates. Our hypothesis is that the proposed *offline NLG approach significantly increases retrieval performance.*

---

[3] The data is publically available from http://data.gov

6.2   Empirical Evaluation Results and Discussion

Indexing. The system was subjected to the generation of search engines for English and German based on the data described above. This paper presents some findings for the query corpus, which was only in English (we leave the evaluation for German for future work). A set of rules were authored, which took about 3 hours for ESGG, and 1.5 hours for TOXICSPILL, respectively. One database was completely modeled, the other one partially (but our method ensures to always obtain a functional system). The result is shown in Table 4.

We carried out the indexing on a desktop PC with Intel CoreDuo PC with two x86 cores (model 6,400) at 2.13 GHz with 3 GB RAM. The method's observed offline indexing run time (Figure 3) and the persistent storage space required for the generated inverted file indices (Figure 4) grow linearly with the number of rules and database rows (since the number of templates is constant) in the offline indexing step, and there is no additional runtime overhead for searching the generated index of surrogate documents.

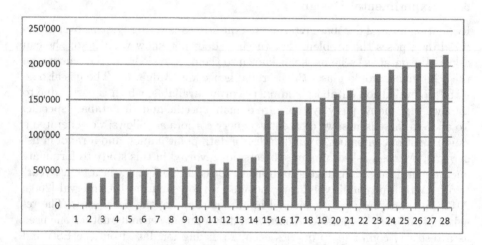

Fig. 3. Scalability of Text Generation Rules: Index Generation Time (in ms) Grows Nearly Linearly With the Number of Rules (ESG database)

Retrieval. The system was tested on the test corpus of 50 test queries[4], half of which ask for information from the ESG database and the other half with information needs from the TOXICSPILL database. The system logs all queries so once the system has been used by real users, more large-scale evaluations can be carried out. We found that the precision at rank one was 80% for the ESG data set and 88% for the the TOXICSPILL data set, amounting to a total of P@1 = 84%. If we look at precision at the top five ranks, precision rises to 92% for both databases. Note that our query corpus is small, so getting

---

[4] Comprising only English examples; the evaluation of German is left for future work.

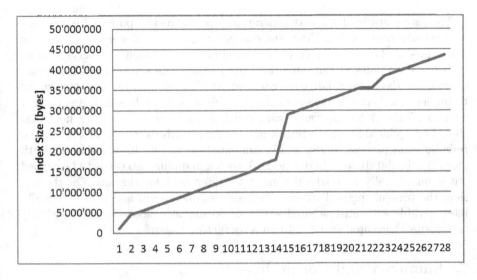

Fig. 4. Scalability of Text Generation Rules: Index Size (in bytes) Grows Nearly Linearly As a Function of the Number of Rules (ESG database)

just a single query wrong leads to a 0.04 decrease, as it happened with one query where the user gave a short form of a company, in a situation where many other companies with similar names exist: in query #29 from the test set, `Metal Finishing lead 2002` the user is seeking information on lead spills of a company called `Metal Finishing Co. Inc.` in the year 2002. When run against the TOXICSPILL database *E-Mesh*, returns the correct result on rank 3 instead of top rank (wrongly placing `Quality Metal Finishing Co.`'s lead emissions at the top). Overall, looking at the top five results for any query, our method ($P@5 = 92\%$) improves over the baseline (64%).

Discussion. Our hypothesis, namely that the proposed offline NLG approach significantly increases retrieval performance, could be confirmed: it holds individually for the ESG dataset, and it holds overall, although we did not observe an improvement for the TOXICSPILL dataset. In future work, we will investigate whether this is caused by a shortage in generation rules.

We also measured the retrieval speed, and observed a mean processing time of 12 ms per query on average.

Table 5. Evaluation Results

	E-Mesh Dev.		Baseline Test		*E-Mesh* Test	
	P@1	P@5	P@1	P@5	P@1	P@5
ESG	0.32	0.40	0.20	0.36	0.80 ▲	0.92 ▲
TOXICSPILL	0.92	0.92	0.92	0.92	0.88	0.92
all	0.62	0.66	**0.56**	**0.64**	**0.84** ▲	**0.92** ▲

The rather static offline text generation process has two potential issues that we have identified so far. First, the concept of surrogate document generation from rows of data sometimes lacks flexibility; an extension to querying more than one data point at the same time (responding to a single natural language query), i.e. the natural language counterpart of a join, is not straight forward to implement. Second, dynamic computations cannot be handled directly. For example, Table 1 has an "age" row, which is an example of a data field that should be updated regularly. Such changing data needs to trigger an incremental re-indexation process. If the age is computed on the fly from a date of birth stored in the database (a better design), such a mapping also needs to be carried out symmetrically at retrieval time if the user asks for the age. On the flip side, the present method trades space for runtime speed, and it is maximally interoperable with unstructured search of document collections, as it basically transforms structured search into an *un*-structured search.

7   Summary and Conclusion

We presented a novel method that uses natural language generation for making structured data searchable. We described its implementation in our search engine *E-Mesh*, which was evaluated on a set of queries against two databases. The method proved to be both fast and offers high accuracy at a reasonable index overhead.

In future work, the method could be extended to provide more complex text generation capabilities. Another possible extension is the combination of data-points from more than one database in a single surrogate document. Our query corpus is limited in size; once logfiles have been gathered after production deployment, the effectiveness can be more systematically assessed. We also plan to evaluate the system performance for other languages (such as German, which we already implemented).

To the best of our knowledge, this is the first paper to suggest how search engines for structured data including numeric data can be "generated" by harnessing NLG techniques to create an inverted file index offline for a set of structured databases, and we offer a way of doing it that easily supports multiple languages. We measured an improvement in precision of the method when retrieving the top five answers for a corpus of 50 queries from a set of two federated databases (performing at $P@5 = 92\%$ compared to a baseline of $P@5 = 64\%$. In the future, we plan to evaluate languages other than English, notably our German rule set.

Acknowledgments. The authors would like to thank Peter Pircher and Khalid al-Kofahi for supporting this work. We thank Jack G. Conrad and Peter Pircher, who contributed to the query corpus in the evaluation. Frank Schilder, Isabelle Moulinier and Tom Zielund, together with the first author, brainstormed the original idea, and provided valuable feedback. We also acknowledge Amanda West, whose Thomson Reuters Environmental Innovation Award inspired this work (and this research is based on our winning entry in that competition).

The work described in this paper was funded by Thomson Reuters Global Resources (TRGR).

References

1. Cafarella, M.J., Halevy, A., Madhavan, J.: Structured data on the Web. Communications of the ACM 54(2), 72–79 (2011)
2. KPMG: KPMG International Survey of Corporate Responsibility Reporting 2008 (2008), http://ec.europa.eu/enterprise/policies/sustainable-business/corporate-social-responsibility/reporting-disclosure/swedish-presidency/files/surveys_and_reports/international_survey_of_csr_reporting_2008_-_kpmg_en.pdf (cited March 3, 2013)
3. Manning, C.D., Raghavan, P., Schütze, H.: An Introduction to Information Retrieval. Cambridge University Press (2008)
4. Leidner, J.L., Bos, J., Dalmas, T., Curran, J.R., Clark, S., Bannard, C.J., Webber, B.L., Steedman, M.: QED: The Edinburgh TREC-2003 question answering system. In: TREC Workshop Notes, pp. 631–635 (2003)
5. Androutsopoulos, I., Ritchie, G.D., Thanisch, P.: Natural language interfaces to databases – an introduction. Natural Language Engineering 1(1), 29–81 (1995)
6. Blunschi, L., Jossen, C., Kossmann, D., Mori, M., Stockinger, K.: Data-thirsty business analysts need SODA: Search over data warehouse. In: Macdonald, C., Ounis, I., Ruthven, I. (eds.) Proceedings of the 20th ACM Conference on Information and Knowledge Management, CIKM 2011, Glasgow, United Kingdom, October 24-28, pp. 2525–2528. ACM (2011)
7. Chen, Y., Wang, W., Liu, Z., Lin, X.: Keyword search on structured and semi-structured data. In: Proceedings of the 35th SIGMOD International Conference on Management of Data, SIGMOD 2009, pp. 1005–1010. ACM, New York (2009)
8. Agrawal, S., Chaudhuri, S., Das, G.: DBXplorer: A system for keyword-based search over relational databases. In: Proceedings of the 18th International Conference on Data Engineering (ICDE), pp. 5–16. IEEE Computer Society, Washington, DC (2002)
9. Bicer, V., Tran, T., Nedkov, R.: Ranking support for keyword search on structured data using relevance models. In: Proceedings of the 20th ACM International Conference on Information and Knowledge Management, CIKM 2011, pp. 1669–1678. ACM, New York (2011)
10. Coffman, J., Weaver, A.C.: Structured data retrieval using cover density ranking. In: Proceedings of the 2nd International Workshop on Keyword Search on Structured Data, KEYS 2010, pp. 1:1–1:6. ACM, New York (2010)
11. Garcia-Alvarado, C., Ordonez, C.: Keyword search across databases and documents. In: Proceedings of the 2nd International Workshop on Keyword Search on Structured Data, KEYS 2010, pp. 2:1–2:6. ACM, New York (2010)
12. Li, G., Ooi, B.C., Feng, J., Wang, J., Zhou, L.: EASE: An effective 3-in-1 keyword search method for unstructured, semi-structured and structured data. In: Proceedings of the 2008 ACM SIGMOD International Conference on Management of Data, SIGMOD 2008, pp. 903–914. ACM, New York (2008)
13. The World Bank Group: Environment (2012), http://data.worldbank.org/topic/environment (cited March 3, 2013)
14. Harmancioglu, N.B., Singh, V.P., Alpaslan, M.N. (eds.): Environmental Data Management. Kluwer Academic Publishers, Norwell (1998)

15. United Nations Environment Programme: Environmental data explorer (2012), http://geodata.grid.unep.ch/ (cited March 3, 2013)
16. Kihn, E., Zhizhin, M., Siquig, R., Redmon, R.: The environmental scenario generator (ESG): A distributed environmental data archive analysis tool. Data Science Journal 3, 10–28 (2004)
17. Bicer, V., Tran, T., Abecker, A., Nedkov, R.: KOIOS: Utilizing semantic search for easy-access and visualization of structured environmental data. In: Aroyo, L., Welty, C., Alani, H., Taylor, J., Bernstein, A., Kagal, L., Noy, N., Blomqvist, E. (eds.) ISWC 2011, Part II. LNCS, vol. 7032, pp. 1–16. Springer, Heidelberg (2011)
18. Ribando, J.M., Bonne, G.: A new quality factor: Finding alpha with ASSET4 ESG data, Research Note (2010)
19. Paiva, D.S.: A survey of applied natural language generation systems. Technical Report ITRI-98-03, University of Brighton (1998)
20. Piwek, P., van Deemter, K.: Constraint-based natural language generation: A survey. Technical report, Open University, Technical Report No. 2006/03 (2006)
21. Reiter, E., Dale, R.: Building Natural Language Generation Systems. Studies in Natural Language Processing. Cambridge University Press (2000)
22. The Apache Foundation: Hibernate Search (2012), http://www.hibernate.org/subprojects/search.html (cited March 3, 2013)
23. U.S. Environmental Protection Agency (EPA), Toxic spill data (2013), http://data.gov (cited March 3, 2013)

A Sample XML Generation Rule

```
<generate-rule>
  <template lang="en">
    <index>company codes</index>
    <index>CO2</index>
    <render>The total CO&lt;sub&gt;2&lt;/sub&gt; emissions
            of ${Company} in ${Year} was ${CO2} tons.</render>
    <index>carbon dioxide equivalents</index>
    <default>I don't know how large were the total CO&lt;sub&gt;2&lt;/sub&gt;
            emissions of ${Company} in ${Year}.</default>
  </template>
  <datareader database="Asset4">
    <param result="CO2">En_En_ER_DP023</param>
  </datareader>
</generate-rule>
```

On Top-k Retrieval for a Family of Non-monotonic Ranking Functions*

Nicolás Madrid[1] and Umberto Straccia[2]

[1] Centre of Excellence IT4Innovations, Division of the University of Ostrava,
Institute for Research and Applications of Fuzzy Modeling, Czech Republic
nicolas.madrid@osu.cz

[2] Istituto di Scienza e Tecnologie dell'Informazione (ISTI - CNR), Pisa, Italy
straccia@isti.cnr.it

Abstract. We presented a top-k algorithm to retrieve tuples according to the order provided by a non-necessarily monotone ranking funtion that belongs to a novel family of functions. The conditions imposed on the ranking functions are related to the values where the maximum score is achieved.

1 Introduction

Usually when users make queries in databases, they are only interested in a subset of answers. Consider just a search for a house in a database according to some preferences about size, location, etc. In such a case, an user usually is not interested in knowing which is the worst house according to his preferences: an answer with simply the ten best houses is good enough. Top-k algorithms deal with that issue and have become an important topic of interest in the last years [1,2,5,6,8,9]. Roughly speaking, the answer of a top-k query is the subset with the k best results.

It is worth mentioning that, a priori, top-k algorithms can be defined on various and diverse frameworks; for instance on *fuzzy logic programming* [9] and on *uncertain databases* [11,12]. However, most approaches of top-k retrieval have been developed on relational databases [3,4]; this paper is not an exception. Hence, for us, an answer of a top-k query is a set of k tuples with the greatest score (according to a ranking function f) among those which satisfy a relation R (called *the joint condition*).

The typical procedure to obtain the answer of a top-k query consists in developing a *threshold algorithm* which computes an upper bound for the scores of tuples non retrieved yet. Usually, threshold algorithms need two requirements: firstly a database sorted in decreasing order with respect to score; and secondly, a monotonic ranking function.

In this paper instead of requiring monotonic mappings, we will consider a most general family of ranking functions. Note that removing the monotonicity

---

* This work was supported by the European Regional Development Fund projects CZ.1.05/1.1.00/02.0070 and CZ.1.07/2.3.00/30.0010.

H.L. Larsen et al. (Eds.): FQAS 2013, LNAI 8132, pp. 507–518, 2013.

as a requirement in ranking functions arises naturally in many contexts. Just consider an user searching for a house who is interested preferably in houses with a size close to $100 \ m^2$. In such a case the function determining how much far is the size of a house from $100 \ m^2$ is obviously not monotonic.

Allowing the use of non monotonic ranking functions in top-k algorithms is a current challenge. To the best of our knowledge, there are only two papers dealing with non-monotonic ranking functions; namely [7] and [10]. In [10], the authors require an indexed-merge structure in the database and the procedure consists in partitioning the domain in sub-domains where the ranking function (a priori arbitrary) is monotonic (or semi-monotonic). On the other hand, [7] defines the top-k procedure by considering *isolines* in a specific family of ranking functions.

The approach described in this paper is the first step of a more general research towards the definition of a top-k algorithm for arbitrary ranking functions. Specifically, in this paper we present a top-k procedure for a family of ranking functions allowing the use of distance functions among others.

In the following, we proceed as follows. In Section 2 we present the properties that a ranking function has to verify. Moreover, we provide some additional properties and methods to construct them. In Section 3 we describe our top-k algorithm and give the theorem of correctness of the procedure. Finally in Section 4 we present the conclusions and address future work.

2   Ranking Functions

A *ranking function* is defined as follows.

Definition 1. *A mapping* $f \colon [0,1]^n \to \mathbb{R}$ *is called a* ranking function *if there exists an element* $(m_1, \dots, m_n) \in [0,1]^n$ *such that for all* $(c_1, \dots, c_n) \in [0,1]^n$ *and all* $i \in \{1, \dots, n\}$, *the mapping defined by:*

$$g_i(x) = f(c_1, \dots, c_{i-1}, x, c_{i+1}, \dots, c_n)$$

is monotonic in $[0, m_i]$ *and antitonic in* $[m_i, 1]$.[1]

Let us explain briefly the condition imposed on ranking functions. The tuple $(m_1, \dots, m_n)$ (called the *best preference* of f) represents the scores associated to the best possible answer ranked by f. Hence, the coefficients $m_1, \dots, m_n$ can be interpreted as a preference for tuples with such scores. Moreover, those coefficients do not represent simply "an overall best preference" but also "*local best preferences*", since the closer the i-th variable to m_i, the greater the value of f; independently of the values in the rest of variables.

The family of ranking functions contains some interesting families as the set of monotonic and antitonic mapping defined from $[0,1]^n$ to $\mathbb{R}$.

Proposition 1. *Any* monotonic mapping *(resp.* antitonic mapping*) is a ranking function.*

---

[1] For the sake of clarity, we recall that a mapping $f \colon [0,1] \to \mathbb{R}$ is monotonic (resp. antitonic) if $x \leq y$ implies $f(x) \leq f(y)$ (resp. $f(x) \geq f(y)$) for all $x, y \in [0,1]$.

The following example shows that our notion of ranking functions can deal, somehow, with distances as well.

Example 1. Let d be the Euclidean distance, then the mapping defined by:

$$f((x_1, x_2)) = \sqrt{2} - d((x_1, x_2), (1/2, 1/2)) = \sqrt{2} - \sqrt{(1/2 - x_1)^2 + (1/2 - x_2)^2}$$

is a ranking function, whose maximal score is reached at $(1/2, 1/2)$.

Actually, the family of distances induced by norms can be considered as ranking functions as shown in the following proposition.

Proposition 2. *Let* $d \colon [0, 1]^n \times [0, 1]^n \to \mathbb{R}^+$ *be a distance induced by a norm and let* $(m_1, \ldots, m_n) \in [0, 1]^n$. *Then the mapping defined by:*

$$f \colon [0, 1]^n \to \mathbb{R}$$
$$f(x_1, \ldots, x_n) = -d((x_1, \ldots, x_n), (m_1, \ldots, m_n))$$

is a ranking function whose maximal score is reached at $(m_1, \ldots, m_n)$.

Proof. To prove that f is a ranking function we have to show that for all $(c_1, \ldots, c_n) \in [0, 1]^n$, each mapping g_i defined by

$$g_i(x) = -d((c_1, \ldots, c_{i-1}, x, c_{i+1}, \ldots, c_n), (m_1, \ldots, m_{i-1}, m_i, m_{i+1}, \ldots, m_n))$$

is monotonic on $[0, m_i]$ and antitonic on $[m_i, 1]$.

Without lost of generality we assume that $i = 1$. Moreover, we use a well known result of metric space theory: if d is a distance induced by a norm, then the closed ball of radius $r > 0$ centered at x:

$$B(x, r) = \{y \in X \colon d(y, x) \le r\}$$

is convex (i.e. all line segment bounded by two points of $B(x, r)$ is contained in $B(x, r)$).

Now, consider $y \le x \le m_1$ and let us show that $g_1(y) \le g_1(x)$; that is

$$-d((y, \ldots, c_n), (m_1, \ldots, m_n)) \le -d((x, \ldots, c_n), (m_1, \ldots, m_n))$$

or equivalently:

$$d((y, \ldots, c_n), (m_1, \ldots, m_n)) \ge d((x, \ldots, c_n), (m_1, \ldots, m_n))$$

Note that

$$d((y, \ldots, c_n), (m_1, \ldots, m_n)) = d((2m_1 - y, \ldots, c_n), (m_1, \ldots, m_n))$$

since:

$$- d((y, \ldots, c_n), (m_1, \ldots, m_n)) = ||(m_1 - y, \ldots, m_2 - c_n)||$$

$$- d((2m_1 - y, \ldots, c_n), (m_1, \ldots, m_n)) = ||(2m_1 - y - m_1, \ldots, m_2 - c_n)|| =$$
$$= ||(m_1 - y, \ldots, m_2 - c_n)||.$$

Now, consider $r = d((y, \ldots, c_n), (m_1, \ldots, m_n))$. Then both $(y, \ldots, c_n)$ and $(2m_1 - y, \ldots, c_n)$ belongs to $B((m_1, \ldots, m_n), r)$. Additionally, note that $(x, \ldots, c_n)$ belongs to the line segment bounded by $(y, \ldots, c_n)$ and $(2m_1 - y, \ldots, c_n)$ since $y \leq x \leq m_1 \leq 2m_1 - y$. Thus, by the result presented above, $(x, \ldots, c_n) \in B((m_1, \ldots, m_n), r)$; or equivalently:

$$d((x, \ldots, c_n), (m_1, \ldots, m_n)) \leq r = d((y, \ldots, c_n), (m_1, \ldots, m_n)) .$$

The previous result allows us to use the idea *"the closer, the better"* in ranking functions. The following result has the aim of facilitating the obtainment of ranking functions.

Proposition 3. *Let $f_1 \colon [0, 1]^n \to [0, 1]$ and $f_2 \colon [0, 1]^k \to [0, 1]$ be two ranking functions and let $g \colon [0, 1]^2 \to [0, 1]$ be a monotonic function. Then the function defined by: $f \colon [0, 1]^n \times [0, 1]^k \to [0, 1]$*

$$f(x, y) = g(f_1(x), f_2(y))$$

is a ranking function. Moreover, if $n = k$ and the best preferences of f_1 and f_2 coincide, then the function defined by: $\overline{f} \colon [0, 1]^n \to [0, 1]$

$$\overline{f}(x) = g(f_1(x), f_2(x))$$

is a ranking function as well.

Proof. The proof is straightforward just taking into account that if m_1 and m_2 denote the best preferences of f_1 and f_2, then (m_1, m_2) is the best preference of f. Moreover, in the case that both best preference coincide, then it is easy to prove that m_1 (or equivalently m_2) is the best preference of $\overline{f}$.

Note that although the proposition above has been defined to deal with two ranking functions, it easily extensible to deal with a numerable amount of ranking functions. Moreover, as a consequence, compositions of ranking functions with monotonic mappings are also ranking functions.

Corollary 1. *Let $f \colon [0, 1]^n \to [0, 1]$ be a ranking functions and let $h \colon [0, 1]^2 \to [0, 1]$ be a monotonic function. Then the function defined by $h(f(x))$ is a ranking function.*

Proof. Just apply Proposition 3 to functions $f_1 = f_2 = f$ and $g = h \circ p_1$; where $p_1 \colon [0, 1]^2 \to [0, 1]$ denotes the projection on the first component.

Example 2. It is easy to show that the mapping provided in Example 1 is a ranking function by using Proposition 2 and Corollary 1. Consider the mappings $f_1 \colon [0, 1]^2 \to \mathbb{R}$ and $f_2 \colon \mathbb{R} \to \mathbb{R}$ given by

$$f_1(x_1, x_2) = -\sqrt{(1/2 - x_1)^2 + (1/2 - x_2)} \qquad f_2(x) = \sqrt{2} + x$$

Note that f_1 is the negated Euclidean distance from the point $(1/2, 1/2)$, so we can assert that f_1 is a ranking function (Proposition 2). On the other hand, f_2 is monotonic, so by Corollary 1, f is a ranking function. Finally, the mapping f given in Example 1 is the composition of f_1 and f_2; i.e. $f(x_1, x_2) = f_2(f_1(x_1, x_2))$.

Example 3. The mapping given by:

$$f(x_1, x_2) = \begin{cases} 2 & \text{if } x \geq 1/2 \text{ and } y \geq 1/2 \\ 2 - \sqrt{(x_1 - 1/2)^2 + (x_1 - 1/2)^2} & \text{otherwise} \end{cases}$$

is a ranking function since is the maximum of the following two ranking functions

$$f_1(x_1, x_2) = \begin{cases} 2 & \text{if } x \geq 1/2 \text{ and } y \geq 1/2 \\ 0 & \text{otherwise} \end{cases} \quad f_2(x_1, x_2) = 2 - \sqrt{(x_1 - 1/2)^2 + (x_1 - 1/2)^2}$$

with the same best preference $(m = (1/2, 1/2))$. That is

$$f(x_1, x_2) = \max\{f_1(x_1, x_2), f_2(x_1, x_2)\} .$$

Below we present some properties of our ranking functions. The first result shows that the best preference is the point where the maximum is reached.

Proposition 4. *Let $f : [0, 1]^n \to \mathbb{R}$ be a ranking function and let $m \in [0, 1]^n$ be its best preference. Then:*

$$\max_{x \in [0,1]^n} f(x) = f(m)$$

The next Lemma is slightly more general than the previous Proposition.

Lemma 1. *Let $f : [0, 1]^n \to \mathbb{R}$ be a ranking function with best preference $(m_1, \ldots, m_n)$. Then for all $c \in [0, 1]$,*

$$\max_{(x_1, \ldots, x_n) \in [0,1]^{n-1}} f(x_1, \ldots, x_{i-1}, c, x_{i+1}, \ldots, x_n) = f(m_1, \ldots, m_{i-1}, c, m_{i+1}, \ldots, m_n) .$$

Proof. Let us show that for all $(a_1, \ldots, a_{i-1}, c, a_{i+1}, \ldots, a_n) \in [0, 1]^n$ we have:

$$f(a_1, \ldots, a_{i-1}, c, a_{i+1}, \ldots, a_n) \leq f(m_1, \ldots, m_{i-1}, c, m_{i+1}, \ldots, m_n) .$$

Note that from the above inequality, the result is straightforward. Let us assume that $a_1 \in [0, m_1]$ (resp. $a_1 \in [m_1, 1]$). By using that f is a ranking function, we have that the mapping defined by $g(x) = f(x, a_2 \ldots, a_{i-1}, c, a_{i+1}, \ldots, a_n)$ is monotonic on $[0, m_1]$ (resp. antitonic on $[m_1, 1]$); thus we have the inequality:

$$f(a_1, \ldots, a_{i-1}, c, a_{i+1}, \ldots, a_n) \leq f(m_1, \ldots, a_{i-1}, c, a_{i+1}, \ldots, a_n) .$$

By using the same reasoning inductively for $a_2, \ldots, a_n$, we obtain eventually:

$$f(a_1, \ldots, a_{i-1}, c, a_{i+1}, \ldots, a_n) \leq f(m_1, \ldots, m_{i-1}, c, m_{i+1}, \ldots, m_n) .$$

The following result plays an important role in the correctness of the top-k algorithm described in the next section.

Lemma 2. *Let $f\colon [0,1]^n \to \mathbb{R}$ be a ranking function with best preference $m = (m_1, \ldots, m_n)$ and let $I = [a_1, b_1] \times \cdots \times [a_n, b_n] \subset [0,1]^n$ be an interval satisfying that $m \in I$. Then for all $x \in [0,1]^n \smallsetminus I$ we have:*

$$f(x) \leq \max \big\{ f(a_1, m_2, \ldots, m_n), f(m_1, a_2, \ldots, m_n), \ldots, f(m_1, m_2, \ldots, a_n),$$
$$f(b_1, m_2, \ldots, m_n), f(m_1, b_2, \ldots, m_n), \ldots, f(m_1, m_2, \ldots, b_n) \big\} \ .$$

Proof. Consider $x = (x_1, \ldots, x_n) \in [0,1]^n \smallsetminus I$, then at least one component x_i has to be either lesser or equal than its respective a_i or greater or equal than its respective b_i. Let us assume without lost of generality that $x_1 \leq a_1$. Then, by Lemma 1:

$$\max_{(y_2, y_3, \ldots, y_n) \in [0,1]^{n-1}} \{ f(x_1, y_2, \ldots, y_n) \} = f(x_1, m_2, \ldots, m_n) \ .$$

As a particular case we obtain that:

$$f(x_1, x_2, \ldots, x_n) \leq f(x_1, m_2, \ldots, m_n) \ .$$

So, thanks to the monotonicity of $f(x, m_2, \ldots, m_n)$ in $[0, a_1] \subseteq [0, m_1]$ and that $x_1 \in [0, a_1]$ we conclude:

$$f(x_1, x_2, \ldots, x_n) \leq f(x_1, m_2, \ldots, m_n) \leq f(a_1, m_2, \ldots, m_n) \ .$$

3   Top-k Retrieval Algorithm

In this section we describe a top-k retrieval algorithm[2] for query answering over relational databases, where tuples are ranked by a function belonging to the family of ranking functions introduced in Section 2. The algorithm presented in this paper is based on [3], which presents a top-k retrieval algorithm for monotonic ranking functions. However, the presence of a more general family of ranking functions than the monotonic functions one requires considering a different tuple retrieval strategy.

3.1   Join Strategy

The join strategy used to retrieve tuples plays an important role in top-k algorithms, since it is crucial for the correctness and response time. Note that if tuples associated to the best preference have not been considered yet, we can not ensure that we have already retrieved the top-1. So, it seems appropriated to check the join condition from the tuples which scores are associated with the best preference instead of from the top. Somehow, the idea underlying in this alternative is: *the closer from the best preference, the better are the results.*

The main drawbacks of this strategy with respect to the strategies starting from the top are two. On the one hand, for each table we must consider now

---

[2] For lack of space, it is not possible to describe formally the top-k problem. So the reader is referred to [3].

two directions to retrieve tuples: up and down (or equivalently decreasing and increasing the scores from the start point). So this increments slightly the storage cost of the procedure. On the other hand, before applying the algorithm, we need to arrive somehow to the tuples associated to the best preference. Just note that the computational cost of this pre-procesing is aceptable ($O(log\ n)$).

The join strategy used in the top-k described in this paper generalizes the "*symmetric Ripple Join strategy*" given in [3]. In the case of making joins with only two tables A and B, the difference between both strategies can be described easily. That is because we can represent graphically how both strategies sweep the cartesian product both table's scores $X.A \times X.B$. Roughly speaking, the difference between both approaches is that whereas [3] sweeps the cartesian plane from a corner, our approach sweeps the cartesian plane from a point allocated, a priori, anywhere.

Figure 1 below represents five steps of the join strategy by considering a "*clockwise movement*". Suppose that the best preference of the weak ranking function f is (a, b). The first iteration simply gets two tuples, one from A and another from B with scores a and b respectively and checks the join condition for both tuples. In the case there is not tuples with scores a in A (resp. b in B), we insert a new tuple with score a (resp.b) in A (resp. in B) non satisfying the join condition with any tuple. The second step consists in considering the tuple of A with the closest score to a between the tuples non considered yet and with a score less or equal than a. So the new tuple considered in A is achieved by decreasing the score. Before to pass to the next step, it is necessary to verify if the new tuple considered in A holds the join condition with the only tuple considered in B. In the third step we retrieve a new tuple of B, as in the previous step, by choosing a tuple decreasingly. In other words, we consider the tuple in B with the closest score to b between the tuples non considered yet and with a score less or equal than b. Additionally, in this step we verify the join condition between all possible combination of tuples considered up to this step. The fourth and fifth steps are similar to the second and third steps respectively but by considering greater scores. That is, firstly it is considered the tuple in A (resp. B) with the closest score to a (resp. b) between the tuples non retrieved yet and with a score greater or equal than a (resp. b); secondly it is verified the join condition for all possible combination between the tuples already considered in B (resp. A).

We provide two remarks to finish this section. Firstly, the Ripple Join can be considered as a particular case of the Clockwise Join; specifically when the best preference coincides with the point $(1, 1)$. Secondly, it is possible to consider asymmetric strategies following the idea underlying in the clockwise join. That is, instead of retrieving tuples one by one (either one of A by one of B or one increasingly by one decreasingly), we can use heuristic to determine a preferential direction.

3.2   Top-k Retrieval Algorithm

The algorithm is divided in two procedures, the main source *Top.k.retrieval* and the subroutine *Get.Next.Tuple*. Briefly, *Get.Next.Tuple* retrieves in each

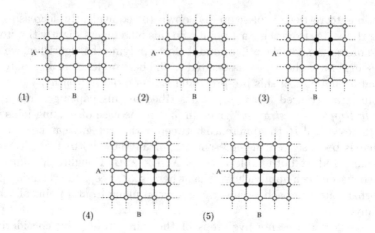

Fig. 1. Our join strategy

step the best result between the tuples non retrieved yet and $Top.k.retrieval$ determines when $Get.Next.Tuple$ can stop of retrieving tuples. To facilitate the presentation, both algorithms are defined and described here by considering only two tables; however it can be generalized to an arbitrary number of tables.

For the sake of clarity we describe the variables appearing in $Top.k.retrieval$:

- $Top.Tuple.List$: in this list we include one by one the top-k results. Actually, this list is the output of the algorithm.
- Q: this list contains temporally each tuple retrieved by using the join condition but non included in $Top.Tuple.List$ yet.
- $firstTuple.A$ and $firstTuple.B$: these Boolean variables are used in the subroutine $Get.Next.Tuple$ to get in buffer the variables $A.init$ and $B.init$; where the value of the best preference is saved.
- tuple: is the answer of $Get.Next.Tuple$. Such variable can be either a tuple belonging to the top-k or the value "NULL" if no more tuple can be obtained by the join condition.

The procedure $Top.k.retrieval$ works as follows. Initially the lists $Top.Tuple.List$ and Q are empty and the Boolean variables $firstTuple.A$ and $firstTuple.B$ are considered to be $true$. In each loop, a call to the procedure $GetNextTuple$ is done. The latter returns the tuple with the best score between the tuples satisfying the join condition and are not already in the list $Top.Tuples.List$. The loop is broken only in two cases: either if we have already retrieved k tuples by using the subroutine $GetNextTuple$ (step 4) or if it is impossible to retrieve more tuples by using the join condition (step 6).

The variables appearing in $GetNextTuple$ and not in $Top.k.retrieval$ are described as follows:

- $Top.Q$: gets the tuple in Q with the best score.

```
Procedure: Top.k.retrieval(A,B,R,k,f)
output: the list Top.Tuples.List with the k best tuples
inputs: A,B: two ranked relational tables
       R: a relation used in the join condition
       k: number of best tuples we are interested in
       f: weak ranking function
init:
    firstTuple.A= true;
    firstTuple.B= true;
    Top.Tuples.List = ∅;
    Q = ∅;
1.   Loop
2.      tuple := GetNextTuple;
3.         include tuple in Top.Tuples.List;
4.         if (|Top.Tuples.List| = k) then
5.            break Loop;
6.         if (tuple = NULL) then
7.            break Loop;
8.      end Loop;
9.   end;
```

Fig. 2. Top.k.retrieval Algorithm

- $A_\uparrow$; $A_\downarrow$; $B_\uparrow$; $B_\downarrow$: are the scores of the respective last tuples seen in each table w.r.t. each direction (see Section 3.1).
- $A.init$; $B.init$: are the scores of the respective first tuple retrieved by each table. Note that those values represent the best preference of the ranking function.
- T: this value is a threshold which upper bounds the score of the rest of tuple non retrieved yet.
- $nextTuple$: this Boolean variable is used to indicate when it is imposible to retrieve more tuple satisfying the join condition.

The procedure $GetNextTuple$ works as follows. Previously to start with the loop, the algorithm checks if there is already any tuple in Q with a score greater or equal than the threshold T; or equivalently if the top-1 of the tuples non contained in $Top.Tuple.List$ belongs to Q. In such a case, is not necessary to activate the loop and the procedure ends (step 5). Otherwise (as in the first call, since Q is empty) the procedure activates a loop. That loop generates tuples till one answer can be returned. Two cases are considered to generate new tuples:

- First of all, if every tuple in tables A and B has been already considered, then no more tuples can be generated. In such case the Boolean variable $nextTuple$ gets the value $false$ and the algorithm goes directly to step 21.
- Otherwise, we consider one new tuple (either from A or from B depending on the joint strategy considered). If this new tuple is the first tuple considered in the table (step 11), then the score of such tuple is one of the coefficients of the best preference; so this value is saved in the variable $I.init$. Anyway, the loop between the steps 15 and 20 recomputes the threshold T, determines all possible join combinations and includes the new tuples in Q decreasingly by using the ranking function f.

Between the steps 20 and 32 the algorithm computes the answer to return to *Top.k.retrieval*. Such answer depends on which case holds:

- If there exists in Q a tuple with score greater or equal than T (step 20) then the answer to return is the tuple with the greatest score in Q.
- If the algorithm has retrieved the whole set of tuples satisfying the joint condition and the list Q is not empty (step 26), the algorithm returns the tuple in Q with the greatest score.
- Finally, if every tuple of A and B has been considered and Q is empty, the algorithm returns "NULL" . This answer means that no more results can be obtained by the join condition, so it is imposible to return the k better results since there is no k tuples satisfying the join condition. With this answer, the main procedure *Top.k.retrieval* breaks its loop and ends.

```
Procedure: GetNextTuple
output: Next tuple to attach in Top.Tuples.List
init: NextTuple=true;
1.    if (Q ≠ ∅) then
2.        tuple = Top.Q
3.        if (tuple.score ≥ T) then
4.            return tuple;
5.        end;
6.    Loop
7.        Determine the next seen tuple, I* (Comment: I* ∈ {A↑, A↓, B↑, B↓});
8.        if (No next seen tuple I* can be considered) then
9.            nextTuple = false ;
10.       else
11.           if (firstTuple.I = true) then
12.               I.init = I*.score
13.               firstTuple.I = false
14.           I*.lastSeen = I*.score
15.           T = max(f(A.init, B↑.lastSeen), f(A.init, B↓.lastSeen),
16.                   f(A↑.lastSeen, B.init), f(A↓.lastSeen, B.init));
17.           determine all possible join combination;
18.           For each valid join combination by using the relation R
19.               compute the score result by using f;
20.               insert each join result in Q and rank them;
21.       if (Q ≠ ∅) then
22.           tuple = Top.Q
23.           if (tuple.score ≥ T) then
24.               remove tuple from Q;
25.               break Loop;
26.           else
27.               if ( nextTuple = false ) then
28.                   remove tuple from Q;
29.                   break Loop;
30.       else
31.           if ( nextTuple = false ) then
32.               tuple = NULL;
33.               break Loop;
34.   end Loop;
35.   return tuple;
36.   end;
```

Fig. 3. GetNextTuple Algorithm

3.3 Correctness of the Algorithm

At follows we attend to the correctness of the algorithm.

Lemma 3. *The algorithm GetNextTuple correctly reports the best join result (according to the ordering provided by f) among the tuples which do not belong to the set Top.Tuple.List and satisfy the join condition.*

Proof. Let x_A and x_B be the scores in A and B associated to the tuple returned by *GetNextTuple*. By definition (step 22) the score of such tuple ($f(x_A, x_B)$) is greater or equal that any score of the tuples belonging to the list Q. Note as well that, by the join strategy, the scores in A (resp. B) associated to any tuple in Q belongs to the interval $[A_\downarrow, A_\uparrow]$ (resp. $[B_\downarrow, B_\uparrow]$).

Let $\overline{x}_A$ and $\overline{x}_B$ be the scores in A and B associated to a tuple satisfying the join condition but non retrieved yet; and therefore non belonging neither to *Top.Tuple.List* nor to Q. Then, by the join strategy, the tuple $(\overline{x}_A, \overline{x}_B)$ belongs to $[0,1]^2 \setminus [A_\downarrow, A_\uparrow] \times [B_\downarrow, B_\uparrow]$. Applying now the Lemma 2, we can assert that the score of such a tuple is upper bounded by:

$$f(\overline{x}_A, \overline{x}_B) \leq \max \left\{ f(A_\downarrow, B.init); f(A_\uparrow, B.init); f(A.init, B_\downarrow); f(A.Init, B_\uparrow) \right\} = T$$

Now, (by step 23) the score of the tuple returned by *GetNextTuple* holds necessarily $f(\overline{x}_A, \overline{x}_B) \leq T \leq f(x_A, x_B)$. In conclusion, the score of the tuple returned by *GetNextTuple* is greater or equal that the score of any tuple in Q and any tuple non retrieved yet; in other words, the score of such a tuple has the greatest score between the join results non belonging to *Top.Tuple.List*.

Theorem 1. *Let A and B be two relational tables, R be a binary relation, k be a natural number and f be a ranking function. Then $Top.k.retrieval(A, B, R, k, f)$ correctly reports the top-k join results ordered by f.*

Proof. The proof comes directly from the Lemma 3, since in each step the loop *GetNextTuple* inserts in *Top.Tuple.List* the tuple with the greatest score among the tuples non belonging to *Top.Tuple.List* and satisfying the join condition.

4 Conclusion and Future Work

We have presented a top-k algorithm to retrieve tuples according to the order provided by a non-necessarily monotone ranking function that belongs to a novel family of functions satisfying some conditions related to the values where the maximum is achieved. For instance, this approach is the first one that allows the use of the mapping $f : [0,1]^2 \to \mathbb{R}$ given by $f((x,y)) = e^y - (x - 0.5)^2$ as a ranking function in top-k retrieval over non-index-merge paradigms.

The ultimate goal of our research is to develop a top-k algorithm for queries ordered by arbitrary mappings. An idea to achieve this goal may be the following. First we may try to develop a method to decompose arbitrary functions in terms of *ranked functions* (under Definition 1). For instance, every distance $d(x, y)$ can be decomposed in the following two ranking functions:

$$f_1(x, y) = \begin{cases} d(x, y) & \text{if } x \leq y \\ 0 & \text{otherwise} \end{cases} \qquad \text{and} \qquad f_2(x, y) = \begin{cases} d(x, y) & \text{if } x \geq y \\ 0 & \text{otherwise} \end{cases}$$

by considering the supremum; i.e $d(x, y) = \sup\{f_1(x, y), f_2(x, y)\}$.

The second step would consist in generalizing the join strategy and the threshold computation according to the decomposition given in the previous step. For instance, considering the decomposition given above for arbitrary distances, the threshold for d would be given in terms of the threshold of f_1 and f_2.

References

1. Badr, M., Vodislav, D.: A general top-k algorithm for web data sources. In: Hameurlain, A., Liddle, S.W., Schewe, K.-D., Zhou, X. (eds.) DEXA 2011, Part I. LNCS, vol. 6860, pp. 379–393. Springer, Heidelberg (2011)
2. He, Z., Lo, E.: Answering why-not questions on top-k queries. IEEE Transactions on Knowledge and Data Engineering 99, 1 (2012)
3. Ilyas, I., Aref, W., Elmagarmid, A.: Supporting top-k join queries in relational databases. The VLDB Journal 13(3), 207–221 (2004)
4. Ilyas, I.F., Beskales, G., Soliman, M.A.: A survey of top-k query processing techniques in relational database systems. ACM Comput. Surv. 40(4), 11:1–11:58 (2008)
5. Luo, Y., Wang, W., Lin, X., Zhou, X., Wang, J., Li, K.: Spark2: Top-k keyword query in relational databases. IEEE Transactions on Knowledge and Data Engineering 23(12), 1763–1780 (2011)
6. Marian, A., Bruno, N., Gravano, L.: Evaluating top-k queries over web-accessible databases. ACM Trans. Database Syst. 29(2), 319–362 (2004)
7. Ranu, S., Singh, A.K.: Answering top-k queries over a mixture of attractive and repulsive dimensions. Proc. VLDB Endowment 5(3), 169–180 (2011)
8. Straccia, U.: Top-k retrieval for ontology mediated access to relational databases. Information Sciences 198, 1–23 (2012)
9. Straccia, U., Madrid, N.: A top-k query answering procedure for fuzzy logic programming. Fuzzy Set and Systems 205, 1–29 (2012)
10. Xin, D., Han, J., Chang, K.C.: Progressive and selective merge: computing top-k with ad-hoc ranking functions. In: Proceedings of the 2007 ACM SIGMOD International Conference on Management of Data, SIGMOD 2007, pp. 103–114. ACM, New York (2007)
11. Xu, C., Wang, Y., Lin, S., Gu, Y., Qiao, J.: Efficient fuzzy top-k query processing over uncertain objects. In: Bringas, P.G., Hameurlain, A., Quirchmayr, G. (eds.) DEXA 2010, Part I. LNCS, vol. 6261, pp. 167–182. Springer, Heidelberg (2010)
12. Yi, K., Li, F., Kollios, G., Srivastava, D.: Efficient processing of top-k queries in uncertain databases with x-relations. IEEE Transactions on Knowledge and Data Engineering 20(12), 1669–1682 (2008)

Using a Stack Decoder for Structured Search

Kien Tjin-Kam-Jet, Dolf Trieschnigg, and Djoerd Hiemstra

University of Twente, Enschede, The Netherlands
{tjinkamj,trieschn,hiemstra}@ewi.utwente.nl

Abstract. We describe a novel and flexible method that translates free-text queries to structured queries for filling out web forms. This can benefit searching in web databases which only allow access to their information through complex web forms. We introduce boosting and discounting heuristics, and use the constraints imposed by a web form to find a solution both efficiently and effectively. Our method is more efficient and shows improved performance over a baseline system.

1 Introduction

Many web pages contain structured information that cannot be indexed by general web search engines like Bing or Google [4]. Web search engines use crawlers to follow hyperlinks and download web pages in order to index these pages, which enables fast keyword search. This crawler architecture has three drawbacks [1, 15]. First, a large part of the web cannot be crawled by simply following hyperlinks. Many pages are hidden behind web forms which cannot be automatically filled out by a crawler. Second, the indices of crawler-based search engines are only a snapshot of the state of the web. Pages containing real-time or highly dynamic information like traffic information or stock information are outdated as soon as they are indexed. Third, most of this information resides in structured databases that allow structured queries, a powerful means of searching. In contrast, putting this information in indices of crawler-based search engines would only allow unstructured keyword queries, a less powerful means of searching.

In this work we alleviate these problems by providing a single free-text search box to search multiple websites through complex web forms. We address the problem of translating a free-text query into a structured query, i.e., key-value pairs accepted by web forms. For instance, the free-text query "acer travelmate at least 4gb" could be mapped to the fields 'brand', 'model' and 'minimum memory' of a shopping website. As results, our system would return forms containing such fields, filled out and ready to be submitted. Note that in order to return results, the system does not need to crawl the web pages behind the forms, it just needs to know how to fill out the form given the free-text query. The problem can be decomposed into a *segmentation* problem of cutting up the free-text query into parts (segments); and a *labeling* problem of actually assigning each segment to the right input field. Our work extends existing segmentation & labeling methods based on HMMs (Hidden Markov Models) [16]. Segmenting is based on whitespace and punctuation characters, and subsequent labeling is based

H.L. Larsen et al. (Eds.): FQAS 2013, LNAI 8132, pp. 519–530, 2013.

on a probabilistic model. Our contributions are as follows. We propose a novel method that incorporates constraint information (see Sect. 3) and segments, labels, and normalizes queries; thereby deriving structured queries. We show that it is beneficial to apply boosting and discounting heuristics; that our method can be applied to a multi-domain, multi-site per domain setting; and, that our method outperforms a well known baseline. *Paper outline:* In Sect. 2, we discuss and compare related work to this work. We then formalize the problem and describe our framework in Sects. 3 and 4. We describe our data in Sect. 5, and our evaluations in Sect. 6. Finally, we round up with our conclusions in Sect. 7.

2   Related Work

Correct **query segmentation in web IR** can substantially improve retrieval results, e.g., grouping 'new' and 'york' as 'new york' can make a big difference. Li et al. [14] argue that supervised methods require expensive labeled data and propose an unsupervised segmentation model that can be trained on click log data. Hagen et al. [7] show that their segmentation algorithm, which uses only raw web n-gram frequencies and Wikipedia titles, is faster than state-of-the-art techniques while having comparable segmentation accuracy. Lastly, Yu and Shi [19] train a CRF (Conditional Random Field [12]) with tokens from a database. They first predict labels for each word in the query, and then segment at each start (S-) label. For example, given the query *Green Mile Tom Hanks* and the predicted labels {[S-MOVIE],[R-MOVIE],[S-ACTOR],[R-ACTOR]}, it is segmented as "Green Mile" and "Tom Hanks".

Query Segmentation & Labeling. The previous example illustrates that CRFs can both indicate segment offsets (e.g., with start/rest labels) and assign entire segments to fields (e.g., ACTOR or MOVIE). However, CRFs need a lot of expensive (manually labeled) training data. To avoid the high costs of manually labeled data, Li et al. [13] used two data sources to train CRFs: a pool of 19K queries labeled by human annotators; and a pool of 70K queries, automatically generated by matching entries from click logs with information from a product listings database. However, the generated queries did not contain all possible labels. Still, the highest performance was obtained when combining the evidence of both sources. In contrast, Kiseleva et al. [10] train multiple CRFs solely on click log data. But unlike manually labeled data, click log data suffers from noise and sparsity. In a follow-up study [11], they did use some manual data (brand synonyms and abbreviations) and artificially expanded their training set aiming to reduce data sparsity. Sarkas et al. [17] propose an unsupervised approach to segment & label web queries. They train an open language model (LM) on tokens derived from a general web log, and attribute LMs on tokens from the structured data residing in tables. They score results using a generative model of the probability of choosing: a set of attributes $T.\mathcal{A}$ from table T, a set of tokens $\mathcal{A}T$ given $T.\mathcal{A}$, and a set of free tokens $\mathcal{F}T$ given the table T. Further, they decide whether a query is intended as a web keyword query, or as a structured search query. DATAMOLD, by Borkar et al. [3], uses nested HMMs to segment &

label short unformatted text into structured records. They modify the Viterbi algorithm [6] to include semantic constraints, restricting it from exploring invalid paths. Since this violates the independence assumption, they re-evaluate a path when some state transition is disallowed by the constraints. Zhang and Clark [20] describe a framework that uses the averaged perceptron algorithm for training and a beam search algorithm (which is essentially, a stack decoder with a small stack) for decoding, and apply it to various syntactic processing tasks, like joint segmentation and POS-tagging. Our approach differs from these approaches in that it uses a stack decoder [2] and incorporates constraint information to prune, boost and discount; it is not purely probabilistic and works without training; and while it does not require, it can benefit from training.

Conclusion. Probabilistic methods like HMMs or CRFs outperform other methods for segmentation & labeling, but require large amounts of expensive training data, while fully unsupervised methods suffer from noisy training data. As a general remark, there is no agreed upon test collection to compare these methods, which makes it hard to determine the best method. That is, if such a conclusion can be made at all, since each method was developed for very specific use cases.

3   Problem Description and Approach

Our query translation problem can be formalized as:

Given a web form and a free-text query, find the intended values and assign the values to their intended fields, under the constraints imposed by the web form.

A *web form* has input *fields*, it only accepts queries as *structured information needs* consisting of a set of field-value assignments, e.g., $F_i = v_i$, for $i = 1 \ldots n$; and, a *free-text query* is an unstructured sequence of characters describing an intended structured information need. Next, we describe the types of constraints, how they can aid free-text to structured query translation, and our approach.

3.1   Hard Constraints

Web forms impose *constraints* that only allow certain combinations of fields and values. Queries satisfying these constraints are *valid*. Otherwise, they are *invalid*.

Mandatory Fields. A web form may require certain fields to be filled out before it can be submitted. For example, it may require either the `make` field, or both the `min` and `max` price fields to be filled out before it can be submitted. Formally, *mandatory field constraints* are propositions of the form: $(F_i) \vee (F_j \wedge F_k) \vee \ldots$, stating that at least one set of fields must be present in the query.

Conditional Fields. While a field may not be mandatory, it may be required if some other field is used. For example, consider a query that contains the text *5 miles near*, which states a radius (near some place). A web form with fields `radius` and `place`, may require that if you fill out `radius`, you must also fill out `place`. Formally, *assertive conditional constraints* are implications of the form: $F_i \rightarrow F_j$, stating that if some field F_i is present, then so must F_j. *Negative* *conditional constraints* are implications of the form: $F_i \rightarrow \neg F_j$, stating that if some field F_i is present, then F_j may and must not also be present.

Field Frequency. We refer to fields that allow only one value as *single-valued* fields and to fields that allow more values as *multi-valued* fields. *Frequency* constraints state that if a field is single-valued, it can be used at most once.

Categories. A category defines a set of values. For example, the category `base color` defines 'red', 'green' and 'blue' as values, while `year` could define numbers between 1970 and 2015 as values. Closed categories have a limited set of values, which are typically stored in a dictionary. Open categories have a limitless set of values, such as the set of real numbers. These are typically modeled by regular expressions. An input field will only accept values of one specific category.

Dependencies. The values allowed for one field may depend on the value of another. For example, if a `make` field has value *Ford*, then `model` may have *Fiesta*, but not *Laguna*. Formally, *dependency constraints* are implications of the form: $F_i \wedge F_j \rightarrow f(\lambda(F_i), \lambda(F_j))$, stating that if two dependent fields F_i and F_j are used, then the function f applied on their values $\lambda(F_i)$ and $\lambda(F_j)$ must be true. Here, f can be any function that takes two values as input and returns a boolean.

3.2   Soft Constraints

Soft constraints indicate which filled out form is more likely, given a valid query.

Patterns. A pattern determines when to assign values to a particular field by detecting field-specific hints that appear just before or after the values of a field's expected category. Formally, a pattern is defined as a 4-tuple {field name, prefixes, category, postfixes}. Prefixes and postfixes denote a set of words which may be empty. For example, consider the query *to New York from Dallas* and assume that *New York* and *Dallas* are values of the category `city`, which can be assigned to the fields: departure or destination. Then, a pattern for the destination field could for example be: {destination, [to], `city`, []}.

Field Order. Ideally, when a query contains a hint for some field F, followed by a value v of the category expected by F, then by all means, assign v to F. In practice however, queries may just contain values, like the query *New York Dallas*. The system would benefit from knowing that a particular field order is more likely than another, e.g., that P(departure, destination) $\geq$ P(destination, departure). We make the Markov assumption and model the probability of a sequence of fields as: $P(F_1, F_2, \ldots, F_n) = \prod_{i=1}^{n} P(F_i | F_{i-2}, F_{i-1})$.

3.3   Approach

Our approach consists of three steps: *a)* **segmenting**, i.e., splitting the free-text query into smaller *segments* ready to be assigned to some field—a segment is a subsequence of the characters of the free-text query. At each character position in the query, we search for known values which are defined by a regular expression or are contained in a dictionary. Our dictionary is based on a Bursttrie [8], but is modified to tolerate spelling errors as long as the first few characters of the search string are error free, and return search completions even if the string being completed has a spelling error. Whenever a value is found, it is added to the segment in which it was found. This process yields a set of segments, each segment containing a list of values, e.g., the segment 'red' can contain the values '4' (a color), and 'red hat' (an operating system name); b) **labeling**, i.e., indicating what to do with a segment value. A label assigned to a segment indicates one of three roles, namely that the segment contains: 1) a value v that will be assigned to some field F; 2) a field name, hinting that the value of an adjacent segment must be assigned to F; or, 3) no useful information. During this process, we also determine an actual segmentation. A segmentation denotes a list of segments such that the whole query can be reconstructed by concatenating each segment from the list. This also implies that the chosen segments may not overlap each other. In Section 4, we discuss how we apply our stack decoder for this labeling task; and c) **normalizing**, i.e., (slightly) rewriting the field value into a format accepted by the form, if necessary. A field has a format in which a value must be specified. For example, a field may require that a time be entered as hh:mm, i.e., two digits for the hour, a colon, and two digits for the minutes. If the query contains a time as *ten to five am*, it should be normalized to 04:50. For normalizing dates and times, we created a separate function. Other normalizations, like when the color *red* actualy has a value 4, or when a word is misspelt, are dealt with using a dictionary.

4   Stack Decoding

Given a free-text query, we first segment it into a set of segments, each segment containing a list of values. Next, we initialize a sorted stack with an empty *path*. A path has a score and a list of labeled segments. We then iteratively decode the query as follows: 1) remove the best path from the stack; 2) look up all segments S that follow immediately after the last segment in the path; 3) for each value in each segment $s \in S$, determine the possible labels and label the segment; 4) for each labeled segment, create a new path and add it to the stack. The process iteratively extends partial paths to become complete paths. When a path is complete, it is removed from the stack and stored as a result for further processing. The decoding stops when the stack is empty, or when some stopping criterion is met (e.g., some max decoding time t has elapsed).

Scoring. A path's score is based on the field values, and on the field order which was discussed in Section 3.2. The score of a value v from some closed category C is initially modeled as a uniform probability of $\frac{1}{|C|}$ for observing v. The score of a numeric value from an open category is determined heuristically: based on the number of digits, it diminishes quadratically such that a 4-digit value gets the highest score, then 3-digit and 5-digit values, and so on. An important issue in stack decoders is the comparability of partial paths [2, 20]. We lower a partial path's score by the number of characters that must yet be processed. This basically estimates for any partial path what the score would be if the whole query was processed. Note that lowering the score too much causes the decoder to proceed in a depth-first search manner instead of best-first search manner.

Pruning. With enough time and memory resources, we could theoretically examine all possible paths, including invalid ones. In practice however, we have little time and resources and need to reduce the time spent on processing invalid paths. Therefore, we prune partial paths that violate the dependency, field frequency, or negative conditional constraints defined in Section 3.1.

Boosting & Discounting. The speed of a stack decoder depend on it repeatedly choosing and expanding the best partial path until it finds the best complete path. The choice is based on fields and values seen so far, without regard for possible further fields and values. This is not always desirable. For example, consider the query *BMW 2000 euro* and a form with three fields: `make`, `year` and `price`. The segment 'BMW' is labeled as `make` and we must now label the segment '2000'. If we only considered segments up to and including '2000', then both labels `year` and `price` would seem fine. However, if we would have looked ahead when labeling '2000' as `year`, we would have known that this label is not likely, therefore we would have lowered the position of this path in the stack. The process of looking ahead and deciding to raise or lower a path's position in the stack is referred to as *boosting* or *discounting*, respectively. We can rank the complete paths by their original scores or by the boosted & discounted scores.

5   Data Used for Evaluation

Our aim was to obtain realistic queries under three conditions: *a*) **multi-domain** search environment. Participants should be able to search in different domains, like travel planning or second hand cars, and get real-time query suggestions; *b*) **multi-site** domains. Each domain should have different sites that may or may not offer the same search functionality. For example, in travel planning, one site might offer bus travel results, while other sites offer train or flight results; and, *c*) **minimal query bias**. Participants should not be persuaded to any kind of information need nor to any structure in which the they can phrase a query.

5.1   Data Acquisition

We setup an online search system covering 3 multi-site domains, and instructed the participants that they could search these domains. We briefly describe the domains and instructions for the participants. The **travel planning** domain has 3 sites, each providing either bus, train, or flight travel information. Instruction: *Find travel advice (for example, a traintrip to someone you know) and rate the result.* The **second hand cars** domain has 5 sites, each having a web form with fields for at least minimum price, maximum price, make, and model. Instruction: *Find cars with specific characteristics (for example, find cars with characteristics like your own car or a car of someone you know) and rate the result.* The **currency exchange** domain has 3 sites, each with a form that has three input fields (*from* currency, *to* currency, and *amount*). Instruction: *Find the exchange rate (of currencies of your choice) and rate the result.*

Participants started with a training session in which they could issue multiple queries in each of the three domains. Whenever a result was clicked on, a box appeared asking to rate the result as either: 'completely wrong', 'iffy', or 'completely right'. After rating a result, the system prompted for the next domain. It is natural to rephrase the query if a system returns no or unsatisfying results. However, if a participant believed that the query could have been answered corectly by the system, he/she could indicate this and optionally describe what kind of results should have been returned. During the training session, participants got acquainted with the system and discovered the search functionality by themselves. After introducing all domains, the participant was asked to conduct 10 different searches and rate at least one result of each search request. As an incentive to continue with the experiment, a score was shown based on, amongst others: the number of queries issued, the number of results rated, and the search functionality[1] discovered so far. Participants could quit whenever they wanted.

5.2   Manual Analysis and Labeling

We manually analysed all submitted queries and specified which forms could return relevant results and how the forms should be filled out. For each form, we compiled a testcorpus specifying the set of field-value assignments for the queries that make sense to the web form. We then measured how much our judgments agreed with those of the participants using the *overlap* between our manually assigned query-result pairs and those of the participants. Overlap is defined as the size of the intersection of the sets of relevant results divided by the size of their union, and has been used by several studies for quantifying the agreement among different annotators [18, 9, 5]. We needed to compile the testcorpora ourselves because: first, participants did not (and were not expected to) find and label all correct results. Second, the system may not have returned any correct results, making it impossible for participants to label all correct results.

5.3   Data Obtained

In total, 47 participants interacted with the system and 23 opted to state their age and gender, resulting in 17 males (age: 19–81, avg. 39) and 6 females (age: 25–41, avg. 30). We analyzed 363 queries, but nearly half were invalid, either missing mandatory fields or asking information that was out of scope. Examples of invalid queries are: *to Amsterdam*; *how long is the Golden Gate bridge*; *kg to pound*; and, *for sale: 15 year old mercedes*. In total, we labeled 194 valid queries containing enough information to fill out a form in our experiment. When multiple forms could be filled out for a given query, we chose the ones in which we could specify most key-value pairs of the query. A summary of the results for the travel planning, currency exchange, and second hand cars domains is shown in Table 1. The rows 'A' to 'K' each correspond to a form in the specified domain and shows: *Q:* the number of queries submitted in that form; *Max:* the maximum

---

[1] Search functionality here means the number of different fields in all clicked results, divided by the total number of fields from all web forms configured in the system.

number of different ways to fill out that form for a single query; *Avg:* the average
number of filled out forms per query; and, *Std.dev:* the standard deviation from
this average. The row 'All' shows the results when aggregating all forms, and
should be interpreted as: 194 queries were submitted in this aggregated form;
there was a query that could be filled out in 19 different ways; there were 2.99
filled out forms per query on average, with a standard deviation of 2.43.

Table 1. Manual labeling results

Travel		Q.	Max.	Avg.	Std.dev.	Cars		Q.	Max.	Avg.	Std.dev.
	A	52	8	1.19	0.99		G	24	2	1.04	0.20
	B	5	5	1.80	1.79		H	59	7	1.39	1.16
	C	12	3	1.25	0.62		I	61	9	1.38	1.29
Currency							J	52	4	1.12	0.51
	D	61	1	1.00	0.00		K	49	3	1.20	0.58
	E	61	2	1.03	0.18	Merged					
	F	62	1	1.00	0.00		All	194	19	2.99	2.43

A result (i.e., a filled out form) denotes a set of field-value pairs. On a result
level, the agreement of our judgments and those of the participants is 0.33, which
is consistent with the "key" agreement reported in [5]. Though it might seem
low, it is a direct result of the strict comparison: one slightly different field value
causes results to disagree completely. If we considered field-value pairs instead,
and averaged the field-value agreement per result, the agreement is 0.68.

6   Evaluating the Stack Decoder

We evaluated our system using the data described in Section 5.3. We investigated
how different stopping criteria, boosting, discounting, and ranking on original or
on boosted scores, affected the decoding time and retrieval performance—which
was measured using MAP (Mean Average Precision [18]). Table 2 lists the 6
stopping criteria that we used. The decoding stopped when: a maximum of r
results was found; or, more than t time elapsed during decoding; or, the next
result's score was lower than some absolute minimum *abs.min*; or, when it was
lower than some mimimum *rel.min* relative to the best result. Further, we tested
two settings for pruning probably irrelevant paths based on the percentage of
the query that was ignored. A path was discarded if more than $j\%$ was ignored
(e.g., due to unknown words). One (fairly strict) setting required the system to
interpret at least 60% of the query, while the other required only 20%.

Table 2. Stopping criteria, sorted by number of results and "strictness"

	Results	Time	Abs. min.	Rel. min.	Ignore %		Results	Time	Abs. min.	Rel. min.	Ignore %
A	10	0.5	-200	-150	40	D	50	45	-200	-150	40
B	10	0.5	-200	-150	80	E	50	45	-200	-150	80
C	10	0.5	-600	-550	80	F	50	45	-600	-550	80

Table 3. Results obtained without training. The headers A–F denote stopping criteria (see Table 2). The leftmost letters B, D, and R denote *boosting*, *discounting*, and *ranking* by original score, respectively. *Time* is the average query decoding time. *Map1* and *Map2* are the MAP of filled out forms, and of segmentation & labeling, respectively.

(a) Evaluation results, averaged over the individual tests per form.

			A			B			C			D			E			F		
B	D	R	MAP1	Time	MAP2	MAP1	Time	MAP2	MAP1	Time	MAP2	MAP1	Time	MAP2	MAP1	Time	MAP2	MAP1	Time	MAP2
-	-	-	0.485	0.05	0.549	0.551	0.04	0.608	0.622	0.07	0.625	0.485	0.05	0.548	0.551	0.05	0.608	0.629	0.31	0.627
0	1	0	0.502	0.04	0.576	0.568	0.04	0.636	0.641	0.07	0.656	0.501	0.05	0.575	0.568	0.05	0.635	0.647	0.30	0.653
1	0	0	0.503	0.04	0.567	0.569	0.04	0.627	0.639	0.07	0.641	0.503	0.05	0.566	0.569	0.05	0.626	0.647	0.16	0.645
1	1	0	0.504	0.04	0.579	0.570	0.04	0.638	0.640	0.07	0.652	0.504	0.04	0.577	0.570	0.04	0.637	0.649	0.15	0.656
0	1	1	0.519	0.04	0.583	0.582	0.04	0.644	0.642	0.07	0.664	0.521	0.05	0.586	0.583	0.04	0.647	0.649	0.31	0.666
1	0	1	0.521	0.04	0.580	0.583	0.04	0.642	0.642	0.07	0.656	0.522	0.05	0.583	0.585	0.05	0.645	0.649	0.16	0.664
1	1	1	0.522	0.04	0.585	0.584	0.04	0.646	0.643	0.07	0.659	0.523	0.04	0.588	0.586	0.05	0.649	0.650	0.16	0.668

(b) Evaluation results of the aggregated web forms.

			A			B			C			D			E			F		
B	D	R	MAP1	Time	MAP2	MAP1	Time	MAP2	MAP1	Time	MAP2	MAP1	Time	MAP2	MAP1	Time	MAP2	MAP1	Time	MAP2
-	-	-	0.414	0.09	0.523	0.444	0.08	0.547	0.442	0.23	0.539	0.446	0.11	0.556	0.475	0.10	0.581	0.504	1.64	0.597
0	1	0	0.424	0.08	0.539	0.454	0.08	0.562	0.453	0.22	0.554	0.455	0.11	0.571	0.484	0.10	0.596	0.516	1.66	0.611
1	0	0	0.427	0.08	0.539	0.463	0.08	0.570	0.461	0.22	0.553	0.462	0.14	0.573	0.495	0.13	0.603	0.514	1.09	0.608
1	1	0	0.428	0.08	0.545	0.464	0.08	0.576	0.464	0.22	0.558	0.461	0.13	0.576	0.493	0.12	0.606	0.517	1.16	0.613
0	1	1	0.402	0.08	0.509	0.434	0.08	0.540	0.439	0.22	0.540	0.417	0.10	0.527	0.452	0.10	0.558	0.479	1.88	0.579
1	0	1	0.407	0.08	0.511	0.440	0.08	0.548	0.447	0.22	0.538	0.424	0.13	0.533	0.457	0.12	0.567	0.477	1.23	0.579
1	1	1	0.408	0.08	0.514	0.441	0.08	0.550	0.449	0.22	0.537	0.422	0.12	0.532	0.456	0.12	0.567	0.478	1.25	0.582

6.1 Untrained and Individual, "per Form" Evaluation

One at a time, we loaded a form's dictionary and constraints and ran its tests. We did not train the system but used a uniform field order distribution[2]. Table 3(a) shows the averaged results of the individual tests, weighted by the number queries per form. The results show that we should not prune "improbable" paths beforehand, i.e., paths with low scores and in which up to 80% of the query is ignored. It also shows that boosting and discounting affects MAP, especially with relatively strict stopping criteria; and that as the criteria relaxes, the effect decreases. This is due to the relatively small search space in the individual tests. The stopping criteria limit the part of the search space can be inspected, and the boosting and discounting try to sneak in as many relevant paths to this limited space as possible. Thus when the stopping criteria are sufficiently relaxed, the effects of boosting and discounting will naturally decrease. For the individual tests, we can conclude that boosting reduces decoding time, and that boosting, discounting, and ranking on original scores yields the best retrieval performance.

6.2 Untrained and Collective, "Aggregated Forms" Evaluation

We collectively loaded all forms into our system. This causes the search space to be much larger, and aside from determining how to fill out a form, the system must also determine which forms to return in the first place. We also aggregated the tests, specifying for each query all forms that should be returned and

---

[2] Except in one form where we manually specified that "departure" fields were more likely followed by "destination" fields, instead of other fields. However, this was done before going online and gathering data, so before we had even seen the test data.

all ways of filling out a form for that query. From the collective evaluation results in Table 3(b), we can conclude that: we should not prune "improbable" paths beforehand, which agrees with the results of the individual tests; Boosting, discounting, and ranking on the boosted & discounted scores yields the best retrieval performance, which contrasts with the individual tests where you should rank on the original scores; Finally, our system effectively brokers over different sites across different domains (e.g., travel planning, currency, second hand cars).

6.3 Baseline Evaluation

To our knowledge, no other system translates free-text queries to filled out forms, normalizes values, and checks against constraints. However, LingPipe[3] is a suitable baseline, as it recognizes named entities by segmenting & labeling text, and is a widely used text processing toolkit. We manually segmented the queries and labeled each segment. Filled out forms naturally correlate with segmented & labeled queries. However, due to normalization and constraint checking, there may not be a valid filled out form even if the query is correctly segmented.

We evaluated both systems on their prediction of which query segments contained field values and what label to assign to each segment. We used 3 data sets to simulate "untrained" up to "fully trained" systems: *set A* contains uniform field transitions and uniform token counts; *set B* contains field transitions from the queries, but uniform token counts; and, *set C* contains both field transitions and token counts taken from the queries. We cross-validated LingPipe using out-of-the-box settings for named entity recognition. In each test, we loaded the dictionary but no regular expressions because they cannot be used together (at least, not out-of-the-box). We cross-validated our system using the parameters from Table 3 that gave the best filled out forms (i.e., with the highest $MAP1$, and lowest time if $MAP1$ is equal). So, for the individual tests we used {criteria=C; B,D,R=1,1,1}, and for the collective tests {criteria=B; B,D,R=1,1,0}.

The segmentation & labeling results are shown in Table 4. Row A denotes results of untrained systems (i.e., they are only "trained" on uniform distributions). Rows B and C denote 5-fold cross validation results of the systems. The collective cross-validations tests are stratified, i.e., 1/5-th of the queries of each form is used in each fold. As expected with no training (row A), LingPipe performs poorly, which constrasts with our untrained system. For now, our system

Table 4. Segmentation & labeling results. Training set A involves no training. In B we train on field transitions, and in C on both field transitions and token counts.

(a) Averaged individual tests.

LingPipe			Our system		
Training set	MAP	Time	Training set	MAP	Time
A	0.302	0.27	A	0.659	0.07
B	0.459	0.04	B	0.717	0.05
C	0.708	0.04	-	-	-

(b) Collective tests.

LingPipe			Our system		
Training set	MAP	Time	Training set	MAP	Time
A	0.117	66.88	A	0.576	0.08
B	0.207	5.16	B	0.629	0.07
C	0.289	5.11	-	-	-

---

[3] Alias-i. 2013. LingPipe 4.1.0. http://alias-i.com/lingpipe (accessed March 1, 2013).

can only train on field transitions (row B), and this already improves performance. Training LingPipe on only field transitions also improves performance; but training on both transitions and token counts (for which it was designed) gives the biggest improvement. Since LingPipe does not know that once it uses labels of one form it cannot use labels of others, it performs very poorly in the collective tests. Then again, it was not developed for such a task.

6.4 Further Discussion

The problem of converting non-structured queries to structured queries goes back as far as 30 years, and solutions were proposed based on heuristics, grammars, and graphs. Due to space limits however, we focussed on probabilistic, state-of-the-art approaches to segmentation & labeling in Sect. 2. In Sect. 3, the form's constraints must be specified manually. Automatic detection of such constraints would be beneficial and warrants further research. Regarding the results, after inspecting a sample of the results we noted that OOV (out-of-vocabulary) words were lowering retrieval performance. Some OOV words can easily be added (e.g., new car models), but others consitute natural language phrases that must be interpreted in context and cannot easily be added. The problem of OOV words must be further researched. Online learning using click log data is potentially the cheapest solution, but comes with several challenges (see Section 5.2). We also noticed that few labels were used for numerical tokens, e.g., a number was often intended as a price, but never as the engine displacement. This makes it easier for LingPipe to guess the right label, as it is ignorant of the actual possible labels for each numerical token and just considers the labels seen during training. Finally, we will extend our system to train on token counts as well (i.e., data from *set C*), which should further improve retrieval results.

7 Conclusion

We introduced a novel and flexible method for translating free-text queries to structured queries for filling out web forms. This enables users to search structured content using free-text queries. In contrast, web search engines struggle to index structured content from web databases, and users cannot enter structured queries in a typical web search engine. Our method consists of three steps: segmenting, labeling, and normalizing. We use the constraints imposed by web forms to prune the search space and apply boosting & discounting heuristics. Our results confirm that our heuristics are effective, reducing decoding time and raising retrieval performance. We also showed that without training, our system outperforms an untrained baseline on the individual and the collective tests. Compared to a trained baseline, our trained system is still better on the individual tests and outperforms the baseline on the collective tests.

Acknowledgment. We thank the anonymous FQAS '13 reviewers for their comments. This research was funded by the Netherlands Organization for Scientific Research, NWO, grant 639.022.809.

References

[1] Baeza-Yates, R., Castillo, C., Junqueira, F., Plachouras, V., Silvestri, F.: Challenges on distributed web retrieval. In: ICDE 2007, pp. 6–20 (April 2007)

[2] Bahl, L.R., Jelinek, F., Mercer, R.L.: A maximum likelihood approach to continuous speech recognition. In: Readings in Speech Recognition, pp. 308–319. Morgan Kaufmann Publishers Inc., San Francisco (1990)

[3] Borkar, V., Deshmukh, K., Sarawagi, S.: Automatic segmentation of text into structured records. In: SIGMOD 2001, pp. 175–186. ACM, New York (2001)

[4] Chang, K.C.-C., He, B., Li, C., Patel, M., Zhang, Z.: Structured databases on the web: observations and implications. SIGMOD Record 33(3), 61–70 (2004)

[5] Demeester, T., Nguyen, D., Trieschnigg, D., Develder, C., Hiemstra, D.: What snippets say about pages in federated web search. In: Hou, Y., Nie, J.-Y., Sun, L., Wang, B., Zhang, P. (eds.) AIRS 2012. LNCS, vol. 7675, pp. 250–261. Springer, Heidelberg (2012)

[6] Forney Jr., G.D.: The viterbi algorithm. Proc. of the IEEE 61(3), 268–278

[7] Hagen, M., Potthast, M., Stein, B., Braeutigam, C.: Query segmentation revisited. In: WWW 2011, pp. 97–106. ACM, New York (2011)

[8] Heinz, S., Zobel, J., Williams, H.E.: Burst tries: a fast, efficient data structure for string keys. In: TOIS 2002, vol. 20(2), pp. 192–223 (2002)

[9] Hiemstra, D., van Leeuwen, D.A.: Creating an information retrieval test corpus for dutch. In: CLIN 2001, Amsterdam, The Netherlands. Language and Computers - Studies in Practical Linguistics, vol. 45, pp. 133–147. Rodopi (2002)

[10] Kiseleva, J., Guo, Q., Agichtein, E., Billsus, D., Chai, W.: Unsupervised query segmentation using click data: preliminary results. In: WWW 2010, pp. 1131–1132. ACM, New York (2010)

[11] Kiseleva, J., Agichtein, E., Billsus, D.: Mining query structure from click data: a case study of product queries. In: CIKM 2011, pp. 2217–2220. ACM, New York (2011)

[12] Lafferty, J.D., McCallum, A., Pereira, F.C.N.: Conditional random fields: Probabilistic models for segmenting and labeling sequence data. In: ICML 2001, San Francisco, CA, USA, pp. 282–289. Morgan Kaufmann Publishers Inc. (2001)

[13] Li, X., Wang, Y.-Y., Acero, A.: Extracting structured information from user queries with semi-supervised conditional random fields. In: SIGIR 2009, pp. 572–579. ACM, New York (2009)

[14] Li, Y., Hsu, B.-J.P., Zhai, C., Wang, K.: Unsupervised query segmentation using clickthrough for information retrieval. In: SIGIR 2011, pp. 285–294. ACM, New York (2011)

[15] Madhavan, J., Ko, D., Kot, L., Ganapathy, V., Rasmussen, A., Halevy, A.: Google's deep web crawl. Proc. VLDB Endow. 1(2), 1241–1252 (2008)

[16] Rabiner, L.R.: A tutorial on hidden markov models and selected applications in speech recognition. Proc. of the IEEE 77(2), 257–286 (1989)

[17] Sarkas, N., Paparizos, S., Tsaparas, P.: Structured annotations of web queries. In: SIGMOD 2010, pp. 771–782. ACM, New York (2010)

[18] Voorhees, E.M.: Variations in relevance judgments and the measurement of retrieval effectiveness. Inf. Processing and Management 36(5), 697–716 (2000)

[19] Yu, X., Shi, H.: Query segmentation using conditional random fields. In: KEYS 2009, pp. 21–26. ACM, New York (2009)

[20] Zhang, Y., Clark, S.: Syntactic processing using the generalized perceptron and beam search. Computational Linguistics 37(1), 105–151 (2011)

On Cosine and Tanimoto Near Duplicates Search among Vectors with Domains Consisting of Zero, a Positive Number and a Negative Number

Marzena Kryszkiewicz

Institute of Computer Science, Warsaw University of Technology
Nowowiejska 15/19, 00-665 Warsaw, Poland
mkr@ii.pw.edu.pl

Abstract. The cosine and Tanimoto similarity measures are widely applied in information retrieval, text and Web mining, data cleaning, chemistry and bio-informatics for finding similar objects, their clustering and classification. Recently, a few very efficient methods were offered to deal with the problem of lossless determination of such objects, especially in large and very high-dimensional data sets. They typically relate to objects that can be represented by (weighted) binary vectors. In this paper, we offer methods suitable for searching vectors with domains consisting of zero, a positive number and a negative number; that is, being a generalization of weighted binary vectors. Our results are not worse than their existing analogs offered for (weighted) binary vectors.

Keywords: the cosine similarity, the Tanimoto similarity, nearest neighbors, near duplicates, exact duplicates, non-zero dimensions, vector's length.

1   Introduction

The cosine and Tanimoto similarity measures are widely applied in information retrieval, text and Web mining, data cleaning, chemistry, biology and bio-informatics for finding similar vectors representing objects [4, 10, 11] as well as for meaningful clustering and classification of objects based on their vectors' representation. In particular, documents are often represented as term frequency vectors or its variants such as *tf_idf* vectors [9]. While approximate lossy search of similar vectors is quite popular [3, 6], competing approaches enabling their lossless search have been offered recently [1, 2, 8, 12]. In particular, lossless search of sufficiently similar vectors is vital for data cleaning [1], as well as for plagiarism discovery. The search of similar vectors should be carried out very efficiently when the task is to be carried out for very large number of vectors (for example, in the case of data cleaning).

In this paper, we focus on efficient lossless search of *cosine and Tanimoto near duplicate vectors* of a given vector *u*; that is, on searching vectors that are similar to a vector *u* in a given degree with respect to the cosine or Tanimoto similarity, respectively. The cosine similarity of vectors is defined as the cosine of the angle between them. Vectors are treated as cosine near duplicates if the angle between them

H.L. Larsen et al. (Eds.): FQAS 2013, LNAI 8132, pp. 531–542, 2013.
© Springer-Verlag Berlin Heidelberg 2013

is small; that is, if its cosine is close to 1. In the case of binary vectors, with dimensions' domains restricted to {0, 1}, the Tanimoto similarity between two vectors equals the ratio of the number of attributes with "1s" shared by both vectors to the number of attributes with "1s" that occur in either vector. In this case, the Tanimoto similarity is equivalent to the Jaccard similarity, which is defined for sets, and equals the ratio of the cardinality of the intersection of the two sets to the cardinality of their union. However, the Tanimoto similarity by definition is computable for any non-zero real-valued vectors, and thus can be regarded as a generalization of the Jaccard similarity. In the literature, one may also meet different definitions of so called "weighted Jaccard similarity". Beneath, we describe briefly two typical definitions of a weighted Jaccard similarity in which it is assumed that a weight $w(e)$ is associated with each element e that may occur in a set. We will also comment their relationships with the Tanimoto similarity. In one of these approaches, a weighted Jaccard similarity is defined for multi-sets as usual Jaccard similarity for unweighted bags obtained from the multi-sets by making $rw(e)$ copies of each element e, where $rw(e)$ is the rounded value of $w(e)$. This definition of a weighted Jaccard similarity can be expressed in terms of the Tanimoto similarity provided each element e will be represented by $rw(e)$ dimensions. The resultant vectors would be binary (non-weighted). Nevertheless, the above definition of a weighted Jaccard similarity was found unsatisfactory in [1]. Another approach to generalizing the Jaccard similarity consists in calculating the ratio of the sum of weights of elements common in two sets to the sum of weights of all elements in their union [12]. This definition of a weighted Jaccard similarity can be expressed equivalently in terms of the Tanimoto similarity provided each vector domain i is two-valued and equals $\{0, \sqrt{w(e_i)}\}$, where e_i is an element corresponding to domain i.

In nineties, first probabilistic algorithms based on the idea of *locality-sensitive hashing* (LSH) were offered for efficient approximate search of near duplicates in very large and high dimensional data sets [3, 6]. Nevertheless, they do not guarantee the identification of all near duplicates [1, 2]. In addition, recently, a few very efficient methods were offered to deal with the problem of lossless determination of near duplicates in the case of large and very high-dimensional data sets that are competitive to LSH methods [1, 2, 12]. They typically relate to objects that can be represented by (weighted) binary vectors [1, 2, 12]. In this paper, however, we consider vectors each domain of which may contain at most three values: zero, a positive value and a negative value. We will call such vectors as *ZPN-vectors*. Beneath, we consider example applications that illustrate usefulness of such vectors.

One possible application of *ZPN*-vectors is the search of documents that cite similar papers in a similar way. If a paper is cited as valuable, it could be graded with a positive value; if it is cited as invaluable, it could be graded with a negative value; if it is not cited, it could be graded with 0. Let us consider another example of using vectors with three-valued dimensions. Some teachers grade answers to test queries in three ways: a positive answer is graded with a positive value, a negative answer is graded with a negative value and lack of an answer is graded with 0. Such grading might discourage students from guessing answers. *ZPN*-vectors may be also used to better describe and analyse the results of medical tests – in order to indicate that a result of a particular medical test is positive, it can be expressed by a positive value that reflects the importance of the test; to indicate that a result of the test is negative, it can be expressed by a negative value. Please note that the restriction to weighted binary domains of the form $\{0, a\}$ would ignore either negative (if $a > 0$) or positive (if $a < 0$) significance of a medical test. ZPN-vectors representations allow us to avoid this shortcoming.

In our paper, we offer and prove how to use knowledge about non-zero dimensions of ZPN-vectors and their lengths for lossless finding of cosine and Tanimoto near duplicates with respect to a given similarity threshold. We also derive specific properties of a special case of near duplicates of ZPN-vectors called *exact duplicates* for the cosine and Tanimoto similarity measures. In fact, our finding concerning exact Tanimoto duplicates is more general and covers the case of real-valued vectors.

Our paper has the following layout. Section 2 provides basic notions used in this paper. In particular, the definitions of the cosine and Tanimoto similarities are recalled here. In Section 3, we derive bounds on lengths of cosine and Tanimoto near duplicate ZPN-vectors, while in Section 4, we offer theoretical results related to using non-zero dimensions for reducing candidates for near duplicate ZPN-vectors. In Section 5, we derive properties of exact duplicates. In Section 6, we present related work. Section 7 summarizes our contribution.

2 Basic Notions

In the paper, we consider n-dimensional vectors. A vector u will be also denoted as $[u_1, ..., u_n]$, where u_i is the value of the i-th dimension of u, $i = 1..n$.

By $NZD(u)$ we denote the set of those dimensions of vector u which have values different from 0; that is,

$$NZD(u) = \{i \in \{1, ..., n\} \mid u_i \neq 0\}.$$

Analogously, by $ZD(u)$ we denote the set of those dimensions of vector u which have zero values; that is,

$$ZD(u) = \{i \in \{1, ..., n\} \mid u_i = 0\}.$$

A *dot product* of vectors u and v is denoted by $u \cdot v$ and is defined as $\sum_{i=1..n} u_i v_i$. One may easily observe that the dot product of vectors u and v that have no common non-zero dimension equals 0.

A *length of vector u* is denoted by $|u|$ and is defined as $\sqrt{u \cdot u}$.

In Table 1, we present an example set of ZPN-vectors that will be used throughout this paper. In Table 2, we provide its alternative sparse representation, where each vector is represented only by its non-zero dimensions and their values. More precisely, each vector is represented by a list of pairs, where the first element of a pair is a non-zero dimension of the vector and the second element – its value. For future use, in Table 2, we also place the information about lengths of the vectors.

Table 1. Dense representation of an example set of *ZPN*-vectors

Id	1	2	3	4	5	6	7	8	9
v1	-3,0	4,0			3,0	5,0	3,0	6,0	
v2	3,0	-2,0				5,0		6,0	
v3			6,0	4,0					
v4		-2,0		4,0		5,0		-5,0	
v5				4,0	-3,0		3,0		
v6			-9,0	4,0					5,0
v7			6,0	4,0					
v8		4,0		4,0					5,0
v9				-2,0	3,0		3,0		5,0
v10		-2,0	-9,0						

Table 2. Sparse representation of the example set of *ZPN*-vectors from Table 1 (extended by the information about vectors' lengths)

Id	(non-zero dimension, value) pairs	length
v1	{(1,-3.0), (2, 4.0), (5, 3.0), (6, 5.0), (7, 3.0), (8, 6.0)}	10.20
v2	{(1, 3.0), (2,-2.0), (6, 5.0), (8, 6.0)}	8.60
v3	{(3, 6.0), (4, 4.0)}	7.21
v4	{(2,-2.0), (4, 4.0), (6, 5.0), (8,-5.0)}	8.37
v5	{(4, 4.0), (5,-3.0), (7, 3.0)}	5.83
v6	{(3,-9.0), (4, 4.0), (9, 5.0)}	11.05
v7	{(3, 6.0), (4, 4.0)}	7.21
v8	{(2, 4.0), (4, 4.0), (9, 5.0)}	7.55
v9	{(4,-2.0), (5, 3.0), (7, 3.0), (9, 5.0)}	6.86
v10	{(2,-2.0), (3,-9.0)}	9.22

The *cosine similarity between* vectors u and v is denoted by $cosSim(u, v)$ and is defined as the cosine of the angle between them; that is,

$$cosSim(u, v) = \frac{u \cdot v}{|u \| v|}.$$

The *Tanimoto similarity* between vectors u and v is denoted by $T(u, v)$ and is defined as follows,

$$T(u, v) = \frac{u \cdot v}{u \cdot u + v \cdot v - u \cdot v}.$$

Clearly, the Tanimoto similarity between two non-zero vectors u and v can be expressed equivalently in terms of their dot product and their lengths as follows:

$$T(u, v) = \frac{u \cdot v}{|u|^2 + |v|^2 - u \cdot v}.$$

For any non-zero vectors u and v, $T(u, v) \in \left[-\frac{1}{3}, 1\right]$ (see [10]).

Obviously, if $u \cdot v = 0$ for non-zero vectors u and v, then $cosSim(u, v) = T(u, v) = 0$. Hence, when looking for vectors v such that $cosSim(u, v) \geq \varepsilon$ or, respectively, $T(u, v) \geq \varepsilon$, where $\varepsilon > 0$, non-zero vectors w that do not have any common non-zero dimension with vector u can be skipped as in their case $cosSim(u, w) = T(u, w) = u \cdot w = 0$. In Example

1, we show how to skip such vectors by means of *inverted indices* [11]. In a simple form, an inverted index stores for a dimension *dim* the list $I(dim)$ of identifiers of all vectors for which *dim* is a non-zero dimension.

Example 1. Let us consider the set of vectors from Table 2 (or Table 1). In Table 3, we present the inverted indices that would be created for this set of vectors. Let us assume that we are to find vectors v similar to vector $u = v1$ such that $cosSim(u, v) \geq \varepsilon$ (or $T(u, v) \geq \varepsilon$) for some positive value of ε. To this end, it is sufficient to compare u only with vectors that have at least one non-zero dimension common with $NZD(u) = \{1, 2, 5, 6, 7, 8\}$ (see Table 2); i.e., with the vectors in the following set (see Table 3):

$$\bigcup \{I(i)|i \in NZD(u)\} = I(1) \cup I(2) \cup I(5) \cup I(6) \cup I(7) \cup I(8)$$
$$= \{v1, v2, v4, v5, v6, v8, v9, v10\}.$$

Hence, the application of inverted indices allowed us to reduce the set of vectors to be evaluated from 10 vectors to 8. □

Table 3. Inverted indices for dimensions of vectors from Table 2 (and Table 1)

dim	I(dim)
1	<v1, v2>
2	<v1, v2, v4, v8, v10>
3	<v3, v6, v7, v10>
4	<v3, v4, v5, v6, v7, v8, v9>
5	<v1, v5, v9>
6	<v1, v2, v4>
7	<v1, v5, v9>
8	<v1, v2, v4>
9	<v6, v8, v9>

Please note that in Example 1 the value of ε has not been taken into account when determining vectors potentially similar to the given vector u.

In this paper, we will propose a more powerful mechanism for reducing the number of candidates for similar vectors than the one presented by means of Example 1; namely, we will derive which subsets among non-zero dimensions of vector u can be used to restrict the number of candidate vectors v that have a chance to meet the similarity condition $cosSim(u, v) \geq \varepsilon$ and, respectively, the similarity condition $T(u, v) \geq \varepsilon$ for a given positive value of ε and will derive bounds on lengths of such vectors. We will also examine more deeply properties of *exact duplicates*, that is, near duplicates found for the similarity threshold $\varepsilon = 1$.

3   Bounds on Lengths of Near Duplicate *ZPN*-Vectors

In this section, we examine a possibility of reducing the search space of candidates for near duplicate *ZPN*-vectors by taking into account their lengths. We will start with an observation (Proposition 1), which we will use later to derive respective theorems.

Let u and v be *ZPN*-vectors and i be any of their dimensions. Let $\{0, a, b\}$, where a is a real positive number and b is a real negative number, be the domain of dimension i. Table 4 presents the values of multiplying i-th dimension of vector u by i-th dimension of vector v ($u_i v_i$), as well as the square of i-th dimension of vector u (u_i^2) and the square of i-th dimension of vector v (v_i^2) for all nine possible combinations of values of i-th dimension of vectors u and v. Based on the results of the multiplications from Table 4, we conclude:

Proposition 1. For any ZPN-*vectors* u and v, the following holds:

a) $\forall_{i \in \{1,\dots,n\}} (u_i v_i \le u_i^2)$.

b) $u \cdot v \le |u|^2$.

Table 4. The result of multiplying any i-th dimension with domain $\{0, a, b\}$, where $a > 0$ and $b < 0$, of a pair of ZPN-vectors u and v

u_i	v_i	$u_i v_i$	u_i^2	v_i^2
0	0	0	0	0
0	a	0	0	a^2
0	b	0	0	b^2
a	0	0	a^2	0
a	a	a^2	a^2	a^2
a	b	ab	a^2	b^2
b	0	0	b^2	0
b	a	ab	b^2	a^2
b	b	b^2	b^2	b^2

Theorem 1. Let u and v be non-zero ZPN-vectors, $cosSim(u, v) \ge \varepsilon$ and $\varepsilon \in (0,1]$. Then:

a) $|v| \in \left[\varepsilon |u|, \dfrac{|u|}{\varepsilon} \right]$.

b) $|v|^2 \in \left[\varepsilon^2 |u|^2, \dfrac{|u|^2}{\varepsilon^2} \right]$.

Proof. Ad a) By Proposition 1b, $u \cdot v \le |u|^2$. Hence, $cosSim(u, v) = \dfrac{u \cdot v}{|u| \, \|v\|} \le \dfrac{|u|^2}{|u| \, \|v\|}$

$= \dfrac{|u|}{|v|}$. As $cosSim(u, v) \ge \varepsilon$, we conclude further that $\varepsilon \le \dfrac{|u|}{|v|}$. Analogously, one may

derive that $\varepsilon \le \dfrac{|v|}{|u|}$. Hence, as $\varepsilon > 0$, we obtain: $|v| \in \left[\varepsilon |u|, \dfrac{|u|}{\varepsilon} \right]$.

Ad b) Follows immediately from Theorem 1a.   □

Example 2. Let us consider vector $u = v1$ from Table 2 and let $\varepsilon = 0.85$. By

Theorem 1a, only vectors the lengths of which belong to the interval $\left[\varepsilon |u|, \dfrac{|u|}{\varepsilon} \right] =$

$\left[0.85 \times 10.20, \dfrac{10.20}{0.85} \right] \approx [8.67, 12.00]$ have a chance to be sought near cosine duplicates of u. Hence, only vectors $v1$, $v6$ and $v10$, which fulfill this length condition, have such a chance.   □

Beneath, we provide and prove Theorem 2 related to bounds on lengths of Tanimoto similar ZPN-vectors.

Theorem 2. Let u and v be non-zero ZPN-vectors, $T(u, v) \geq \varepsilon$ and $\varepsilon \in (0,1]$. Then:

a) $|v|^2 \in \left[\varepsilon |u|^2, \dfrac{|u|^2}{\varepsilon} \right]$.

b) $|v| \in \left[\sqrt{\varepsilon} |u|, \dfrac{|u|}{\sqrt{\varepsilon}} \right]$.

Proof. Ad a) By Proposition 1b, $u \cdot v \leq |u|^2$ and $|u|^2 - u \cdot v \geq 0$. Hence, $T(u, v)$

$= \dfrac{u \cdot v}{|u|^2 + |v|^2 - u \cdot v} \leq \dfrac{|u|^2}{|v|^2}$. As $T(u, v) \geq \varepsilon$, we conclude that $\varepsilon \leq \dfrac{|u|^2}{|v|^2}$. Analogously,

one may derive that $\varepsilon \leq \dfrac{|v|^2}{|u|^2}$. Hence, as $\varepsilon > 0$, we get: $|v|^2 \in \left[\varepsilon |u|^2, \dfrac{|u|^2}{\varepsilon} \right]$.

Ad b) Follows immediately from Theorem 2a.   □

Clearly, Theorem 2b, which concerns the Tanimoto similarity, specifies a more restrictive condition on candidates for near duplicate vectors than analogous Theorem 1a, which concerns the cosine similarity.

Property 1. Let $\varepsilon \in (0,1]$, u be a non-zero ZPN-vector and D be a set of non-zero ZPN-vectors. Then:

$$\left\{ v \in D \middle| |v| \in \left[\sqrt{\varepsilon} |u|, \frac{|u|}{\sqrt{\varepsilon}} \right] \right\} \subseteq \left\{ v \in D \middle| |v| \in \left[\varepsilon |u|, \frac{|u|}{\varepsilon} \right] \right\}.$$

One may easily note that if u is a ZPN-vector whose each dimension has domain $\{0, 1, -1\}$, then $|u|^2 = |NZD(u)|$. This observation allows us to obtain the following corollary from Theorem 1 and Theorem 2, respectively, which refers to the number of non-zero dimensions of similar vectors.

Corollary 1. Let u and v be non-zero ZPN-vectors whose each dimension has domain $\{0, 1, -1\}$ and $\varepsilon \in (0,1]$. Then:

a) If $cosSim(u, v) \geq \varepsilon$, then $|NZD(v)| \in \left[\varepsilon^2 |NZD(u)|, \dfrac{|NZD(u)|}{\varepsilon^2} \right]$.

b) If $T(u, v) \geq \varepsilon$, then $|NZD(v)| \in \left[\varepsilon |NZD(u)|, \dfrac{|NZD(u)|}{\varepsilon} \right]$.

4   Employing Non-zero Dimensions for Determining Near Duplicate *ZPN*-Vectors

In this section, we offer and prove theorems related to using non-zero dimensions for searching near duplicate *ZPN*-vectors. Let us start with Theorem 3 concerning cosine (dis)similar vectors.

Theorem 3. Let u and v be non-zero *ZPN*-vectors, $J \subseteq NZD(u)$, $J \subseteq ZD(v)$ and $\varepsilon \in (0,1]$. Then:

a)   If $cosSim(u, v) \geq \varepsilon$, then $\sum_{i \in J} u_i^2 \leq (1 - \varepsilon^2)| u |^2$.

b)   If $\sum_{i \in J} u_i^2 > (1 - \varepsilon^2)| u |^2$, then $cosSim(u,v) < \varepsilon$.

c) $(1 - \varepsilon^2) \in [0,1)$.

Proof. Ad a) Let $cosSim(u, v) \geq \varepsilon$. Since $J \subseteq ZD(v)$, then $\sum_{i \in J} u_i v_i = 0$. Therefore,

$$cosSim(u, v) \quad = \quad \frac{u \cdot v}{|u \| v|} = \frac{\sum_{i \in \{1,..,n\} \setminus J} u_i v_i}{|u \| v|}. \quad \text{We} \quad \text{also} \quad \text{note}$$

$\sum_{i \in \{1,..,n\} \setminus J} u_i v_i \leq \sum_{i \in \{1,..,n\} \setminus J} u_i^2$ (by Proposition 1a) and $| v | \geq \varepsilon| u |$ (by Theorem 1a).   Thus,

$$cosSim(u,v) \leq \frac{\sum_{i \in \{1,..,n\} \setminus J} u_i^2}{\varepsilon| u |^2} = \frac{\sum_{i \in \{1,..,n\}} u_i^2 - \sum_{i \in J} u_i^2}{\varepsilon| u |^2} = \frac{| u |^2 - \sum_{i \in J} u_i^2}{\varepsilon| u |^2}. \quad \text{So,}$$

$0 < \varepsilon \leq cosSim(u, v) \leq \dfrac{| u |^2 - \sum_{i \in J} u_i^2}{\varepsilon| u |^2}$. Hence, $\sum_{i \in J} u_i^2 \leq (1 - \varepsilon^2)| u |^2$.

Ad b) Follows from Theorem 3a.
Ad c) Trivial.   □

In the beneath example, we illustrate the usefulness of Theorem 3b for reducing the number of vectors that should be considered as candidates for near duplicates.

Example 3. Let us consider vector $u = v1$ from Table 2. $NZD(u) = \{1, 2, 5, 6, 7, 8\}$ (see Table 2). Let $J = \{8\}$ (so, $J \subseteq NZD(u)$) and $\varepsilon = 0.85$. We note that $\sum_{i \in J} u_i^2 = u_8^2 = 6.0^2$ and $(1 - \varepsilon^2)| u |^2 = (1 - 0.85^2) \times 10.20^2 = 28.8711$. Hence, $\sum_{i \in J} u_i^2 > (1 - \varepsilon^2)| u |^2$. Thus, by Theorem 3b, all vectors in Table 2 for which all dimensions in J are zero dimensions (here: J contains only dimension 8) are guaranteed not to be sufficiently similar to vector u for $\varepsilon = 0.85$. Hence, only the remaining vectors; that is, $\bigcup\{I(i) | i \in J\} = I(8) = \{v1, v2, v4\}$ (please, see Table 3 for $I(8)$), have a chance to be sought near duplicates of vector u.

Please observe that for the same vector u and the same similarity threshold ε, we found a different set of candidates for cosine duplicates of u in Example 2, where we applied bounds on lengths (according to Theorem 1a). There we found the following vectors: $v1$, $v6$, $v10$ as candidates for near duplicates of u. The application of both non-zero dimensions and the bounds on lengths of near duplicate ZPN-vectors allows us to restrict the number of candidates even further to $\{v1, v2, v4\} \cap \{v1, v6, v10\} = \{v1\}$. □

In Theorem 4, we formulate properties of Tanimoto (dis)similar ZPN-vectors.

Theorem 4. Let u and v be non-zero ZPN-vectors, $J \subseteq NZD(u)$, $J \subseteq ZD(v)$ and $\varepsilon \in (0,1]$. Then:

a) If $T(u,v) \geq \varepsilon$, then $\sum_{i \in J} u_i^2 \leq (1-\varepsilon)|u|^2$.

b) If $\sum_{i \in J} u_i^2 > (1-\varepsilon)|u|^2$, then $T(u,v) < \varepsilon$.

c) $(1-\varepsilon) \in [0,1)$.

Proof. Ad a) Let $T(u, v) \geq \varepsilon$. Since $J \subseteq ZD(v)$, then $\sum_{i \in J} u_i v_i = 0$. Thus, $T(u, v) =$

$$\frac{u \cdot v}{|u|^2 + |v|^2 - u \cdot v} = \frac{\sum_{i \in \{1,\dots,n\} \setminus J} u_i v_i}{|u|^2 + |v|^2 - u \cdot v}.$$ We also note $\sum_{i \in \{1,\dots,n\} \setminus J} u_i v_i \leq$

$\sum_{i \in \{1,\dots,n\} \setminus J} u_i^2$ (by Proposition 1a) and $|v|^2 - u \cdot v \geq 0$ (by Proposition 1b). Hence,

$$T(u, v) \leq \frac{\sum_{i \in \{1,\dots,n\} \setminus J} u_i^2}{|u|^2} = \frac{\sum_{i \in \{1,\dots,n\}} u_i^2 - \sum_{i \in J} u_i^2}{|u|^2} = \frac{|u|^2 - \sum_{i \in J} u_i^2}{|u|^2}$$ So,

$$0 < \varepsilon \leq T(u, v) \leq \frac{|u|^2 - \sum_{i \in J} u_i^2}{|u|^2}.$$ Therefore, $\sum_{i \in J} u_i^2 \leq (1-\varepsilon)|u|^2$.

Ad b) Follows from Theorem 4a.
Ad c) Trivial. □

Please note that the threshold $(1-\varepsilon^2)|u|^2$ obtained in Theorem 3 for the cosine similarity is greater than or equal to the threshold $(1-\varepsilon)|u|^2$ obtained in Theorem 4 for the Tanimoto similarity for $\varepsilon \in (0,1]$ and any vector u.

Property 2. Let $\varepsilon \in (0,1]$ and u be a vector. Then:

$$(1-\varepsilon)|u|^2 \leq (1-\varepsilon)(1+\varepsilon)|u|^2 = (1-\varepsilon^2)|u|^2 .$$

Taking into account that $|u|^2 = |NZD(u)|$ and $\sum_{i \in J} u_i^2 = |J|$ if u is a ZPN-vector whose each dimension has domain $\{0, 1, -1\}$ and $J \subseteq NZD(u)$, we obtain Corollary 2 from Theorem 3b and Theorem 4b, respectively.

Corollary 2. Let u and v be non-zero ZPN-vectors whose each dimension has domain $\{0, 1, -1\}$, $J \subseteq NZD(u)$, $J \subseteq ZD(v)$ and $\varepsilon \in (0,1]$. Then:

a)   If $|J| > (1 - \varepsilon^2)|NZD(u)|$, then $cosSim(u, v) < \varepsilon$.

b)   If $|J| > (1 - \varepsilon)|NZD(u)|$, then $T(u, v) < \varepsilon$.

5   Searching Exact Duplicate ZPN-Vectors

In this section, we examine properties of vectors maximally similar with regard to the cosine and Tanimoto measures; that is, for the similarity threshold $\varepsilon = 1$.

Theorem 5. Let u and v be non-zero ZPN-vectors. Then:

$$(cosSim(u, v) = 1) \Leftrightarrow (u = v).$$

Proof. ($\Rightarrow$) Let $cosSim(u, v) = 1$. Then, $\angle(u, v) = 0°$; that is, $u = kv$ for some $k > 0$. Hence, $\forall_{i=1,...,n} u_i = kv_i$ for some $k > 0$. In the case of ZPN-vectors, this means that for each domain i, u_i and v_i have identical value, which is either positive or negative or equal to 0. Thus, $cosSim(u, v) = 1$ implies that $\forall_{i=1,...,n} u_i = v_i$; that is, $u = v$.

($\Rightarrow$) Trivial. $\square$

Lemma 2. Let u and v be non-zero vectors and $h = \dfrac{|u|}{|v|} + \dfrac{|v|}{|u|}$. Then:

a) $h \geq 2$.

b) $T(u, v) = \dfrac{cosSim(u, v)}{h - cosSim(u, v)}$.

Proof. Ad a) Follows from the fact that $(|u| - |v|)^2 \geq 0$.

Ad   b)   Follows   from   the   definition   of $T(u, v)$ and   the   fact   that $u \cdot v = |u| \| v | cosSim(u, v)$. $\square$

Theorem 6. Let u and v be non-zero vectors. Then:

$$(T(u, v) = 1) \Leftrightarrow (u = v).$$

Proof. By Lemma 2, $T(u,v) = \dfrac{cosSim(u, v)}{h - cosSim(u, v)}$, where $h = \dfrac{|u|}{|v|} + \dfrac{|v|}{|u|}$ and $h \geq 2$.

Thus, $(T(u,v) = 1) \Leftrightarrow ((2cosSim(u, v) = h) \wedge (2 \geq 2cosSim(u, v)) \wedge (h \geq 2)) \Leftrightarrow ((2 = 2cosSim(u, v) \wedge (h = 2)) \Leftrightarrow ((cosSim(u, v) = 1) \wedge (|u| = |v|) \Leftrightarrow (u = v).$ $\square$

In conclusion, cosine similar ZPN-vectors as well as Tanimoto similar ZPN-vectors are exact duplicates if and only if they are equal.

6   Related Work

Most research on efficient lossless search of near duplicates in large high-dimensional data sets is carried out under assumption that objects are represented by (weighted) binary sets or (multi-)sets that can be implemented as such vectors. Very efficient methods to search near duplicates in such data were offered and verified experimentally in [1, 2, 12]. In [12], the authors offered methods for using non-zero dimensions in searching near duplicate (weighted) binary vectors that are equivalent to Theorem 3b and Theorem 4b. In [1, 2, 7], the authors offered bounds on lengths of near duplicate (weighted) binary vectors that are the same as the bounds specified in Theorem 1 and/or Theorem 2. Our contribution can be regarded as a generalization of those results in that we derived equally strong mechanisms for using non-zero dimensions and lengths of vectors for finding near duplicates for ZPN-vectors, which are more general than weighted binary vectors.

It should be mentioned, however, that [2] offers also a method for discovering near duplicates among positive real valued vectors. Namely, it was shown in [2] that if $cosSim(u, v) \geq \varepsilon > 0$ for positive real-valued non-zero vectors u and v, then $|NZD(v)| \geq \dfrac{\varepsilon |u|}{\max\{u_i \mid i = 1,...,n\}}$. On the other hand, we derived in [8] that if $T(u, v) \geq \varepsilon > 0$ for real-valued non-zero vectors u and v, then $|v| \in \left[\dfrac{1}{\alpha}|u|, \alpha |u| \right]$, where

$\alpha = \dfrac{1}{2}\left(\left(1 + \dfrac{1}{\varepsilon}\right) + \sqrt{\left(1 + \dfrac{1}{\varepsilon}\right)^2 - 4} \right)$. The bounds on lengths of near duplicate ZPN-

vectors specified in Theorem 2b are more precise than those obtained for real-valued vectors in [8] (see Proposition 2).

Proposition 2. Let u be a non-zero vector and $\varepsilon \in (0, 1]$. Then:

$$\left[\sqrt{\varepsilon}|u|, \dfrac{|u|}{\sqrt{\varepsilon}} \right] \subseteq \left[\dfrac{1}{\alpha}|u|, \alpha |u| \right], \text{ where } \alpha = \dfrac{1}{2}\left(\left(1 + \dfrac{1}{\varepsilon}\right) + \sqrt{\left(1 + \dfrac{1}{\varepsilon}\right)^2 - 4} \right).$$

Proof. $\left(\sqrt{\varepsilon} - 1\right)^2 \geq 0$. Hence, $\dfrac{1}{\sqrt{\varepsilon}} \leq \dfrac{\varepsilon + 1}{2\varepsilon} = \dfrac{1}{2}\left(1 + \dfrac{1}{\varepsilon}\right) \leq \alpha$.   □

Table 5. Coefficients determining up to how many times the lengths of similar vectors can be longer/shorter than the length of a given vector

ε	angle [°] $arccos(\varepsilon)$	cosSim (Theorem 1a; ZPN-vectors) $\dfrac{1}{\varepsilon}$	Tanimoto (Theorem 2b; ZPN-vectors) $\dfrac{1}{\sqrt{\varepsilon}}$	Tanimoto ([8]; real-valued vectors) α
1.000	0.0°	1.00	1.00	1.00
0.999	2.6°	1.00	1.00	1.03
0.99	8.1°	1.01	1.01	1.11
0.97	14.1°	1.03	1.02	1.19
0.95	18.2°	1.05	1.03	1.26
0.90	25.8°	1.11	1.05	1.39
0.85	31.8°	1.18	1.08	1.52
0.80	36.9°	1.25	1.12	1.64

Table 5 provides a comparison of the bounds on lengths of near cosine and Tanimoto duplicates in the case of *ZPN*-vectors and the bounds on lengths of near Tanimoto duplicates in the case of real-valued vectors for different values of the similarity threshold ε. As follows from Table 5, near duplicate *ZPN*-vectors have very similar lengths.

7   Summary

In this paper, we have offered and proved how to use knowledge about: 1) non-zero dimensions of *ZPN*-vectors and 2) their lengths for finding cosine and Tanimoto similar *ZPN*-vectors with respect to a given similarity threshold. By means of an example, we have shown that these two types of the derived filtering conditions on candidates for near duplicate *ZPN*-vectors are complementary and can be used together to obtain a better reduction ratio. Our results are not worse than the corresponding ones proposed for objects that can be represented by (weighted) binary vectors, which are a particular case of *ZPN*-vectors. We have also shown that exact cosine duplicates of a *ZPN*-vector u are all and only vectors which are equal to u. In addition, we have shown that exact Tanimoto duplicates of a non-zero vector u are all and only vectors which are equal to u even in the case of real-valued vectors. In the future, we plan to continue our work on efficient search of near duplicates for other similarity measures (see e.g. [5] for a survey of similarity measures) as well.

References

1. Arasu, A., Ganti, V., Kaushik, R.: Efficient exact set-similarity joins. In: Proc. of VLDB 2006. ACM (2006)
2. Bayardo, R.J., Ma, Y., Srikant, R.: Scaling up all pairs similarity search. In: Proc. of WWW 2007, pp. 131–140. ACM (2007)
3. Broder, A.Z., Glassman, S.C., Manasse, M.S., Zweig, G.: Syntactic Clustering of the Web. Computer Networks 29(8-13) 1157–1166 (1997)
4. Chaudhuri, S., Ganti, V., Kaushik, R.L.: A primitive operator for similarity joins in data cleaning. In: Proceedings of ICDE 2006. IEEE Computer Society (2006)
5. De Baets, B., De Meyer, H., Naessens, H.: A class of rational cardinality-based similarity measures. J. Comput. Appl. Math. 132, 51–69 (2001)
6. Gionis, A., Indyk, P., Motwani, R.: Similarity Search in High Dimensions via hashing. In: Proc. of VLDB 1999, pp. 518–529 (1999)
7. Kryszkiewicz, M.: Efficient Determination of Binary Non-Negative Vector Neighbors with Regard to Cosine Similarity. In: Jiang, H., Ding, W., Ali, M., Wu, X. (eds.) IEA/AIE 2012. LNCS (LNAI), vol. 7345, pp. 48–57. Springer, Heidelberg (2012)
8. Kryszkiewicz, M.: Bounds on Lengths of Real Valued Vectors Similar with Regard to the Tanimoto Similarity. In: Selamat, A., Nguyen, N.T., Haron, H. (eds.) ACIIDS 2013, Part I. LNCS, vol. 7802, pp. 445–454. Springer, Heidelberg (2013)
9. Salton, G., Wong, A., Yang, C.S.: A vector space model for automatic indexing. Communications of the ACM 18(11), 613–620 (1975)
10. Willett, P., Barnard, J.M., Downs, G.M.: Chemical similarity searching. J. Chem. Inf. Comput. Sci. 38(6), 983–996 (1998)
11. Witten, I.H., Moffat, A., Bell, T.C.: Managing Gigabytes: Compressing and Indexing Documents and Images. Morgan Kaufmann (1999)
12. Xiao, C., Wang, W., Lin, X., Yu, J.X.: Efficient similarity joins for near duplicate detection. In: Proc. of WWW Conference, pp. 131–140 (2008)

L2RLab: Integrated Experimenter Environment for Learning to Rank

Óscar J. Alejo[1], Juan M. Fernández-Luna[2],
Juan F. Huete[2], and Eleazar Moreno-Cerrud[1]

[1] Informatic Faculty, University of Cienfuegos,
Cienfuegos, Cuba
{alejo,ii200829}@ucf.edu.cu
[2] Departamento de Ciencias de la Computación e I.A.,
E.T.S.I. Informática y de Telecomunicación. CITIC-UGR,
Universidad de Granada, Granada, Spain
{jmfluna,jhg}@decsai.ugr.es

Abstract. L2RLab is a development environment that lets us to integrate all the stages to develop, evaluate, compare and analyze the performance of new learning-to-rank models. It contains tools for individual and multiple pre-processed of the data collections, it also lets us to study the influence of the features in the ranking, the format conversion (e.g., Weka's .ARFF) and visualization. This software facilitates the comparison between two or more methods taking as parameters the performance achieved in the ranking, also includes functionalities for the statistical analysis on the query-level precision of the algorithm proposed regarding to those referenced in the literature. The study of the learning curves' behavior of the different methods is another feature of the tool. L2RLab is programmed in java and is designed as a tool oriented to the extensibility, therefore, the addition of new functionalities is an easy task. L2RLab has an easy-to-use interface that avoids the reprogramming of the applications for our experiments. Basically, L2RLab is structured by two main modules: the visual application and a framework that facilitates the inclusion of the new algorithms and the performance measures developed by the researcher.

Keywords: Information Retrieval, Learning to Rank, Experimentation Tool, Data analysis tool.

1 Introduction

The interest for the Learning to Rank (L2R) task has not been in the air for that long, it began about 15 years ago. Specifically, since 2005, a large number of studies have been conducted on L2R [1–3] and its application to Information Retrieval (IR) [4].

Nowadays, the majority of the research works seek the development of new L2R models that require mandatorily a high level of experimentation, for their conception and refinement.

H.L. Larsen et al. (Eds.): FQAS 2013, LNAI 8132, pp. 543–554, 2013.

Normally, all the experiments are carried out using different data collections and evaluation measures. Besides, fixing the ideal combination of parameters of a model for a data collection is a complex task.

On the other hand, to prove that a specific proposal is superior to others (speaking of performance, computational time, etc.) it is necessary to compare them in similar situations, (e.g., hardware requirement, executions, data collections, evaluation measures, others), which is not always that easy. This is mainly because the algorithm varies in structure, programming language, etc.; and in most of the cases the resources implemented by the other authors are inaccessible.

In the latest years the scientific community has created different tools for the machine learning task that can be adapted to assist the researchers within this area of the Artificial Intelligence.

Weka [5] is an example of it. Weka is a vast collection of machine learning algorithms developed by the Waikato University (New Zealand) implemented in java and using the GPL[1] license. Although Weka has the tools needed to carry out transformations on the data, classification tasks, regression, clustering, association and visualization, it does not contemplate an appropriate structure for the L2R task. In general, results more feasible and less expensive design a new tool, than to modify the structure of Weka to include functionalities that facilitate the L2R.

On the other hand, there are professional applications as MatLab[2]. The MatLab high-performance language for technical computing integrates computation, visualization, and programming in an easy-to-use environment.

It could be mentioned the Java Data Mining (JDM) too. JDM is a standard Java API for developing data mining applications and tools. JDM defines an object model and Java API for data mining objects and processes. JDM enables applications to integrate data mining technology for developing predictive analytic applications and tools.

However, none of these two alternatives are developed specifically for the L2R, that's why if the user decides to use them; he must spend a long time in its adaptation and usage. Even more, some of them are not free according to GNU license, which is the case for example of MatLab.

For the current topic, we have LETOR and RankLib, as more specific proposals.

In LETOR[3] official web site (**LE**arning **TO R**ank for Information Retrieval by Microsoft Research), we can find a complete schema of experimentation to facilitate the L2R task. There are also data collections published, the results of some of the main L2R methods and a few evaluation tools programmed in Perl.

---

[1] GNU Public License. http://www.gnu.org/copyleft/gpl.html

[2] Available in: http://www.mathworks.com

[3] Available in: http://research.microsoft.com/en-us/um/beijing/projects/letor/

RankLib[4] is a library of learning to rank algorithms. Currently eight popular algorithms have been implemented in its last version. It also implements many retrieval metrics as well as provides many ways to carry out evaluation. However, such elements in both cases are not enough to carry out pre-processing tasks, data collections analysis, comparison, statistical analysis and to study deeply the behavior of new L2R models.

To solve the above mentioned necessity, in this work we propose the Learning-to-Rank Laboratory (L2RLab), which is a new tool that has as main purpose to assist and facilitate the researcher labor in the experimentation process with L2R models and data collections.

L2RLab is programmed in java and is designed as a tool oriented to the extensibility, therefore, the addition of new functionalities is an easy task.

Basically, L2RLab is structured by two main modules: the visual application and a framework that facilitates the inclusion of the new algorithms and the performance measures developed by the researcher.

The system interfaces allow us to set and control the experiments execution. The results of these executions are visualized both numerically and graphically.

Also, L2RLab allows us to carry out transformations on the data collections, to evaluate and compare the performance of two or more L2R algorithms, to study the learning curves' behaviors, and to carry out statistical analysis, among others tasks.

With the goal of explaining properly the usage of this tool, the rest of the work is organized as following.

In section 2 the technical aspects of L2RLab are described, meanwhile in section 3 the principal functionalities are explained. In section 4, a case of study is analyzed. Finally, in Section 5 we conclude this work and point out some directions for future research.

2   Technical Aspects of L2RLab

L2RLab was programmed under the Java technology, which is a high level language developed by Sun Microsystems[5]. So this proposed tool is architecture independent and works in any platform on which exists a Java virtual machine.

As mentioned in the introduction, L2RLab is composed by two main components: a framework for the creation and inclusion of L2R algorithm and performance measures, and a visual application to control the experiments execution and to visualize the obtained results. That is why the user just has to worry about the programming of his proposals. In the following subsections the details of these two components will be explained.

2.1   Framework

An Object Oriented Framework (OOF) is the reusable design of a system that describes how should itself break down in a group of objects that interact which

---

[4] Available in: `https://sourceforge.net/p/lemur/code/2535/tree/RankLib/tags/release-2.1/bin/RankLib.jar`

[5] Available in: `http://www.sun.com/`

each other. Different to a simple software architecture, an OOF is expressed in a programming language and it is based on the domain of a specific problem [6].

Basically, a OOF is composed by two types of main elements: hot points and frozen points. The hot points represent expandable source through abstract classes and interfaces, while the frozen points are functionalities that cannot be changed for the final user and that define the problem domain logic (in this case, experimentation with L2R models). So, with the idea of supplying the researcher with the needed facilities in the proposal and problems programming, it was implemented an OOF, and its classes are shown in Figure 1.

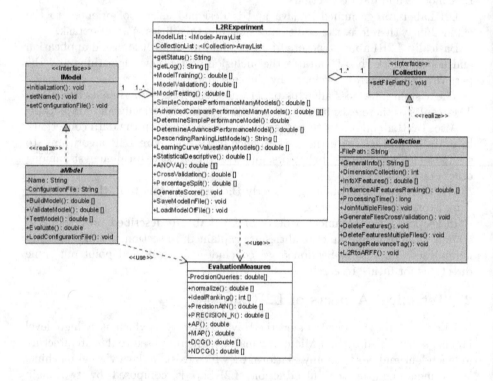

Fig. 1. Framework classes diagram included in L2RLab - Some methods, attributes and parameters are excluded for a better comprehension -.

The classes with a more intense tonality represent the hot points (e.g. *aModel*, *aCollection*, etc.), while the lighter ones are the frozen points. In this diagram other methods and attributes are excluded for a better comprehension.

Note that an experiment, defined by the *L2RExperiment* class, consists of one or many L2R models and one or many data collections; where this class is the one in charge of carrying out and controlling the many L2R tasks.

There is also a class named *EvaluationMeasures*, that gathers a collection of evaluation measures, in charge of quantify the performance of the learning models concerning to the data collections used.

The inclusion of the interfaces gives more freedom to the user in his proposals when programming, while the abstract classes reduces the implementation time due to fact that they contain some functionalities. Then the user can extend the framework taking as initial point the interfaces or the abstract classes already implemented. The *L2RExperiment* class is in charge of finally carrying out the L2R task.

2.2   Visual Application

The visual application developed is intuitive and simple. Through 3 modules visually connected from the same main window, it is possible to access the different system functionalities.

The system modules are: (1) Collection Data Pre-processing, (2) Models Training and Testing, and (3) Model analysis and comparison.

The main class of the application communicates with the users code through the framework, specifically with the *L2RExperiment* class. Each L2R model is represented by a class implemented by the user, which reference to a configuration file loaded (see *LoadConfigurationFile* method) when the application starts.

Besides the parameters definitions of the learning algorithm, in this file it is included the name of the class with a brief description of it. Each class defines by the user gets through its constructor the values of the parameters for the algorithm, like others specific for the executions. One of the advantages that the inclusion of configuration files as the ones used gives is that each class could be associated to several different files, with the consequent derivation of several instances or scenarios for each learning algorithm.

3   Main Functionalities

In the following subsections the main functionalities of the L2RLab are explained, particularly the visual part.

3.1   Datasets Pre-processing

The datasets pre-processing module allows carrying out several operations upon one or many datasets and data collections.

The format L2RLab supports for the collections analysis and processing is the standard L2R format, where each row is a query-document pair. The first column contains the relevance label of this pair, the second column contains the query id, the following columns contains the features, and the end of the row is comment about the pair, including id of the document. The larger the relevance label, the more relevant the query-document pair. Here it is one example row from the MQ2007 dataset:

---

2 qid:10032 1:0.056537 2:0.000000 3:0.666667 4:1.000000 5:0.067138 ... 45:0.000000 46:0.076923
#docid = GX029-35-5894638 inc = 0.0119881192468859 prob = 0.139842

It is important to mention that the values added next to the features (e.g., #docid, inc and prob), do not influence in the correct datasets pre-processing.

Knowing this and once the data collection is selected according to the **File - Open** option, it is shown in the corresponding window the basic information (Figure 2), it means, the collection amount of queries, irrelevant documents and the amount of relevant documents separated according to the relevance category they belong to.

It is also shown; the amount of retrieved documents for each one of the queries; and for each one of the attributes that compose the data, a statistical summary is detailed with: the range of the data, the arithmetic mean and the standard deviation.

It can also be determined, the influence of the features in the ranking; in this action we can check first the time that will take the operation to be completed (according to the collection and our computer hardware characteristics).

If we execute this functionality we will obtain a graphic, where it is reflected how influential each feature in the ranking is, it also facilitates us in a new window each one of the features descendingly ordered according to its strength to achieve a better ranking.

Once the general information of one or more collections is known, we can carry out other individual tasks or group tasks, as example, the union and modification of data files, the clustering of datasets starting from general collections, the format conversion, among others.

L2RLab gives the possibility to the user of specifying names to the resulting files of each operation, in case the user does not want to name the new file, the application uses an agreement of intuitive names for each action. On the other hand, the tool lets us to save in a new file the modifications and eliminations carried out on the collections, keeping unaltered the original data and guaranteeing the reliability of a new process of pre-processing. Another of the common characteristics for all functionalities in each one of the interfaces is to use a progress bar and to notify the state and result of the different operations. The *Log* button lets us to access this information.

The modification of one or more data files can be carried out through different options that facilitate the individual or by group elimination, in queries, features, and documents. For the case of the relevance judgments, they can be substituted by new labels.

The tool is also able to generate multiple data files in a controlled way, starting from the information of a main file, where all the queries belong to a certain data collection. Firstly, we have the option of creating a file for each query of the data collection. In a second option it is possible to generate the three classic files for the experimentation (training, validation and testing), when specifying which percentage of data of the main file, will correspond to each one of these new

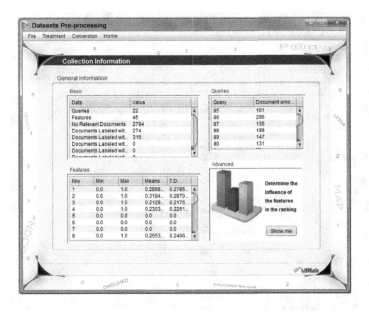

Fig. 2. Screen of the module #1 that shows specific information about a data collection

files. Finally, the researcher, also has the option to decide which queries of the main file, will be included in the conformation of the training, validation and testing files. In these cases, we should be careful to avoid a data fracture. In general, we recommend to use the experimental outline (5-fold-cross validation) proposed in LETOR for each data collection.

On the other hand, this module offers a simple interface that allows us to select and to convert a dataset from (.L2R) to the well-known format of Weka (.arff). In the transformed file, the features will be considered as attributes, where the missing features will be labeled with the "?" character; and on the other hand, the relevance judgments will be the classes.

3.2 Models Training and Testing

As we can see in the Figure 3 this module allows us to take the control of the training and evaluation of new L2R models.

In a first moment, the L2R algorithm, the training set and the evaluation measure used to build the ranking function is selected.

The tool then, allows us to specify different options to carry out the evaluation of the learned model: (1) to use the whole training set, (2) to select a new dataset, (3) to carry out n-fold cross validation, where the n value will be fixed by the researcher, and finally, (4) to specify what percent of the training set will be used for the evaluation.

Finished the phases of training and testing, it is visualized in the results area, the learned model, the computational time used to build the model and

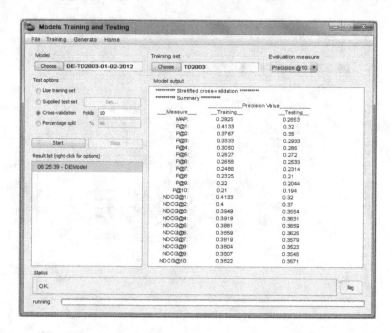

Fig. 3. Screen of the module #2 that allows us to carry out trainings to evaluate the behavior of the L2R models

a summary of the performances achieved by the algorithm for the evaluation measures used in the Learning to Rank for Information Retrieval, such as, Mean Average Precision (MAP) [7], Normalized Discounted Cumulative Gain (NDCG) [8] and Precision at n (P@n) [7]. For these last two measures the precision is shown in the first 10 positions of the ranking. None of the tests carried out to the model, are not eliminated, but rather are registered in the results list - located in the left inferior part -. This option has the advantage of allowing the consultation of the previous results, while other tests are executed. These results also, are stored in independent text files inside the *Results* folder of the main directory of L2RLab. The inferior panel of this screen presents information of the state of the current execution of the algorithm and by clicking the *Log* button we can access the system history of events for this module.

Another of the functionalities of this module is that a scores file can be created from a L2R model and a dataset. These scores represent how good the ranking function considers a certain recovered document with regard to the corresponding question. These scores files can be used to determine the performance of the method evaluated at the query level.

3.3 Model Analysis and Comparison

The third module of L2RLab seeks to analyze and to compare the performances of the different L2R methods, which have to be interesting to the researcher.

We can configure our own comparison scheme to determine the performance achieved by each learning method as for precision in the ranking. We choose then, the methods that we will confront, the evaluation measure and the corresponding dataset. The final results are visualized graphically, where the performances obtained by each method at the dataset level and for each specific query are shown. The evaluation at the query level is vital to determine how robust and stable it is the model learned by each algorithm.

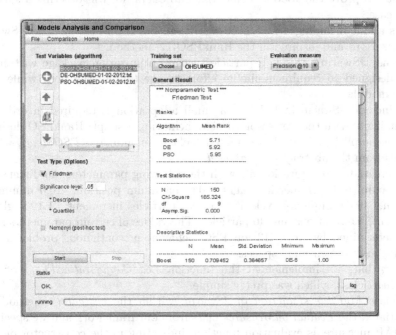

Fig. 4. Screen of the module #3 that lets us to carry out nonparametric tests, considering the query-level performances of the L2R methods

Another of the functional options of this module is lead to facilitate the study of the learning curves behavior of each L2R method. The resulting graph gives us the possibility to observe in n-iterations how the learning of each algorithm behaves, while it constructs its ranking function. This study is essential to understand the operation of a learning method and to be able to know its capacity of adaptation under diverse conditions of training.

On the other hand, L2RLab lets us carry out nonparametric tests. Specifically, in the screen of Figure 4, an entire outline is shown for a variance analysis of second way (two way ANOVA) [9]. For this purpose, the Friedman test can be used [10]. If the null-hypothesis is rejected, we can proceed with a post-hoc test. We recommend Nemenyi test [11], this test is similar to the Tukey test for ANOVA and is used when all algorithms are compared to each other.

This model was applied because is adjusted well to compare different algorithms on the same group of queries; all that which allowed to carry out a comparative study of the performances of each algorithm with a high level of truthfulness and accuracy. Concluded the analysis, a complete report of the ranks, test statistics, and descriptive statistics is shown, besides the pairwise comparisons for each L2R algorithm.

4 Study Case

In order to prove the goodness that L2RLab offers, we will describe a specific study case.

It is about including a new L2R algorithm, based on PSO (Particle Swarm Optimization) and inspired in the RankPSO[12] method.

We use JSwarm[6], this is a particle swarm optimization package written in Java; designed to require minimum effort to use (out of the box) while also highly modular.

To include JSwarm in L2RLab a small adaptation is required in order to communicate with the framework design, achieving a simple RankPSO design.

To accomplish this, we have created a class denominated *PSOModel*, which extends from the abstract class *aModel*.

We created a configuration file with the following parameters: 45 dimension, true maximize, 0.721 inertia weight, 30.0 maximum position, -10.0 minimum position, 40.0 minimum velocity, 1.193 particle increment, 1.193 global increment, 50 iterations and 40 particles. The values of certain parameters have been determined by work [13] on PSO behavior in continuous problems and complex multidimensional spaces.

After that, we implemented the function *LoadConfigurationFile()* of our *PSOModel* class, which was pretty simple.

In class *MyFitnessFunction* of JSwarm, within the *evaluate()* method we instantiated the *EvaluationMeasures* class from our framework and we specified the MAP measure as evaluation function, indicating in the constructor of this class that this function should be maximized.

Then, in the *Evaluate()* function of the *PSOModel*, we set as result of this function the output of the *evaluate()* method that is in class *MyFitnessFunction*.

In a different site, in the main method *BuildModel()* of our model, we created an instance of the *Swarm* class from the used package, we passed it the parameters read from the configuration file and we generically defined the parameter training set and then invoked the *evolve()* method. For the *TestModel()* method case, we just specified as parameter the testing set and the evaluation measure; this function uses directly the *EvaluationMeasures* class.

Once this link between classes is finished, we used the interface shown in figure 3, where we specified each one of the options needed to carry out the experiment (training and testing).

It is important to highlight that when we select the training set, a *OHSUMEDCollection* class is created that extends from the abstract class

---

[6] Available in:
http://en.sourceforge.jp/frs/g_redir.php?m=jaist&f=%2Fjswarm-pso%2Fjswarm-pso%2Fjswarm_pso_1_2%2Fjswarm-pso.jar

aCollection, an updating of the *MyParticle* class from JSwarm is also carried out, by specifying in the constructor that a 45-dimensional particle (due to the amount of features extracted from OHSUMED correlates with this value) must be created.

Following the experimental settings of LETOR, we carried out a 5-fold-cross-validation and the final results were the next ones: MAP 0.4525, P@1 0.6718, P@2 0.6192, P@3 0.5926, P@4 0.5929, P@5 0.5786, P@6 0.5577, P@7 0.5470, P@8 0.5330, P@9 0.5211, P@10 0.5058, NDCG@1 0.5584, NDCG@2 0.4940, NDCG@3 0.4824, NDCG@4 0.4818, NDCG@5 0.4784, NDCG@6 0.4707, NDCG@7 0.4675, NDCG@8 0.4607, NDCG@9 0.4586 and NDCG@10 0.4539.

This experimental schema let us carry out direct comparisons with the referenced L2R methods in the literature, which performance have been published in LETOR. Finally, we conclude that for all the considered evaluation measure, the precision values achieved by this method [12] are highly competitive (the best and second best result), both for the 10 first positions of the ranking and general average.

Our future works will be intended to include new algorithms and evaluation measures, with the objective of obtaining a new framework with the main exponents of the state-of-the-art L2R. Promoting the comparison and statistical analysis between the new proposals, as well as the saving of time in the coding and result visualization. The future availability of this tool will be of free access.

5   Conclusions and Future Work

In this paper, we have proposed an integrated experimenter environment for learning to rank called L2RLab. This tool lets us to integrate all the stages to develop, evaluate, compare and analyze the performance of the new learning-to-rank models.

The main advantages are the following: gives a practical tool to assist the experimentation process with L2R models; the code reuse and implemented resources are another possibility; greater speed (time saving) and dependability (errors reduction) in the experimentation process; the researcher will be able to concentrate on the learning-to-rank task, without having to worry about technical aspects in the experiments design. Also, as the technologies used to develop the L2RLab are free, therefore it is not necessary to pay licenses.

On the other hand, we described a case of study with the goal of illustrating one typical workflow in L2RLab, when adding and evaluating new algorithms.

As future work, we plan to address the following issues:

- To incorporate to L2RLab the main state-of-the-art algorithms in the learning-to-rank field.
- To add to L2RLab the possibility to carry out new statistical tests that facilitate the analysis and the comparison between L2R models.
- To provide a new framework for data analysis.
- To improve the system's help, incorporating a detailed description of the procedures in each module and for each L2R method.

Acknowledgment. This paper has been supported by the Spanish Consejería de Innovación, Ciencia y Empresa de la Junta de Andalucía and the Ministerio de Ciencia e Innovación under the projects P09-TIC-4526 and TIN2011-28538-C02-02, respectively.

References

1. Joachims, T.: A support vector method for multivariate performance measures. In: ICML 22, pp. 377–384 (2005)
2. Valizadegan, H., Jin, R., Zhang, R., Mao, J.: Learning to Rank by Optimizing NDCG Measure. In: Proceedings of NIPS 23, pp. 1883–1891 (2009)
3. Wu, J., Yang, Z., Lin, Y., Lin, H., Ye, Z., Xu, K.: Learning to Rank Using Query-Level Regression. In: SIGIR 2011: Proceedings of the 34th Annual International ACM SIGIR Conference on Research and Development in Information Retrieval, Beijing, China, pp. 1091–1092 (2011)
4. Xu, J., Li, H.: Adarank: a boosting algorithm for information retrieval. In: SIGIR 2007: Proceedings of the 30th Annual International ACM SIGIR Conference on Research and Development in Information Retrieval, Amsterdam, Netherlands, pp. 391–398 (2007)
5. Witten, I.H., Frank, E.: Data Mining: Practical Machine Learning Tools and Techniques, 2nd edn., p. 560. Morgan Kaufmann Publishers, San Francisco (2005)
6. Fayad, M.E., Schmidt, D.C., Johnson, R.E.: Building application frameworks: Object oriented foundations of framework design. Application Frameworks, ch., 1st edn., p. 638 (1999)
7. Liu, T.-Y., Xu, J., Qin, T., Xiong, W., Li, H.: Letor: Benchmark dataset for research on learning to rank for information retrieval. In: Proceedings of SIGIR 2007 Workshop on Learning to Rank for Information Retrieval (2007)
8. Järvelin, K., Kekäläinen, J.: Cumulated gain-based evaluation of ir techniques. ACM Transactions on Information Systems 20(4), 422–446 (2002)
9. Cárdenas, J.: Análisis Univariado. Capítulo # 6.- Análisis de varianza multifactorial. In: C.-A.d.v.m.D.J.R.C.P.F.d.E (eds.) UCLV, Facultad de Economía, UCLV (2003)
10. Friedman, M.: A comparison of alternative tests of significance for the problem of m rankings. Annals of Mathematical Statistics 11, 86–92 (1940)
11. Nemenyi, P.B.: Distribution-free multiple comparisons. PhD thesis, Princeton University (1963)
12. Alejo, O.J., Fernández, J.M., Huete, J.F.: RankPSO: a new L2R algorithm based on Particle Swarm Optimization. Waitin for being published in Journal of Multiple-Valued Logic and Soft Computing (JMVLSC) (2013)
13. Clerc, M., Kennedy, J.: The particle swarm - explosion, stability, and convergence in a multidimensional complex space. IEEE Transactions Evolutionary Computation 6(1), 58–73 (2002)

Heuristic Classifier Chains
for Multi-label Classification

Tomasz Kajdanowicz and Przemyslaw Kazienko

Wroclaw University of Technology, Wroclaw, Poland
Faculty of Computer Science and Management, Institute of Informatics
{tomasz.kajdanowicz,przemyslaw.kazienko}@pwr.wroc.pl

Abstract. Multi-label classification, in opposite to conventional classification, assumes that each data instance may be associated with more
than one labels simultaneously. Multi-label learning methods take advantage of dependencies between labels, but this implies greater learning
computational complexity.

The paper considers Classifier Chain multi-label classification method,
which in original form is fast, but assumes the order of labels in the chain.
This leads to propagation of inference errors down the chain. On the
other hand recent Bayes-optimal method, Probabilistic Classifier Chain,
overcomes this drawback, but is computationally intractable. In order
to find the trade off solution it is presented a novel heuristic approach
for finding appropriate label order in chain. It is demonstrated that the
method obtains competitive overall accuracy and is also tractable to
higher-dimensional data.

1   Introduction

Assuming that $\mathcal{X} \in \mathbb{R}^d$ is a d-dimensional real input space and $\mathcal{L} = \{\lambda_1, \lambda_2, \ldots, \lambda_l\}$ is a finite set of class labels, where l denotes the number of class labels,
in multi-label classification setting an instance $x \in \mathcal{X}$ can be associated with
a subset of labels $\Lambda \in \mathcal{L}$ that refers to set of relevant labels for x. Given a
training data set of size n drawn identically and independently from an unknown
probability distribution on $\mathcal{X} \times 2^{\mathcal{L}}$, namely:

$$\{(x_1, \Lambda_1), (x_2, \Lambda_2), \ldots, (x_n, \Lambda_n),\} \tag{1}$$

the aim of multi-label classification is to train a classifier $\Psi(x) : \mathcal{X} \to 2^{\mathcal{L}}$ that
provide generalization abilities over these instances. In other words, multi-label
classification is a mapping from an input $x \in \mathcal{X}$ to an output $\Lambda \in 2^{\mathcal{L}}$.

The paper brings closer view on the Classifier Chain (CC) multi-label classification method (Section 2). There are discussed origins and motivations that
influenced the research and creation of the method, its general idea with appropriate algorithms for training and testing as well as complexity assessment.
It is highlighted, that the method may perform inaccurate classification due to
the fact that it assumes particular order of generalization in chain. In order to

H.L. Larsen et al. (Eds.): FQAS 2013, LNAI 8132, pp. 555–566, 2013.
© Springer-Verlag Berlin Heidelberg 2013

deal with that problem it is proposed in the paper the extension to CC that can protect it from inappropriate generalization (Section 3). The proposed Heuristic Classifier Chain Method with Ordered Training utilizes three distinct proposals, formalized as heuristics, in the searching process of label factorization. The results of empirical evaluation of the proposed extension of CC is contained in Section 4 and concluded in Section 5.

2   Backgroud of Classifier Chain Method for Multi-label Classification

Classifier Chain multi-label classification method, which was a result of the original work initially introduced in parallel in [1] and [2], was motivated by a need of new multi-label classification method which is able to model the dependencies that exist between labels in multi-label data and achieve it with limited time complexity. The initial point of the research was a solution in which each label generalization was treated as independent binary classification problem (Binary Relevance). That approach presents reasonable complexity, but does not model label dependencies. The general idea of the new method was to assume capturing label dependencies by decomposition of the multi-label problem into l number of binary classification problems, where l denotes the number of distinct labels in the label-set $\mathcal{L}$.

The origins of classifier chain method can be found in the early work on the idea of stacked generalization [3]. The classification organized in the stacked approach assumes to accomplish two stages in learning and inference. The first one, so called *level-0-model*, learns k base classifiers in k-cross-fold validation manner and infers k classification results for each data instance using models learnt on each fold. Thus, the first stage of learning from population of instances x returns the $(\Phi_1(x_i), \Phi_2(x_i), \ldots, \Phi_k(x_i))$ classifiers result for each instance i. In the second stage - learning of *level-1-model* is performed using original input values x augmented with the results from the first stage, namely a new model is learnt on $\{(x_i, \Phi_1(x_i), \Phi_2(x_i), \ldots, \Phi_k(x_i))\}$, $i = 1, 2, \ldots, n$ data. The output of inference in *level-1-model* is the final classification result.

What is worth mentioning, in [4] it was shown that successful stacked generalization requires using output class probabilities rather than class predictions.

However, stacked generalization by its nature is not able to be applied directly to multi-label classification since it was originally proposed to enhance the accuracy in single-label classification. The direct adaptation of the method to multi-label classification was proposed in [5]. The authors proposed using stacking while learning each of labels in Binary Relevance scheme. The original data set was split into k disjoint parts and in each of them l binary classifiers were learning the relevance of labels independently, one classifier per label (level-0). In this setting each label was generalized by the k classifiers. The level-1 classification model was learning from the original data augmented by k classification results obtained in level-0. It can be concluded that the solution presented in [5] introduced a Binary Relevance Stacking. In other words the idea was based on simple bootstrapping scheme that was generalized by ensemble of possibly diverse binary classifiers. Its output was utilized to extend the input space for final decision making.

Similar concept of input space augmentation was applied in Classifier Chain. However, the information that expands the input space as well as the scheme of classification process are different.

2.1 Description of Classifier Chain Method

The Classifier Chain (CC) method transforms the multi-label classification problem to $|\mathcal{L}|$ binary classification problems, one for each label from $\mathcal{L}$. Therefore, CC method may be treated as binary relevance (BR) model. However it differs from BR on the attribute space used in each binary model. The CC method utilizes extended learning attribute space with the 0/1 outputs indicating label relevance obtained from partially generalized output label space.

Similarly to common supervised learning techniques, the CC method performs two phases: training, that results with learned multi-label chain of classifiers and testing, which infer the output value for previously unseen instances. Training and testing phases of the CC method are presented in Algorithm 1 and Algorithm 2.

The Ψ multi-label Classifier Chain is formed of $|\mathcal{L}|$ base classifiers, namely $\Psi = (\Phi_1, \ldots, \Phi_{|\mathcal{L}|})$. As it can be observed in line 7 of Algorithm 1 each base classifier Φ_i corresponds to binary classification of ith label. Each base classifier Φ_i is trained using the attribute space augmented by all previously accomplished classifications in the chain. Following the Algorithm 1 the input x may be composed of any variable type (binary, numerical or categorical values). However the chained attributes $(y_i, \ldots, y_i)$ are always binary. The graphical notation of the dependencies that are modelled by Classifier Chain method is presented in Figure 1. It can be observed that generalization of λ_i label is obtained from learning on initial attributes of data instance x and all labels $\lambda_1, \ldots, \lambda_{i-1}$.

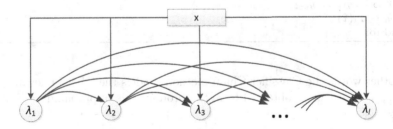

Fig. 1. The learning scheme of Classifier Chain method

In order to obtain the classification results from classifier chain Ψ a straightforward process is performed. Each base classifier Φ_i from the chain $(\Phi_1, \ldots, \Phi_{|\mathcal{L}|})$ provides binary prediction of ith label's relevance. As presented in lines 1 to 4 of Algorithm 2 the classification propagates along the chain using input attribute space augmented by all previously obtained classification results in the chain. Hence in the chain the information of label relevance is passed between following classifiers. It allows CC to model the dependencies in the label space and overcome the problem of inability of label correlation modelling known in binary relevance. Although the additional attributes constitute only a part of attribute space, if there exist a strong correlation, the following base classifiers may be provided with relatively valuable input and result with better classification accuracy.

Algorithm 1. The training phase of Classifier Chain multi-label classification method.

Input: $\mathcal{D}^{tr}$, Φ untrained base classifier
Output: $\Psi = (\Phi_1, \ldots, \Phi_{|\mathcal{L}|})$ trained classifiers chain
  1: **for** $i = 1, \ldots, |\mathcal{L}|$ **do**
  2:    $\hat{D}_i^{tr} \leftarrow \{\}$
  3:    **for all** $(x, y) \in \mathcal{D}^{tr}$ **do**
  4:       $\hat{x} \leftarrow [x, y_1, \ldots, y_{i-1}]$
  5:       $\hat{D}_i^{tr} \leftarrow \hat{D}_i^{tr} \cup (\hat{x}, y_i)$
  6:    **end for**
  7:    train Φ_i to predict y_i
  8: **end for**

Algorithm 2. The testing phase of Classifier Chain multi-label classification method.

Input: $(x, y) \in \mathcal{D}^{ts}$, $\Psi = (\Phi_1, \ldots, \Phi_{|\mathcal{L}|})$ trained classifiers chain
Output: $\hat{y}$ predicted output
  1: **for** $i = 1, \ldots, |\mathcal{L}|$ **do**
  2:    $\hat{x} \leftarrow [x, \hat{y}_1, \ldots, \hat{y}_{i-1}]$
  3:    $\hat{y}_i \leftarrow \Phi_i(\hat{x})$
  4: **end for**

In other words the Classifier Chain method models the labels' relevance $y = (y_1, \ldots, y_l)$ that can be obtained from the product rule of probability:

$$P(y|x) = P(y_1|x) \cdot \prod_{i=2}^{l} P(y_i|x, y_1, \ldots, y_{i-1}) \tag{2}$$

The estimation of joint distribution of labels is possible through realization of l functions $f_i(\cdot)$ on augmented input space $\mathcal{X} \times \{0, 1\}^{i-1}$ with $(y_1, \ldots, y_{i-1})$ additional attributes. If $f_i(\cdot)$ is interpreted as a probabilistic classifier resulting with probability that $y_i = 1$, it can be written:

$$P(y|x) = f_1(x) \cdot \prod_{i=2}^{l} f_i(x, y_1, \ldots, y_{i-1}) \tag{3}$$

According to Equation 3 we can observe that the order of the output variables y_i used in the product part may be important. For instance the label y_2 estimated in the second partial product may not be as good argument for f_3 as y_4 in order to obtain accurate generalization for label y_3. This feature of the method may lead to inaccurate prediction. Nevertheless there are at least three ways how to deal with it. First, the easiest solution to this problem is to use some a priori knowledge on the dependencies between the labels and model the order according to it. This is probably the most obvious method but usually it does not hold and the a priori knowledge on label dependencies is not often available. This can be appointed in modelling of problems described by hierarchical structure of labels. The second idea of modelling dependencies in the method is to reformulate the Classifier Chain enabling it to take all possible dependencies between labels. This can be formulated as presented in the work of [6] in the idea called *Probabilistic Classifier Chain (PCC)*. First difference depicts the output shape of each of classifiers in the chain - whereas CC uses $\{0, 1\}$ indicators of label relevance and is a deterministic approximation the PCC method is able to handle continuous scores. The key change to original CC recalling the work presented in [7] and [6] is in the way the probability of each label combination $y = (y_1, \ldots, y_m)$ is obtained. Please indicate that PCC accomplishes the estimation as in Equation 4.

$$P(y|x) = \prod_{i=1}^{l} f_i(x, y_1, \ldots, y_{i-1}, \ldots, y_{i+1}, \ldots, y_l) \tag{4}$$

We can see that the multi-label problem reformulated as in Equation 4 is much more flexible than in 3. However it is much more complicated and in practical applications might not be possible to obtain.

The problem of learning order in Classifier Chain method might be now seen as a problem of searching a right path in a tree that denotes a proper learning order. Nodes in this tree denote labels $\lambda \in \mathcal{L}$. According to presented Algorithm 1 and 2 CC accomplishes only single path in this tree. This might result in labelling with not necessarily the highest accuracy. However as long as CC searches only for single path of length m in the aforementioned tree, the PCC method requires to check 2^m paths. This works for CC method and makes it much more applicable, even for data sets larger than 15 labels, which was identified as a limit for PCC [6]. Searching a proper path in such a tree may be a third solution to the problem. The proposal for Classifier Chain Method with Ordered Training is provided in following Section 3.

2.2 Complexity of Classifier Chain Method

The computational complexity of Classifier Chain method depends outright on the size of label set $|\mathcal{L}|$. It can be quantified as $O(|\mathcal{L}| \times f(d + |\mathcal{L}|, N))$, where $f(d + |\mathcal{L}|, N)$ denotes the complexity of underlying base classifier performing on data set with d attributes and N instances. Therefore the CC's complexity is comparable to original BR's complexity $O(|\mathcal{L}| \times f(d, N))$ and is incurred by penalty of handling $|\mathcal{L}|$ additional attributes. In practice the difference in running time of CC and BR tends to be small as long as $|\mathcal{L}| < d$, which is normally expected in multi-label classification.

The Classifier Chain method does not manifest ability to be parallelized in the training phase. However CC may be easily serialized in such a way that at each time only a single binary classification is required to be handled in memory. This constitute a clear advantage of the method over other methods based on a single large model.

3 Heuristic Classifier Chain Method with Ordered Training

Needless to say the classifier chain method may provide poor estimation of the true distribution $P(y|x)$. In overall the method performs generalization in sequential manner, step by step for each of the labels. Obtained label's relevance is then utilized as an additional input for next labels classification. Therefore, the order in which labels are generalized may have significant impact on the overall classification accuracy of the whole label set [8, 9].

As a complete overview over the space of all possible labels' orderings is exponentially complex, for instance as it is proposed in Probabilistic Classifier Chain [6], there should be a method providing reasonable ordering with acceptable computational complexity. In order to provide the ordering of the training for classifier chain, three distinct proposals, formalized as heuristics in the searching process, are proposed.

3.1 Description of Proposed Ordering Schemes

As the order in which chain of classifiers is constructed may influence classification accuracy the classifier chain method needs to be enriched with a step allowing selection of label order that constitutes the chain of classifiers. In the Algorithm 3 this is accomplished by calculation of ordering OR in line 1. It is assumed that the decision on the order is performed externally to the training of base classifiers but equally good it can be accomplished dynamically inside the training using momentary measurements such as classification accuracy, hamming loss or other.

Searching for the appropriate order of training steps can be presented as a tree searching problem. In such representation nodes denote particular labels from the label set. Each level of the tree indicates appropriate position in the training order, e.g. the first level of the tree indicates the first position in the chain, whereas each next level corresponds to the following positions in the chain.

The complete overview of the whole tree is highly complex and therefore a less complex method providing approximate, good solution is required. In order to provide such mechanism, at each of non-leading-to-leaf nodes, the decision on

Algorithm 3. The training phase of Classifier Chain with ordering multi-label classification method.

Input: $\mathcal{D}^{tr}$, Φ untrained base classifier, Υ ordering method
Output: $\Psi = (\Phi_1, \ldots, \Phi_{|\mathcal{L}|})$ trained classifiers chain
1: calculate training order $OR = <l_1, \ldots, l_{|\mathcal{L}|}>$ using Υ
2: **for** $i = 1, \ldots, |OR|$ **do**
3:    $\hat{D}_i^{tr} \leftarrow \{\}$
4:    **for all** $(x, y) \in \mathcal{D}^{tr}$ **do**
5:       $\hat{x} \leftarrow [x, y_{l_1)}, \ldots, y_{l_{(i-1)}}]$
6:       $\hat{D}_i^{tr} \leftarrow \hat{D}_i^{tr} \cup (\hat{x}, y_{l_i})$
7:    **end for**
8:    train Φ_i to predict y_{l_i}
9: **end for**

Algorithm 4. The testing phase of Classifier Chain with ordering multi-label classification method.

Input: $(x, y) \in \mathcal{D}^{ts}$, $\Psi = (\Phi_1, \ldots, \Phi_{|\mathcal{L}|})$ trained classifiers chain, $OR = <l_1, \ldots, l_{|\mathcal{L}|}>$ training order
Output: $\hat{y}$ predicted output
1: **for** $i = 1, \ldots, |OR|$ **do**
2:    $\hat{x} \leftarrow [x, \hat{y}_{l_1}, \ldots, \hat{y}_{l_{i-1}}]$
3:    $\hat{y}_i \leftarrow \Phi_i(\hat{x})$
4: **end for**

which branch to explore next should be taken. Three heuristics providing such decisions in the searching process are proposed below.

The first proposed heuristic $H1$ is constructed on the basis of classification error minimization. The heuristic determines which label should be taken in consideration in the next step of training process performed in chain. It is done by assessment of the classification error of base classifier for all remaining labels in output space that has not been chained yet. The label that is classified with the smallest error is taken as the next one in the chain. This approach seems to search for the best single-label classification which in general, after complete chain is formed, may not result in the best training order. However the $H1$ heuristic tries to minimize the propagation of error in conditional probability estimation which in CC method is based on part of labels obtained from chain.

In Figure 2(a) it is presented a toy example of how $H1$ works incorporated in the Classifier Chain method. The process starts with separate evaluation of four distinct labels $\{\lambda_1, \lambda_2, \lambda_3, \lambda_4\}$. Depicted by triangles, base classifiers (see Figure 2(a)) are trained on input space $\mathcal{X}$ and evaluated on the testing data set and the one with the best accuracy (dark triangle) is selected to be a first classifier in the chain. Therefore the first modelled label in the chain is λ_2. In next step for all 3 remaining labels $\{\lambda_1, \lambda_3, \lambda_4\}$ again the evaluation of base classifier is performed. However this time the classifiers are trained on augmented input space $\mathcal{X}$ by the λ_2 relevance indicator - a new input attribute. We can see that the best accuracy

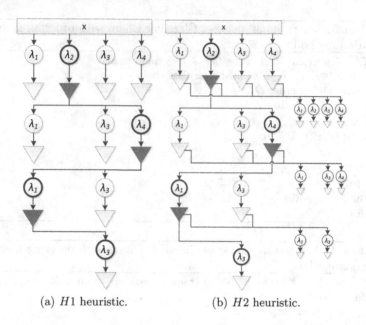

(a) $H1$ heuristic.   (b) $H2$ heuristic.

Fig. 2. Illustration of heuristic approaches, which discovers the order of generalisation in Heuristic Classifier Chain algorithm

was obtained in λ_4 generalization. The process continues until whole label space is covered.

Another proposed heuristic $H2$ provides similarly the decision of exploration on the basis of minimal classification error. The label that should be modelled in the chain is obtained by evaluation of classification error obtained in the second-level descendant node. It means that classification error needs to be calculated for all 2-element permutations of labels. For instance, discovering the label to be chosen on the level 1 (first label modelled in the chain), the heuristic accedes all descendant nodes (level 2) and computes the classification error using input attributes enriched with the first chosen label. When the element at level 1 provides the best training ability resulting in the smallest average error for local classification of all descendants (at level 2), it is then selected as a next label to be modelled in the chain.

In Figure 2(b) it is presented an example of $H2$ used in the Classifier Chain method. It can be noticed that the process starts with evaluation of accuracy obtained by CC for each permutation of four distinct labels $\{\lambda_1, \lambda_2, \lambda_3, \lambda_4\}$. λ_2 is elected as a first label to be used in final training of Classifier Chain because it resulted with the smallest average classification error for all permutations of the shape $(\lambda_2, \cdot)$ obtained in CC. Then similarly the next label is chosen, but the input space is augmented as in $H1$.

Third order construction method for classification in classifier chain $H3$ is realized likewise the $H2$. Similarly to $H2$, the label that should be modelled in the chain is obtained by evaluation of classification error obtained in the second level descendant node. However, the decision on the node representing a label to be chosen as the next one in training is selected based on mean classification accuracy for the pair of elements: parent(ascendant) and a child (descendant). The node which has minimal average classification error with any of its descendant nodes is chosen then.

3.2 Complexity

The computational complexity of classifier chain method accompanied by an ordering calculation can be considered as standard complexity of CC with additional computational cost of order generation. In case of $H1$ order calculation method modelling the output space of $|\mathcal{L}|$ labels requires to perform $\frac{|\mathcal{L}|^2+|\mathcal{L}|}{2} - 1$ evaluations to propose an ordering solution. For instance considering multi-label classification with ten labels $H1$ heuristics requires ten error evaluations to determine the first label in the training order. The second label will be chosen from nine remaining and so on, up to the tenth label, together 54 evaluations.

The complexity of classifier chain method accompanied by $H1$ ordering heuristics will arise to $O((\frac{|\mathcal{L}|^2+3|\mathcal{L}|}{2} - 1) \times f(d+|\mathcal{L}|, N))$, where $f(d+|\mathcal{L}|, N)$ denotes the complexity of underlying base classifier performing on data set with d attributes and N instances.

In case of $H2$ heuristics computing the ordering for chain of $\mathcal{L}$ length requires $\frac{\mathcal{L}^3-\mathcal{L}}{3} - 1$ evaluations. Again, the computational complexity of CC grows to $O((\frac{|\mathcal{L}|^3+2|\mathcal{L}|}{3} - 1) \times f(d + |\mathcal{L}|, N))$.

The most complex ordering calculation - $H3$ heuristic - requires $\frac{\mathcal{L}^3}{3}+\frac{\mathcal{L}^2}{2}+\frac{\mathcal{L}}{6}-1$ evaluations for the label set of $|\mathcal{L}|$ elements in order to provide the ordering solution. The total complexity of CC and $H3$ is quantified by $O((\frac{\mathcal{L}^3}{3} + \frac{\mathcal{L}^2}{2} + \frac{7\mathcal{L}}{6} - 1) \times f(d + |\mathcal{L}|, N))$.

4 Experiments

The main objective of the performed experiments was to evaluate predictive performance of the Classifier Chain method that uses proposed heuristics $H1$, $H2$, $H3$ in comparison to random Classifier Chain. The main attention in the experiment was concentrated on the evaluation of three proposed heuristics deriving distinct ordering schemes. In the experiment there were examined Hamming Loss (HL) and Classification Accuracy (CA) separately for six distinct data sets.

4.1 Datasets

In order to evaluate and compare all proposed approaches, the experiments were carried out on six distinct data sets, with assumed train–test split, from four diverse application domains: semantic scene analysis, bioinformatics, music categorization and text processing. The image data set *scene* [10] semantically indexes still scenes. The biological data set *yeast* [11] concerns micro-array expressions and phylogenetic profiles for genes classification. The music data set *emotions* [12], in turn, contains data about songs categorized into one or more classes of emotions. The *medical* [13] data set is based on the Computational Medicine Center's 2007 Medical Natural Language Processing Challenge and contains clinical free text reports labelled with disease codes. Another data set, *enron*, is based on annotated email messages exchanged between Enron Corporation employees. The last data set, *genbase* [14] refers to protein classification.

4.2 Evaluation Measures

The first examined measure, Hamming Loss HL, is defined as:

$$HL = \frac{1}{N} \sum_{i=1}^{N} \frac{Y_i \triangle F(x_i)}{|Y_i|} \tag{5}$$

where: N is the total number of instances x in the test set; Y_i denotes actual (real) list of labels for instance x_i, $F(x_i)$ is a sequence of labels predicted by multi-label classifier for instance x_i and $\triangle$ stands for the symmetric difference of two vectors, which is the vector-theoretic equivalent of the exclusive disjunction in Boolean logic.

The second evaluation measure is Classification Accuracy CA defined as:

$$CA = \frac{1}{N} \sum_{i=1}^{N} I(Y_i = F(x_i)) \tag{6}$$

where: N, Y_i, $F(x_i)$ have the same meaning as in Eq. 5, $I(true) = 1$ and $I(false) = 0$.

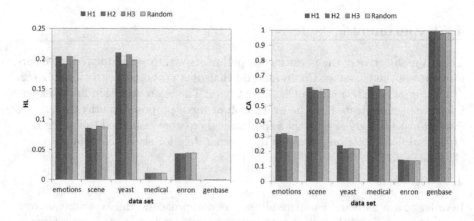

Fig. 3. Hamming Loss (HL) and Classification Accuracy (CA) results obtained by Heuristic Classifier Chain using H1, H2 and H3 ordering schemes for distinct data sets

Measure CA provides very strict evaluation as it requires the predicted set of labels to be an exact match of the true set of labels.

4.3 Results

As it is presented in Figure 3 $H1$, $H2$ and $H3$ ordering heuristics applied to CC revealed different predictive abilities. It can be observed that among all examined data sets $H2$ provided the best HL and comparative to others CA results. Moreover $H2$ provides slightly better results in comparison to CC with random generalization order.

5 Conclusions

The proposal for extension of Classifier Chain multi-label classification method was presented in the paper. Three distinct heuristics for generalization order were proposed and examined in terms of predictive accuracy. In general it can be concluded that by providing the ordering schemes to Classifier Chain it is possible to obtain better accuracy than in random ordering. The proposed heuristics are able to overcome the propagation of inference errors down the chain and are tractable to higher-dimensional data in comparison to Bayes-optimal Probabilistic Classifier Chain method.

References

[1] Kajdanowicz, T., Kazienko, P.: Hybrid repayment prediction for debt portfolio. In: Nguyen, N.T., Kowalczyk, R., Chen, S.-M. (eds.) ICCCI 2009. LNCS, vol. 5796, pp. 850–857. Springer, Heidelberg (2009)

[2] Read, J., Pfahringer, B., Holmes, G., Frank, E.: Classifier chains for multi-label classification. Machine Learning 85(3), 333–359 (2011)

[3] Wolpert, D.H.: Stacked generalization. Neural Networks 5, 241–259 (1992)

[4] Ting, K.M., Witten, I.H.: Issues in stacked generalization. Journal of Artificial Intelligence Research 10, 271–289 (1999)

[5] Tsoumakas, G., Dimou, A., Spyromitros, E., Mezaris, V., Kompatsiaris, I., Vlahavas, I.: Correlation-based pruning of stacked binary relevance models for multi-label learning. In: Proceedings of 1st International Workshop on Learning from Multi-Label Data, MLD 2009, at the European Conference on Machine Learning and Principles and Practice of Knowledge Discovery in Databases, pp. 101–116 (2009)

[6] Dembczynski, K., Cheng, W., Hullermeier, E.: Bayes optimal multilabel classification via probabilistic classifier chains. In: Proceedings of the 27th International Conference on Machine Learning, ICML 2010, Haifa, Israel, pp. 279–286. Omnipress (June 2010)

[7] Dembczynski, K., Waegeman, W., Cheng, W., Hullermeier, E.: On label dependence in multi-label classification. In: Workshop Proceedings of Learning from Multi-Label Data, Haifa, Israel, pp. 5–12 (June 2010)

[8] Kajdanowicz, T., Kazienko, P.: Structured output element ordering in boosting-based classification. In: Corchado, E., Kurzyński, M., Woźniak, M. (eds.) HAIS 2011, Part II. LNCS, vol. 6679, pp. 221–228. Springer, Heidelberg (2011)

[9] Kajdanowicz, T., Kazienko, P.: Learning and inference order in structured output elements classification. In: Pan, J.-S., Chen, S.-M., Nguyen, N.T. (eds.) ACIIDS 2012, Part I. LNCS, vol. 7196, pp. 301–309. Springer, Heidelberg (2012)

[10] Boutell, M.R., Luo, J., Shen, X., Brown, C.M.: Learning multi-label scene classi-fication. Pattern Recognition 37(9), 1757–1771 (2004)

[11] Elisseeff, A., Weston, J.: A kernel method for multi-labelled classification. In: Ad-vances in Neural Information Processing Systems. Neural Information Processing Systems, pp. 681–687. MIT Press (2001)

[12] Trohidis, K., Tsoumakas, G., Kalliris, G., Vlahavas, I.: Multilabel classification of music into emotions. In: Proceedings of 9th International Conference on Music Information Retrieval, ISMIR 2008, Philadelphia, PA, USA, pp. 325–330 (2008)

[13] Pestian, J., Brew, C., Matykiewicz, P., Hovermale, D., Johnson, N., Bretonnel Cohen, K., Duch, W.: A shared task involving multi-label classification of clinical free text. In: Proceedings of the Workshop on BioNLP 2007: Biological, Transla-tional, and Clinical Language Processing. Association of Computational Linguis-tics (2007)

[14] Diplaris, S., Tsoumakas, G., Mitkas, P., Vlahavas, I.: Protein classification with multiple algorithms. In: Bozanis, P., Houstis, E.N. (eds.) PCI 2005. LNCS, vol. 3746, pp. 448–456. Springer, Heidelberg (2005)

Weighting Component Models by Predicting from Data Streams Using Ensembles of Genetic Fuzzy Systems

Bogdan Trawiński[1], Tadeusz Lasota[2], Magdalena Smętek[1], and Grzegorz Trawiński[3]

[1] Wrocław University of Technology, Institute of Informatics,
Wybrzeże Wyspiańskiego 27, 50-370 Wrocław, Poland
[2] Wrocław University of Environmental and Life Sciences, Dept. of Spatial Management
ul. Norwida 25/27, 50-375 Wrocław, Poland
[3] Wrocław University of Technology, Faculty of Electronics,
Wybrzeże S. Wyspiańskiego 27, 50-370 Wrocław, Poland
{magdalena.smetek,bogdan.trawinski}@pwr.wroc.pl,
tadeusz.lasota@up.wroc.pl, grzegorz.trawinsky@gmail.com

Abstract. Our recently proposed method to predict from a data stream of real estate sales transactions based on ensembles of genetic fuzzy systems was extended to include weighting component models. The method consists in incremental expanding an ensemble by models built over successive chunks of a data stream. The predicted prices of residential premises computed by aged component models for current data are updated according to a trend function reflecting the changes of the market. The impact of different techniques of weighting component models on the accuracy of an ensemble was compared in the paper. Three techniques of weighting component models were proposed: proportional to their estimated accuracy, time of ageing, and dependent on property market fluctuations.

Keywords: genetic fuzzy systems, data stream, sliding windows, ensembles, weighting component models, trend functions, property valuation.

1 Introduction

The paper reports the extension of our study on predicting from a data stream of real estate sales transactions using ensembles of genetic fuzzy systems [18], [19], [20]. Working out the methods of processing data streams poses a considerable challenge because they require taking into account memory limitations, short processing times, and single scans of arriving data. Many strategies and techniques for mining data streams have been devised. Gaber in his recent overview paper categorizes them into four main groups: two-phase techniques, Hoeffding bound-based, symbolic approximation-based, and granularity-based ones [7]. Much effort is devoted to the issue of concept drift which occurs when data distributions and definitions of target classes change over time [6], [16], [24], [25]. Among the instantly growing methods of handling concept drift in data streams Tsymbal distinguishes three basic approaches, namely instance selection, instance weighting, and ensemble learning

H.L. Larsen et al. (Eds.): FQAS 2013, LNAI 8132, pp. 567–578, 2013.

[22]. The latter has been systematically overviewed in [12], [17]. In adaptive ensembles, component models are generated from sequential blocks of training instances. When a new block arrives, models are examined and then discarded or modified based on the results of the evaluation. Several methods have been proposed for that, e.g. accuracy weighted ensembles [23] and accuracy updated ensembles [3]. In [1], [2] Bifet et al. proposed two bagging methods to process concept drift in a data stream: ASHT Bagging using trees of different sizes, and ADWIN Bagging employing a change detector to decide when to discard underperforming ensemble members.

The goal of our research we have been conducting recently is to apply a non-incremental genetic fuzzy systems (GFSs) to build reliable predictive models from a data stream. The approach was inspired by the observation of a real estate market of in one big Polish city in recent years when it experienced a violent fluctuations of residential premises prices. Our method consists in the utilization of aged models to compose ensembles and correction of the output provided by component models by means of trend functions reflecting the changes of prices in the market over time. In this paper we propose three methods of weighting the members of an ensemble: proportional to their estimated accuracy, time of ageing, and dependent on property market fluctuations.

2   Motivation and GFS Ensemble Approach

The approach based on fuzzy logic is especially suitable for property valuation because professional appraisers are forced to use many, very often inconsistent and imprecise sources of information, and their familiarity with a real estate market and the land where properties are located is frequently incomplete. Moreover, they have to consider various price drivers and complex interrelation among them. The appraisers should make on-site inspection to estimate qualitative attributes of a given property as well as its neighbourhood. They have also to assess such subjective factors as location attractiveness and current trend and vogue. So, their estimations are to a great extent subjective and are based on uncertain knowledge, experience, and intuition rather than on objective data.

So, the appraisers should be supported by automated valuation systems which often incorporate data driven models for premises valuation developed employing sales comparison method. The data driven models, considered in the paper, were generated using real-world data on sales transactions taken from a cadastral system and a public registry of real estate transactions. So far, we have investigated several methods to construct regression models to assist with real estate appraisal based on fuzzy approach: i.e. genetic fuzzy systems as both single models [11] and ensembles built using various resampling techniques [10], [14], but in this case the whole datasets had to be available before the process of training models started. All property prices were updated to be uniform in a given point of time. Good performance revealed evolving fuzzy models applied to cadastral data [13], [15].

The outline of the *GFS* ensemble approach to predict from a data stream is depicted in Fig. 1. The data stream is partitioned into data chunks according to the periods of a constant length t_c. Each time interval determines the shift of a sliding time window which comprises training data to create *GFS* models. The window is shifted step by step of a period t_s in the course of time. The length of the sliding window t_w is equal to the multiple of t_c so that $t_w=jt_c$, where $j=1,2,3,...$. The window determines the scope of training data to generate from scratch a property valuation model, in our case GFS_i. It is assumed that the models generated over a given training dataset is valid for the next interval which specifies the scope for a test dataset. Similarly, the interval t_t which delineates a test dataset is equal to the multiple of t_c so that $t_t=kt_c$, where $k=1,2,3,...$. The sliding window is shifted step by step of a period t_s in the course of time, and likewise, the interval t_s is equal to the multiple of t_c so that $t_s=lt_c$, where $l=1,2,3,...$.

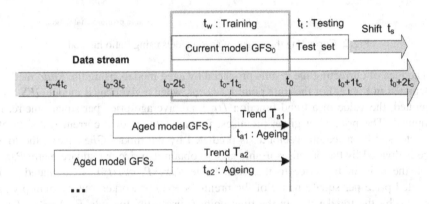

Fig. 1. GFS ensemble approach to predict from a data stream

We consider in Fig. 1 a point of time t_0 at which the current model GFS_0 was built over data that came in between time t_0-2t_c and t_0. The models created earlier, i.e. GFS_1, GFS_2, etc. have aged gradually and in consequence their accuracy has deteriorated. However, they are neither discarded nor restructured but utilized to compose an ensemble so that the current test dataset is applied to each component GFS_i. However, in order to compensate ageing, their output produced for the current test dataset is updated using trend functions $T(t)$. As the functions to model the trends of price changes the polynomials of the degree from one to five were employed [18]. The trends were determined over two time periods: shorter and longer ones. The shorter periods encompassed the length of a sliding window plus model ageing intervals, i.e. t_w plus, t_{ai} for a given aged model GFS_i. In turn, the longer periods took into account all data since the beginning of the stream.

We proposed two different methods of updating the prices of premises according to the trends of the value changes over time. The first one based on the difference between a price and a trend value in a given time point and we called it the *Delta* method. In turn, the second technique utilized the ratio of the price to the trend value and it was named the *Ratio* method of price correction [18]. In this paper we utilize only the latter method.

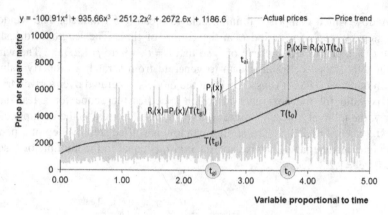

$y = -100.91x^4 + 935.66x^3 - 2512.2x^2 + 2672.6x + 1186.6$ ····· Actual prices —— Price trend

Fig. 2. Correcting the output of aged models using Ratio method

The idea of correcting the results produced by aged models using the *Ratio* method is depicted in Fig. 2. For the time point $t_{gi}=t_0-t_{ai}$, when a given aged model GFS_i was generated, the value of a trend function $T(t_{gi})$, i.e. average price per square metre, is computed. The price of a given premises, i.e. an instance of a current test dataset, characterised by a feature vector x, is predicted by the model GFS_i. Next, the total price is divided by the premises usable area to obtain its price per square metre $P_i(x)$. Then, the ratio of the price to the trend value $R_i(x)=P_i(x)/T(t_{gi})$ is calculated. The corrected price per square metre of the premises $\dot{P}_i(x)$ is worked out by multiplying this ratio by the trend value in the time point t_0 using the formula $\dot{P}_i(x) =R_i(x)T(t_0)$, where $T(t_0)$ is the value of a trend function in t_0. Finally, the corrected price per square metre $\dot{P}_i(x)$ is converted into the corrected total price of the premises by multiplying it by the premises usable area.

3   Proposed Methods for Component Model Weighting

In our previous works we have used the arithmetic mean as an ensemble output averaging function assuming that each component model contributes equally to the final average. However, it seems to be more adequate to weight ensemble members according to their expected accuracy, stability, dependability, etc. Using the weighted mean shown in Formula 1 we assure that components with a high weight contribute more to the weighted mean than do elements with a low weight. In this formula Err_i denotes the accuracy of individual model expressed in terms of its error over a test set, e.g. mean square error, and $\overline{Err}_{ens}$ stands for the accuracy of the whole ensemble.

$$\overline{Err}_{ens} = \frac{\sum_{i=1}^{N} w_i Err_i}{\sum_{i=1}^{N} w_i} \tag{1}$$

Three following methods of weighting models embraced by ensembles have been proposed in our study, namely weights proportional to model ageing time (denoted by wA), weights proportional to the predictive accuracy estimated for a given model (*wP*), and weights dependent on fluctuations of real estate market during a model generation (*wF*). The denotation used in the description of proposed weighting methods is illustrated in Fig. 3.

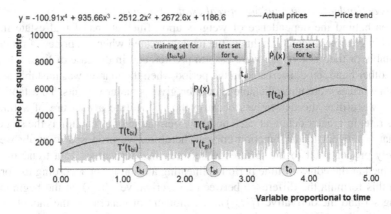

Fig. 3. Denotation used in the description of proposed weighting methods

Weights proportional to model ageing time (wA)
The first method of weighting ensemble members assumes that the models created earlier age gradually and in consequence their accuracy deteriorates. Therefore, we proposed to apply weights reversely proportional to ageing time according to Formula (2). In this formula the difference between time point of consideration t_0 and time of generation of a given model t_{gi} determines the ageing period t_{ai} expressed e.g. in months. In turn, the constant c can take real values from 0 to 1 and controls the strength of differentiating individual models. The values closer to 1 the bigger differences among weights and conversely the values closer to 0 the smaller discrepancy.

$$w_{Ai} = \frac{1}{1 + c\left(t_0 - t_{gi}\right)} = \frac{1}{1 + c\, t_{ai}} \tag{2}$$

Weights proportional to estimated model predictive accuracy (wP)
The second method of weighting models presumes that the models which reveal better predictive accuracy should acquire greater weights. Therefore, we proposed to assign weights reversely proportional to error level produced by individual models (Err_i) in terms of mean absolute error (*MAE*) or root mean square error (*RMSE*) according to Formula (3). In this formula $Err_i(t_{gi})$ stands for error value of *i-th* model estimated at the moment of its generation t_{gi}. Moreover, Err_i is normalized by dividing it by the value of the trend function $T_i(t_{gi})$ determined for the *i-th* model. In turn, the constant c takes real values from the bracket 0 to 1 and differentiates individual models. The values of c closer to 1 the bigger differences among weights and conversely the values closer to 0 the smaller discrepancy.

$$w_{Pi} = \cfrac{1}{1 + c \cfrac{Err_i(t_{gi})}{T_i(t_{gi})}} \tag{3}$$

Weights dependent on fluctuations of real estate market (wF)

The idea behind the dependence of weights upon fluctuations of real estate market consists in granting bigger weights to the models created when the prices of properties were stable or their rise or fall were easy to predict, e.g. in the case of linear changes. On the other hand, models created in the period when the market was unstable and the real estate prices were unpredictable should be given smaller weights. To establish the weights we utilize the derivatives because they constitute a measure of a function change rate, in greater detail, at each point the derivative of a function is the slope of a line that is tangent to the function curve. Therefore, we proposed to apply weights reversely proportional to the difference between the derivatives of a trend function determined at the edges of time period delineating a training set according to Formula (4). In this formula the difference between the derivative $T_i'(t_{bi})$ at the beginning of training set and the derivative $T_i'(t_{gi})$ at the moment of generating the model reflects the fluctuation rate. We take the absolute value of this difference to be able to consider both growth and fall of the trend function. In turn, the constant c taken from the bracket of real values 0 to 1 is used to control the diversity among the models. The values of c closer to 1 the bigger differences among weights and conversely the values closer to 0 the smaller discrepancy.

$$w_{Fi} = \frac{1}{1 + c\, Abs(T_i'(t_{bi}) - T_i'(t_{gi}))} \tag{4}$$

For comparison, ensembles comprising component models without weights, i.e. with all weights equal to one (denoted by *w1*), were also used in experiments.

4   Experimental Setup and Results

The investigation was conducted with our experimental system implemented in Matlab devoted to carry out research into machine learning algorithms using various resampling and multi-model methods for regression problems. We have recently extended our system to include the functions of building ensembles over a data stream and weighting their component models. The trends are modelled using the Matlab function *polyfit(x,y,n)*, which finds the coefficients of a polynomial $p(x)$ of degree n that fits the y data by minimizing the sum of the squares of the deviations of the data from the model (least-squares fit).

Real-world dataset used in experiments was drawn from an unrefined dataset containing above 100 000 records referring to residential premises transactions accomplished in one Polish big city with the population of 640 000 within 14 years from 1998 to 2011. In this period the majority of transactions were made with non-market prices when the council was selling flats to their current tenants on preferential terms. First of all, transactional records referring to residential premises sold at market prices were selected. Then, the dataset was confined to sales transaction data of residential premises (apartments) where the land was leased on terms of perpetual usufruct. The other transactions of premises with the ownership of the land were omitted due to the conviction of professional appraisers stating that the land ownership and lease affect substantially the prices of apartments and therefore they should be used separately for sales comparison valuation methods. The final dataset counted 9795 samples. Due to the fact we possessed the exact date of each transaction we were able to order all instances in the dataset by time, so that it can be regarded as a data stream. Four following attributes were pointed out as main price drivers by professional appraisers: usable area of a flat (*Area*), age of a building construction (*Age*), number of storeys in the building (Storeys), the distance of the building from the city centre (*Centre*), in turn, price of premises (*Price*) was the output variable. In order to characterize quantitatively the data stream the sizes of one-year datasets are given in Table 1.

Table 1. Number of instances in one-year datasets

1998	1999	2000	2001	2002	2003	2004
446	646	554	626	573	790	774
2005	**2006**	**2007**	**2008**	**2009**	**2010**	**2011**
740	776	442	734	821	1296	577

The property valuation models were built from scratch by genetic fuzzy systems over chunks of data stream determined by a sliding window which was 12 months long. The parameters of the architecture of fuzzy systems as well as genetic algorithms are listed in Table 2. Similar designs are described in [4], [11].

Table 2. Parameters of GFS used in experiments

Fuzzy system	Genetic Algorithm
Type of fuzzy system: Mamdani	Chromosome: rule base and mf, real-coded
No. of input variables: 4	Population size: 100
Type of membership functions (mf): triangular	Fitness function: MSE
No. of input mf: 3	Selection function: tournament
No. of output mf: 5	Tournament size: 4
No. of rules: 15	Elite count: 2
AND operator: prod	Crossover fraction: 0.8
Implication operator: prod	Crossover function: two point
Aggregation operator: probor	Mutation function: custom
Defuzzyfication method: centroid	No. of generations: 100

The results of evaluating experiments were considered within three periods of time, namely 1) 2002-2004, 2) 2005-2007, and 3) 2008-2010 marked with grey shades in Fig. 4. In the first period a modest rise of real estate prices just before Poland entered the European Union (EU) could be observed, in the second one the prices of residential premises were increasing rapidly during the worldwide real estate bubble. The third one corresponds to a modest growth of prices after the bubble burst and during the global financial crisis. This period was characterized by unstable real estate market and great fluctuations of prices due to nervous behaviour of both buyers and sellers. The trend of premises price changes over 14 years from 1998 to 2011, shown in Fig. 4, can be modelled by the polynomial function of degree four.

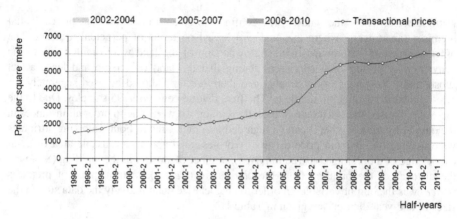

Fig. 4. Change trend of average transactional prices per square metre over time

This study is the continuation of our recent works on predicting from a data stream using the GFS ensembles [18], [19], [20]. So far we have investigated among others the impact of ensemble size and different trend functions on the accuracy of single and ensemble fuzzy models. Moreover, we proved the usefulness of ensemble approach incorporating the correction of individual component model output. Based on the results of our previous study, we were able to determine following parameters of our first series of experiments including two phases: generating single GFS models and building GFS ensembles.

Generating single GFS models
- Set the length of the sliding window to 12 months, $t_w = 12$.
- Set the starting point of the sliding window, i.e. its right edge, to 2000-01-01 and the terminating point to 2010-12-01.
- Set the shift of the sliding window to 1 month, $t_s = 1$.
- Move the window from starting point to terminating point with the step $t_s = 1$.
- At each stage generate a GFS from scratch over a training set delineated by the window. In total 108 GFSs were built.

Building GFS ensembles
- Select periods characterizing different fluctuation of real estate market, i.e. 2002-2004, 2005-2007, and 2008-2010.
- For each three-year period built 36 ensembles composed of 24 ageing GFSs each. An ensemble is created in the way described in Section 2 with shift equal to one month, $t_s = 1$.
- Take test sets actual for each t_0 over a period of 3 months, $t_t = 3$.
- Compute the output of individual GFSs and update it using trend functions of degree four determined over time periods from the beginning of the stream.
- Derive weights for each model using formulas (2), (3), or (4).
- As the output of individual ensembles compute the weighted means applying the derived weights and additionally determine the arithmetic mean.
- Conduct statistical analysis of the results obtained.

The analysis of the results was performed using statistical methodology including nonparametric tests followed by post-hoc procedures designed especially for multiple $N \times N$ comparisons [5], [8], [9], [21]. The idea behind statistical methods applied to analyse the results of experiments was as follows. The commonly used paired tests i.e. parametric t-test and its nonparametric alternative Wilcoxon signed rank tests are not appropriate for multiple comparisons due to the so called family-wise error. The proposed routine starts with the nonparametric Friedman test, which detect the presence of differences among all algorithms compared. After the null-hypotheses have been rejected the post-hoc procedures should be applied in order to point out the particular pairs of algorithms which produce differences. For $N \times N$ comparisons nonparametric Nemenyi's, Holm's, Shaffer's, and Bergamnn-Hommel's procedures are recommended.

The goal of the statistical analysis was to compare the accuracy of *GFSs* ensembles created using the aforementioned techniques and to conduct nonparametric tests of statistical significance within three periods 2002-2004, 2005-2007, and 2008-2010 characterized by different fluctuations of real estate market. Within each of these periods 36 values of *MAE* computed using weighting methods *wA*, *wP*, *wF*, and *w1*, respectively, constituted the points of observation. The Friedman test performed in respect of *MAE* values of the ensembles showed that there were significant differences between some ensembles. Average ranks of individual models are shown in Table 3, where the lower rank value the better model. In the period 2002-2004 the ranks are not statistically insignificant so that we were not justified to proceed to the non-parametric post-hoc procedures. In the next two periods 2005-2007 and 2008-2010 the *wP* and *wF* were in the first two places whereas for all years *wP* took the first and *wF* the last position.

Table 3. Average rank positions of weighting methods produced by Friedman tests

2002-2004 p-value=0.47529		2005-2007 p-value=0.00000		2008-2010 p-value=0.0000		All years p-value=0.0000	
Rank	Weighting	Rank	Weighting	Rank	Weighting	Rank	Weighting
2.31	wP	1.53	wP	1.75	wA	1.96	wP
2.39	w1	2.14	wF	2.06	wP	2.45	wA
2.56	wA	3.06	wA	2.61	w1	2.76	w1
2.75	wF	3.28	w1	3.58	wF	2.82	wF

Adjusted *p-values* for Nemenyi's, Holm's, Shaffer's, and Bergmann-Hommel's post-hoc procedures for $N \times N$ comparisons for all possible pairs of algorithms are shown in Tables 4, 5, and 6 for the 2005-2007, 2008-2010, and all year periods, respectively. For comparison, the results of paired Wilcoxon tests are placed in all tables. The *p-values* indicating the statistically significant differences between given pairs of algorithms are marked with italics. The significance level considered for the null hypothesis rejection was 0.05. Following main observations could be done based on the most powerful Shaffer's, and Bergmann-Hommel's post-hoc procedures. For the period 2005-2007 both weighting methods *wP* and *wF* lead to significantly better performance than the other methods but there is not any significant difference between them. For the period 2008-2010 both weighting methods *wA* and *wP* provide significantly better performance than the other methods but one, however, there is not any significant difference between them. For all periods together the weighting method *wP* yields significantly better results than any other method, in turn, there are not significant differences among *wA*, *wF*, and *w1* weighting methods.

Table 4. Adjusted p-values for $N \times N$ comparisons of weighting methods over 2005-2007

Meth. vs Meth.	pWilcox	pNeme	pHolm	pShaf	pBerg
wP vs w1	0.000000	5.32E-08	5.32E-08	5.32E-08	5.32E-08
wP vs wA	0.000284	3.09E-06	2.57E-06	1.54E-06	1.54E-06
wF vs w1	0.000001	0.001092	7.28E-04	5.46E-04	5.46E-04
wF vs wA	0.000731	0.015548	0.007774	0.007774	0.002591
wP vs wF	0.001351	0.267658	0.089219	0.089219	0.089219
wA vs w1	0.001211	1.000000	0.465209	0.465209	0.465209

Table 5. Adjusted p-values for $N \times N$ comparisons of weighting methods over 2008-2010

Meth. vs Meth.	pWilcox	pNeme	pHolm	pShaf	pBerg
wA vs wF	0.000041	1.01E-08	1.01E-08	1.01E-08	1.01E-08
wP vs wF	0.000001	3.09E-06	2.57E-06	1.54E-06	1.54E-06
w1 vs wF	0.000000	0.008388	0.005592	0.004194	0.002796
wA vs w1	0.000173	0.027938	0.013969	0.013969	0.013969
wP vs w1	0.000066	0.407335	0.135778	0.135778	0.067889
wA vs wP	0.000433	1.000000	0.315302	0.315302	0.315302

Table 6. Adjusted p-values for $N \times N$ comparisons of weighting methods for all years

Meth. vs Meth.	pWilcox	pNeme	pHolm	pShaf	pBerg
wP vs wF	0.000000	5.71E-06	5.71E-06	5.71E-06	5.71E-06
wP vs w1	0.000000	3.50E-05	2.91E-05	1.75E-05	1.75E-05
wP vs wA	0.887873	0.031300	0.020866	0.015650	0.010433
w1 vs wF	0.020658	1.000000	0.712178	0.712178	0.712178
wA vs wF	0.271165	0.210090	0.105045	0.105045	0.105045
wA vs w1	0.465694	0.491942	0.163981	0.163981	0.105045

5   Conclusions and Future Work

In the paper we reported our further study of methods to predict from a data stream of real estate sales transactions based on ensembles of genetic fuzzy systems. Our ensemble approach consists in incremental expanding an ensemble by models built from scratch over successive chunks of a data stream determined by a sliding window. The predicted prices of residential premises computed by aged component models for current data are updated according to trend functions which model the changes of the market. In the paper we proposed three methods for weighting component models of ensembles created to predict from a data stream of real estate sales transactions. The methods derive weights which are reversely proportional to model ageing time, reversely proportional to the expected errors produced by the models, and are dependent on fluctuations of real estate market.

We conducted intensive evaluating experiments using real-world data taken from cadastral systems. They consisted in generating ensembles for 108 points of time and then comparing the impact of proposed weighting methods on the ensemble accuracy using nonparametric tests of statistical significance adequate for multiple comparisons. The time points fell in to three periods of time characterized by different rates of premises price growth and different stability of real estate market.

For the time period 1) 2002-2004, by the moderate price growth and relative stability of the market, there were no statistically significant differences among the methods. In turn, for the time period 2) 2005-2007, when the process were soaring due to the real estate bubble, the methods which derived the weights proportional to expected model accuracy and reflecting the fluctuations of the market lead to significantly better performance than the other methods. And finally, for the time period 3) 2008-2010, characterized by the market instability due to the global financial crisis, the method determining weights proportional to model ageing time provided the best results. However, the comparison over all periods together pointed out as the best solution the method which consisted in the estimation of a model accuracy.

The study opens the area for our further research into the selection of the best parameters of weighted ensembles including ensemble pruning as well as the impact of the constant c. So far, we have treated an models as black boxes we will make an attempt to explore the intrinsic structure of component models, i.e. their knowledge and rule bases, as well as their generation efficiency, interpretability, the problems of overfitting and outliers. We also intend to conduct experiments employing as base learning algorithms other methods capable of learning from concept drifts such as: decision trees, recurrent neural networks, support vector regression, etc. Moreover, we plan to compare the outcome produced by proposed genetic fuzzy models with human based predictions.

Acknowledgments. This paper was partially supported by the Polish National Science Centre under grant no. N N516 483840.

References

1. Bifet, A., Holmes, G., Pfahringer, B., Gavaldà, R.: Improving Adaptive Bagging Methods for Evolving Data Streams. In: Zhou, Z.-H., Washio, T. (eds.) ACML 2009. LNCS (LNAI), vol. 5828, pp. 23–37. Springer, Heidelberg (2009)
2. Bifet, A., Holmes, G., Pfahringer, B., Kirkby, R., Gavalda, R.: New ensemble methods for evolving data streams. In: Elder IV, J.F., et al. (eds.) KDD 2009, pp. 139–148. ACM Press, New York (2009)
3. Brzeziński, D., Stefanowski, J.: Accuracy Updated Ensemble for Data Streams with Concept Drift. In: Corchado, E., Kurzyński, M., Woźniak, M. (eds.) HAIS 2011, Part II. LNCS (LNAI), vol. 6679, pp. 155–163. Springer, Heidelberg (2011)
4. Cordón, O., Herrera, F.: A Two-Stage Evolutionary Process for Designing TSK Fuzzy Rule-Based Systems. IEEE Tr. on Sys., Man and Cyber., Part B 29(6), 703–715 (1999)
5. Demšar, J.: Statistical comparisons of classifiers over multiple data sets. Journal of Machine Learning Research 7, 1–30 (2006)
6. Elwell, R., Polikar, R.: Incremental Learning of Concept Drift in Nonstationary Environments. IEEE Transactions on Neural Networks 22(10), 1517–1531 (2011)
7. Gaber, M.M.: Advances in data stream mining. Wiley Interdisciplinary Reviews: Data Mining and Knowledge Discovery 2(1), 79–85 (2012)
8. García, S., Herrera, F.: An Extension on "Statistical Comparisons of Classifiers over Multiple Data Sets" for all Pairwise Comparisons. Journal of Machine Learning Research 9, 2677–2694 (2008)
9. Graczyk, M., Lasota, T., Telec, Z., Trawiński, B.: Nonparametric Statistical Analysis of Machine Learning Algorithms for Regression Problems. In: Setchi, R., Jordanov, I., Howlett, R.J., Jain, L.C. (eds.) KES 2010, Part I. LNCS (LNAI), vol. 6276, pp. 111–120. Springer, Heidelberg (2010)

10. Kempa, O., Lasota, T., Telec, Z., Trawiński, B.: Investigation of bagging ensembles of genetic neural networks and fuzzy systems for real estate appraisal. In: Nguyen, N.T., Kim, C.-G., Janiak, A. (eds.) ACIIDS 2011, Part II. LNCS (LNAI), vol. 6592, pp. 323–332. Springer, Heidelberg (2011)

11. Król, D., Lasota, T., Trawiński, B., Trawiński, K.: Investigation of Evolutionary Optimization Methods of TSK Fuzzy Model for Real Estate Appraisal. International Journal of Hybrid Intelligent Systems 5(3), 111–128 (2008)

12. Kuncheva, L.I.: Classifier ensembles for changing environments. In: Roli, F., Kittler, J., Windeatt, T. (eds.) MCS 2004. LNCS, vol. 3077, pp. 1–15. Springer, Heidelberg (2004)

13. Lasota, T., Telec, Z., Trawiński, B., Trawiński, K.: Investigation of the eTS Evolving Fuzzy Systems Applied to Real Estate Appraisal. Journal of Multiple-Valued Logic and Soft Computing 17(2-3), 229–253 (2011)

14. Lasota, T., Telec, Z., Trawiński, G., Trawiński, B.: Empirical Comparison of Resampling Methods Using Genetic Fuzzy Systems for a Regression Problem. In: Yin, H., Wang, W., Rayward-Smith, V. (eds.) IDEAL 2011. LNCS, vol. 6936, pp. 17–24. Springer, Heidelberg (2011)

15. Lughofer, E., Trawiński, B., Trawiński, K., Kempa, O., Lasota, T.: On Employing Fuzzy Modeling Algorithms for the Valuation of Residential Premises. Information Sciences 181, 5123–5142 (2011)

16. Maloof, M.A., Michalski, R.S.: Incremental learning with partial instance memory. Artificial Intelligence 154(1-2), 95–126 (2004)

17. Minku, L.L., White, A.P., Yao, X.: The Impact of Diversity on Online Ensemble Learning in the Presence of Concept Drift. IEEE Transactions on Knowledge and Data Engineering 22(5), 730–742 (2010)

18. Trawiński, B.: Evolutionary fuzzy system ensemble approach to model real estate market based on data stream exploration. Journal for Universal Computer Science 19(4), 539–562 (2013)

19. Trawiński, B., Lasota, T., Smętek, M., Trawiński, G.: An Analysis of Change Trends by Predicting from a Data Stream Using Genetic Fuzzy Systems. In: Nguyen, N.-T., Hoang, K., Jędrzejowicz, P. (eds.) ICCCI 2012, Part I. LNCS (LNAI), vol. 7653, pp. 220–229. Springer, Heidelberg (2012)

20. Trawiński, B., Lasota, T., Smętek, M., Trawiński, G.: An Attempt to Employ Genetic Fuzzy Systems to Predict from a Data Stream of Premises Transactions. In: Hüllermeier, E., Link, S., Fober, T., Seeger, B. (eds.) SUM 2012. LNCS, vol. 7520, pp. 127–140. Springer, Heidelberg (2012)

21. Trawiński, B., Smętek, M., Telec, Z., Lasota, T.: Nonparametric Statistical Analysis for Multiple Comparison of Machine Learning Regression Algorithms. International Journal of Applied Mathematics and Computer Science 22(4), 867–881 (2012)

22. Tsymbal, A.: The problem of concept drift: Definitions and related work. Technical Report. Department of Computer Science, Trinity College, Dublin (2004)

23. Wang, H., Fan, W., Yu, P.S., Han, J.: Mining concept-drifting data streams using ensemble classifiers. In: Getoor, L., et al. (eds.) KDD 2003, pp. 226–235. ACM Press, New York (2003)

24. Widmer, G., Kubat, M.: Learning in the presence of concept drift and hidden contexts. Machine Learning 23, 69–101 (1996)

25. Zliobaite, I.: Learning under Concept Drift: an Overview. Technical Report. Faculty of Mathematics and Informatics, Vilnius University, Vilnius (2009)

Weighted Aging Classifier Ensemble for the Incremental Drifted Data Streams

Michał Woźniak, Andrzej Kasprzak, and Piotr Cal

Department of Systems and Computer Networks
Wroclaw University of Technology
Wyb. Wyspianskiego 27, 50-370 Wroclaw, Poland
{michal.wozniak,andrzej.kasprzak,piotr.cal}@pwr.wroc.pl

Abstract. Evolving systems are recently focus of intense research because for most of the real problems we can observe that the parameters of the decision tasks should adapt to new conditions. In classification such a problem is usually called concept drift. The paper deals with the data stream classification where we assume that the concept drift is sudden but its rapidity is limited. To deal with this problem we propose a new algorithm called Weighted Aging Ensemble (WAE), which is able to adapt to changes of classification model parameters. The method is inspired by well-known algorithm Accuracy Weighted Ensemble (AWE) which allows to change the line-up of a classifier ensemble, but the proposed method incudes two important modifications: (i) classifier weights depend on the individual classifier accuracies and time they have been spending in the ensemble, (ii) individual classifier are chosen to the ensemble on the basis on the non-pairwise diversity measure. The proposed method was evaluated on the basis of computer experiments which were carried out on SEA dataset. The obtained results encourage us to continue the work on the proposed concept.

Keywords: machine learning, classifier ensemble, data stream, concept drift, incremental learning, forgetting.

1 Introduction

The market-leading companies realize that smart analytic tools which are capable to analyze collected, fast-growing data could lead to business success. Therefore they desire to exploit strength of machine learning techniques to extract hidden, valuable knowledge from the huge databases. One of the most promising directions of that research is classification task, which is widely used in computer security (e.g. designing intrusion detection/prevention systems IDS/IPS), medicine, finance (e.g., fraud detection or credit approval), or trade. Designing such solutions we should take into consideration that in the modern world the most of the data arrive continuously and it causes that smart analytic tools should respect this nature and be able to interpret so-called data streams. Unfortunately most of the traditional methods of classifier design do not take into consideration that:

H.L. Larsen et al. (Eds.): FQAS 2013, LNAI 8132, pp. 579–588, 2013.

- the statistical dependencies between the observations of a given objects and their classifications could change,
- data can come flooding in the analyzer what causes that it is impossible to label all records.

This work focuses on the first problem called *concept drift* [17] and it comes in many forms, depending on the type of change. In general, the following approaches can be considered to deal with the mentioned above problem

- Rebuilding a classification model if new data becomes available, which is very expensive and impossible from a practical point of view, especially if the concept drift occurs rapidly.
- Detecting concept changes in new data and if these changes are sufficiently "significant", then rebuilding the classifier.
- Adopting an incremental learning algorithm for the classification model.

We will concentrate on the last proposition. Adapting the learner is a part of an incremental learning [10]. The model is either updated (e.g., neural networks) or needs to be partially or completely rebuilt (as CVFDT algorithm [4]). Usually we assume that the data stream is given in a form of data chunks (windows). When dealing with the sliding window the main question is how to adjust the window size. On the one hand, a shorter window allows focusing on the emerging context, though data may not be representative for a longer lasting context. On the other hand, a wider window may result in mixing the instances representing different contexts. Therefore, certain advanced algorithms adjust the window size dynamically depending on the detected state (e.g., FLORA2 [17]) or algorithms can use multiple windows [9]. One of the important group of algorithms dedicated to stream classification exploits strength of ensemble systems, which work pretty well in static environments [8], because according to "no free lunch theorem" [18] there is not a single classifier that is suitable for all the tasks, since each of them has its own domain of competence. A strategy for generating the classifier ensemble should guarantee its diversity improvement therefore let us enumerate the main propositions how to get a desirable committee:

- The individual classifiers could be train on different datasets, because we hope that classifiers trained on different inputs would be complementary.
- The individual classifiers can use the selected features only.
- Usually it could be easy to decompose the classification problem into simpler ones solved by the individual classifier. The key problem of such approach is how to recover the whole set of possible classes.
- The last and intuitive method is to use individual classifiers trained on different models or different versions of models.

It has been shown that a collective decision can increase classification accuracy because the knowledge that is distributed among the classifiers may be more comprehensive [14]. Usually, a diversity may refer to the classifier model, the feature set, or the instances used in training, but in a case of data stream classification diversity can also refer to the context, but the problem how the diversity of the classifier ensemble should be measured still remains.

Several strategies are possible for a data stream classification:

1. Dynamic combiners, where individual classifiers are trained in advance and their relevance to the current context is evaluated dynamically while processing subsequent data. The level of contribution to the final decision is directly proportional to the relevance [5]. The drawback of this approach is that all contexts must be available in advance; emergence of new unknown contexts may result in a lack of experts.
2. Updating the ensemble members, where each ensemble consists of a set of online classifiers that are updated incrementally based on the incoming data [2].
3. Dynamic changing line-up of ensemble e.g., individual classifiers are evaluated dynamically and the worst one is replaced by a new one trained on the most recent data.

Among the most popular ensemble approaches, the following are worth noting: the Streaming Ensemble Algorithm (SEA) [15] or the Accuracy Weighted Ensemble (AWE)[16]. Both algorithms keep a fixed-size set of classifiers. Incoming data are collected in data chunks, which are used to train new classifiers. All the classifiers are evaluated on the basis of their accuracy and the worst one in the committee is replaced by a new one if the latter has higher accuracy. The SEA uses a majority voting strategy, whereas the AWE uses the more advanced weighted voting strategy. A similar formula for decision making is implemented in the Dynamic Weighted Majority (DWM) algorithm [7].

In this work we propose the dynamic ensemble model called WAE (Weighted Aging Ensemble) which can modify line-up of the classifier committee on the basis of diversity measure. Additionally the decision about object's label is made according to weighted voting, where weight of a given classifier depends on its accuracy and time spending in an ensemble. The detailed description of WAE is presented in the next section. Then we present preliminary results of computer experiments which were carried out on SEA dataset and seem to confirm usefulness of proposed algorithm. The last section concludes our research.

2 Algorithm

We assume that the classified data stream is given in a form of data chunks denotes as $\mathcal{DS}_k$, where k is the chunk index. The concept drift could appear in the incoming data chunks. We do not detect it, but we try to construct self-adapting classifier ensemble. Therefore on the basis of the each chunk one individual is trained and we check if it could form valuable ensemble with the previously trained models. In our algorithm we propose to use the Generalized Diversity (denoted as $\mathcal{GD}$) proposed by Partridge and Krzanowski [11] to assess all possible ensembles and to choose the best one. $\mathcal{GD}$ returns the maximum values in the case of failure of one classifier is accompanied by correct classification by the other one and minimum diversity occurs when failure of one classifier is accompanied by failure of the other.

$$GD(\Pi) = 1 - \frac{\sum_{i=1}^{L} \frac{i(i-1)p_i}{L(L-1)}}{\sum_{i=1}^{L} \frac{ip_i}{L}} \qquad (1)$$

where L is the cardinality of the classifier pool (number of individual classifiers) and p_i stands for the probability that i randomly chosen classifiers from Π will fail on randomly chosen example.

Lets $P_a(\Psi_i)$ denotes frequency of correct classification of classifier Ψ_i and $itter(\Psi_i)$ stands for number of iterations which Ψ_i has been spent in the ensemble. We propose to establish the classifier's weight $w(\Psi_i)$ according to the following formulae

$$w(\Psi_i) = \frac{P_a(\Psi_i)}{\sqrt{itter(\Psi_i)}} \qquad (2)$$

This proposition of classifier aging has its root in object weighting algorithms where an instance weight is usually inversely proportional to the time that has passed since the instance was read [6] and Accuracy Weighted Ensemble (AWE)[16], but the proposed method called Weighted Aging Ensemble (WAE) incudes two important modifications:

1. classifier weights depend on the individual classifier accuracies and time they have been spending in the ensemble,
2. individual classifier are chosen to the ensemble on the basis on the non-pairwise diversity measure.

The WAE pseudocode is presented in Alg.1.

3   Experimental Investigations

The aims of the experiment were to assess if the proposed method of weighting and aging individual classifiers in the ensemble is valuable proposition compared with the methods which do not include aging or weighting techniques.

3.1   Set-Up

All experiments were carried out on the SEA dataset describes in [15]. Each object belongs to the on of two classes and is described by 3 numeric attributes with value between 0 and 10, but only two of them are relevant. Object belongs to class 1 (TRUE) if $arg_1 + arg_2 < \phi$ and to class 2 (FALSE) if $arg_1 + arg_2 \geq \phi$. ϕ is a threshold between two classes, so different thresholds correspond to different concepts (models).Thus, all generated dataset is linearly separable, but we add 5% noise, which means that class label for some samples is changed, with expected value equal to 0. The number of objects, noise and the set of concepts are set by user. We simulated drift by instant random model change.

Algorithm 1. Weighted Aging Ensemble (WAE)

Require: input data stream, data chunk size, classifier training procedure, ensemble
    size L
  1: $i := 1$
  2: **repeat**
  3:    collect new data chunk DS_i
  4:    train classifier Ψ_i on the basis of DS_i
  5:    add Ψ_i to the classifier ensemble Π
  6:    **if** $i > L$ **then**
  7:        $\Psi_{k+1} = \Psi_i$
  8:        $\Pi_t = \emptyset$
  9:        $GD_t = 0$
10:        **for** $j = 1$ to $L + 1$ **do**
11:            **if** $\mathcal{GD}(\Pi \backslash \Psi_i)$ (calculated according to (1)) $> GD_t$ **then**
12:                $\Pi_t = \Pi \backslash \Psi_i$
13:            **end if**
14:        **end for**
15:        $\Pi = \Pi_t$
16:    **end if**
17:    $w := 0$
18:    **for** $j = 1$ to L **do**
19:        calculate $w(\Psi_i)$ according to (2)
20:        $w := w + w(\Psi_i)$
21:    **end for**
22:    **for** $j = 1$ to L **do**
23:        $w(\Psi_i) := \frac{w(\Psi_i)}{w}$
24:    **end for**
25:    $i := i + 1$
26: **until** end of the input data stream

For each of the experiments we decided to form homogenous ensemble i.e., ensemble which consists of the classifier using the same model. We repeated experiments for Naive Bayes, decision tree trained by C4.5 [13], and SVM with polynomial kernel trained by the sequential minimal optimization method (SMO) [12].

During each of the experiment we tried to evaluate dependency between data chunk sizes (which were fixed on 50, 100, 150, 200) and overall classifier quality (accuracy and standard deviation) for the following ensembles:

1. $w0a0$ - an ensemble using majority voting without aging.
2. $w1a0$ - an ensemble using weighted voting without aging, where weight assigned to a given classifier is inversely proportional to its accuracy.
3. $w1a1$ - an ensemble using weighted voting with aging, where weight assigned to a given classifier is calculated according to (2).

Method of ensemble pruning was the same for each ensemble and presented in Alg.1. The only difference was line 19 of the pseudocode what was previously described. All experiments were carried out in the Java environment using Weka classifiers [3].

3.2 Results

The results of experiments are presented in Fig.1-6. Fig. 1-3 show the accuracies of the tested ensembles for a chosen experiment. Unfortunately, because of the space limit we are not able to presents all extensive results, but they are available on demand from corresponding author. Fig.4-6 present overall accuracy and standard deviation for the tested methods and how they depend on data chunk size.

3.3 Discussion

On the basis of presented results we can formulate several observations. It does not surprise us that quality improvements for all tested method according to increasing data chunk size. Usually the WAE outperformed others, but the differences are quite small and only in the case of ensemble built on the basis of Naive Bayes classifiers the differences are statistical significant (t-test) [1] i.e., differences among different chunk sizes. The observation is useful because the bigger size of data chunk means that effort dedicated to building new models is smaller because they are being built rarely.

Another interesting observation is that the standard deviation is smaller for bigger data chunk and usually standard deviation of WAE is smallest among all

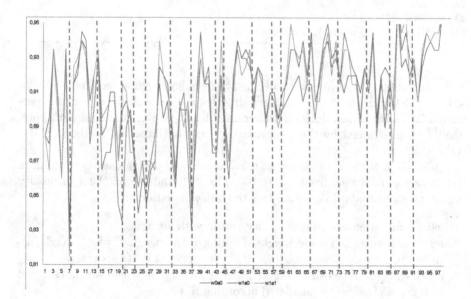

Fig. 1. Classification accuracy of the ensembles consist of Naive Bayes classifiers for the chunk size = 200. Vertical dotted lines indicate concept drift appearances.

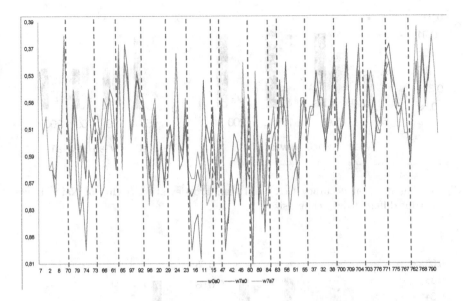

Fig. 2. Classification accuracy of the ensembles consist of C4.5 (decision tree) classifiers for the chunk size = 150. Vertical dotted lines indicate concept drift appearances.

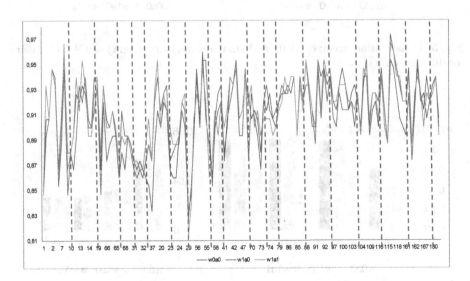

Fig. 3. Classification accuracy of the ensembles consist of SVM classifiers for the chunk size = 150. Vertical dotted lines indicate concept drift appearances.

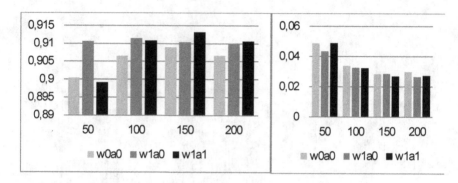

Fig. 4. Classification accuracy (left) and standard deviation (right) of Naive Bayes classifier for different data chunk sizes

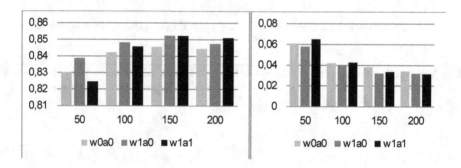

Fig. 5. Classification accuracy (left) and standard deviation (right) of C4.5 classifier for different data chunk sizes

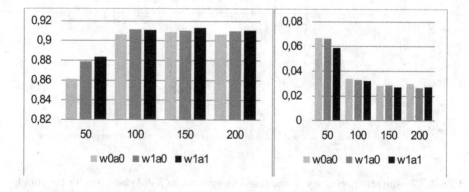

Fig. 6. Classification accuracy (left) and standard deviation (right) of SVM classifier for different data chunk sizes

tested methods. It means that the concept drift appearances have the weakest impact on the WAE accuracy.

We realize that the scope of the experiments we carried out is limited and derived remarks are limited to the tested methods and one dataset only. In this case formulating general conclusions is very risky, but the preliminary results are quite promising, therefore we would like to continue the work on WAE in the future.

4 Conclusions

The paper presented the original classifier for data stream classification tasks. Proposed WAE algorithm uses dynamic classifier ensemble i.e., its line-up is formed when new data chunk is come and the decision which classifier is chosen to the ensemble is made on the basis of General Diversity (diversity measure). The decision about object's label is made according to weighted voting where weight assigned to a given classifier depends on its accuracy (proportional) and how long the classifier participates in the ensemble (inversely proportional). The experiments conformed that proposed method can adapt to changing concept returning stable classifier. We would like to emphasize that we presented preliminary study on WAE which is a starting point for the future research. In the near future we are going to:

- carry out experiments on the wider number of datasets,
- evaluate WAE's behavior for more sharp sudden concept drift,
- evaluate usefulness of the other diversity measures for WAE's classifier ensemble pruning,
- assess more sophisticated combination rules based on support functions of individual classifiers,
- check if training set of different classifier model on the basis of new data chunk could have an impact on WAE's quality, because such an approach will lead to the more diverse heterogenous classifier ensemble.

Acknowledgment. The work was supported by the statutory funds of the Department of Systems and Computer Networks, Wroclaw University of Technology and by the Polish National Science Center under a grant N N519 650440 for the period 2011-2014.

References

1. Alpaydin, E.: Introduction to Machine Learning, 2nd edn. The MIT Press (2010)
2. Bifet, A., Holmes, G., Pfahringer, B., Read, J., Kranen, P., Kremer, H., Jansen, T., Seidl, T.: Moa: a real-time analytics open source framework. In: Gunopulos, D., Hofmann, T., Malerba, D., Vazirgiannis, M. (eds.) ECML PKDD 2011, Part III. LNCS, vol. 6913, pp. 617–620. Springer, Heidelberg (2011)
3. Hall, M., Frank, E., Holmes, G., Pfahringer, B., Reutemann, P., Witten, I.H.: The weka data mining software: an update. SIGKDD Explor. Newsl. 11(1), 10–18 (2009)

4. Hulten, G., Spencer, L., Domingos, P.: Mining time-changing data streams. In: Proceedings of the Seventh ACM SIGKDD International Conference on Knowledge Discovery and Data Mining, pp. 97–106 (2001)
5. Jacobs, R.A., Jordan, M.I., Nowlan, S.J., Hinton, G.E.: Adaptive mixtures of local experts. Neural Comput. 3, 79–87 (1991)
6. Klinkenberg, R., Renz, I.: Adaptive information filtering: Learning in the presence of concept drifts, pp. 33–40 (1998)
7. Kolter, J.Z., Maloof, M.A.: Dynamic weighted majority: a new ensemble method for tracking concept drift. In: Third IEEE International Conference on Data Mining, ICDM 2003, pp. 123–130 (November 2003)
8. Kuncheva, L.I.: Combining Pattern Classifiers: Methods and Algorithms. Wiley-Interscience (2004)
9. Lazarescu, M.M., Venkatesh, S., Bui, H.H.: Using multiple windows to track concept drift. Intell. Data Anal. 8(1), 29–59 (2004)
10. Muhlbaier, M.D., Topalis, A., Polikar, R.: Learn^{++}.nc: Combining ensemble of classifiers with dynamically weighted consult-and-vote for efficient incremental learning of new classes. IEEE Transactions on Neural Networks 20(1), 152–168 (2009)
11. Partridge, D., Krzanowski, W.: Software diversity: practical statistics for its measurement and exploitation. Information and Software Technology 39(10), 707–717 (1997)
12. Platt, J.C.: Fast training of support vector machines using sequential minimal optimization. In: Advances in Kernel Methods, pp. 185–208. MIT Press, Cambridge (1999)
13. Quinlan, J.R.: C4.5: Programs for Machine Learning. Morgan Kaufmann Series in Machine Learning. Morgan Kaufmann Publishers (1993)
14. Shipp, C.A., Kuncheva, L.: Relationships between combination methods and measures of diversity in combining classifiers. Information Fusion 3(2), 135–148 (2002)
15. Nick Street, W., Kim, Y.: A streaming ensemble algorithm (sea) for large-scale classification. In: Proceedings of the Seventh ACM SIGKDD International Conference on Knowledge Discovery and Data Mining, KDD 2001, pp. 377–382. ACM, New York (2001)
16. Wang, H., Fan, W., Yu, P.S., Han, J.: Mining concept-drifting data streams using ensemble classifiers. In: Proceedings of the Ninth ACM SIGKDD International Conference on Knowledge Discovery and Data Mining, KDD 2003, pp. 226–235. ACM, New York (2003)
17. Widmer, G., Kubat, M.: Learning in the presence of concept drift and hidden contexts. Mach. Learn. 23(1), 69–101 (1996)
18. Wolpert, D.H.: The supervised learning no-free-lunch theorems. In: Proc. 6th Online World Conference on Soft Computing in Industrial Applications, pp. 25–42 (2001)

An Analysis of Change Trends by Predicting from a Data Stream Using Neural Networks

Zbigniew Telec[1], Tadeusz Lasota[2], Bogdan Trawiński[1], and Grzegorz Trawiński[3]

[1] Wrocław University of Technology, Institute of Informatics,
Wybrzeże Wyspiańskiego 27, 50-370 Wrocław, Poland
[2] Wrocław University of Environmental and Life Sciences, Dept. of Spatial Management
ul. Norwida 25/27, 50-375 Wrocław, Poland
[3] Wrocław University of Technology, Faculty of Electronics,
Wybrzeże S. Wyspiańskiego 27, 50-370 Wrocław, Poland
{zbigniew.telec,bogdan.trawinski}@pwr.wroc.pl,
tadeusz.lasota@up.wroc.pl, grzegorz.trawinsky@gmail.com

Abstract. A method to predict from a data stream of real estate sales transactions based on ensembles of artificial neural networks was proposed. The approach consists in incremental expanding an ensemble by models built over successive chunks of a data stream. The predicted prices of residential premises computed by aged component models for current data are updated according to a trend function reflecting the changes of the market. The impact of different trend functions on the accuracy of ensemble neural models was investigated in the paper. The results indicate it is necessary to make selection of correcting functions appropriate to the nature of market changes.

Keywords: neural networks, data stream, sliding windows, ensembles, predictive models, trend functions, property valuation.

1 Introduction

Processing data streams demands elaboration of specific methods because they should take into account memory limitations, short processing times, and single scans of incoming data. Many strategies and techniques for mining data streams have been devised. Gaber in his recent overview paper categorizes them into four main groups: two-phase techniques, Hoeffding bound-based, symbolic approximation-based, and granularity-based ones [7]. Much effort is devoted to the issue of concept drift which occurs when data distributions and definitions of target classes change over time [6], [21], [29], [30]. Among the instantly growing methods of handling concept drift in data streams Tsymbal distinguishes three basic approaches, namely instance selection, instance weighting, and ensemble learning [27]. The latter has been systematically overviewed in [14], [22]. In adaptive ensembles, component models are generated from sequential blocks of training instances. When a new block arrives, models are examined and then discarded or modified based on the results of the evaluation. Several methods have been proposed for that, e.g. accuracy weighted ensembles [28] and accuracy updated ensembles [4]. In [2], [3] Bifet et al. proposed two bagging

H.L. Larsen et al. (Eds.): FQAS 2013, LNAI 8132, pp. 589–600, 2013.

methods to process concept drift in a data stream: ASHT Bagging using trees of different sizes, and ADWIN Bagging employing a change detector to decide when to discard underperforming ensemble members.

This study is the continuation of our recent works on predicting from a data stream using the genetic fuzzy system ensembles [23], [24], [25]; we spread our approach over artificial neural networks. Our approach was inspired by the observation of a real estate market of in one big Polish city in recent years when it experienced a violent fluctuations of residential premises prices. Our method consists in the utilization of aged models to compose ensembles and correction of the output provided by component models by means of trend functions reflecting the changes of prices in the market over time. So far we have investigated among others the impact of ensemble size and different trend functions on the accuracy of single and ensemble fuzzy models. Moreover, we proved the usefulness of ensemble approach incorporating the correction of individual component model output.

The goal of research reported in this paper is to apply a non-incremental artificial neural networks (ANNs) to build reliable predictive models from a data stream. In this paper we explore multilayer perceptron within the above mentioned framework.

2   Motivation and ANN Ensemble Approach

Professional appraisers are forced to use many, very often inconsistent and imprecise sources of information, and their familiarity with a real estate market and the land where properties are located is frequently incomplete. Moreover, they have to consider various price drivers and complex interrelation among them. The appraisers should make on-site inspection to estimate qualitative attributes of a given property as well as its neighbourhood. They have also to assess such subjective factors as location attractiveness and current trend and vogue. So, their estimations are to a great extent subjective and are based on uncertain knowledge, experience, and intuition rather than on objective data. So, the appraisers should be supported by automated valuation systems which often incorporate data driven models for premises valuation developed employing sales comparison method. The data driven models, considered in the paper, were generated using real-world data on sales transactions taken from a cadastral system and a public registry of real estate transactions.

So far, we have investigated several methods to construct regression models to assist with real estate appraisal based on neural networks. Our earlier works focused on exploring single models built by means of neural networks [12], [18], [19] and next on ensembles created using various resampling techniques [1], [11], [13], [20]. Finally, we constructed property valuation models using such more advanced fusion and ensemble techniques as mixture of experts [9], [15], random subspaces, random forests and rotation forests [16], [17]. The methods utilized in these investigations required that the whole datasets had to be available before the process of training models started.

The outline of the *ANN* ensemble approach to predict from a data stream is depicted in Fig. 1. The data stream is partitioned into data chunks according to the periods of a constant length t_c. Each time interval determines the shift of a sliding time window which comprises training data to create *ANN* models. The window is shifted step by step of a period t_s in the course of time. The length of the sliding window t_w is equal to the multiple of t_c so that $t_w=jt_c$, where $j=1,2,3,...$. The window determines the scope of training data to generate from scratch a property valuation model, in our case ANN_i. It is assumed that the models generated over a given training dataset is valid for the next interval which specifies the scope for a test dataset. Similarly, the interval t_t which delineates a test dataset is equal to the multiple of t_c so that $t_t=kt_c$, where $k=1,2,3,...$. The sliding window is shifted step by step of a period t_s in the course of time, and likewise, the interval t_s is equal to the multiple of t_c so that $t_s=lt_c$, where $l=1,2,3,...$.

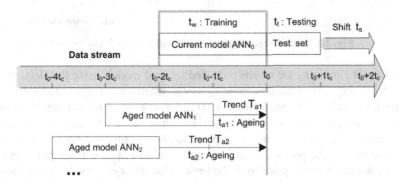

Fig. 1. ANN ensemble approach to predict from a data stream

We consider in Fig. 1 a point of time t_0 at which the current model ANN_0 was built over data that came in between time t_0-2t_c and t_0. The models created earlier, i.e. ANN_1, ANN_2, etc. have aged gradually and in consequence their accuracy has deteriorated. However, they are neither discarded nor restructured but utilized to compose an ensemble so that the current test dataset is applied to each component ANN_i. However, in order to compensate ageing, their output produced for the current test dataset is updated using trend functions $T(t)$. As the functions to model the trends of price changes the polynomials of the degree from one to five were employed, denoted in the rest of the paper by *T1*, *T2*,..,*T5*, respectively. The trends were determined over two time periods: shorter and longer ones. The shorter periods encompassed the length of a sliding window plus model ageing intervals, i.e. t_w plus, t_{ai} for a given aged model ANN_i. In turn, the longer periods took into account all data since the beginning of the stream. Hence, the shorter periods are denoted by *Age* and the longer periods are marked by *Beg* in the symbols of methods used in tables presenting the experimental results further on in the paper. In order to be concise, in remaining text of the paper we will call the former *Age Trends* and the latter *Beg Trends*.

Moreover, we proposed two different methods of updating the prices of premises according to the trends of the value changes over time. The first one based on the difference between a price and a trend value in a given time point and we called it the *Delta* method. In turn, the second technique utilized the ratio of the price to the trend value and it was named the *Ratio* method of price correction [22]. In the present paper we utilize only the former method.

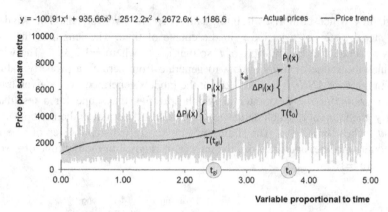

Fig. 2. Correcting the output of aged models using *Delta* method

The idea of correcting the results produced by aged models using the *Delta* method is depicted in Fig. 2. For the time point $t_{gi}=t_0-t_{ai}$, when a given aged model ANN_i was generated, the value of a trend function $T(t_{gi})$, i.e. average price per square metre, is computed. The price of a given premises, i.e. an instance of a current test dataset, characterised by a feature vector x, is predicted by the model ANN_i. Next, the total price is divided by the premises usable area to obtain its price per square metre $P_i(x)$. Then, the deviation of the price from the trend value $\Delta P_i(x)=P_i(x)-T(t_{gi})$ is calculated. The corrected price per square metre of the premises $\dot{P}_i(x)$ is worked out by adding this deviation to the trend value in the time point t_0 using the formula $\dot{P}_i(x)=\Delta P_i(x)+T(t_0)$, where $T(t_0)$ is the value of a trend function in t_0. Finally, the corrected price per square metre $\dot{P}_i(x)$ is converted into the corrected total price of the premises by multiplying it by the premises usable area.

3   Experimental Setup

The investigation was conducted with our experimental system implemented in Matlab environment. The system was designed to carry out research into machine learning algorithms using various resampling methods and constructing and evaluating ensemble models for regression problems. We have recently extended our system to include the functions of building ensembles over a data stream and weighting their component models. The trends are modelled using the Matlab function *polyfit(x,y,n)*, which finds the coefficients of a polynomial $p(x)$ of degree n that fits the y data by minimizing the sum of the squares of the deviations of the data from the model (least-squares fit).

Real-world dataset used in experiments was drawn from an unrefined dataset containing above 100 000 records referring to residential premises transactions accomplished in one Polish big city with the population of 640 000 within 14 years from 1998 to 2011. In this period the majority of transactions were made with non-market prices when the council was selling flats to their current tenants on preferential terms. First of all, transactional records referring to residential premises sold at market prices were selected. Then, the dataset was confined to sales transaction data of residential premises (apartments) where the land was leased on terms of perpetual usufruct. The other transactions of premises with the ownership of the land were omitted due to the conviction of professional appraisers stating that the land ownership and lease affect substantially the prices of apartments and therefore they should be used separately for sales comparison valuation methods. The final dataset counted 9795 samples. Due to the fact we possessed the exact date of each transaction we were able to order all instances in the dataset by time, so that it can be regarded as a data stream. Four following attributes were pointed out as main price drivers by professional appraisers: usable area of a flat (*Area*), age of a building construction (*Age*), number of storeys in the building (Storeys), the distance of the building from the city centre (*Centre*), in turn, price of premises (*Price*) was the output variable. In order to characterize quantitatively the data stream the sizes of one-year datasets are given in Table 1.

Table 1. Number of instances in one-year datasets

1998	1999	2000	2001	2002	2003	2004
446	646	554	626	573	790	774
2005	**2006**	**2007**	**2008**	**2009**	**2010**	**2011**
740	776	442	734	821	1296	577

The property valuation models were built from scratch by artificial neural networks over chunks of data stream determined by a sliding window which was 12 months long. The multilayer perceptron Matlab function *mlp* was used with four inputs, a three neuron hidden layer and one output. The number of epochs to learn each *mlp* network was equal to 100. As the performance measure the root mean square error (*RMSE*) was used.

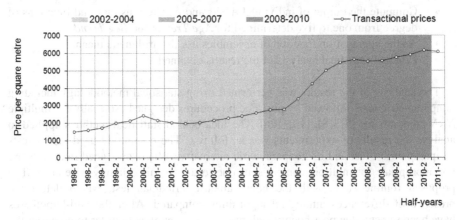

Fig. 3. Change trend of average transactional prices per square metre over time

The results of evaluating experiments were considered within three periods of time, namely 1) 2002-2004, 2) 2005-2007, and 3) 2008-2010 marked with grey shades in Fig. 3. In the first period a modest rise of real estate prices just before Poland entered the European Union (EU) could be observed, in the second one the prices of residential premises were increasing rapidly during the worldwide real estate bubble. The third one corresponds to a modest growth of prices after the bubble burst and during the global financial crisis. This period was characterized by unstable real estate market and great fluctuations of prices due to nervous behaviour of both buyers and sellers. The trend of premises price changes over 14 years from 1998 to 2011, shown in Fig. 3, can be modelled by the polynomial function of degree four.

Based on the results of our previous study, we were able to determine following parameters of our experiments including two phases: generating single *ANN* models and building *ANN* ensembles.

Generating single ANN models
- Set the length of the sliding window to 12 months, $t_w = 12$.
- Set the starting point of the sliding window, i.e. its right edge, to 2000-01-01 and the terminating point to 2010-12-01.
- Set the shift of the sliding window to 1 month, $t_s = 1$.
- Move the window from starting point to terminating point with the step $t_s = 1$.
- At each stage generate a *ANN* from scratch over a training set delineated by the window. In total 108 *ANNs* were built.

Building ANN ensembles
- Select periods characterizing different fluctuation of real estate market, i.e. 2002-2004, 2005-2007, and 2008-2010.
- For each three-year period built 36 ensembles composed of 24 ageing *ANNs* each. An ensemble is created in the way described in Section 2 with the shift equal to one month, $t_s = 1$.
- Take test sets actual for each t_0 over a period of 3 months, $t_t = 3$.
- Compute the output of individual *ANNs* and update it using trend functions of degree from one to five determined for *Age Trends* and *Beg Trends*.
- As the aggregation function of ensembles use the arithmetic mean.
- Conduct statistical analysis of the results obtained.

The analysis of the results was performed using statistical methodology including nonparametric tests followed by post-hoc procedures designed especially for multiple $N \times N$ comparisons [5], [8], [10], [26]. The idea behind statistical methods applied to analyse the results of experiments was as follows. The commonly used paired tests i.e. parametric t-test and its nonparametric alternative Wilcoxon signed rank tests are not appropriate for multiple comparisons due to the so called family-wise error. The proposed routine starts with the nonparametric Friedman test, which detect the presence of differences among all algorithms compared. After the null-hypotheses have been rejected the post-hoc procedures should be applied in order to point out the particular pairs of algorithms which produce differences. For $N \times N$ comparisons nonparametric Nemenyi's, Holm's, Shaffer's, and Bergamnn-Hommel's procedures are recommended.

4   Statistical Analysis of Results

The goal of the statistical analysis was to compare the accuracy of the ensembles composed of *ANN* models which output was updated using trend functions *T1*, *T2*, *T3*, *T4*, and *T5* for *Age Trends* and *Beg Trends*. Additionally, the results not updated with any trend function were utilized for comparison, they were denoted by *noT*. Nonparametric tests of statistical significance were conducted within three periods 2002-2004, 2005-2007, and 2008-2010. Within each of these periods 36 values of *RMSE*, determined for each trend function, constituted the points of observation. Finally, similar analysis was done for all years together.

The Friedman test performed in respect of *RMSE* values of the ensembles showed that there were significant differences among ensembles within each time period considered. Average ranks of individual models for *Age Trends* and *Beg Trends* are shown in Tables 2 and 3, respectively, where the lower rank value the better model. For *Age Trends* in can be noticed in Table 2 that *noT* was in the last place within all periods but one. In turn, the differences in positions taken by other ensembles were rather slight. This observation was confirmed by the analysis of the number of null-hypotheses rejected by the post-hoc procedures (see Table 4). For three periods of time the number was equal to five and meant that statistically significant differences occurred only between *noT* and other ensembles. Therefore, the results provided by the post-hoc procedures for Age Trends are not presented here in detail.

Table 2. Average rank positions of *Age Trends* produced by Friedman tests

2002-2004		2005-2007		2008-2010		All years	
Rank	Trend	Rank	Trend	Rank	Trend	Rank	Trend
2.61	AgeT5	2.56	AgeT2	2.61	AgeT1	2.93	AgeT2
3.00	AgeT4	2.78	AgeT3	3.08	AgeT2	2.98	AgeT5
3.08	AgeT3	2.92	AgeT4	3.08	AgeT4	3.00	AgeT4
3.14	AgeT2	3.19	AgeT5	3.14	AgeT5	3.15	AgeT3
4.47	noT	3.56	AgeT1	3.58	AgeT3	3.62	AgeT1
4.69	AgeT1	6.00	noT	5.50	noT	5.32	noT

Table 3. Average rank positions of *Beg Trends* produced by Friedman tests

2002-2004		2005-2007		2008-2010		All years	
Rank	Trend	Rank	Trend	Rank	Trend	Rank	Trend
2.11	BegT5	1.86	BegT4	1.94	BegT3	2.65	BegT4
2.44	BegT4	1.97	BegT5	2.00	BegT2	2.78	BegT5
3.00	BegT1	2.17	BegT3	3.64	BegT4	2.81	BegT3
4.06	noT	4.03	BegT2	3.78	BegT1	3.70	BegT2
4.31	BegT3	4.97	BegT1	4.25	BegT5	3.92	BegT1
5.08	BegT2	6.00	noT	5.39	noT	5.15	noT

Table 4. Number of hypotheses rejected by post-hoc procedures out of 15 possible ones

Trends	2002-2004	2005-2007	2008-2010	All years
Age Trends	8	5	5	5
Beg Trends	8	10	11	11

For *Beg Trends* it can be observed in Table 3 that *AgeT4* and *AgeT5* were in the first two places in all periods but 2008-2010. In turn, *noT* took the last positions in all periods but 2002-2004. It is shown in Table 4 that the most powerful Shaffer's, and Bergmann-Hommel's post-hoc procedures rejected from 8 to 11 null-hypotheses out of 15 possible pairs of ensembles.

Adjusted *p-values* for Nemenyi's, Holm's, Shaffer's, and Bergmann-Hommel's post-hoc procedures for *N×N* comparisons for all possible 15 pairs of ensembles for *Beg Trends* are shown in Tables 5, 6, 7, and 8 for the 2002-2004, 2005-2007, 2008-2010, and *All year* periods, respectively. For comparison, the results of paired Wilcoxon tests are placed in all tables. The *p-values* indicating the statistically significant differences between given pairs of ensembles are marked with italics. The significance level considered for the null hypothesis rejection was 0.05. Following main observations could be done based on the most powerful Shaffer's, and Bergmann-Hommel's post-hoc procedures.

For the period 2002-2004 *BegT4* and *BegT5* ensembles revealed significantly better performance than *BegT2, BegT3, and noT* models. There were no significant differences among the *BegT1*, *BegT4*, and Beg*T5* ensembles as well as among *BegT2, BegT3, and noT* models.

For the period 2005-2007 *BegT3*, *BegT4* and *BegT5* ensembles produced significantly better results than Beg*T1, BegT2, and noT* models. No significant differences among *BegT3*, *BegT4,* and *BegT5* were shown.

For the period 2008-2010 *BegT2* and *BegT3* ensembles provided significantly better performance than other models. No significant differences were shown between *BegT2* and *BegT3* as well as among *BegT1*, *BegT4* and *BegT5* ensembles. In turn, noT yielded significantly worse results than any other model.

For *All years* together the *BegT3*, *BegT4* and *BegT5* ensembles revealed significantly better performance than other models. No significant differences were discovered among *BegT3*, *BegT4* and *BegT5* as well as between *BegT1* and *BegT2* ensembles. In turn, *noT* led to significantly worse performance than the other ensembles.

For all considered periods of time the paired Wilcoxon test allowed for rejection of a greater number of null hypotheses than post-hoc procedures did.

Table 5. Adjusted p-values for *N×N* comparisons of *Beg Trends* methods over 2002-2004

Meth. vs Meth.	pWilcox	pNeme	pHolm	pShaf	pBerg
BegT2 vs BegT5	*0.000002*	*2.37E-10*	*2.37E-10*	*2.37E-10*	*2.37E-10*
BegT2 vs BegT4	*0.000002*	*3.26E-08*	*3.04E-08*	*2.17E-08*	*2.17E-08*
BegT3 vs BegT5	*0.000000*	*9.71E-06*	*8.42E-06*	*6.47E-06*	*6.47E-06*
BegT1 vs BegT2	*0.000001*	*3.46E-05*	*2.77E-05*	*2.31E-05*	*1.61E-05*
noT vs BegT5	*0.000173*	*1.55E-04*	*1.14E-04*	*1.04E-04*	*7.25E-05*
BegT3 vs BegT4	*0.000000*	*3.65E-04*	*2.44E-04*	*2.44E-04*	*1.46E-04*
noT vs BegT4	*0.000185*	*0.003878*	*0.002327*	*0.001810*	*0.001034*
BegT1 vs BegT3	0.875162	*0.046038*	*0.024554*	*0.021484*	*0.012277*
noT vs BegT1	*0.000009*	0.250140	0.116732	0.116732	*0.050028*
noT vs BegT2	*0.000111*	0.296474	0.118589	0.118589	0.118589
BegT1 vs BegT5	*0.001084*	0.657297	0.219099	0.175279	0.175279
BegT2 vs BegT3	*0.000341*	1.000000	0.311040	0.311040	0.175279
BegT1 vs BegT4	*0.001146*	1.000000	0.623136	0.623136	0.415424
BegT4 vs BegT5	*0.031371*	1.000000	0.899384	0.899384	0.899384
noT vs BegT3	0.109051	1.000000	0.899384	0.899384	0.899384

Table 6. Adjusted p-values for $N \times N$ comparisons of *Beg Trends* methods over 2005-2007

Meth. vs Meth.	pWilcox	pNeme	pHolm	pShaf	pBerg
noT vs BegT4	0.000000	9.34E-20	9.34E-20	9.34E-20	9.34E-20
noT vs BegT5	0.000000	9.89E-19	9.23E-19	6.59E-19	6.59E-19
noT vs BegT3	0.000000	5.29E-17	4.58E-17	3.52E-17	2.47E-17
BegT1 vs BegT4	0.000000	2.58E-11	2.07E-11	1.72E-11	1.72E-11
BegT1 vs BegT5	0.000000	1.53E-10	1.12E-10	1.02E-10	6.13E-11
BegT1 vs BegT3	0.000000	2.98E-09	1.99E-09	1.99E-09	7.94E-10
BegT2 vs BegT4	0.000000	1.34E-05	8.05E-06	6.26E-06	6.26E-06
BegT2 vs BegT5	0.000000	4.71E-05	2.51E-05	2.20E-05	1.26E-05
noT vs BegT2	0.000000	1.16E-04	5.41E-05	5.41E-05	4.64E-05
BegT2 vs BegT3	0.000000	3.65E-04	1.46E-04	1.46E-04	7.31E-05
noT vs BegT1	0.000000	0.296474	0.098825	0.079060	0.079060
BegT1 vs BegT2	0.000000	0.483145	0.128839	0.128839	0.128839
BegT3 vs BegT4	0.171683	1.000000	1.000000	1.000000	1.000000
BegT3 vs BegT5	0.186940	1.000000	1.000000	1.000000	1.000000
BegT4 vs BegT5	0.648673	1.000000	1.000000	1.000000	1.000000

Table 7. Adjusted p-values for $N \times N$ comparisons of *Beg Trends* methods over 2008-2010

Meth. vs Meth.	pWilcox	pNeme	pHolm	pShaf	pBerg
noT vs BegT3	0.000000	8.49E-14	8.49E-14	8.49E-14	8.49E-14
noT vs BegT2	0.000000	2.29E-13	2.14E-13	1.53E-13	1.53E-13
BegT3 vs BegT5	0.000011	2.56E-06	2.22E-06	1.71E-06	1.71E-06
BegT2 vs BegT5	0.000731	5.03E-06	4.02E-06	3.35E-06	2.01E-06
BegT1 vs BegT3	0.000000	4.82E-04	3.54E-04	3.22E-04	2.25E-04
BegT1 vs BegT2	0.000000	8.31E-04	5.54E-04	5.54E-04	2.25E-04
noT vs BegT4	0.000080	0.001084	6.51E-04	5.54E-04	5.06E-04
BegT3 vs BegT4	0.000016	0.001826	9.74E-04	8.52E-04	7.30E-04
BegT2 vs BegT4	0.001146	0.003028	0.001413	0.001413	8.08E-04
noT vs BegT1	0.000000	0.003878	0.001551	0.001551	0.001034
noT vs BegT5	0.000408	0.147021	0.049007	0.039206	0.029404
BegT4 vs BegT5	0.001351	1.000000	0.663147	0.663147	0.663147
BegT1 vs BegT5	0.509354	1.000000	0.852644	0.852644	0.663147
BegT1 vs BegT4	0.648673	1.000000	1.000000	1.000000	1.000000
BegT2 vs BegT3	0.813698	1.000000	1.000000	1.000000	1.000000

Table 8. Adjusted p-values for $N \times N$ comparisons of *Beg Trends* methods over All years

Meth. vs Meth.	pWilcox	pNeme	pHolm	pShaf	pBerg
noT vs BegT4	0.000000	1.39E-21	1.39E-21	1.39E-21	1.39E-21
noT vs BegT5	0.000000	1.91E-19	1.78E-19	1.27E-19	1.27E-19
noT vs BegT3	0.000000	5.29E-19	4.59E-19	3.53E-19	2.47E-19
noT vs BegT2	0.000000	2.10E-07	1.68E-07	1.40E-07	9.78E-08
BegT1 vs BegT4	0.000000	9.41E-06	6.90E-06	6.27E-06	6.27E-06
noT vs BegT1	0.000000	1.98E-05	1.32E-05	1.32E-05	9.22E-06
BegT1 vs BegT5	0.000002	1.15E-04	6.93E-05	5.39E-05	4.62E-05
BegT1 vs BegT3	0.000000	1.91E-04	1.02E-04	8.92E-05	5.10E-05
BegT2 vs BegT4	0.000064	5.07E-04	2.37E-04	2.37E-04	2.03E-04
BegT2 vs BegT5	0.000137	0.004138	0.001655	0.001655	0.000828
BegT2 vs BegT3	0.000000	0,006284	0.002095	0.001676	0.000838
BegT1 vs BegT2	0.000006	1.000000	1.000000	1.000000	1.000000
BegT4 vs BegT5	0.034175	1.000000	1.000000	1.000000	1.000000
BegT3 vs BegT4	0.766223	1.000000	1.000000	1.000000	1.000000
BegT3 vs BegT5	0.892718	1.000000	1.000000	1.000000	1.000000

5   Conclusions and Future Work

In the paper we reported our study of the method to predict from a data stream of real estate sales transactions based on ensembles of artificial neural networks, namely multilayer perceptrons. Our ensemble approach consists in incremental expanding an ensemble by models built from scratch over successive chunks of a data stream determined by a sliding window. The predicted prices of residential premises computed by aged component models for current data are updated according to trend functions which model the changes of the market. The impact of different trend functions on the accuracy of ensemble neural models was investigated in the paper.

We conducted first series of evaluating experiments using real-world data taken from cadastral systems. They consisted in generating ensembles for 108 points of time and then comparing their predictive accuracy using nonparametric tests of statistical significance adequate for multiple comparisons. As the functions to model the trends of price changes the polynomials of degree from one to five were employed. The trends were determined over two time periods: shorter and longer ones. The shorter periods encompassed the length of a sliding window plus model ageing time whereas the longer ones took into account all data since the beginning of the stream. The method of correcting the output of component models was based on the difference between a predicted price and a trend value in a given time point.

We analysed the ensemble performance within three periods of time characterized by different rates of premises price growth and different stability of real estate market. The results we obtained are not consistent which indicates it is necessary to make selection of correcting functions appropriate to the nature of market changes.

For the time period 1) 2002-2004, by the moderate price growth and relative stability of the market, *BegT4* and *BegT5* led to significantly better performance than the other models. In turn, for the time period 2) 2005-2007, when the process were soaring due to the real estate bubble, *BegT3* and again *BegT4* and *BegT5* revealed the best performance. And finally, for the time period 3) 2008-2010, characterized by the market instability due to the global financial crisis, *BegT2* and *BegT3* provided the best results. For *All years* together the ensembles composed of aged models which output was corrected with polynomial functions of higher degrees, i.e. *BegT3*, *BegT4* and *BegT5* yielded the best accuracy.

The other conclusions are as follows. The trends determined over shorter time periods led to the rejection of the smaller number of null-hypotheses than the trends over longer periods did. The paired Wilcoxon test can lead to over-optimistic decisions compared to post-hoc nonparametric procedures adequate for multiple comparisons, because it allowed for the rejection of a greater number of null hypotheses.

The study opens the area for our further research into the selection of the best parameters of the proposed method including the number of aged models encompassed by an ensemble as well as the selection of the degree of a trend function adequate for the dynamics of a given time period. We also intend to conduct experiments employing as base learning algorithms other methods capable of learning from concept drifts such as: decision trees, recurrent neural networks, support vector regression, etc. Moreover, we plan to compare the outcome produced by proposed ensembles with human based predictions.

Acknowledgments. This paper was partially supported by the statutory funds of the Institute of Informatics, Wrocław University of Technology and the Polish National Science Centre under grant no. N N516 483840.

References

1. Bańczyk, K., Kempa, O., Lasota, T., Trawiński, B.: Empirical Comparison of Bagging Ensembles Created Using Weak Learners for a Regression Problem. In: Nguyen, N.T., Kim, C.-G., Janiak, A. (eds.) ACIIDS 2011, Part II. LNCS (LNAI), vol. 6592, pp. 312–322. Springer, Heidelberg (2011)
2. Bifet, A., Holmes, G., Pfahringer, B., Gavaldà, R.: Improving Adaptive Bagging Methods for Evolving Data Streams. In: Zhou, Z.-H., Washio, T. (eds.) ACML 2009. LNCS (LNAI), vol. 5828, pp. 23–37. Springer, Heidelberg (2009)
3. Bifet, A., Holmes, G., Pfahringer, B., Kirkby, R., Gavalda, R.: New ensemble methods for evolving data streams. In: Elder IV, J.F., et al. (eds.) KDD 2009, pp. 139–148. ACM Press, New York (2009)
4. Brzeziński, D., Stefanowski, J.: Accuracy Updated Ensemble for Data Streams with Concept Drift. In: Corchado, E., Kurzyński, M., Woźniak, M. (eds.) HAIS 2011, Part II. LNCS (LNAI), vol. 6679, pp. 155–163. Springer, Heidelberg (2011)
5. Demšar, J.: Statistical comparisons of classifiers over multiple data sets. Journal of Machine Learning Research 7, 1–30 (2006)
6. Elwell, R., Polikar, R.: Incremental Learning of Concept Drift in Nonstationary Environments. IEEE Transactions on Neural Networks 22(10), 1517–1531 (2011)
7. Gaber, M.M.: Advances in data stream mining. Wiley Interdisciplinary Reviews: Data Mining and Knowledge Discovery 2(1), 79–85 (2012)
8. García, S., Herrera, F.: An Extension on "Statistical Comparisons of Classifiers over Multiple Data Sets" for all Pairwise Comparisons. Journal of Machine Learning Research 9, 2677–2694 (2008)
9. Graczyk, M., Lasota, T., Telec, Z., Trawiński, B.: Application of mixture of experts to construct real estate appraisal models. In: Graña Romay, M., Corchado, E., Garcia Sebastian, M.T. (eds.) HAIS 2010, Part I. LNCS (LNAI), vol. 6076, pp. 581–589. Springer, Heidelberg (2010)
10. Graczyk, M., Lasota, T., Telec, Z., Trawiński, B.: Nonparametric Statistical Analysis of Machine Learning Algorithms for Regression Problems. In: Setchi, R., Jordanov, I., Howlett, R.J., Jain, L.C. (eds.) KES 2010, Part I. LNCS (LNAI), vol. 6276, pp. 111–120. Springer, Heidelberg (2010)
11. Graczyk, M., Lasota, T., Trawiński, B., Trawiński, K.: Comparison of Bagging, Boosting and Stacking Ensembles Applied to Real Estate Appraisal. In: Nguyen, N.T., Le, M.T., Świątek, J. (eds.) ACIIDS 2010. LNCS (LNAI), vol. 5991, pp. 340–350. Springer, Heidelberg (2010)
12. Graczyk, M., Lasota, T., Trawiński, B.: Comparative Analysis of Premises Valuation Models Using KEEL, RapidMiner, and WEKA. In: Nguyen, N.T., Kowalczyk, R., Chen, S.-M. (eds.) ICCCI 2009. LNCS (LNAI), vol. 5796, pp. 800–812. Springer, Heidelberg (2009)
13. Kempa, O., Lasota, T., Telec, Z., Trawiński, B.: Investigation of bagging ensembles of genetic neural networks and fuzzy systems for real estate appraisal. In: Nguyen, N.T., Kim, C.-G., Janiak, A. (eds.) ACIIDS 2011, Part II. LNCS (LNAI), vol. 6592, pp. 323–332. Springer, Heidelberg (2011)

14. Kuncheva, L.I.: Classifier ensembles for changing environments. In: Roli, F., Kittler, J., Windeatt, T. (eds.) MCS 2004. LNCS, vol. 3077, pp. 1–15. Springer, Heidelberg (2004)

15. Lasota, T., Londzin, B., Trawiński, B., Telec, Z.: Investigation of Mixture of Experts Applied to Residential Premises Valuation. In: Selamat, A., Nguyen, N.T., Haron, H. (eds.) ACIIDS 2013, Part II. LNCS, vol. 7803, pp. 225–235. Springer, Heidelberg (2013)

16. Lasota, T., Łuczak, T., Trawiński, B.: Investigation of Random Subspace and Random Forest Methods Applied to Property Valuation Data. In: Jędrzejowicz, P., Nguyen, N.T., Hoang, K. (eds.) ICCCI 2011, Part I. LNCS, vol. 6922, pp. 142–151. Springer, Heidelberg (2011)

17. Lasota, T., Łuczak, T., Trawiński, B.: Investigation of Rotation Forest Method Applied to Property Price Prediction. In: Rutkowski, L., Korytkowski, M., Scherer, R., Tadeusiewicz, R., Zadeh, L.A., Zurada, J.M. (eds.) ICAISC 2012, Part I. LNCS, vol. 7267, pp. 403–411. Springer, Heidelberg (2012)

18. Lasota, T., Makos, M., Trawiński, B.: Comparative Analysis of Neural Network Models for Premises Valuation using SAS Enterprise Miner. In: Nguyen, N.T., et al. (eds.) New Challenges in Computational Collective Intelligence. SCI, vol. 244, pp. 337–348. Springer, Berlin (2009)

19. Lasota, T., Mazurkiewicz, J., Trawiński, B., Trawiński, K.: Comparison of Data Driven Models for the Validation of Residential Premises using KEEL. International Journal of Hybrid Intelligent Systems 7(1), 3–16 (2010)

20. Lasota, T., Telec, Z., Trawiński, G., Trawiński, B.: Empirical Comparison of Resampling Methods Using Genetic Neural Networks for a Regression Problem. In: Corchado, E., Kurzyński, M., Woźniak, M. (eds.) HAIS 2011, Part II. LNCS (LNAI), vol. 6679, pp. 213–220. Springer, Heidelberg (2011)

21. Maloof, M.A., Michalski, R.S.: Incremental learning with partial instance memory. Artificial Intelligence 154(1-2), 95–126 (2004)

22. Minku, L.L., White, A.P., Yao, X.: The Impact of Diversity on Online Ensemble Learning in the Presence of Concept Drift. IEEE Transactions on Knowledge and Data Engineering 22(5), 730–742 (2010)

23. Trawiński, B.: Evolutionary fuzzy system ensemble approach to model real estate market based on data stream exploration. Journal of Universal Computer Science 19(4), 539–562 (2013)

24. Trawiński, B., Lasota, T., Smętek, M., Trawiński, G.: An Analysis of Change Trends by Predicting from a Data Stream Using Genetic Fuzzy Systems. In: Nguyen, N.-T., Hoang, K., Jędrzejowicz, P. (eds.) ICCCI 2012, Part I. LNCS (LNAI), vol. 7653, pp. 220–229. Springer, Heidelberg (2012)

25. Trawiński, B., Lasota, T., Smętek, M., Trawiński, G.: An Attempt to Employ Genetic Fuzzy Systems to Predict from a Data Stream of Premises Transactions. In: Hüllermeier, E., Link, S., Fober, T., Seeger, B. (eds.) SUM 2012. LNCS (LNAI), vol. 7520, pp. 127–140. Springer, Heidelberg (2012)

26. Trawiński, B., Smętek, M., Telec, Z., Lasota, T.: Nonparametric Statistical Analysis for Multiple Comparison of Machine Learning Regression Algorithms. International Journal of Applied Mathematics and Computer Science 22(4), 867–881 (2012)

27. Tsymbal, A.: The problem of concept drift: Definitions and related work. Technical Report. Department of Computer Science, Trinity College, Dublin (2004)

28. Wang, H., Fan, W., Yu, P.S., Han, J.: Mining concept-drifting data streams using ensemble classifiers. In: Getoor, L., et al. (eds.) KDD 2003, pp. 226–235. ACM Press, New York (2003)

29. Widmer, G., Kubat, M.: Learning in the presence of concept drift and hidden contexts. Machine Learning 23, 69–101 (1996)

30. Zliobaite, I.: Learning under Concept Drift: an Overview. Technical Report. Faculty of Mathematics and Informatics, Vilnius University, Vilnius (2009)

An Autocompletion Mechanism for Enriched Keyword Queries to RDF Data Sources

Grégory Smits[1], Olivier Pivert[1], Hélène Jaudoin[1], and François Paulus[2]

[1] IRISA – University of Rennes 1
Technopole Anticipa 22305 Lannion Cedex, France
{gregory.smits,olivier.pivert,helene.jaudoin}@irisa.fr
[2] SEMSOFT
80, Avenue Des Buttes De Coesmes 35700 Rennes, France
francois.paulus@semsoft-corp.com

Abstract. This article introduces a novel keyword query paradigm for end users in order to retrieve precise answers from semantic data sources. Contrary to existing approaches, connectors corresponding to linking words or verbal structures from natural languages are used inside queries to specify the meaning of each keyword, thus leading to a complete and explicit definition of the intent of the search. An example of such a query is *name of person at the head of company and author of article about "business intelligence"*. In order to help users formulate such connected keywords queries and to translate them into SPARQL, an interactive mechanism based on autocompletion has been developed, which is presented in this article.

1 Introduction

Social networks have boosted the need for efficient and intuitive query interfaces to access large scale knowledge graphs whose semantics is defined by means of ontologies. The buzz around the launches of Google Knowledge Graph[1] and Facebook Graph Search[2] clearly illustrates how crucial that issue is. Before getting a great deal of media attention, this issue has been largely addressed by the scientific community where interesting approaches have been developed to propose keyword-based access to structured data stored in XML documents [5], relational databases [16] or ontologies [19]. Classical queries based on simple unordered lists of keywords may be highly ambiguous and their interpretation wrt. a data structure may lead to several concurrent analyses, and consequently to erroneous answers. To let users clearly express the meaning of their search, some works have been dedicated to the enrichment of the expressivity of query interfaces especially using natural language processing techniques [6]. In the spirit of the approaches proposed in [14,12], the approach presented in this article is half-way between Natural Language (NL) queries and classical keyword queries.

---

[1] http://www.google.com/insidesearch/features/search/knowledge.html
[2] https://www.facebook.com/about/graphsearch

H.L. Larsen et al. (Eds.): FQAS 2013, LNAI 8132, pp. 601–612, 2013.

In between NL queries that are difficult to analyze due to the high diversity of linguistic structures and classic keyword queries that are difficult to interpret due to the lack of information about the meaning of each keyword, a pseudo-formal model of keyword queries is introduced that relies on grammatical connectives that clearly express how keywords have to be linked. These connectives correspond to linking words (conjunctions, prepositions) or verbal structures that are used to express the meaning of the keyword they precede.

The use of connectives to link keywords together facilitates the translation of the keyword query into a formal language as SPARQL. An example of such a connected keyword query is: *name* **of** *person* **at the head of** *company* **and author of** *article* **about** *"business intelligence"* where one clearly remarks that the connectives in bold disambiguate the keywords they link. Keywords and connectives being attached to elements of the searchable ontology, an autocompletion mechanism is proposed to help users navigate in the ontology and interactively define connected keyword queries. Developed in a context of data mediation guided by a domain ontology, this approach is particularly relevant for native apps to access distributed heterogeneous data sources over the Web. Contributions of this work are twofold: i) the formalization of connected keyword queries addressed to a domain ontology and ii) the proposition of an efficient and suitable autocompletion system to help users make the most of this novel keyword query system and to support the translation of keyword queries into SPARQL queries. The rest of the article is organized as follows. Section 2 introduces the context of this work, whereas Section 3 gives a formalization of connected keyword queries wrt. to an ontology. Section 4 points out the crucial role of the autocompletion strategy that acts as the unique interface to the searchable data sources. Before concluding and drawing some perspectives for future works in Section 6, Section 5 compares the proposed approach with existing ones.

2   Context and Data Modeling

2.1   Applicative Context

The keyword query approach presented in this article has been developed on top of an integration system called AGGREGO server developed by the company SEMSOFT[3]. As shown in Figure 1, AGGREGO server relies on a mediation layer associated with a domain ontology corresponding to the searchable ontology and adaptors to make the integration of distributed and heterogeneous data sources easier [17]. SPARQL queries submitted to this mediation layer are rewritten in terms of views [13] describing the schemas of the searchable data sources.

This context of data mediation introduces interesting particularities but also strong constraints that have influenced the development of the keyword query approach presented in this paper. First, one takes advantage of the fact that AGGREGO server uses a mediation layer and a reference domain ontology to give access to distributed and semantically heterogeneous data sources. Thus,

---

[3] http://semsoft-corp.com

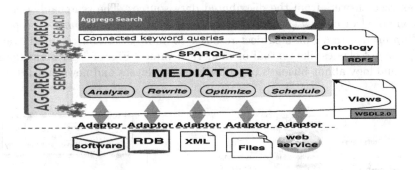

Fig. 1. Architecture of AGGREGO toolbox

queries addressed to this mediation system are defined according to the reference ontology. Then, contrary to existing keyword query approaches that rely on an index of the complete searchable vocabulary, data is not directly accessible in such a mediation architecture. This is why an adapted autocompletion strategy has been defined as explained in Section 4. The query strategy is thus particularly suitable to design a lightweight native app for tablet computers as only the domain ontology has to be stored on the device as well as the code of the query interface.

Figure 2 shows how the proposed keyword query strategy may be used to extract exact answers from very large distributed data sources like Twitter, Linkedin, BFM, Coface Services, Infolegale, etc.

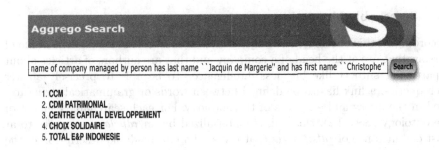

Fig. 2. AGGREGO SEARCH on top of AGGREGO SERVER

2.2   Data Structure

The searchable distributed data sources are defined by means of a central RDFS ontology. SPARQL queries submitted to the aforementioned mediation system exploit this central ontology and a view-based rewriting strategy is then used to extract answers from the distributed data sources. The searchable ontology is composed of classes (*rdf:class*) that are used to divide *resources*, properties (*rdfs:property*) that link a subject resource to an object resource, and instances (*rdf:type*) of classes like textual or numerical values. Figure 3 gives an example of an ontology about business organizations, employees and news articles.

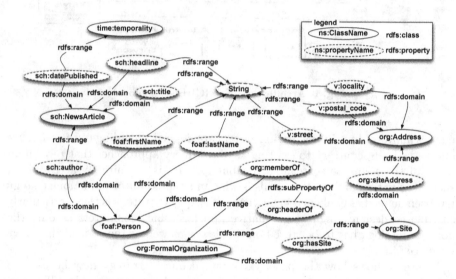

Fig. 3. Extract of an ontology

2.3   Searchable Vocabulary

According to this ontology, one may be interested in retrieving e.g. the name of persons that are at the head of a company and that are authors of articles about a particular subject like business intelligence. To be able to precisely answer such queries, a link has to be defined between words or grammatical structures used in the query and elements of the ontology. For each searchable element of the ontology (Sec. 2.2), this link is materialized by an *rdfs:label* property to at least one instance of *rdfs:Literal* that gives a human readable description of the concerned resource.

The searchable vocabulary that may be used inside so-called connected keyword queries are thus composed of labels associated with classes, properties and instances of the ontology. An *rdf:Property* is defined by its name, its domain and range, and expresses a link between a subject resource taken from its domain and an object resource taken from its range. Three *rdfs:label* properties are attached to each searchable property to obtain a complete human description: one attached to the property itself to describe its meaning, and two others attached respectively to its domain and range.

Thus, textual literals linked by *rdfs:label* properties to searchable elements of the ontology form the query vocabulary that can be used inside connected keyword queries. These labels are *a priori* defined by an expert so as to be concatenated in order to form an explicit human readable description, as close as possible to natural language, of a path in the ontology. Thus, the searchable vocabulary that can be used inside connected keyword queries is composed of:

- labels of classes hereafter called *CLASSNAME*,
- labels of properties hereafter called *PROPNAME*,
- labels of property ranges hereafter called *RANCON*,
- labels of property domains hereafter called *DOMCON*,
- *rdfs:Literal* instances hereafter called *VALUES*,
- and additional logical connectives like *and* and some tool words like *of* that make the query more natural and explicit (see Section 3.1).

To illustrate this use of labels in connected keyword queries, let us consider the query introduced at the beginning of this section whose aim is to retrieve the name of persons that are at the head of a company and that are authors of articles about business intelligence. Figure 4 illustrates how labels have been associated with the elements of the ontology introduced in Figure 3. For the sake of clarity, only *rdfs:label* properties concerned by the query are informally specified in rectangles. Using these labels, the connected keyword query version of the example query is: *name of person at the head of company and author of article about "business intelligence"*.

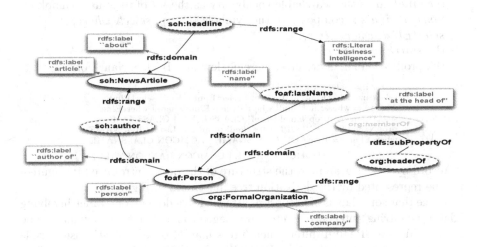

Fig. 4. Graph view of the query *name of person at the head of company and author of article about "business intelligence"*

3   Structured Keyword Queries and SPARQL Translation

3.1   Query Structure

The final goal of the presented approach is to translate a user-defined keyword query into a SPARQL query. A SPARQL query is composed of a projection part introduced by the keyword SELECT and a selection part introduced by the keyword WHERE. The projection clause is used to declare the variables on which matching patterns defined in the selection clause are applied when querying the graph. For example, the keyword query illustrated in Figure 4 is translated into SPARQL as follows:

```
SELECT DISTINCT ?name
WHERE {
    ?name rdf:type foaf:lastName.
    ?person rdf:type foaf:Person. ?person foaf:name ?name.
    ?person org:headerOf ?comp. ?person sch:author ?art.
    ?art rdf:type sch:NewsArticle. ?art sch:headline ?head.
    FILTER regex(?head, "business intelligence") }
```

In our model, it is assumed that keyword queries are also composed of a first projection part where the expected information is specified, and a second optional selection part where filtering criteria are defined. A context free grammar has been defined to determine the patterns that may compose a valid connected keyword query. This grammar $G = (terminals, nonTerminals, startSymbol, rules)$ is defined as follows:

- *terminals* uses the searchable vocabulary as the set of terminal symbols,
- *nonTerminals* is composed of the symbols {*query, select, where, selectElmt, whereElmt*},
- the *startSymbol* is *query*,
- the production *rules* are given hereinbelow in a Backus Naur Form.

```
query :: = select where | select
select ::= selectElmt 'and' select | selectElmt select | selectElmt
where ::= whereElmt 'and' where | whereElmt where | whereElmt
selectElmt ::= propNameList 'of' CLASSNAME | CLASSNAME
propNameList ::= PROPNAME 'and' propNameList | PROPNAME
whereElmt ::= RANCON CLASSNAME | DOMCON CLASSNAME
              | DOMCON VALUE | DOMCON RANCON VALUE
```

According to this grammar, the structure of the query introduced in Figure 4 may be represented by the derivation tree illustrated in Figure 5.

Notice that some labels of properties attached to a domain or a range involving a datatype (string, integer, real, etc.) are tagged to indicate that they have to be interpreted as a SPARQL filter. Such filters may also be explicitly used inside a selection statement of the type *DOMCON RANCON VALUE* as in: *has title contains "SPARQL"*, where *has title* may be a label attached to the domain of property *sch:title* and *contains* is attached to its range.

The current grammar covers a limited number of query patterns only. As explained in Section 4, this is not a problem in practice as this grammar is used

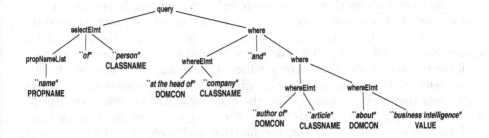

Fig. 5. Derivation tree of the query from Fig. 4

to guide an autocompletion system only and not to syntactically validate the structure of freely-typed user queries.

4 Autocompletion System

As illustrated in Figure 7, the query interface of the approach presented in this article is composed of a single field. Currently, this field is not completely freely editable by the user, but may only be used with an autocompletion mechanism that plays a crucial role as it: i) helps users define connected keyword queries whose structure is covered by the grammar, and ii) builds its SPARQL translation on-the-fly.

4.1 Autocompletion of Structured Queries

Compared with classical autocompletion systems embedded in search engines, suggested completions are relevant regarding not only the first letters typed by the user but also regarding the structure of the query whose construction is in progress. Suggestions made by the autocompletion system are guided by the grammar detailed in Section 3.1 that is internally represented as a directed labelled graph (cf. Figure 6).

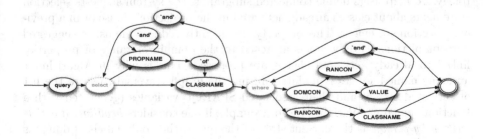

Fig. 6. Graph representation of the grammar

Algorithm 1 shows that the autocompletion system simply performs a kind of breadth-first-traversal of the graph representation of the grammar until the final node is reached, which corresponds to the submission of the query by the user. Starting with the initial node *query*, the autocompletion system suggests all the vocabulary elements corresponding to the category of the nodes directly connected to the current node. In Algorithm 1, the list of categories of vocabulary elements that may be appended to the query after the current node, say N, is given by the function $nextCat(N)$. Then, for each possible next category of vocabulary element, say *cat*, the function $getVocab(cat, actClass)$ returns the vocabulary elements to suggest, where the role of the second parameter is explained just below.

curNod ← *query*;
actClass = [];
while *curNod* ≠ *finalNode* **do**
  completions ← [];
  C ← nextCat(curNod);
  foreach C *as* c **do**
    | completions ← completions ∪ getVocab(c, actClass);
  end
  display(completions);
  curNod ← read userSelectedNode;
  if *curNod.category* = *CLASSNAME* and *curNod.resource* ∉ *activatedClass*
  then
    | actClass ← actClass ∪ (curNod.resource, newVariable());
  end
end

Algorithm 1. Autocompletion process

One considers SPARQL queries with joins between selection statements only, i.e. pairwise sharing of a common variable, and not queries involving cartesian products [2] since such queries would not make sense in the context considered. Thus, a valid SPARQL query corresponds to a connected subgraph of the ontology. To help users define connected subgraphs, the system suggests selection statements about classes already activated in the projection clause or in a previous selection statement. This property significantly reduces the list of suggested autocompletions as only labels attached to the domain or range of properties linked to already activated classes are proposed. This is why in Algorithm 1 an array named actClass is updated during the graph traversal to store the list of activated classes and their associated SPARQL variables (generated with a function named newVariable()). For example, if one considers *headline of article written by person* as the current state of the query, then only labels of domains of properties linked to the activated classes *sch:NewsArticle* and *foaf:Person* are suggested as illustrated in Figure 7.

As explained in Section 2, this keyword query approach is implemented in a particular context of mediated-access to distributed data sources. As it is impossible to index all the data (i.e. values) of the different sources, the autocompletion system only suggests vocabulary elements of a reference domain ontology. Thus, if a selection statement involving a value is being proposed, the user is invited to type this value as a quoted string. In the example illustrated in Figure 4, the label *about* associated with the domain of the property *sch:headline*, which is itself linked to the class *sch:article*, introduces a value. This is why a quoted string is opened directly after the selection by the user of the keyword *about* to let him/her type a description of the headline he/she is interested in, e.g. "business intelligence".

To be able to determine which activated class a selection statement is related to, two solutions may be envisaged. The first one is to impose non ambiguous labels, i.e. each searchable element of the ontology is associated with a discriminative label. The second one is to handle ambiguous labels and to display additional information in case of ambiguities between suggested completions in order to explain which element an ambiguous suggestion is related to. If one considers an ambiguous label of property, e.g. *name*, attached to two different classes, e.g. *foaf:Person* and *org:FormalOrganization*, then instead of suggesting the property label *name* twice, complete projection statements are displayed as *name <of Person>* and *name <of Company>*. This last point of disambiguation is all the more important when it concerns selection statements starting with ambiguous labels of domain properties whose range is linked to a string element. In order to precisely transcribe the sense of the keyword query into SPARQL, it is indeed mandatory to attach each selection statement to its right variable.

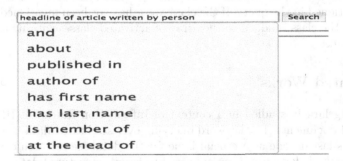

Fig. 7. Suggestion of relevant completions

4.2 On-the-Fly SPARQL Query Construction

During query construction using the autocompletion system (Algorithm 1), a translation of the keyword query into SPARQL is performed on-the-fly. As soon as a projection statement is completed by the user, the *SELECT* clause of the target SPARQL query is completed. A translation rule is indeed associated with each of the two covered projection statements. For the first one (*propNameList PROJCON CLASSNAME*), a variable is created for each property enumerated in the *propNameList* element as well as a link to a variable representing the class explicitly defined as the *CLASSNAME* element of the rule. Variables representing classes involved in the query are stored in the already mentioned array of activated classes named *actClass*. For the second type of covered projection statement that contains the name of a class only, one completes this statement with all the properties linked to this class. For example, the projection statement described only by *article* is completed with *title and published date and headline and author of article* that is translated into:

```
SELECT DISTINCT ?title ?published_data ?headline ?author
WHERE {
    ?title rdf:type sch:title. ?headline rdf:type sch:headline.
    ?published rdf:type sch:datePublished. ?author rdf:type sch:author.
    ?article rdf:type sch:NewsArticle.
    ?article sch:title ?title. ?article sch:datePublished ?published.
    ?article sch:headline ?headline. ?article sch:author ?author.
```

As selection statements concern classes already mentioned in the query and referenced in the array of activated classes, their translation into SPARQL is straightforward. For example, if a class say Person is associated with the variable ?person in the array of activated classes, then the selection statement *author of article* corresponding to the pattern *DOMCON (author of) CLASSNAME (article)* is translated into:

```
    ?article ref:type sch:NewsArticle. ?person sch:author ?article.
```

where ?person is the variable associated with the class Person that has to be previously mentioned in the query. If the class Article has not been mentioned so far in the query, the class is added to the array of activated classes with its associated variable ?article .

5   Related Works

After being largely studied in a context of Information Retrieval (IR) for unstructured documents [11], keyword querying over structured data has emerged during this last decade as a crucial issue for the database research community. IR-style methods have first been extended to handle structured data containing both textual, numerical and categorical values [7,10]. Whatever the model used to structure the data: XML documents [5,4], relational databases [15,9,16] or ontologies [19,8], all the researchers that tackled this problem agree that in order to efficiently return relevant answers, the interpretation process of a keyword query has to take into account the data structure and not only string similarities and terms frequency or entropy. Most of the contributions to keyword search over structured data consider that a query is composed of an unstructured enumeration of keywords referring values and metadata of the searchable data source. To get a better expressiveness, Pound et al. [14] introduces a pseudo-formal query language to let users define the exact meaning of their search, whereas Bergamaschi et al. [1] take into account the order of the keywords as a indicator of their relative importance. In terms of expressivity, the approach detailed in this article shares some similarity with the STRUCT system described in [12]. However, instead of being interpreted with basic Natural Language Processing (NLP) tools, keyword queries are interactively constructed and interpreted using an autocompletion system. As in [3], one considers that autocompletion systems are less intrusive than visual tools like [18] and thus more likely to be adopted by end users. Contrary to approaches based on NLP [12,6], the autocompletion system proposed in this article guarantees that all the constructed queries are semantically valid wrt. the ontology and return exact answers.

6 Conclusion and Perspectives

This article introduces a novel autocompletion-based interactive strategy to help users manipulate searchable elements of an ontology to finally obtain a semantically and syntactically valid SPARQL query. The aforementioned query construction guide relies first on a matching between searchable elements of the ontology and keyword-based human readable descriptions. Then, a formalization as a context-free grammar of the SPARQL queries handled by the system is used to control the autocompletion system and to display relevant suggestions only.

Currently, this grammar handles basic query structures only. Even if in practice this limited coverage is sufficient to precisely retrieve some information from an ontology, a perspective for future works is to increase the coverage of the grammar with more elaborated query structures like nested-queries. We also intend to perform a qualitative assessment of the system usability and its answers relevance through experiments with real users. To let users freely type their query with more expressivity and without using the autocompletion system, a NLP chain composed of lexical and syntactic analyzers is under development.

References

1. Bergamaschi, S., Domnori, E., Guerra, F., Trillo Lado, R., Velegrakis, Y.: Keyword search over relational databases: a metadata approach. In: Proceedings of the 2011 ACM SIGMOD International Conference on Management of Data, pp. 565–576 (2011)
2. Cyganiak, R.: A relational algebra for sparql. Digital Media Systems Laboratory HP Laboratories Bristol. HPL-2005-170 (2005)
3. Fan, J., Li, G., Zhou, L.: Interactive SQL query suggestion: Making databases user-friendly. In: 27th International Conference on Data Engineering, pp. 351–362 (April 2011)
4. Florescu, D., Kossmann, D., Manolescu, I.: Integrating keyword search into XML query processing. Computer Networks 33(1-6), 119–135 (2000)
5. Guo, L., Shao, F., Botev, C., Shanmugasundaram, J.: XRANK: ranked keyword search over xml documents. In: Proc. of the 2003 ACM SIGMOD International Conference on Management of Data, pp. 16–27 (2003)

6. Heinecke, J., Toumani, F.: A natural language mediation system for e-commerce applications: an ontology-based approach. In: Proc. of 2nd International Semantic Web Conference (2003)

7. Hristidis, V., Gravano, L., Papakonstantinou, Y.: Efficient IR-style keyword search over relational databases. In: Proceedings of the 29th International Conference on Very Large Data Bases, vol. 29, pp. 850–861. VLDB Endowment (2003)

8. Lei, Y., Uren, V.S., Motta, E.: Semsearch: A search engine for the semantic web. In: Staab, S., Svátek, V. (eds.) EKAW 2006. LNCS (LNAI), vol. 4248, pp. 238–245. Springer, Heidelberg (2006)

9. Li, G., Ooi, B.C., Feng, J., Wang, J., Zhou, L.: EASE: an effective 3-in-1 keyword search method for unstructured, semi-structured and structured data. In: Proceedings of the 2008 ACM SIGMOD International Conference on Management of Data, pp. 903–914 (2008)

10. Liu, F., Yu, C., Meng, W., Chowdhury, A.: Effective keyword search in relational databases. In: Proceedings of the 2006 ACM SIGMOD International Conference on Management of Data, pp. 563–574 (2006)

11. Manning, C.D., Raghavan, P., Schütze, H.: Introduction to Information Retrieval. Cambridge University Press (2008)

12. Patil, R., Chen, Z.: Struct: Incorporating contextual information for english query search on relational databases. In: Proc. of the 3rd Workshop KEYS (2012)

13. Pottinger, R., Halevy, A.: Minicon: A scalable algorithm for answering queries using views. The VLDB Journal 10(2-3), 182–198 (2001), http://dl.acm.org/citation.cfm?id=767141.767146

14. Pound, J., Ilyas, I.F., Weddell, G.: Expressive and flexible access to web-extracted data: a keyword-based structured query language. In: Proceedings of the 2010 ACM SIGMOD International Conference on Management of Data, pp. 423–434 (2010)

15. Qin, L., Yu, J.X., Chang, L.: Keyword search in databases: the power of RDBMS. In: Proceedings of the 35th SIGMOD International Conference on Management of Data, pp. 681–694 (2009)

16. Simitsis, A., Koutrika, G., Ioannidis, Y.: Précis: from unstructured keywords as queries to structured databases as answers. The VLDB Journal 17, 117–149 (2008)

17. Wiederhold, G.: Mediators in the architecture of future information systems. Computer 25(3), 38–49 (1992)

18. Zenz, G., Zhou, X., Minack, E., Siberski, W., Nejdl, W.: From keywords to semantic queries—incremental query construction on the semantic web. Web Semantics: Science, Services and Agents on the World Wide Web 7(3), 166–176 (2009)

19. Zhou, Q., et al.: SPARK: Adapting keyword query to semantic search. In: Aberer, K., Choi, K.-S., Noy, N., Allemang, D., Lee, K.-I., Nixon, L.J.B., Golbeck, J., Mika, P., Maynard, D., Mizoguchi, R., Schreiber, G., Cudré-Mauroux, P. (eds.) ASWC 2007 and ISWC 2007. LNCS, vol. 4825, pp. 694–707. Springer, Heidelberg (2007)

Querying Sentiment Development over Time

Troels Andreasen, Henning Christiansen, and Christian Theil Have

Research Group PLIS: Programming, Logic and Intelligent Systems
Dept. of Communication, Business and Information Technologies
Roskilde University, Denmark
{troels,henning,cth}@ruc.dk

Abstract. A new language is introduced for describing hypotheses about fluctuations of measurable properties in streams of timestamped data, and as prime example, we consider trends of emotions in the constantly flowing stream of Twitter messages. The language, called EMOEPISODES, has a precise semantics that measures how well a hypothesis characterizes a given time interval; the semantics is parameterized so it can be adjusted to different views of the data. EMOEPISODES is extended to a query language with variables standing for unknown topics and emotions, and the query-answering mechanism will return instantiations for topics and emotions as well as time intervals that provide the largest deflections in this measurement. Experiments are performed on a selection of Twitter data to demonstrates the usefulness of the approach.

1 Introduction

Social media are becoming more and more popular in humans' daily lives and networking on social media is an increasingly important social activity. With the evolving social networks and increased media activity the already massive amount of media data is rapidly growing and so is the potential value of these data as sources for analysis of structure, relationships and structural trends. There is also a huge potential for the derivation of valuable and reliable indications of trends in development of opinions and sentiments from the immense messaging on social media, but there is still a lack of systematic approaches for these purposes. However, messaging is target for various new and promising approaches to data mining and sentiment analysis and this area is undergoing a rapid development. The present paper emphasizes a new direction for this kind of analysis where opinions on topics can be investigated, not only for appearance but also for development over time.

We introduce a language, EMOEPISODES, for hypothesis testing and querying. Hypotheses about sentiments and emotions and their development over time can be formulated in the language. The language allows for querying on emotions about specified topics, for instance "fear of asteroids" and the degree to which an emotion is expressed about a given topic. An emotion about a given topic can be queried for a specific period, or the period can be left open to investigate peeks. In addition, consecutive sequences of periods can be enclosed in queries

H.L. Larsen et al. (Eds.): FQAS 2013, LNAI 8132, pp. 613–624, 2013.
© Springer-Verlag Berlin Heidelberg 2013

to investigate development of emotions over time, for instance to query peeks in shifts from negative to positive attitudes about a given topic. A query can focus on a specific topic, such as "how did the fear of astroids develop during Christmas" or refer to correlated topics, such as "did the School shooting influence the opinion against Obama's policy". We present a general and flexible language for formulating hypothesis – and thus queries – about detailed patterns of the evaluation of emotions over time. This is in contrast to earlier efforts which have mostly been concerned with measuring specific trends in social media data.

The validity of a given hypothesis in some time interval is measured by a satisfaction degree which is a number in the unit interval, and the grading is applied for ranking of best matches for a given hypothesis. Specifically, we consider here streams of timestamped messages in social media and hypotheses about fluctuations of emotions related to given topics.

Emotions are characterized and measured for some fixed **time granule** adapted to the application at hand. For applications in literature, one year or one decade may be chosen, for historical studies perhaps centuries. For the application chosen here – evolving emotions in social media – we have chosen one day as the granule. This choice abstracts away variations caused by the rotation of the earth and the uneven distribution of users around the globe, but allows to characterize relatively fast changes. In our current implementation, we arbitrarily chose 24h periods starting at 00:00:00 GMT+1, i.e., CET without Summer Time. Within each time granule we characterize topics by **level of emotion** where emotion is a classification of expression and level can be chosen from a finite set of fuzzy linguistic expressions.

Different applications may require different ways of combining satisfaction degrees for the constituents of complex hypotheses and as indicated we may combine topics and evaluate emotions as they appear simultaneously as well as consecutively over time. We discuss aspects of **aggregation** of constituent emotions and introduce a parameterizable function to adapt the different kinds of aggregations needed as well as for user preference.

The present paper is structured as follows. In section 2 we introduce the language EMOEPISODES and in section 3 we describe the chosen semantics for our application at hand – mining trends in social media. In section 4 we describe experiments and evaluation based on an implementation of the language and on an application on Twitter data from about 1.5 month from late December to early February 2013. In section 5 we discuss related work and finally in section 6 we conclude.

2   A Proposal for an Emotional Episode Language, EMOEPISODES

In the following, a *time point* refers to a specific time granule having a fixed position along a time line, and a *time interval* is a contiguous set of time points. The symbol $\mathcal{TI}$ refers to all such time intervals; below we use letter d to refer to time intervals (as t will be used for topics; below).

2.1   The Basic Emotional Episode Language

An arbitrary set of *topics* $\mathcal{TI}$ is assumed, e.g., $\mathcal{T} = \{\mathsf{Xmas, love, beer, asteroids,}$ $\ldots\}$ and set of *emotions*, e.g, $\mathcal{E} = \mathsf{fear, happiness, anger, sadness}, \ldots\}$. As mentioned above, the *level* of an emotion is described by a finite set of symbols $\mathcal{L}$, ordered by magnitude; we use for our main example high > medium > low, but more or less (at least two) steps may be used.

An atomic emotion *statement* for a given topic associates an emotion and a level to that topic. Example:

asteroids: fear(high)

The data semantics is given by a satisfaction degree function $SD : \mathcal{AS} \times \mathcal{TI} \rightarrow [0; 1]$ where $\mathcal{AS}$ is the set of such atomic statements, i.e., a measurement of how well a given statement characterizes a given time interval.[1]

A *compound* statement consists of a collection of atomic ones written with curly brackets; when more than one atomic statements concerns the same topic, we may group these also by curly brackets. Examples:

{asteroids: fear(high), doomsday: fear(high)}
asteroids: {fear(medium), excitement(high)}

The second example is a shorthand for a compound consisting of two atomic statements about asteroids.

A compound statement is understood as a sort of conjunction, i.e., all contained atomic statements influence the satisfaction of the compound; $\mathcal{S}$ will denote the set of all statements, with $\mathcal{AS} \subset \mathcal{S}$. The satisfaction degree function is extended to go all statements, $SD : \mathcal{S} \times \mathcal{TI} \rightarrow [0; 1]$, as follows where $\bigotimes^{com}$ is an aggregation function $[0; 1]^* \rightarrow [0; 1]$.

$$SD(\{\phi_1, \ldots, \phi_n\}, d) = \bigotimes_{i=1..n}^{com} SD(\phi_i, d)$$

A *scene* is a statement with an associated *time constraint*. Examples:

asteroids:fear(high)[5 days]
asteroids:fear(high)[> 5 days]
asteroids:fear(high)[2 days, from 2013-02-15]
asteroids:fear(high)[2013-02-15]
asteroids: {fear(medium), excitement(high)}[> 5 days]

The detailed language for time constraints is not specified further; we assume a natural definition of whether a given time interval is matched by a constraint. For example, [5 days] matches any interval of exactly 5 days. A *time assignment* for a scene is any interval of days that satisfies its time constraint, so, e.g.,

---

[1] As it appears, SD measured over an interval is not a mere aggregation of SD for each granule in the interval; this provides a freedom to let SD measure, say, relative frequencies over the entire interval.

the interval consisting of the days from 2013-02-14 to 2013-02-18 can be an assignment for asteroids:fear(high)[C], where, say, C is '5 days', '> 2 days' or '5 days, after 2012-31-12'.

Finally, an *episode* is a sequence of consecutive scenes, indicated by semicolon. Example:

> asteroids:fear(medium)[5 days] ;
>   {meteorites:fear(high), doomsday:fear(high)}[2 days, after 2013-02-15]

A time assignment for an episode is a sequence of assignments of consecutive intervals for each of its scenes; it is also referred to as a *match*. An episode is called *inconsistent* if no possible match exists, otherwise it is *consistent*. In order to allow 'holes' in episodes, we introduce the empty statement {} having $SD(\{\}, -) = e$ where e is a neutral element (not specified further).[2]

The satisfaction degree of an episode with respect to a time assignment is an aggregation $\bigotimes^{eps}$ of the satisfaction degrees for the individual scenes in their respective, assigned time interval. A *best* match in a given context is a match with the highest satisfaction degree.

Notice as a consequence of our definitions that the satisfaction degree of, say asteroids:fear(medium)[5 days], in a given time interval is independent of the emotions for asteroids in the days before or after the chosen period.

2.2   EMOEPISODES as a Query Language

The semantics of EMOEPISODES can produce a measure of the satisfaction degree of a given episode for a specific time assignment, and this is the basis on which we can form a query language. Here we discuss how EMOEPISODES can be used as a query language; actual experiments performed on Twitter messages are shown in section 4, below.

It gives good sense to query with a given episode for *the* best match within a larger time interval, i.e., for the best time assignment together with its satisfaction degree. Such a query mechanism can be extended with an *a priori* defined threshold and only report a match better than this threshold.

While for many other information retrieval tasks, it is relevant to ask for a sorted list of the k best matches, this is more dubious for episode matching due to overlapping matches. To see this, consider the following episode

> asteroids: fear(medium)[> 10 days] ; asteroids: fear(high)[> 10 days]

and a data set in which the frequency of observations of asteroids: fear grows slowly over a period of 1000 days. Here we may expect one optimal match about two thirds into the data, and it will be surrounded by a cloud of near optimal matches. It will be even worse in case asteroids: fear varies in very long waves, having peaks of different magnitudes. Here the sorted list approach may likely

---

[2] A proper formalization of the empty statement would require an extension of the domain for satisfaction degree to $[0; 1] \cup \{e\}$ and a specification of how the different aggregation operators treats e. These details are not interesting and left out.

report only the highest peak and the cloud around it, and having the second and third highest peaks outside the k best matches shown to the user, which we do not consider to be desirable.

If graphical output is an option, the best way to present the result of a query for an episode E may be as a curve showing, for each time point t, the satisfaction degree for an occurrence of E starting at t. When textual or symbolic output is expected, it may be suggested to report islands (defined as contiguous unions of matches better that a threshold) together with the best match in each island (the peaks), sorted according to the latter.

We can extend the EMOEPISODES language for queries involving variables that stands for unknown parts of an episode. The expected answers to a query with variables is a list of alternative instantiations of the variables, sorted according to the satisfaction degrees for their respective best match in the data set. In an interactive query system, a mouse click may, for each such instantiation, open a window with a satisfaction degree curve of a list of islands and peaks as explained above.

In order to obtain efficient query-answering algorithms, we allow only variables in positions where a topic, an emotion or a degree is expected (and not for, say, entire statements or time constraints).

EMOEPISODES as a query language can also be used for giving alert when certain patterns are observed. A journalist may, for example, want a report whenever the attitude towards the topic president changes. He can do this by setting up a tenant that sends a report whenever one of the following two queries have a match in the current data stream.

president: X(high)[3 days] ; president: X(low)[1 day]
president: X(low)[3 days] ; president: X(high)[1 day]

The three day interval indicates a certain stability, but the journalist allows only one day to see a possible shift, so he may be the first to write about it if it happens to be interesting.

3 EMOEPISODES Semantics for Mining Trends on Twitter

In this section we show two different semantics for EMOEPISODES by different choice of data semantics, both considered relevant for the sample applications, mining trends in Twitter data. We suggest also relevant choices for aggregation operators, that fit with both data semantics.

The data semantics for atomic scenes $(t : e(\ell), d)$, with t being a topic, e an emotion, ℓ a level, and d a time interval, is defined in terms of the number of messages tagged with t and classified as e in d. For scenes with compound statements, $t_1 : e_1(\ell_1), t_2 : e_2(\ell_2), \ldots$ the aggregation is based on counting messages tagged with all of $t_1, t_2, \ldots$.

3.1   Atomic Statement Semantics

Let T be the set of topics and D the set of duration specifications (time intervals). Apart from marking some messages with detected emotions, currently among $E = \{anger, disgust, fear, joy, sadness, surprise\}$, the given prototype also provides a sentiment classification for each message from $S = \{negative, neutral, positive\}$. We want to cover both by the language, but since each of these is a disjoint classification of messages (where E is partial and S is complete), we generalize to calculate satisfaction degrees for a specific classification among a given set $\mathcal{C}$ of classifications, where $\mathcal{C} = \{E, S\}$. For a classification $C \in \mathcal{C}$ and $d \in D$, we define $\delta_C(d)$ as the set of all messages during d that are classified by C, while, for topic $t \in T$ and class $c \in C$, $\delta_C(t, d)$ and $\delta_C(t : c, d)$ denotes the set of all messages during d on topic t and the set of all messages during d on topic t classified as c respectively.

Based on cardinalities of these sets, we define the relative satisfaction R_C of a statement $\phi = (t : c)$ during d for class (emotion or sentiment) $c \in C$ (with $C \in \mathcal{C} = \{E, S\}$) by

$$R_C(\phi, d) = R_C(t : c, d) = \frac{|\delta_C(t : c, d)|}{|\delta_C(t, d)|}, \quad c \in C$$

and measure the satisfaction degree of sentences by way of compliance with simple fuzzy linguistic terms $low, medium,$ and $high$ over relative satisfaction. Membership functions for these terms are defined individually for each classification such that the membership function for $medium$ is symmetric around $1/n$ where $n = |C|$. Figure 1(a) shows definitions of relative satisfaction level terms $low, medium, high$ for classifications E and figure 1(b) for classifier S (notice that $|E| = 6$ and $|S| = 3$). For the relative satisfaction level $level \in \{low, medium, high\}$ the satisfaction degree of a statement $\phi = (t : e)$ during d is defined by:

$$SD(\phi, d) = SD(t : c[level], d) = \mu_{level}(R_C(t : c, d)), \quad c \in C$$

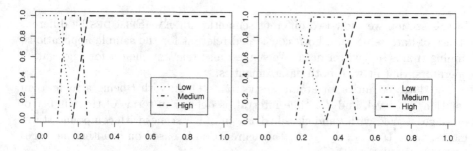

Fig. 1. Membership functions μ_{level} for fuzzy linguistic terms over relative satisfaction with $level \in \{low, medium, high\}$. Shown in (a) for classifier E and in (b) for classifier S (6 and 3 classes respectively).

3.2   Elitist and Populist Data Semantics

Queries are evaluated over time intervals and satisfaction for an individual scene enclosed in a query is based on the fraction of tweets, in the given time interval, satisfying the scene. We consider two data semantics that differ in the way this fraction is derived: elitist and populist data semantics.

When applying **elitist semantics** $SD(t : c[level], d)$ is measured relative those tweets that refer to topic t during d. This corresponds to applying the relative satisfaction as introduced above, that is:

$$R(\phi, d) = R(t : c, d) = \frac{|\delta(t : c, d)|}{|\delta(t, d)|}$$

An alternative is to apply **populist semantics** where $SD(t : c[level], d)$ rather is measured relative to all tweets during d, thus based on a relative satisfaction, which is:

$$R(\phi, d) = R(t : c, d) = \frac{|\delta(t : c, d)|}{|\delta(d)|}$$

As indicated elsewhere we have chosen the elitist for our preliminary experiments. However the choice of semantics could be left to the user as a preference parameter.

3.3   Compound Statement and Episode Semantics

To complete the semantics of the language we must specify aggregation principles for compound statements (aggregating multiple simultaneous statements for a given time interval) as well as for episodes (aggregating statements over continuous time intervals). Intuitively both aggregations should reflect a conjunction of the statements, but we need to take into account the graded satisfaction of statements so an obvious choice is to go for a single, but parametrizable, graded averaging aggregation function. We adopt the Order Weighted Averaging (OWA) function [13] and introduce a simplification of the parameterization of this – scaling with a single parameter a class of averaging functions with min and max as extremes. OWA aggregates n values $a_1, \cdots, a_n$ by means of an ordering vector $W = [w_1, ..., w_n]$, applying w_1 to the highest value among $a_1, \cdots, a_n$, w_2 to next highest value, etc. Thus OWA is defined by:

$$F_W(a_1, \cdots, a_n) = \sum_{i=1}^{n} w_i b_i; \quad w_i \in [0, 1]; \quad \sum_{i=1}^{n} w_i = 1$$

where b_i is the i'th largest among $a_1, \cdots, a_n$ and $b_1, \cdots, b_n$ is thus the descending ordering of the values $a_1, \cdots, a_n$. By modifying W we can obtain different aggregations, for instance, $W = [1, 0, 0, \cdots]$ corresponds to the maximum, $W = [1/n, 1/n, \cdots]$ becomes the average, and $W = [0, 0, \cdots, 1]$ the minimum. Order weights can, independent of n, be modeled by an increasing function $K : [0, 1] \to [0, 1]$ such that:

$$w_i = K\left(\frac{i}{n}\right) - K\left(\frac{i-1}{n}\right)$$

Assuming $K(0) = 0$ and $K(1) = 1$, aggregations such as max can be modeled by $K(x) = 1$ for $x > 0$, min by $K(x) = 0$ for $x < 1$, $average$ by $K(x) = x$ and for instance a restrictive aggregation (closer to max that min) by $K(x) = x^3$. Using this principle of prescribing weights by increasing functions, order weight specification can be further simplified using a single parameter $\beta \in [0,1]$, as in:

$$K(x) = G_\beta(x) = \begin{cases} 0 & \text{for } \beta = 0 \\ x^{\frac{1}{\beta}-1} & \text{for } \beta > 0 \end{cases}$$

where values $0, 1$ and 0.5 for β corresponds to min, max and $average$ respectively, while values closer to zero corresponds to more restrictive aggregations and values closer to 1, to less.

OWA aggregation conveniently adapts intuitive definitions of "linguistic quantifiers". Using the single-parameter approach above we can define $EXISTS$ by $G_\beta(x)$ with $\beta = 1$ and $FOR\ ALL$ with $\beta = 0$, while quantifiers such as $A\ FEW$, $SOME$, $MOST$, and $ALLMOST\ ALL$ can be modeled by β-values such as 0.8, 0.5, 0.2 and 0.05 respectively.

While compound statement as well as episode aggregation intuitively are conjunctive, we consider the former as indicating more restrictive quantification than the latter. We chose, for application in the prototype the $MOST$ aggregation, setting the corresponding β-value to 0.2.

4   Experiments and Evaluation

We illustrate applications of our query language to the social network Twitter using our prototype implementation of the query language.

The implementation is realized in Prolog and R (bridged through the R..eal Prolog library [2]). The system has a grounding component which for each variable and flexible duration generates all possible *query variants* in which query variables are replaced by possible values and durations are given as a fixed number of dates. We refer to a query variant and its associated score as a *match*. A query result set contains all matches to a query sorted by their score. Query evaluation is a recursive procedure which uses fuzzy membership functions for atomic statements and Order Weighted Averaging for compound statements as described in section 3. A search for a match of a specific episode runs in time linear in the size of the data. The number of variables in a query may have drastic influence on the number of query variants and hence on time complexity, but may be controlled by heuristic score cut-offs. Furthermore, both the independent scoring of query variants and the recursive matching procedure is well-suited for parallelization using a map-reduce strategy [5].

To test our implementation we have gathered almost 500GB of data from Twitter[3] over a period of about one month (From December 23, 2012 to February

---

[3] A tweet including meta-data takes about 1 kilobyte.

7, 2013). The data were collected by monitoring the sample firehose [12] – a
service provided by Twitter which gives access to a random subset of Twitter
messages (tweets) as they are produced in realtime. Twitter also provide a search
interface, but unlike the firehose, the returned data are not purely random but
are filtered based on criteria known only by Twitter. Each tweet is provided in
a JSON format, which in addition to the tweet text, provides metadata such as
the language of the of tweet. We consider only tweets for which the language is
indicated to be English.

For each tweet collected, we identify topics and perform sentiment analysis
using the **sentiment** package for R [6]. We utilize Twitters hashtags as topic
classifications, which has the advantage that we do not have to decide on a
prospective set of topics in advance. Hashtags are words in a tweet that are
prefixed with the # character and serves as a way for users to volunteer a
sort of topic classifications of their tweets. For our purposes, it is not sensible
to consider all hashtags since some are so infrequent that it is not possible to
measure a trend within our time limit. We consider only hashtags which occurs
in at least 500 of the collected tweets (corresponding to an average of at least
ten tweets per day). In our dataset, only 3494 out of 2045318 unique hashtags
occur in at least 500 messages.

4.1   Example 1: Stock Prices and Surprising Events

As a case to illustrate the query language, we are interested in finding sudden
events involving a company which either angers or pleases the public. We use the
company Apple (tag #apple) in this example. Reactions to sudden events may
involve an expression of surprise as well as an indication of attitude towards the
event. To characterize attitude we consider joy and anger rather than positive
or negative sentiment because these are more extreme emotions and more likely
to pertain to surprising events. We expect that sudden events of these kinds can
have an effect on the stock price of a company.

We detect such events with EMOEPISODES using the following two queries:

Q_1: apple:{surprise(high), joy(high)}[$\geq$ 1 day]
Q_2: apple:{surprise(high), anger(high)}[$\geq$ 1 day]

The first query matches scenes of arbitrary duration for which surprise and
joy for #apple is high and the second matches scenes for which surprise and
anger is high. For each of these two queries we collect all matches.

The matches to a query are likely to overlap. For instance, a match with a
period of three days may be overlapped by (similar) matches with periods of
more than or fewer than three days, but covering some of the same three day
period. Figure 2 shows curves of the maximal scores of any match to Q_1 and
Q_2 together with a normalized stock prize. Even if the stock price curve is not
a perfect match to either of the query curves, it is easy to spot correlations.

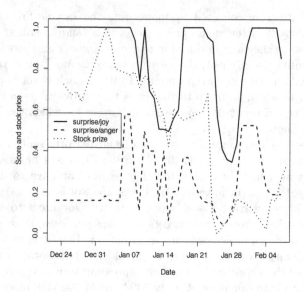

Fig. 2. The solid curve displays the maximal score for query Q_1 (surprise and joy) and the dashed curve displays the maximal score for query Q_2 (surprise and anger). The dotted curve displays Apples stock price normalized to the range $[0, 1]$.

4.2   Example 2: Changes of Attitude

The previous example does not make use of variables which are useful to reason about unspecified tags, sentiments and emotions. We consider here matching topics for which a change in attitude has occurred. In particular, we consider a change from a positive attitude to a negative attitude.

Q_3: X:positive(high)[10 days] ; X:negative(high)[10 days]

Four tags achieve perfect score for this query. The most interesting of these is the tag #14jan ($\approx$ Jan 14–Feb 2) which refer to a march in Pakistan led by Canadian-Pakistanian Tahir-ul-Qadri in protest of the government. While the protest march led to an agreement of reforms signed by the government, the negative sentiments period (Jan 24–Feb 2) contains many critical tweets about the dual nationality of Tahir-ul-Qadri and about his agenda.

The rest are non-topical tags: #emblemfollowspree ($\approx$ Jan 10–Jan 29), #twitterhastaughtme ($\approx$ 29 Dec–17 Jan), #dontbemadatmebecause ($\approx$ 29 Dec–18 Jan). These are Twitter-specific tags for which there is a high volume of these tweets for a relatively short duration (trending tags). Tweets using one of these tags seem to use a sarcastic tone, which explain negative sentiment classifications. In the first few days of these trends, they receive a lot of positive attention. This is apparent from tweets using the tag to comment on the tag/trend to indicate, e.g., that it is funny. This explains a great deal of positive sentiment classifications.

5   Related Work

Our terminology is inspired by the seminal work of [7] who suggested a way to define episodes in sequences of discrete events (from a finite alphabet of such) and gave algorithms to search for a sort of association rules among such episodes. Before that, [1] described algorithms for mining frequent, sequential patterns in a transaction database. See a recent survey [8] on later work inspired by [1,7].

Our work differs from the referenced work in that we present 1) a logical language EMOEPISODES for specifying scenes and episodes sequences, and 2) these episodes refer to measurements over large sets of timestamped objects (exemplified by tweet messages) with associated multi-dimensional features (exemplified by hash tags and emotions), rather that a finite alphabet. Furthermore, our episodes can be parameterized in arbitrary ways, we can include very general time constraints, and we can search for (ranked lists of) instances of the parameters that provides the best match. Our declarative episode specification language has a well-defined, graduated truth semantics, that is parameterized in a way that allows different interpretations of the data.

An SQL-based query language for specifying *search* for sequential patterns of simple events is introduced in [11]. The queries inherit the generality of SQL, but examples in the paper gives an impression that formulation of queries may be a non-trivial task. There is no account for a priority between different answers, although this may possibly be encoded with aggregate functions and SORT BY in SQL. A generalization of the query language of [11] to handle episodes specified in EMOEPISODES does not seem feasible, as all details of the underlying data semantics and aggregations would need to be unfolded in an SQL style within each query.

Sentiment analysis of Twitter messages over time has previously been demonstrate to correlate with public opinion measured by Gallup polls in [9] which, however, only measures positive/negative sentiment. Another study correlates tweet sentiments and emotions to socio-economic phenomena [4]. Correlation of the sentiment of Twitter messages to stock prices is studied in [3], though not in relation to surprising events. In contrast to the specific nature of these studies, EMOEPISODES provides the flexibility to adapt to a variety of use-cases.

6   Conclusions

We have presented a language for querying and mining the development of sentiments and emotions over time and illustrated its use with Twitter data. The language gives the ability to answer questions on how and when opinions change. It is also useful as a data mining tool to discover sequential patterns of sentiments and emotions associated to topics which may not be known in advance. Being able to answer queries as those supported by our language can have important implications for, e.g., social, socio-economical and political science as well as for market analytics. As a monitoring tool it can discover unexpected events and be used to alert media and decision makers to events worthy of attention. To our knowledge, a query language with these capabilities has not been seen before.

Our implementation can be improved in a number of ways, particularly with regard to the method of sentiment analysis and topic classification. Using hashtags as topic classifications is convenient but it is possibly ambiguous and may be insufficient if tags are non-topical or missing. The use of NLP techniques and an ontology of tag meaning, i.e., MOAT [10], can be used mitigate the issue. Similarly, it would be more accurate to classify sentiment in reference to topics rather than classifying the overall sentiment of a tweet.

Social media data is a multi-faceted, global phenomena which take many forms. Besides the temporal aspect, social media data also contain a geographical dimension which provide relevant information to include in a sentiment query language. In addition, integrating different sources of temporal data such as, e.g., stock prices and news reports, may further increase utility of the language.

References

1. Agrawal, R., Srikant, R.: Mining sequential patterns. In: Yu, P.S., Chen, A.L.P. (eds.) ICDE, pp. 3–14. IEEE Computer Society (1995)
2. Angelopoulos, N., Costa, V.S., Azevedo, J., Wielemaker, J., Camacho, R., Wessels, L.: Integrative functional statistics in logic programming. In: Proc. of Practical Aspects of Declarative Languages, vol. 7752 (2013)
3. Bollen, J., Mao, H., Zeng, X.: Twitter mood predicts the stock market. Journal of Computational Science 2(1), 1–8 (2011)
4. Bollen, J., Pepe, A., Mao, H.: Modeling public mood and emotion: Twitter sentiment and socio-economic phenomena. In: Proceedings of the Fifth International AAAI Conference on Weblogs and Social Media, pp. 450–453 (2011)
5. Dean, J., Ghemawat, S.: Mapreduce: simplified data processing on large clusters. Communications of the ACM 51(1), 107–113 (2008)
6. Jurka, T.P.: sentiment: Tools for Sentiment Analysis. Version 0.2, http://github.com/timjurka/sentiment
7. Mannila, H., Toivonen, H., Verkamo, A.I.: Discovery of frequent episodes in event sequences. Data Min. Knowl. Discov. 1(3), 259–289 (1997)
8. Mooney, C.H., Roddick, J.F.: Sequential pattern mining – approaches and algorithms. ACM Comput. Surv. 45(2), 19:1–19:39 (2013)
9. O'Connor, B., Balasubramanyan, R., Routledge, B.R., Smith, N.A.: From tweets to polls: Linking text sentiment to public opinion time series. In: Proceedings of the International AAAI Conference on Weblogs and Social Media, pp. 122–129 (2010)
10. Passant, A., Laublet, P.: Meaning of a tag: A collaborative approach to bridge the gap between tagging and linked data. In: Bizer, C., Heath, T., Idehen, K., Berners-Lee, T. (eds.) LDOW. CEUR Workshop Proceedings, vol. 369. CEUR-WS.org (2008)
11. Sadri, R., Zaniolo, C., Zarkesh, A.M., Adibi, J.: Optimization of sequence queries in database systems. In: Buneman, P. (ed.) PODS. ACM (2001)
12. Twitter. The streaming APIs, https://dev.twitter.com/docs/streaming-apis
13. Yager, R.: On ordered weighted averaging aggregation operators in multicriteria decisionmaking. IEEE Transactions on Systems, Man and Cybernetics 18(1), 183–190 (1988)

SuDoC: Semi-unsupervised Classification of Text Document Opinions Using a Few Labeled Examples and Clustering

František Dařena and Jan Žižka

Department of Informatics, FBE, Mendel University in Brno
Zemědělská 1, 613 00 Brno, Czech Republic
{frantisek.darena,jan.zizka}@mendelu.cz

Abstract. The presented novel procedure named *SuDoC* – or Semi-unsupervised Document Classification – provides an alternative method to standard clustering techniques when it is necessary to separate a very large set of textual instances into groups that represent the text-document semantics. Unlike the conventional clustering, SuDoC proceeds from an initial small set of typical specimen that can be created manually and which provides the necessary bias for generating appropriate classes. SuDoC starts with a higher number of generated clusters and – to avoid over-fitting – reiteratively decreases their quantity, increasing the resulting classification generality. The unlabeled instances are automatically labeled according to their similarity to the defined labeled samples, thus reaching higher classification accuracy in the future. The results of the presented strengthened clustering procedure are demonstrated using a real-world data set represented by hotel guests' unstructured reviews written in natural language.

Keywords: document labeling, clustering, small sample sets, text mining, natural language.

1 Introduction

Managing a very large set of relevant textual documents can provide valuable knowledge as many successful text-mining applications demonstrate [3,4,19]. A typical text-mining task is based on classification or prediction. Having a sufficient number of labeled instances, a user can employ them as training samples for adjusting specific parameters of a chosen classification algorithm: *supervised learning* [17]. Usually, the more textual documents representing individual classes are available, the better knowledge can be revealed, as the practice shows. Unfortunately, not always the respective labels that determine a document category are available. If a set of collected document samples is not too large, the supplementary information in the form of labels can be appended manually by a human expert. Naturally, for very extensive and potentially highly valuable collections of unlabeled documents (maybe hundreds of thousands or millions),

H.L. Larsen et al. (Eds.): FQAS 2013, LNAI 8132, pp. 625–636, 2013.

the manual method is impracticable. In such a situation, the labeling can be automatically created with the help of clustering algorithms: *unsupervised learning* [17]. However, because a clustering method has less initial information available, the synthetic, machine-controlled division of the set of unlabeled documents between different categories (clusters) can be often imperfect resulting typically in a lower classification accuracy, purity, precision, and so like.

The *semi-supervised learning* procedure represents a trade-off between the supervised and unsupervised learning. It works initially with usually a small set of instances with known labels, which positively supports the supplementary labeling of a much larger remainder of unlabeled instances. The semi-supervised learning can employ several algorithms as *self-training, co-training, McLachlan, expectation-maximization, boosting, label propagation*, and others – an interested reader can find a good summary in [1].

This article suggests using a small set of labeled instances for improving the clustering process, too. However, because the presented method is not one of the typical semi-supervised learning approaches, here it is called *semi-unsupervised learning*, or *SuDoC – Semi-unsupervised Document Classification*, making reference to its similarity to the principles of the semi-supervised methods: Using a small set of beforehand labeled instances as bias in favor of improved automatic supplementary labeling. Here, the beforehand labeled instances play a role known as *training samples* and the following text uses this term.

At the beginning, a lot of unlabeled instances are biased (by a small number of training samples) to create a high number of clusters because such clusters are characterized in that they have higher purity. (An extreme – but here not used – case could be initiating clusters having just one instance: perfect purity, but no generality.) Then, reiteratively, the following steps decrease the quantity of clusters and propagate the labels through fewer larger clusters to still unlabeled instances, which eventually represent classes. The goal is therefore to create groups from a lot of instances, which provide improved classification functionality.

The method presented in this paper aims at providing an alternative possibility how to deal with one of the typical problems comprised by the missing labels for classification. In the following sections, the article describes the suggested clustering improvement procedure, the experiments carried out with a large real-world data collection from the Internet, and the results and their interpretation as well as a comparison with some selected typical classifiers.

2   Classification Based on a Small Number of Examples

In order to evaluate the performance of different classification methods, the set of all available instances – represented by positive and negative textual unstructured reviews of hotel service customers – was divided into two parts: a training and testing set. Labeled reviews from the training set were used for classification of the documents in the testing set. For the classification, several well-known standard supervised techniques were used. However, in this case, the number of reviews in the training set was very low (not more than 20 reviews) compared to the size of the testing set. Then, the results of those classifiers were compared to the suggested novel approach based on "strengthened" clustering described further.

Because all the reviews in the testing set were labeled too, it was possible to evaluate the quality of classification as well as the SuDoC process. There exist many different metrics applied to evaluation of a classifier performance. The presented experiments used *Accuracy, Precision, Recall*, and *F-measure* that are commonly accepted for estimating the classification performance evaluation measures [18]. In addition, the performed experiments were repeated 100 times in order to exclude the influence of randomness.

2.1 Training Samples Selection

From all textual reviews that were available (1,245 reviews), a small subset containing maximally 20 reviews was created using random selection from the whole assemblage of available documents. The reviews in the selection were then carefully manually labeled as negative or positive by the authors according to the real content from the human-like semantic point of view. When it was impossible to clearly determine whether a review was positive or negative, the document was simply excluded from the set of labeled samples – certain reviews contained both some positive and negative points of the used accommodation service, not being plainly distinctive. Thus, in the experiments, the subsets contained from 14 to 20 examples (in most cases typically 18 or 19). The sample sets, containing two classes, were relatively balanced, that is, they contained almost the same numbers of positive and negative review samples.

2.2 Using Common Supervised Techniques

To evaluate a possible contribution of SuDoC to processing of the data described in detail in the section 3, the first group of experiments assessed outputs of supervised machine-learning algorithms implemented in the popular data-mining system WEKA [10]. The various algorithms included: J48 [16], Naïve Bayes [12], Logistic Regression [5], Support Vector Machines [15], K-star [6], Instance based learning [2], and J-Rip [7]. During the experiments, default parameters of the mentioned classifiers were used.

2.3 SuDoC – Semi-unsupervised Document Classification

The presented SuDoC approach is grounded on the assumption that it is possible to label all mutually similar instances in a certain group (cluster) using the exploration of just a few of them. The more homogeneous the group is, the smaller amount of instances must be studied. In an ideal situation where the instances are perfectly (or very closely) similar (i.e., they belong to one class according to their sentiment or topic), it is sufficient to examine just one of them in order to label all remaining ones. In a situation that is not perfect (the group contains instances from different classes), the selection of an instance from a minority class might inevitably cause a high error rate of the classification. In order to prevent this situation, more instances as examples should be selected and the label that has the majority among the labels assigned to them can be used for the rest of the cluster, see Fig. 1. If the heterogeneity of the cluster in such a situation is not too high, it is possible to label all instances in the group with a user's acceptable accuracy.

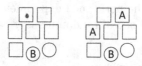

Fig. 1. Two identical clusters containing instances of two classes – class A (squares) and class B (circles); some of the instances are initially labeled (marked with a letter representing the assigned label of the document); remaining instances in the clusters are assigned to the same class that prevail within the randomly selected instances. Left – all instances are assigned to the class B with error 75% (6 from 8); right – all instances are assigned to the class A according the majority class of initially labeled instances with error 25% (2 from 8).

To be successful when using the approach demonstrated above, it is necessary to have *not too many* groups of instances with *sufficient homogeneity*. Such a state can be achieved through a process known as *clustering*. Clustering, as the most common form of unsupervised learning, enables automatic grouping of unlabeled instances into subsets called clusters according to the mutual (dis)similarity of the instances. The instances in each subset are more similar to each other than to instances in other subsets. During assigning the instances to the clusters, a particular clustering criterion function defined over the entire clustering solution is minimized or maximized.

The quality of the clustering solution is frequently measured by *purity, entropy, mutual information,* or *F-measure* [9]. Purity measures the extent to which each cluster contains instances from primarily one class. In a perfect clustering solution the clusters contain instances from only a single class, i.e., purity equals to 1 [20]. The smaller the clusters are, the higher their homogeneity generally is. In the situation when each cluster contains just one instance, the homogeneity is naturally perfect. This is, however, a quite useless solution lacking any generality as it is over-fitted just to the available instances while the goal is to categorize the instances that would appear in the future. Any solution containing fewer clusters with lower heterogeneity is thus better. It is therefore necessary to find a solution with a low number of clusters having a sufficient quality from a user view.

In the experiments, the clustering process used the software package *Cluto* version 2.1.2 [23]. This free software provides various different clustering methods working with several criterion functions and similarity measures, and it is very suitable for operating on very large datasets [13]. Cluto's criterion functions that are optimized during the clustering process operate either with *vector representation* (internal, external, and hybrid functions) of the objects to be clustered, or with *graph-based representation* (graph-based functions). Internal criterion functions are defined over the instances that are parts of each cluster and do not take into account the instances assigned to different clusters. External criterion functions derive the clustering solution from the difference between individual clusters. Various clustering criterion functions can be also combined to define a set of hybrid criterion functions that simultaneously optimize the individual criterion functions [20].

During the experiments, the following parameters of the clustering were set:

- similarity function: cosine similarity,
- clustering method: k-means (Cluto's specific variation), which iteratively adapts the initial randomly generated k cluster centroids' positions,
- criterion function optimized during clustering process: *H2* (hybrid).

Other clustering parameters of Cluto remained set to their default values, i.e., *Number of iterations*: 10, *Number of trials*: 10, *Cluster selection*: best, and *Row model*: none. The above parameterization was chosen based on the results of the previous experiments published in [22].

As it was mentioned above, sufficiently high quality (acceptable for a user) of clusters is essential for the success of classification based on a few *typical* examples. The relation between the number of clusters and the quality of the clusters was demonstrated in an experiment based on clustering documents represented by the below mentioned customer reviews. Table 1 contains information about the quality of clustering solutions for different numbers of clusters for the given specific data. The quality is measured by *Purity* which is defined as

$$Purity = \sum_{i=1}^{k} \frac{|C_i|}{|D|} \cdot \frac{1}{|C_i|} \max_j(|C_i|_{class=j}),$$

where k is the number of clusters, $|C_i|$ is the size o the i-th cluster, $|D|$ is the number of instances to be clustered, and $|C_i|_{class=j}$ is the number of instances of class j in cluster C_i.

Table 1. Quality of clustering solutions with different numbers of clusters measured by the *Purity* criterion

Number of clusters	1	2	20	50	100	200	300	500	750	1000
Purity	0.53	0.83	0.84	0.86	0.86	0.87	0.88	0.88	0.92	0.97

It is obvious that by labeling the documents in clusters according to a randomly selected document in each cluster, we might achieve about the 90% accuracy when there are 750 clusters created and when at least one document in each cluster is labeled. However, this number is too high for manual labeling. Having, for example, 20 clusters, the accuracy of the same process would be significantly lower (84% or less), also because the chance that a document of a minority class in a cluster might be randomly selected (this is a consequence of lower quality of the clustering solution). The following steps focus on achieving higher accuracy of such a classification with just a small number of specimen documents that need to be manually labeled in advance.

Such a small number, N_i, of documents that has to be manually labeled is the same as described in the section 2.1 in order to compare the presented procedure with commonly used classification techniques.

These initially labeled documents are used to label the remaining documents in the clusters they belong to. When the number of initially labeled documents, N_i, is low compared to the whole number of clusters (a low N_i is desired), we might achieve higher classification accuracy but not all documents will be classified. When N_i is close to the number of clusters, i.e., the number of clusters is low (having a low N_i), we achieve lower accuracy but all or almost all documents will be labeled, see Fig. 2 for a model example representing this situation.

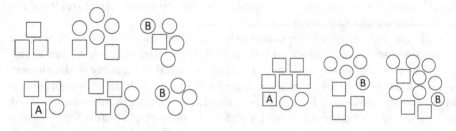

Fig. 2. Two clustering solutions of the same document collection. Squares represent documents of class A and circles documents of class B; documents marked with letters A or B are initially labeled documents. Remaining documents in the clusters are later labeled according to the three initially labeled ones. Left – error rate of such process is about 15% but not all documents are labeled (because three clusters do not contain any initially labeled documents); right – error rate is about 30% but all documents are labeled.

As told above, a higher number of clusters is better for achieving higher classification accuracy. However, the problem of having many unclassified documents must be eliminated. We therefore propose spreading the labels (assigned with higher accuracy in a clustering solution having a higher number of clusters, see Fig. 3) within a clustering solution having a lower number of clusters (see Fig. 4). Thanks to a few initially labeled documents, we can have much more labeled instances available, with a sufficient accuracy (Fig. 3). These labeled documents can be later used for labeling other unlabeled documents that are in other clusters at this moment. Thus, a different clustering solution is needed. In order to ultimately label all document instances, such a solution must have a smaller number of clusters as illustrated in Fig. 4.

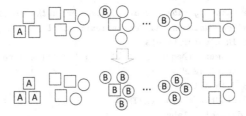

Fig. 3. Top – a few randomly selected examples are labeled. Bottom – labels of the examples are used to label the remaining instances in the clusters; some of the clusters contain instances without labels because the number of selected instances is much smaller than the number of clusters. However, the classification accuracy is higher.

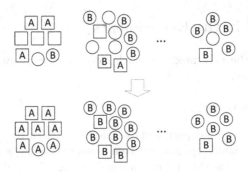

Fig. 4. A different clustering solution of the same collection as in Fig. 3. Top – distribution of the documents with labels that were assigned in the previous step. Bottom – the assigned labels are used to label remaining instances in the clusters. All or most of the clusters contain labeled instances.

The following pseudocode recapitulates the SuDoC algorithm:

```
/* manual labeling of document samples */
FOR EACH sample IN samples DO
    DISPLAY sample.text
    GET sample.label
END FOR

/* numbers of clusters in iterations are given as algorithm parameters */
FOR EACH nc IN number_of_clusters_in_iterations DO
    /* creating nc cluters from all documents, assigning a cluster
       number to every document */
    CREATE_CLUSTERS(all_documents, nc)

    /* counting the occurences of assigned labels in individual clusters */
    FOR EACH sample IN samples DO
        FIND d IN all_documents WHERE d = sample
        ADD sample.label TO cluster_labels[ d.cluster ]
    END FOR
```

```
/* assigning the majority label to all documents in the clusters
   where at least one document has an assigned label */
FOR EACH cluster IN cluster_labels
    FIND label with max. frequency IN cluster_labels[ cluster ]
    IF one label found THEN
        FOR EACH d IN all_documents DO
            IF d.cluster = cluster THEN
                d.label = label
            END IF
        END FOR
    END IF
END FOR
END FOR
```

3   Data Used in the Experiments

In order to verify the presented approach, several experiments with real-world data were carried out.

3.1   Data Characteristics

The text data used in the experiments was a subset from the data described in [21], containing customers' opinions written in many languages of several millions of hotel guests who – via the on-line Internet service – booked accommodations in many different hotels and countries all over the world. The subset used in our experiments contained 1,245 both positive and negative opinions related to one particular hotel. The data characteristics: *minimal review length* = 1 word, *maximal review length* = 262 words, *average review length* = 30.5 words, *standard deviation* = 35.7 words.

The reviews were always labeled as either positive or negative and this labeling was performed carefully. However, there were several entries that were originally categorized obviously wrongly as the consequence of their authors' errors. For some of the reviews, it was also not possible to determine the opinion polarity without knowing the context. For example, the review "Nothing!" was labeled as positive because it was an answer to a question: "What did you not like about the hotel?" However, this review might be perceived as negative when it would have been an answer to a question: "What did you like about the hotel?" Without knowing such a question (context), one could not decide whether the review was positive or negative.

The reviews were written only by people who made their reservation through the web and who really stayed in the hotel, thus based on their real experience. The reviews were often written quite formally, but most of them embodied all deficiencies typical for texts written in natural languages (i.e., mistyping, transposed letters, missing letters, grammar errors, and so like).

3.2   Data Preprocessing

To be able to use supervised and unsupervised machine learning techniques, the data must be transformed to a representation suitable for selected algorithms. In our experiments, the words in the documents were selected as meaningful units (terms) of the texts. A big advantage of such a word-based representation is its simplicity and the straightforward process of creation [11]. Each of the documents was therefore simply transformed into a bag-of-words, a sequence of words where the ordering was irrelevant. Every document was then represented by a numerical vector where individual dimensions were the words the values of which represented the weights of individual attributes (the words) of the text.

The procedure used the known *tf-idf* (Term Frequency-Inverse Document Frequency, see, e.g., [14]) weighting scheme that usually provided better results than a representation not employing global *idf* weights [22].

The quality of the document vector representation could be increased in several ways, e.g., by using n-grams, adding some semantics, removing very frequent or very infrequent words, eliminating stop words, stemming, but often with only marginal effects [8]. During the preprocessing phase, none of the mentioned techniques was performed.

4   Results

The following section summarizes the results of the experiments described in the previous sections. The non-trivial process of the SuDoC algorithm consists of two major steps, that is:

- preparing clustering solutions of the entire data to be labeled with different numbers of clusters, and
- spreading the labels from initially labeled document instances within at least two clustering solutions, starting with a clustering solution with a higher number of clusters and continuing with one or more clustering solutions having gradually fewer clusters.

The results of both of these tasks could be influenced by several parameter settings. The parameters of the clustering process were already examined in [22] and the settings that were found to provide the best clustering solutions were applied to the presented experiments.

In the second step, the number of clustering solutions used for spreading the known labels and the number of clusters in each of these solutions needed to be determined. In the initial experiments, it was revealed that using just two clustering solutions was quite sufficient. This finding enabled to achieve significantly better results when applying SuDoC than using the well known classifiers, with a lower number of computations than in the case of using three or more clustering solutions. Having more than 1,000 reviews to be automatically labeled, the best results were achieved when the first clustering solution contained a hundred of clusters and the second contained five clusters (100/5). When the numbers of clusters in the two clustering solutions were chosen differently, e.g., 100/10, or 200/10, the classification performance measures reached slightly worse values.

Table 2 contains averaged values of the chosen performance metrics of the experiments with different classifiers – commonly used classifiers on the top of the table, and SuDoC with tree different settings at the bottom (see also Fig. 5).

Table 2. Classification performance metrics for the used classifiers. The presented values are average values obtained from 100 experiments. *Acc* represents Accuracy, *Prec* Precision, *Rec* Recall, and *F* F-measure for the corresponding classes: + for positive reviews and − for negative ones. The numbers as 100/5 following SuDoC stand for the number of clusters in two used clustering solutions.

Classifier	Acc	$Prec^+$	$Prec^-$	Rec^+	Rec^-	F^+	F^-
J48	0.638	0.670	0.647	0.677	0.593	0.650	0.589
Naïve Bayes	0.694	0.699	0.715	0.762	0.619	0.720	0.648
Logistic Regression	0.719	0.736	0.722	0.747	0.688	0.733	0.694
Support Vector Machines	0.706	0.722	0.730	0.751	0.655	0.719	0.669
K-star	0.648	0.659	0.698	0.749	0.534	0.677	0.563
Instance based learning	0.659	0.700	0.704	0.710	0.602	0.668	0.590
J-Rip	0.594	0.652	0.619	0.603	0.583	0.627	0.562
SuDoC(100/5)	0.788	0.865	0.754	0.729	0.851	0.779	0.783
SuDoC(100/10)	0.758	0.823	0.736	0.706	0.810	0.742	0.757
SuDoC(200/10)	0.762	0.816	0.741	0.724	0.799	0.753	0.755

From the presented results, it is obvious that for the given specific task and processed data SuDoC significantly outperformed the commonly used classifiers.

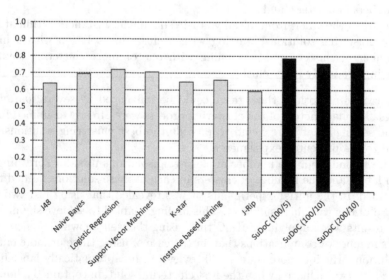

Fig. 5. Accuracy for the used classifiers and SuDoC algorithm

5   Conclusions

This paper presents a novel approach to labeling unknown document instances based on a small number of initially labeled examples. The described approach, called *SuDoC* (Semi-unsupervised Document Classification), can be used as an alternative to commonly used well-known classifiers, such as the Naïve Bayes classifier, Decision Trees, Support Vector Machines, and others. SuDoC's main idea is grounded on using a small number of specimen – a limited set of manually labeled instances representing considered classes.

This set is used for biasing the unlabeled instances so that they get automatically appropriate labels, thus creating classes supporting the future classification or prediction. The classes are generated reiteratively, from a larger number of smaller, less general clusters to a lower quantity of bigger, more general ones. Such a procedure demonstrated better results than applying traditional training of classification algorithms using the limited number of training samples.

The presented results, based on 100 experimental runs for each of the 10 algorithms, their initial conditions and settings, demonstrate that it is possible to achieve better values of the chosen classification performance metrics when using the SuDoC algorithm unlike the traditional clustering procedures.

The future work is going to focus on determining the number of used clustering solutions and the numbers of clusters in each of such solutions which are major aspects of the SuDoC procedure. This will include a large number of experimental runs with the data used in the presented experiments as well as with some different data, and a thorough analysis and comparison of the results. The process will also involve a large amount of manual labeling in order to arrive at representative outcomes. The SuDoC algorithm will be also used for processing documents in different natural languages.

Acknowledgments. This work was supported by the research grants of the Czech Ministry of Education VZ MSM No. 6215648904 and IGA of the Mendel University in Brno No. 4/2013.

References

1. Abney, S.P.: Semisupervised Learning for Computational Linguistics. Chapman & Hall/CRC (2008)
2. Aha, D., Kibler, D.: Instance-based learning algorithms. Machine Learning 6, 37–66 (1991)
3. Berry, M.W., Kogan, J. (eds.): Text Mining: Applications and Theory. John Wiley & Sons (2010)
4. Van Britsom, D., Bronselaer, A., De Tré, G.: Concept Identification in Constructing Multi-Document Summarizations. In: Greco, S., Bouchon-Meunier, B., Coletti, G., Fedrizzi, M., Matarazzo, B., Yager, R.R., et al. (eds.) IPMU 2012, Part II. CCIS, vol. 298, pp. 276–284. Springer, Heidelberg (2012)
5. le Cessie, S., van Houwelingen, J.C.: Ridge Estimators in Logistic Regression. Applied Statistics 41, 191–201 (1992)

6. Cleary, J.G., Trigg, L.E.: K*: An Instance-based Learner Using an Entropic Distance Measure. In: 12th International Conference on Machine Learning, pp. 108–114 (1995)
7. Cohen, W.W.: Fast Effective Rule Induction. In: Twelfth International Conference on Machine Learning, pp. 115–123 (1995)
8. Figueiredo, F., Rocha, L., Couto, T., Salles, T., Goncalves, M.A., Meira, W.: Word co-occurrence features for text classification. Information Systems 36, 843–858 (2011)
9. Ghosh, J., Strehl, A.: Similarity-Based Text Clustering: A Comparative Study. In: Grouping Multidimensional Data, pp. 73–97. Springer, Berlin (2006)
10. Hall, M., et al.: The WEKA Data Mining Software: An Update. SIGKDD Explorations 11, 10–18 (2009)
11. Joachims, T.: Learning to classify text using support vector machines. Kluwer Academic Publishers (2002)
12. John, G.H., Langley, P.: Estimating Continuous Distributions in Bayesian Classifiers. In: Eleventh Conference on Uncertainty in Artificial Intelligence, pp. 338–345 (1995)
13. Karypis, G.: Cluto: A Clustering Toolkit. Technical report, University of Minnesota (2003)
14. Nie, J.Y.: Cross-Language Information Retrieval. Synthesis Lectures on Human Language Technologies 3, 1–125 (2010)
15. Platt, J.: Fast Training of Support Vector Machines using Sequential Minimal Optimization. In: Schoelkopf, B., Burges, C., Smola, A. (eds.) Advances in Kernel Methods – Support Vector Learning (1998)
16. Quinlan, R.: C4.5: Programs for Machine Learning. Morgan Kaufmann Publishers, San Mateo (1993)
17. Russel, S., Norvig, P.: Artificial Intelligence: A Modern Approach. Pearson Education, Upper Saddle River (2010)
18. Sokolova, M., Japkowicz, N., Szpakowicz, S.: Beyond Accuracy, F-Score and ROC: A Family of Discriminant Measures for Performance Evaluation. In: Sattar, A., Kang, B.-H. (eds.) AI 2006. LNCS (LNAI), vol. 4304, pp. 1015–1021. Springer, Heidelberg (2006)
19. Weiss, S.M., Indurkhya, N., Zhang, T., Damerau, F.J.: Text Mining: Predictive Methods for Analyzing Unstructured Information. Springer, New York (2010)
20. Zhao, Y., Karypis, K.: Criterion Functions for Document Clustering: Experiments and Analysis. Technical report, University of Minnesota (2003)
21. Žižka, J., Dařena, F.: Mining Significant Words from Customer Opinions Written in Different Natural Languages. In: Habernal, I., Matoušek, V. (eds.) TSD 2011. LNCS, vol. 6836, pp. 211–218. Springer, Heidelberg (2011)
22. Žižka, J., Burda, K., Dařena, F.: Mining Opinion-Clusters from Very Large Unstructured Real-World Textual Data. In: Ramsay, A., Agre, G. (eds.) AIMSA 2012. LNCS, vol. 7557, pp. 38–47. Springer, Heidelberg (2012)
23. http://glaros.dtc.umn.edu/gkhome/cluto/cluto/download/ (March 2013)

Efficient Visualization of Folksonomies
Based on «Intersectors»

A. Mouakher[1], S. Heymann[2], S. Ben Yahia[1], and B. Le Grand[3]

[1] University of Tunis El Manar, Faculty of Sciences of Tunis, Tunis, Tunisia
sadok.benyahia@fst.rnu.tn
[2] LIP6, CNRS, Université Pierre et Marie Curie, Paris, France
sebastien.heymann@lip6.fr
[3] CRI, Université Paris 1 Panthéon - Sorbonne, Paris, France
benedicte.le-grand@univ-paris1.fr

Abstract. Social bookmarking systems have recently received an increasing attention in both academic and industrial communities. This success is owed to their ease of use that relies on a simple intuitive process, allowing their users to label diverse resources with freely chosen keywords aka *tags*. The obtained collections are known under the nickname of *Folksonomy*. In this paper, we introduce a new approach dedicated to the visualization of large folksonomies, based on the "intersecting" minimal transversals. The main thrust of such an approach is the proposal of a reduced set of "key" nodes of the folksonomy from which the remaining nodes would be faithfully retrieved. Thus, the user could navigate in the folksonomy through a folding/unfolding process.

1 Introduction

Social bookmarking tools are rapidly emerging on the Web as it can be witnessed by the overwhelming number of participants. Indeed, within the last years, social software on the Web, such as FLICKR[1], DEL.ICIO.US[2], BIBSONOMY[3] to cite but a few, has received a tremendous impact with regard to hundreds of millions of users. A key factor to the success of social software tools in the Web is their grass-roots approach to sharing information between a broad community or people (folks) allowing them to browse and to search tags attached to information resources. The result of this collaborative tagging activity in the systems has led to user-generated classifications called folksonomies. However, during folksonomy exploration, it is difficult for a user with a limited subjective knowledge to find related resources which may represent best his/her current interests [1]. Dedicated literature witnesses a wealthy work about improving the quality of queries based on folksonomies. Hence, one stream of research has attempted to refine query results using clustering and tag clouds techniques. However, the relations derived by these techniques are only taxonomical relationships, such as hyperonym relations, and are not based on meanings. Indeed, relying on statistical association or co-occurrence of tags, clustering techniques group the search results into

---

[1] http://www.flickr.com
[2] http://www.delicious.com
[3] http://www.bibsonomy.org

H.L. Larsen et al. (Eds.): FQAS 2013, LNAI 8132, pp. 637–648, 2013.
© Springer-Verlag Berlin Heidelberg 2013

several subsets and recommend related resources based on selected tags. On the other hand, tag cloud, which contains very general terms such as "computer" or "picture", is a somewhat rough approach to organizing tags. It shows a subset of frequently used tags which sizes reflect their frequencies. Actually, the overwhelming size and density of the induced structure is an actual hamper towards a fluent visualisation and comprehension by the user [2]. The aim of this paper is to palliate this weakness by introducing a novel approach of folksonomy visualisation based on "intersector" nodes. The main thrust of this approach relies in the efficient detection of a reduced number of user nodes that allows to faithfully retrieve the remaining nodes. These particular nodes called ambassadors, diffusors, multimembers [3–5] have recently received a renewed of attention within viral marketing. Thus, a user only has to face a very reduced number of nodes even for very large graphs. Indeed, experiments carried out in [5] showed that the number of "intersector" nodes is around twenty even for very large folksonomies. Interestingly enough, from a theoretical point of view, we show that these intersectors are the smallest minimal transversals in terms of size. Roughly speaking, these intersectors are the first reachable minimal transversals if the search space is swept in breadth first manner. The main thrust of the introduced approach is the swift detection of these intersectors and also their reduced number with respect to the set of all nodes.

Standing within a perspective of folding/unfolding and in compliance with Shneiderman's mantra [6] "*Overview first, then details on demand*", a user does not have to face a hazy dense folksonomy. Indeed, (s)he has to start exploring a reduced and manageable fraction of the graph, composed of intersector nodes. The exploration process is based on the unfolding operation that, by clicking on an intersector node, allows to smoothly and gradually unveil the covered and connected nodes. The remainder of this paper is organized as follows. In section 2, we outline the preliminary notions on folksonomies. Then, we scrutinize the works to the dedicated visualization of folksonomies as well as works that focus on the quality of tagging in folksonomies. Section 3 is dedicated to a thorough description of our folksonomy visualization approach based on intersector nodes. Section 4 provides experimental results to illustrate the proposed approach. The last section concludes this paper and opens perspectives for future work.

2   Related Work

Before presenting our approach, we provide a simplified definition of some concepts used throughout in this paper, namely, folksonomy and triadic concept which is an adapted formalism for folksonomies as explained in the following.

2.1   Basic Concepts on Folksonomies

A *folksonomy* describes users, resources, tags, and allows users to arbitrary assign tags to resources. Similarly to [7], we define a *folksonomy* as follows:

Definition 1. (FOLKSONOMY) *A folksonomy is a set of tuples* $\mathcal{F} = (\mathcal{U}, \mathcal{T}, \mathcal{R}, \mathcal{P})$, *where* $\mathcal{U}$ *is a finite set of users,* $\mathcal{T}$ *is a finite set of tags,* $\mathcal{R}$ *is a finite set of resources, and* $\mathcal{P} \subseteq \mathcal{U} \times \mathcal{R} \times \mathcal{T}$ *is a ternary relation and each* $p \subseteq \mathcal{P}$ *can be represented as a triple:*

$$p = \{(u, r, t) \mid u \in \mathcal{U}, r \in \mathcal{R}, t \in T\}.$$

which describes the assignment of the tag t by the user u to the resource r.

A *folksonomy* is also called *"Social Tagging"*, a process which enables users to add, annotate, edit tags to share resources. In other terms, it consists in a web 2.0 support for the classification of resources. Indeed, the annotation of resources (by tags) facilitates the sharing and then the information retrieval. In the following, we recall the main definitions related to triadic concepts, that exactly mimic the structure of a *folksonomy*.

Definition 2. (TRIADIC CONTEXT) *[8] A triadic extraction context (or a triadic context for short) is a quadruple* $\mathcal{K} = (\mathcal{E}, \mathcal{I}, C, \mathcal{Y})$*, where* $\mathcal{E}$*,* $\mathcal{I}$ *and* C *are sets, and* $\mathcal{Y}$ *is a ternary relation between* $\mathcal{E}$*,* $\mathcal{I}$ *and* C*, i.e.,* $\mathcal{Y} \subseteq \mathcal{E} \times \mathcal{I} \times C$*. Elements of* $\mathcal{E}$*,* $\mathcal{I}$ *and* C *are respectively called objects, attributes, and links and* $(e, i, c) \in \mathcal{Y}$*, underlies that the object e has the attribute i according to the link c.*

Example 1. *An example of a triadic context for a folksonomy* $\mathcal{F}$ *is depicted by Table 1 with* $\mathcal{U} = \{u_1, \ldots, u_8\}$*,* $T = \{t_1, \ldots, t_6\}$ *and* $\mathcal{R} = \{r_1, r_2, r_3\}$*. Each cross represents a ternary relation between a user from* $\mathcal{U}$*, a tag from* T *and a resource from* $\mathcal{R}$*, i.e., a user has tagged a particular resource with a particular tag.*

Table 1. A triadic context/*folksonomy*

U/R-T	t_1	t_2	t_3	t_4	t_5	t_6	t_1	t_2	t_3	t_4	t_5	t_6	t_1	t_2	t_3	t_4	t_5	t_6
			r_1						r_2						r_3			
u_1	×	×					×	×										
u_2	×	×	×	×			×	×	×	×								
u_3			×	×					×	×			×	×				×
u_4					×								×	×				
u_5	×	×					×	×					×	×				
u_6				×											×	×	×	
u_7			×	×					×	×					×	×	×	×
u_8															×	×	×	

Formally, a tri-concept is defined as follows:

Definition 3. ((FREQUENT) TRIADIC CONCEPT) *A triadic concept (or a tri-concept for short) of a triadic context* $\mathcal{K} = (\mathcal{E}, \mathcal{I}, C, \mathcal{Y})$ *is a triple* (U, T, R) *with* $U \subseteq \mathcal{E}$*,* $T \subseteq C$*, and* $R \subseteq \mathcal{I}$ *with* $U \times T \times R \subseteq \mathcal{Y}$ *such that the triple* (U, T, R) *is maximal,i.e., for* $U_1 \subseteq U$*,* $T_1 \subseteq T$ *and* $R_1 \subseteq R$ *with* $U_1 \times T_1 \times R_1 \subseteq \mathcal{Y}$*, the containments* $U \subseteq U_1$*,* $T \subseteq T_1$*, and* $R \subseteq R_1$ *always imply* $(U, T, R) = (U_1, T_1, R_1)$*. A tri-concept is said to be frequent whenever it is frequent tri-set. The set of all tri-concepts of* $\mathcal{K}$ *is equal to* $TC_{\mathcal{K}} = \{TC_i \mid TC_i = (U, T, R) \in \mathcal{Y}$ *is a tri-concept,* $i = 1 \ldots n\}$*.*

For a tri-concept (U, T, R), the sets U, R and T are respectively called **Extent**, **Intent**, and **Modus** of the tri-concept (U, T, R).

The main challenge, that would face a user, is the navigation within the hazy graph plotting a large folksonomy. In the following subsection, we review the pioneering approaches that paid attention to the assessment of the tagging quality within folksonomies.

2.2 Tag's Quality

Related literature witnesses a wealthy work about tag quality measures for folksonomies, *e.g*, [9–12] to cite but a few. Among them, the well-adapted to our approach are those proposed by [11] to evaluate tag's relevance. In the following, we briefly define these measures that will be used in the remainder.

1. (HIGH FREQUENCY TAGS) : for a given folksonomy, the frequency of a given tag for the same resource is considered as a factor to judge its relevance. So, for each tagged resource, we order the tags and count the frequency of each distinct tag. Tags with higher frequencies are considered better in terms of quality.

2. (TAG AGREEMENT) : for a given resource x, this metric is defined as the set of tags that are selected by most users who have tagged resource x, i.e., $Agr(t_{x,y}) = \frac{Freq(t_{x,y})}{\sum |u_i|}$. We first determine the frequency of each unique tag. Then, we calculate the number of users who have tagged each resource. The tag agreement is consequently calculated by dividing the tag frequency by the number of users that have tagged a resource. When all users agree on a certain tag, this number should be equal to 1. The closer to 0, the lesser the agreement on that particular tag is.

3. (TF-IRF) : This metric is derived from the well known Term Frequency Inverse Document Frequency "TF-IDF". The latter is a common metric in the domain of automatic indexing for finding good descriptive keywords for a document. In fact, this measure evaluates the importance of a term in a given document. When selecting the appropriate keywords for a certain document, the TF-IDF formula takes the intra as well as inter document frequency of keywords into account. The higher the TF-IDF weight, the more valuable the keyword. Some adjustments have been made on this measure to evaluate tags. Indeed, textual information or documents have been excluded from TF-IDF formula since tagged resources are not always textual in a folksonomy (e.g. an mp3 audio file). Thus, [11] proposed the equation given below to compute the TF-IRF weight for a certain tag annotated to a resource: $TF - IRF(t_{x,y}) = \frac{Freq(t_{x,y})}{T_y} * \log(\frac{\sum |r_i|}{r_x})$, where T_y = total number of tags for resource y and r_x = sum of resources that have tag t_x.

In the following, we will briefly present some related works on folksonomies's visualization as well as their practical applications.

2.3 Visualizing Folksonomies as Graphs

Intensive studies have been made about efficient visualization of large folksonomies.

At first, this tripartite hypergraph of users, tags and resources was represented by a *"tag cloud"*. In this respect, [13] established a comparative study about performance of different models of *"tag cloud"* for visual exploration. Nevertheless, [14] pointed out that this representation does not provide any information about tags's relationships and recommended graph visualization. Thus, the authors of [2] highlighted that the efficiency of such a visualization is higher with a simplification and even a reduction of the quantity of information to be provided to the user. In this respect, [15] introduced projection methods in order to uncover the structure of the folksonomy in less complex ones. The visualization of such structures heavily relies on an assessment of the correlation of similar nodes. Also worth of mention, the project *TagGraph*[4] that tackled the issue of visualization. In the latter, the author has studied the Flickr photography dataset. The provided tool actually acts as an explorer that retrieves the most recent photos that have been indexed by tags matching the one provided by the user. In the same trend, the *Tag Galaxy*[5] tool explores the Flickr dataset as a galaxy composed of planets metaphoring the tags. The authors of [14] also proposed a technique, FOLKSO-VIZ, for automatically deriving semantic relations between tags and for visualizing the tags and their relations. Indeed, they apply various rules and models based on Wikipedia corpus to find the equivalence, subsumption and similarity relations for their approach. FOLKSOVIZ manages the display of the discovered semantical relation between tags in order to provide as intuitive as possible to the user with a visualization of the folksonomy. In [16], the authors introduced the FOLKVIEW tool for the dynamic visualization of a folksonomy by using a multi-agent approach. Thus, the provided tool has been able to provide personalized views of the folksonomy to the user. In the same context, authors in [17] advocated an interaction model that allows users to remotely browse the immediate context graph around a specific node of interest. To do so, they adapted the basic concept of degree of interest from tree to graphs to tackle the complexity of large and dense graphs.

According to the different surveyed approaches, it is a sighting fact that the major problem of folksonomies's visualization stands the inability to display large amount of data. Indeed, the size and density of large folksonomies is steadily growing since user can freely tag different resources. Thus, the main challenge of folksonomy visualization is to provide a reduction strategy to obtain manageable structures [2] Trickily reducing such a structure seems to be a *sine qua non* condition. Furthermore, most of literature works on folksonomies visualization have neglected relationships between users and has been only interested in relations between tags. The visualization approach that we introduce in the remainder aims at tackling these thriving compound challenges.

3   VIF: VIsualization of Large Folksonomies

An efficient visualization usually follows the three steps indicated by the Schneiderman's visual design guidelines [6]. The author recommended the following scheduling tasks *"Overview first, zoom and filter, then details on demand."* In fact, grasping a huge amount of data can very often encourage the user to perform a superficial exploration rather than an in-depth one. That's why a visualization tool has to provide an overview of the information, which permits a direct access to overall knowledge and interconnections within the knowledge space. There are many cases where the user is simply not interested in a global view of the whole graph, but interested in solving a particular concrete task on the graph instead [17]. Once the user focuses on a portion of data, a smooth user-zooming capability should be performed to obtain more details. Similarly, the authors of [18] state that an efficient visualization has to fulfil the three following factors: (*i*) Global overview of the system; (*ii*) Details on demand; and (*iii*) Interaction

---

[4] http://taggraph.com/
[5] http://taggalaxy.de/

with user. In the following, we present a detailed description of VIF, the new graph-based approach of VIsualizing Folksonomies. Let us remind that the main idea of our new approach is to provide a simpler view of the graph by focusing on specific nodes. In fact, displaying an overview of the whole graph is not always relevant to the user. So, we propose to present a reduced set of "key" nodes from which the remaining nodes would be faithfully retrieved through a folding/unfolding process. As shown in figure 1, the "VIF" approach operates in two steps:

1. **Extraction of the tri-concepts** : the set of extracted tri-concepts which are a loss-less concise representation of the whole folksonomy.For this task, we opted for the TRICONS algorithm [19] which is a generate-and-test algorithm. The latter takes as input a *folksonomy* $\mathcal{F}$ as well as three user-defined thresholds : $minsup_u$, $minsup_t$ and $minsup_r$. The TRICONS algorithm outputs the set of all frequent TCs that fulfil these aforementioned thresholds. Standing on the fact that tri-generators are the smallest elements within an equivalence class, so their detection/traversal is largely eased. TRICONS operates in a level-wise manner and heavily relies on the order ideal shape of tri-Minimal Generator family towards achieving better performances.

   Example 2. *With respect to the* folksonomy *illustrated by table 1,* TC_1=({u_1, u_2, u_5}, {t_1, t_2}, {r_1, r_2}) *is a tri-concept of* $\mathcal{F}$, *i.e, is the maximal set of tags and resources shared between* u_1, u_2 *and* u_5. *In addition, the 5 other extracted tri-concepts are:* TC_2=({u_2, u_3, u_7}, {t_3, t_4}, {r_1, r_2}), TC_3=({u_4, u_6}, {t_5}, {r_1}), TC_4=({u_3, u_4, u_5}, {t_1, t_2}, {r_3}), TC_5=({u_6, u_7, u_8}, {t_3, t_4, t_5}, {r_3}) TC_6=({u_3, u_7}, {t_6}, {r_3}).

2. **Extraction of the minimal transversal "Intersectors"**: As explained in section, it operates on the tri-concepts extracted in the previous step. Given the fact that we are searching for particular elements which are strongly connected with most possible nodes in the graph, minimal transversal's notion provides an ideal frame-work to localize these "intersectors". Thus, we have to perform a projection on the "User" dimension. Doing so, we obtain an hypergraph from which we extract the "intersector" nodes.

Indeed, a social network can be defined as a set of entities interconnected with one another [20]. The entities are generally actors or organizations but can also be web pages, scientific articles, films, *etc.* The relations depict their interactions. The analysis of the strategic position that some actors may have in the network and their identifi-cation are mainly studied in the field of sociology notably with measures like degree, closeness, betweenness, designed to this end. The community structure of the network has also been considered [21, 22]. In this respect, we propose to define the notion of "Intersector" in a graph which is based on the concepts of hypergraph and minimal transversal.

In the context of hypergraphs, a recent trend put the focus on determining "key ac-tors" in the underlying community structure. Indeed, in the dedicated literature, such ac-tors are also called as *mediators, ambassadors, experts* depending on their position, and consequently depending on their role in the network [23]. Among these actors whose

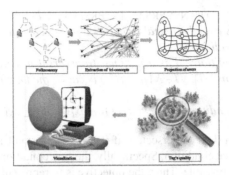

Fig. 1. The VIF approach at a glance

identification has received a lot of attention, also appears the influencer who can be defined as an actor who has the ability to influence the behaviour or opinion of the other members in the social network [24]. Thus, in this respect, we have to focus on minimal transversal "Intersector" defined as follows:

Definition 4. HYPERGRAPH *[25] Let* $H = (\mathcal{X}, \xi)$ *with* $\mathcal{X} = \{x_1, x_2, \ldots, x_n\}$ *a finite set of elements and* $\xi = \{e_1, e_2, \ldots, e_m\}$ *a family of subsets of* $\mathcal{X}$. *H is a hypergraph on* $\mathcal{X}$ *if : (i)* $e_i \neq \emptyset, i \in \{1, \ldots, m\}$ *and (ii)* $\bigcup_{i=1,\ldots,m} e_i = \mathcal{X}$.

The elements of $\mathcal{X}$ are called *vertices* of the hypergraph and they correspond to entities. Elements of ξ, called *hyperedges* of the hypergraph, correspond to communities. Figure 2 depicts a hypergraph $H = (\mathcal{X}, \xi)$ such that $\mathcal{X} = \{1, 2, 3, 4, 5, 6, 7, 8\}$ and $\xi = \{\{1, 2\}, \{2, 3, 7\}, \{3, 4, 5\}, \{4, 6\}, \{6, 7, 8\}, \{7\}\}$. From this hypergraph, we have $TC=(\{u_1, u_2, u_5\}, \{u_2, u_3, u_7\}, \{u_4, u_6\}, \{u_3, u_4, u_5\}, \{u_6, u_7, u_8\}, \{u_3, u_7\})$.

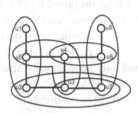

Fig. 2. Hypergraph related to the set of extracted tri-concepts from the folksonomy given by table 1

Table 2. Extraction context corresponding to the hypergraph depicted by figure 2

	u_1	u_2	u_3	u_4	u_5	u_6	u_7	u_8
T_1	×	×			×			
T_2		×	×				×	
T_3				×		×		
T_4			×	×	×			
T_5						×	×	×
T_6			×				×	

Definition 5. MINIMAL TRANSVERSAL *[25] Let a hypergraph* $H = (\mathcal{X}, \xi)$ *where* $\mathcal{X}$ *is the set of vertices and* $\xi = \{e_1, e_2, \ldots, e_m\}$ *is the set of hyperedges.* $T \subset \mathcal{X}$ *is a transversal of H if* $T \bigcap e_i \neq \emptyset \ \forall i = 1, \ldots, m$. γ_H *denotes the set of the transversals defined on* $H : \gamma_H = \{T \subset \mathcal{X} | T \bigcap e_i \neq \emptyset \ \forall i = 1, \ldots, m\}$. *A transversal T of a hypergraph is called* minimal *if* $\nexists T_1 \subset T$ *s.t.* $T_1 \in \gamma_H$. *In the following, we denote* $\mathcal{M}_H$, *the set of minimal transversals of H.*

Definition 6. MINIMAL TRANSVERSAL INTERSECTOR *Let* $H = (\mathcal{X}, \xi)$, *a hypergraph and* $I \subset \mathcal{X}$. *I is called minimal transversal intersector, denoted* TMI, *if I fulfils these two following conditions :*

   *1. (**Minimality Condition**): I is a minimal transversal in cardinality terms:* $|I| = \tau(H)$ *where* $\tau(H) = Min \{|T|, \forall T \in \mathcal{M}_H\}$. $\tau(H)$ *is called the minimal level of transversality.*

   *2. (**Maximal quality Condition**): Relevance(I)=max{Quality_tag(tag(I)) s.t.* $|I| = \tau(H)$*}*

A set of nodes is a TMI "Intersector" if its size is as small as possible and if it maximizes the relevance condition. Specifically, the first condition aims at keeping the set of TMI as small as possible. Thus, the objective is to represent all the communities with a minimum of nodes. The second condition, called *maximal quality*, takes into account the fact that several TMI can belong to a same community. In such a case, the aim is to favour the elements which belong also to the communities with more relevant tags. It is noteworthy that Jelassi et al. [5] defined the notion of multi-member in a social network by choosing minimal transversal that fulfils the maximal connectivity constraint.

The aim of the approach that we introduce, is to minimize as far as possible the number of nodes of a graph with a maximal quality in terms of tags. Even though several algorithms have been presented in the literature for the extraction of a minimal transversal of a hypergraph, our aim is a little bit different. Indeed, our intersectors are a special class of the set of minimal transversals of the induced hypergraph. Roughly speaking, these Intersectors are the smallest minimal transversals in size terms, i.e., the first reachable minimal transversals if we sweep the search space in levelwise manner. The main thrust of the introduced approach is the swift detection of these gateways as well as their reduced number with respect to the set of all nodes. Once detected, these intersectors will be connected to each other as a complete graph. Then, each intersector will be directly connected to the peers that it belongs to the same hyperedges as them.

In our approach, we used the TMD-MINER algorithm proposed by [5]. The latter provides very interesting performances compared to those obtained by the pioneering algorithms of the literature. Indeed, it swiftly identifies the minimimal transversality's degree, i.e., the level k where we can find TMIs. Then, once having at hand the value of k, the algorithm directly jumps in the search space and only generates all the candidates which their degree is equal to k without uselessly generating candidates with lower sizes. After that, from these all k candidates, only candidates fulfilling the minimality condition of the definition 6. These intersector candidates are then filtered with respect to maximal tagging quality condition. This filtering allows at the end to select our intersectors. In our experimental study, detailed in the next section, we have focused on an illustrative example to explain our approach with more details.

4   Illustrative Example and Experimental Results

To illustrate our approach, let us consider the folksonomy depicted by Table 1. First of all, we extract the cover of tri-concepts from the folksonomy. Thus, we apply the TRI-CONS algorithm with $minsup_u = 2$, $minsup_t = 1$ and $minsup_r = 1$ as thresholds.

Hence, we obtain the hypergraph of the Figure 2. The latter can be also represented by table 2 where hyperedges T_i are the projection of users in the different tri-concepts.

The following step is to localize for the smallest smallest minimal transversals in size terms. Such minimal transversals are flagged as candidate TMIs. According to our example, we can extract the following candidate intersectors: $\{u_1u_4u_7\}$, $\{u_2u_4u_7\}$ and $\{u_4u_5u_7\}$.

Then, for each candidate, we compute the quality of tags and keep those corresponding to the most relevant. Indeed, a user who belongs to a given TMI, has tagged one or many resources in the folksonomy. Only relevant set of tags is retained. In fact, the evaluation of tags is done through the RELEVANCE Function which averages the metrics Agr and TF-IRF defined previously in section 2.2. Table 3 contains the set of all the minimal transversals that may be drawn with corresponding relevance tags. Let's take for example the candidate $\{u_1u_4u_7\}$. The user u_1 has tagged the resource r_1 with $\{t_1, t_2\}$ and the resource r_2 with tags $\{t_1, t_2\}$. The user u_4 has tagged the resource r_1 with $\{t_5\}$ and the resource r_3 with tags $\{t_1, t_2\}$. On the other hand, the user u_7 has tagged the resource r_1 with $\{t_3, t_4\}$, the resource r_2 with the same tags and the resource r_3 with tags $\{t_3, t_4, t_5, t_6\}$. The relevance of this candidate $\{u_1u_4u_7\}$ is the sum of the relevance of all this set of tags. The same applies for the other candidates. Finally, we retain the TMI "Intersector" $\{u_4, u_5, u_7\}$ as it corresponds to the best quality of tags.

Table 3. Process of computing TMI

TMI	{Tags}	Relevance	TMI	{Tags}	Relevance	TMI	{Tags}	Relevance
$u_1u_4u_7$	$\{t_1,t_2\}\to r_1$	0.19	$u_2u_4u_7$	$\{t_1,t_2,t_3,t_4\}\to r_1$	0.06	$u_4u_5u_7$	$\{t_1,t_2\}\to r_1$	0.19
	$\{t_1,t_2\}\to r_2$	0.19		$\{t_1,t_2,t_3,t_4\}\to r_2$	0.06		$\{t_1,t_2\}\to r_2$	0.19
	$\{t_5\}\to r_1$	0.16		$\{t_5\}\to r_1$	0.16		$\{t_1,t_2\}\to r_3$	0.19
	$\{t_1,t_2\}\to r_3$	0.19		$\{t_1,t_2\}\to r_3$	0.19		$\{t_5\}\to r_1$	0.16
	$\{t_3,t_4\}\to r_1$	0.19		$\{t_3,t_4\}\to r_1$	0.19		$\{t_1,t_2\}\to r_3$	0.19
	$\{t_3,t_4\}\to r_2$	0.19		$\{t_3,t_4\}\to r_2$	0.19		$\{t_3,t_4\}\to r_1$	0.19
	$\{t_3,t_4,t_5,t_6\}\to r_3$	0.06		$\{t_3,t_4,t_5,t_6\}\to r_3$	0.06		$\{t_3,t_4\}\to r_2$	0.19
							$\{t_3,t_4,t_5,t_6\}\to r_3$	0.06
		1.17			0.91			**1.36**

We display the TMI "Intersectors" with *Linkurious*[6], a Web-based application for searching and visualizing property graph databases, supported by *Linkurious* SAS. Property graph databases are a specific kind of database made for storing and querying large graphs of millions of nodes, relationships and properties associated to them. The search result of *Linkurious* displays intersectors to the graph visualization, then the user can add the neighbors of the displayed nodes to the visualization on demand, enabling a local exploration of the underlying complete graph. The property graph database we use is a *Neo4j*[7] instance. Neo4j is an open source Java technology supported by Neo Technology Inc. As highlighted in figure 3 we can see the TMI "Intersector" extracted from the folksonomy considered in this example. The nodes displayed in this step are $\{u_4, u_5, u_7\}$. User can easily get specific information and browse through this small set of nodes. *Linkurious* allows to display specific relationships between nodes as illustrated in figure 4. Thus, user can extract useful information and control the complexity of the generated visualization.

---

[6] http://linkurio.us/

[7] http://www.neo4j.org/

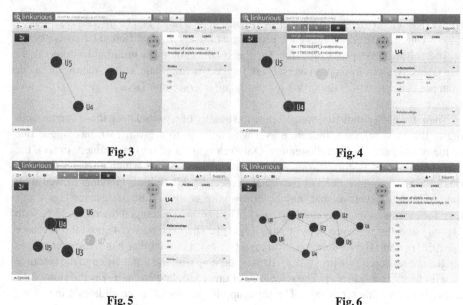

Fig. 3   Fig. 4

Fig. 5   Fig. 6

He/she has also the possibility to more explore nodes that are "touched" by this transversal and expand the graph by an unfolding process, as illustrated in figure 5. The latter shows that by clicking on node u_4, its neighbors become visible namely u_3 and u_6. When user clicks on all the nodes of the TMI "Intersector", we can obtain, as it shown in 6, all the nodes of the folksonomy. An unfolding process can also be done on each node to hide its neighbors.

During our experimental evaluation, we considered the following folksonomies:
1-DEL.ICIO.US[8]: the social bookmarks system DEL.ICIO.US is a service of social bookmarking that offers its users the opportunity to share their favourite web pages. The dataset[9] contains all the bookmarks added in January 2007 to the site *http://delicious. com*. The recovery process has approximately $494,636$ bookmarks that have been published by $54,915$ users through $64,968$ tags of $129,220$ resources.
2- MOVIELENS[10]: it is a film rating system MOVIELENS. This dataset[11] contains explicit evaluations of films. The first folksonomy includes 1 million ratings from 1 to 5 stars, made by about $6,000$ users, and the second consists in $100,000$ ratings provided by 943 users on 1682 movies, between September 1997 and April 1998.

By varying the $minsupp_u$ and as shown by Table 4, we obtained three datasets from the DEL.ICIO.US folksonomy (denoted DELI1, DELI2 and DELI3) and three datasets from the MOVIELENS folksonomy (denoted MOVLENS1, MOVLENS2 and MOVLENS3). In the hypergraph associated to each dataset, the vertices correspond to the users and the hyperedges to the communities where a community represents a

---

[8] www.delicious.com
[9] Freely downloadable at http://data.dai-labor.de/corpus/delicious/.
[10] www.movielens.umn.edu
[11] Freely available to the public[12].

subset of users who have shared a same subset of resources with the same subset of tags. The aim of this experimental validation is to show that even for large folksonomies the number of intesectors remains manageable. The corresponding visualisation snapshots are omitted due to lack of available space. At a glance, we note that the lower the value of $minsupp_u$, the higher the number of TMIs is, and the larger their size, reaching 21 for DELI3 and 20 for MOVLENS3. For the dataset DELI1, we extracted 10 TMIs of size 6, i.e., composed of 6 vertices. This means that we have to find at least 6 users to represent all the communities. A scrutiny of these TMIs, shows that 4 of them are specially active ones, namely the nodes 10, 47, 77 and 78, since they belong to all the extracted TMIs from this dataset.

Table 4. Characteristics of the considered folksonomies, where $:minsupp_u$: minimum number of vertices within a community; $\#Hyper$: number of hyperedges; $\#MT$: number of all minimal transversals; $\#TMI$: number of TMIs; $|TMI|$: size of a TMI in terms of number of vertices

	$minsupp_u$	$|\mathcal{X}|$	$\#Hyper$	$\#MT$	$\#TMI$	$|TMI|$
DELI1	5	119	91	52	10	6
DELI2	3	165	157	1800	78	13
DELI3	2	248	179	8976	201	21
MOVLENS1	5	88	80	108	1	6
MOVLENS2	3	143	246	172	3	12
MOVLENS3	2	196	501	306	26	20

5 Conclusion

The main thrust of such an approach stands in the definition of a reduced set of nodes from which the remaining nodes would be faithfully retrieved. We have illustrated this approach on 2 large datasets and showed an example of intersectors visualization. In the near future, we plan to study in depth to influence of the quality tagging on the popularity of the intersector nodes and setting a evaluation protocole of the proposed visualization interface.

References

1. Trabelsi, C., Jrad, A., Ben, S.: Yahia: Bridging folksonomies and domain ontologies: Getting out non-taxonomic relations. In: Proc. of the 2010 IEEE Intl. Conference on Data Mining Workshops, ICDMW 2010, pp. 369–379. IEEE Computer Society, Washington, DC (2010)
2. Lohmann, S., Díaz, P.: Representing and visualizing folksonomies as graphs: a reference model. In: Proc. of the Intl. Working Conference on Advanced Visual Interfaces, AVI 2012, pp. 729–732. ACM, New York (2012)
3. Scripps, J., Tan, P.N., Esfahanian, A.H.: Node roles and community structure in networks. In: Proc. of the 1st Workshop on Web Mining and Social Network Analysis (SNA-KDD 2007), San José, California, pp. 26–35 (2007)
4. Opsahl, T., Hogan, B.: Growth mechanisms in continuously-observed networks: Communication in a facebook-like community. CoRR (2010)

5. Jelassi, N., Largeron, C., Ben Yahia, S.: TMD-Miner: Une nouvelle approche pour la détection des diffuseurs dans un système communautaire. In: Actes de la 12eme Conférence Intl.e Francophone EGC, Bordeaux, France, pp. 423–428 (2012)
6. Shneiderman, B.: The eyes have it: A task by data type taxonomy for information visualization. In: Press, I.C.S. (ed.) Proc. IEEE Symposium on Visual Languages, Boulder, Colorado, pp. 336–343 (1996)
7. Hotho, A., Jäschke, R., Schmitz, C., Stumme, G.: Information retrieval in folksonomies: Search and ranking. In: Sure, Y., Domingue, J. (eds.) ESWC 2006. LNCS, vol. 4011, pp. 411–426. Springer, Heidelberg (2006)
8. Ganter, B., Wille, R.: Formal Concept Analysis. Springer, Heidelberg (1999)
9. Sen, S., Harper, M.F.M., Lapitz, A., Riedl, J.: The quest for quality tags. In: Proc. of the 2007 Intl. ACM Conference on Supporting Group Work, pp. 361–370. ACM (2007)
10. Krestel, R., Chen, L.: The art of tagging: Measuring the quality of tags. In: Domingue, J., Anutariya, C. (eds.) ASWC 2008. LNCS, vol. 5367, pp. 257–271. Springer, Heidelberg (2008)
11. Damme, C., Hepp, M., Coenen, T.: Quality Metrics for Tags of Broad Folksonomies. In: Proc. of I-semantics 2008, Graz, Austria (2008)
12. Gu, X., Wang, X., Li, R., Wen, K., Yang, Y., Xiao, W.: Measuring social tag confidence: is it a good or bad tag? In: Wang, H., Li, S., Oyama, S., Hu, X., Qian, T. (eds.) WAIM 2011. LNCS, vol. 6897, pp. 94–105. Springer, Heidelberg (2011)
13. Lohmann, S., Ziegler, J., Tetzlaff, L.: Comparison of tag cloud layouts: Task-related performance and visual exploration. In: Gross, T., Gulliksen, J., Kotzé, P., Oestreicher, L., Palanque, P., Prates, R.O., Winckler, M. (eds.) INTERACT 2009, Part I. LNCS, vol. 5726, pp. 392–404. Springer, Heidelberg (2009)
14. Kangpyo, L., Hyunwoo, K., Hyopil, S., Hyoung-Joo, K.: Folksoviz: A semantic relation-based folksonomy visualization using the wikipedia corpus. In: Proc. of the 10th Intl. Conference ACIS, pp. 24–29. IEEE Computer Society, Washington, DC (2009)
15. Lambiotte, R., Ausloos, M.: Collaborative tagging as a tripartite network. In: Alexandrov, V.N., van Albada, G.D., Sloot, P.M.A., Dongarra, J. (eds.) ICCS 2006. LNCS, vol. 3993, pp. 1114–1117. Springer, Heidelberg (2006)
16. Dattolo, A., Pitassi, E.: Visualizing and managing folksonomies. In: Proc. of the Workshop on Semantic Adaptive Social Web, Girona, Spain, pp. 6–14. Springer (2011)
17. Ham, F.V., Perer, A.: "Search, show context, expand on demand": Supporting large graph exploration with degree-of-interest. IEEE Trans. Vis. Comput. Graph. 15, 953–960 (2009)
18. Le Grand, B.: Extraction d'information et visualisation de systèmes complexes sémantiquement structurés. Doctorat d'université, Paris. Université Pierre et Marie Curie (Décembre 2001)
19. Trabelsi, C., Jelassi, N., Ben Yahia, S.: Scalable mining of frequent tri-concepts from folksonomies. In: Tan, P.-N., Chawla, S., Ho, C.K., Bailey, J. (eds.) PAKDD 2012, Part II. LNCS, vol. 7302, pp. 231–242. Springer, Heidelberg (2012)
20. Wasserman, S., Faust, K.: Social Network Analysis, methods and application (1994)
21. Scripps, J., Tan, P.N., Esfahanian, A.H.: Exploration of link structure and community-based node roles in network analysis. In: Proc. of the 7th IEEE Intl. Conference on Data Mining (ICDM 2007), pp. 649–654 (2007)
22. Forestier, M., Stavrianou, A., Velcin, J., Zighed, D.A.: Roles in social networks: Methodologies and research issues. Web Intelligence and Agent Systems 10, 117–133 (2012)
23. Borgatti, S.P., Everett, M.G.: Notions of Position in Social Network Analysis. Sociological Methodology 22, 1–35 (1992)
24. Anagnostopoulos, A., Kumar, R., Mahdian, M.: Influence and correlation in social networks. In: Proc. of the 14th ACM SIGKDD Intl. Conference, pp. 7–15. ACM, New York (2008)
25. Berge, C.: Hypergraphs: Combinatorics of finite sets, p. 256 (1989)

Ukrainian WordNet:
Creation and Filling

Anatoly Anisimov, Oleksandr Marchenko, Andrey Nikonenko,
Elena Porkhun, and Volodymyr Taranukha

Faculty of Cybernetics, Taras Shevchenko National University of Kyiv, Ukraine
ava@unicyb.kiev.ua, rozenkrans@yandex.ua, andrey.nikonenko@gmail.com,
{ellen2003,taranukha}@ukr.net

Abstract. This paper deals with the process of developing a lexical semantic database for Ukrainian language – UkrWordNet. The architecture
of the developed system is described in detail. The data storing structure and mechanisms of access to knowledge are reviewed along with the
internal logic of the system and some key software modules. The article
is also concerned with the research and development of automated techniques of UkrWordNet Semantic Network replenishment and extension.

Keywords: Information Extraction, WordNet, Wikipedia, Knowledge
Representation, Ontologies.

1 Introduction

For the last decade, the engineering and development of knowledge bases have
been traditionally considered as one of the most important areas in artificial
intelligence. Knowledge bases are the foundation for creating a variety of intelligent information systems. They range from expert systems to natural language
processing (NLP) systems. This period has seen a impressive step forward in
the knowledge bases development and evolution. It is impossible to imagine
constructing a "smart" program that effectively addresses complex information
problems in real time without using the knowledge base technology. One of the
most widely used semantic knowledge resources is the lexical database WordNet
[1]. Focusing on new possibilities of NLP, which appeared after the development
of the WordNet, different countries created projects for developing their own national WordNets. These lexical semantic nets rely on the Princeton WordNet as
a base, but they are filled with NL knowledge of specific languages. The largest
Europian WordNet-building project for national languages is EuroWordNet [2]
that combines knowledge bases for Dutch, Spanish, Italian, German, French,
Czech and Estonian languages. A number of national projects are concerned
with building lexical databases similar to WordNet for Hungarian,Turkish, Arabic, Tamil, Chinese, Korean, Russian and other languages, as well as global
projects for association of national lexical semantic bases into a single resource,
for example, BalkaNet.

H.L. Larsen et al. (Eds.): FQAS 2013, LNAI 8132, pp. 649–660, 2013.

This paper highlights the main results of a project to create Ukrainian lexical-semantic knowledge base UkrWordNet (UWN). The project has been going on for several years and made significant progress in localization of WordNet lexical nodes, construction and filling UWN own database structures, creating tools, which automate filling knowledge base structures. A layered software system allowing analysis, updating and editing knowledge bases has been developed. A number of applications realizing for visualization of knowledge base content and offering a convenient interface to access data have also been developed.

The first work on creating the Ukrainian WordNet was conducted in mid-2000. The experience gained from the project indicated that it was necessary to systematize and automate the process of adoption, extension and translation. To address these needs a new model of knowledge representation was suggested. A number of custom-built tools for working with semantic lexical database content were created. Also a method of automatic linking Ukrainian Wikipedia articles to the hierarchical network of UWN as new nodes was implemented.

2   System Architecture

Development of any information system begins with formulating goals, creating a requirements list and choosing the architecture that is able to satisfy the mentioned requirements. The idea behind the UWN project was to unite available information about the language and algorithms that would process it into a single online system. This is fundamentally useful for complex linguistic methods of NL text analysis. Such an approach can also significantly improve the speed of data processing.

The main task of the project was to create a platform that would combine the semantic knowledge base, the morphological knowledge base and logic to handle the knowledge. The platform in question was intended to work online, it has to include a set of tools that work with bases content, provide opportunities for joint work allowed for elaborating linguistic applications. Therefore, the development of a proper system design was one of the project key objectives. The following demands were put forward:

- online access to the data;
- high speed data access;
- ease of editing and updating the data;
- access to the data via a special utility and in a "manual" mode;
- a flexible storage structure that could be easily changed or extended;
- automatic data conversion after changes in a metadata structure;
- support of connectivity between concepts in a multilingual lexical database;
- data access security;
- built-in control over data changes;
- the ability to combine data and algorithms at the same place;
- sustainable and simultaneous work with multiple users.

Detailed analysis showed that the two-level client-server architecture is consistent best with the demanded requirements. The main advantage of the two-level architecture over the three-level one is the ability to deploy the server logic on the same network node with DB. Oracle was used for developing the system. The Oracle Database XE was selected, which is available under license [3]. The Oracle database operates both as an application server and a database. High reliability, quality, speed and extensive internal language PL/SQL ensured the popularity of this database among mobile operators, stock exchanges and banks.

The Main Structural Elements. The main structural elements to accommodate logic and data in Oracle databases are schemas. Each schema can contain tables with data, modules with program logic, individual procedures and functions, user-defined data types, triggers and scheduled processes. In addition, each schema can have its own policy for security and access to other schema facilities.

The assignment of key blocks of the UWN logic and data storage is described below:

- *ua_guest* – schema is used to connect clients to the database. As a single junction point of external systems connection to UWN we created a schema *ua_guest*, which serves as a buffer zone for security and allows connecting any user or a system to UWN. This schema has no any objects, as it has no access rights to objects of other schemas. The only exceptions are interfaces of access to UWN security;
- *ua_security* – schema is responsible for authentication of applications that connect to UWN. It grants access rights according to the software application profile. Also, this schema serves as a single access point to the UWN internal interfaces, i.e. the server logic that performs data processing and modification of semantic and morphological bases. Additionally, this schema contains mechanisms for logging access attempts. The data include time, IP-address and other characteristics of a client machine together with a list of commands that the application performs;
- *ua_ontology* – schema stores information about content of the Ukrainian WordNet. It also contains logic for concepts manipulation. Operations include search for synsets, retrieval of a synset information, modification of the synset data, receiving and setting special features of synsets, retrieval information about relationships between the concepts and making changes in those data. Logic for some applications is also included into this schema;
- *ua_alg* – schema is used for storing a variety of semantic algorithms and the server logic of linguistic applications as well. It also accommodates software and algorithmic parts for linguistic research (e.g., software implementations of methods for measuring the degree of semantic relatedness);
- *ua_morphology* – schema stores morphological dictionary of the Ukrainian language and provides mechanisms for access to it. Spelling check-up and correction algorithms are also deployed in this schema.

Knowledge Representation. Ontology structure development (or constructing an ontological knowledge representation language) is quite a complex and
lengthy process, so, instead of developing our own ontology format, we took the
format typical to the WordNet family. The knowledge is presented in the form
of concepts (synsets) and connections between them. On the physical level the
data are located in a special tabular database structure.

Key Logic Modules. The algorithmic and logical component system consists
of several types of modules:

1. Rights management and access to system objects.
2. Logging user login/logout and a list of operations performed.
3. Recording changes.
4. Modules that work with the database content.
5. Modules of the linguistic logic.
6. Subsystem modules that contain the server logic for client applications.

The structure and use of these modules can be presented in detail as follows:

1. Module *security_pkg* deals with access rights management of client applications. It analyzes each function calls, for which a client tries to get access.
   Any client application that establishes a connection to the server through
   the technological UWN account *ua_guest*, must specify what the subsystem it is (ontoeditor, ontocorrector, etc.). After successful identification of
   an application by the server, the client gets the right to run functions in
   accordance with the access profile of this application type.
2. Module *run* is responsible for access to internal system objects. It includes
   interfaces to internal packages and system procedures. In fact, the module
   run is the single point of access to all server logic deployed inside UWN.
   The subsystem of client authentication and issuing rights for access to the
   server logic has two types of public interfaces. These are accessible for any
   client that may connect to the server through the *ua_guest* account.
3. The module *security_pkg* logs connections of a client machine to the server
   as well as logs a sequence of actions taken. The record of connecting a new
   application is created immediately after successful registration. Information
   about illegal access attempts or unregistered types of applications is collected
   and analysed in the special security system deployed at the administrative
   schema.
   The subsystem *security_pkg* that checks access rights of a client application
   to the server functions also logs the sequence of all actions taken. This type
   of logging is mainly used for testing purposes and can be configured to track
   only certain commands or subsystems.
4. Unlike the modules that provide control, security and access, whose logic is
   concentrated in one place, the changes logging mechanism is of a decentralized nature, with its elements being placed in different schemas. Every node
   of the system that contains important data has its own logging mechanism. A
   key role in the logging mechanism is performed by triggers and data change

logs that provide tracking time of changes, user who made them, and some other parameters. Due to the sizable changelogs generated in multiuser work, logs are used only for the main data.

5. Modules for manipulating each WordNet are placed in the same schema as the data. In general, there are several types of modules in the system that contain the server logic with high complexity degrees of implementation. These modules contain functions for finding, removing, and editing data. This type of a module is a layer between clients and UWN, their main goal is to ensure the correctness of work done with database. Another common module type is modules that implement the entire subsystem for the UWN. Also systems to analyze linguistic data characteristics and experiments can be implemented as individual modules.

6. Linguistic logic modules are deployed in the *ua_alg* schema. It represents a separate class of auxiliary means for various phases of intelligent analysis (i.e. word sense disambiguation, semantic relatedness measure, etc.). A typical example is the module that contains the software implementation of various methods of measuring the semantic similarity degree.

7. Some modules are used to implement the server side of a specific applications that solve a computational linguistic tasks. These modules contain all the logic required to interact between UWN and the end-user. They are deployed in *ua_alg* schema or in some especially dedicated schemas.

Access Model and Data Protection. The main problem in the development of popular client-server systems is data protection against unauthorized access. Also it is important to ensure collaboration of a large number of users working with the same data, alongside the issues of allocation of server resources to clients. In UWN, all applications use the same access model. The basic principles rely on making it impossible to establish direct access to the data and fulfilling requirements to use intermediate specialized interfaces.

The mechanism of interaction between the client applications and the UWN server part is shown in detail. Here description is given in terms of the new *get/set* interface, *get* access model in the old interface is slightly different.

1. The first step is to connect client application to the server part of UWN using the technological *ua_guest* account. This provides access to the *ua_security* schema for further identification.

2. The second step is calling authentication procedure on the *ua_security* schema and to register client application in the system using *security_pkg* module. After successfully passing the registration, a client gets the right to work with UWN objects according to the profile of its applications.

3. Upon successfully executing the second step, the client application gets the right to perform a certain subset of commands in the *run* module. Because the *run* module is an interface to access internal UWN objects, then the application gets the right to call procedures and functions from other schemas of the system.

4. The fourth step is the direct work with the client application. These are calls of certain linguistic features and interaction with the server-side of logic.

3   Methods of Automated Filling and Extension of the Ukrainian WordNet

A number of methods are used to automate the process of filling the Ukrainian WordNet (UWN):

- Localisation knowledge of the English lexical database WordNet to the realities of Ukrainian language through semantic translation of synsets and further heuristic post-processing of translation results;
- Filling the UkrWordNet with new concepts by integrating knowledge from electronic encyclopedias. It is done through creating new nodes that semantically corresponds to Ukrainian Wikipedia articles and joining these nodes to the UkrWordNet.

At the first stage of the Ukrainian WordNet development adaption and translation of the English synsets into Ukrainian has been done. Translation was carried out using electronic dictionaries and using available automatic translation means. To ensure high data quality the obtained Ukrainian synsets undergo manual checking and editing.

Due to the lack of relevant concepts in semantic fields of English language, the problem of incorporating new original Ukrainian synsets that do not have direct analogues in the WordNet arises. The task of updating the UkrWordNet database consists of two steps. First, one must find a suitable synset in the WordNet taxonomy, against which Ukrainian synset can be defined as a descendant. Second, one needs to fill new semantic links of different types for this new synset. While the second stage is problematic in terms of automating the process, the first step is much more promising.

4   Methods to Bind Wikipedia Articles to WordNet Nodes

The development of algorithms for automation of knowledge extraction and filling semantic knowledge bases is very weighty today. The most abundant resources to gain knowledge are electronic encyclopedias, including Wikipedia [4]. Therefore, a number of research projects are aimed at developing means for formalization of knowledge in Wikipedia and integrating them into different kinds of knowledge bases: BabelNet [5], DBpedia [6], WikiTax2WordNet [7], WordNet++ [8] etc. This direction of data integration (from Wikipedia to WordNet hierarchical taxonomy) looks most promising. The WordNet has a high quality structure of hyponymic relations. Wikipedia being a public project with rather ad hoc connections contains cycles and errors. WordNet contains about 120 000 concept nodes. This is not enough for qualitative analysis and text processing in arbitrary topics. At the same time currently Ukrainian Wikipedia contains approximately 500,000 articles, each of them can be considered as a notion. Integrating these notions (with the names-titles of corresponding articles) as new nodes into WordNet by attaching them to the most semantically close synsets may allow producing semantic knowledge base with WordNet hierarchical quality and size of 500 000 concepts. It is likely to become the most powerful structured source of knowledge in the Ukrainian language today.

The vast majority of methods linking Wikipedia articles as new nodes into WordNet uses different measures of semantic similarity of an article text to the lexemes of a certain synset and its glossary.

These algorithms take into account both lexical intersection similarity measure suggested by Lesk [9] as well as relational position of an article, i.e. the correspondence between an article and its links to other articles and a synset and its links to other synsets, which also have to be similar. A good example of such methods is shown in [5] and [10].

5   A New Method for Binding Wikipedia Articles as Nodes to WordNet

A new measure of semantic similarity using latent semantic analysis [11] was proposed to improve the quality of bindings Wikipedia articles to semantically nearest WordNet nodes. This method determines the similarity of the meanings of words and texts using statistical calculations on a large text corpus of Ukrainian Wikipedia.

Construction of TD-matrix via Wikipedia Processing. The first step is the sequential processing of Ukrainian Wikipedia articles to build an *articles ×word* matrix which contains the frequency of word usages in texts of Wikipedia articles. The category *words* can contain phrases alongside with words, e.g. "Trafalgar Square" or "Champs Elysees". The Ukrainian language requires morphological analysis as preprocessing to produce normal word forms – lemmas. Commonly used words have to be excluded from consideration while producing *articles × word* matrix. One can use a stop-list or TF-IDF statistics to remove those words.

The position of a word in a Wikipedia article is a very important factor. When the word is used in the first sentence of an article – it is the part of the definition of the concept described in the article. Therefore, the word has the highest priority and receives a bonus score when its frequency is calculated and *articles × word* matrix is being composed. Let it be 3 for each occurrence in this position. When the word is used in the first paragraph of the text starting with the second sentence, it is also a sign of a high priority. The word is scored 2 for each appearance in this position. All other words in the text are scored 1 for each occurrence.

Decompositions of the TD-matrix. The LSA algorithm is applied to the resulting *articles × word* matrix. The algorithm decomposes the matrix into product of two matrices. The first one represents Wikipedia articles mapping into k-dimensional latent semantic space of topics. The second one represents words mapping into k-dimensional latent semantic space of topics. Column vector in these matrices of a reduced rank k allows fast computation of semantic similarity between words, texts, words and texts, by means of calculating a scalar product of corresponding vectors. Matrix singular decomposition or non-negative matrix factorization (NMF) is usually used to this purpose. For NMF the method by Lee and Seung [12], also known as the multiplicative update rules algorithm, is used. Frobenius norm is exploited as the basis of the cost function.

To facilitate calculation of μ_{sem} semantic similarity measure, one needs to process Ukrainian Wikipedia, generate matrix V ($articles \times words$), factorize it into matrices W and H, where $V = WH$. Then one can calculate μ_{sem} between two objects: between two words as the scalar product of the vector-columns of H, between two articles as the scalar product of the vector-rows of W, between a word and a text of an article as a scalar product of the corresponding row W and the transposed column H. The μ_{sem} measure is a key tool for binding Ukrainian Wikipedia articles to the WordNet nodes.

Binding Ukrainian Wikipedia Articles to WordNet Nodes. The algorithm is presented in the form of a pseudocode. Ukrainian WordNet and Ukrainian Wikipedia are the input of the algorithm.

The main procedure of binding the i-th article of Ukrainian Wikipedia to a nearest WordNet synset looks as follows:

```
Procedure Linking (i:longint;var Sset:SynsetOfWordNet;
                  var MeasureSem:double);
Begin
    Take i-th article of Wikipedia and vector row W[i];
    Find in WordNet all synsets containing the article name;
    Fork (A, B, C):
        If (there is only one synset Sset) then
            Begin (A)
                Calculate μsem between W[i] and each H[t] where
                t is a word of synset Sset.
                Calculate mean:
```

$$\mu_{sem-average} = \frac{\sum_{word_t \in SSet} \mu_{sem}(W[i], H[t])}{count(word_t \in SSet)}$$

```
                If (μsem-average > Thrshld)
                /*Thrshld is obtained experimentally*/
                then Bind the i-th article to the synset Sset.
                MeasureSem is set to μsem-average.
            End(A);
```

```
If (there is a set of synsets {Ssets}, |{Ssets}|>1, each
synset of {Ssets} contains the article name) then
   Begin(B)
        For j =1 to |{Ssets}| do begin
        Calculate μsem between W[i] and each H[t] where
        t is a word of synset Ssets[j]. Calculate mean:
```

$$\mu_{sem-average} = \frac{\sum\limits_{word_t \in SSets[j]} \mu_{sem}(W[i], H[t])}{count(word_t \in SSets[j])}$$

```
        Save correspondence between synset Ssets[j] and
        its μsem-average
        end;
        Select a synset Sset from {Ssets} with highest value
        of μsem-average.
        Bind the i-th article to selected Sset.
        MeasureSem is set to highest μsem-average.
   End(B);
If (no such synset found) then
   Begin(C)
        Take i-th article links from "Categories"
        and obtain list of corresponding articles {Plinks};
        Set variable Best = 0;
        Set variable pbest = None;
        For p in {Plinks} do begin
            If (p is binded to some WordNet synset Sset1) then
            Calculate semantic similarity
            MeasureSem1 = μsem-average between p
            and corresponding synset Sset1
            else call Linking(p,sset1,MeasureSem1);
            Save synset Sset1 that is nearest to p and its
            value of semantic similarity MeasureSem1:
            If (MeasureSem1 > Best) then begin
            Best = MeasureSem1; pbest = p; Sset1best = Sset1
            end;
        end;
     /*This way the article pbest and the corresponding
        synset Sset1best with best similarity is found. Then
        algorithm launches recursive search function FindBest
        to find descendant of synset Sset1best with best
        value of μsem-average to i-th article.*/
     Call FindBest(i, Sset1best, 0, BestSynset, BestValue);
     If (BestValue > 0) then begin
        Sset = BestSynset; MeasureSem = BestValue;
     end;
     Bind i-th Wikipedia article as new node to the
```

```
        BestSynset synset;
    End(C);
End of Linking;

Function FindBest(i:longint,Sset:SynsetOfWordNet,Cut:double;
                var BestSynset:SynsetOfWordNet,BestValue:double)
Begin
    Get set W of all words from Sset; Calculate M:
```

$$M = \frac{\sum\limits_{w_t \in SSet} \mu_{sem}(W[i], H[t])}{count(w_t \in SSet)}$$

```
    If (M < Cut) then Return;
    else begin BestSynset = Sset; BestValue = M; end;
    Using Hyponymy relation get set Sons of all immediate
    descendants of Sset;
    For Son in Sons do begin
        FindBest(i,Son,M,BestSynset1,BestValue1);
        If (BestValue1 > BestValue) then begin
            BestValue = BestValue1; BestSynset = BestSynset1
        end;
    end;
End of Function FindBest;
```

Incorporating new original Ukrainian synsets that have no direct equivalents in the English WordNet corresponds to case C in the algorithm. In case C the method processes an article of Ukrainian Wikipedia, which name-title is not contained in lexicon of UWN – so no synset represented this new notion.

It should be noted that proposed method with measuring semantic similarity of the new nodes to the existing ones successfuly establishes only the hyponymy/hyperonymy relations to the new created synsets. Development of extended special methods for analysis and establishing other types of WordNet relations (meronymy, holonymy, etc.) to the new created nodes is priority direction of further research in the project of the UWN.

6   Experiments

Testing methodology. Testing was performed on a network of Ukrainian nouns in the WordNet database. As it was mentioned above, all English original synsets were replaced by their Ukrainian counterparts. After manual post-processing the Ukrainian WordNet network of nouns contains more than 82,000 synsets and about 145,000 nouns in the lexicon of the system.

Matrix V *articles* × *words* relies on two types of articles from Ukrainian Wikipedia. The first type consists of all articles whose names are contained in the lexicon of the Ukrainian WordNet (multiword names of Ukrainian Wikipedia articles are also contained in UWN lexicon). The second type consists of 50,000 articles that are "lower" in the hierarchy of Ukrainian Wikipedia categories comparing to the articles of the first type. These 50,000 articles were selected at random with normal distribution in different parts and at different levels of the Ukrainian Wikipedia categorical hierarchy. Then matrix V was factorized as it was proposed by Lee and Seung[12]. Factorization was significantly accelerated with the help of GPU-paralleling technology, as described in [13]. As noted above, the algorithm of binding the i-th Ukrainian Wikipedia article to semantically nearest synset in the WordNet is structurally broken down into 3 cases:

(A) One article corresponds to one synset – a definite case;
(B) One article meets several synsets – an ambiguous case (polysemy);
(C) There is no synset containing the name of the i-th article of Ukrainian Wikipedia. One has to find the most semantically similar synset and bind an article to it as a descendant (we call it an uncertain case).

For the purpose of testing 50,000 articles have been processed. They were bounded to some synsets in the WordNet. In the course of binding, logs were made about its results. Every ninth case selected uniformly was logged alongside with the note of a binding type – A, B or C. Then, the log sequence was divided into samples A, B, and C respectively.

Each sample was evaluated by experts for accuracy of bindings (precision), completeness (recall) and, accordingly, F1-measure. The summary of quality rating of the presented algorithm is shown in Table 1.

Table 1. Summary of Wikipedia pages and WordNet synsets binding quality

	A	B	C
	a definite case	polysemy	uncertain case
precision	95,70%	89.43%	85.65%
recall	91,67%	81.86%	77.19%
F1	93,64%	85,48%	81,20%

For comparison, the method of binding pages of Wikipedia to WordNet synsets, described in [10] has the score of 91.1% (in general) and 83.9% (in cases of polysemy) [10]. The paper [5] describes the methodology of mapping a set of articles Wikipedia to a set of WordNet concepts. It shows P = 81.9, R = 77.5, F1 = 79.6. Compared with previously proposed methods, our method showed better performance.

7   Conclusions

This article describes some of the important aspects of creating an Ukrainian lexical semantic network based on the world-famous WordNet. The first part of the paper describes the architecture of the system and its main structural elements: the model of knowledge representation, the key logic modules, the access model and methods of information security and data protection. The system objectives set and its reflection into the list of requirements is shown. Processes and protocols of external users interaction with database are described. The second part of the paper concerns research and development of methods for automating constructions and filling the semantic knowledge base UkrWordNet. The method for creating new nodes generated from Ukrainian Wikipedia articles and binding them to synsets of the UkrWordNet was developed. It allows great extension of UkrWordNet semantic fields. Experiments showed high performance comparing to the best existing methods for binding Wikipedia articles to WordNet synsets.

References

1. Miller, G., Beckwith, R., Fellbaum, C., Gross, D., Miller, K.: Introduction to Word-Net: An On-line Lexical Database,
   http://wordnetcode.princeton.edu/5papers.pdf
2. Vossen, P., Diez-Orzas, P., Peters, W.: Multilingual Design of EuroWordNet,
   http://acl.ldc.upenn.edu/W/W97/W97-0801.pdf
3. Oracle Technology Network Developer License,
   http://www.oracle.com/technetwork/licenses/xe-license-152020.html
4. Wikipedia, http://www.wikipedia.org
5. Navigli, R., Ponzetto, S.P.: BabelNet: Building a Very Large Multilingual Semantic Network. In: Proc. of the 48th Annual Meeting of the Association for Computational Linguistics (ACL 2010), Uppsala, Sweden, pp. 216–225 (2010)
6. Bizer, C., Lehmann, J., Kobilarov, G., Auer, S., Becker, C., Cyganiak, R., Hellmann, S.: DBpedia – A Crystallization Point for the Web of Data. Web Semantics 7(3), 154–165 (2009)
7. Navigli, R., Ponzetto, S.P.: Large-Scale Taxonomy Mapping for Restructuring and Integrating Wikipedia. In: Proc. of the 21st International Joint Conference on Artificial Intelligence (IJCAI 2009), Pasadena, California, pp. 2083–2088 (2009)
8. Navigli, R., Ponzetto, S.P.: Knowledge-rich Word Sense Disambiguation Rivaling Supervised Systems. In: Proc. of the 48th Annual Meeting of the Association for Computational Linguistics (ACL 2010), pp. 1522–1531 (2010)
9. Lesk, M.: Automatic Sense Disambiguation Using Machine Readable Dictionaries: How to Tell a Pine Cone from an Ice Cream Cone. In: SIGDOC 1986: Proceedings of the 5th Annual International Conference on Systems Documentation, pp. 24–26. ACM, New York (1986)
10. Ruiz-Casado, M., Alfonseca, E., Castells, P.: Automatic Assignment of Wikipedia Encyclopedic Entries to WordNet Synsets. In: Szczepaniak, P.S., Kacprzyk, J., Niewiadomski, A. (eds.) AWIC 2005. LNCS (LNAI), vol. 3528, pp. 380–386. Springer, Heidelberg (2005)
11. Deerwester, S., Dumais, S.T., Furnas, G.W., Landauer, T.K., Harshman, R.: Indexing by Latent Semantic Analysis. Journal of the American Society for Information Science 41(6), 391–407 (1990)
12. Lee, D.D., Seung, H.S.: Algorithms for Non-Negative Matrix Factorization. NIPS (2000), http://hebb.mit.edu/people/seung/papers/nmfconverge.pdf
13. Kysenko, V., Rupp, K., Marchenko, O., Selberherr, S., Anisimov, A.: GPU-Accelerated Non-negative Matrix Factorization for Text Mining. In: Bouma, G., Ittoo, A., Métais, E., Wortmann, H. (eds.) NLDB 2012. LNCS, vol. 7337, pp. 158–163. Springer, Heidelberg (2012)

Linguistic Patterns
for Encyclopaedic Information Extraction[*]

Jesús Cardeñosa, Miguel Ángel de la Villa, and Carolina Gallardo

Validation & Business Applications Research Group
Universidad Politécnica de Madrid
{carde,mvilla,carolina}@opera.dia.fi.upm.es

Abstract. Information extraction has almost always focused on extracting retrievable data from a text. Approaches that manage to extract elaborated information have seldom been devised. Through the use of interlingua-type language-independent contents representation, the semantic relations of the contents can be used to search a set of information concerning a particular entity. This way, the person asking a question to find out something about a city or a person, for example, would have to know no more than the name to be used to run a search. This approach is very promising as the person asking the question does not have to know what type of information he or she can request from a documentary source. Our work targets the goal of, given a user's query, providing a complete report about such topic or event, composed of what we consider encyclopaedic knowledge. We describe the origins of this research and the followed procedure, as well as an illustrative case of this on-going research.

Keywords: Information extraction, Linguistic patterns, UNL.

1   Introduction

The most advanced information extraction systems are capable of searching for very specific fact-related information, entities or events ([1], [2], [3]) (hereafter referred to as entity), and, very especially, information that can be "queried" using "factoid" questions, that is, questions that start with What, Where, Who, etc... ([4], [5], [6], [7], [8]).

This type of information is extremely useful. But, for this to work, the answer has effectively to be in the documentary source and the person asking the question has to be very clear about what he or she wants to know. This approach very often does not reflect the position of the person who wants to find out something about a specific fact, event or entity. In other words, very specific questions can only be formulated by someone who is rather well acquainted with the documentary source who is merely looking for a missing data item.

Our current approach aims to satisfy a more common need of someone who is looking for specific information that is not confined to just a single data item. The goal is to somehow emulate encyclopaedic search. If, for work or pleasure, we wanted

---

[*] The production of this article has been sponsored by DAIL-Software S.L. www.dail-software.com

H.L. Larsen et al. (Eds.): FQAS 2013, LNAI 8132, pp. 661–670, 2013.

to find out something about Picasso, for example, we would look up Picasso in the encyclopaedia, where we would find an entry of variable length stating when and where he was born, when and where he died, what are his most famous works, etc. All this information could be clustered in or dispersed across a documentary source, and anyone asking about each individual information item would, as mentioned above, have to know that the source contains that information.

If, on top of this, we had a scenario where the information is not only disperse but also written in different languages, the knowledge or data about an event, fact or entity could only be fused if the knowledge representation were language independent. The work that we describe in this paper represents a further step of Question Answering systems towards systems able to search for information, gather it and generate reports.

In this article, we set out a preliminary approach to this problem, based on a language-independent contents representation. For this representation, we chose UNL [9]. UNL is able to represent written content as an acyclic directed graph on which matching and directed search functions are defined.

In the following sections we will explain first the background of this work (Section 2). Then we will explain why we chose UNL as a contents representation mechanism, and describe this language's semantic properties and how it can be applied for our purposes (Section 3). Section 4 describes the experiment planning, its results and results analysis.

2   Previous Work

In [10] it is explained how specific information could be found in document bases encoded in UNL. User's "factoid" questions (formulated in his/her native language) are transformed into language independent queries, specifically designed to operate on UNL expressions. These queries take the form of unconnected directed graphs, where a relation (the so-called "main relation") pointed to an unknown node representing the answer to the question. The different possible matches of queries in the UNL document base lead to different possible answers, selecting finally the one whose corresponding match form the smallest-size graph (Fig. 1).

It is also observed that UNL expressions containing answers use to contain additional information about the subject of the questions. This additional information is not captured because it has not been asked for, and it can be found in the vicinity of the answer node.

Therefore, if the scope of the query is extended to a limited number of nodes around an entity, much more information about that entity is obtained. This would be the inverse process of searching answers to questions, that is, to obtain all queries that can yield results in the document base about a particular entity, capturing very diverse information, and thus freeing the "questioner" of the obligation to know precisely what to ask, enabling formulations like "tell me all you know about..."

Question: *"Why was Aubert awarded with Camèré prize?"*
Sentence: *"Aubert was awarded the Camèré prize in 1934 by the Academy of Sciences for a new type of movable dam".*

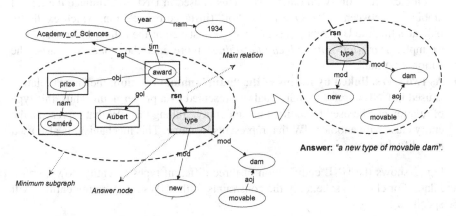

Fig. 1. Answer extraction from UNL expression

Queries obtained by this method could be compiled and generalized by entity type, forming pattern libraries that could be used to draft a very concrete, very specific report about a particular entity.

This approach has driven our recent research and empowered the information extraction system to do much more than simple data extraction. The direction to take now is to find data linked through consistent semantic relations, and beyond, to be able to interpret this kind of information.

3   Linguistic Patterns for Information Extraction

UNL was originally designed as an interlingua capable of unambiguously representing the meaning of a text. Semantic nets were chosen as the formalism for this purpose. Basically, a UNL expression is a directed graph, where the nodes contain the words of the sentence that is to be represented, whereas the arcs indicate how those words are related in the sentence through thematic roles.

The underlying principles are, therefore, purely linguistic. To some extent, this detaches UNL from the more common knowledge representation formalisms [11].

One of the important and more immediately exploitable features is the universality and non-ambiguity of the UNL vocabulary: each of its elements, called UWs (universal words), represents a single concept, and its notation is also fixed, that is, it does not inflect according to gender, number, aspect, etc. These categories are considered as circumstantial traits of the sentence and are expressed in UNL through node attributes. This way, a UNL document can be searched for a particular concept or UW with 100% accuracy, without having to run a more thorough syntactic and/or semantic analysis (no disambiguation is required) [12].

Another key property is that the entities are specified inside the expressions. In UNL, there are two possible codifications of an entity:

- **As a single UW**: the actual codification of the UWs specifies certain, generally, ontological, relations with other UWs. This is used in UNL to eliminate the lexical ambiguity between UWs. One is the "iof" (instance of) relation, which explicitly indicates that the UW represents an entity or instance of the following concept. For example, *Paris(iof>national_capital>thing)* represents the concept "Paris, the nation's capital".
- **As two UWs, linked by means of the "nam" (name) relation**: the "nam" relation is used in UNL to refer to individuals represented by a proper noun. Thus, this type of relation can be used to codify a named entity, linking the UW that represents the entity class with another UW that represents its name. This mechanism is also used to codify dates.

Fig. 2 shows the UNL codification of three different types of entity: town, cruise and date (For clarity's sake, only the constraints of the UWs representing entities will be specified):

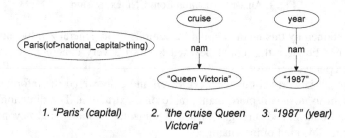

1. *"Paris" (capital)*   2. *"the cruise Queen*   3. *"1987" (year)*
                                Victoria"

Fig. 2. Different ways of codifying entities in UNL

For example, the UNL expression shown in Fig. 3 contains a subnet (highlighted in bold) that codifies factual information about the entity "Hunter Rose", specifically, "Hunter Rose was born in Toledo".

The remainder of the sentence ("had his early education there") could also be said to contain other factual information about Hunter Rose ("he was educated in Toledo"). To gather this datum, however, we first had to deduce that the UW "there" refers to the city of Toledo. This then requires a previous step of linguistic inference.

Such generally context-dependent deductions are beyond the objectives of this research, which focuses exclusively on factual information explicitly expressed inside the UNL expressions and through the very UWs that they contain, as mentioned earlier.

On the other hand, it will not always be possible to find subnets that represent factual information in the vicinity of the entities, meaning that we will need some criteria for detecting and selecting these structures. After running several experiments, we found that these subnets usually have the following characteristics:

- They quite often appear across different UNL expressions.
- They depend on the entity type.

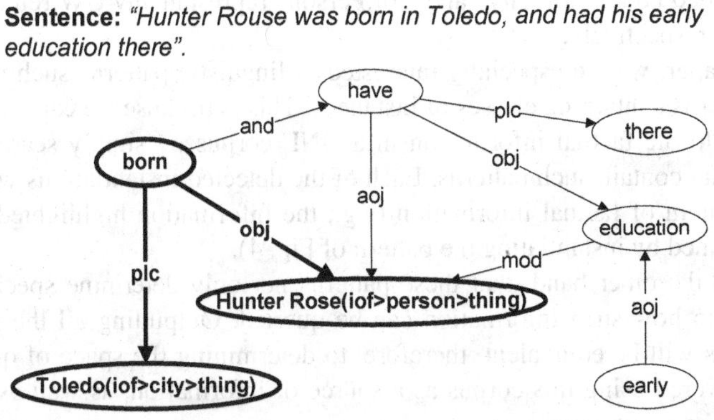

Sentence: *"Hunter Rouse was born in Toledo, and had his early education there".*

have

and

born

plc

there

obj

aoj

obj

plc

mod

education

Hunter Rose(iof>person>thing)

aoj

Toledo(iof>city>thing)

early

Fig. 3. Factual information inside a UNL expression (in bold)

- They contain full information.
- They are semantically coherent.
- They often relate several entities.

The fact that the form and composition of these subnets are related to the type of entity (or entities) that they contain also suggests that they can be generalized to form what we will term *linguistic patterns*.

A linguistic pattern in UNL is a specific way to relate particular UW classes within an expression. It will be depicted in the same manner as a UNL graph, where the UW classes will be denoted by means of box-shaped nodes. Taking into account this casuistry, we can define the concept of linguistic pattern more formally as:

> *A linguistic pattern is the conjunction of semantic relations (in our case, those from the UNL model) and nodes, representing the terms of a domain and linked by the semantic relations, that allows for the expression of events and situations in a semantically coherent way.*

These linguistic patterns, that may vary according to the application domain, allows for searching concrete facts in order to make up an information report based on facts related to what we generically called encyclopaedic knowledge.

Having **observed** that the structure highlighted in Fig. 3 is quite often repeated for entities like person and city, we can generalize it by defining the linguistic pattern shown in Fig. 4.

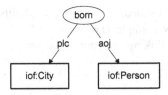

Fig. 4. Linguistic pattern relating persons and cities

The labelled nodes "iof:City" and "iof:Person" represent any UW referring to cities and people, respectively.

In this paper, we are especially interested in linguistic patterns such as the above, that is, patterns containing classes of instances. This is because we can use this type of patterns to locate factual information in a UNL corpus by simply searching the expressions that contain such patterns. Each of the detected instantiations will constitute a different item of factual information (e.g., the information highlighted in Figure 3 can be obtained by instantiating the pattern of Fig. 4).

Note, on the other hand, that these patterns not only determine specific information, but also how such information can be queried. Outputting all the patterns in a UNL corpus will be equivalent, therefore, to determining the space of questions that can be answered using this corpus as a source of information, as well as the specific answers to these questions.

This latter idea is extremely useful for information extraction, because a system capable of locating linguistic patterns related to entities in UNL documents would be able to visualize all the data that those documents contain about a particular entity for the user.

4   Experimental Results: A Case of Study

We ran an experiment on a real UNL document aimed at discovering such linguistic patterns as mentioned above.

The chosen database was the UNL codification of several articles belonging to the EOLSS collection [13]. Specifically, we selected the article "Biographies of eminent water resources personalities", composed of 601 UNL expressions and a total of 2534 UWs, for the tests.

This article is a compilation of a series of biographies of several historical figures linked to the use of water resources. It is perfect for our purposes, as it contains a lot of factual information (dates, people, cities, countries, etc.).

In the pursuit of our goal, we applied a two-step method:

1. Identify the entities that are referenced in the article.
2. Discover the linguistic patterns that determine factual information about the above entities.

4.1   Entity Identification

The entities can be easily located, as they are explicitly codified as such in the UNL expressions, either by a single UW with the "iof" (instance of) constraint or through the "nam" relation linking the entity type with its name (for named entities and dates).

After inspecting the expressions in the article according to these rules, we found a total of 351 different entities of 42 different types. Table 1 shows the count and classification by type (only the most frequent entries are shown):

Table 1. Entity types referenced in article

Entity type	Count
Dates	130
Persons	83
Cities	44
Rivers	13
European countries	9
National capitals	...
North American countries	...
...	...
Total	**351**

Clearly, the distribution conforms to the characteristics of the article: the date is the most frequent type of entity (130), followed by people (83). This is followed by several locations (cities and countries), together with rivers, whose frequency is significant because the biographies contained in the article refer to people related to water resources.

4.2 Pattern Discovery

Basically, the idea is to find repeated structures for particular entity types by building the respective patterns from the generalization of these structures.

To design the procedure, we first had to state some hypotheses on what the UNL nets would have to be like to be considered as valid structures for representing factual information. These hypotheses are related to their characterization in section 3.

In the following we will list each of the steps of the procedure, illustrated by means of an example.

Step 1. Choose two different types of entities.

For our example, we will select the most common types of entity in the document: "date" and "person".

Step 2. Search the corpus for the smallest-sized UNL structures that relate the two types of entities.

To do this step, we will have to run through all the expressions of the document in search of those that contain both types of entities. For each located expression, we will have to find the smallest-sized subgraph that contains each pair of entities of different types. Fig. 5 illustrates this idea by means of a diagram of expression containing a single pair of entities:

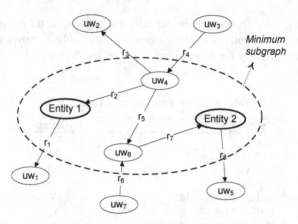

Fig. 5. Minimum subgraph containing a pair of entities

When we have run through all the expressions, we will have output the set formed by all the smallest UNL nets in the corpus that relate the two entity types.

In this example, we found a total of 93 expressions in which the person and date entities appear together. From these, we obtained 105 minimum subgraphs.

Step 3. Build patterns from the detected subgraphs.

For each subgraph output in Step 2, the entities are replaced by nodes that represent their types, outputting a pattern. In this case, we will substitute the nodes that contain specific person and date entities by nodes that represent their classes: "iof:Person" and "iof:Date", respectively. Fig. 6 shows a real size-3 pattern, which is output this way and was found in the article.

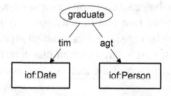

Fig. 6. Linguistic pattern relating persons and dates

Step 4. Select and complete the most frequent patterns in the corpus.

The selected patterns will have to be completed with the aim of codifying sentences with the meaning closest to the context of the corpus. To do this, we will add relations to their internal UWs (that is, to the nodes that do not represent entity classes). These will be the relations necessary to vest the UWs with the argument structure (set of relations) that they normally have in the expressions containing the pattern. Each new relation (argument) will mean adding a new node, which any UW will be able to instantiate.

In this example, the most common pattern is shown in Fig. 6 (appears a total of five times).

To complete this pattern, we will examine the expressions containing the pattern with the aim of determining the argument structures for the internal UW "graduate". For the case of "graduate", we find "tim", "gol" and "agt".

Thus, we will have to add the relation "gol" to the pattern, linking its internal UW to a new node that will act as a free variable.

This completes the process, outputting the pattern shown in Fig. 7.

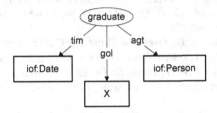

Fig. 7. Completed linguistic pattern

4.3   Factual Data Extraction

Using the pattern output above, we can locate factual information in the document by searching all its instantiations. In the following, we show the sentences generated by instantiating the "graduate" pattern:

- "Howard Penman graduated in Physics in 1930."
- "Koch graduated MD in 1866."
- "William Unwin graduated BSc in 1861."
- "Hunter Rouse graduated in Civil Engineering in 1929."
- "Robert Sellin graduated in Civil Engineering in 1955."

5   Conclusions

The small case described illustrates a new approach to information extraction systems. Users of systems like this would not be bound to be acquainted with the contents of the documentary source to guarantee that they are not asking meaningless questions. If the documentary repository contains any information about the entity concerned, this information will be served to the user in the target language as the UNL graph is able to generate outputs in any language. This is an exhaustive approach, and the intelligent part of systems like these would be to determine the patterns. This should not be a major challenge, as there is a limited range of questions that users could ask. These early experiments proved to be very promising for retrieving full and editable contents rather than mere data.

References

1. Balog, K., Serdyukov, P., de Vries, A.P.: Overview of the TREC 2011 Entity Track. In: Proceedings of the 20th Text Retrieval Conference (TREC 2011), Gaithersburg, Maryland (2012)

2. Cardeñosa, J., Tovar, E.: Intelligent knowledge extraction from the Web. International Journal of Uncertainty, Fuzziness and Knowledge Based Systems 11(supp. 1), 117–134 (2003), doi:10.1142/S0218488503002302

3. Li, G., Deng, D., Feng, J.: Faerie: efficient filtering algorithms for approximate dictionary-based entity extraction. In: Proceedings of the 2011 ACM SIGMOD International Conference on Management of Data (SIGMOD 2011), pp. 529–540. ACM, New York (2011), doi:10.1145/1989323.1989379

4. Dang, H.T., Lin, J., Kelly, D.: Overview of the TREC 2006 question answering track. In: Proceedings of the 15th Text Retrieval Conference, Gaithersburg, Maryland (2006)

5. Heie, M.H., Whittaker, E.W.D., Furui, S.: Question answering using statistical language modelling. Computer Speech & Language 26(3), 193–209 (2012), doi:10.1016/j.csl.2011.11.001

6. López, V., Fernández, M., Motta, E., Stieler, N.: PowerAqua: Supporting users in querying and exploring the Semantic Web. Semantic Web Journal 3(3), 249–265 (2011), doi:10.3233/SW-2011-0030

7. Sagara, T., Hagiwara, M.: Natural language neural network and its application to question-answering system. In: Proceedings of the 2012 International Joint Conference on Neural Networks (IJCNN), Brisbane, Australia, pp. 1–7 (2012), doi:10.1109/IJCNN.2012.6252553

8. Unger, C., Bühmann, L., Lehmann, J., Ngomo, A.C.N., Gerber, D., Cimiano, P.: Template-based question answering over RDF data. In: Proceedings of the21st International Conference on World Wide Web (WWW 2012), pp. 639–648. ACM Press, New York (2012), doi:10.1145/2187836.2187923

9. Uchida, H.: The Universal Networking Language (UNL) Specifications, version 2005 (edition 2006) (2006), http://www.undl.org/unlsys/unl/unl2005-e2006

10. Cardeñosa, J., Gallardo, C., de la Villa, M.A.: Interlingual information extraction as a solution for multilingual QA systems. In: Andreasen, T., Yager, R.R., Bulskov, H., Christiansen, H., Larsen, H.L. (eds.) FQAS 2009. LNCS, vol. 5822, pp. 500–511. Springer, Heidelberg (2009), doi:10.1007/978-3-642-04957-6_43

11. Bateman, J., Magnini, B., Fabris, G.: The generalized upper model Knowledge Base: Organization and use. In: Mars, N.J.I. (ed.) Towards Very Large Knowledge Bases: Knowledge Building and Knowledge Sharing, pp. 60–72. IOS Press, Amsterdam (1995)

12. Boguslavsky, I., Cardeñosa, J., Gallardo, C.: A novel approach to creating disambiguated multilingual dictionaries. Applied Linguistics 30(1), 70–92 (2009), doi:10.1093/applin/amn036

13. EOLSS, Encyclopedia of Life Support Systems (2002), http://www.eolss.net/

Contextualization and Personalization of Queries to Knowledge Bases Using Spreading Activation

Ana B. Pelegrina[1], Maria J. Martin-Bautista[2], and Pamela Faber[1]

[1] Department of Translation and Interpreting
University of Granada
{abpelegrina,pfaber}@ugr.es
[2] Department of Computer Science and Artificial Intelligence
University of Granada
mbautis@decsai.ugr.es

Abstract. Most taxonomies and thesauri offer their users a huge amount of structured data. However, this volume of data is often excessive, and, thus does not fulfill the needs of the users, who are trying to find specific information related to a certain concept. While there are techniques that may partially alleviate this problem (e.g. visual representation of the data), some of the effects of the information overload persist. This paper proposes a four-step mechanism for personalization and knowledge extraction, derived from the information about users' activities stored in their profiles. More precisely, the system extracts contextualization from the users' profiles by using a spreading activation algorithm. The preliminary results of this approach are presented in this paper.

1 Introduction

In the world today, storage, processing and networking technologies have advanced significantly. Users thus have at their disposal huge amounts of structured data regarding almost any topic or field of knowledge. Such information repositories are known as Knowledge Bases (henceforth KBs) and often take the form of thesauri, taxonomies or ontologies. Well known examples of KBs include BabelNet[1], DBPedia [2], WordNet [3,4], and EcoLexicon [5]. However, one of the constraining factors when working with certain repositories is that users may find it difficult to retrieve the information that they are seeking from the vast amount of data available. Therefore, it is necessary to provide them with effective tools to browse and query the data.

Software personalization is a commonly used mechanism to adapt content to user needs and goals. The automatized personalization of software systems is not a trivial task and requires the use of one or more techniques to capture interests of potential users and to translate these preferences into the software. This can be achieved with the use of different techniques such as clustering [6,7,8], fuzzy logic [9,10], data mining [11,10,12,13], and spreading activation [14,15].

Spreading activation is used as a search method for associative networks. Although its roots lie in psychology, semantics processing and linguistics [16,17],

H.L. Larsen et al. (Eds.): FQAS 2013, LNAI 8132, pp. 671–682, 2013.

it has been rapidly adopted in the computer sciences field as a way to simulate a 'human-like' search for information or documents (see Sect. 2 and Sect. 3 for a detailed description of *spreading activation* and its applications).

A simple way to offer users a clear understanding of the data and its structure is to provide them with a visual representation of its content. Nevertheless, given the size of such databases, this visualization often includes too much information to be processed at one glance. This could confuse the user and hinder his/her search efforts. A possible solution for this problem is to personalize the visualization, and show only the entities that are relevant to the user's preferences.

Additionally, a user who is accessing and querying the data may possess a certain degree of specialized knowledge. This can be reflected in the user's interactions with the system (e.g. a geologist will query the system with related terms within the domain of expertise, thus reflecting a knowledge of geology). Generally speaking, the explicit elicitation of this kind of knowledge requires the effort and time of experts, which may be costly. However, alternatively, this knowledge could be implicitly extracted and added to the knowledge base without interfering with the user's activities. This would have the advantage of bypassing expert consultation and avoiding explicit knowledge elicitation.

In this paper, we propose the use of spreading activation as a means to customize the search tools and visualization of knowledge bases. Furthermore, our objective is to combine the information provided by the users through their interactions with the system and the results given by the spreading activation algorithm to extract implicit expert knowledge from the users. The result is a simple, non-intrusive, and implicit mechanism to extract contextualization from users and include it in the user profiles. This contextualization may be later used as a way to provide users with a recommender system [18] to guide non-experts who access the system. This concept is applied to and exemplified in the EcoLexicon knowledge base.

This paper is organized as follows: In Sect. 2, we provide a background on spreading activation. Section 3 discusses the related works and an overview of our proposal. The proposed approach and its implementation are described in Sect. 4. The preliminary results of the application of the technique over a subset of the knowledge base are shown in Sect. 5. Finally, conclusions and future work are presented in Sect. 6.

2   Spreading Activation

The spreading activation theory attempts to model how humans represent and retrieve their knowledge. This theory claims that knowledge in the human mind is represented as nodes and links between nodes in the manner of a semantic network of concepts. It also proposes that the human recall process starts with the activation of a set of concepts (or nodes), which spreads to the neighboring nodes in a series of pulses in a decreasing gradient [17].

For spreading activation, each node in the network has an activation value and each edge has a weight. These values can be initialized to certain values that represent certain properties of the information in the network. For instance, the initial weight of the edges can be set to the strength of the relationship between the connected nodes.

The search process begins with a set of source nodes (e.g. the search terms), which are updated to a new *activation* value. This activation spreads through the nodes that are directly connected to the initial nodes in a series of pulses (or iterations). For each activated node, the activation value is calculated by using (1) and (2) [19]. More specifically, the initial input value I_j for the node j is computed by adding the output value (O_i) of the nodes directly connected to it, which are weighted by the weight $w_{i,j}$ of the links between those nodes. The output, or activation, value O_j is calculated by applying the activation function, f, to the input value I_j. The process finishes once one of the stop conditions (e.g., number of iterations) has been reached.

$$I_j = \sum_i O_i w_{i,j} \tag{1}$$

$$O_j = f(I_j) \tag{2}$$

This technique can be constrained by a set of restrictions (briefly outlined below) that limit the spreading activation process in order to obtain results of relevance. These restrictions thus compensate for some of the disadvantages of pure spreading activation (e.g. saturation of the network produced by the activation of all the nodes in the network) [19].

- **Distance**: the spreading process should stop when it reaches a node whose distance from the initial nodes is greater than a predeterminate value.
- **Path**: activation should spread following preferred paths. This can be achieved adding weight to the relationships between nodes.
- **Activation**: a threshold function can be used to restrict the spreading of the activation of the nodes.
- **Fanout**:the spreading process should stop when it reaches a node that is connected to a large number of nodes.

Furthermore, more restrictions can be added to the models such as the number of activated nodes, number of pulses or the execution time [20], in order to adapt spreading activation to some particular factors that might arose when applying this technique to some specific problems.

3 Related Work

Spreading activation is widely applied in a broad variety of fields in Computer Science, such as information and document retrieval [19,21], computing metrics for information retrieval [22], Web search [23,14,24], collaborative recommendation systems [25,26], data analysis [27], and data visualization [28,29].

This widespread application can be explained by the following: (i) the simplicity and customizability of the algorithm; (ii) the widespread use of the data structure required by the technique, which makes it easy to apply spreading activation to existing network-based knowledge bases; (iii) the 'human-like' results of this technique, though not optimal, may be viewed by users as the most relevant.

Spreading activation has been proposed as a way to extract user preferences or interests in a number of research studies, such as [30,15,14,24,25]. However, none of them is focused on gaining an understanding of the knowledge itself. The goal of this work is not only to personalize the search results but to also provide a way to extract the domain knowledge of the users and apply it in order to attain a deeper understanding of the underlying knowledge structure (e.g., the domains in which can be divided).

The application of spreading activation to data visualizations has been proposed in [28] and [29]. VisLink [28] is a method to interactively explore visualizations and the relationships between them. This method includes spreading activation as means to add analytical power to VisLink. Kuß et al. [29] propose a method based on spreading activation to generate visualizations for a neuroanatomical atlas. Unlike our proposal, neither of these research pieces take advantage of the results of past spreading activation executions in order to improve the personalization of the systems (i.e., they only use spreading activation as a search algorithm with no memory of past searches).

4   Proposal

This research has two objectives. The first is to use the spreading activation technique to personalize search, browsing, and visualization tools for large amounts os structored data. The second is to combine the information provided by the users through their interactions with the system and the results provided by the spreading activation algorithm, and employ them to extract the implicit expert knowledge of the users. For example, we can use the elements searched by the user, the domains that the user is skilled in (e.g., geology) and the search result to automatically classify the concepts included in the knowledge base in different domains. This provides an implicit mechanism to extract contextualization from users. The process is composed of four steps, as shown in Fig. 1:

1. **User profiles generation.** Based on each user's search queries and data browsing, his/her user profile is constructed (see Sect. 4.1).
2. **Contextual information extraction.** Based on the user profile information, a spreading activation algorithm is applied to the data in order to extract the contextualizations. This paper is focused on this step.
3. **Integration of the contextualization in the knowledge base.** Finally, the information regarding the contextualization is incorporated into the knowledge base.
4. **Application of the contextualization.** When searching or visualizing the data, the contextualization obtained and processed in steps 2 and 3 will be employed to customize the searching, browsing and visualization. The new user activities generate new information, and then return to step one.

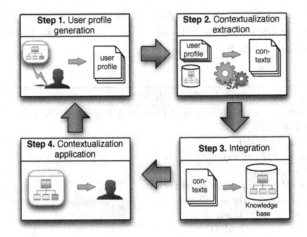

Fig. 1. The steps involved in the proposed mechanism

4.1   User Profiles

The user profiles store user preferences and provide the basis for the personalization of the search and for browsing the data and the recommender system. A variation of the simple profile presented in [31] will be used: *"A group of concepts extracted from the user's activity and deemed interesting by him/her"*. These concepts are extracted from the user's interaction with the system (e.g., searching and browsing activities) and stored in the user profiles as a list of items included in the system (e.g. [*water, concrete, cement*]). When a user first starts accessing the system, the contextualization will only be provided by the constrained spreading activation algorithm and the system will not supply the new user with any customization. Once the initial information about the user has been collected, the system will provide him/her with personalization as well as contextualization.

4.2   Contextual Information Extraction

In order to extract the contextual information implicitly stored in the user profiles, an spreading activation algorithm is applied. This algorithm will activate the concepts that are strongly related to the concepts included in the user profile, defining a context based on the specialized knowledge of the user. In the future, the collected information may be put to use in a recommender system. This system will serve as a guide to new users of the system, and will suggest terms, relations and searches based on their preferences, their interaction with the system, and the extracted experience of past users.

Spreading Activation Implementation. The spreading activation technique has been implemented by using Algorithm 1. The inputs of the algorithm are the entities included in the user profile and the output is a vector of activation values, one for each node in the network. The parameters in the algorithm and their values are given below:

Algorithm 1. Spreading Activation Algorithm

---

procedure SPREADINGACTIVATION($NODES, VERTEX, ORIGIN, F, D$)
   $toFire \leftarrow ORIGIN$
   while $toFire \neq \varnothing$ **do**
     for all $i \in toFire$ **do**
      toFire.remove(i)
      for all $j \in adjacents(i)$ **do**
       $A_j, O_j \leftarrow A_j + DO_i w_{i,j}$
       if $O_j > F$ **then**
        toFire.add(j)
       else
        $O_j \leftarrow 0$
       end if
      end for
     end for
   end while
end procedure

- Activation threshold (F): 0.05 and decay factor (D): 0.75. Both parameters have been adjusted to the conditions of the experimental setup presented in Sect 5. These conditions are a few starting nodes and a small graph.
- Activation function (see (3) and (4)): First, the activation value (A_j) for the node j is calculated using its previous activation value (A_{j-1}), the output value of node i (O_i), the weight of the edge between nodes i and j ($w_{i,j}$), and the decay factor (D). Subsequently, the output value (O_j) to be used in the next iteration, is computed by using the threshold constraint (F).
- Edge weight: in order to reduce de influence of highly connected nodes (see Sect. 2), the weight of a node is calculate following (5): where $w_{i,j}$ is the weight of the edge; $degree(i)$ is the number of edges that leave the node i; and α is a constant that limits the fan-out effect.

$$A_j = A_{j-1} + O_i D w_{i,j} \qquad (3)$$

$$O_j = \begin{cases} 0 & \text{if } A_j < F \\ A_j & \text{otherwise} \end{cases} \qquad (4)$$

$$w_{i,j} = \begin{cases} 1 & \text{if } degree(i) = 1 \\ \alpha \frac{1}{degree(i)} & \text{if } degree(i) > 1 \end{cases} \qquad (5)$$

5   Experimental Example

This proposal was evaluated with the EcoLexicon knowledge base [5], a thesaurus for the specialized domain of the Environment. It is the result of hundreds of hours of work of experts in the areas of translation, terminology, and environmental sciences. This knowledge base includes more than 4,000 concepts, 10,000 terms in six languages, as well as the annotated relations between the aforementioned concepts and terms. EcoLexicon also provides a visual tool (available at: http://ecolexicon.ugr.es) that enables the browsing and searching of its content.

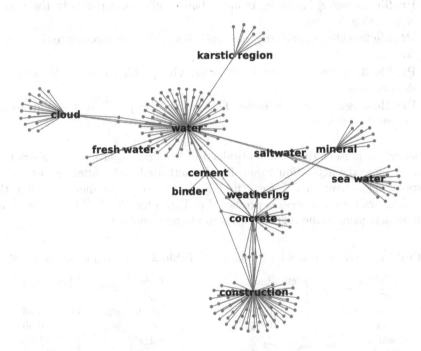

Fig. 2. Database subset (for readability, only some labels are displayed)

In order to address the issue of information overload in EcoLexicon, a contextualization based on the classification of the instances of relations in different domains is presented in [32]. However, these contexts are rigidly defined and require the explicit effort of various experts to define and implement them. The aim of this section is to present an alternative to these contextual domains by means of user-generated contexts using spreading activation. Based on user interactions with EcoLexicon (i.e., search queries), the system builds user profiles that are used as a starting point for the spreading activation process. The initial results of this approach are presented in this paper.

First, we extract a subset of the data in the knowledge base related to the concept 'WATER', which generates a set of concepts and relationships between them. Afterwards, the Networkx library [33] is used to build a graph-based representation of the subset. The resulting network contains 201 nodes (concepts) and 220 links between nodes (relations between two concepts) and as shown in Fig. 2.

After building the network, four user profiles were selected. Each profile includes a set of search terms (though the user profiles can be constructed by also using other user interactions with the system). These profiles are the starting nodes for the spreading activation process. As shown below, each profile corresponds to an area of expertise:

- **Profile 1**: *water, concrete, cement.* This profile corresponds to the *coastal engineering* domain.
- **Profile 2**: *water, weathering, mineral.* This profile corresponds to the *geology* domain.
- **Profile 3**: *water, seawater, fresh water.* This profile corresponds to the *hydrology* domain.
- **Profile 4**: *water, matter, cloud.* This profile This profile corresponds to the *meteorology* domain.

Below, we present the results obtained in the first experiments. Tables 1, 2, 3 and 4 show the activation values for the activated nodes after the execution of spreading activation for every profile. The state of the network after the spreading activation process is presented in Fig. 5 for *Profile 1* (only the nodes with an activation value bigger than zero are represented).

Table 1. Activated nodes for Profile 1

Node	Activation
sand	0.90
karstic region	0.26
rock	0.06
water	1.00
binder	0.45
conglomerate	0.36
region	0.23
soil component	0.26
dissolved oxygen	0.26
weathering	0.26
matter	0.51
placer mine	0.06
water action	0.26
erosion	0.26
cement	1.00
aggregate	0.81
mineral	0.09
intrusion	0.26
concrete	1.00

Table 2. Activated nodes for Profile 2

Node	Activation
sand	0.49
karstic region	0.26
rock	0.60
water	1.00
conglomerate	0.13
region	0.23
soil component	0.26
dissolved oxygen	0.26
weathering	1.00
matter	1.00
placer mine	0.60
water action	0.26
erosion	0.26
aggregate	0.13
mineral	1.00
intrusion	0.26
concrete	0.36

Table 3. Activated nodes for Profile 3

Node	Activation
sea water	1.00
salt marsh	0.54
sand	0.13
karstic region	0.26
rock	0.06
water	1.00
conglomerate	0.03
region	0.23
soil component	0.26
dissolved oxygen	0.26
fresh water	1.00
weathering	0.26
matter	0.18
placer mine	0.06
saltwater	0.90
water action	0.26
erosion	0.26
aggregate	0.03
sandripple	0.90
mineral	0.09
intrusion	0.26
salt pan	0.54
concrete	0.09

Table 4. Activated nodes for Profile 4

Node	Activation
atmosphere	0.36
low atmospheric pressure	0.36
sand	0.13
karstic region	0.26
hydrometeor	0.36
cloud	1.00
rock	0.06
water	1.00
conglomerate	0.03
region	0.23
soil component	0.26
dissolved oxygen	0.26
weathering	0.26
glory	0.36
matter	1.00
placer mine	0.06
water action	0.26
erosion	0.26
aggregate	0.03
mineral	0.09
intrusion	0.26
concrete	0.09

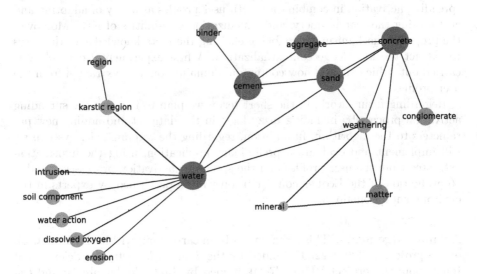

Fig. 3. Activated nodes after the spreading activation process for Profile 1. The size of each node is proportional to its activation value.

We performed a comparison between the generated contextualizations and the domains proposed by the experts in [32]. The results for profile 1 are presented in Table 5. As can be seen, the contextualization coincides with the experts domain (*coastal engineering*) in the 84% of the concepts, using only a small user profile of three terms.

Table 5. Comparison between the generated contextualization and the domains proposed by the experts for *Profile 1*

Node	Present in domain	Node	Present in domain
sand	Yes	weathering	Yes
karstic region	No	matter	Yes
rock	Yes	placer mine	Yes
water	Yes	water action	No
binder	Yes	erosion	Yes
conglomerate	Yes	aggregate	Yes
region	Yes	cement	Yes
soil component	Yes	mineral	Yes
dissolved oxygen	Yes	intrusion	No

6   Conclusions and Future Work

As the quality and quantity of large data sets steadily increase, there is a growing need for more efficient means of adapting the browsing and searching of these data sets to user preferences. In response to this issue, we propose the use of spreading activation in combination with user profiles as a way of adapting and customizing the search, query, and visualization capabilities of KBs. Moreover, the proposed mechanism is capable of eliciting the meta-knowledge of the users to extract the knowledge contextualization. A first experimental example was carried out, which showed how contextual domains could be extracted from the user profiles.

Regarding future work, in the short term, we plan to improve the spreading activation process by including fuzzy logic in the data set and adding new parameters to the algorithm. In addition, regarding the recommender system, we will implement and evaluate a preliminary application, using the information extracted from the user profiles and the spreading activation processes. Finally, an evaluation of the EcoLexicon extension will be performed by experts in the environmental domain

Acknowledgements. This research has been carried out within the framework of the project FFI2011-22397, funded by the Spanish Ministry for Science and Innovation; the project P11-TIC7460, funded by Junta de Andalucía; and the project FP7-SEC-2012-312651, funded from the European Union in the Seventh Framework Programme [FP7/2007-2013] under grant agreement No 312651.

References

1. Navigli, R., Ponzetto, S.P.: Babelnet: Building a very large multilingual semantic network. In: Proceedings of the 48th Annual Meeting of the Association for Computational Linguistics, pp. 216–225. Association for Computational Linguistics (2010)
2. Auer, S., et al.: Dbpedia: A nucleus for a web of open data. In: Aberer, K., Choi, K.-S., Noy, N., Allemang, D., Lee, K.-I., Nixon, L.J.B., Golbeck, J., Mika, P., Maynard, D., Mizoguchi, R., Schreiber, G., Cudré-Mauroux, P. (eds.) ASWC 2007 and ISWC 2007. LNCS, vol. 4825, pp. 722–735. Springer, Heidelberg (2007)
3. Fellbaum, C.: Wordnet. In: Theory and Applications of Ontology: Computer Applications, pp. 231–243 (2010)
4. Miller, G.A., Beckwith, R., Fellbaum, C., Gross, D., Miller, K.J.: Introduction to wordnet: An on-line lexical database*. International Journal of Lexicography 3(4), 235–244 (1990)
5. Reimerink, A., Faber, P.: Ecolexicon: A frame-based knowledge base for the environment. In: Proceedings of the International Conference Towards eEnvironment, pp. 25–27 (2009)
6. Ferragina, P., Gulli, A.: A personalized search engine based on web-snippet hierarchical clustering. Software: Practice and Experience 38(2), 189–225 (2008)
7. Han, L., Chen, G.: A fuzzy clustering method of construction of ontology-based user profiles. Advances in Engineering Software 40(7), 535–540 (2009)
8. Leung, K.T., Ng, W., Lee, D.L.: Personalized concept-based clustering of search engine queries. IEEE Transactions on Knowledge and Data Engineering 20(11), 1505–1518 (2008)
9. Widyantoro, D., Yen, J.: Using fuzzy ontology for query refinement in a personalized abstract search engine. In: Joint 9th IFSA World Congress and 20th NAFIPS International Conference, vol. 1, pp. 610–615 (July 2001)
10. Kim, K.J., Cho, S.B.: Personalized mining of web documents using link structures and fuzzy concept networks. Applied Soft Computing 7(1), 398–410 (2007)
11. Duong, T.H., Uddin, M.N., Li, D., Jo, G.S.: A Collaborative Ontology-Based User Profiles System. In: Nguyen, N.T., Kowalczyk, R., Chen, S.-M. (eds.) ICCCI 2009. LNCS, vol. 5796, pp. 540–552. Springer, Heidelberg (2009)
12. Mobasher, B., Cooley, R., Srivastava, J.: Automatic personalization based on web usage mining. Communications of the ACM 43(8), 142–151 (2000)
13. Mulvenna, M.D., Anand, S.S., Büchner, A.G.: Personalization on the net using web mining: introduction. Communications of the ACM 43(8), 122–125 (2000)
14. Jiang, X., Tan, A.H.: Learning and inferencing in user ontology for personalized Semantic Web search. Information Sciences 179(16), 2794–2808 (2009)
15. Katifori, A., Vassilakis, C., Dix, A.: Ontologies and the brain: Using spreading activation through ontologies to support personal interaction. Cognitive Systems Research 11(1), 25–41 (2010)
16. Anderson, J.R.: A spreading activation theory of memory. Journal of Verbal Learning and Verbal Behavior 22(3), 261–295 (1983)
17. Collins, A.M., Loftus, E.F.: A spreading-activation theory of semantic processing. Psychological Review 82(6), 407 (1975)
18. Resnick, P., Varian, H.R.: Recommender systems. Communications of the ACM 40(3), 56–58 (1997)
19. Crestani, F.: Application of spreading activation techniques in information retrieval. Artificial Intelligence Review (1997)

20. Alvarez, J.M., Polo, L., Jimenez, W., Abella, P., Labra, J.E.: Application of the spreading activation technique for recommending concepts of well-known ontologies in medical systems. In: Proceedings of the 2nd ACM Conference on Bioinformatics, Computational Biology and Biomedicine, BCB 2011, p. 626 (2011)
21. Preece, S.E.: Spreading activation network model for information retrieval. Dissertation Abstracts International Part B: Science and Engineering (Diss. Abst. Int. Pt. B- Sci. & Eng.) 42(9) (1982)
22. Gouws, S., Rooyen, G.J.V., Engelbrecht, H.A.: Measuring conceptual similarity by spreading activation over wikipedia's hyperlink structure, 46–54 (August 2010)
23. Crestani, F., Lee, P.L.: Searching the web by constrained spreading activation. Information Processing & Management 36(4), 585–605 (2000)
24. Sieg, A., Mobasher, B., Burke, R.: Ontological User Profiles for Representing Context in Web Search. In: 2007 IEEE/WIC/ACM International Conferences on Web Intelligence and Intelligent Agent Technology - Workshops, pp. 91–94. IEEE (November 2007)
25. Sieg, A., Mobasher, B., Burke, R.: Improving the effectiveness of collaborative recommendation with ontology-based user profiles. In: Proceedings of the 1st International Workshop on Information Heterogeneity and Fusion in Recommender Systems, HetRec 2010, pp. 39–46. ACM Press, New York (2010)
26. Sieg, A., Mobasher, B., Burke, R.: Ontology-Based Collaborative Recommendation. In: ITWP (2010)
27. Teufl, P., Payer, U., Parycek, P.: Automated analysis of e-participation data by utilizing associative networks, spreading activation and unsupervised learning. In: Macintosh, A., Tambouris, E. (eds.) ePart 2009. LNCS, vol. 5694, pp. 139–150. Springer, Heidelberg (2009)
28. Collins, C., Carpendale, S.: Vislink: Revealing relationships amongst visualizations. IEEE Transactions on Visualization and Computer Graphics 13(6), 1192–1199 (2007)
29. Kuß, A., Prohaska, S., Meyer, B., Rybak, J., Hege, H.C.: Ontology-based visualization of hierarchical neuroanatomical structures. In: Proc. Vis. Comp. Biomed., pp. 177–184 (2008)
30. Eyharabide, V., Amandi, A.: Ontology-based user profile learning. Applied Intelligence 36(4), 857–869 (2011)
31. Martín-Bautista, M.J., Kraft, D.H., Vila, M., Chen, J., Cruz, J.: User profiles and fuzzy logic for web retrieval issues. Soft Computing-A Fusion of Foundations, Methodologies and Applications 6(5), 365–372 (2002)
32. León Araúz, P., Magaña Redondo, P.: Ecolexicon: contextualizing an environmental ontology. In: Proceedings of the Terminology and Knowledge Engineering (TKE) Conference 2010, pp. 341–355 (2010)
33. Hagberg, A.A., Schult, D.A., Swart, P.J.: Exploring network structure, dynamics, and function using networkx. In: Varoquaux, G., Vaught, T., Millman, J. (eds.) Proceedings of the 7th Python in Science Conference, Pasadena, CA, USA, pp. 11–15 (2008)

Utilizing Annotated Wikipedia Article Titles to Improve a Rule-Based Named Entity Recognizer for Turkish

Dilek Küçük

Electrical Power Technologies Group
TÜBİTAK Energy Institute
06800 Ankara, Turkey
dilek.kucuk@tubitak.gov.tr

Abstract. Named entity recognition is one of the information extraction tasks which aims to identify named entities such as person/location/ organization names along with some numeric and temporal expressions in free natural language texts. In this study, we target at named entity recognition from Turkish texts on which information extraction research is considerably rare compared to other well-studied languages. The effects of utilizing annotated Wikipedia article titles to enrich the lexical resources of a rule-based named entity recognizer for Turkish are discussed after evaluating the enriched named entity recognizer against its initial version. The evaluation results demonstrate that the presented extension improves the recognition performance on different text genres, particularly on historical and financial news text sets for which the initial recognizer has not been engineered for. The current study is significant as it is the first study to address the utilization of Wikipedia articles as an information source to improve named entity recognition on Turkish texts.

1 Introduction

Information extraction (IE) from textual data is an important natural language processing task which gains increasing research attention mainly due to the need to automatically process and analyze textual documents available on the Web in several different languages. Named entity recognition (NER) is a well studied IE subtask targeting at the extraction of the names of people, organizations, and locations, along with some numeric and temporal expressions from free natural language texts [1].

NER studies generally range from manually engineered rule-based systems to learning-based and statistical systems [1,2,3]. While the former systems, which make use of gazetteers and other lexical resources, do not require annotated corpora, they achieve low success rates when tested on diverse text genres, particularly on genres different from the initial target domain of these systems. The learning/statistical systems are freed from this problem since they can be trained

H.L. Larsen et al. (Eds.): FQAS 2013, LNAI 8132, pp. 683–691, 2013.

on new domains provided that the required annotated training corpora are available [2]. Nevertheless, building annotated corpora manually is a highly costly task and the latter systems need an additional training phase on such corpora in order to be applicable to new domains. Decision trees [4], Bayesian learning [5], hidden Markov models (HMMs) [6], relational learning (such as inductive logic programming) [5], support vector machines (SVM) [7,8], and conditional random fields (CRF) [9] are among the most prominent learning/statistical techniques used for IE tasks including NER.

On well-studied languages such as English, several different NER studies with high performance rates have been conducted and reported in the literature while NER research on other languages including Turkish is quite rare. Considering the NER research conducted on Turkish texts, to the best of our knowledge, the first study on the topic is the language-independent named entity recognizer reported in [10] where the recognizer is evaluated on Turkish texts along with other texts in Romanian, English, Greek, and Hindi. The statistical name tagger proposed in [11] is an HMM based statistical system [11] to extract person, location, and organization names in Turkish texts. In [12], the first rule-based NER system for Turkish is described. This system utilizes a set of lexical resources and patterns bases to extract person, location, and organization names as well as money/percentage and date/time expressions from Turkish news texts [12]. This system is turned into a hybrid recognizer with a capability to extend its lexical resources by extracting high-confidence named entities from annotated corpora through rote learning and the ultimate hybrid system [13] is shown to outperform its rule-based predecessor. The aforementioned rule-based recognizer has also been successfully integrated into automatic and semi-automatic semantic video annotation systems for Turkish news videos [14,15] where named entities are automatically extracted from video texts to be used as semantic annotations for the videos. A system utilizing CRF and a set of morphological features is presented in [16] where the author argues that CRF provides advantages over HMMs during NER in Turkish. Finally, a rule learning system for the NER task in Turkish texts is presented in [17].

In this study, we annotate Wikipedia article titles with named entity tags and utilize this annotated data to enrich the resources employed by the rule-based NER system presented in [12]. After this extension to the recognizer, we have evaluated the performance of this extended recognizer against its predecessor on diverse text genres to observe the effects of this annotated data to named entity recognition performance and discussed the evaluation results. As Wikipedia articles are known to constitute a plausible information source for different knowledge-based applications, the effects of their utilization to improve a rule-based NER system is a significant contribution to the literature especially for Turkish texts for which NER studies are still limited in number and scope. The rest of the paper is organized as follows: In Section 2, the process of annotating Wikipedia article titles in Turkish is presented, Section 3 is devoted to the evaluation results of the ultimate NER system, corresponding to the original rule-based system enriched with the Wikipedia annotations, against its predecessor and discussion of the results. Finally, Section 4 concludes the paper together with future research directions to pursue.

2   Annotation and Utilization of Wikipedia Article Titles

Wikipedia[1] is known to be an important semantic information source and several text processing applications benefit from the available textual data in Wikipedia and its existing structure [18]. Regarding the NER task, there are several studies ranging from proposals of automatic generation of gazetteers or annotations from Wikipedia such as [19], [20] to those that target at NER on Wikipedia itself [21].

In this study, we consider utilizing Wikipedia article titles in Turkish to enhance a rule-based named entity recognizer [12], as a considerable proportion of these titles belongs to a named entity category, usually a person, location, or organization name, among a diverse set of categories. In addition to this, these titles include prominent named entities in other languages as well, such as the names of foreign politicians, artists, and organizations. Hence, Wikipedia article titles stand as a plausible information source to extend the resources of any rule-based recognizer.

Our initial named entity recognizer for Turkish [12] is a rule-based system that has been manually engineered for news texts. The recognizer employs a set of lexical resources and pattern bases to identify named entities of type person, location, and organization names, in addition to date/time and percent/money expressions [12]. These resources are illustrated in Figure 1 as excerpted from [12]. Although the recognizer, in its default settings, does not make use of the capitalization information during the recognition process, it can also be made to execute utilizing the capitalization clue.

We mainly target at the Wikipedia article titles in Turkish to arrive at new sets of lexical resources to be integrated into the recognizer. We have used the Turkish Wikipedia database dump dated 12 July 2012 which contains about

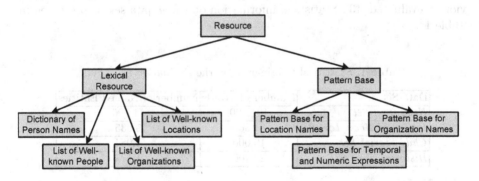

Fig. 1. The Taxonomy of the Resources Employed by the Rule-Based Named Entity Recognizer for Turkish [12]

---

[1] http://en.wikipedia.org/wiki/Main_Page

388,500 article titles after removing some empty and non alphanumeric titles. We have divided this set into 20 distinct subsets (the first 19 of them containing about 20,000 and the last one containing 8,500 titles) and considered only the first subset for manual annotation, as automatic annotation of the remaining 19 subsets using the annotated first subset can be considered as a plausible future work.

The subset of the database dump considered for manual annotation contains 20,000 article titles, with a total of about 34,592 tokens, which is annotated using the named entity annotation interface presented in [12]. This interface enables human annotators to annotate the corresponding entities of type person, location, organization names; date, time, percent and money expressions in an input text with the corresponding named entity tags of ENAMEX, TIMEX, and NUMEX as utilized in the MUC conference series[2]. The annotation process has taken a total of about 11 hours for the author and resulted in the annotation of 7,098 named entities.

The annotated 7,098 named entities are checked against the existing lexical resources of the initial named entity recognizer [12] and those annotated entities which already exist in these resources are removed. After this post-processing stage, the number of the resulting annotated entities drops down to 4,631; 3,069 of them being person names, 974 of them being location names, and finally 588 of them being organization names. These new named entities are added to the corresponding lexical resources employed by the initial recognizer according to their types to arrive at the extended named entity recognizer for Turkish.

3   Evaluation of the Extended Named Entity Recognizer and Discussion of the Results

The extended form of the named entity recognizer is evaluated on the set of four distinct data sets on which the original rule-based recognizer has been previously evaluated [13]. Statistical information on these data sets is presented in Table 1[3].

Table 1. Statistical Information on the Evaluation Data Sets

Data Set	Number of Words	Number of Named Entities
News Text Set	101,700	11,206
Financial News Text Set	84,300	5,635
Child Stories Set	19,000	1,084
Historical Text Set	20,100	1,173

---

[2] http://www-nlpir.nist.gov/related_projects/muc/

[3] It should be noted that only person and organization names are annotated and hence considered on financial news text set which accounts for the comparatively lower number of named entities in this data set.

The news text data set is compiled from METU Turkish corpus [22], financial news texts are from a Turkish news provider, Anadolu Agency[4], child stories set comprises two stories by an author, and finally historical text data set corresponds to the first three chapters of a book about Turkish history.

The evaluation is performed based on the commonly employed metrics of precision (P), recall (R), and F-Measure (F), calculated as follows:

$$Precision = (Correct + 0.5 * Partial)/(Correct + Spurious + 0.5 * Partial)$$
$$Recall = (Correct + 0.5 * Partial)/(Correct + Missing + 0.5 * Partial)$$
$$F-Measure = (2 * Precision * Recall)/(Precision + Recall)$$

It should be noted that the above metrics give half credit to partial extractions where the type of a named entity extracted by the recognizer is correct but its span is not correct (either missing some of the required tokens or including excessive tokens) [13], mainly following the corresponding metric definitions in [23].

The evaluation results of the initial and extended rule-based recognizers, when the capitalization feature is turned off, on the data sets are presented in Table 2. The evaluation results of the recognizers when the capitalization feature is turned on (i.e., when the capitalization information is utilized by the recognizers), are provided in Table 3. In both tables, F-Measure rates are presented in boldface to illustrate the differences in the performance of the recognizers.

Table 2. Evaluation Results of the Initial and Extended Rule-Based Named Entity Recognizers (Capitalization Feature is Turned Off)

Data Set	Rule-Based Recognizer			Extended Rule-Based Recognizer		
	P (%)	R (%)	**F (%)**	P (%)	R (%)	**F (%)**
News Text Set	85.15	83.23	**84.18**	84.82	83.86	**84.34**
Financial News Text Set	71.70	50.36	**59.17**	71.61	51.72	**60.06**
Child Stories Set	72.67	76.81	**74.68**	70.67	77.28	**73.83**
Historical Text Set	53.57	70.70	**60.96**	54.86	75.12	**63.41**

Table 3. Evaluation Results of the Initial and Extended Rule-Based Named Entity Recognizers (Capitalization Feature is Turned On)

Data Set	Rule-Based Recognizer			Extended Rule-Based Recognizer		
	P (%)	R (%)	**F (%)**	P (%)	R (%)	**F (%)**
News Text Set	93.41	83.12	**87.96**	92.99	83.78	**88.14**
Financial News Text Set	79.02	50.14	**61.35**	78.76	51.50	**62.28**
Child Stories Set	87.50	76.47	**81.61**	86.36	76.84	**81.32**
Historical Text Set	79.38	70.34	**74.59**	80.28	74.87	**77.48**

Before comparing the two recognizer versions, it should be noted that the original rule-based named entity recognizer suffers from considerable performance

---

[4] http://www.aa.com.tr

degradations for the three text genres of financial news, child stories, and historical texts. This situation is usually termed as *the porting problem*, as the recognizer achieves low performance rates when ported to domains other than its original target domain, which is the news text genre for the recognizer under consideration [12].

From the performance rates given in Table 2 and Table 3, the following conclusions can be drawn:

– In both evaluation settings (whether capitalization feature is turned off or on), the extended recognizer achieves better F-Measure rates than the initial recognizer for all text genres, excluding the child stories set. Considering the child stories set, the named entities included do not seem to be compatible with the common information available on Wikipedia and hence, the new entries obtained from Wikipedia lead to more false positives than true positives. This is in fact a plausible consequence since the child stories set employed does not cover many well-known person/location/organization names. Nevertheless, further evaluations on other child stories of larger sizes are imperative to derive more generic conclusions.

– Since Wikipedia article titles are utilized to extend the lexical resources of the initial recognizer, the recall rates of the extended recognizer are better than that of the initial recognizer for all text genres considered, as expected. The highest increase in recall is observed on historical text set and next, on financial text set while the increases in recall on news text and child stories sets are comparatively lower. The small increase in recall for the news text data set can be attributed to the fact that the initial recognizer (i.e, its resources) has already been engineered for news texts and as previously pointed out, the child stories set does not actually include many named entities that can be recognized using the common information available on Wikipedia which accounts for the small recall increase for this data set. On the other hand, the lexical resources of the initial recognizer lack several necessary historical named entities (like past empires, their emperors, etc.) and well-known financial institutions (i.e., organization names) to help recognize the named entities in the historical and financial text sets while the Wikipedia article titles annotated and added to the lexical resources of the extended recognizer have led to correct recognition of several of such named entities in the corresponding text sets. In other words, the newly added sets of entries obtained from Wikipedia, to a certain degree, have alleviated the effects of the porting problem for these two text genres and paved the way for further extension of the resources of the recognizer.

– The original recognizer suffers from performance degradations due to some foreign named entities available in the evaluation data sets (especially news text and financial news texts sets) utilized during evaluation. The extended recognizer is able to recognize some of these foreign named entities since Wikipedia article titles in Turkish include prominent foreign entities which has also resulted in a slight increase in the recall rates of the extended recognizer.

– Since a total of 4,631 new entries are added to the resources of the original recognizer, some of these newly added entries has led to a slight decrease in the precision rates of the extended recognizer due to few false positives, for all text genres excluding the historical text set.

To summarize, the utilization of annotated Wikipedia article titles improves the performance of the recognizer although a random subset of size one-twentieth of the whole Turkish Wikipedia database is considered. The improvement is especially significant for the text genres of historical and financial news texts as the Wikipedia article titles cover many prominent financial and historical named entities both in Turkish and in foreign languages.

To the best of our knowledge, the current study is significant as it is the first study to address the utilization of Turkish Wikipedia articles for improving named entity recognition on Turkish texts. The findings of the current study create a firm basis to consider utilizing all Wikipedia articles titles and texts in an automated manner to substantially improve named entity recognition performance on Turkish texts.

4 Conclusion

Wikipedia articles are known to be utilized in several text processing tasks including named entity recognition. In this study, we investigate the effects of annotated Wikipedia article titles to a rule-based named entity recognizer for Turkish news texts. A subset of Turkish Wikipedia database is considered and the annotated article titles in this subset are added to the lexical resources of the recognizer to arrive at an extended recognizer. This ultimate recognizer is evaluated against its predecessor on four different text genres and the evaluation results demonstrate that extending the resources of the recognizer with annotated Wikipedia article titles usually improves its performance, especially on historical and financial text sets on which the initial recognizer suffers from the porting problem. The current study is significant as it is the first study to address the utilization of Wikipedia to improve named entity recognition in Turkish considering the fact that information extraction research on Turkish texts is quite insufficient.

Future work includes implementation of approaches for automatic processing and annotation of all Wikipedia article titles and article contents to improve named entity recognition on Turkish texts, in-depth evaluation of these approaches, and comparison of the approaches with the related work.

References

1. Nadeau, D., Sekine, S.: A survey of named entity recognition and classification. Lingvistica Investigationes 30(1), 3–26 (2007)
2. Grishman, R.: Information extraction. In: Mitkov, R. (ed.) The Oxford Handbook of Computational Linguistics. Oxford University Press (2003)

3. Turmo, J., Ageno, A., Catala, N.: Adaptive information extraction. ACM Computing Surveys 38(2), 1–47 (2006)
4. Sekine, S., Grishman, R., Shinnou, H.: A decision tree method for finding and classifying names in Japanese texts. In: Proceedings of the Sixth Workshop on Very Large Corpora (1998)
5. Freitag, D.: Machine learning for information extraction in informal domains. Machine Learning 39(2-3), 169–202 (2000)
6. Bikel, D.M., Miller, S., Schwartz, R., Weischedel, R.: Nymble: a high-performance learning name-finder. In: Proceedings of the Fifth Conference on Applied Natural Language Processing, pp. 194–201 (1997)
7. Li, Y., Bontcheva, K., Cunningham, H.: Adapting SVM for Data Sparseness and Imbalance: A Case Study on Information Extraction. Natural Language Engineering 15(2), 241–271 (2009)
8. Mayfield, J., McNamee, P., Piatko, C.: Named entity recognition using hundreds of thousands of features. In: Proceedings of the Seventh Conference on Natural language Learning (CONLL) at HLT-NAACL, pp. 184–187 (2003)
9. McCallum, A., Li, W.: Early results for named entity recognition with conditional random fields, feature induction and web-enhanced lexicons. In: Proceedings of the Seventh Conference on Natural Language Learning, pp. 188–191 (2003)
10. Cucerzan, S., Yarowsky, D.: Language independent named entity recognition combining morphological and contextual evidence. In: Proceedings of the Joint SIG-DAT Conference on Empirical Methods in Natural Language Processing and Very Large Corpora, pp. 90–99 (1999)
11. Tür, G., Hakkani-Tür, D., Oflazer, K.: A statistical information extraction system for Turkish. Natural Language Engineering 9(2), 181–210 (2003)
12. Küçük, D., Yazıcı, A.: Named entity recognition experiments on Turkish texts. In: Andreasen, T., Yager, R.R., Bulskov, H., Christiansen, H., Larsen, H.L. (eds.) FQAS 2009. LNCS, vol. 5822, pp. 524–535. Springer, Heidelberg (2009)
13. Küçük, D., Yazıcı, A.: A hybrid named entity recognizer for Turkish. Expert Systems with Applications 39(3), 2733–2742 (2012)
14. Küçük, D., Yazıcı, A.: Exploiting information extraction techniques for automatic semantic video indexing with an application to Turkish news videos. Knowledge-Based Systems 24(6), 844–857 (2011)
15. Küçük, D., Yazıcı, A.: A semi-automatic text-based semantic video annotation system for Turkish facilitating multilingual retrieval. Expert Systems with Applications 40(9), 3398–3411 (2013)
16. Yeniterzi, R.: Exploiting morphology in Turkish named entity recognition system. In: Proceedings of the ACL Student Session, pp. 105–110 (2011)
17. Tatar, S., Çicekli, İ.: Automatic rule learning exploiting morphological features for named entity recognition in Turkish. Journal of Information Science 37(2), 137–151 (2011)
18. Medelyan, O., Milne, D.N., Legg, C., Witten, I.H.: Mining meaning from Wikipedia. International Journal of Human-Computer Studies 67(9), 716–754 (2009)
19. Toral, A., Munoz, R.: A proposal to automatically build and maintain gazetteers for named entity recognition by using Wikipedia. In: Proceedings of the Conference of the European Chapter of the Association for Computational Linguistics (EACL) (2006)
20. Nothman, J., Curran, J.R., Murphy, T.: Transforming Wikipedia into named entity training data. In: Proceedings of the Australasian Language Technology Association Workshop, pp. 124–132 (2008)

21. Balasuriya, D., Ringland, N., Nothman, J., Murphy, T., Curran, J.R.: Named entity recognition in Wikipedia. In: Proceedings of the Workshop on The People's Web Meets NLP: Collaboratively Constructed Semantic Resources, pp. 10–18 (2009)
22. Say, B., Zeyrek, D., Oflazer, K., Özge, U.: Development of a corpus and a treebank for present-day written Turkish. In: Proceedings of the 11th International Conference of Turkish Linguistics (ICTL) (2002)
23. Maynard, D., Tablan, V., Ursu, C., Cunningham, H., Wilks, Y.: Named entity recognition from diverse text types. In: Proceedings of the Conference on Recent Advances in Natural Language Processing (RANLP) (2001)

Author Index

Akaichi, Jalel 40
Akhgar, Babak 124
Alejo, Óscar José 543
Alzebdi, Mohammedsharaf 317
Andreasen, Troels 1, 613
Andrews, Simon 124
Anisimov, Anatoly 649
Anker Jensen, Per 1
Appelgren Lara, Gloria 112
Arcaini, Paolo 352
Aryadinata, Yogi S. 437
Atanassov, Krassimir 317

Baazaoui-Zghal, Hajer 100
Barranco, Carlos D. 447
Barro, S. 269
Bayoudhi, Amine 67
Benghazi, Kawtar 471
Ben Yahia, S. 637
Bermúdez, Jesús 13
Berzal, Fernando 293
Besbes, Ghada 100
Billiet, Christophe 401
Bordogna, Gloria 352
Bouchon-Meunier, Bernadette 257
Bronselaer, Antoon 233, 364
Bugarin, A. 269
Bulskov, Henrik 1

Cal, Piotr 579
Cardeñosa, Jesús 661
Castelltort, Arnaud 437
Castillo-Ortega, R. 245
Chamorro-Martínez, J. 198
Chountas, Panagiotis 317
Christiansen, Henning 613
Ciecierski, Konrad 328
Cubero, Juan-Carlos 293

Dařena, František 625
de la Villa, Miguel Ángel 661
Delgado, Miguel 57, 112, 143, 305
De Tré, Guy 233, 364, 401
Dubois, Didier 376
Dujmović, Jozo 364

Espín, Vanesa 471

Faber, Pamela 671
Fajardo, Waldo 305
Fernández-Luna, Juan M. 543
Fischer Nilsson, Jørgen 1
Furfaro, Angelo 340

Gallardo, Carolina 661
Garrido, E. 91
Gerdes, Anne 155
Ghorbel, Hatem 67
González, Marta 13
Groccia, Maria Carmela 340

Hadrich Belguith, Lamia 67
Heymann, S. 637
Hiemstra, Djoerd 519
Hirsch, Laurence 124
Huete, Juan F. 543
Hurtado, María V. 471

Illarramendi, Arantza 13

Jaudoin, Hélène 425, 601
Julián-Iranzo, Pascual 413

Kajdanowicz, Tomasz 555
Kamkova, Darya 495
Kasprzak, Andrzej 579
Kazienko, Przemyslaw 555
Kryszkiewicz, Marzena 531
Küçük, Dilek 683
Kupka, Jiří 209

Labbadi, Wissem 40
Larsen, Henrik Legind 155
Lasota, Tadeusz 567, 589
Lassen, Tine 1
Laurent, Anne 437
Le Grand, B. 637
Leidner, Jochen L. 495
Lesot, Marie-Jeanne 257
Liétard, Ludovic 389
Loukanova, Roussanka 164

Madrid, Nicolás   507
Marchenko, Oleksandr   649
Marín, Nicolás   112, 245
Martin-Bautista, Maria J.   143, 671
Martínez-Jiménez, P.   198
Medina, Juan Miguel   447
Mena, Eduardo   13
Molina, Carlos   176, 245
Molina-Solana, Miguel   57, 305
Moreno, Antonio   100
Moreno-Cerrud, Eleazar   543
Moreno-García, Juan   221
Morrissey, Joan   79
Mouakher, A.   637
Moyse, Gilles   257
Murinová, Petra   186

Nielandt, Joachim   364
Nikonenko, Andrey   649
Noguera, Manuel   471
Novák, Vilém   186, 209

Olivas, J.A.   91
O'Riordan, Colm   459
Ortiz-Arroyo, Daniel   134

Pasi, Gabriella   459
Paulus, François   601
Pelegrina, Ana B.   671
Peña Yañez, Carmen   176
Peska, Ladislav   483
Pivert, Olivier   425, 601
Pons, José Enrique   401, 447
Pons, Olga   401, 447
Porkhun, Elena   649
Prade, Henri   376
Prados de Reyes, Miguel   176
Prados-Suárez, Belen   176
Przybyszewski, Andrzej W.   328
Puente, C.   91

Quesada, Luis   293

Ramos-Soto, A.   269
Raś, Zbigniew W.   328
Rocacher, Daniel   389

Rombo, Simona E.   340
Romero, Francisco P.   221
Ros, María   57
Rouces, Jacobo   155
Rubio-Manzano, Clemente   413
Ruiz, M. Dolores   143

Sala, Michel   437
Salazar-Santis, Esteban   413
Sánchez, Daniel   143, 198, 245
Sanchez-Valdes, Daniel   281
San Martín-Villarroel, Eduardo   413
Smętek, Magdalena   567
Smits, Grégory   425, 601
Soto-Hidalgo, J.M.   198
Stedmon, Alex   124
Štěpnička, Martin   209
Sterlacchini, Simone   352
Straccia, Umberto   507
Stuhr, Magnus   25

Taboada, J.   269
Tamani, Nouredine   389
Taranukha, Volodymyr   649
Telec, Zbigniew   589
Theil Have, Christian   613
Tjin-Kam-Jet, Kien   519
Torre-Bastida, Ana I.   13
Touazi, Fayçal   376
Trawiński, Bogdan   567, 589
Trawiński, Grzegorz   567, 589
Trieschnigg, Dolf   519
Trivino, Gracian   281

Van Britsom, Daan   233
Veres, Csaba   25
Vojtas, Peter   483

Woźniak, Michał   579

Yates, Simeon   124
Younus, Arjumand   459

Zhao, Ruoxuan   79
Žižka, Jan   625